SEED DISPERSAL

Theory and its Application in a Changing World

SEED DISPERSAL

Theory and its Application in a Changing World

Edited by

Andrew J. Dennis

Eugene W. Schupp

Ronda J. Green

and

David A. Westcott

www.cabi.org

CSIRO

Rainforest CRC

CABI is a trading name of CAB International

CABI Head Office
Nosworthy Way
Wallingford
Oxfordshire OX10 8DE
UK

Tel: +44 (0)1491 832111
Fax: +44 (0)1491 833508
E-mail: cabi@cabi.org
Website: www.cabi.org

CABI North American Office
875 Massachusetts Avenue
7th Floor
Cambridge, MA 02139
USA

Tel: +1 617 395 4056
Fax: +1 617 354 6875
E-mail: cabi-nao@cabi.org

A catalogue record for this book is available from the British Library, London, UK.

Library of Congress Cataloging-in-Publication Data

Seed dispersal : theory and its application in a changing world / edited by Andrew J. Dennis ... [*et al.*].
p. cm.
Includes bibliographical references and index.
ISBN 978-1-84593-165-0 (alk. paper)
1. Seeds--Dispersal--Congresses. I. Dennis, Andrew J. II. Title.
QK929.S443 2007
581.4′67--dc22

2007002294

ISBN 978 1 84593 165 0

Produced and typeset by Columns Design Ltd, Reading, UK
Printed and bound in the UK by Biddles Ltd, King's Lynn

Contents

Contributors

Frederick R. Adler, *Department of Mathematics and Department of Biology, 257 South 1400 East, University of Utah, Salt Lake City, UT 84112, USA. adler@math.utah.edu*

Ofir Altstein, *Movement Ecology Laboratory, Department of Evolution, Systematics and Ecology, The Hebrew University of Jerusalem, Givat Ram, Jerusalem 91904, Israel. ofiralt@pob.huji.ac.il*

A. Elizabeth Arnold, *Division of Plant Pathology and Microbiology, Department of Plant Sciences, University of Arizona, Tucson, AZ 85721, USA. arnold@ag.arizona.edu*

Juliann E. Aukema, *Zoology Department, North Carolina State University, Raleigh, NC 27695, USA. jaukema@alumni.brown.edu*

John G. Blake, *Department of Biology, University of Missouri–St Louis, One University Blvd, St Louis, MO 63121, USA. blakej@msx.umsl.edu*

Pedro G. Blendinger, *LIEY–CONICET, Universidad Nacional de Tucumán, CC 34, 4107 Yerba Buena, Tucumán, Argentina. blendinger@birdecology.com.ar*

Katrin Böhning-Gaese, *Institut für Zoologie, Abt. V, Johannes Gutenberg-Universität Mainz, Becherweg 13, 55099 Mainz, Germany. boehning@uni-mainz.de*

Judith L. Bronstein, *Department of Ecology and Evolutionary Biology, University of Arizona, Tucson, AZ 85721, USA. judieb@email.arizona.edu*

Tomás A. Carlo, *Biology Department, University of Washington, Box 351800, Seattle, WA 98115, USA. dendrophthora@gmail.com*

John E. Carlson, *School of Forest Resources, Pennsylvania State University, University Park, PA 16802, USA. jec16@psu.edu*

Connie J. Clark, *School of Natural Resources and Environment and Department of Zoology, University of Florida, Gainesville, FL 32611, USA. cclark@zoo.ufl.edu*

Thomas A. Contreras, *Department of Biology, Wilkes University, Wilkes-Barre, PA 18766, USA. Present Address: Department of Biology, Washington and Jefferson College, Washington, PA 15301, USA. tcontreras@washjeff.edu*

Richard T. Corlett, *Department of Ecology and Biodiversity, The University of Hong Kong, Pokfulam Road, Hong Kong, China. corlett@hku.hk*

James W. Dalling, *Department of Plant Biology, University of Illinois, Urbana, IL 61801, USA. Smithsonian Tropical Research Institute, Apartado 2072, Balboa, Panamá. dallingj@life.uiuc.edu*

Stuart J. Davies, *Center for Tropical Forest Science–Arnold Arboretum Asia Program, Harvard University, 22 Divinity Avenue, Cambridge, MA 02138, USA. sdavies@oeb.harvard.edu*

Andrew J. Dennis, *CSIRO Sustainable Ecosystems and the Rainforest CRC, PO Box 37, Herberton 4887, Australia. Aviceda.Creek@bigpond.com*

Camila I. Donatti, *Department of Biological Sciences, Stanford University, CA 94305, USA. Laboratório de Biologia da Conservação, Grupo de Fenologia e Dispersão de Sementes, Departamento de Ecologia, Universidade Estadual Paulista (UNESP), C.P. 199, 13506-900 Rio Claro, SP, Brazil. cdonatti@stanford.edu*

Pierre-Michel Forget, *Muséum National d'Histoire Naturelle, Département Ecologie et Gestion de la Biodiversité, UMR 7179 CNRS-MNHN, 4 av. du Petit Château, F-91800 Brunoy, France. pmf@mnhn.fr*

Sérgio Furtado dos Reis, *Departamento de Parasitologia, I.B., Universidade Estadual de Campinas (UNICAMP), 13083-970, Campinas, SP, Brazil. sfreis@unicamp.br*

Mauro Galetti, *Laboratório de Biologia da Conservação, Grupo de Fenologia e Dispersão de Sementes, Departamento de Ecologia, Universidade Estadual Paulista (UNESP), C.P. 199, 13506-900 Rio Claro, SP, Brazil. mgaletti@rc.unesp.br*

Rachel E. Gallery, *Department of Plant Biology, University of Illinois, Urbana, IL 61801, USA. Present address: Department of Zoology, University of Oxford, South Parks Road, Oxford OX1 3PS, UK. rachel.gallery@gmail.com*

Catherine A. Gehring, *Department of Biological Sciences, Northern Arizona University, Flagstaff, AZ 86011, USA. Catherine.Gehring@nau.edu*

Héctor Godínez-Alvarez, *UBIPRO, FES Iztacala, UNAM, Av. de los Barrios 1, Los Reyes Iztacala, Tlalnepantla 54090, Edo. de México, Ap. Postal 314, México. hgodinez@campus.iztacala.unam.mx*

Ronda J. Green, *Faculty of Environmental Sciences, Griffith University, Nathan, Qld 4111, Australia. platypuscorner@bigpond.com*

Paulo Guimarães, *Daitan Labs, Galleria Office, Bloco 4, CJ 444, Campinas, SP, Brazil. prguima@daitanlabs.com*

Paulo R. Guimarães Jr, *Programa de Pós-Graduação em Ecologia, Universidade Estadual de Campinas (UNICAMP), 13083-970, Campinas, SP, Brazil. Integrative Ecology Group, Estación Biológica de Doñana, CSIC, Pabellón del Perú, Avda. M. Luisa S/N, E-41013, Sevilla, Spain. prguima@ifi.unicamp.br*

David C. Haak, *Department of Biology, Box 351800, University of Washington, Seattle, WA 98195-1800, USA. haakd@u.washington.edu*

Britta Denise Hardesty, *University of Georgia, Department of Plant Biology, Athens, GA 30602 USA. Present address: CSIRO Sustainable Ecosystems, PO Box 780, Atherton 4883, Australia. Smithsonian Tropical Research Institute, Apartado 2072, Balboa, Ancón, Republic of Panama. Denise.Hardesty@csiro.au*

Kimberly M. Holbrook, *Department of Biology and International Center for Tropical Ecology, University of Missouri–St. Louis, One University Blvd, St Louis, MO 63121-4499, USA. kholbrook@umsl.edu*

Ido Izhaki, *Department of Biology, University of Haifa at Oranim, 36006 Tivon, Israel. izhaki@research.haifa.ac.il*

Patrick A. Jansen, *Community and Conservation Ecology, University of Groningen, PO Box 14, 9750 AA Haren, The Netherlands. p.a.jansen@rug.nl*

Pedro Jordano, *Integrative Ecology Group, Estación Biológica de Doñana, CSIC, Pabellón del Perú, Avda. M. Luisa S/N, E-41013, Sevilla, Spain. jordano@ebd.csic.es*

Shumpei Kitamura, *Department of Microbiology, Faculty of Science, Mahidol University, Rama 6 Road, Bangkok 10400, Thailand. kshumpei@wg8.so-net.ne.jp*

Charles Kwit, *Savannah River Ecology Laboratory, PO Drawer E, Aiken, SC 29802, USA. Present address: Department of Botany, Miami University, Oxford, OH 45056, USA. kwitc@muohio.edu*

Joanna E. Lambert, *Departments of Anthropology and Zoology, 5240 W.H. Sewell Building, University of Wisconsin–Madison, Madison, WI 53706, USA. jelambert@wisc.edu*

Douglas J. Levey, *Department of Biology, Box 351800, Univ. of Washington, Seattle, WA 98195-1800 USA. Present: Department of Zoology, PO 118525, University of Florida, Gainesville, FL 32611-8525, USA. dlevey@zoo.ufl.edu*

Bette A. Loiselle, *Department of Biology, University of Missouri–St Louis, One University Blvd, St Louis, MO 63121, USA. loiselle@umsl.edu*

Alon Lotan, *Department of Evolutionary and Environmental Biology, University of Haifa, Haifa 31905, Israel. alonlotan99@yahoo.com*

Susan J. Mazer, *Department of Ecology, Evolution and Marine Biology, University of California, Santa Barbara, CA 93106, USA. mazer@lifesci.ucsb.edu*

Amy B. McEuen, *Department of Biology, Wilkes University, Wilkes-Barre, PA 18766, USA. Present Address: Department of Biology, University of Illinois, Springfield, Springfield, Il, 62703, USA. mceuen.amy@uis.edu*

Juan M. Morales, *The Statistical Laboratory, Centre for Mathematical Sciences, Wilberforce Road, Cambridge CB3 0WB, UK. Present Address: Laboratorio ECOTONO, Universidad Nacional del Comahue, Quintral 1250, (8400). jmorales05@gmail.com*

Helene C. Muller-Landau, *Department of Ecology, Evolution and Behavior, University of Minnesota, 1987 Upper Buford Circle, Saint Paul, MN 55108, USA. hmuller@umn.edu*

Ran Nathan, *Movement Ecology Laboratory, Department of Evolution, Systematics and Ecology, The Hebrew University of Jerusalem, Givat Ram, Jerusalem 91904, Israel. rnathan@cc.huji.ac.il*

Jens M. Olesen, *Institute of Biological Sciences, University of Aarhus, Ny Munkegade Building 1540, DK-8000 Aarhus C, Denmark. jens.olesen@biology.au.dk*

Marco Aurélio Pizo, *Universidade do Vale do Rio dos Sinos (UNISINOS), Ciências da Saúde, Av. Unisinos, 950, 93022-900, São Leopoldo, RS, Brazil. mapizo@unisinos.br*

Matthew D. Potts, *Department of Biology and Center for Ecosystem Science and Policy, University of Miami, Coral Gables, FL 33124-0421, USA. potts@bio.miami.edu*

John R. Poulsen, *Department of Zoology, University of Florida, Gainesville, FL 32611, USA. jpoulsen@zoo.ufl.edu*

Alastair W. Robertson, *Ecology, Institute of Natural Resources, Massey University, Private Bag 11222, Palmerston North, New Zealand. A.W.Robertson@massey.ac.nz*

Javier Rodríguez-Pérez, *Institut Mediterrani d'Estudis Avançats (CSIC-UIB), C/Miquel Marqués 21, 07190 Esporles, Mallorca, Balearic Islands, Spain. jrodriguez@uib.es*

Sabrina E. Russo, *Center for Tropical Forest Science–Arnold Arboretum Asia Program, Harvard University, 22 Divinity Avenue, Cambridge, MA 02138, USA. Present address: School of Biological Sciences, University of Nebraska, Lincoln, Nebraska, USA. srusso2@unl.edu*

Thomas B. Ryder, *Department of Biology, University of Missouri–St Louis, One University Blvd, St Louis, MO 63121, USA. pipridae@umsl.edu*

H. Martin Schaefer, *Faculty of Biology, University of Freiburg, Department of Evolutionary Ecology, Hauptstr. 1, 79104 Freiburg, Germany. martin.schaefer@biologie.uni-freibur*

Veronika Schaefer, *Institute of Biology 1, Hauptstrasse 1, 79104 Freiburg, Germany. Faculty of Biology, University of Freiburg, Department of Evolutionary Ecology, Hauptstr. 1, 79104 Freiburg, Germany. veronika.schaefer@biologie.uni-freiburg.de*

Eugene W. Schupp, *Department of Wildland Resources, and The Ecology Center, Utah State University, Logan, UT 84322-5230, USA. schupp@cc.usu.edu*

Wesley R. Silva, *Departamento de Zoologia, I.B., Universidade Estadual de Campinas (UNICAMP), 13083-970, Campinas, SP, Brazil. wesley@unicamp.br*

Peter D. Smallwood, *Department of Biology, University of Richmond, Richmond, VA 23173, USA. psmallwo@richmond.edu*

Orr Spiegel, *Movement Ecology Laboratory, Department of Evolution, Systematics and Ecology, The Hebrew University of Jerusalem, Givat Ram, Jerusalem 91904, Israel. ors@pob.huji.ac.il*

Michael A. Steele, *Department of Biology, Wilkes University, Wilkes-Barre, PA 18766, USA. michael.steele@wilkes.edu*

Pablo R. Stevenson, *IDPAS, SUNY at Stony Brook, NY, USA. Present address: Dept. Ciencias Biológicas, Universidad de Los Andes, Cr. 1 No. 18A-10, Bogotá, Colombia. pstevens@uniandes.edu.co*

Sylvester Tan, *Sarawak Forestry Corporation, Kuching, Sarawak, Malaysia. sylvestertks@sarawakforestry.com*

William B. Terzaghi, *Department of Biology, Wilkes University, Wilkes-Barre, PA 18766, USA. terzaghi@wilkes.edu*

Joshua J. Tewksbury, *Department of Biology, Box 351800, University of Washington, Seattle, WA 98195-1800, USA. tewksjj@u.washington.edu*

Tad C. Theimer, *Department of Biological Sciences, Northern Arizona University, Flagstaff, AZ 86011, USA. Tad.Theimer@nau.edu*

Anna Traveset, *Institut Mediterrani d'Estudis Avançats (CSIC-UIB), C/Miquel Marqués 21, 07190 Esporles, Mallorca, Balearic Islands, Spain. atraveset@uib.es*

Ella Tsahar, *Department of Biology, Technion, Israel Institute of Technology, Haifa 32000, Israel. Elat@techunix.technion.ac.il*

Susan A. Turner, *Savannah River Ecology Laboratory, PO Drawer E, Aiken, SC 29802, USA. Present address: Department of Biological Sciences, Greenville Technical College, Greenville, SC 29605, USA. sturner@gmail.com*

Alfredo Valido, *Integrative Ecology Group, Estación Biológica de Doñana, CSIC, Avda. Mª Luisa s/n. Pabellón del Perú, 41013 Sevilla, Spain. avalido@ebd.csic.es*

David A. Westcott, *CSIRO Sustainable Ecosystems and the Rainforest CRC, PO Box 780, Atherton 4883, Australia. David.Westcott@csiro.au*

Brett T. Wolfe, *Department of Plant Biology, University of Illinois, Urbana, IL 61801, USA. btwolfe@gmail.com*

S. Joseph Wright, *Smithsonian Tropical Research Institute, Apartado 0843-03092, Balboa, Ancón, Republic of Panama. WRIGHTJ@si.edu*

Preface

Many researchers working in the field of seed dispersal have recognized that, periodically, the field has begun to lose cohesion and has faced the risk of fraying into a tangle of disconnected directions. This book is the fourth in a series of edited volumes that have appeared at long intervals and have marked the empirical and theoretical position of the field. Our aim, like that of the editors of previous volumes, has been to present recent advances, provide reviews of the state of knowledge and to chart future directions; to provide a reference point in a burgeoning literature. Like its predecessors (Estrada and Fleming, 1986; Fleming and Estrada, 1993; Levey *et al.*, 2002) it has grown from an international symposium/workshop on frugivores and seed dispersal. Most of the chapters in this book are derived from invited presentations given at the *Fourth International Symposium/Workshop on Frugivores and Seed Dispersal* held in Brisbane, Australia, in 2005. A few were invited additions, selected to complement the existing chapters to make a more thorough reference.

The study of seed dispersal and of the plants and animals involved has been developing, and maturing, at an impressive rate. While many concepts and methods used in the past have maintained their currency, many fresh approaches are providing new perspectives and directions. The combination has lead to a suite of exciting new questions, new answers to old questions, new approaches, and a reinvigoration of the field. And they point the way for greater progress in the future.

We are proud to be able to present a book that demonstrates the vigour and vibrancy of the field. Without losing touch with its theoretical and empirical past, seed dispersal research has in recent years undergone a renaissance and has surged forward. From a pure science perspective, these are exciting times. But this is also a time when it is becoming ever more important that our knowledge is applied in an effort to meet the growing environmental challenges that threaten natural systems. The chapters in this volume show that research into the ecology of seed dispersal is focusing on these issues and identifying and applying aspects of research to meet these challenges. Although there is much more to be done, this promises to be one of the growth areas in the field.

The previous three volumes (Estrada and Fleming, 1986; Fleming and Estrada, 1993; Levey *et al.*, 2002) have each represented landmarks in the field and have had enormous impact. We sincerely hope that this fourth volume will live up to the benchmark set by its predecessors and that it serves not only as a reference but also as a source of scientific inspiration and insight.

References

Estrada, A. and Fleming, T.H. (eds) (1986) *Frugivores and Seed Dispersal.* Dr. W. Junk, Dordrecht, The Netherlands.

Fleming, T.H. and Estrada, A. (eds) (1993) *Frugivory and Seed Dispersal: Ecological and Evolutionary Aspects.* Kluwer Academic, Dordrecht, The Netherlands.

Levey, D.J., Silva, W.R. and Galetti, M. (eds) (2002) *Seed Dispersal and Frugivory: Ecology, Evolution and Conservation*

Acknowledgements

We sincerely thank all the authors of chapters for their efforts, advice, diligence and patience. Thanks also to the numerous reviewers who assisted in vetting and improving the content and flow of the chapters; your efforts are very much appreciated. We would also like to thank those who assisted with the organization of the conference from which this book has grown: Pedro Jordano, Mauro Galetti, Doug Levey, Carla Catterall, Cath Moran, Jenny Marsden, Darren Green, Shannon Hogan, David Snow, William T. Cooper, Matt Bradford, Cassia Camillo, Norbert Cordeiro, Jennifer Cramer, Elaine Hooper, Palitha Jayasekara, Kara LeFevre, Adam McKeown, Steve McKenna, Wendy Neilan, Ruhyat Partasasmita, Soumya Prasad and Wangworn Sankamethawee. Many people helped in big and small ways, and should there be anyone we have forgotten to include here we beg your forgiveness. We gratefully acknowledge the financial support of Griffith University, CSIRO Sustainable Ecosystems, the Rainforest Cooperative Research Centre, Araucaria Ecotours, National Science Foundation (USA: OISE No. 0456402 to E.W.S.), Earthwatch Institute, the Ecology Center at Utah State University, Nokomis Editions and CSIRO Publishing. We thank William T. Cooper for the use of his artwork on the cover. A.J.D. would like to thank his wife, Diana, and three children, Mitchell, Jordan and Marni for their support and patience throughout the editing of this book, which was done almost exclusively in 'family time'. A.J.D. would also like to thank his parent, June and Paul, for travelling thousands of kilometres to look after the grandkids in the final stages. E.W.S. thanks his life partner, Janis Boettinger, for the continual support and inspiration that makes life, as well as work, wonderful and productive. D.A.W. would like to thank Frederieke Kroon for her support.

I Frugivores and Frugivory

ANDREW J. DENNIS

Introduction

The process of seed dispersal frequently begins with a fruit interacting with a frugivore, and this interaction is a fitting place to begin a book on seed dispersal. The interactions between animals, fruits and their seeds dominate plant reproduction in most habitats. Although we often think of tropical forests, even many desert plants produce fleshy fruit designed to attract animal dispersers, and these plants are frequently abundant species (see Bronstein *et al.*, Chapter 7). Of all the potential vectors for dispersing seeds, animals are the most susceptible to disturbance by humans (see Part IV), making it essential that the process of seed dispersal mediated by animals is understood to the fullest possible extent in our rapidly changing world.

The dawn of research into seed dispersal probably began with naturalists being fascinated by colourful fruits, sometimes flamboyant animals, and the interactions between the two. These interactions have caught the attention of people interested in a range of animal behaviours such as food choice, diet composition, nutrients, and foraging behaviour. However, more than most aspects of animal behaviour, the consumption of fruits and seeds by animals has consequences that can reach far beyond the immediate players. Many animals with non-fruit diets can have impacts on the demography of their food items or can influence community composition (e.g. Augustine and McNaughton, 1998) but few have such complicated and potentially far-reaching consequences as animals that eat fruit (see, e.g. Terborgh *et al.*, 2002).

The act of an animal eating a fruit can influence the survival, dormancy and germination of a seed and the subsequent probability of survival and growth rate for a seedling (see Traveset *et al.*, Chapter 4). The immediate influence an animal has over a seed is determined by how the seed is handled and the kind of animal doing the handling (e.g. Schupp, 1993). There are four key areas where a frugivore's handling can have significant impacts on the immediate outcomes for a seed. They include the removal of flesh, scarification of the seed coat, other materials deposited with the seed in the dung, and the number and combination of seeds deposited (see Traveset *et al.*, Chapter 4). Combined, these influences can result in a complex array of outcomes, all without considering the location of deposition, which has a whole new suite of influences (see Part III; Schupp, Chapter 21).

While an individual animal can influence an individual seed's fate, assemblages of animals can influence the recruitment of plants and their populations (see Part II). This influence extends to community composition and dynamics and to the evolution of plant traits at individual (see Levey *et al.*, Chapter 2; Schaefer and Schaefer, Chapter 3), regional and continental scales (see Forget *et al.*, Chapter 1; Donatti *et al.*, Chapter 5). Plants and animals have co-evolved in a number of ways. It is common for many species of animals to interact with a single plant (see Dennis and Westcott, Chapter 9), which means that selective pressures and the co-evolution of fruits and their consumers have been diffuse (e.g. Herrera, 1985). However, it is not just the combination of animals consuming a given fruit at any one time or place that influences its evolution, but a combination of past and present selective pressures, and those exerted in different locations (Forget *et al.*, Chapter 1; Donatti *et al.*, Chapter 5).

Despite the diffuse nature of selective pressures, frugivory and seed dispersal have strongly influenced the traits of fruits. For example, colour and bloom on fruits act as visual communication to mutualists and antagonists. Colours and blooms that contrast best with the background appear to have been selected so that fruits can attract those seed-dispersing animals that provide advantages to the plants (see Schaefer and Schaefer, Chapter 3). Clearly, such simple communication is not the full story, and a range of other more subtle messages remain to be fully explored (see Schaefer and Schaefer, Chapter 3). Among these are chemical messages and controls on animal behaviour, which frequently serve complex functions, trading off the benefits of attracting animal dispersers and of deterring pathogens and seed-eating animals (see Levey *et al.*, Chapter 2).

The results of selection can vary from place to place and in subtle and complex ways. For example, the traits of fruits selected by lizards differ considerably between locations. This may be due to different groups of lizards being dominant at different locations, despite lizards seeming to be a uniform group of dispersers throughout the world (see Valido and Olesen, Chapter 6). This pattern is also reflected in patterns of seed size and shape when analysed at a community level and compared between floras on different continents. In this case, differences in animal disperser assemblages are not always reflected, as expected, in the shapes and sizes of seeds (see Forget *et al.*, Chapter 1).

This diffuse co-evolution, brought about by the diversity of the animals potentially interacting with any one plant leads to complex interaction networks which can have high levels of resilience at community and individual levels (see, e.g. Donatti *et al.*, Chapter 5; Silva *et al.*, Chapter 26). This resilience can be brought about by redundancy in the system, where the loss of one or two dispersers or even whole functional groups of dispersers does not mean the total loss of service to a plant (see Loiselle *et al.*, Chapter 8). Ecological redundancy in the system further complicates our ability to fully understand and predict the outcomes of disruption to the process. In some instances, the ability to persist without dispersal or a range of biotic and abiotic dispersal strategies can mean survival until lost dispersers can be replaced by new ones in different locations or times (see Forget *et al.*, Chapter 1; Donatti *et al.*, Chapter 5). However, as this diversity of dispersal agents and ecological redundancy erodes through human impacts, the system's resilience may also decline (see Part IV).

Because of these complexities, the study of frugivores and the fruits they eat has quickly gone beyond the study of animals and their food resources into the true study of a mutualism – seed dispersal – and all of its consequences. While seed dispersal has developed a range of directions and theories, its foundations still rest on understanding the interaction between fruits and frugivores. This highlights the critical importance of a sound understanding of the natural history of these interactions. Without observing the patterns of food choice and the behaviours of animals after feeding, it will be impossible to navigate the complexities of this process and apply our understanding to managing the sustainability of the earth's ecosystems.

This first section includes some excellent chapters dealing with frugivores and frugivory. It does not attempt to cover all aspects of this broad field but instead provides some novel insights and comprehensive reviews, filling in some gaps and consolidating some lines of investigation. It provides new springboards for continuing research and understanding that can be applied in solutions to ecological problems.

References

Augustine, D.J. and McNaughton, S.J. (1998) Ungulate effects on the functional species composition of plant communities: herbivore selectivity and plant tolerance. *Journal of Wildlife Management* 62, 1165–1183.

Herrera, C.M. (1985) Determinants of plant–animal coevolution: the case of mutualistic dispersal of seeds by vertebrates. *Oikos* 44, 132–141.

Schupp, E.W. (1993) Quantity, quality and the effectiveness of seed dispersal by animals. *Vegetatio* 107/108, 15–29.

Terborgh, J., Pitman, N., Silman, M., Schichter, H. and Nunez, V.P. (2002) Maintenance of tree diversity in tropical forests. In: Levey, D.J., Silva, W.R. and Galetti, M. (eds) *Seed Dispersal and Frugivory: Ecology, Evolution and Conservation.* CAB International, Wallingford, UK, pp. 1–17.

1 Seed Allometry and Disperser Assemblages in Tropical Rainforests: A Comparison of Four Floras on Different Continents

P.-M. Forget, A.J. Dennis, S.J. Mazer, P.A. Jansen, S. Kitamura, J.E. Lambert and D.A. Westcott

Introduction

A wide range of independent factors can exert selection on plants, influencing the evolution of both fruit and seed size and shape (e.g. Herrera, 1992; Mazer and Wheelwright, 1993; Jordano, 1995a; Lord, 2004). Seed size comes under a range of selection pressures from plant traits such as growth form, seedling size, leaf area and shade tolerance. For example, species with large seeds are usually better adapted to the low-light conditions of mature tropical rainforests (Salisbury, 1974; Foster, 1986; Kelly, 1993; Westoby *et al.*, 1996; Moles *et al.*, 2005a,b). Fruit traits such as secondary compounds (see Levey *et al.*, Chapter 2, this volume) or colour (see Schaefer and Schaefer, Chapter 3, this volume), as well as frugivore traits such as body size, gape width and gut length (Charles-Dominique *et al.*, 1981) may influence the handling and treatment of fruits and seeds and therefore also exert selective pressure on their size (e.g. Jordano, 1995a,b; Rey *et al.*, 1997; Alcantara and Rey, 2003). A number of studies have examined fruit characteristics (Lord, 2004, and references therein) and the size and/or gape width of animals swallowing or regurgitating them (Gautier-Hion *et al.*, 1985; Kitamura *et al.*, 2002) but few have dealt with seed size and shape in relation to endozoochory. Theoretically, as with fruit characteristics, seed size and shape are likely to be selected and constrained by the size of the seed-dispersing animals that swallow them (see, e.g. Tiffney, 1984; Westoby *et al.*, 1996), as well as by the predators that handle them (e.g. Mack, 1998; Dalling and Harms, 1999; Moles and Drake, 1999).

Just as egg allometry is related to birds' body size (Huxley, 1932; Calder, 1979), seed allometry has probably been affected by selective pressures exerted by dispersers via the transportation and passage of seeds through the body (Lord, 2004). In particular, the diameter of a frugivore's intestinal tube places an upper limit on the size of seeds that it may successfully disperse. In general, the larger the disperser, the bigger the fruit or seed it can handle, swallow and disperse via gut transit (Gautier-Hion *et al.*, 1985). For example, independent of growth form, Peruvian plants dispersed predominantly by mammals have larger seeds than those dispersed mostly by birds, reflecting the greater average size of dispersing mammals compared with birds (Kelly, 1995).

Tropical plants tend to have larger fruits and seeds than those of temperate floras, and most are consumed and dispersed primarily by frugivorous or granivorous animals rather than by abiotic vectors (Howe, 1986). Tropical frugivores, ranging in size from tiny ants and dung beetles to massive tapirs, elephants and cassowaries, disperse plant diaspores, ranging in size from 0.1 g to several hundred grams in weight. Rainforest fruits are largest in Africa, where the largest known frugivores occur (Mack, 1993; but see Dubost, 1968, 1987). However, few comparative studies of seed size and allometry have been conducted on a large biogeographic scale. Assuming that differences in mean body size exist between Old and New World frugivores, Mack (1993) predicted corresponding differences between fruits and seeds of these regions and suggested that selection for large seed size should be less constrained in the Old World than in the New World. Indeed, he found evidence that fruits were larger in the Palaeotropics than in the Neotropics, but he could only assume that there was a corresponding difference in seeds. In general, as seed size increases, the number of frugivores likely to disperse them by endozoochory decreases due to the physical constraints associated with transit (e.g. Pratt and Stiles, 1985; Hamann and Curio, 1999; Githiru *et al.*, 2002). However, seeds can change shape as well as size. Lord (2004) found that New Zealand fruit became more ellipsoid (elongate) with increasing size, allowing passage through the gut of smaller frugivores than would be possible with spherical fruits.

If the passage of seeds is constrained by gut diameter, then the elongation ratio of a seed (the ratio of length to width), in combination with the relative sizes of seed and frugivore, will determine whether it is swallowed and defecated or regurgitated or handled without swallowing. Therefore, selection on seed size and elongation imposed by disperser size may be dissipated by disperser behaviour. For example, small frugivores such as scatter-hoarding rodents may sometimes effectively disperse very large seeds by exozoochory (*sensu* Lobova and Mori, 2005; see also Forget and Vander Wall, 2001); seeds that might otherwise be dispersed by endozoochorous megafauna, either extinct or extant (Janzen and Martin, 1982; see Donatti *et al.*, Chapter 5, this volume).

Current patterns of fruit and seed allometry may represent the product of past selection pressures, and the fruits and seeds of extant plants could reflect selection exerted by now extinct dispersers and seed predators (Janzen and Martin, 1982; Eriksson *et al.*, 2000b; see Donatti *et al.*, Chapter 5, this volume). In addition, trans-oceanic dispersal of often large-seeded species (Dick *et al.*, 2003; Pennington and Dick, 2004; de

Queiroz, 2005) could also have led to colonization of islands and isolated continents where the original co-evolved fauna or alternative endozoochorous seed dispersers are absent. Both scenarios would lead to inconsistency between the frequency distributions of the sizes and shapes of fruits and seeds of extant plants and that of the body mass of the frugivores that consume and disperse them (see Donatti *et al.*, Chapter 5, this volume).

Given the myriad selection pressures on seed size, both past and present, can we detect an effect of dispersers on seed size? Does present-day animal size and gut diameter reflect seed size and shape? Paraphrasing Traveset *et al.* (2004), who focused on fruit traits, if frugivores have exerted major selection pressures on seed characteristics in recent history, then frugivore communities that differ in body size in different regions should produce different arrays of seed traits, despite similarities in geographical origin and phylogenetic relatedness. Alternatively, as stated by Kelly (1995), under the exaptation hypothesis, are current seed sizes and shapes the outcome of selection by a now-extinct fauna, which nevertheless remain adapted for dispersal by current frugivores? One way to examine these questions is to compare the distribution of seed sizes and shapes in communities from different continents with different suites of dispersers and different evolutionary histories.

Previous comparisons of tropical fruits have considered geographically restricted floras (e.g. Gabon – Gautier-Hion *et al.*, 1985; Costa Rica – Mazer and Wheelwright, 1993; Thailand – Peru – Kelly, 1995; Kitamura *et al.*, 2002) or floras in two regions (e.g. the Neotropics and Palaeotropics – Mack, 1993; see Corlett and Primack, 2006). The few existing intercontinental comparisons consider temperate floras (e.g. Herrera, 1992; Lord *et al.*, 1995; Edwards and Westoby, 1997; Lord, 2004; Traveset *et al.*, 2004, and references therein) and tropical-to-subtropical forest comparisons (Lord *et al.*, 1997). A worldwide cross-continental comparison of seed size and shape has never been attempted. If disperser size and gut transit capabilities influence seed size and shape then we might predict that the characteristics of a disperser assemblage may influence the general trends in these characters for a particular flora. Given the very different evolutionary histories of the world's tropical rainforest regions (see Corlett and Primack, 2006) and rapid selection for seed size when dispersers are lacking (see Kelly *et al.*, 2004), one might expect that because of long isolation and the recent disappearance of megafauna (see Donatti *et al.*, Chapter 5, this volume), seeds should be smaller and more elongate in tropical America, Asia and Australia than in Africa.

The goals of this chapter are to investigate whether geographically separated plant communities inhabited by frugivores with distinct frequency distributions for body size differ with respect to mean seed size or mean seed elongation. This assumes that endozoochorous dispersal exerts the strongest pressure on seed size and shape. However, exozoochory may relax the limitations on seed size and shape that endozoochory imposes, weakening selection to favour smaller or more elongate seeds in large-seeded species and allowing smaller-bodied frugivores to disperse and select for large and round seeds that were initially adapted to endozoochory by larger now-extinct frugivores (Janzen and Martin, 1982). This could result in weaker patterns of difference between communities with large dispersers compared with communities with small dispersers. Despite this, we predict that a correspondence between seed size and shape and body size of dispersers, driven by constraints imposed by the passage of seeds through the gut, will be evident when comparing communities from different continents. This should result in:

1. Seeds being smaller and showing greater elongation on continents with smaller-sized dispersing faunas.
2. Plant communities occupied by large dispersers (e.g. megafauna) showing less evidence of constraints on seed diameter relative to those occupied by relatively small frugivores.
3. Guilds of dispersers (e.g. birds, bats, other mammals) differentially affecting selection on the set of seeds that they disperse, resulting in distinct seed size distributions among guilds on different continents.

To test these predictions, we compiled data on the shapes and sizes of seeds consumed and dispersed by frugivores in four neotropical and palaeotropical rainforests. We analysed the relationship between seed width and length, and we compared the mean size and shape of seeds across continents for whole communities and within groups of plants dispersed by particular guilds of animals.

Methods

Data compilation

We compiled a database on tropical plants with seeds known to be consumed and dispersed by frugivores from four rainforest floras on different continents: the Guianas (e.g. Guyane – also known as French Guiana; Suriname – formerly known as Dutch Guiana); Central Africa (e.g. Ivory Coast, Gabon, Cameroun, Uganda); Southern Thailand; and the wet tropics of Australia in Queensland (see Appendices 1 and 2). In some instances, we used data from reports in other countries when information was not available or complete for a region. For example, data for French Guiana included observations of spider monkeys from Suriname (van Roosmalen, 1985b) and Cuvier's toucans from Ecuador (K. Holbrook, Missouri, 2005, personal communication). Much of the data on seed size was obtained from botanical collections at four research stations: Arataye-Nouragues, French Guiana (Belbenoit *et al.*, 2001; Bongers *et al.*, 2001); Makokou, Gabon (Hladik *et al.*, 1987; A. Hladik, MNHN, Brunoy, 2005, personal communication); Khao Yai National Park, Thailand (Kitamura *et al.*, 2002); and the central Wet Tropics Bioregion, Australia (Dennis *et al.*, 2005; A.J. Dennis and D.A. Westcott, 2005, unpublished results). We also obtained seed size data from the literature on frugivore gut content, including primary literature (Authors, 1961–2005; Jansen-Jacobs *et al.*, 1985–2003; van Roosmalen, 1985a; Vivien and Faure, 1996a,b; Mori *et al.*, 2002), from a thesis (Sabatier, 1983), from the Internet (van Roosmalen and van Roosmalen-Blijenberg, 2003; Mori and Lobova, 2005), and from personal communications (see Acknowledgements). In most instances, the specimens measured were from species whose consumers were known (see references in Appendices 1 and 2). Consumption of a fruit or seed was not treated as necessarily indicating successful dispersal; 'consumption' can also result in seeds being dropped and/or killed, since many seed dispersers are also seed predators (e.g. scatter-hoarding rodents, primates with cheek pouches).

Whenever possible, we took measurements (length and width to the nearest millimetre) of at least 5–10 seeds per species (range 1–30). For each species (including morphotypes), we calculated the mean length and width of a seed. Length and width included the two longest dimensions in a two-dimensional space, which are the major constraints to oral handling and gut transit by consumers (Mazer and Wheelwright, 1993). The third dimension, thickness or depth of seeds, was not used because: (i) it is rarely mentioned in flora or studies; (ii) it often equals width; and (iii) it rarely (if ever) constrains oral handling or gut transit. Data were included only for those records that were identified to species level or included referenced morphospecies. This was particularly important for studies relying on gut content, where identification was often uncertain (e.g. Dubost, 1984; Erard *et al.*, 1989a; Henry *et al.*, 2000). Nomenclature was revised according to the International Plant Name Index (Anonymous, 2004). Our data set is not exhaustive, but instead is a sample of plant species that may be consumed by each animal inhabiting a given forest. Therefore, it is representative of the sizes of seeds harvested and potentially transported by animals.

We extracted two subsets of the original database. In the first, which we named the Plant Data Base (PDB), we pooled all the data from frugivore diets within each continent in order to compare rainforest floras. In the second, the Animal Data Base (ADB), we pooled all species according to their occurrence in the diets of members of different frugivore groups (e.g. volant birds, arboreal mammals, ruminants/ungulates). Therefore, in the ADB each individual species of plant may be listed more than once if it is dispersed by more than one group of frugivores. Most of our frugivore groups were based on high-level taxonomy (e.g. Order or higher; see Table 1.5). However, some groups were less taxonomically constrained. For example, our megafauna group includes the tapir (*Tapirus terrestris*; Tapiridae), cassowary (*Casuarius casuarius*; Casuaridae) and elephant (*Loxodonta africana*; Elephantidae), which we consider comparable, as they all swallow very large items and pass them intact. This differentiates the group from ruminants/ungulates, which either spit out or chew large seeds (e.g. Feer *et al.*, 2001). In addition, all arboreal mammals are pooled. This group includes, and is dominated by, scansorial and arboreal primates, but also includes carnivores and marsupials.

In this chapter we present a summary of seed traits across continents based on species in the diets of neo- and palaeotropical animals. These animals range in size from a weight of several grams (manakins) to several tonnes (elephant). Body mass data are from Gautier-Hion *et al.* (1999) and C. Erard (MNHN, 2005, personal communication) for Central Africa; Emmons and Feer (1990) and a few other sources (see Appendix 2) for the Guianas; Kitamura *et al.* (2002) for Thailand; and A.J. Dennis and D.A. Westcott (unpublished data), Strahan (1995); Baker *et al.* (1997) and Dennis (1997) for Australia. Values for body mass are the average for males and females. Mean weights for all species in our samples, and within guilds, were compared between locations using log-transformed values and ANOVA but excluding animals weighing >100 kg from the whole-community analysis. Animals of this size, including *Gorilla gorilla* (Hominidae), *Loxodonta africana*, *Tapirus terrestris*, *Cervus unicolor* (Cervidae) and *Elephas maximus*, were excluded due to their disproportionate impact on the means.

Analyses

Relationship between size and elongation in seeds

There are two ways to investigate the allometry of seed size across continents. The first approach is based on an ecological or 'static' perspective, examining the bivariate relationship between seed size and shape among extant species in which seeds are dispersed by animals. In this approach, each phenotype represents an independent data point, regardless of its relationship to other species in the data set. This approach asks the question: Do seed width and length increase proportionally or is there evidence that seed diameter is constrained relative to changes in seed length? For these 'cross-species' analyses, we regressed seed width against seed length using log-transformed data (to improve normality) and standardized, major-axis regressions, with the software utility SMATR (Warton *et al.*, 2006, http://www.bio.mq.edu.au/ecology/SMATR). When the resulting slope is <1.0 and the 95% confidence interval does not overlap with 1.0, the relationship is one of 'negative allometry', indicating that seed width increases more slowly than seed length. Negative allometry supports the hypothesis that large seeds have evolved greater elongation in response to constraints imposed by their dispersers.

The second approach is based on an evolutionary perspective and incorporates the phylogenetic relatedness of plants in a given community. It examines the correlated evolution of length and width among frugivore-dispersed seeds. This approach also asks: 'Do seed width and length increase proportionally?' but uses phylogenetically independent contrasts (PICs; Felsenstein, 1985; Harvey and Pagel, 1991) to examine this relationship without the complication of particular families or genera driving observed patterns. This approach necessarily results in fewer data points than the ecological or static approach because only cases in which closely related taxa have diverged will contribute a 'contrast' to the data set. We expect that if seed length and width evolve at the same rate, a 50% increase in length would be accompanied by a 50% increase in width: an isometric relationship. However, if gut diameter places an upper limit on the width of seeds that can be swallowed and dispersed, then we would expect that an increase in length would be accompanied by a proportionally smaller increase in width.

To use this approach, we compiled data by identifying evolutionary divergences between pairs of closely related taxa (measured as phenotypic differences). These divergences were identified after mapping trait values on a cladogram which represented the phylogenetic relationships between co-occurring species of plants. We used *Phylomatic* (Webb and Donoghue, 2005) to construct pruned cladograms for the angiosperm communities in each of our four study regions. The cladograms were pruned (or incomplete) because they only contained data on species that were known to occur in the diets of animals in each area. They represent relationships among the included taxa based on the dynamic, family-level phylogeny available online (Stevens, 2004), a hypothesized relationship between species. Given that the precise phylogenetic relationships among many species and genera within families are unknown, confamilial genera and congeneric species are often represented as polytomies on the resulting tree. Nevertheless, among many of the included taxa, phenotypic divergences between pairs of confamilial and/or congeneric taxa were identified.

We used the 'Analysis of Traits' module (version 3.0; Ackerly, 2004) in *Phylocom* version 3.22 (Webb *et al.*, 2004) to identify phylogenetically independent contrasts within each phylogenetic tree. The values of each independent contrast for seed length and width were calculated at each node where the pair of descendent taxa differed in either trait. The trait values of internal nodes were estimated using a weighting procedure that assumes that traits evolve by Brownian motion (Felsenstein, 1985). The treatment of polytomies is described by Ackerly (2004).

To maintain consistency in the bivariate relationship, we consistently subtracted the phenotypic value of the taxon or node with the shorter seed(s) from that with the longer seed(s), which yields contrasts for seed length that are always positive. The corresponding contrast for seed width could, in theory, be either positive or negative. Generally, though, where seed length increased between two taxa or nodes, seed width did as well.

Once divergences were identified and contrasts calculated, we regressed seed width against length using log-transformed data (to improve normality) and standardized, major-axis regressions (using SMATR) and fitting regression lines forced through the origin (Garland *et al.*, 1992; Warton *et al.*, 2006). As with the previous analysis, a slope significantly <1.0 suggests that evolutionary changes in seed width are constrained relative to those in seed length, and that larger seeds are more elongate than smaller ones.

Seed attributes across rainforests and frugivore guilds

Using ANOVAS, we tested whether the mean seed length and the width : length ratio (W:L estimates roundness) across all species differed between rainforests and frugivore guilds on different continents. Following Mazer and Wheelwright (1993) and Mack (1993), we used one dimension and one index to analyse size and shape of seeds harvested by animals. They were: (i) length (L; mm), which is an indication of seed size and is highly correlated with width (handling constraint); and (ii) the W:L ratio, a measure of roundness and gut transit constraints. Values of W:L range between 0 and 1; when width equals length, W:L is 1 and the seed is spherical. The greater the deviation of W:L from 1, the greater the elongation of a seed. Seed length was log-transformed and W:L ratios were square-root-transformed for ANOVA. Species were nested within genus and family for each continent (see Appendix 1).

Results

We compiled seed size and shape data (PDB) for 1558 species and morphospecies belonging to 650 genera and 119 families which were consumed by frugivores in four neo- and palaeotropical rainforests (Appendix 1). We also compiled records of plant species with seeds consumed by a total of 154 species in the four rainforests (Appendix 2).

Disperser sizes on different continents

The disperser assemblages sampled in this analysis differed between continents in the distributions of weights (Fig. 1.1). Australia had a much higher frequency of small-sized dispersers than the other locations. The Guianas had a more even distribution of body sizes up to about 100 kg but lacked very large species. Central Africa and Thailand were the most similar. Australia had a mean disperser weight statistically less than that of Central Africa (Fig. 1.2; $F_{(3, 143)} = 6.92$, $P < 0.001$, Tukey HSD) but a trend was also evident, with Central Africa > the Guianas ~ Thailand > Australia.

Birds sampled from Africa and Guiana were heavier than those from Thailand and Australia ($F_{(3, 71)} = 7.81$, $P < 0.001$). Bats were lighter in the Neotropics (Guiana) than in parts of the Palaeotropics (Africa and Australia; $F_{(2, 10)} = 7.57$, $P < 0.01$). Similarly, primates were lighter in the Neotropics than in parts of the Palaeotropics (Africa and Thailand), although this was only marginally significant ($F_{(2, 19)} = 3.42$, $P = 0.054$). Rodent and ruminant/ungulate groups did not differ between regions ($F_{(3, 19)} = 2.15$, $P = 0.13$ and $F_{(2, 7)} = 0.39$, $P = 0.69$, respectively). The largest dispersers were elephants in Africa and Thailand, tapir in the Guianas and cassowary in Australia.

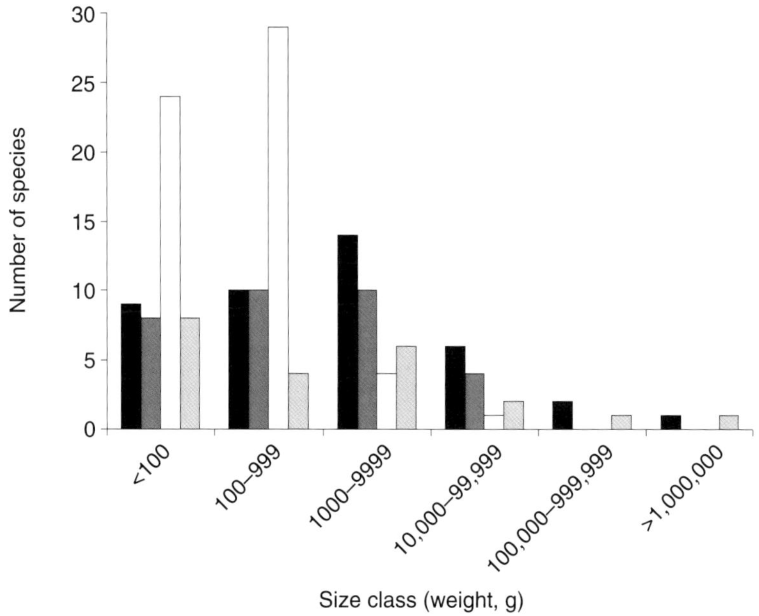

Fig. 1.1. The number of species of disperser known to consume fruits and seeds in different size classes (weight, g) in our databases for Central Africa (black bars), the Guianas (dark grey bars), Australia's wet tropics (white bars) and Thailand (light grey bars).

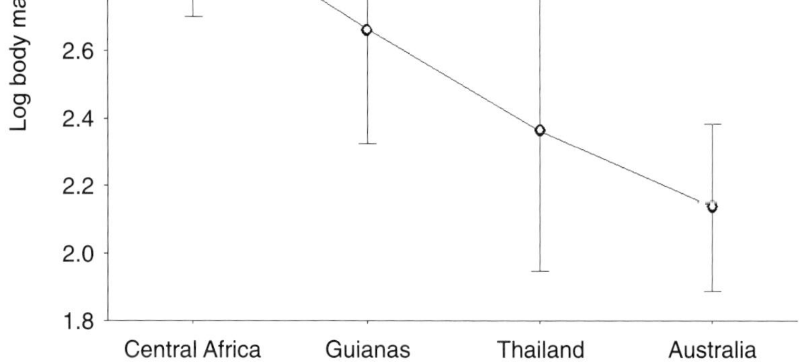

Fig. 1.2. Mean body mass (log-transformed) of disperser communities for our sample floras in Central Africa, the Guianas, Thailand and Australia, excluding species over 100 kg.

Relationship between size and elongation in seeds

The relationship between seed width and seed length differed among floras, and the differences depended on the analysis conducted. The cross-species analyses were more likely to detect evidence of negative allometry than were the PIC analyses (Table 1.1; Figs 1.3 and 1.4).

Among the cross-species analyses, the slope of the regression of log(seed width) on log(seed length) was significantly <1.0 in Central Africa and in Thailand (Fig. 1.3; Table 1.1): longer-seeded species were more elongate than shorter-seeded species. In contrast, seed width increased significantly more rapidly than seed length in the Guianan flora (the slope was significantly >1.0), meaning that large seeds were more spherical than small seeds. In Australia's wet tropical flora, width and length increased isometrically, so seeds retained a similar shape independent of size.

PIC analyses revealed similar patterns, although fewer results were significant. Large seeds were more elongate than small seeds only for species from Thailand (Table 1.1; Fig. 1.4; $s = 0.942$; 95% CI 0.890–0.997). In the three other regions, seed length increased either isometrically with width (Central Africa and Australia) or evolutionary increases in seed length were accompanied by disproportionately large increases in seed width (the Guianas).

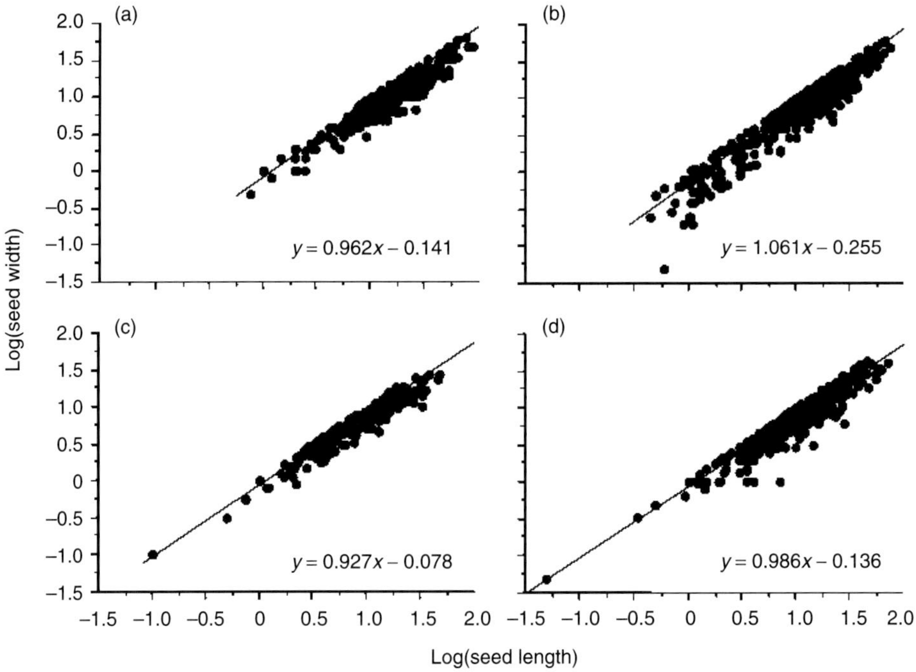

Fig. 1.3. Standardized major axis regressions of the cross-species analysis for log(seed length) versus log(seed width) in tropical rainforest seeds from (a) Central Africa, (b) the Guianas, (c) Thailand and (d) Australia.

Table 1.1. Relationship between seed width and seed length for species of plants dispersed by animals in four rainforests on different continents. Summary of the standardized major-axis regressions of the cross-species means (intercept unconstrained; see Methods) and the phylogenetically independent contrasts (PICs; intercept forced through origin) for log(seed width) versus log(seed length). Slopes with 95% confidence intervals that remain below 1.0 indicate that larger seeds are more elongate than smaller seeds. $P < 0.05$ indicates regressions in which the slope is significantly different from 1.0 and these have a slope summary in bold type.

Continent	n*	Slope	Slope summary	95% CI	R^2	Intercept (CI)	P value of $H_o : s = 1$
Cross-species							
Central Africa	304	0.962	**$s < 1$**	0.932–0.993	0.922	−0.141 (−0.177 – −0.104)	0.0167
Guianas	556	1.061	**$s > 1$**	1.035–1.088	0.913	−0.255 (−0.284 – −0.226)	< 0.0001
Thailand	250	0.927	**$s < 1$**	0.901–0.952	0.95	−0.078 (−0.102 – −0.054)	< 0.0001
Queensland	442	0.986	$s = 1$	0.959–1.012	0.918	−0.136 (−0.163 – −0.108)	0.2893
PICs							
Central Africa	122	1.001	$s = 1$	0.941–1.065	0.881	0	< 0.0001
Guianas	202	1.092	**$s > 1$**	1.032–1.155	0.836	0	0.0025
Thailand	122	0.942	**$s < 1$**	0.890–0.997	0.90	0	0.0416
Queensland	174	1.001	$s = 1$	0.946–1.059	0.86	0	0.9810

* Number of species for cross-species regression, number of contrasts for PIC-based regression.

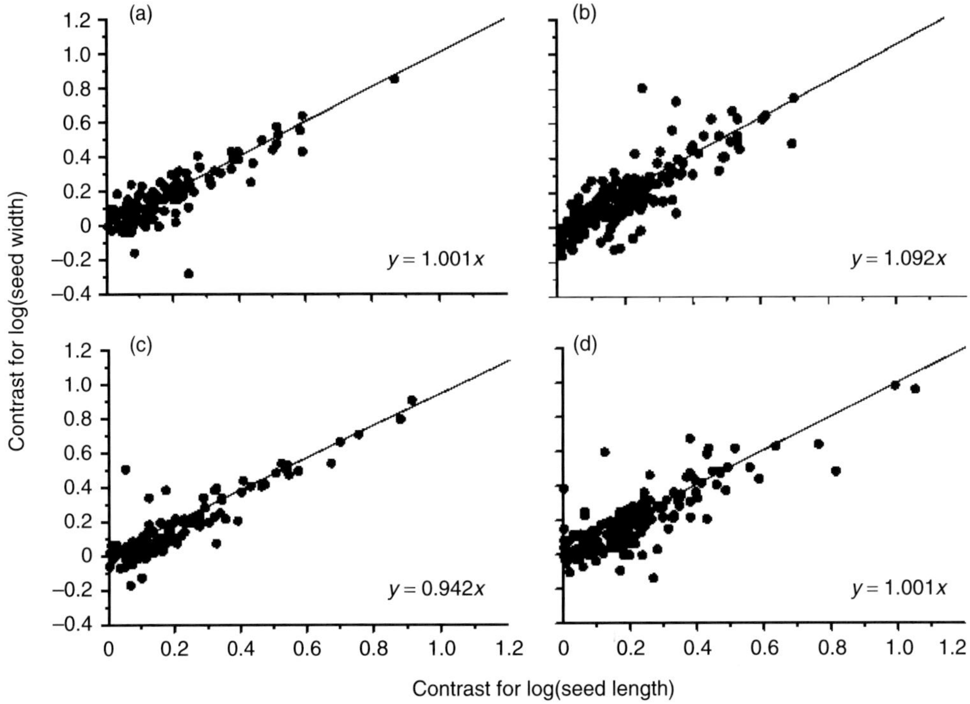

Fig. 1.4. Standardized major axis regressions for the phylogenetically independent contrasts (PICs) of log(seed length) versus log(seed width) in tropical rainforest seeds from (a) Central Africa, (b) the Guianas, (c) Thailand and (d) Australia.

Seed attributes across rainforests on different continents

Length of seeds consumed by frugivores differed significantly among rainforests ($F_{(3,\,1180)} = 22.114$, $P < 0.001$). Species in Central Africa had significantly longer seeds (mean 19 mm) than species in the Guianas (14.9 mm), Australia (12.8 mm) and Thailand (10.1 mm) (Fig. 1.5a; post hoc Bonferroni test: Central Africa > Guianas = Australia > Thailand; see grand mean for all families in Table 1.2). Seed roundness was also significantly different among continents ($F_{(3,\,1180)} = 15.302$, $P < 0.001$); species in Thailand and Australia had significantly rounder seeds than species in Central Africa and the Guianas (Fig. 1.5b; see grand mean in Table 1.2).

Following Mack (1993), we also compared seed traits across pantropical plant families (Table 1.2). Overall, eight of 12 families showed significant ($P < 0.05$) or marginally significant ($P < 0.1$) differences in seed size across rainforests and continents (Table 1.3). For instance, seeds in the families Annonaceae and Sapotaceae had significantly longer seeds in Central Africa than in the Guianas, and Arecaceae seeds were longer in Central Africa than in Australia. In contrast, Moraceae and Rubiaceae had

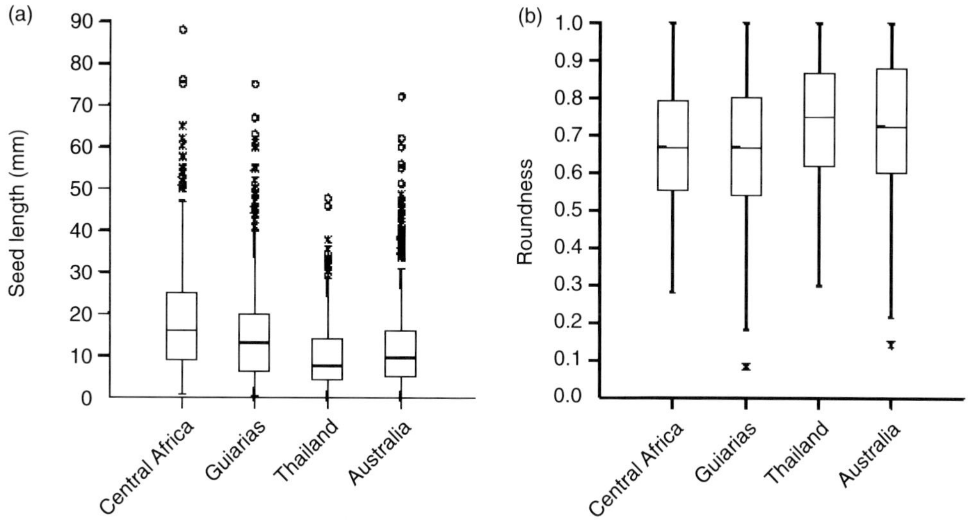

Fig. 1.5. Comparative length (a) and roundness (b) of seeds consumed and dispersed by frugivores in four rainforests from different continents. Box plot showing the median of the sample: each box shows the range within which the central 50% of the values fall, with the box edges at the first and third quartiles. Values between the inner and outer fences are plotted with asterisks, and outside values are plotted with empty circles.

longer seeds in the Guianas than in Central Africa, and Euphorbiaceae were smaller in Australia than in the three other continents. Anacardiaceae, Lauraceae, Meliaceae and Myristicaceae did not differ across continents (all $P > 0.05$ for both L and W:L). Clusiaceae had rounder seeds in Australia than in the Guianas ($P = 0.026$) but no other family differed in seed shape among continents (all W:L < 0.6).

Seed attributes across frugivore guilds

Bats

Fruits consumed by neotropical bats fall into two classes for which seed size and shape are not equally constrained: (i) those with seeds that are swallowed and defecated (endozoochory); or (ii) those carried away in the mouth (exozoochory). As an example, in the Guianas, endozoochorous seeds average 2 × 1.3 mm (L range 0.5–7 mm, $n = 71$), while exozoochorous seeds average 25.3 × 17 mm (L range 6.0–63 mm, $n = 83$). We found no difference in seed length or shape among any of the three continents tested for either of these two classes (no data available for Thailand; all $P > 0.452$; Table 1.4). We then pooled the data across classes and retested; again, neither size nor shape differed across continents (Table 1.5). The species dispersed by bats ranged from those with minute seeds such as *Clidemia octona* (Melastomataceae; 0.5 × 0.3 mm) and

Table 1.2. Seed dimensions and shape for species of plant from four rainforests on four continents for each of 12 pantropical families and the grand mean for all families (CA – Central Africa; GA – the Guianas; TH – Thailand; AU – Australia's wet tropics). Values with the same superscript letters indicate those values that were not significantly different in Bonferroni post hoc tests when P was significant ($P < 0.05$).

	Mean length (mm)					Average roundness (W:L)					No. spp.			
Plant family	CA	GA	TH	AU	*P*	CA	GU	TH	AU	*P*	CA	GU	TH	AU
Anacardiaceae	24.3	19.7	23.1	13.9	0.500	0.665	0.708	0.719	0.818	0.104	11	4	5	4
Annonaceae	17.1^{a}	12.1^{b}	15.1^{ab}	14.4^{b}	0.019	0.588	0.655	0.674	0.685	0.169	30	21	14	8
Arecacaeae	31.4^{a}	20.9^{a}	20.0^{ab}	13.0^{a}	< 0.001	0.760	0.781	0.704	0.794	0.967	6	16	3	13
Clusiaceae	27.1	13.8	25.1	22.2	0.378	0.627^{ab}	0.509^{a}	0.511^{ac}	0.784^{bc}	0.029	7	17	2	4
Euphorbiaceae	12.6^{a}	13.0^{a}	5.6^{b}	10.4^{a}	< 0.001	0.774	0.710	0.818	0.808	0.065	23	14	18	15
Lauraceae	35.8	17.0	14.7	22.2	0.273	0.610	0.739	0.668	0.745	0.257	3	12	15	53
Meliaceae	14.1	20.7	18.9	16.4	0.082	0.675	0.662	0.737	0.636	0.645	7	10	10	16
Moraceae	4.2	6.7^{b}	3.2^{a}	1.9^{a}	< 0.001	0.920	0.804	0.864	0.695	0.195	11	28	21	25
Myristicaceae	24.2	21.6	21.3	17.3	0.596	0.665	0.711	0.718	0.802	0.405	5	17	2	2
Rubiaceae	4.0^{a}	7.5^{b}	5.3^{a}	4.8^{a}	< 0.001	0.754	0.744	0.665	0.658	0.707	17	14	26	11
Sapindaceae	19.8^{a}	12.6^{ab}	14.8^{ab}	12.9^{b}	0.011	0.569^{a}	0.764^{b}	0.768^{b}	0.715^{b}	0.044	8	13	7	25
Sapotaceae	26.6^{a}	19.4^{b}	16.5^{ab}	20.7^{ab}	0.052	0.575	0.571	0.635	0.610	0.771	21	41	2	15
Grand mean	20.2^{a}	12.7^{b}	12.0^{c}	11.8^{b}	< 0.001	0.682^{a}	0.652^{a}	0.732^{b}	0.747^{b}	< 0.001	54	70	61	75

Table 1.3. A comparison of seed length and shape across rainforest floras on different continents with genera nested within family. Significant differences among continents are indicated with Tukey post hoc probabilities between brackets (CA – Central Africa; GU – Guianas; TH – Thailand; AU – Australia's wet tropics).

		Length (mm)			Roundness		
Family*	*df*	*F*	*P*	Observed trends	*F*	*P*	Observed trends
Anacardiaceae	3,8	0.859	0.500		2.864	0.104	
Annonaceae	3,31	3.835	0.019	CA > GU (0.032)	1.794	0.169	
Arecaceae[a]	3,12	12.821	< 0.001	CA = GU > AU (0.001)	0.085	0.967	
Clusiaceae[a]	2,13	1.051	0.378		4.134	0.029	GU < AU (0.026)
Euphorbiaceae[a]	3,42	13.231	< 0.001	CA = GU = AU > TH (0.01)	2.741	0.065	
Lauraceae[a,b]	2,65	1.326	0.273		1.387	0.257	
Meliaceae	3,25	2.502	0.082		0.562	0.645	
Moraceae[a]	3,59	12.997	< 0.001	GU > CA = TH = AU (0.04)	1.618	0.195	
Myristicaceae	3,16	0.647	0.596		1.031	0.405	
Rubiaceae[a]	3,28	16.882	< 0.001	GU > CA = AS = AU (< 0.001)	0.468	0.707	
Sapindaceae	3,21	3.032	0.052	CA > AU (0.042)	3.204	0.044	CA < GU (0.056)
Sapotaceae[a,b]	3,52	4.061	0.011	CA > GU (0.011)	0.376	0.771	

* Significant effect ($P < 0.05$) of Genus nested within Family and Continent for [a] Length and [b] Roundness.

Table 1.4. Seed size and shape for species of plant consumed by different frugivore guilds, for each of four rainforests on different continents, with seed treatment and body mass of the frugivores.

Group	Rainforest/ animal	Treatment*	Body mass (g)	Mean length (mm)	SD	Min.	Max.	Mean width (mm)	SD	Min.	Max.	Average roundness (W:L ratio)	No. spp.
Bats	Central Africa	C – Sw	250–300	18.2	19.4	1.0	88.0	13.1	14.0	0.8	65.0	0.720	37
	Guianas	C – Sw	15–70	13.2	14.4	0.5	63.0	8.8	10.0	0.2	40.0	0.668	155
	Australia	C – Sw	700	13.3	13.4	0.5	60.0	9.2	9.2	0.5	35.0	0.698	49
Volant birds	Central Africa	R – Sw	250–1,300	18.0	8.3	3.4	40.7	11.3	4.8	3.1	23.6	0.663	53
	Guianas	P – Sw	13–3,000	12.9	7.9	0.5	45.0	8.7	5.0	0.3	28.0	0.701	191
	Thailand	R – Sw	<50–3,000	9.1	7.5	0.1	35.6	6.1	4.4	0.1	20.3	0.744	151
	Australia	P – R – Sw	10–2,270	9.2	7.2	0.1	40.4	6.5	5.3	0.1	35.0	0.737	311
Arboreal mammals	Central Africa	P – Sw – Sp	1,100–175,000	18.7	14.2	0.8	88.0	12.0	9.9	0.5	60.8	0.673	222
	Guianas	P – Sw	300–10,500	14.9	9.4	0.5	67.0	9.5	6.4	0.2	60.0	0.652	326
	Thailand	P – Sw – Sp	4,000–9,000	12.8	9.1	0.1	47.8	8.2	5.2	0.1	27.7	0.715	86
Rodents	Central Africa	Ca – P	20–1,000	23.8	15.6	1.0	76.2	16.0	11.5	1.0	60.8	0.686	111
	Guianas	Ca – P	35–3,500	24.7	15.5	1.3	75.0	16.9	11.8	0.5	60.0	0.684	124
	Thailand	Ca – P	300–1,400	13.9	10.2	0.1	47.8	9.0	6.1	0.1	27.7	0.718	103
	Australia	Ca – P	80–650	22.3	13.6	1.5	56.0	16.8	11.4	1.0	41.9	0.734	56
Terrestrial marsupials	Australia	C – Ca – P – Sw	520–4,500	26.1	14.6	2.1	62.1	20.1	11.6	1.0	42.3	0.772	56
Ruminants/ ungulates	Central Africa	P – Sp – Sw	4,900–68,000	21.8	14.1	1.0	76.2	14.5	10.1	1.0	51.4	0.679	123
	Guianas	P – Sp – Sw	16,000–44,000	16.7	9.7	1.3	45.5	11.0	6.3	0.5	35.0	0.675	91
	Thailand	P – Sp – Sw	24,000–222,500	17.5	10.6	0.1	47.8	10.6	6.1	0.1	27.7	0.658	46
Megafauna	African elephant	P – Sw	5,000,000	26.6	17.4	0.8	88.0	17.2	11.7	0.5	51.4	0.667	81
	Brazilian tapir	Sw	250,000	18.3	13.7	1.5	55.0	12.1	9.5	0.9	45.0	0.668	30
	Asian elephant	Sw	4,000,000	17.9	11.2	1.0	30.2	11.3	7.4	1.0	22.9	0.673	9
	Cassowary	Sw	60,000	16.3	13.1	0.1	72.2	11.6	9.5	0.1	41.7	0.732	228

*Seed treatment: C – carried; Ca – cached; P – predated; R – regurgitated; Sw –swallowed; Sp – spat.

Table 1.5. Results of ANOVA showing differences in seed length and shape among rainforests on different continents for species of plants consumed by different groups of animals (species nested within genus and family). Significant differences are indicated in parentheses with probabilities from Tukey post hoc tests; CA – Central Africa; GU – Guianas; TH – Thailand; AU – Australia's wet tropics.

		Length (mm)			Roundness		
Guild	*df*	*F*	*P*	Contrast	*F*	*P*	Contrast
Bats	2,213	1.828	0.163		1.475	0.231	
Birds	3,541	22.923	0.000	CA > GU > TH = AU (< 0.004)	4.901	0.02	CA = GU < TH =AU (0.045)
Arboreal mammals	2,441	4.634	0.010	CA > TH (0.007)	2.664	0.071	
Rodents	3,293	16.707	< 0.001	CA = GU = AU > TH (< 0.001)	2.241	0.084	
Ruminants/ungulates	2,180	1.488	0.233		0.234	0.791	
Megafauna	2,345	10.032	< 0.001	CA > AU (< 0.001)	5.136	0.005	CA < AU (0.013)

Pothomorphe peltata (Piperaceae; 0.5 × 0.5 mm) up to large stones such as *Dipteryx odorata* (Fabaceae; 60 × 35 mm) and *Caryocar villosum* (Caryocaraceae; 55 × 39 mm) in the Guianas, and *Balanites wilsoniana* (Zygophyllaceae; 88 × 47 mm) in Central Africa. Shape ranged from seeds as round as *Andira coriacea* (Fabaceae; W:L ratio = 1.0) to elongated ones such as those of *Dipteryx* spp. (Fabaceae; W:L ratio = 0.57).

Volant birds

In contrast to bats, and despite comparable body mass range (Table 1.4), both the length and shape of seeds consumed by birds differed significantly across continents (Table 1.5). Species consumed by birds had larger seeds in Central Africa than in the Guianas, and these were in turn larger than those of Thailand and Australia (Table 1.5). Seeds in Central Africa and the Guianas were more slender than those consumed by birds in Australia and Thailand. In Central Africa, the largest seed was dispersed by hornbills (*Annonidium mannii*, Annonaceae; 41 × 24 mm); in Ecuador by toucans, *Rhamphastos cuvieri* (*Iryanthera grandis*, Myristicaceae; 45 × 28 mm). In Thailand and Australia, *Canarium euphyllum* (Burseraceae; 36 × 17 mm) and *Endiandra sideroxylon* (Lauraceae; 40 × 22 mm), respectively, had the largest dimensions for seeds dispersed by volant birds. A seed length of approx. 40 mm appears to reflect an upper threshold size limit for dispersal by volant birds.

Arboreal mammals

Species of plants with fruits consumed by this group had larger seeds in Central Africa (18.7 mm in length) than in the Guianas (14.9 mm) and Thailand (12.8 mm), although only the difference between Thailand and Central Africa was significant ($P = 0.007$; Table 1.5). Seeds from fruit consumed by arboreal mammals in Thailand were marginally rounder than those in Central Africa and the Guianas ($P = 0.07$; Table 1.5). It appears that even the largest scansorial frugivores, such as *Pan paniscus* (Hominidae) and *Gorilla gorilla*, do not swallow large seeds (> 30–40 mm). Instead, they extract the pulp and reject the seeds of fruits with large seeds or nuts such as *Baillonella toxisperma* (Sapotaceae; 58 × 34 mm), *Panda oleosa* (Pandaceae; 54 × 38 mm), *Diospyros manii* (Ebenaceae; 52 × 23 mm) and *Parinari holstii* (Chrysobalanaceae; 52 × 42 mm). The largest-seeded species dispersed via endozoochory and defecated by arboreal mammals were *Platonia insignis* (Clusiaceae; 42 × 24 mm), *Gnetum* spp. (Gnetaceae; 40 × 20 mm) and *Dussia discolor* (Fabaceae; 36 × 21 mm) in the Guianas. In contrast, large seeds of *Elaeagnus latifolia* (Elaeagnaceae; 33 × 10 mm) and *Beilschmiedia maingayi* (Lauraceae; 32 × 14 mm) in Thailand and a wide range of species with seed length 20–42 mm in Central Africa are spat out after processing. The contrast in seed shape and size across neo- and palaeotropical primates will be further analysed in another study (P.-M. Forget *et al.*, 2006, unpublished results).

Rodents

Large rodents are well documented as seed dispersers in the Guianas and Australia (Forget and Vander Wall, 2001; Theimer, 2001), but little is known about the role of large rodents as seed dispersers in Central Africa (but see Ewer, 1965; Emmons, 1980) and Thailand (but see Yasuda *et al.*, 2000; Kitamura *et al.*, 2004). Because rodents hoard the same species they consume, we consider them seed dispersers despite the frequent high cost of seed loss to the plants. Seeds taken by rodents appeared to be smaller in Thailand (L = 14 mm) than in the three other continents (L > 22 mm; $P < 0.001$), which may be an artefact of the smaller mean seed size in Thailand. In Central Africa, the range of seed sizes harvested by rodents overlapped with that of seeds handled by primates, fruit bats and elephants (Tables 1.4 and 1.5). Seeds taken by rodents in Central Africa and the Guianas were very similar in shape but marginally more elongate than those in Australia and Thailand ($P = 0.08$; Table 1.5). In the Guianas, dasyproctid rodents are dispersers of large seeds such as *Parinari montana* (75 × 48 mm), *Couepia bracteosa* (Chrysobalanaceae; 63 × 40 mm), *Dipteryx odorata* (60 × 35 mm), *Attalea maripa* (Arecaceae; 55 × 28 mm) and *Caryocar villosum* (Caryocaraceae; 55 × 39 mm). In Central Africa, a large terrestrial squirrel *Epixerus ebii* (Sciuridae; Emmons, 1980) and the nocturnal rodent *Cricetomys eminii* (Muridae; Debroux, 1988) may disperse seeds of comparable size, such as *Baillonella toxisperma*, *Panda oleosa*, *Parinari holstii* (52 × 42 mm) and *Irvingia grandifolia* (Irvingiaceae; 45 × 22 mm; Emmons, 1982; Debroux, 1998). In Australia, *Endiandra xanthocarpa* (56 × 29 mm), *Beilschmiedia bancroftii* (49 × 42 mm) and *Syzygium gustavioides* (Myrtaceae; 47 × 42 mm) are among the largest-seeded species scatter-hoarded by rodents such as *Uromys caudimaculatus* (Muridae). In Thailand, the largest seeds taken by rodents overlap with those dispersed by primates.

Ground-dwelling marsupials

Overall, seeds swallowed or carried by ground-dwelling marsupials in Australia were large (mean = 26 × 20 mm) and round (Table 1.4). Data for this group were dominated by a single species, the musky rat-kangaroo, *Hypsiprymnodon moschatus* (Hypsiprymnodontidae), which carries and disperses relatively large seeds (mean = 22.6 × 18.5 mm; $n = 48$). These animals scatter-hoard fruits and seeds and play a comparable ecological role to that of large rodents in dispersing large-seeded species that sometimes lack alternative dispersal modes, although usually with less seed consumption than large rodents (Dennis, 2003).

Ruminants/Ungulates

Seeds consumed by ruminants in the Palaeotropics and by ungulates in the Neotropics usually do not survive passage of the digestive tract. However, seeds with hard integuments may be spat out and dispersed. For this guild of consumers, there was no significant difference in seed allometry across three rainforests (Tables 1.4 and 1.5).

Modern megafauna

Finally, we independently considered the three largest ground-dwelling frugivores: the tapir, *T. terrestris*, in the Guianas (insufficient data for Thailand), the elephant, *L. africana*, in Central Africa (insufficient data for Thailand, but see Williams, 1978), and the cassowary, *C. casuarius*, in Australia. Along with the rhinoceros (Dinerstein and Wemmer, 1988; Dinerstein, 1991), these are the only species likely to play a role as megafauna seed dispersers (*sensu* Janzen and Martin, 1982). Seeds dispersed by *L. africana* were larger and more slender than seeds dispersed by *T. terrestris* or *C. casuarius* (Tables 1.4 and 1.5; Fig. 1.6). In Central Africa, the largest elephant-dispersed seeds were *Balanites wilsoniana*, *Tieghemella africana* (Sapotaceae; 65 × 35 mm), *Detarium macrocarpum*, *Baillonella toxisperma*, *Maranthes glabra* (Chrysobalanaceae; 55 × 33 mm), *Panda oleosa* and *Raphia leptobothrys* (Arecaceae; 54 × 34 mm), some of which were also listed as dispersed by bats, primates and rodents. Similarly, many seeds dispersed by *T. terrestris* in the Guianas, including large palm nuts such as *Attalea maripa* (55 × 28 mm), *Mauritia flexuosa* (Arecaceae; 45 × 45 mm), *Astrocaryum paramaca* (Arecaceae; 33 × 19 mm) and *Jessenia bataua* (Arecaceae; 32 × 20 mm), were also

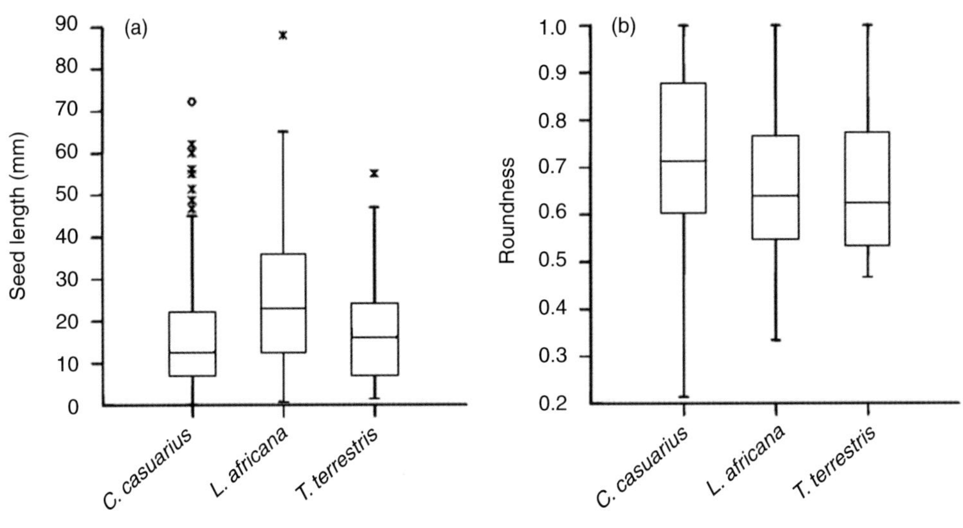

Fig. 1.6. Comparative length (a) and roundness (b) of seeds consumed and dispersed by three megafauna species (the cassowary, *C. casuarius*, in Australia; the elephant, *L. africana*, in Africa; the tapir, *T. terrestris*, in the Guianas) in three rain forests on different continents. Box plot showing the median of the sample: each box shows the range within which the central 50% of the values fall, with the box edges at the first and third quartiles. Values between the inner and outer fences are plotted with asterisks, and outside values are plotted with empty circles.

dispersed by scatter-hoarding Dasyproctid rodents. In Australia, species dispersed by *C. casuarius*, such as *Endiandra microneura* (62 × 33 mm) and *Elaeocarpus stellaris* (Elaeocarpaceae; 51 × 38 mm), on the one hand, and *Beilschmeidia bancroftii* (49 × 42 mm) on the other, are also scatter-hoarded by *H. moschatus* and *U. caudimaculatus*, respectively, while *Terminalia catappa* (Combretaceae; 60 × 35 mm) appears to be dispersed by both *C. casuarius* and the flying fox, *Pteropus* spp.

Discussion

The flora with the largest mean seed size, that of Central Africa, exhibited negative allometry, suggesting that the larger-seeded species were constrained with respect to their diameters (cross-species analysis; see Table 1.1; Fig. 1.1). This is surprising given the availability of megafauna in Africa, as well as the large size of its bats, primates and ruminants. However, while large seeds are more elongate today, there was no evidence that selective pressures created elongate seeds; the PIC analysis demonstrated that seed shape remained similar between small and large seeds within taxonomic groups (Table 1.1; Fig. 1.4). This suggests that taxa with elongate seeds encountered no constraints as larger seed sizes were selected for.

The sampled flora of southern Thailand, on the other hand, exhibited negative allometry in both static and evolutionary contexts (Table 1.1; Figs 1.3 and 1.4), suggesting strong constraints on seed shape both present and past. In addition, Thailand had a relatively small mean seed size despite the availability of large volant birds, primates and, potentially, elephants and rhinoceros as seed dispersers (Table 1.4). However, volant birds, primates and rodents in Thailand were all observed to consume relatively small-seeded species, so collectively they may place an upper limit on the size of seeds likely to be successfully dispersed. Together, these results suggest selective pressures limiting the enlargement of seeds and restricting increases in seed size to a single dimension. The strength of the pressure necessary to create this effect depends on the concurrent pressures selecting for large seed size. Conditions that reduced selection for large seeds could allow a weak selective pressure for small or elongate seeds to have an impact.

In contrast to Thailand, the Guianas demonstrated positive allometry in both analyses (Table 1.1; Figs 1.3 and 1.4); seed width increased more than length as seeds became larger, and large seeds were relatively rounder than small ones. This occurred even though seeds in general were more elongate in the Guianas than in Thailand. This result suggests that large seeds in the Guianas do not rely on passage through a gut for dispersal, or if they do, the animals in question impose less constraints on seed width than animals that eat smaller fruit do.

In Australia's wet tropics, the relationship between seed length and width remained isometric for both analyses, suggesting little or no selective constraints on seed shape despite Australia having the smallest dispersers. While mean seed size was smaller than in Africa and South America, Australian seeds were rounder. This may be due to large seeds being carried rather than swallowed and passed through the gut. Alternatively, when swallowed and dispersed, the disperser did not or does not constrain the size of the seeds. It may be worth mentioning that rodents and marsupials in Australia and rodents in the Guianas handle quite large-seeded species (comparable to those handled by rodents in Central Africa), potentially releasing them from selection based on gut-width limitations.

The lack of concordance between seed characteristics and frugivore sizes on different continents is not easily explained. We initially predicted that as the size of animals in disperser assemblages decreased, fruit size should also decrease (see below), and that large fruit should be more elongate and show evidence of selection for elongation. This was not clearly demonstrated in our results, suggesting that selective pressures exerted by disperser communities cannot be analysed at the broad level attempted here, or that seeds may not respond to selection in the manners predicted. The optimum fruit shape and size may differ among floras and it may be affected by the relative abundances of different frugivore guilds, an issue that we were unable to examine with the current data. One productive way of evaluating this hypothesis more carefully would be to examine cross-species and evolutionary relationships between seed width and length within each frugivore group and continent. In this way, one may be able to detect differences among floras in the pattern or strength of selection imposed by individual guilds, which may be obscured when all of the guilds are pooled.

Contrast in seed size and shape and frugivore guilds across four rainforests

In this comparison of floras on different continents, we found that seeds were larger (longer) in Central African rainforests than in other rainforests (Table 1.2). Thus, we confirm the previous assumption by Mack (1993) that seed size is greatest in Africa, the continent with the largest frugivores. Although this was the overall trend, these differences were not demonstrated by all pantropical families (Tables 1.2 and 1.3). In addition, the trend across the four floras was not a reflection of the largest frugivores but was instead a reflection of the community mean excluding the largest species (Fig. 1.2; Table 1.2). Seed shape was relatively comparable across continents, although seeds were more elongate in Central Africa and the Guianas than in Thailand or Australia (Fig. 1.5).

Among birds and megafauna, we found correspondence between the size of seeds and the size of dispersers across the four rainforests (see Results and Table 1.5) However, we also found that small frugivores such as bats and scatter-hoarding rodents or marsupials dispersed seeds as large as those dispersed by megafauna (Table 1.4), regardless of whether megafauna were present in a community. Thus, we conclude there is no clear relationship between the size of the seeds and the size of the frugivores dispersing them.

We assume that after the break-up of the continents, the floras of different places experienced different selective pressures based, in part, on differing disperser assemblages on each continent. Selective pressures have continued to change with the recent extinction (i.e. 10,000–40,000 years ago) of a variety of megafauna in many locations (see, e.g. Donatti *et al.*, Chapter 5, this volume). This may have led to the disappearance of some plants, but others are now dispersed by smaller animals that filled the gap or continued to disperse the same seeds after megafauna declined. The use of new tools has shown that radiation of plants throughout the tropics might be more recent than previously thought; many species may have recently colonized new continents after their seeds floated long distances across oceans (de Queiroz, 2005). Therefore, fruit and seed attributes that might have evolved in one rainforest, may be pre-adapted or exapted to other seed dispersers in another rainforest lacking the original fauna.

Together, the historical changes in dispersing animals within an area and the spread of plants to new areas creates mechanisms for the disjunction between seed attributes and their dispersing fauna. We have collated a large number of examples of seeds that appear primarily adapted to endozoochory (elongate) but are also dispersed through exozoochory by small animals, as well as fruit that are large (but not elongate) that are dispersed both by large and small animals (Table 1.6). For example, on Polillo Island in the Philippines, the gigantic lizards, *Varanus olivaceus* (up to 180 cm, 10 kg; Varanidae) swallow and disperse the large seeds of *Canarium ovatum* (54 × 21; Burseraceae) (Bennett, 2005; D. Bennett, Brisbane, 2005, personal communication), which resemble palm nuts. In the Neotropics, the large *Attalea* palm nuts, *Canarium*-like seed, are primarily dispersed by tapirs (Fragoso, 1997; Fragoso *et al.*, 2003), secondarily dispersed by agoutis *Dasyprocta leporina* and depredated by various vertebrates (Forget *et al.*, 1994, Silvius and Fragoso, 2003). In Central Africa, *Raphia leptobothrys*, a species related to *Attalea* (Pennington and Dick, 2004), is dispersed by elephant and the giant rat *Cricetomys emini* (Debroux, 1988) and possibly by terrestrial squirrels that scatter-hoard (Ewer, 1965). In New Guinea, large-seeded *Meliaceae* (*Aglaia* spp.) are dispersed by cassowary and occasionally by large fruit bats (Mack, 1995). In Thailand, seeds in another related *Aglaia* species are dispersed by hornbills and cached by scatter-hoarding rodents (Kitamura *et al.*, 2004; see also Becker and Wong, 1985). These parallel observations from different rainforests suggest the possibility that many of these seeds might have once been dispersed by giant reptiles (today represented by giant *Varanus*) now extinct, and that large size in seeds (i.e. > 25–35 mm L) and elongated shape is exapted to their current dispersal agents.

There is no broad correspondence between the size of seeds and the body mass of current dispersers, which may be because the same (or similar) seeds might be ingested whole and defecated (e.g. *Varanus*, tapir, elephant), regurgitated (hornbill), spat out (primates), scatter-hoarded

Table 1.6. Examples of large-seeded plants from pantropical families dispersed by endozoochores and/or exozoochores in four rainforests on different continents (CA – Central Africa; GU – Guianas; S-EA – South-east Asia; AU – Australia's wet tropics).

	Species	Family	L × W (mm) or weight (g)	Endozoochores (swallower, spitter)	Exozoochores (carrier)	References
CA	*Beilschmiedia mannii*	Lauraceae	50 × 33	*Pan troglodytes*, *Cercopithecus cephus*	?	Gautier-Hion *et al*., 1985
	B. obscura	Lauraceae	33 × 15	*Ceratogymna atrata*, *C. cylindricus*	?	Gautier-Hion *et al*., 1985
	Canarium sweinfurthii	Burseraceae	35 × 16	*Loxodonta africana*	?	Alexandre, 1978
	Carapa spp.	Meliaceae	30 × 25	*L. africana*	*Cricetomys* spp.	Debroux, 1988
	Dacryodes buttnueri	Burseraceae	34 × 15	Hornbills, Primates	?	Gautier-Hion *et al*., 1985; Ebigbo, 2005
	Parinari curatellifolia	Chrysobalanaceae	35 × 25	*L. africana* ?	*Eidolon helvum*	Ebigbo, 2005
	P. holstii	Chrysobalanaceae	52 × 42	*L. africana*, Primates	*E. helvum*	Gautier-Hion *et al*., 1985
	Raphia leptobothrys	Arecaceae	54 × 34	*L. africana*	?	Alexandre, 1978
	Symphonia globulifera	Clusiaceae	25 × 12	Hornbills, Primates	?	Gautier-Hion *et al*., 1985
	Spondias mombin	Anacardiaceae	32 × 20	? unknown	*Eidolon helvum*	Okon, 1974
GU	*Attalea maripa*	Arecaceae	55 × 28	*T. terrestris*	Dasyproctidae	Forget *et al*., 1994; Fragoso, 1997; Fragoso *et al*., 2003; Silvius and Fragoso, 2003
	Carapa spp.	Meliaceae	30–35 × 20–25	? extinct	Dasyproctidae	Forget, 1996; P.-M. Forget and P.A. Jansen, personal observation
	Mauritia flexuosa	Arecaceae	45 × 45	*T. terrestris*	Dasyproctidae	van Roosmalen, 1985a
	Parinari campestris	Chrysobalanaceae	40 × 26	? extinct	*Artibeus* spp.	Charles-Dominique, 1986; van Roosmalen, 1985a
	P. exelsa	Chrysobalanaceae	32 × 21	? extinct	*Artibeus* spp.	Charles-Dominique, 1986; cf. van Roosmalen, 1985a
	P. montana	Chrysobalanaceae	75 × 48	? extinct	Dasyproctidae	P.-M. Forget, personal observation
	Symphonia globulifera	Clusiaceae	20 × 14	*T. terrestris*	*Rhynophylla pumilio*, *Artibeus* spp.	Charles-Dominique, 1986; Henry *et al*., 2000
	Spondias mombin	Anacardiaceae	32 × 20	*T. terrestris, Mazama* spp., *Ateles paniscus*, *Alouatta seniculus*	Bats	Guillotin *et al*., 1994; Henry *et al*., 2000; Janzen, 1985; van Roosmalen, 1985b

S-EA	*Aglaia harmsiana*	Meliaceae	17.9 diam.	*Varanus olivaceus*		Auffenberg, 1988
	A. spectabilis	Meliaceae	30–46 × 17–28	Hornbills	*Maxomys surifer*	Kitamura *et al.*, 2004
	Aglaia sp.	Meliaceae	25–45 × 15–27	*Anthracoceros malayanus*	*Callosciurus prevostii*	Becker and Wong, 1985
	Beilschmiedia maingayi	Lauraceae	32 × 14	*Ducula badia*, Hornbills, *Hylobates* spp., *Macaca nemestrina*	Rodents	Kitamura *et al.*, 2002
	Canarium hirsutum	Burseraceae	26 × 20	*V. olivaceus*	?	Auffenberg, 1988; D. Bennett, personal communication
	C. indicum	Burseraceae	30–60 × 20–30	*Cynopterus sphinx*	?	Docters van Leeuwen, 1935
	C. ovatum	Burseraceae	54 × 21	*V. olivaceus*	?	Auffenberg, 1988; D. Bennett, personal communication
	C. euphyllum	Burseraceae	36 × 17	Hornbills, *Ducula badia*, *Cervus unicolor*, *Muntiacus muntjak*	*Menetes berdmorei*	Kitamura *et al.*, 2006
	Spondias spinnata	Anacardiaceae	25.8 diam.	*Varanus olivaceus*	?	Auffenberg, 1988
AU	*Beilschmiedia bancroftii*	Lauraceae	49 × 42	*Casuarius casuarius*	*Hypsiprimnodon moschatus*, *Uromys caudimaculatus*	Theimer, 2001
	Entiandra microneura	Lauraceae	62 × 33	*C. casuarius*	*H. moschatus*, *U. caudimaculatus*	Stocker and Irvine, 1983; Dennis, 2003
	E. xanthocarpa	Lauraceae	56 × 29	*C. casuarius*	*H. moschatus*, *U. caudimaculatus*	Stocker and Irvine, 1983; Dennis, 2003
	Aglaia cf. *flavida*	Meliaceae	118 g	*Casuarius bennettii*	*Dobsonia moluccensis*	Mack, 1995

(rodents, *Hypsiprymnodon*) or dispersed by abiotic means. This combination of dispersal modes and vectors means that selective pressures on seed size and shape could vary considerably among different places and times.

Another example of such exaptation is given by *Symphonia globulifera* (Clusiaceae), which evolved in the Palaeotropics and then later dispersed by drifting on oceanic currents to colonize the Neotropics (Pennington and Dick, 2004). The palaeotropical form is known in Central Africa from fossils as old as 45 MYBP (Dick *et al.*, 2003), and is now dispersed by hornbills (regurgitated), primates (spat out; Table 1.6) and cephalophes (swallowed; *Cephalophus monticola* and *Hyemoschus aquaticus*). In the Neotropics, where it arrived some 15 MYBP, *S. globulifera* it is now dispersed by endozoochory through tapirs and by exozoochory through bats and later consumed by granivores such as *Tinamou major*, small rodents, primates (*Pitheca pithecia*) and deer. While *S. globulifera* do not differ in the size and shape of their seeds across continents, there are two forms that differ in seed size in the Guianas: one small (10–20 mm in length) and the other large (>30 mm). The small-seeded form grows on hills and would benefit from a wide array of vertebrate dispersal services, perhaps constraining seed size. The large-seeded form grows in flooded and swamp forests and benefits from both vertebrate and water dispersal, perhaps relieving the constraints on seed size. Both forms are known to be dispersed by bats and tapir. The debate remains open.

Conclusions

Is there a relationship between seed size and shape and the size of the animals that disperse them? Differences in the mean size of seeds in plants from four rainforests on different continents reflected the mean size of the animals handling them. However, the relationship was a tenuous one and, when examined in more detail, the expected or straightforward patterns broke down. Large seeds from our African sample were elongate relative to smaller ones, though we expected few constraints on the shape of large seeds there. Large seeds in Thailand were also relatively elongate and showed strong selective pressure to become elongate as seed size increased. Given that they had the second smallest mean disperser size, this is not surprising. However, Australia, with the smallest mean disperser size, had seeds that did not show a change in shape as a function of seed size, nor any selective pressure to do so. In contrast, seeds from the Guianas, with the second largest mean disperser weight, had large seeds that were rounder than small ones and demonstrated strong selection for this to occur. These patterns indicate that different trends in seed allometry do occur at the broad community scales investigated here, but also that the evolution of seed size cannot easily be unravelled by examining patterns at this broad community scale. Why patterns in seed size and shape should be so different on different continents is an interesting question requiring further investigation. Clearly, different animal behaviours, such as carrying and dispersing seeds rather than swallowing them, can exert opposing selection on seed size and shape. For example, the largest seeds in the Guianas may no longer be swallowed but only dispersed externally, releasing them from selection for elongation, whereas in Africa the largest seeds may still be swallowed as well as carried, resulting in continued selection for elongate seeds. To unravel the reasons for these patterns further, we need more data on the diets, behaviours and relative contributions of dispersers.

Acknowledgements

We are very grateful to Ian J. Wright (Macquarie University), who kindly conducted the standardized major axis regression analyses reported here. We thank numerous colleagues for sharing data on seed size, fruit consumption and seed dispersal or predation by animals, especially Françoise Bayart, Daniel Bennett, Richard Corlett, Njikoha Ebigbo, Christian Erard, François Feer, Mauro Galetti, Annette Hladik, Kimberley Holbrook, Tatyana Lobova, Kim McConkey, Marco Pizo and Ken Whitney. We also thank Scott Mori for checking the taxonomy of the Lecythidaceae. Finally, we are indebted to Andrew Mack and Dave Kelly for their comments and suggestions, which improved the manuscript. P.A.J. was supported by the Netherlands Organization for the Advancement of Tropical Research (grant W85-239). Such a compilation would not have been possible without the dedication of the many students and researchers who collected and preserved the fruit and the seeds observed to be consumed by their study animals in the four rainforests – we thank them all.

References

Ackerly, D. (2004) Analysis of traits (AOT), version 3.0: a module of *Phylocom*, version 3.22 [available at http://www.phylodiversity.net/phylocom].

Alcantara, J.M. and Rey, P.J. (2003) Conflicting selection pressures on seed size: evolutionary ecology of fruit size in a bird-dispersed tree, *Olea europaea*. *Journal of Evolutionary Biology* 16, 1168–1176.

Alexandre, D.Y. (1978) Le rôle disséminateur des éléphants en forêt de Taï, Côte d'Ivoire. *La Terre et la Vie* 32, 47–72.

Anonymous (2004) *The International Plant Names Index*. [available at http://www.ipni.org/index.html].

Auffenberg, W. (1988) *Gray's Monitor Lizard.* University of Florida Press, Gainesville, Florida.

Authors (1961–2005) *Flore du Gabon.* Muséum National d'Histoire Naturelle, Paris.

Baker, G.B., Dettmann, E.B., Scotney, B.T., Hardy, L.J. and Drynan, D.A.D. (1997) *Report on the Australian Bird and Bat Banding Scheme, 1995–1996.* Environment Australia, Canberra.

Becker, P. and Wong, M. (1985) Seed dispersal, seed predation, and juvenile mortality of *Aglaia* sp. (Meliaceae) in lowland Dipterocarp rainforest. *Biotropica* 17, 230–237.

Belbenoit, P., Poncy, O., Sabatier, D., Prévost, M.-F., Riéra, B., Blanc, P., Larpin, D. and Sarthou, C. (2001) Appendix 1: Floristic checklist of the Nouragues area. In: Bongers, F., Charles-Dominique, P., Forget, P.-M. and Thery, M. (eds) *Nouragues: Dynamics and Plant–Animal Interactions in a Neotropical Rainforest.* Kluwer Academic, Dordrecht, The Netherlands, pp. 301–341.

Bennett, D. (2005) *Polillo Butaan Project: 2002–2005 Summary: The Ecology of* Varanus olivaceus *on Polillo Island and Implications for Other Giant Frugivorous Lizards in the Phillipines.* Report to Mampam Conservation, UK [available at http://www.mampam.com/gfl/bennettgiantfruitylizardsdec05.pdf].

Bongers, F., Charles-Dominique, P., Forget, P.-M. and Théry, M. (2001) *Nouragues: Dynamics and Plant–Animal Interactions in a Neotropical Rainforest.* Kluwer Academic, Dordrecht, The Netherlands.

Calder, W.A.I. (1979) The kiwi and the egg design: evolution as a package deal. *Bioscience* 29, 461–467.

Charles-Dominique, P. (1986) Inter-relations between frugivorous vertebrates and pioneer plants: *Cecropia*, birds and bats in French Guyana. In: Estrada, A. and Fleming, T.H. (eds) *Frugivores and Seed Dispersal.* Dr. W. Junk, Dordrecht, The Netherlands, pp. 119–135.

Charles-Dominique, P., Atramentowicz, M., Charles-Dominique, M., Gérard, H., Hladik, A., Hladik, C.-M. and Prévost, M.-F. (1981) Les mammifères frugivores arboricoles nocturnes d'une forêt guyanaise: inter-relations plantes-animaux. *Revue d'Ecologie (Terre et Vie)* 35, 341–435.

Corlett, R.T. and Primack, R.B. (2006) Tropical rainforests and the need for cross-continental comparisons. *Trends in Ecology and Evolution* 21, 104–110.

Dalling, J.W. and Harms, K.E. (1999) Damage tolerance and cotyledonary resource use in the tropical tree *Gustavia superba. Oikos* 85, 257–264.

de Queiroz, A. (2005) The resurrection of oceanic dispersal in historical biogeography. *Trends in Ecology and Evolution* 20, 68–73.

Debroux, L. (1988) L'aménagement des forêts tropicales fondé sur la gestion des populations d'arbres: l'exemple du moabi (*Baillonella toxisperma* Pierre) dans la forêt du Dja, Cameroun. PhD Sciences Agronomiques et Ingénieurie Biologique. Faculté Universitaire des Sciences Agronomiques de Gembloux, Gembloux, Belgium.

Dennis, A.J. (1997) Musky rat-kangaroos, *Hypsiprymnodon moschatus*: cursorial frugivores in Australia's wet tropical rain forests. PhD thesis, James Cook University of North Queensland, Townsville, Australia.

Dennis, A.J. (2003) Scatter-hoarding by musky rat-kangaroos, *Hypsiprymnodon moschatus*, a tropical rain-forest marsupial from Australia: implications for seed dispersal. *Journal of Tropical Ecology* 19, 619–627.

Dennis, A.J., Lipsett-Moore, G.J., Harrington, G., Collins, E. and Westcott, D. (2005) Seed predation, seed dispersal and habitat fragmentation: does context make a difference in tropical Australia? In: Forget, P.-M., Lambert, J., Hulme, P. and Vander Wall, S.B. (eds) *Seed Fate: Predation, Dispersal and Seedling Establishment.* CAB International, Wallingford, UK, pp. 117–135.

Dick, C.W., Abdul-Salim, K. and Bermingham, E. (2003) Molecular systematic analysis reveals cryptic tertiary diversification of a widespread tropical rain forest tree. *American Naturalist* 162, 691–703.

Dinerstein, E. (1991) Seed dispersal by greater one-horned rhinoceros (*Rhinoceros unicornis*) and the flora of *Rhinoceros* latrines. *Mammalia* 55, 355–362.

Dinerstein, E. and Wemmer, C.M. (1988) Fruits rhinoceros eat: dispersal of *Trewia nudiflora* (Euphorbiaceae) in lowland Nepal. *Ecology* 69, 1768–1774.

Docters van Leeuwen, W. (1935) The dispersal of plants by fruit-eating bats. *Gardens Bulletin, Straits Settlements* 9, 58–63.

Dubost, G. (1968) Les niches écologiques des forêts tropicales sud-américaines et africaines, sources de convergences remarquables entre rongeurs et artiodactyles. *La Terre et la Vie* 1, 3–28.

Dubost, G. (1984) Comparison of the diets of frugivorous forest ruminants of Gabon. *Journal of Mammalogy* 65, 298–316.

Dubost, G. (1987) Une analyse écologique de deux faunes de mammifères forestiers tropicaux. *Mammalia* 51, 415–436.

Ebigbo, N.M. (2005) The role of flying foxes as seed dispersers on the vegetation dynamics in a west African forest–savannah mosaic in Côte d'Ivoire. Ulm University, Ulm, Germany.

Edwards, W. and Westoby, M. (1997) Reserve mass and dispersal investment in relation to geographic range of plant species: phylogenetically independent contrasts. *Journal of Biogeography* 23, 329–338.

Emmons, L.H. (1980) Ecology and resource partitioning among nine species of African rain forest squirrels. *Ecological Monograph* 50, 31–54.

Emmons, L.H. (1982) Ecology of *Proechimys* (Rodentia, echimidae) in south-eastern Peru. *Tropical Ecology* 23, 280–290.

Emmons, L.H. and Feer, F. (1990) *Neotropical rainforest mammals*. University of Chicago Press, Chicago, Illinois.

Erard, C., Théry, M. and Sabatier, D. (1989a) Régime alimentaire de *Rupicola rupicola* (Cotingidae) en Guyane française: relations avec la frugivorie et la zoochorie. *Revue d'Ecologie (Terre et Vie)* 44, 47–74.

Eriksson, O., Friis, E.M., Pedersen, K.R. and Crane, P.R. (2000b) Seed size and dispersal systems of early cretaceous angiosperms from Famalicao, Portugal. *International Journal of Plant Sciences* 161, 319–329.

Ewer, R.F. (1965) Food burying in the African ground squirrel, *Xerus erythropus* (E. Geoff.). *Zeitschrift für Tierpsychologie* 22, 321–327.

Feer, F., Henry, O., Forget, P.-M. and Gayot, M. (2001) Frugivory and seed dispersal by terrestrial mammals. In: Bongers, F., Charles-Dominique, P., Forget, P.-M. and Théry, M. (eds) *Nouragues: Dynamics and Plant–Animal Interactions in a Neotropical Rainforest.* Kluwer Academic, Dordrecht, The Netherlands, pp. 227–232.

Felsenstein, J. (1985) Phylogenies and the comparative method. *American Naturalist* 125, 1–15.

Forget, P.-M. (1996) Removal of seeds of *Carapa procera* (Meliaceae) by rodents and their fate in rainforest in French Guiana. *Journal of Tropical Ecology* 12, 751–761.

Forget, P.M. and Vander Wall, S.B. (2001) Scatter-hoarding rodents and marsupials: convergent evolution on diverging continents. *Trends in Ecology and Evolution* 16, 65–67.

Forget, P.M., Munoz, E. and Leigh, E.G., Jr (1994) Predation by rodents and bruchid beetles on seeds of *Scheelea* palms on Barro Colorado Island, Panama. *Biotropica* 26, 420–426.

Foster, S.A. (1986) On the adaptative value of large seeds for tropical moist forest trees: a review. *Botanical Review* 52.

Fragoso, J.M.V. (1997) Tapir-generated seed shadows: scale-dependent patchiness in the Amazon rain forest. *Journal of Ecology* 85, 519–529.

Fragoso, J.M.V., Silvius, K.M. and Correa, J. (2003) Long-distance seed dispersal by tapirs increases seed survival and aggregates tropical trees. *Ecology* 83, 1998–2006.

Garland, T.J., Harvey, P.H. and Ives, A.R. (1992) Procedures for the analysis of comparative data using phylogenetically independent contrasts. *Systematic Biology* 41, 18–32.

Gautier-Hion, A., Colyn, M. and Gautier, J.-P. (1999) *Histoire naturelle des Primates d'Afrique Centrale*, Libreville, Gabon.

Gautier-Hion, A., Duplantier, J.-M., Quris, R., Feer, F., Sourd, C., Decoux, J., Dubost, G., Emmons, L., Erard, C., Hecketsweiler, P., Moungazi, A., Roussilhon, C. and Thiollay, J.-M. (1985) Fruit characters as a basis of fruit choice, and seed dispersal by a tropical forest vertebrate community. *Oecologia* 65, 324–337.

Githiru, M., Lens, L., Bennur, L.A. and Ogol, C.P.K.O. (2002) Effects of site and fruit size on the composition of avian frugivore assemblages in a fragmented Afrotropical forest. *Oikos* 96, 320–330.

Guillotin, M., Dubost, G. and Sabatier, D. (1994) Food choice and food composition among three major primate species of French Guiana. *Journal of Zoology of London* 233, 551–579.

Hamann, A. and Curio, E. (1999) Interactions among frugivores and fleshy fruit trees in a Philippine submontane rainforest. *Conservation Biology* 13, 766–773

Harvey, P.H. and Pagel, M.D. (1991) *The Comparative Method in Evolutionary Biology.* Oxford University Press, Oxford, UK.

Henry, O., Feer, F. and Sabatier, D. (2000) Diet of the lowland tapir (*Tapirus terrestris* L.) in French Guiana. *Biotropica* 32, 364–368.

Herrera, C.M. (1992) Interspecific variation in fruit shape: allometry, phylogeny, and adaptation to dispersal agents. *Ecology* 73, 1832–1841.

Hladik, A., Dubost, G., Ezavin, O., Posso, P. and Hadley, M. (1987) *Makokou, Gabon. Une station de recherche en écologie forestière tropicale.* IRET/ECOTROP (CNRS)/ UNESCO, Paris.

Howe, H.F. (1986) Seed dispersal by fruit-eating birds and mammals. In: Murray, D.R. (ed.) *Seed Dispersal.* Academic Press, Sydney, Australia, pp. 123–189.

Huxley, J. (1932) *Problems of Relative Growth* (1972 reprint). Dove Publications, New York.

Jansen-Jacobs, M.J., ter Steege, H. and Zijlstra, G. (1985–2003) *Flora of the Guianas.* Nationaal Herbarium Nederland, Utrecht, The Netherlands.

Janzen, D.H. (1985) *Spondias mombin* is culturally deprived in megafauna-free forest. *Journal of Tropical Ecology* 1, 131–155.

Janzen, D.H. and Martin, P.S. (1982) Neotropical anachronisms: the fruits the gomphotheres ate. *Science* 215, 19–27.

Jordano, P. (1995a) Angiosperm fleshy fruits and seed dispersers: a comparative analysis of adaptation and constraints in plant–animal interactions. *American Naturalist* 145, 163–191.

Jordano, P. (1995b) Frugivore-mediated selection on fruit and seed size: birds and St. Lucie's cherry, *Prunus mahaleb. Ecology* 76, 2627–2639.

Kelly, C.K. (1993) Seed size and establishment conditions in tropical trees. *Oecologia* 94, 356–360.

Kelly, C.K. (1995) Seed size in tropical trees: a comparative study of factors affecting seed size in Peruvian angiosperms. *Oecologia* 102, 377–388.

Kelly, D., Ladley, J.J. and Robertson, A.W. (2004) Is dispersal easier than pollination? Two tests in New Zealand Loranthaceae. *New Zealand Journal of Botany* 42, 89–103.

Kitamura, S., Shunsuke, S., Yumoto, T., Poonswad, P., Chuailua, P., Plongmai, K., Noma, N., Maruhashi, T. and Suckasam, C. (2004) Dispersal of *Aglaia spectabilis*, a large-seeded tree species in a moist evergreen forest in Thailand. *Journal of Tropical Ecology* 20, 421–427.

Kitamura, S., Suzuki, S., Yumoto, T., Poonswad, P., Chuailua, P., Plongmai, K., Maruhashi, T., Noma, N. and Suckasam, C. (2006) Dispersal of *Canarium euphyllum* (Burseraceae), a large-seeded tree species, in a moist evergreen forest in Thailand. *Journal of Tropical Ecology* 22, 137–146.

Kitamura, S., Yumoto, T., Poonswad, P., Chuailua, P., Plongmai, K., Maruhashi, T. and Noma, N. (2002) Interactions between fleshy fruits and frugivores in a tropical seasonal forest in Thailand. *Oecologia* 133, 559–572.

Lobova, T.A. and Mori, S.A. (2005) *Checklist of Species Dispersed by Bats in Central French Guiana* [available at http://nybg.org/botany/tlobova/mori/batsplants/Checklist/Checklist_Home.htm]. The New York Botanical Garden, New York.

Lord, J., Egan, J., Clifford, T., Jurado, E., Leishman, M., Willimas, D. and Westoby, M. (1997) Larger seeds in tropical floras: consistent patterns independent of growth form and dispersal mode. *Journal of Biogeography* 24, 205–211.

Lord, J., Westoby, M. and Leishman, M. (1995) Seed size and phylogeny in six temperate floras: constraint, niche conservatism, and adaptation. *American Naturalist* 146, 349–364.

Lord, J.M. (2004) Frugivore gape size and the evolution of fruit size and shape in southern hemisphere floras. *Austral Ecology* 29, 430–436.

Mack, A.L. (1993) The sizes of vertebrate-dispersed fruits: a neotropical-paleotropical comparison. *American Naturalist* 142, 840–856.

Mack, A.L. (1995) Distance and non-randomness of seed dispersal by the dwarf cassowary *Cassarius bennetti*. *Ecography* 18, 286–295.

Mack, A.L. (1998) An advantage of large seed size: tolerating rather than succumbing to seed predators. *Biotropica* 30, 604–608.

Mazer, S.J. and Wheelwright, N.T. (1993) Fruit size and shape: allometry at different taxonomic levels in bird-dispersed plants. *Evolutionary Ecology* 7, 556–575.

Moles, A.T. and Drake, D.R. (1999) Post-dispersal seed predation on eleven large-seeded species from the New Zealand flora: a preliminary study in secondary forest. *New Zealand Journal of Botany* 37, 679–685.

Moles, A.T., Ackerly, D.D., Webb, C.O., Tweddle, J.C., Dickie, J.B., Pitman, A.J. and Westoby, M. (2005a) Factors that shape seed mass evolution. *Proceedings of the National Academy of Sciences of the USA* 102, 10540–10544.

Moles, A.T., Ackerly, D.D., Webb, C.O., Tweddle, J.C., Dickie, J.B. and Westoby, M. (2005b) A brief history of seed size. *Science* 307, 576–580.

Mori, S. and Lobova, T. (2005) *Bat/Plant Interactions in the Neotropics*. New York Botanical Garden.

Mori, S.A., Cremers, G., Gracie, C.A., de Granville, J.J., Heald, S.V., Hoff, M. and Mitchell, J.D. (2002) *Guide to the Vascular Plants of French Guiana. Part 2. Dicotyledons.* Memoirs 76. New York Botanical Garden, New York.

Okon, E.E. (1974) Fruit bats at Ife: their roosting and food preferences (Ife Fruit Bat Project No. 2). *Nigerian Field* 39, 33–40.

Pennington, R.T. and Dick, C.W. (2004) The role of immigrants in the assembly of the South American rainforest tree flora. *Philosophical Transactions of the Royal Society London B* 359, 1611–1622.

Pratt, T.K. and Stiles, E.W. (1985) The influence of fruit size and structure on composition of frugivore assemblages in New Guinea. *Biotropica* 17, 314–321.

Rey, P.J., Guttierrez, J.E., Alcantara, J.M. and Valera, F. (1997) Fruit size in wild olives: implications for avian seed dispersal. *Functional Ecology* 11, 611–618.

Sabatier, D. (1983) Fructification and dissémination en forêt guyanaise: l'exemple de quelques espèces ligneuses. Thèse de doctorat de l'Université USTL, Montpellier, France.

Salisbury, E. (1974) Seed size and mass in relation to environment. *Proceedings of the Royal Society London B* 186, 83–88.

Silvius, K.M. and Fragoso, J.M.V. (2003) Red-rumped agouti (*Dasyprocta leporina*) home range use in an Amazonian forest: implications for the aggregated distribution of forest trees. *Biotropica* 35, 74–83.

Stevens, P.F. (2004) *Angiosperm Phylogeny Website*, Version 5, May 2004 [available at http://www.mobot.org/MOBOT/research/APweb/].

Stocker, G.C. and Irvine, A.K. (1983) Seed dispersal by cassowaries *Casuarius casuarius* in north Queensland's rainforests. *Biotropica* 15, 170–176.

Strahan, R. (ed.) (1995) *The Mammals of Australia*. Reed Books/Australian Museum, Chatswood, Australia.

Theimer, T.C. (2001) Seed scatterhoarding by white-tailed rats: consequences for seedling recruitment by an Australian rain forest tree. *Journal of Tropical Ecology* 17, 177–189.

Tiffney, B.H. (1984) Seed size, dispersal, syndromes, and the rise of the angiosperms: evidence and hypothesis. *Annals of the Missouri Botanical Garden* 71, 551–576.

Traveset, A., Willson, M.E. and Verdu, M. (2004) Characteristics of fleshy fruits in southeast Alaska: phylogenetic comparison with fruits from Illinois. *Ecography* 27, 41–48.

van Roosmalen, M.G.M. (1985a) *Fruits of the Guiana Flora.* Institute of Systematic Botany, Utrecht University, Wageningen, The Netherlands.

van Roosmalen, M.G.M. (1985b) Habitat preferences, diet, feeding strategy and social organization of the black spider monkey (*Ateles paniscus* Linnaeus 1758) in Surinam. *Acta Amazonica* 15 (Suppl.), 1–238.

van Roosmalen, M.G.M. and van Roosmalen-Blijenberg, B. (2003) *Fruits of the Amazon Flora.* Departamento de Botânica, INPA, Brazil.

Vivien, J. and Faure, J.J. (1996a) *Arbres des forêts denses d'Afrique centrale. Espèces du Cameroun. Agence de Coopération Culturelle et Technique, Paris.*

Vivien, J. and Faure, J.J. (1996b) *Fruitiers sauvages d'Afrique. Espèces du Cameroun.* Editions NGUILA-KEROU, Clohars Carnoet.

Warton, D.I., Wright, I.J., Falster, D.S. and Westoby, M. (2006) Bivariate line-fitting methods for allometry. *Biological Reviews* 81, 259–291.

Webb, C.O. and Donoghue, M.J. (2005) Phylomatic: tree assembly for applied phylogenetics. *Molecular Ecology Notes* 5, 181–183.

Webb, C.O., Ackerly, D.D. and Kembel, S. (2004) *Phylocom: Software for the Analysis of Community Phylogenetic Structure and Character Evolution* [available at http://www.phylodiversity.net/phylocom/].

Westoby, M., Leishman, M. and Lord, M. (1996) Comparative ecology of seed size and dispersal. *Philosophical Transactions of the Royal Society London B* 351, 1309–1318.

Williams, K.D. (1978) Aspects of the ecology and behaviour of the Malayan Tapir (*Tapirus indicus* Desmarest) in the National Park of West Malaysia. MSc thesis, Michigan State University, East Lansing, Michigan.

Yasuda, M., Miura, S. and Hussein, N.A. (2000) Evidence for hoarding behaviour in terrestrial rodents in Pasoh Forest reserve, a Malaysian lowland rain forest. *Journal of Tropical Forest Science* 12, 164–173.

2 Evolutionary Ecology of Secondary Compounds in Ripe Fruit: Case Studies with Capsaicin and Emodin

D.J. Levey, J.J. Tewksbury, I. Izhaki, E. Tsahar and D.C. Haak

Introduction

Vertebrate-consumed fruits should be characterized by features that facilitate seed dispersal – features that increase attractiveness to any organism capable of ingesting and defecating viable seeds. Yet, as any inquisitive child with outdoor experience can certainly attest, most wild fruits appear delicious and taste bad. Astringency is the rule, not an exception (Herrera, 1982). It is equally puzzling that most species of birds and mammals rely only partially on fruits or not at all. Because wild fruits are 'designed' to be eaten, they should embody the proverbial free lunch. Why do they so often remain seemingly untouched for long periods (McCarty *et al.*, 2002; Tang *et al.*, 2005)?

A potential explanation emerges when one adopts a broad view of fruits and frugivores (Janzen, 1977; Cipollini and Levey, 1997a; Levey and Martínez del Rio, 2001). Fruits are more than packages of seeds and nutritious pulp, and frugivores are more diverse than those that can be easily seen. In particular, secondary compounds are a universal constituent of fruits, and microbes are ubiquitous frugivores that are generally detrimental to plant fitness (because they do not disperse seeds). To understand the ecology of fruits and vertebrate frugivores, one must view fruit–frugivore interactions as an evolutionary triad between fruiting plants, their mutualists (vertebrate frugivores), and their predators and parasites, including seed predators, insects and microbial frugivores (Herrera, 1982, 2002; Cipollini, 2000; Jordano, 2000; Tewksbury, 2002; Levey, 2004). Secondary compounds are thought to mediate many of these interactions (Cipollini, 2000; Jordano, 2000; Herrera, 2002).

We believe that the key to understanding the functional significance of any secondary compound in fruits is to examine its role in an integrative manner. This requires knowledge of its biosynthetic pathway, the costs and benefits of its production, and the impact it has on the complete spectrum of organisms that interact with the plant, including mutualists and antagonists. Effects should not be viewed in isolation. Ideally, studies of the evolutionary ecology of secondary compounds should be intraspecific, because individual variance in the trait of interest can be more easily separated from variation in other, unrelated, traits and because natural selection acts on individuals.

These points are not new (Berenbaum, 1995a; Wink and Schimmer, 1999; Cipollini, 2000). We raise them because progress in the field of fruit secondary metabolites continues to be slow and the potential for significant discoveries seems high – we know of no ecological study that has rigorously explored fruit secondary compounds and *not* found an impact of some kind. These impacts, however, are bewilderingly broad in scope, including effects on microbes, arthropods, vertebrate seed dispersers and vertebrate seed predators (Cipollini, 2000). In addition, the literature is largely a hodgepodge of studies about the effects of a particular compound on a particular organism, often without any ecological or evolutionary context.

In this chapter, we synthesize ongoing and past work on two prominent secondary metabolites in ripe fruit, emodin and capsaicinoids. Our goal is to illustrate the importance of adopting an integrative approach in deciphering the ecological role and evolutionary context of these compounds. To place our review into a broad context, we start with a short critique of how secondary compounds have been viewed by plant ecologists. We then describe the biosynthetic pathways of emodin and capsaicinoids, where they occur in plant tissue, and how their occurrence is linked to abiotic conditions. Next, we shift our attention to biotic interactions, organizing our discussion around three sequential phases of plant life-history: pre-dispersal, dispersal, and post-dispersal. Finally, we draw attention to the similarities and differences between secondary compounds in fruits and in other plant tissues. Our unifying theme is that an unusually wide array of selection pressures on fruits and seeds has led to an unusually diverse set of functions for secondary compounds in ripe fruits.

Terminology and Function of Plant Secondary Compounds

Traditionally, plant metabolites have been divided into those with a direct role in primary metabolism (i.e. growth and development) and those without such a role (i.e. 'secondary'; Whittaker and Feeny, 1971; Stamp, 2003a). In this framework, the classification of secondary compounds is simple: if a metabolite is not in a major and essential pathway – photosynthesis, respiration or uptake – it is a secondary compound. Most of the theory that developed around this dichotomy assumed a defensive role for secondary compounds. This assumption persists, despite much controversy about the evolutionary ecology of secondary compounds (Berenbaum, 1995a; Hamilton *et al.*, 2001; Koricheva, 2002; Lerdau and Coley, 2002; Stamp, 2003a,b).

For the purpose of this chapter, we focus attention on the tenuous distinction between what is 'primary' and what is 'secondary' in plant metabolism. It is now widely accepted that production processes of primary and secondary metabolites are often linked via biosynthetic pathways; they are not as independent as their classification implies. This linkage may lead to trade-offs between growth, defence and reproduction. For example, there can be a negative correlation between growth and production of some tannins, suggesting that tannins are sometimes produced at the expense of growth (Coley, 1986; Sagers and Coley, 1995). More fundamentally, primary compounds can function as secondary compounds by playing a role in plant defence (e.g. citric acid; Cipollini and Stiles, 1992) and secondary compounds can have non-defensive functions (e.g. protection from UV; Dixon and Paiva, 1995).

The important point is that many plant compounds can perform 'primary' *and* 'secondary' functions, depending on when and where they are produced (Herms and Mattson, 1992; Berenbaum, 1995b). In fruits, their production is often tightly controlled, especially during the ripening process (Lund and Bohlman, 2006). This view of secondary compounds in fruits has important implications for the evolution of fruits, as we will explain towards the end of this chapter.

Biosynthesis of Emodin and Capsaicin

Emodin

Emodin belongs to the anthraquinones, a group of more than 170 natural phenolic compounds that comprise the largest group of natural quinones (Thomson, 1987, 1997; Harborne *et al.*, 1999). There are two distinct biosynthetic pathways leading to anthraquinones in higher plants: the acetate–malonate pathway in Polygonaceae, Leguminosae and Rhamnaceae, and the O-succinylbenzoic acid pathway in the Bignoniaceae and Verbenaceae (Evans, 1996; Dewick, 1998; Harborne *et al.*, 1999). The basic chemical structure of anthraquinone is an anthracene ring (tricyclic aromatic) with two ketone groups in positions C9 and C10. In plants, anthraquinones are mostly present as sugar derivatives (glycosides) but the free forms (aglycones) are widely distributed as well (Thomson, 1987, 1997; Harborne *et al.*, 1999). Among the commonest naturally occurring anthraquinone aglycones in higher plants are emodin, rhein, chrysophanol, aloe-emodin and physcion (Evans, 1996; Harborne *et al.*, 1999). Several biochemical pathways that transform one anthraquinone to another have been discovered. For example, chrysophanol is synthesized in plants by dehydroxylation of emodin, an enzymatic conversion that is mediated by NADPH (Anderson *et al.*, 1988). It also appears that physcion is derived from emodin (Thomson, 1997). The anthraquinone glycosides are formed when one or more sugar molecules, mostly glucose or rhamnose, are bound to the aglycone by a β-glycoside linkage to the hydroxyl group at position C8 (in the case of glucose) or the one at C6 (in the case of rhamnose) (Dewick, 1998). Among the most common emodin-related glycosides are emodin-8-glucose, frangulin and glucofrangulin (Harborne *et al.*, 1999).

Within-species variation among the chemical and morphological traits of ripe fruits has been studied in *Rhamnus alaternus* (Rhamnaceae; Izhaki *et al.*, 2002). Variation in chemical traits of its fruit (emodin, macronutrients and minerals) was typically much higher than variation in morphological traits (e.g. fruit size). This discrepancy may be due to differences in environmental conditions between microsites that imposed greater variability on fruit nutrient composition than on fruit morphological traits, and by lower selective pressure by birds on fruit chemical traits than on morphological traits. Alternatively, it may be explained by differences in measurement error between chemical and physical traits.

Izhaki (2002a) also found that emodin concentration in fruits was positively correlated with concentrations of non-structural carbohydrates and negatively correlated with lipid concentration. Such interrelationships between chemical constituents in fruit may indicate synergistic and antagonistic interactions between these constituents, some of which might also be explained by plant–environment relations. Emodin levels were also found to depend on season and light intensity, perhaps reflecting a trade-off between plant development and defence (Paneitz and Westendorf, 1999). However, no study has explored the effect of abiotic conditions on emodin concentration in fruit.

Capsaicinoids

The secondary metabolite capsaicin (8-methyl-*N*-vanillyl-6-nonenamide), along with a series of homologous alkyl vanillylamides called capsaicinoids, are responsible for the pungent ('hot') flavour of chilli fruit (Iwai *et al.*, 1979; Kawada *et al.*, 1985; Govindarajan, 1986; Cordell and Araujo, 1993). Capsaicinoids are unique to the genus *Capsicum* (Solanaceae; Govindarajan *et al.*, 1987; Govindarajan and Sathyanarayana, 1991; Bosland, 1994), and they are produced only in the fruit (Fugiwake *et al.*, 1982; Suzuki and Iwai, 1984). Thus, their adaptive function cannot be ascribed to interactions in other plant tissues. In addition, because capsaicinoids are both odourless and colourless (Iwai *et al.*, 1979; Kawada *et al.*, 1985; Cordell and Araujo, 1993), they are unlikely to function as signalling compounds, and thus any adaptive function should be tightly linked to direct defence of the developing embryos. Capsaicinoids are produced during fruit maturation, after elongation is complete (Estrada *et al.*, 1999), and are the result of a condensation reaction between the end products of two distichous pathways, the phenylpropanoid pathway and the fatty-acid biosynthesis pathway. The phenylpropanoid pathway contributes the aromatic head-group, while the fatty acid pathway provides the acyl moiety (Sukrasno and Yeoman, 1993). The phenylpropanoid pathway produces compounds encompassing a large range of structural and secondary bioactive plant chemicals, all of which are unique to plants and bacteria (Kessler and Baldwin, 2002). The initial conversions of phenylalanine to phenolic compounds such as cinnamates, coumerates, tannins, saponifiable cell-wall phenolics, monolignols, flavonoids and caffeic acids is a part of the phenylpropanoid pathway common to all angiosperms. However, the conversion from ferulic acid to capsaicin is known to occur only in the genus *Capsicum* (Sukrasno and Yeoman, 1993; Dixon and Paiva, 1995). Starting with labelled phenylalanine, Hall and Yeoman (1991) found that a mere 1% of precursor was incorporated into capsaicin, while nearly 50% was incorporated into 'lignin-like' substances. One of the most important products of phenylpropanoid metabolism is a class of compounds, monolignols, which polymerize to form lignin, a structural component of plant cell walls. Importantly, lignin and capsaicin share common precursors in developing fruit; thus it appears that carbon can be allocated either to lignin production or capsaicin production but not both. It remains unclear whether or when these precursors may be limiting. If they are in short supply, a trade-off in allocation is set up between investment towards physical defence (lignin) and chemical defence (capsaicin). An important implication is that one type of defence cannot be understood without considering the other type of defence, and that the genetic and physiological mechanisms controlling production of capsaicin hold the key to revealing where and when capsaicin is produced.

Recent investigations support the hypothesis that capsaicinoid production is a monophyletic, derived trait. *Capsicum lanceolatum*, *C. rhomboideum*, *C. lycianthoides* and *C. geminifolium* are non-pungent and appear basal to all pungent taxa (Walsh and Hoot, 2001; L. Bohs, Utah, 2006, personal communication). Furthermore, a single quantitative trait locus, *cap*, accounts for 34–38% of the phenotypic variation in capsaicinoid content (Blum *et al.*, 2002). The only comprehensive theory advanced to explain the origin of pungency in *Capsicum* centres on an initial radiation from the arid, high-elevation, interior valleys of Bolivia (McLeod *et al.*, 1982). The proposed ancestral species in this radiation was a form of *C. chacoense*. More recent work using nuclear and chloroplast DNA supports a radiation of pungent taxa from dry mountainous regions, either in Bolivia or Peru (Walsh and Hoot, 2001). This is based on the earliest-branching pungent species (*C. eximium* and *C. cardinasii*) and inconclusive bootstrap values separating the baccatum clade (*C. chacoense* and *C. baccatum*). Finally, the discovery of a natural polymorphism for fruit pungency in an accession of *C. chacoense* from south-eastern Bolivia (1959 collection by Paul Smith, USDA #PI260433, polymorphism discovered by P. Bosland, New Mexico), along with the discovery of wild polymorphic populations in two additional species, *C. baccatum* and *C. eximium* (Tewksbury *et al.*, 2006) demonstrates that this genus does not always have pungent fruit. While early collectors have noted rare encounters with non-pungent plants (D'Arcy and Eshbaugh, 1974), details of this polymorphism have never been examined. The discovery and description of zones of polymorphism for capsaicinoid production may be a first step in understanding the origin and adaptive significance of capsaicinoids in wild fruit, and their role in the radiation of the genus.

We conclude this section by summarizing three important differences in the occurrence of emodin and capsaicin. Taken together, they strongly suggest that emodin is a more primitive (derived early) compound with generalized effects, and capsaicin is more recently derived with more restricted effects.

1. Emodin is found throughout the plants that contain it, whereas capsaicin occurs only in fruit and on seeds (Iwai *et al.*, 1979; Izhaki, 2002a). This difference suggests that the ecological functions of capsaicin may be more tightly linked to the activity of frugivores and seed predators than those of emodin.
2. Emodin is produced in a wide variety of plant families and even by some fungi, whereas capsaicin appears restricted to a single genus of Solanaceae, *Capsicum* (Bosland, 1998; Izhaki, 2002a), again suggesting a more specialized role of capsaicin.
3. Although concentrations of both emodin and capsaicin can vary widely among nearby plants, this variation is much more extreme for capsaicin: to our knowledge, polymorphism for emodin production does not occur in any species, whereas it has been found in at least three wild species of *Capsicum* (Tewksbury *et al.*, 2006).

Because these three differences between capsaicin and emodin will be reflected throughout the following sections on ecological functions, they can be viewed as an organizing framework. We caution, however, that our discussion is restricted to only capsaicin and emodin, and that they may or may not be representative of other secondary compounds with similar characteristics.

Pre-dispersal Effects

Without chemical protection prior to dispersal, fruits and seeds are especially vulnerable to pathogens and predators because they occur in large numbers for a relatively long time in a predictable place, and are both nutritious and moist. This pre-dispersal phase includes both ripe and unripe fruit. Removal or damage to unripe fruit is obviously detrimental to plant fitness. Removal of ripe fruit, on the other hand, may be either detrimental or beneficial, depending upon whether the seed is killed or dispersed after removal. In this section, we consider effects of emodin and capsaicin on all organisms that reduce consumption of fruit by seed dispersers.

Concentrations of emodin are highest in unripe fruits (Tsahar *et al.*, 2002), while concentrations of capsaicinoids are highest in ripe fruits (Estrada *et al.*, 2000). The reduction in emodin concentration with fruit ripening is typical of fruit secondary metabolites, suggesting a generalized defensive role and potential conflicts with beneficial consumers. By contrast, the increase in capsaicinoid concentration with fruit ripening seen in *Capsicum* suggests a more specialized role for these chemicals – one that is focused explicitly on fruit consumption. If emodin and capsaicinoids were simply a by-product of a plant's general metabolic processes (Ehrlen and Eriksson, 1993), one would not expect their concentrations to change markedly during fruit maturation, nor that these changes would be so common among other secondary compounds and species.

Protection of fruit and seeds prior to seed dispersal is critical before and after the fruit ripens, but trade-offs and selective pressures shift during fruit ripening. For example, early in fruit development, herbivores will often consume fruit, creating significant fitness costs (Hulme and Benkman, 2002), and chemicals that deter herbivory in other plant parts are often found in unripe fruits (Ehrlen and Eriksson, 1993). During fruit maturation, chemicals that deter beneficial seed dispersers are detrimental, and most plant secondary metabolites rapidly decrease in concentration as fruits ripen (Cipollini and Levey, 1997a). However, in almost all fruits there is a period of vulnerability after fruit maturation but before dispersal (Howe, 1977; Thompson and Willson, 1979), during which pre-dispersal seed predation by invertebrates and microbes can destroy fruit (Hulme and Benkman, 2002). Protecting ripe fruit may be more difficult for a plant than protecting unripe fruit, as chemicals involved in the protection of ripe fruit may also harm legitimate dispersers, necessitating more selective fruit chemistry (Herrera, 1982; Cipollini and Levey, 1997a; Tewksbury, 2002; Tsahar *et al.*, 2002; Schaefer *et al.*, 2003). Chemical protection of unripe fruit might thus be directed primarily at herbivores, while chemical protection of ripe fruit may be most often directed against microbial and invertebrate attack, as these organisms thrive on the same nutritional qualities that attract legitimate dispersers – sugars, lipids, carbohydrates and proteins (Herrera, 1982). However, considerable overlap is expected, as consumers and fruit chemistry often cross the boundary between ripe and unripe fruit. Indeed, in the two systems we examine here, both emodin and capsaicin appear to deter a broad spectrum of organisms, including herbivores and microbes.

Emodin

Emodin has larvicidal activity against dipterans (Yang *et al.*, 2003) and inhibits growth of some protozoa (Wang, 1993), bacteria (Wang and Chung, 1997) and fungi (Kim *et al.*, 2004). Among vertebrates, emodin strongly deters food consumption by yellow-vented bulbuls (*Pycnonotus xanthopygos*; Pycnonotidae), house sparrows (*Passer domesticus*; Passeridae), and white-footed mice (*Peromyscus leucopus*; Cricetidae) (Sherburne, 1972; Tsahar *et al.*, 2002) and can be lethal when force-fed to redwing blackbirds (*Agelaius phoeniceus*; Icteridae), European starlings (*Sturnus vulgaris*; Sturnidae) and one-day-old cockerels (*Gallus domesticus*; Phasianidae) (Wells *et al.*, 1975; Schafer *et al.*, 1983). Unfortunately, many of these studies are fundamentally limited by a lack of ecological context. Most were not done in the field and effects were tested using unusually high concentrations of emodin on organisms that rarely or never encounter emodin in nature. A notable exception is a study that examined natural variation in emodin concentration in *Rhamnus alaternus* fruit and reported that plants with more emodin suffered lower pre-dispersal seed damage by insects (Tsahar *et al.*, 2002). However, it is unclear whether variation in emodin or some other factor generated the differences in seed damage. Thus, the most we can conclude about emodin's function during the pre-dispersal phase is that it has great potential for thwarting attacks on unripe fruit and that this potential is general across a wide range of taxa.

Capsaicinoids

The functional role of capsaicinoids in unripe and ripening fruits is clearer than that of emodin. Like emodin, capsaicin inhibits growth of many types of microbes and invertebrates *in vivo* and under laboratory conditions (Debkirtaniya *et al.*, 1980; Cichewicz and Thorpe, 1996; Oh *et al.*, 2004; Zeyrek, 2005). However, the polymorphism for pungency in some species of wild chilli allows one to move beyond these type of studies and examine the effects of capsaicin in an ecologically relevant setting. In Bolivia, ripe fruits on wild pungent and non-pungent *Capsicum chacoense* plants were monitored for fungal infection (J.J. Tewksbury *et al.*, 2006, unpublished results). After 45 days, no fruits on pungent plants showed signs of fungal attack, whereas 12% of fruits on non-pungent plants did so. This experiment does not rule out effects of other compounds which might vary between pungent and non-pungent plants, but it strongly suggests an anti-fungal role for capsaicinoids. A similar experiment using pungent and non-pungent varieties of *C. chacoense* also demonstrated that capsaicin deters consumption of chilli fruits and seeds by small rodents in the field and in the laboratory (Tewksbury *et al.*, 1999; Tewksbury and Nabhan, 2001). This is noteworthy because rodents are generally seed predators. Interestingly, capsaicin does not strongly or consistently deter consumption of foliage by large herbivorous mammals (Andelt *et al.*, 1994; Wagner and Nolte, 2000; Santilli *et al.*, 2004), which are more likely than small rodents to defecate ingested seeds in a viable condition (Janzen, 1971).

Dispersal Effects

For the purposes of this review, we define dispersal as the phase between the time when a fruit is removed from a plant by a seed disperser and when its seeds are eventually defecated, dropped or regurgitated. We include fruit selection by seed dispersers, as it ultimately determines which fruits are consumed.

Emodin

Birds and mammals appear universally sensitive to variation in emodin concentration, although this sensitivity is not equal among taxa (Izhaki, 2002a). For example, yellow-vented bulbuls (*Pycnonotus xanthopygos*) in feeding trials distinguished between diets containing 0%, 0.001% and 0.002% emodin, always preferring the diet with the lowest concentration (Tsahar *et al.*, 2002). House sparrows (*Passer domesticus*) were less sensitive than bulbuls. They did not distinguish between 0.001% and 0.002% diets but did between 0.005% and 0.01%. Functionally, this means that bulbuls are sensitive to the natural range of emodin variation in ripe fruits (0.001–0.002%), whereas house sparrows are only sensitive to the range of variation between ripe and unripe fruits (0.001–0.01%) (Tsahar *et al.*, 2002). Sherburne (1972) also reported sensitivity of vertebrates to emodin: American robins (*Turdus migratorius*) and white-footed mice (*Peromyscus leucopus*) tended to avoid ripe fruits coated with emodin but readily consumed uncoated (control) fruits and very rarely consumed unripe fruits, which contain the highest concentrations of emodin.

Despite the widespread detrimental effect of emodin on fruit consumption by vertebrates, ripe fruits that contain emodin are frequently consumed by a wide variety of seed-dispersers (Herrera, 1984; Izhaki and Safriel, 1985; Izhaki, 2002b; Tsahar *et al.*, 2002), suggesting that the net benefits of fruit consumption somehow outweigh the negative effects of emodin. In particular, the nutritional reward of carbohydrates, lipids and proteins in fruit pulp can be viewed as more important to consumers than the presence of co-occurring secondary compounds (Cipollini and Levey, 1997b). This does not mean that consumers are blind to natural variation in compounds such as emodin. Tsahar *et al.* (2002) found that removal of *Rhamnus alaternus* fruits was significantly higher on plants with lower concentrations of emodin in one of two years of study.

Consumption of emodin has physiological effects on frugivores, which probably influence seed fate and hence plant fitness. Domestic chicks force-fed 3.7–37 mg/kg and American robins force-fed approximately 0.07–70 mg/kg developed severe diarrhoea (Sherburne, 1972; Wells *et al.*, 1975). In contrast, yellow-vented bulbuls feeding voluntarily on a fruit-based diet containing 0.01% (wet mass) of emodin consumed as much of the diet and in a similar pattern to birds on a control diet but had longer intervals between defecations, strongly suggesting a costive effect of emodin (Tsahar *et al.*, 2003). At concentrations of 0.001%, 0.005% and 0.01%, emodin generally increased digestive efficiency of dry matter, nitrogen, fat and organic remains (mostly carbohydrates and protein) by approximately 5–10%. Because this increase could not be attributed to differences in consumption rates, it was likely to have been due to the slowing of transit time through the gut, allowing for more complete assimilation of digesta (Afik and Karasov, 1995). From a bird's perspective, higher digestive efficiencies are presumably beneficial. From a plant's perspective, the costive effects of emodin are probably beneficial because they increase the time between defecations and potentially the distance that seeds are dispersed away from the parent plant and the distance between where seeds are deposited in sequential defecations. Longer gut retention times may also increase the probability that seeds will be thoroughly cleaned of pulp, which is important in species whose seeds require removal of pulp in order to germinate (Barnea *et al.*, 1991; Traveset *et al.*, Chapter 4, this volume).

Capsaicinoids

Birds are the primary consumers and dispersers of wild chillies (Tewksbury and Nabhan, 2001; Levey *et al.*, 2006). Capsaicin has no effect on the feeding behaviour of captive birds (Norman *et al.*, 1992; Tewksbury and Nabhan, 2001). To determine whether this lack of discrimination also occurs in natural settings, we trimmed all fruit from five pungent and eight non-pungent *Capsicum chacoense* plants and wired in their place cut branches with 10–11 ripe fruit from either pungent or non-pungent plants. We monitored fruit removal on branches from pungent and non-pungent plants, and replaced branches where removal had occurred with new branches of the other pungency treatment (i.e. a branch bearing pungent fruit would follow one bearing non-pungent fruit). We found no difference in removal rates of pungent and non-pungent fruits, suggesting that the presence of capsaicin in ripe chilli fruits has no cost to the plant in terms of fruit preference by seed-dispersing birds (Fig. 2.1).

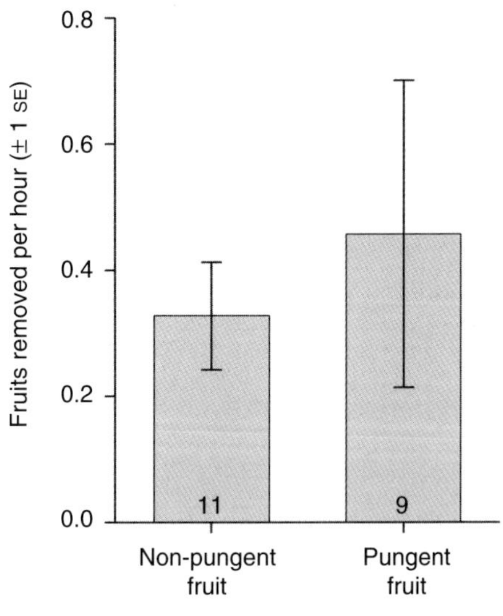

Fig. 2.1. Rate of fruit removal (per hour) for non-pungent and pungent *C. chacoense* fruit in Bolivia. Branches with ten fruits per branch were wired onto chilli bushes where all natural fruit had been removed, and these branches were observed to determine when fruits were removed. Sample size, provided at the base of each bar, is the number of branches monitored.

Once ingested, capsaicin affects the gut processing of seeds. To determine the nature of this effect we studied the most common dispersers of chillies in Bolivia (small-billed elaenia, *Elaenia parvirostris*; Tyrannidae) and in Arizona (curve-billed thrasher, *Toxostoma curvirostre*; Mimidae). Individuals of each species were fed non-pungent chillies injected with a solution containing capsaicin or a control solution, and the time of all defecations was recorded until the last seeds were voided (J.J. Tewksbury *et al.*, 2006, unpublished results). For both species, capsaicin increased seed retention times by an average of 15–20% compared with controls.

Longer retention times induced by capsaicin probably mean that *Capsicum* seeds are dispersed more widely by birds, which would be beneficial to parent plants. Such a benefit may be countered by post-dispersal impacts to the seeds as a result of longer exposure to digestive enzymes and physical processing in the gut (see below). Longer retention times may benefit avian dispersers if they increase assimilation of nutrients in fruit pulp, as happens with emodin; but we did not explore that possibility.

Post-dispersal Effects

Although the post-dispersal phase technically includes all processes from the time of seed deposition to the time of fruit production by the next generation of plants, we restrict the scope of our review to what happens to seeds before germination.

Emodin

The effects of emodin on post-dispersal seed fate are unexplored in species that naturally produce emodin. All that is known comes from experiments on sunflowers (*Helianthus annuus*; Asteraceae) and maize (*Zea mays*; Poaceae), in which application of emodin inhibits germination by 98% and 55–76%, respectively (Hasan, 1998). Seeds of two species of *Rhamnus* will not germinate unless freed of pulp (Gourley and Howell, 1984; Barnea *et al.*, 1991). Emodin is a common constituent of *Rhamnus* fruit pulp but whether it accounts for inhibition of germination is unclear. Likewise, nothing is known about emodin effects on ecologically relevant species of post-dispersal seed predators and pathogens.

Capsaicin

Pungency is most likely to affect seed fate through changes in germination and in vulnerability to seed predators and pathogens. Effects on germination are tied to the differences, described above, in gut retention time. In *C. chacoense* there is no difference between pungent and non-pungent fruits in total percentage of germinated seeds; seeds with and without a coating of capsaicin have equal probabilities of germinating (J.J. Tewksbury *et al.*, 2006, unpublished results). However, seeds from pungent and non-pungent fruits differ substantially in how time in the gut affects seed viability. Seeds from non-pungent plants are not affected by long periods in the gut, whereas seeds from pungent plants decrease from 80% viability if defecated within 30 min to 45% when defecated after 3 h (J.J. Tewksbury *et al.*, 2006, unpublished results). We are currently exploring the reason for the greater susceptibility of pungent seeds to long gut-transit times. Given that capsaicinoids and lignin share a biosynthetic pathway, we hypothesize that pungent plants gain chemical protection from capsaicinoids at the cost of reduced physical protection by lignin. Less lignin in seed coats may mean a greater susceptibility of seeds to physical and chemical breakdown during gut passage.

Capsaicinoids appear to have little influence on post-dispersal predation of *Capsicum* seeds. In Bolivia we coated naturally non-pungent *C. chacoense* seeds with capsaicin and compared their removal to control seeds (naturally non-pungent *C. chacoense* seeds coated with ethanol, the carrier we used for capsaicin). We found no difference in removal rates (J.J. Tewksbury *et al.*, 2006, unpublished results). However, the presence of capsaicin appears to reduce the extent of fungal (*Fusarium* spp.) infection of seeds; infection levels are higher in non-pungent than in pungent *C. chacoense* seeds. This effect is complicated by the fact that passage of seeds through a bird's gut reduces both the amount of capsaicin on seeds and the level of fungal infection, at least temporarily (J.J. Tewksbury *et al.*, 2006, unpublished results).

Discussion

Our take-home message is that fruit secondary compounds have multiple functions on multiple taxa over multiple time scales (Table 2.1). We restricted our review to emodin and capsaicin because no other fruit secondary compounds have been as thoroughly studied in a wide variety of ecological contexts, except perhaps glycoalkaloids (Cipollini and Levey, 1997b,c; Levey and Cipollini, 1998; Wahaj *et al.*, 1998). Despite our fairly narrow scope, the broad spectrum of effects we have described is likely to be typical of fruit secondary metabolites (Wink, 1998; Wink and Schimmer, 1999; Cipollini, 2000). Essentially, all secondary compounds examined in an ecologically relevant way have been reported to have at least some effects on fruit or seed consumers (Cipollini and Stiles, 1991, 1992, 1993; Murray *et al.*, 1994; Bairlein, 1996; Cipollini and Levey, 1997a,b,c; Levey and Cipollini, 1998; Wahaj *et al.*, 1998; Stanley and Lill, 2001; Tewksbury and Nabhan, 2001; Witmer, 2001; Tsahar *et al.*, 2002, 2003; Schaefer *et al.*, 2003; Tsang and Corlett, 2005). Studies that report no effects are rare; we are aware of only one: Struempf and Martínez del Rio (1999) found no effect of amygdalin on food consumption by cedar waxwings (*Bombycilla cedrorum*; Bombycillidae).

Table 2.1. Properties and functions of emodin and capsaicin in fruits before, during and after seed dispersal.

Plant life-history phase	Property or possible function	Emodin	Capsaicin
Pre-dispersal	Higher concentration in unripe fruits	+	+
	Protect fruits and seeds by deterring consumption by:		
	vertebrates	+	+
	invertebrates	+	0
	fungi	+	+
Dispersal	Detectable by dispersing birds	+	0
	Increase gut retention time	+ (Yellow-vented bulbul)	+ (Curve-billed thrasher and Short-billed elaenia)
		− (Chicken and American robin)	
Post-dispersal	Seed viability after bird ingestion	Not studied	− pungent (*C. chacoense*) 0 non-pungent (*C. chacoense*) + pungent (*C. annuum*)
	Protect seeds from predation by:		
	rodents	Not studied	+ (*C. annuum*)
	fungi	Not studied	+ (*C. chacoense*)
	ants	Not studied	0 (*C. chacoense*)

+ = positive effect; − = negative effect; 0 = no effect. Study species are listed in parentheses.

An important corollary is that a narrow exploration of the role of a secondary compound is practically guaranteed to yield an equally narrow and probably misleading interpretation. For example, a study on capsaicin's effect on retention time of seeds in bird guts would lead one to conclude that slower passage through the gut would benefit *Capsicum* plants by ensuring wider dispersal. An examination of seed predation, however, would no doubt reveal that seeds held longer in the gut are stripped of capsaicin and more prone to seed predation than seeds not passed through the gut or those passed more rapidly. In addition, seeds held for >3 h in the gut suffer a reduction in the percentage germinating, presumably due to damage by digestive processes. From the plant's perspective, the *net effect* of a secondary compound is the relevant currency of natural selection. The important point is that this net effect is impossible to gauge without a multifaceted examination of potential effects.

The example with capsaicin, gut passage and germination raises another issue that is often overlooked – metabolic trade-offs. Most studies on secondary metabolites focus on potential benefits, with little recognition of costs. Capsaicin is produced via the phenylpropanoid pathway, which is significant because lignin is also produced by it, and therefore capsaicin and lignin compete for precursors (Sukrasno and Yeoman, 1993). This results in a potential trade-off between physical protection (lignin) and chemical protection (capsaicin), and suggests a mechanistic explanation of the apparently fragile nature of pungent seeds reflected by their lack of germination success after exposure to long gut-transit times. Similarly, emodin concentration in *Rhamnus alaternus* fruits is negatively correlated with amounts of neutral detergent fibre and acid detergent fibre, both of which contain lignin, and other types of fibre (Izhaki *et al.*, 2002). Unfortunately, we cannot understand such trade-offs without a better knowledge of how fruit secondary compounds are created and changed during ripening; fruits and their constituents are highly dynamic. Thus, research on fruit secondary metabolites needs to acknowledge the complex interplay between biosynthetic pathways, genetic regulatory mechanisms, and environmental cues during ripening (Berenbaum, 1995; Lund and Bohlman, 2006).

Although not universal (Cipollini *et al.*, 2004), the kinds of trade-offs described above may be common, because the phenylpropanoid pathway is responsible for the production of practically all fruit secondary metabolites (Dixon and Paiva, 1995). The extent to which metabolic trade-offs determine concentrations of secondary metabolites in plants is under debate, with the outcome largely depending on how limiting metabolic precursors are under different conditions and the relative costs of producing different metabolites (Stamp, 2003a). To our knowledge, only two studies have experimentally examined the relationship between the availability of a potentially limiting precursor and the concentration of secondary compounds in wild fruits. Based on a greenhouse experiment with a desert plant, *Ochradenus baccatus* (Recedaceae), glucosinolate concentration in fruit pulp was positively affected by the sulphur content in soil and was negatively related to water conditions (Lotan, 2005). On the other hand, *Solanum carolinense* (Solanaceae) fruits on plants grown under two nitrogen and two water treatments showed essentially no variation in concentrations of secondary compounds (Cipollini *et al.*, 2004), suggesting that the types of trade-offs we propose for products of the phenylpropanoid pathway are not universal.

Why do fruit secondary metabolites so often have multiple functions?

Multifunctionality of fruit secondary metabolites should not come as a surprise or be viewed as extraordinary. The reason extends beyond previous explanations of multifunctionality, which are based on the simple premise that a compound with more than one benefit will be favoured over a compound with a single benefit (Wink, 1998; Wink and Schimmer, 1999; Cipollini, 2000). That premise begs the question of *how* such variation in function may arise in the first place. What unique characteristics of fruit underlie multifunctionality?

We suggest that fruits are an unusually effective generator of bioactive chemicals. Through their seeds, they link one generation to another and therefore play a central role in determining the fitness of fruiting plants. More fundamentally, because fruit secondary metabolites occur in both pulp and seeds, their impacts span the period from fruit inception through seed dispersal and germination. In species with seed dormancy, this period can be longer than the lifespan of the parent plant. In all species, it includes exposure to practically all types of organisms that typically interact with plants. These organisms include microbes, arthropods and vertebrates. Furthermore, they also encompass mutualists, predators and pathogens. Secondary metabolites mediate interactions with all of them. This enormous variation in type of interaction is essentially an incubator for compounds with multiple functions across diverse taxa of fruit and seed consumers.

In contrast to fruits, leaves generally experience a single type of biotic interaction: herbivory. Any organism that consumes leaf tissue is typically detrimental to plant fitness, whereas organisms that consume fruits may be either beneficial or detrimental. Thus, secondary compounds in leaves have a much narrower scope of selection pressure than those in fruits. In flowers, secondary compounds have the potential to mediate interactions with mutualists (pollinators) and antagonists (nectar robbers, herbivores) in a way similar to what we propose for fruits. Some secondary compounds that occur in nectar are known to deter at least one species of nectar robber, while stimulating nectar consumption by pollinators (Irwin *et al.*, 2004; Tadmor-Melamed *et al.*, 2004; Singaravelan *et al.*, 2005, 2006).

This perspective of fruit secondary metabolites aligns with an unconventional view of fruit pulp. Mack (2000) proposed that fruit pulp originated as a medium in which protective compounds and structures could encase seeds, essentially adding a layer of chemical and physical protection over whatever physical protection was already provided by the seed coat. The evolution of frugivory resulted from vertebrates that specialized in consuming these types of seeds, and gradually developed a more gentle treatment of seeds as plants developed other types of pulp constituents that were more accessible and nutritious to those vertebrates. In much the same way, insect pollination is thought to have derived from insects that initially preyed upon pollen and gradually switched to feeding on nectar as it became available. This view should force a re-evaluation of the distinction between secondary compounds and nutritional rewards in fruit pulp. Carbohydrates, lipids and proteins are typically considered ecologically and evolutionarily distinct from secondary metabolites, yet they occur only because other secondary metabolites created the opportunity in what we view as an evolutionary incubator for such compounds. Furthermore, they fit the classical definition of a secondary metabolite as a compound with no primary (i.e. physiological) function.

More fundamentally, this confusion over categorization requires one to re-examine the usefulness of the term 'secondary metabolite'. Categorizing plant metabolites as primary and secondary can be useful in whole-plant physiological studies where growth is defined as primary and everything else is considered secondary (most notably reproduction, where scents and colours are mediated by secondary metabolites). These categories become conflicted and even irrelevant when a broader ecological or evolutionary view is taken. Major adaptive shifts can be mediated by chemicals for which no 'primary' purpose is known (Bradshaw and Schemske, 2003), but the fitness advantage conferred through the production of the chemical is no less important. Thus, as our understanding of the role of compounds associated with increasing fitness becomes more complete, the false construct of categorizing these compounds as primary or secondary may fall from favour. Indeed, some investigators have begun using the term 'specialized' metabolite, rather than secondary metabolite, to describe compounds that appear to have a relatively narrow function within plant tissues.

If one accepts the premise that fruit secondary metabolites have diverse and multiple functions, the question of how to characterize compounds such as lipids and capsaicinoids has an easy answer: lipids and capsaicinoids are mechanistically different but share some functional similarity as mediators of interactions between plants and their biotic environment. Thus, the primary importance of secondary compounds in fruit pulp may be that essentially *all* compounds in ripe fruit should be given equal billing and regarded, collectively, as primary determinants of fruit fate, seed fate, and ultimately the fitness of fruiting plants.

Acknowledgements

Research on capsaicin was supported by the National Science Foundation (DEB-0129168) and the National Geographic Society Committee for Research and Exploration (7190-02). For logistic support in Bolivia, we thank the Wildlife Conservation Society, Fundación Amigos de la Naturaleza, Museo de Historia Natural Noel Kempf Mercado, and Capitania del Alto y Bajo Izozog (CABI). In the field, Dan Cariveau, Robert Dobbs, Meribeth Huzinga, Carlos Manchego, Haley Morris, Don Odon, Uco Sapag and Melissa Simon helped collect data. Research on emodin was supported by the International Arid Lands Consortium (No. 03R-25), and by the Israel Science Foundation (ISF-600/03). We thank two anonymous reviewers and A. Dennis for helpful suggestions on improving the manuscript.

References

Afik, D. and Karasov, W.H. (1995) The trade-offs between digestion rate and efficiency in warblers and their ecological implications. *Ecology* 76, 2247–2257.

Andelt, W.F., Burnham, K.P. and Baker, D.L. (1994) Effectiveness of capsaicin and Bitrex repellents for deterring browsing by captive Mule Deer. *Journal of Wildlife Management* 58, 330–334.

Anderson, J.A., Bor-Kang, L., Williams, H.J. and Scott, A.I. (1988) Deoxygenation of phenolic natural products: enzymatic conversion of emodin to chrysophanol. *Journal of the American Chemical Society* 110, 1623–1624.

Bairlein, F. (1996) Fruit-eating in birds and its nutritional consequences. *Comparative Biochemistry and Physiology* 113A, 215–224.

Barnea, A., Yom-Tov, Y. and Friedman, J. (1991) Does ingestion by birds affect seed germination. *Functional Ecology* 5, 394–402.

Berenbaum, M.R. (1995a) The chemistry of defense: theory and practice. *Proceedings of the National Academy of Sciences* 92, 2–8.

Berenbaum, M.R. (1995b) Turnabout is fair play: secondary roles for primary metabolites. *Journal of Chemical Ecology* 21, 125–139.

Blum, E., Liu, K., Mazourek, M., Yoo, E.Y., Jahn, M. and Paran, I. (2002) Molecular mapping of the C locus for presence of pungency in *Capsicum*. *Genome* 45, 702–705.

Bosland, P.W. (1994) Chiles: history, cultivation, and uses. In: Charalambous, G. (ed.) *Spices, Herbs and Edible Fungi.* Elsevier, Amsterdam, pp. 347–366.

Bosland, P.W. (1998) Chiles: a gift from a fiery God. *HortScience* 34, 809–811.

Bradshaw, H.D. and Schemske, D.W. (2003) Allele substitution at a flower colour locus produces a pollinator shift in monkeyflowers. *Nature* 426, 176–178.

Cichewicz, R.H. and Thorpe, P.A. (1996) The antimicrobial properties of chile peppers (*Capsicum* species) and their uses in Mayan medicine. *Journal of Ethnopharmacology* 2, 61–70.

Cipollini, M.L. (2000) Secondary metabolites of vertebrate-dispersed fruits: evidence for adaptive functions. *Revista Chilena de Hitoria Natural* 73, 421–440.

Cipollini, M.L. and Levey, D.J. (1997a) Secondary metabolites of fleshy vertebrate-dispersed fruits: adaptive hypotheses and implications for seed dispersal. *American Naturalist* 150, 346–372.

Cipollini, M.L. and Levey, D.J. (1997b) Antifungal activity of *Solanum* fruit glycoalkaloids: implications for frugivory and seed dispersal. *Ecology* 78, 799–809.

Cipollini, M.L. and Levey, D.J. (1997c) Why are some fruits toxic? Glycoalkaloids in *Solanum* and fruit choice by vertebrates. *Ecology* 78, 782–798.

Cipollini, M.L. and Stiles, E.W. (1991) Antifungal activity of ripe ericaceous fruits: phenolic-acid interactions and palatability for dispersers. *Biochemical Systematics and Ecology* 20, 501–514.

Cipollini, M.L. and Stiles, E.W. (1992) Relative risks of microbial rot for fleshy fruits: significance with respect to dispersal and selection for secondary defense. *Advances in Ecological Research* 23, 35–91.

Cipollini, M.L. and Stiles, E.W. (1993) Fruit rot, antifungal defense, and palatability of fleshy fruits for frugivorous birds. *Ecology* 74, 751–762.

Cipollini, M.L., Paulk, E., Mink, K., Vaughn, K. and Fischer, T. (2004) Defense tradeoffs in fleshy fruits: effects of resource variation on growth, reproduction, and fruit secondary chemistry in *Solanum carolinense. Journal of Chemical Ecology* 30, 1–17.

Coley, P.D. (1986) Costs and benefits of defence by tannins in a neotropical tree. *Oecologica* 70, 352–368.

Cordell, G.A. and Araujo, O.E. (1993) Capsaicin: identification, nomenclature, and pharmacotherapy. *Annals of Pharmacotherapy* 27, 330–6.

D'arcy, W.G. and Eshbaugh, W.H. (1974) New World peppers (Capsicum–Solanaceae) north of Columbia: a resume. *Baileya* 19, 93–105.

Debkirtaniya, S., Ghosh, M.R., Mitra, S.R., Adityachaudhury, N. and Chatterjee, A. (1980) Note on the insecticidal properties of the fruits of chilli. *Indian Journal of Agricultural Sciences* 50, 510–512.

Dewick, P.M. (1998) *Medicinal Natural Products: A Biosynthetic Approach.* John Wiley and Sons, New York.

Dixon, R.A. and Paiva, N.L. (1995) Stress-induced phenylpropanoid metabolism. *Plant and Cell* 7, 1085–1097.

Ehrlen, J. and Eriksson, O. (1993) Toxicity in fleshy fruits: a non-adaptive trait? *Oikos* 66, 107–113.

Estrada, B., Pomar, F., Diaz, J., Merino, F., Bernal, M.A. (1999) Pungency level in fruits of the Padron pepper with different water supply. *Scientia Horticulturae* 81, 385–396.

Estrada, B., Bernal, M.A., Diaz, J., Pomar, F. and Merino, F. (2000) Fruit development in *Capsicum annuum*: changes in capsaicin, lignin, free phenolics, and peroxidase patterns. *Journal of Agricultural and Food Chemistry* 48, 6234–6239.

Evans, W.C. (1996) *Trease & Evans' Pharmacognosy*, 4th edn. W.B. Saunders Company, London.

Fujiwake, H., Suzuki, T. and Iwai, K. (1982) Formation and metabolism of the pungent principle of *Capsicum* fruits. 11. Capsaicinoid formation in the protoplast from the placenta of *Capsicum* fruits. *Agricultural and Biological Chemistry* 46, 2591–2592.

Gourley, L.C. and Howell, E. (1984) Factors in buckthorn invasion documented: control measure checked (Wisconsin). *Restoration and Management Notes* 2, 87.

Govindarajan, V.S. (1986) Capsicum: production, technology, chemistry, and quality. 3. Chemistry of the color, aroma, and pungency stimuli. *CRC Critical Reviews in Food Science and Nutrition* 24, 245–355.

Govindarajan, V.S. and Sathyanarayana, M.N. (1991) Capsicum: production, technology, chemistry, and quality. 5. Impact on physiology, pharmacology, nutrition, and metabolism – structure, pungency, pain, and desensitization sequences. *Critical Reviews in Food Science and Nutrition* 29, 435–474.

Govindarajan, V.S., Rajalakshmi, D. and Chand, N. (1987) Capsicum: production, technology, chemistry, and quality. 4. Evaluation of quality. *CRC* Critical *Reviews in Food Science and Nutrition* 25, 185–282.

Hall, R.D. and Yeoman, M.M. (1991) The influence of intracellular pools of phenylalanine derivatives upon the synthesis of capsaicin by immobilized cell-cultures of the chilie pepper *Capsicum frutescens. Planta* 185, 72–80.

Hamilton, J.G., Zangerl, A.R., DeLucia, E.H. and Berenbaum, M.R. (2001) The carbon-nutrient balance hypothesis: its rise and fall. *Ecology Letters* 4, 86–95.

Harborne, J.B., Baxter, H. and Moss, G.P. (1999) *Phytochemical Dictionary. A Handbook of Bioactive Compounds from Plants.* Taylor and Francis, London.

Hasan, H.A.H. (1998) Studies on toxigenic fungi in roasted foodstuff (salted seed) and halotolerant activity of emodin-producing *Aspergillus wentii*. *Folia Microbiologica* 43, 383–391.

Herms, D.A. and Mattson, W.J. (1992) The dilemma of plants: to grow or defend. *Quarterly Review of Biology* 67, 283–335.

Herrera, C.M. (1982) Defense of ripe fruits from pests: its significance in relation to plant–disperser interactions. *American Naturalist* 120, 218–241.

Herrera, C.M. (1984) A study of avian frugivores, bird-dispersed plants, and their interactions in Mediterranean scrublands. *Ecological Monographs* 54, 1–20.

Herrera, C.M. (2002) Seed dispersal by vertebrates. In: Herrera, C.M. and Pellmyr, O. (eds) *Plant–Animal Interactions: An Evolutionary Approach*. Blackwell Science, Oxford, pp. 185–210.

Howe, H.F. (1977) Bird activity and seed dispersal of a tropical wet forest tree. *Ecology* 58, 539–50.

Hulme, P.E. and Benkman, C.W. (2002) Granivory. In: Herrera, C. and Pellmyr, O. (eds) *Plant–Animal Interactions: An Evolutionary Approach*. Blackwell Science, Oxford, pp. 132–156.

Irwin, R.E., Adler, L.S. and Brody, A.K. (2004) The dual role of floral traits: pollinator attraction and plant defense. *Ecology* 85, 1503–1511.

Iwai, K., Suzuki, T. and Fujiwake, H. (1979) Formation and accumulation of pungent principle of hot pepper fruits, Capsaicin and its analogs, in *Capsicum-annuun var annuun* cv. *karayatsubusa* at different growth-stages after flowering. *Agricultural and Biological Chemistry* 43, 2493–2498.

Izhaki, I. (2002a) Emodin: a secondary metabolite with multiple ecological functions in higher plants. *New Phytologist* 155, 205–217.

Izhaki, I. (2002b) The role of fruit traits in determining fruit removal in East Mediterranean ecosystems. In: Levey, D.J., Silva, W.R. and Galetti, M. (eds) *Seed Dispersal and Frugivory: Ecology, Evolution, and Conservation*. CAB International, Wallingford, UK, pp. 161–175.

Izhaki, I. and Safriel, U.N. (1985) Why do fleshy-fruited plants of the Mediterranean scrub intercept fall-passage but not spring-passage of seed-dispersing migratory birds? *Oecologia* 67, 40–43.

Izhaki, I., Tsahar, E., Paluy, I. and Friedman, J. (2002) Within-population variation and interrelationships between morphology, nutritional content, and secondary compounds of *Rhamnus alaternus* fruits. *New Phytologist* 156, 217–223.

Janzen, D.H. (1971) Seed predation by animals. *Annual Review of Ecology and Systematics* 2, 465–492.

Janzen, D.H. (1977) Why fruits rot, seeds mould, and meat spoils. *American Naturalist* 111, 691–713.

Jordano, P. (2000) Fruits and frugivory. In: Fenner, M. (ed.) *Seeds: The Ecology of Regeneration in Natural Plant Communities*. CAB International, Wallingford, UK, pp. 105–151.

Kawada, T., Watanabe, T., Katsura, K., Takami, H. and Iwai, K. (1985) Formation and metabolism of pungent principle of *Capsicum* fruits. 15. Microdetermination of capsaicin by high-performance liquid-chromatography with electrochemical detection. *Journal of Chromatography* 329, 99–105.

Kessler, A. and Baldwin, I.T. (2002) Plant responses to insect herbivory: the emerging molecular analysis. *Annual Review of Plant Biology* 53, 299–328.

Kim, Y.M., Lee, C.H. and Lee, H.S. (2004) Anthraquinones isolated from *Cassia tora* (Leguminosae) seed show an antifungal property against phytopathogenic fungi. *Journal of Agricultural and Food Chemistry* 52, 6096–6100.

Koricheva, J. (2002) The carbon-nutrient balance hypothesis is dead: long live the carbon-nutrient balance hypothesis? *Oikos* 98, 537–539.

Lerdau, M. and Coley, P.D. (2002) Benefits of the carbon-nutrient balance hypothesis. *Oikos* 98, 534–536.

Levey, D.J. (2004) The evolutionary ecology of ethanol production and alcoholism. *Integrative and Comparative Biology* 44, 284–289.

Levey, D.J. and Cipollini, M.L. (1998) A glycoalkaloid in ripe fruit deters consumption by cedar waxwings. *Auk* 115, 359–367.

Levey, D.J. and Martínez del Rio, C. (2001) It takes guts (and more) to eat fruit: lessons from avian nutritional ecology. *Auk* 118, 819–831.

Levey, D.J., Tewksbury, J.J., Cipollini, M.L. and Carlo, T.A. (2006) A field test of the directed deterrence hypothesis in two species of wild chili. *Oecologia* 150, 61–68.

Lotan, A. (2005) The impact of environmental conditions on the morphology, nutritional value and secondary metabolites in *Ochradenus baccatus* fruits. MSc thesis, University of Haifa, Israel.

Lund, S.T. and Bohlman, J. (2006) The molecular basis for wine grape quality: a volatile subject. *Science* 311, 804–805.

Mack, A.L. (2000) Did fleshy fruit evolve as a defence against seed loss rather than as a dispersal mechanism? *Journal of Biosciences* 25, 93–97.

McCarty, J., Levey, D.J., Greenberg, C. and Sargent, S. (2002) Spatial and temporal variation in fruit use by wildlife in a forested landscape. *Forest Ecology and Management* 164, 277–291.

McLeod, M.J., Guttman, S.I. and Eshbaugh, W.H. (1982) Early evolution of chile peppers (*Capsicum*). *Economic Botany* 36, 361–386.

Murray, K.G., Russell, S., Picone, C.M., Winnett-Murray, K., Sherwood, W. and Kuhlmann, M.L. (1994) Fruit laxatives and seed passage rates in frugivores: consequences for plant reproductive success. *Ecology* 75, 989–994.

Norman, D.M., Mason, J.R. and Clark, L. (1992) Capsaicin effects on consumption of food by cedar waxwings and house finches. *Wilson Bulletin* 104, 549–551.

Oh, N.-S., Shin, D.-B., In, M.-J., Chang, Y.-I. and Han, M. (2004) Effects of capsaicin on the growth and ethanol production of *Zygosaccharomyces rouxii* KFY80 isolated from Gochujang (fermented hot pepper paste). *Food Science and Biotechnology* 13, 749–753.

Paneitz, A. and Westendorf, J. (1999) Anthranoid contents of rhubarb (*Rheum undulatum* L.) and other *Rheum* species and their toxicological relevance. *European Food Research and Technology* 210, 97–101.

Sagers, C.L. and Coley, P.D. (1995) Benefits and costs of defence in a neotropical shrub. *Ecology* 76, 1835–1843.

Santilli, F., Mori, L. and Galardi, L. (2004) Evaluation of three repellents for the prevention of damage to olive seedlings by deer. *European Journal of Wildlife Research* 50, 85–89.

Schaefer, H.M., Schmidt, V. and Winkler, H. (2003) Testing the defence trade-off hypothesis: how contents of nutrients and secondary compounds affect fruit removal. *Oikos* 102, 318–329.

Schafer, E.W., Bowles, W.A. and Hurlbut, J. (1983) The acute oral toxicity, repellency, and hazard potential of 998 chemicals to one or more species of wild and domestic birds. *Archives of Environmental Contamination and Toxicology* 12, 355–382.

Sherburne, J.A. (1972) Effect of seasonal changes in the abundance and chemistry of the fleshy fruits of northeastern woody shrubs on patterns of exploitation by frugivorous birds. PhD thesis, Cornell University, Ithaca, New York.

Singaravelan, N., Ne'eman, G., Inbar, M. and Izhaki, I. (2005) Feeding responses of free-flying honeybees to secondary compounds mimicking floral nectar. *Journal of Chemical Ecology* 31, 2791–2804.

Singaravelan, N., Inbar, M., Ne'eman, G., Distl, M., Wink, M. and Izhaki, I. (2006) The effects of nectar-nicotine on colony fitness of caged honeybees. *Journal of Chemical Ecology* 32, 49–58.

Stamp, N. (2003a) Out of the quagmire of plant defence hypotheses. *Quarterly Review of Biology* 78, 23–55.

Stamp, N. (2003b) Theory of plant defensive level: example of process and pitfalls in development of ecological theory. *Oikos* 102, 672–678.

Stanley, M.C. and Lill, A. (2001) Response of silvereyes (*Zosterops lateralis*) to dietary tannins: the paradox of secondary metabolites in ripe fruit. *Australian Journal of Zoology* 49, 633–640.

Struempf, H.M. and Martínez del Rio, C. (1999) Amygdalin, a cyanogenic glycoside in ripe fruit, does not deter consumption by cedar waxwings. *Auk* 116, 749–758.

Sukrasno, N. and Yeoman, M.M. (1993) Phenylpropanoid metabolism during growth and development of Capsicum frutescens fruits. *Phytochemistry* 32, 839–844.

Suzuki, T. and Iwai, K. (1984) Constituents of red pepper species: chemistry, biochemistry, pharmacology, and food science of the pungent principle of *Capsicum* species. In: Brossi, A. (ed.) *The Alkaloids: Chemistry and Pharmacology*. Academic Press, Orlando, Florida, pp. 227–299.

Tadmor-Melamed, H., Markman, S., Arieli, A., Distl, M., Wink, M. and Izhaki, I. (2004) Limited ability of Palestine sunbirds (*Nectarinia osea*) to cope with pyridine alkaloids in nectar of tree tobacco (*Nicotiana glauca*). *Functional Ecology* 18, 844–850.

Tang, A.M.C., Corlett, R.T. and Hyde, K.D. (2005) The persistence of ripe fleshy fruits in the presence and absence of frugivores. *Oecologia* 142, 232–237.

Tewksbury, J.J. (2002) Fruits, frugivores and the evolutionary arms race. *New Phytologist* 156, 137–139.

Tewksbury, J.J. and Nabhan, G.P. (2001) Directed deterrence by capsaicin in chillies. *Nature* 412, 403–404.

Tewksbury, J.J., Nabhan, G.P., Norman, D.M., Suzan, H., Toxill, J. and Donovan, J. (1999) *In situ* conservation of wild chiles and their biotic associates. *Conservation Biology* 13, 98–107.

Tewksbury, J.J., Manchego, C., Haak, D. and Levey, D.J. (2006) Where did the chili get its spice? Biogeography of capsaicinoid production in ancestral wild chili species. *Journal of Chemical Ecology* 32, 547–564.

Thompson, J.N. and Willson, M.F. (1979) Evolution of temperate fruit–bird interactions: phenological strategies. *Evolution* 33, 973–982.

Thomson, R.H. (1987) *Naturally Occurring Quinines. III. Recent Advances.* Chapman and Hall, London.

Thomson, R.H. (1997) *Naturally Occurring Quinines. IV. Recent Advances.* Blackie Academic and Professional, Chapman and Hall, London.

Tsahar, E., Friedman, J. and Izhaki, I. (2002) Impact of fruit removal and seed predation of a secondary metabolite, emodin, in *Rhamnus alaternus* fruit pulp. *Oikos* 99, 290–299.

Tsahar, E., Friedman, J. and Izhaki, I. (2003) Secondary metabolite emodin increases food assimilation efficiency of yellow-vented bulbuls (*Pycnonotus xanthopygos*). *Auk* 120, 411–417.

Tsang, A.C.W. and Corlett, R.T. (2005) Reproductive biology of the *Ilex* species (Aquifoliaceae) in Hong Kong, China. *Canadian Journal of Botany* 83, 1645–1654.

Wagner, K.K. and Nolte, D.L. (2000) Evaluation of Hot Sauce® as a repellent for forest animals. *Wildlife Society Bulletin* 28, 76–83.

Wahaj, S.A., Levey, D.J., Sanders, A.K. and Cipollini, M.L. (1998) Control of gut retention time by secondary metabolites in ripe *Solanum* fruits. *Ecology* 79, 2309–2319.

Walsh, B.M. and Hoot, S.B. (2001) Phylogenetic relationships of *Capsicum* (Solanaceae) using DNA sequences from two noncoding regions: the chloroplast atpB-rbcL spacer region and nuclear *waxy* introns. *International Journal of Plant Sciences* 162, 1409–1418.

Wang, H.H. (1993) Antitrichomonal action of emodin in mice. *Journal of Ethnopharmacology* 40, 111–116.

Wang, H.H. and Chung, J.G. (1997) Emodin-induced inhibition of growth and DNA damage in the *Heliobacter pylori. Current Microbiology* 35, 262–266.

Wells, J.M., Cole, R.J. and Kirksey, J.W. (1975) Emodin, a toxic metabolite of *Aspergillus wentii* isolated from weevil-damaged chestnuts. *Applied Microbiology* 30, 26–28.

Whittaker, R.H. and Feeny, P.P. (1971) Allelochemics: chemical interactions between species. *Science* 171, 757–770.

Wink, M. (1998) Modes of action of alkaloids. In: Roberts, M.F. and Wink, M. (eds) *Alkaloids. Biochemistry, Ecology, and Medicinal Applications.* Plenum Press, New York, pp. 301–326.

Wink, M. and Schimmer, O. (1999) Modes of action of defensive secondary metabolites. In: Wink, M. (ed.) *Function of Plant Secondary Metabolites and their Exploitation in Biotechnology.* Academic Press, Sheffield, UK, pp. 1–16.

Witmer, M.C. (2001) Nutritional interactions and fruit removal: springtime consumption of *Viburnum opulus* fruits by cedar waxwings. *Ecology* 82, 3120–3130.

Yang, Y.C., Lim, M.Y. and Lee, H.S. (2003) Emodin isolated from *Cassia obtusifolia* (Leguminosae) seed shows larvicidal activity against three mosquito species. *Journal of Agricultural and Food Chemistry* 51, 7629–7631.

Zeyrek, F.Y. (2005) *In vitro* activity of capsaicin against *Helicobacter pylori. Annals of Microbiology* 55, 125–127.

3 The Evolution of Visual Fruit Signals: Concepts and Constraints

H.M. Schaefer and V. Schaefer

Introduction

The diversity and beauty of fruit and flower coloration have attracted humans long before the beginning of scientific thought, as seen in the writings of the Ancient Greeks. Soon after Darwin's groundbreaking work on natural selection, researchers hypothesized that the colours of fruits and the visual systems of the animals that commonly feed on them might have mutually influenced each other (see Allen, 1879; Kerner, 1895). In 1879, Allen wrote:

> How fruit began to acquire those brilliant tints is not difficult to see. ... This tendency [to acquire non-green coloration] would ... be increased by natural selection in those seeds, which like fruits-proper, derive benefits from the observation of animals.

In a reply to this work, Alfred Wallace (1879) stated:

> ... the evidence that fruits in general have been modified both in edibility and attractiveness [e.g. colour] in relation to the animals which feed upon them, is by no means so clear as in the case of flowers.

Given the impressive advances in evolutionary ecology over the last 130 years, it is perhaps surprising that Wallace's cautionary statement still holds true and that the evolutionary and ecological processes responsible for fruit colour diversity are not understood (Willson and Whelan, 1990; Burns and Dalen, 2002).

Much of the research on fruit coloration in the last century adopted an adaptive framework. This framework seems intuitively plausible given that fruits dispersed by diurnal vertebrates typically change colour during ripening and thereby become more conspicuous, at least to the human eye. However, recent studies have emphasized that seed dispersers are not the sole selective pressure acting upon fruit coloration. In this chapter we consider the different selective factors (Fig. 3.1) and outline recent developments in the field of communication between fruiting plants and visually oriented animals. Rather than viewing fruit colours simply as colours, we view them as signals. This shift of emphasis offers more general insights into the selective pressures between plants and their frugivorous consumers. Many of the conclusions reached in this chapter may also apply to visual signals in other plant–animal interactions (pollination, herbivory). However, the evolutionary ecology of other signals, for example fruit odours, might differ greatly from the perspective given below.

Approaches to fruit colour diversity

Until recently, two problems have hindered our understanding of the evolution of fruit coloration. First, objective colour measurements allowing for colour quantification independent of a specific type of consumer vision (e.g. human) were not available. Researchers were forced to categorize colour according to human perception, which is a poor indicator of the perception of other seed disperser groups because of distinct differences in their visual capacities (Burkhardt, 1982). Second, the factors determining colour vision in non-human seed dispersers were largely unknown. For these reasons, colour was rarely included in rigorous analyses on fruit traits that tested, and often rejected, the adaptationist view on fruit–frugivore interactions (e.g. Herrera, 1992; Jordano, 1995).

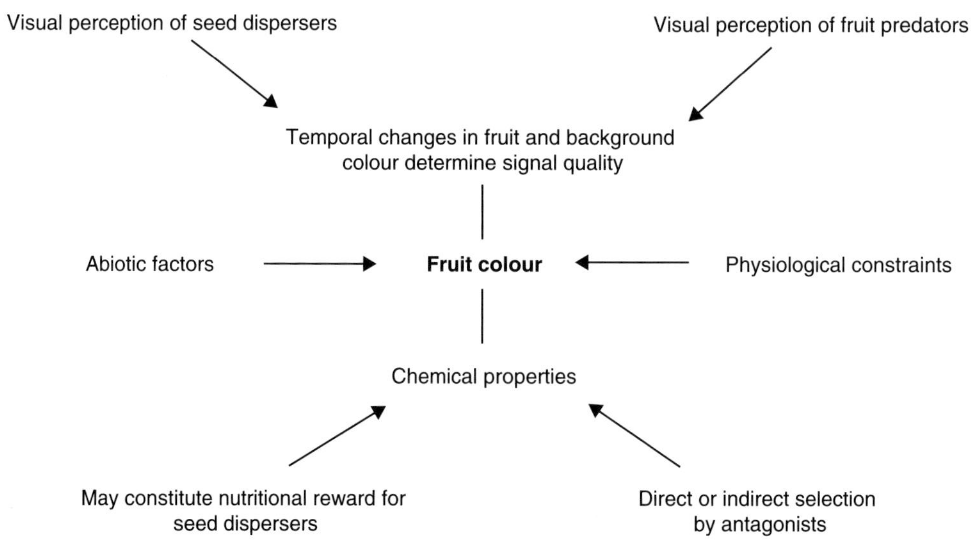

Fig. 3.1. Schematic illustration of the potential selective pressures acting on the different properties of fruit colours.

In the past, studies on fruit coloration were often influenced by the disperser-syndrome hypothesis, which posits that a range of fruit characters, including fruit colour, are associated with a specific type of seed disperser (e.g. McKey, 1975; Howe and Estabrook, 1977). Applying the hypothesis to fruit colour, its core prediction is that the colour vision of consumer groups and fruit coloration are adaptively fine-tuned. Prior to the development of tools for objective colour measurements in the last 15 years, tests of this prediction were mostly restricted to evaluating whether patterns of fruit colour diversity in a given habitat or region matched those of colour selection by representative species of the main seed disperser groups. Intriguingly, studies yielded conflicting results between relatively robust patterns of fruit colour diversity in different habitats (Fig. 3.2) and inconsistent colour selection by most species of seed dispersers in controlled experiments (for a review, see Willson and Whelan, 1990). In particular, researchers noted that temperate and tropical bird-dispersed fruits are predominantly red or black, whereas fruits dispersed by primates are generally orange, brown or yellow (Janson, 1983; Wheelwright and Janson, 1985; Willson *et al.*, 1989). Accounting for phylogenetic effects, a study concluded that colour was the best indicator of nine fruit traits for differential food selection of primates and birds in South African and Malagasy tree communities (Voigt *et al.*, 2004).

Although numerous studies have tried to link the prevalence of red and black bird-dispersed fruits to colour selection by birds, consistent colour

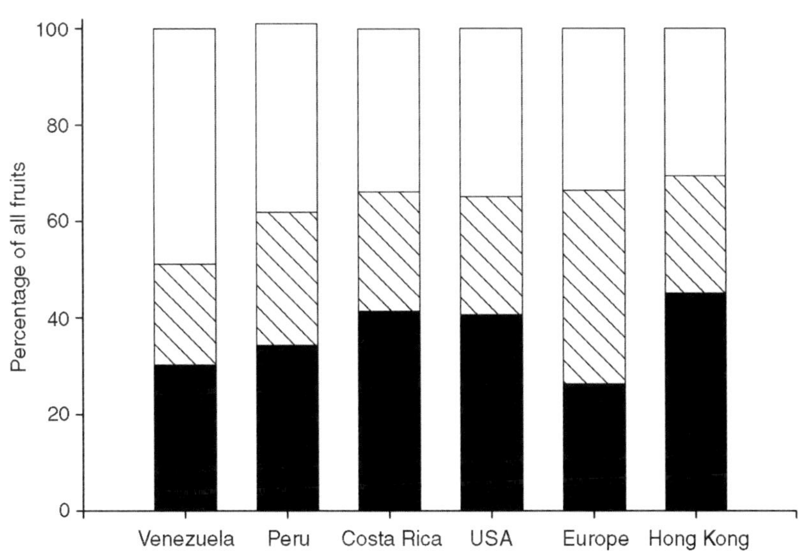

Fig. 3.2. The colour diversity of bird-dispersed fruits is similar across different habitats and latitudes. Note the prevalence of black (black bars) and red fruits (hatched bars) relative to other fruit colours (white bars). Reference sources: Turcek (1963), Long (1971), Wheelwright and Janson (1985), Corlett (1996) and Schmidt *et al.* (2004).

preferences of birds were found only in few species (e.g. Puckey *et al.*, 1996; Siitari *et al.*, 1999; Whitney, 2005). Instead, the emerging picture has been that most species of birds exhibit inconsistent and transient colour choices, with high variability within and between individuals (Table 3.1; Willson *et al.*, 1990; Willson and Comet, 1993; Traveset and Willson, 1998; Schmidt *et al.*, 2004). Similarly, there are age-specific differences in colour preferences. For example, the unlearned bias for red fruits in juvenile blackcaps (*Sylvia atricapilla*; Sylvidae) is lost with experience and is not present in adult birds (Schmidt and Schaefer, 2004), whereas a preference for UV-reflecting fruits in redwings (*Turdus iliacus*; Turdidae) is present in adult birds, but not in juveniles (Siitari *et al.*, 1999; Honkavaara *et al.*, 2004). The observed flexibility in fruit-colour choices is consistent with the large variety of fruits that many tropical and migratory temperate frugivorous species consume over extended periods; for example, fruits from more than 100 species of plants were consumed by Salvin's curassow (*Mitu salvini*; Cracidae) over the course of 14 months (Santamaría and Franco, 2000). In summary, avian preferences for certain hues are apparently not very strong and are thus unlikely to explain the patterns of global colour diversity in bird-dispersed fruits. We suggest that the relationship between fruit colours and consumer type (see previous paragraph) is instead explained by a different mechanism: the likelihood of fruit detection.

Fruit Colours as Signals

Fruit colours meet the criteria of classical signals, which are often defined as structures that increase the fitness of the sender by altering the behaviour of other organisms (Maynard-Smith and Harper, 1995). Natural selection shapes signal evolution by increasing the efficiency of information transfer between organisms (Otte, 1974). In general, the information transfer depends on the detectability and the reliability of a signal in indicating its message. The detectability of a visual signal is determined by its contrast against the background, the visual conditions during signalling and by the visual perception of the animal receiving the signal. Long before the rise of signal theory in modern evolutionary ecology, its essence had been captured by Kerner (1895), who assumed that those fruit colours that are most effective in attracting seed dispersers are most conspicuous: that is, most easily distinguished from the background. His hypothesis, later termed the 'contrast hypothesis' (Schmidt *et al.*, 2004) is consistent with the main tenet of signal theory – that detectability is one of the two main factors driving the evolution of signals (Schluter and Price, 1993).

Supporting the contrast hypothesis, various studies have documented that red and black, the most common colours of bird-dispersed fruits, contrast more strongly against the background than other fruit colours (Lee *et al.*, 1994; Schmidt *et al.*, 2004). Unfortunately, none of the studies assessed contrasts according to avian vision but used the human-based CIE system (Lee *et al.*, 1994) or the consumer-independent Euclidean distance

Table 3.1. Experiments investigating avian fruit colour choices have often yielded inconsistent or transient colour choices. Consistent preferences for red fruits were found in some species of bird. The term 'frugivorous birds' is used if colour was not examined on individual species but on an entire community of frugivorous birds.

Species	Fruit type	Colour	Preference	Reference
Frugivorous birds	Wild	UV-blue, black	None	Willson and Whelan, 1989
Dumetella carolinesis and two *Catharus* species	Artificial	Red, blue, yellow, black	Inconsistent	Willson *et al.*, 1990
Corvus caurinus	Wild and artificial	Red, blue, green, yellow	None in young, inconsistent in adult	Willson and Comet, 1993
Zosterops lateralis	Wild and artificial	Red, yellow, white	Red preferred	Puckey *et al.*, 1996
Frugivorous birds	Wild	Red, orange	Red > orange	Gervais *et al.*, 1999
Zosterops lateralis	Wild	Red, white	Red > white	Giles and Lill, 1999
Turdus iliacus	Wild	UV-blue, black	UV-blue > black	Siitari *et al.*, 1999
Neotropical frugivores	Wild	UV-blue, blue without UV	UV-blue > blue	Altshuler, 2001
Frugivorous birds	Artificial	Red, black, white	Red > black > white	Alves-Costa and Lopes, 2001
Turdus iliacus	Wild	UV-blue, black, red, white	Juvenile: UV-blue = black = red > white; Adult: UV-blue > red	Honkavaara *et al.*, 2004
Sylvia atricapilla	Artificial	Red, orange, white, yellow, green, blue	None in adult, red preference in juveniles	Schmidt and Schaefer, 2004
Four neotropical tanagers	Artificial	Red, orange, white, yellow, green, blue	None	Schmidt *et al.*, 2004
Frugivorous birds	Wild	Yellow, orange, red	Inconsistent	Whitney, 2005

between fruit and background spectra (Schmidt *et al.*, 2004). Further support for the contrast hypothesis stems from behavioural data: four species of frugivorous birds preferred more contrasting fruit displays regardless of the colours that created the contrasts (Schmidt *et al.*, 2004). Thus, if seed dispersers select fruits mainly on the basis of their detectability, then the contrasts between fruit and background (leaves, bark, and structures associated with the fruit display), and how this contrast relates to the visual perception of the animal, are more important, from an evolutionary perspective, than the colour of a fruit per se.

In general, the evolution of signal design is thought to be a compromise between efficient signalling to the intended recipients and inconspicuousness to unintended recipients that may use the signal to the sender's disadvantage (Endler, 2000). Consequently, if we can quantify the detectability of fruit signals according to the perception of various fruit-consumer groups, we can derive strong predictions based on signal theory on how seed dispersers influenced the evolution of fruit colour. The task of detecting fruits against their background might, in turn, have influenced the perceptual tuning of colour vision in some consumer groups, such as primates (Osorio and Vorobyev, 1996). For example, because the interaction between fruiting plants and seed dispersers is a mutualistic one, both partners would benefit from an increase in the efficiency of information transfer either by using fruit colours that are easy to detect or by evolving visual sensitivities that allow for efficient detection of fruits against a predominantly green background. It is important to note, however, that animals' ability to perceive colour is a conservative trait and that visual abilities are often not tightly linked to a species' particular lifestyle (Briscoe and Chittka, 2001). Furthermore, animals use vision for tasks other than foraging. This includes identifying potential mates and predators, which in turn reduces the possibility that a specific task, such as finding fruits, fine-tunes the receptors used for perceiving fruit colours.

Methodology for assessing signals

Using a spectrometer, one can quantify contrast, and hence the detectability of a fruit's signal. The spectrometer measures the reflectance of surfaces, such as fruit skin and the objects of the background (e.g. leaves and any other structures against which fruits are presented), resulting in reflectance spectra (Fig. 3.3). Reflectance spectra are measured and illustrated relative to standard white and black references. Additionally, the spectrometer can measure the irradiance, which characterizes the spectral composition of the ambient light at the place where the fruit is presented and where signalling takes place. Considering ambient light is particularly important because its spectral composition may alter the colour perception of the signal recipient. If the distance between observer and target is short, the influence of the air through which the spectrum travels can usually be ignored. In this case the irradiance and reflectance spectra are combined over the entire visible range of a given recipient, and the resulting radiance spectrum (Fig. 3.4)

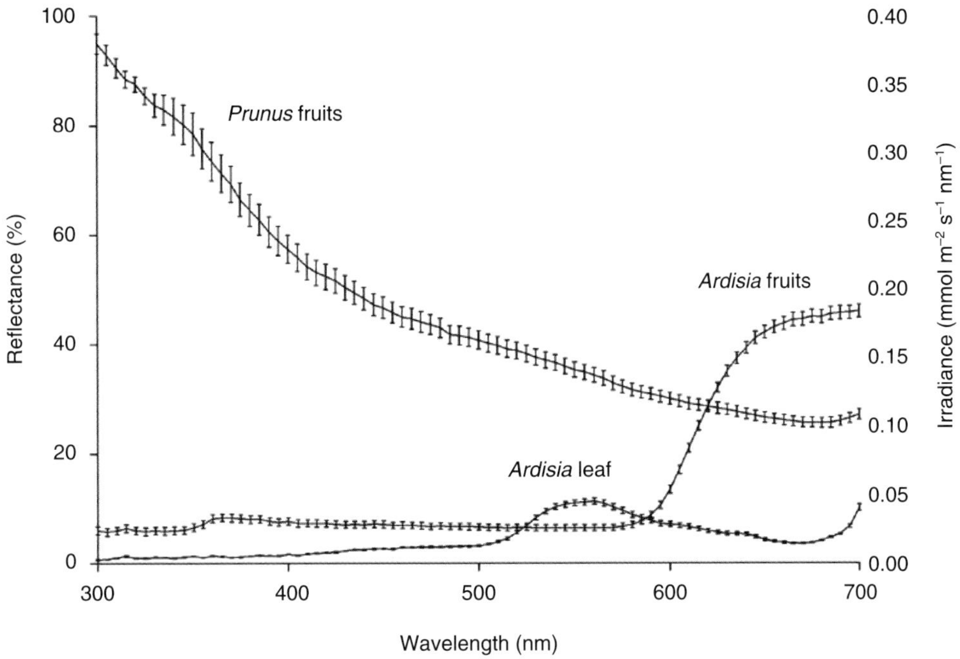

Fig. 3.3. Mean reflectance spectra ± SE are shown of 20 red fruits (*Ardisia crenata*; Myrsinaceae), 20 UV-reflecting fruits (*Prunus spinosus*; Rosaceae) and ten *A. crenata* leaves over the visible range of birds. Using irradiance illustrates the spectral composition of ambient light in the understorey of forests, measured vertically 2 m above the ground.

represents the physical raw data that enter the eye of the consumer (see Endler, 1990; Endler and Mielke, 2005). Before the radiance spectrum can be processed by the photoreceptors in the retina, it has to pass various structures of the eye (lens, macular pigments), which again alter the spectrum (Regan *et al.*, 2001). While spectrometer measurements specify only the radiance spectrum, current eye models (Vorobyev and Osorio, 1998) account for the properties of the visual system of different animal taxa.

In parallel with the use of spectrometry for analysing visual signals, knowledge on the visual perception of seed disperser taxa has increased considerably in recent years (Osorio and Vorobyev, 1996; Vorobyev and Osorio, 1998; Hart, 2001; Regan *et al.*, 2001). The most important technique in investigating the spectral sensitivities of different animals has been microspectrophotometry – the application of standard spectrometry to microscopic samples of the retina (Hart, 2001). These data, together with measurements of the absorbance of oil droplets (in birds) and ocular filters, allow prediction of the spectral sensitivities of different animal taxa (Kelber *et al.*, 2003). Colour quantification and knowledge of consumer perception represent powerful tools which can cast new light on concepts such as the disperser-syndrome hypothesis and Allen's (1879) paradigm of co-evolution, both of which assume that fruit colours are signals that are matched to the perception of the recipient.

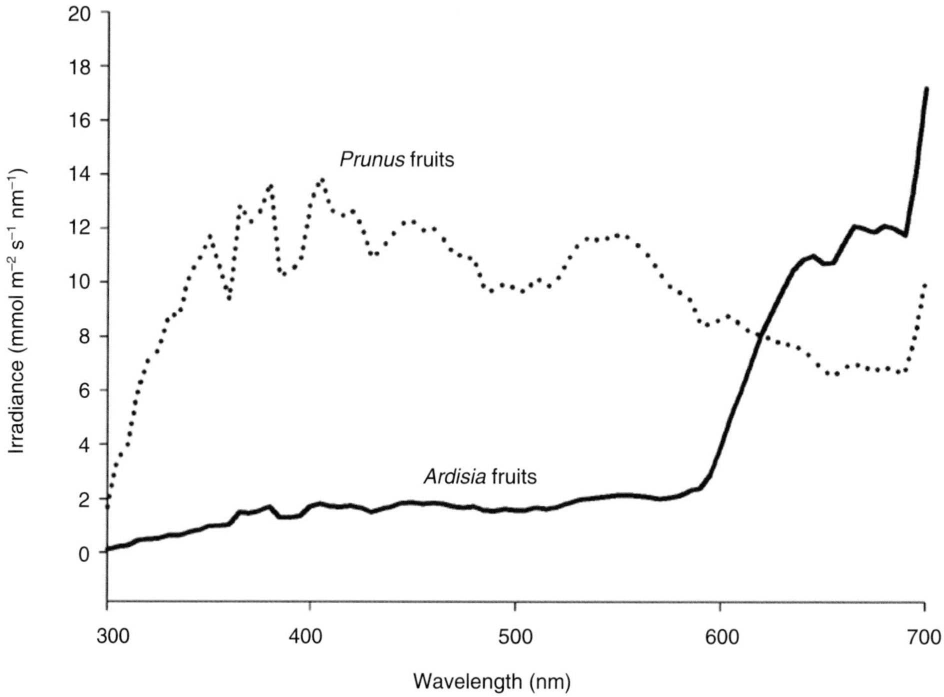

Fig. 3.4. Radiance spectra reaching the eye of the observer are a function of the illuminant and the reflectance spectra (for both, see Fig. 3.2). Note that the UV reflectance of *Prunus spinosus* fruits is degraded more strongly in the forest light than is the colour of red *Ardisia crenata* fruits.

Co-evolution Between Fruit Signals and Consumer Vision?

Primates

The interplay between consumer vision and fruit signals is best understood in primates with human-like colour vision. This is partly because of their similarity to humans but, more important, in order to understand the origins of colour vision in humans, much research activity has focused on the relationship between fruit signals, the extent of frugivory, and the visual capacities of our closest relatives. A particularly intriguing study system is the ecology of colour vision in neotropical primates. Some individuals within a species are dichromatic (two receptor types used for colour vision) and others trichromatic (three receptor types used for colour vision) (see Surridge *et al.*, 2003). Because dichromacy is the ancestral state, studies have focused on the relative advantage of di- and trichromacy for fruit detection. Theoretical studies revealed that trichromacy (present in all Old World monkeys) is advantageous for finding fruits among foliage (Osorio and Vorobyev, 1996), but recent experiments on different species yielded mixed results (Dominy *et al.*, 2003; Smith *et al.*, 2003), suggesting that the difference may not be as clear-cut as once thought.

Studying the properties of visual perception in six species of Old World primates and measuring the fruit colours in an African vegetation community, Sumner and Mollon (2000) found that the spectral positions of the receptors used for colour vision are optimized for finding fruits against a background of foliage. While the visual capacities and the fruit colour diversity in African and neotropical forests are consistent with the paradigm of co-evolution between primates and fruiting plants, they do not require it (Sumner and Mollon, 2000; Regan *et al.*, 2001). This is because the primate visual system is optimized for finding any target that is not a mature leaf among foliage and might thus have evolved to distinguish young nutritive (often reddish) leaves, rather than fruits, from other leaves (Dominy and Lucas, 2001).

Birds

Shifting focus from the fruit consumer as the recipient of fruit signals to the plant as the sender, we can ask whether the detectability of a signal is the most important factor driving fruit colour evolution. If detectability is the driving force, we expect plants to have optimized the contrasts between fruits and leaves. At first glance, such an analysis seems trivial, as all leaves are predominantly coloured by chlorophyll, leaving little variance in leaf coloration. However, carotenoids and anthocyanins also contribute to leaf coloration, and researchers have noted large interspecific differences between leaf colours and have searched for adaptive explanations for this phenomenon (Lev-Yadun *et al.*, 2004). According to a model of the avian eye (Vorobyev, 2003), bird-dispersed fruits (n = 130) do not contrast more strongly against leaves of their own species compared with those of other bird-dispersed species (Schaefer *et al.*, 2007). This result demonstrates that signal optimization has not occurred against their own leaves, which form the predominant background for fruits of most species. However, it would be premature to reject hypotheses of co-evolution on these grounds. Contrasts between fruits and leaves (both their own and those from other species) might be strong enough to reliably warrant detection, thus making further signal optimization unlikely.

Signal messages

Analysing fruit colours only according to their detectability oversimplifies the complexities of signalling systems. For example, in a rainforest community in southern Venezuela, plants employed different signal strategies: more strongly contrasting red and black fruits did not indicate the composition of fruit pulp, whereas less-contrasting fruit colours correlated with protein, carbohydrate and condensed tannin content

(Schaefer and Schmidt, 2004). This suggests a trade-off between the detectability of a fruit colour and its reliability in transmitting a message that is more specific than simply indicating the maturation stage of a fruit. By signalling nutritive rewards, especially rare ones, less-conspicuous fruits may attract consumers, which preferentially consume these rewards. For example, yellow and orange fruits were found to have high protein contents; a scarce resource for frugivores in a rainforest (Schaefer and Schmidt, 2004). Whether such trade-offs between the design and content of fruit signals are widespread or restricted to rainforest plant communities, where rare species or those that bear fruits only irregularly would benefit from this signalling strategy, is a new and challenging research direction.

Background structures

The detectability of fruit displays can be increased not only by the colour of a fruit but also the contrasts created by secondary structures (i.e. non-green coloured bracts, stems and unripe fruits that are associated with the fruit display). That secondary structures enhance the conspicuousness of fruit displays is a long-standing hypothesis (e.g. Stiles, 1982; Willson and Melampy, 1983). Supporting evidence stems from an experiment where painting the naturally red bracts of *Lonicera involucrata* (Caprifoliaceae) green reduced fruit removal by birds relative to the red-painted control (Burns and Dalen, 2002). Again, if we measure the contrasts between the secondary structures, fruits and the remaining background, we can model the signal perception according to the visual properties of fruit consumers. Analysing secondary structures with a model of avian vision documented that these structures do indeed increase the contrasts of the entire fruit display as perceived by the avian eye (Schaefer *et al.*, 2007). Thus, secondary structures represent a way of overcoming potential trade-offs between signal detectability and signal content in fruit skin by decoupling the conspicuousness of the fruit display from the presentation of the fruit itself.

Although the signals sent by plants are relatively constant when compared to the signals sent by animals, they are by no means fixed. The most obvious changes in fruit signals occur during ripening, concomitant with changes in fruit pulp composition. Most fruits are green, and thus relatively cryptic among foliage, when unripe. They increase their conspicuousness through colour changes associated with ripening. Once a fruit is ripe, the human eye does not perceive changes in the colour of the fruit. Even if we assume that the colours of ripe fruits are essentially unchanging for all consumers, the signal may change. This is because, at least in long-persisting fruits, the background does change over time. These changes are particularly pronounced in species from seasonal habitats; for example, those that produce ripe fruits prior to or during the time of autumnal leaf colour changes. It has long been hypothesized that the prevalence of temperate black fruits (relative to red fruits) in autumn represents an adaptation to provide strong contrasts against both green and red autumnal foliage (Stiles, 1982; Burns and Dalen, 2002; but see Willson and Hoppes, 1986).

Further changes in the background occur when leaves are shed in autumn and long-persisting fruits are no longer displayed against leaves but against the predominantly brown colours of bark. For example, the fruits of many *Callicarpa* spp. (Verbenaceae) are pink and commonly displayed directly against the foliage. Because these species have similar fruit and leaf colours, their signals are perceived similarly by birds. However, the fruits of *Callicarpa americana* persist late in winter, long after leaves have been shed (H.M. Schaefer, Florida, 2004, personal observation). Under these conditions, the fruits are displayed against the branches of the shrub and are more conspicuous for birds than when displayed against the leaves (Fig. 3.5). The differences in the contrasts are more pronounced for passerine birds, whose peak absorbance of the shortwave receptor is shifted towards the UV, than for most non-passerine birds (Schaefer *et al.*, 2007). This higher conspicuousness is due to stronger

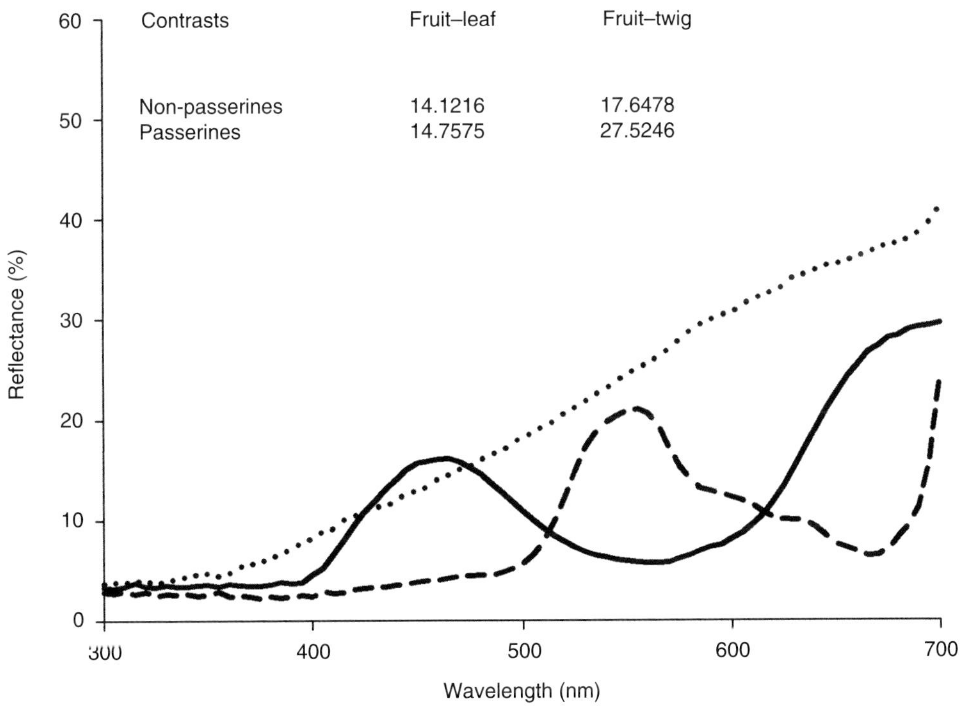

Fig. 3.5. Mean reflectance spectra of fruits (solid line), leaves (dashed line) and branches (dotted line) of *Callicarpa americana* (Verbenaceae). The unit for the magnitude of contrasts are just noticeable differences (jnds), which define the threshold of discrimination at 1 jnd. Contrast calculations follow Schaefer *et al.* (2007).

contrasts in a spectral area of high sensitivity in the typical passerine eye. The increase in conspicuousness is less apparent for most non-passerine birds, which are characterized by lower spectral sensitivities in that area, owing to differences in the spectral position and a greater overlap of the photoreceptor types (Schaefer *et al.*, 2007). This example demonstrates that for an evaluation of signals we need to account for variability in the background structures as well as for the physiological basis of colour vision between, and even within, consumer groups. Moreover, these results also highlight the fact that changes in the background, which often occur unrelated to fruit–frugivore interactions (e.g. shedding of the leaves), may have profound implications for the effectiveness of a given fruit signal, thereby restricting plants' potential for adaptive signalling. An emerging lesson from these examples is that the signalling system between plants and animals is more dynamic than currently reflected in most studies. This conclusion probably also pertains to olfactory plant signals (see, e.g. Ayasse *et al.*, 2000).

So far, we have described fruit colours in their function as signals to fruit consumers. However, apart from signalling, the pigments that impart colour to fruits and flowers might also serve other physiological and ecological roles.

Plant Pigments and Defence

Warning signals

As previously mentioned, signal evolution is thought to be a compromise between efficient signalling to intended recipients (here seed dispersers) and inconspicuous signalling to unintended recipients (fruit and seed predators) (Endler, 2000). Plants might overcome this conflict if their signals consist of colours that are less conspicuous to antagonists than to mutualists. Such a communication has been termed a 'private communication channel' and has been suggested for flower–pollinator interactions (Bradshaw and Schemske, 2003). Whether fruit colours represent private communication channels has not yet been studied. The variation in the visual perception of different seed disperser groups (e.g. birds and mammals) and of common seed predators (e.g. rodents and insects) suggests that they all perceive fruit signals differently, which is a prerequisite for the evolution of private communication channels. The visual system of birds, for example, extends into the ultraviolet wavelengths, which are invisible to humans and other primates, whereas the colour vision of birds and many insects differs in the red part of the spectrum. Also, animals differ in the number and peak absorbance of light receptors used for colour vision: four in birds, three or two in mammals, and three or four in insects (Kelber *et al.*, 2003; Schaefer *et al.*, 2004). We must emphasize, however, that the complexities of fruit–frugivore interactions are probably important constraints for the evolution of private communication channels in fruit signals. The two most important seed disperser groups, birds and primates, comprise not only legitimate seed dispersers but also seed predators (Jordano and Schupp, 2000; Regan *et al.*, 2001). Moreover, the traditional dichotomy of seed dispersers and seed predators is often unjustified, as the same animal might act as seed predator for one species of plant and as seed disperser for another, and some even do both for the same species (Norconk *et al.*, 1998; Regan *et al.*, 2001).

Rather than escaping fruit predators by producing cryptic fruits, fruit colour might act as a warning signal in those species that indicate their defensive strength visually, as suggested by the correlation between condensed tannins and fruit coloration (Schaefer and Schmidt, 2004). Unfortunately, owing to the long-standing focus on seed dispersers, there is a dearth of experiments investigating whether warning signals indicating fruit defence are widespread. However, it is important to note that visual warning signals do not discourage some of the most detrimental fruit predators – fungi and microbes – which are colour-blind. Concerning these agents, the evolution of chemical defences is more important than visual signals to deter potential predators.

Correlated selection on fruit colour

Fruit predators might select fruit colours not only directly but also indirectly through correlated selection on other fruit characters that are linked to the synthesis of pigments. Recent work has documented that insects can exert correlated selection on fruit and flower colour alleles. *Acacia ligulata* (Mimosaceae) is polymorphic for fruit colour, and insects were found to feed differentially on unripe fruit morphs, prior to the development of fruit coloration (Whitney and Stanton, 2004). The authors concluded that pleiotropic effects of fruit colour alleles are responsible for these effects. Similarly, in the flowers of *Raphanus sativus* (Cruciferae), colour morphs correlated differently with the contents of deterrent secondary compounds affecting herbivore survival (Irwin *et al.*, 2003). Owing to these effects, herbivores seem to balance the selective pressures of pollinators on floral colour, and both herbivores and pollinators appear to maintain the colour polymorphism.

The mechanisms responsible for correlated selection in *Acacia* fruits are currently unknown, but consideration of the biosynthetic pathway of pigment production might be revealing. For example, anthocyanins, which are among the most ubiquitous fruit pigments (see Kong *et al.*, 2003) share common precursors with a number of defensive compounds, such as tannins and phenols, which are produced by the phenylpropanoid pathway (Fineblum and Rausher, 1997). Because the production of anthocyanins and defensive compounds is dynamically linked by a common biosynthetic pathway, the expression of anthocyanins is thought to be linked with the synthesis of these defensive compounds in leaves (Schaefer and Rolshausen, 2006). In this case, correlated selection in response to defensive compounds that are linked to anthocyanins might, over evolutionary time, become direct selection against anthocyanins. This is because in plant tissue these pigments are often the most visible sign of the presence of other secondary compounds, which are often colourless themselves.

While this hypothesis, termed the 'defence indication hypothesis' (Schaefer and Rolshausen, 2006), might apply to flowers and leaves (Nozzolillo *et al.*, 1990; Irwin *et al.*, 2003), its applicability to fruits, which in contrast to flowers and leaves are made to be consumed, is currently unknown. This is largely because changes in the profiles of secondary compounds in fruit pulp and skin have rarely been characterized for natural fruits. This lack of data is impeding our understanding of the effects of correlated selection on fruit colour evolution. At any rate, the co-evolutionary process between plants as the senders of fruit signals and seed dispersers as the intended recipients is probably constrained, not necessarily by phylogenetic effects, which are considered to be weak (Voigt *et al.*, 2004), but by the selective pressures exerted directly or indirectly by fruit predators. This conjecture suggests that the traditional paradigm of fruit colours as products of co-evolutionary processes between plants and seed dispersers is probably outdated. Investigating the role of such ecological constraints on the evolution of the communication systems between plants and animals is a particularly interesting emerging research direction.

Lastly, surprisingly little is known on how fruit colours interact with abiotic factors that might contribute to fruit colour evolution (Willson and Whelan, 1990). This long-standing hypothesis is well known in horticultural studies. Particularly well studied are the effects of sun exposure on the development of the waxy UV-reflecting bloom of grapes (*Vitis* spp.; Vitaceae) (Rosenquist and Morrison, 1989). Edaphic factors might also contribute to fruit colour polymorphisms. For example, the orange fruit morph of salmonberry (*Rubus spectabilis*; Rosaceae) germinated faster than the red fruit morph on certain soils (Traveset and Willson, 1998). In a different species, Whitney and Lister (2004) suggested that clinal patterns in fruit colour polymorphism along a gradient of decreasing temperatures and increasing rainfall in Australia could be related to climatic factors. A proximate mechanism underlying such patterns might be that seeds of fruit colour morphs germinate differently. For instance, seeds from blue *Myrtus communis* (Myrtaceae) fruits germinated faster than those from white fruits, although this difference disappeared after several weeks (Traveset *et al.*, 2001). The mechanism responsible for this effect is currently unknown, which reflects our limited knowledge on the importance of abiotic factors.

In temperate fruits, UV reflection is mostly attained through epicuticular waxes that lay on top of the fruit skin (Burkhardt, 1982). In some regions 40–61% of fruits reflect UV, either due to a waxy bloom, as in temperate species, or due to pigments, as in tropical species (Willson and Whelan, 1989; Altshuler, 2001). Because birds are sensitive to UV light and often use UV-reflecting signals in communication, they are seen as the most likely selective agents on UV-reflecting fruits. However, two field studies on the effects of UV coloration in fruits yielded contradictory results, with strong preferences for UV signals being found in only one study (Willson and Whelan, 1989; Altshuler, 2001). It is important to note that the epicuticular waxes in temperate fruits often develop in unripe fruits, which presumably do not signal to seed dispersers. Interestingly, removal of the bloom in unripe fruits led to a higher mortality prior to fruit maturation compared with control fruits whose waxy bloom remained intact (Wilcoxon test, $n = 10$, $z = -2.05$, $P < 0.05$; Fig. 3.6). The higher mortality of unripe fruits was probably due to desiccation, although handling effects cannot be excluded. Consistent with the conjecture that fruits without the waxy bloom had higher desiccation risks is the fact that cuticle and wax layers of domestic grapes increase proportionately to sun exposure (Rosenquist and Morrison, 1989) and that epicuticular waxes are generally thought to protect leaves against desiccation. Two possible mechanisms may underlie reduced evaporation by an epicuticular wax layer. First, epicuticular waxes thicken the fruit skin and provide a hydrophobic barrier, thereby lowering the water exchange between fruit and environment. Second, the much higher overall reflectance caused by the waxy layer may lower fruit temperature relative to fruits without the layer, leading to lower water loss. This is an example of the physiological role of fruit coloration that has rarely been considered in ecological studies, although physiological constraints are likely to influence fruit colour evolution.

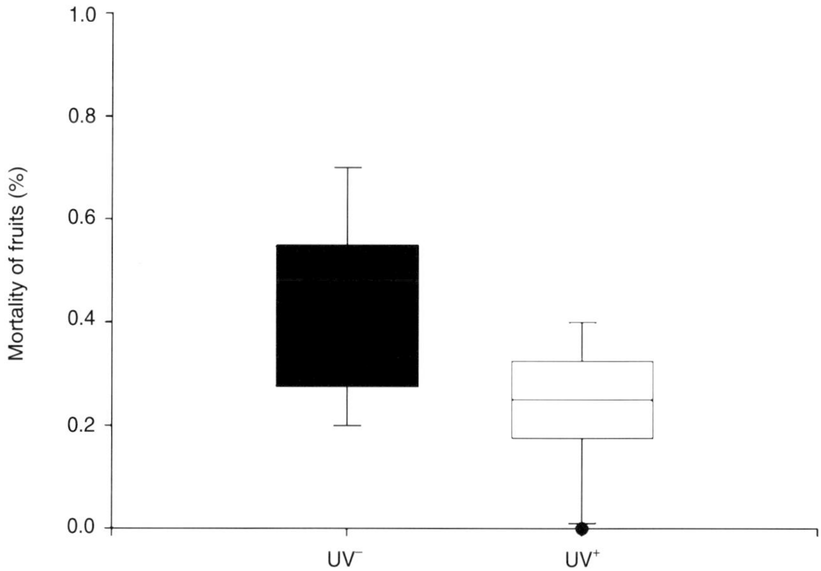

Fig. 3.6. Green unripe blackthorn (*Prunus spinosus*; Rosaceae) fruits suffer a higher mortality without UV-reflecting epicuticular waxes (UV^-) than control green unripe fruits with UV-reflecting waxes (UV^+). Fruits without UV-reflecting waxes had a lower probability of becoming ripe.

Perspective

Similar to the evolution of other fruit traits, the evolution of fruit colours is not shaped by tight co-evolution between plants and seed dispersers, but by diffuse interactions between plants, antagonists and mutualists, as well as by abiotic factors and physiological constraints (Fig. 3.6). We argue that researchers today are better equipped to tease apart the relative influence of each of these factors than they were a few decades ago. Based on the framework of signal theory, we can develop strong predictions on how seed dispersers select fruit colours. The two main predictions say that seed dispersers should prefer the most detectable and/or the most reliable signals in terms of indicating fruit contents. A future research direction will be to determine whether there are trade-offs between these two signal components. This will be an important finding not only for fruit–frugivore interactions but also for developing the framework of biological signals. For fruit predators, the two main hypotheses state that antagonists select fruit colours directly (crypsis or deterrent function of pigments) or indirectly through correlated selection on genes linked to fruit colour expression. Finally, evidence that fruit colours interact with abiotic factors is limited to few species, and encompassing hypotheses to explain such effects are lacking. Integrating these three approaches will enable us to attain a comprehensive picture of the selective pressures and the constraints on the evolution of signals in the communication between disparate organisms.

Acknowledgements

We thank the organizers of the *Fourth International Symposium/Workshop on Frugivores and Seed Dispersal* for hosting such a stimulating conference and for inviting us to contribute our research. We are indebted to Doug Levey for many invaluable suggestions on fruit signalling in general. We thank Doug Levey, Anna Traveset and Andrew Dennis for numerous helpful comments that greatly improved the earlier version of the manuscript. H.M.S. was sponsored by a DFG grant, SCHA 1008/3-1, and V.S. by a DFG grant, SCHM 2113/1-1.

References

Allen, G. (1879) *The Colour-Sense: Its Origin and Development.* Trübner and Co, London.

Altshuler, D. (2001) Ultraviolet reflectance in fruits, ambient light composition and fruit removal in a tropical forest. *Evolutionary Ecology Research* 3, 767–778.

Alves-Costa, C.P. and Lopes, A.V. (2001) Using artificial fruits to evaluate fruit selection by birds in the field. *Biotropica* 33, 713–717.

Ayasse, M., Schiestl, F.P., Paulus, H.F., Lofstedt, C., Hansson, B., Ibarra, F. and Francke, W. (2000) Evolution of reproductive strategies in the sexually deceptive orchid *Ophrys sphegodes*: how does flower-specific variation of odor signals influence reproductive success? *Evolution* 54, 1995–2006.

Bradshaw, H.D. and Schemske, W. (2003) Allele substitution at a flower colour locus produces a pollinator shift in monkeyflowers. *Nature* 426, 176–178.

Briscoe, A.D. and Chittka, L. (2001) The evolution of color vision in insects. *Annual Review of Entomology* 46, 471–570.

Burkhardt, D. (1982) Birds, berries and UV. *Naturwissenschaften* 69, 153–157.

Burns, K.C. and Dalen, J.L. (2002) Foliage color contrasts and adaptive fruit color variation in a bird-dispersed plant community. *Oikos* 96, 463–469.

Corlett, R.T. (1996) Characteristics of vertebrate-dispersed fruits in Hong Kong. *Journal of Tropical Ecology* 12, 819–833.

Dominy, N.J. and Lucas, P.W. (2001) Ecological importance of trichromatic vision to primates. *Nature* 410, 363–365.

Dominy, N.J., Garber, P.A., Bicca-Marques, J.C. and Azevedo-Lopes, M.A.D.O. (2003) Do female tamarins use visual cues to detect fruit rewards more successfully than do males? *Animal Behaviour* 66, 829–837.

Endler, J.A. (1990) On the measurement and classification of color in studies of animal color patterns. *Biological Journal of the Linnean Society* 41, 315–352.

Endler, J.A. (2000) Evolutionary implications of the interaction between animal signals and the environment. In: Espmark, Y., Amundsen, T. and Rosenqvist, G. (eds) *Animal Signals*. Tapir Academic Press, Trondheim, Norway, pp. 11–46.

Endler, J.A. and Mielke, P.W. (2005) Comparing entire colour patterns as birds see them. *Biological Journal of the Linnean Society* 86, 405–431.

Fineblum, W.L. and Rausher, M.D. (1997) Do floral pigmentation genes also influence resistance to enemies? The W locus in *Ipomoea pupurea*. *Ecology* 78, 1646–1654.

Gervais, J.A., Noon, B.R. and Willson, M.F. (1999) Avian selection of the color-dimorphic fruits of salmonberry, *Rubus spectabilis*: a field experiment. *Oikos* 84, 77–86.

Giles, S. and Lill, A. (1999) The effect of fruit abundance, conspicuousness and sugar concentration on fruit colour choice by captive silvereyes. *Ethology, Ecology and Evolution* 11, 229–242.

Hart, N.S. (2001) The visual ecology of avian photoreceptors. *Progress in Retinal and Eye Research* 20, 675–703.

Herrera, C.M. (1992) Interspecific variation in fruit shape, allometry, phylogeny and adaptation to dispersal agents. *Ecology* 73, 1832–1841.

Honkavaara, J., Siitari, H. and Viitala, J. (2004) Fruit colour preferences of redwings (*Turdus iliacus*): experiments with hand-raised juveniles and wild-caught adults. *Ethology* 110, 445–457.

Howe, H.F. and Estabrook, G.F. (1977) Intraspecific competition for avian dispersers in tropical trees. *American Naturalist*, 111, 817–832.

Irwin, R.E., Strauss, S.Y., Storz, S., Emerson, A. and Guibert, G. (2003) The role of herbivores in the maintenance of a flower color polymorphism in wild radish. *Ecology* 84, 1733–1743.

Janson, C.H. (1983) Adaptation of fruit morphology to dispersal agents in a neotropical forest. *Science* 219, 187–189.

Jordano, P. (1995) Angiosperm fleshy fruits and seed dispersers: a comparative analysis of adaptation and constraints in plant–animal interactions. *American Naturalist* 145, 163–191.

Jordano, P. and Schupp, E.W. (2000) Seed disperser effectiveness: the quantity component and patterns of seed rain for *Prunus mahaleb*. *Ecological Monographs* 70, 591–615.

Kelber, A., Vorobyev, M. and Osorio, D. (2003) Animal colour vision: behavioural tests and physiological concepts. *Biological Reviews* 78, 81–118.

Kerner, A. (1895) *The Natural History of Plants: Their Forms, Growth, Reproduction and Distribution.* Holt, London.

Kong, J.-M., Chia, L.-S., Goh, N.-K., Chia, T.-F. and Brouillard, R. (2003) Analysis and biological activities of anthocyanins. *Phytochemistry* 64, 923–933.

Lee, W.G., Weatherall, I.L. and Wilson, J.B. (1994) Fruit conspicuousness in some New Zealand *Coprosma* (Rubiaceae) species. *Oikos* 69, 87–94.

Lev-Yadun, S., Dafni, A., Flaishman, A., Inbar, M., Izhaki, I., Katzir, G. and Ne'eman, G. (2004) Plant coloration undermines herbivorous insect camouflage. *BioEssays* 26, 1126–1130.

Long, R.W. (1971) *A Flora of Tropical Florida.* University of Miami Press, Coral Gables, Florida.

Maynard-Smith, J. and Harper, D.G.C. (1995) Animal signals: models and terminology. *Journal of Theoretical Biology* 177, 305–311.

McKey, D. (1975) The ecology of coevolved seed dispersal systems. In: Gilbert, L.E. and Raven, P.H. (eds) *Coevolution of Animals and Plants.* University of Texas Press, Austin, Texax, pp. 159–191.

Norconk, M.A., Grafton, B.W. and Conklin-Brittain, N.L. (1998) Seed dispersal by neotropical seed predators. *American Journal of Primatology* 45, 103–126.

Nozzolillo, C., Isabelle, P. and Das, G. (1990) Seasonal changes in the phenolic constituents of jack pine seedlings (*Pinus banksiana*) in relation to the purpling phenomenon. *Canadian Journal of Botany* 68, 2010–2017.

Osorio, D. and Vorobyev, M. (1996) Colour vision as an adaptation to frugivory in primates. *Proceedings of the Royal Society London B* 263, 593–599.

Otte, D. (1974) Effects and functions in the evolution of signaling systems. *Annual Review of Ecology and Systematics* 5, 385–417.

Puckey, H.L., Lill, A. and O'Dowd, D.J. (1996) Fruit color choices of captive silvereyes (*Zosterops lateralis*). *Condor* 98, 780–790.

Regan, B.C., Julliot, C., Simmen, B., Vienot, F., Charles-Dominique, P. and Mollon, J.D. (2001) Fruits, foliage and the evolution of primate colour vision. *Philosophical Transactions of the Royal Society London B* 356, 229–283.

Rosenquist, J.K. and Morrison, J.C. (1989) Some factors affecting cuticle and wax accumulation on grape berries. *American Journal of Enology and Viticulture* 40, 241–244.

Santamaría, M. and Franco, A.M. (2000) Frugivory of Salvin's Curassow in a rainforest of the Colombian Amazon. *Wilson Bulletin* 112, 473–481.

Schaefer, H.M. and Rolshausen, G. (2006) Plants on red alert: do insects pay attention? *BioEssays* 28, 65–71.

Schaefer, H.M. and Schmidt, V. (2004) Detectability and content as opposing signal characteristics in fruits. *Proceedings of the Royal Society London B* 271, S370–S373.

Schaefer, H.M., Schaefer, V. and Levey, D.J. (2004) How plant–animal interactions signal new insights in communication. *Trends in Ecology and Evolution* 19, 577–584.

Schaefer, H.M., Schaefer, V. and Vorobyev, M. (2007) Are fruit colors adapted to consumer vision and birds equally efficient in detecting colorful signals? *American Naturalist* 169, S159–S169.

Schluter, D. and Price, T. (1993) Honesty, perception and population divergence in sexually selected traits. *Proceedings of the Royal Society London B* 253, 117–122.

Schmidt, V. and Schaefer, H.M. (2004) Unlearned preference for red may facilitate recognition of palatable food in young omnivorous birds. *Evolutionary Ecology Research* 6, 919–925.

Schmidt, V., Schaefer, H.M. and Winkler, H. (2004) Conspicuousness, not colour as foraging cue in plant–animal signalling. *Oikos* 106, 551–557.

Siitari, H., Honkavaara, J. and Viitala, J. (1999) Ultraviolet reflection of berries attracts foraging birds: a laboratory study with redwings (*Turdus iliacue*) and bilberries (*Vaccinium myrtillus*). *Proceedings of the Royal Society London B* 266, 2125–2129.

Smith, A.C., Buchanan-Smith, H.M., Surridge, A.K., Osorio, D. and Mundy, N.I. (2003) The effect of colour vision status on the detection and selection of fruits by tamarins (*Saguinus* spp.). *Journal of Experimental Biology* 206, 3159–3165.

Stiles, E.W. (1982) Fruit flags: two hypotheses. *American Naturalist* 120, 500–509.

Sumner, P. and Mollon, J.D. (2000) Catarrhine photopigments are optimized for detecting targets against a foliage background. *Journal of Experimental Biology* 203, 1963–1986.

Surridge, A.K., Osorio, D. and Mundy, N.I. (2003) Evolution and selection of trichromatic vision in primates. *Trends in Ecology and Evolution* 18, 198–205.

Traveset, A. and Willson, M.F. (1998) Ecology of the fruit-colour polymorphism in *Rubus spectabilis*. *Evolutionary Ecology* 12, 331–345.

Traveset, A., Riera, N. and Mas, R.E. (2001) Ecology of fruit-colour polymorphism in *Myrtus communis* and differential effects of birds and mammals on seed germination and seedling growth. *Journal of Ecology* 89, 749–760.

Turcek, F.J. (1963) *Color Preference in Fruit- and Seed-Eating Birds: Proceedings of the XIII International Ornithological Congress*. American Ornithological Union, Ithaca, New York, pp. 285–292.

Voigt, F.A., Bleher, B., Fietz, J., Ganzhorn, J.U., Schwab, D. and Böhning-Gaese, K. (2004) A comparison of morphological and chemical fruit traits between two sites with different frugivore assemblages. *Oecologia* 141, 94–104.

Vorobyev, M. (2003) Coloured oil droplets enhance colour discrimination. *Proceedings of the Royal Society London B* 270, 1255–1261.

Vorobyev, M. and Osorio, D. (1998) Receptor noise as a determinant of colour thresholds. *Proceedings of the Royal Society London B* 265, 351–358.

Wallace, A.R. (1879) Colour in nature. *Nature* 19, 580–581.

Wheelwright, N.T. and Janson, C.H. (1985) Colors of fruit displays of bird dispersed plants in two tropical forests. *American Naturalist* 126, 777–799.

Whitney, K.D. (2005) Linking frugivores to the dynamics of a fruit color polymorphism. *American Journal of Botany* 92, 859–867.

Whitney, K.D. and Lister, C.E. (2004) Fruit colour polymorphism in *Acacia ligulata*: seed and seedling performance, clinal patterns, and chemical variation. *Evolutionary Ecology* 18, 165–186.

Whitney, K.D. and Stanton, M.L. (2004) Insect seed predators as novel agents of selection on fruit color. *Ecology* 2004, 2153–2160.

Willson, M.F. and Comet, T.A. (1993) Food choices by northwestern Crows: experiments with captive, free-ranging, and hand-raised birds. *Condor* 95, 596–615.

Willson, M.F. and Hoppes, W.G. (1986) Foliar 'flags' for avian frugivores: signal or serendipity? In: Estrada, A. and Fleming, T.H. (eds) *Frugivores and Seed Dispersal*. Dr. W. Junk Publishers, Dordrecht, The Netherlands, pp. 55–69.

Willson, M.F. and Melampy, M.N. (1983) The effect of bicolored fruit displays on fruit removal by avian frugivores. *Oikos* 41, 27–31.

Willson, M.F. and Whelan, C.J. (1989) Ultraviolet reflectance of fruits of vertebrate-dispersed plants. *Oikos* 55, 341–348.

Willson, M.F. and Whelan, C.J. (1990) The evolution of fruit color in fleshy-fruited plants. *American Naturalist* 136, 790–809.

Willson, M.F., Irvine, A.K. and Walsh, N.G. (1989) Vertebrate dispersal syndromes in some Australian and New Zealand plant communities, with geographic comparisons. *Biotropica* 21, 133–147.

Willson, M.F., Graff, D.A. and Whelan, C.J. (1990) Color preferences of frugivorous birds in relation to the colors of fleshy fruits. *Condor* 92, 545–555.

4 A Review on the Role of Endozoochory in Seed Germination

A. Traveset, A.W. Robertson and J. Rodríguez-Pérez

Introduction

A large fraction of living plants produce fruits that attract animals (64% and 27% of gymnosperm and angiosperm lineages, respectively; Herrera and Pellmyr, 2002). In consuming them, animals can spread the seeds to more or less distant sites from the parent plant, thus contributing to plant regeneration and colonization of new sites (Willson and Traveset, 2000, and references therein). For millions of years both the pulp and seeds of fruit have been subjected to selective pressures exerted by frugivores and granivores. Fruit consumers, specifically, may show preferences for fruit traits such as size, shape, chemical composition and others, and have specific morphologies and physiologies of the digestive tract that affect the survival probability of the ingested seeds in different ways (Herrera and Pellmyr, 2002, and references therein). Seed traits are under further selective pressures imposed by a number of biotic (antagonistic fungi, insects and microbes; Cipollini and Levey, 1997) and abiotic factors (e.g. light, temperature, rainfall; Holl, 1999; Leishman *et al.*, 2000) that influence the dormancy period, germination time and/or future seedling growth, which can ultimately determine germination and seedling success (Verdú and Traveset, 2005, and references therein). Therefore, the specific seed traits in an endozoochorous plant are a result of the combined selection imposed by frugivores, granivores and other biotic and abiotic factors.

In this chapter, we evaluate the role of endozoochory on seed germination. We review the information gathered on germination patterns in experiments aimed at examining the effect of a seed's passage through a frugivore's gut, and give further directions on methods for future studies. Fruit treatment in the disperser's digestive tract (which includes pulp separation from seeds and treatment of the seeds) can determine the capacity of seeds to germinate and, thus, is one of the components of disperser effectiveness that may be crucial for the population dynamics of many fleshy-fruited plant species. The nature of the relationships between fruit chemistry and morphology and the type of disperser is crucial to understanding the co-evolution of plant–frugivore interactions (e.g. Murray *et al.*, 1994; Cipollini and Levey, 1997; Wahaj *et al.*, 1998; Regan *et al.*, 2001; Tewksbury and Nabhan, 2001; Wenny, 2001; Stanley and Lill, 2002b; Alcántara and Rey, 2003).

Biology of the System: From Fruit Swallowing to Seed Fate

Frugivores that swallow fruits (subsequently referred to as frugivores) act differentially on seeds in several ways during ingestion, gut-processing and defecation. Plants, in turn, can be expected to evolve physical and chemical adaptations in fruits and seeds to direct dispersal and to maximize the proportion of seeds that are successfully dispersed and established (Table 4.1). Below we briefly outline the issues surrounding four main components of the process and follow this with detailed reviews in each area.

Removal of pulp surrounding the seed(s)

Fruit pulp often contains germination inhibitors that can block biochemical pathways of germination (e.g. Evenari, 1949; Mayer and Poljakoff-Mayber, 1982; Cipollini and Levey, 1997, and references therein; Meyer and Witmer, 1998). Moreover, the pulp can decrease and even preclude germination by

Table 4.1. Aspects of fruit processing during endozoochory likely to influence the germination of seeds when they are subsequently deposited.

	Pulp removal	Gut passage	Deposition pattern
Frugivore attributes	All frugivores adequate except fruit peckers.	Large frugivores, especially mammals, have lengthy passage times and are chemically and mechanically harsh. Birds tend to be more gentle and pass seeds faster. Diet may have important effects.	Regurgitators tend to drop seeds singly but often close to the maternal tree. Large defecators produce dung with large numbers of seeds and high rates of seed mixing Small frugivores deposit smallnumbers of seeds with less seed-mixing.
Fruit chemistry	Many plants appear to have germination inhibitors in the pulp that must be removed before germination will occur.	Passage rates may be influenced by secondary metabolites that act either as laxatives or constipating agents.	Seeds may contain allelochemicals that inhibit the germination of competitors within the same faecal clump.

altering the seeds' microenvironment (e.g. osmotic pressure and light regime; Mayer and Poljakoff-Mayber, 1982; Meyer and Witmer, 1998; Samuels and Levey, 2005). The removal of fruit pulp and consequent release of the seed from this kind of inhibition (the deinhibition effect – Robertson *et al.*, 2006) has not received as much attention as scarification (the mechanical and chemical processing of the seed coat or endocarp), but the fact that seeds are more or less freed of pulp in the frugivore's gut is part of the seed passage process and thus needs to be considered in future studies, especially when the fate of uneaten fruits is considered (Samuels and Levey, 2005; Robertson *et al.*, 2006).

Mechanical and/or chemical scarification of the seed coat or endocarp

The effect of scarification is usually thought to depend on retention time in the digestive tract and on the type of food ingested along with seeds (Barnea *et al.*, 1990; Traveset, 1998). This mechanism (rather than deinhibition) is the most often invoked when comparing seed germination patterns between ingested and uningested seeds of the same species (Samuels and Levey, 2005). Despite this, we have very little direct empirical evidence of such alteration of the seed coat or endocarp structure.

Effect of the faecal material in which seeds are embedded at the time of deposition

The residues of whatever food items a frugivore has been eating can influence the microenvironment following defecation and may have a fertilizing effect on seedling growth (e.g. Dinerstein and Wemmer, 1988; Malo and Suárez, 1995; Traveset *et al.*, 2001a; Cosyns *et al.*, 2005). In some circumstances, however, the presence of faeces might reduce germination success by facilitating fungal and/or bacterial growth (Meyer and Witmer, 1998). In other cases, faecal material can protect seeds from attack of species-specific parasites and predators (Fragoso *et al.*, 2003). The rather sparse studies on this subject have reported that the material the dispersed seeds are embedded in is another factor that needs to be considered when evaluating the influence of frugivores on the final fate of seeds of fleshy-fruited plants.

The number and combination of seeds in the faeces

Seed-mixing may also influence germination and/or seedling growth. The number of seeds defecated in a dropping can be very relevant to the future establishment success of a plant. For instance, a mechanism of autotoxicity might explain the inverse correlation between seed abundance and germination success found by Barnea *et al.* (1992) for *Solanum luteum* and

Morus nigra. The presence of a large quantity of seeds in a dropping can also influence both post-dispersal seed predation (e.g. Janzen, 1986; Schupp, 1990; Traveset, 1990; Bermejo *et al.*, 1998) and intra- and interspecific competition (e.g. Loiselle, 1990). In addition, when different seed species are defecated in the same dropping, it is possible that one or more may exhibit allelopathic effects that inhibit the germination of the other species (Mayer and Poljakoff-Mayber, 1982; Cipollini and Levey, 1997).

Effect of frugivores separating pulp from seeds (deinhibition)

In drupes and berries, seeds are surrounded by a layer of flesh or pulp that offers a reward to frugivores but may also have other functions such as the protection of the seeds and the prevention of germination while still on the plant (Evenari, 1949; Mayer and Poljakoff-Mayber, 1982). This inhibition can take several forms. The inhibition may be induced by high osmotic pressure caused by the high sugar content of ripe fruits (Samuels and Levey, 2005), by light-blocking pigmentation that prevents enough light reaching the seedlings to stimulate germination (a possible reason why many ripe fruits are darkly pigmented; Cipollini and Levey, 1997), or from secondary metabolites that directly inhibit seed germination (Evenari, 1949; Cipollini and Levey, 1997). Although in most cases the exact mechanism is not known, many laboratory studies have confirmed that seeds often fail to germinate in Petri dishes when in intact fruit, while they readily germinate when hand-cleaned. For example, 60% of the 46 woody species that are dispersed by birds in New Zealand showed poor germination (< 20%) in fruits. However, for most of these species, germination of hand-cleaned seeds was > 90% (Kelly *et al.*, 2004). A recent review (Robertson *et al.*, 2006) showed that more than half of the species tested for deinhibition showed strong effects of removing the pulp on seed germination, though in most cases tests were conducted in Petri dishes, which appear to produce artificially large effects compared with tests conducted in the field (Rodríguez-Pérez *et al.*, 2005; Robertson *et al.*, 2006; see discussion below). To determine the extent to which the release of seeds from the pulp is a critical step in allowing germination, it is important that these tests are also conducted in field conditions, where chemical leaching and biological processing are quite different from those in the laboratory.

In a few cases, there is direct evidence of chemical inhibition of germination from constituents in the fruit pulp. Yagihashi *et al.* (1998) found that while seeds of *Sorbus commixta* (Rosaceae) that were hand-cleaned or that had been ingested and defecated by dusky thrushes (*Turdus naumanni*; Turdidae) germinated equally well in the laboratory, seeds that were either within intact fruits or extracted and treated with a 1% solution of pulp juice, failed to germinate at all. This suggests that a strong chemical which inhibits germination occurs in the pulp. The active component(s) was (were) not identified in this case. Richmond and Ghisalberti (1994) found a similarly strong inhibition of seed germination in *Eremophila maculata* (Myoporaceae) from extracts of the fruit wall of *E. maculata* and *E. racemosa*. The active ingredient in this case was tentatively identified as an aromatic glycoside. Wahaj *et al.* (1998) fed artificial agar 'fruits' containing *Solanum americanum* (Solanaceae) seeds and low concentrations of two naturally occurring *S. americanum* glycoalkaloids to cedar waxwings (*Bombycilla cedrorum*; Bombycillidae). The collected seeds from these 'fruits' showed reduced germination rates compared with seeds collected following defecation from agar 'fruits' that contained no glycoalkaloids. This suggests that these glycoalkaloids can maintain biological activity as germination inhibitors even after gut passage, and that they may have additional roles as allelopathic substances acting on other potentially competing seeds following deposition in faecal clumps. We will return to this idea when we consider deposition patterns.

Effect of frugivores on seed scarification

The seed coat is, to a variable extent, mechanically and/or chemically scarified when passing through the digestive tract of a frugivore. The level of scarification depends upon: (i) the species of frugivore that ingests the fruit; and (ii) the intrinsic traits of the fruits and seeds. The frugivores' morphological and physiological traits, as well as retention times in the gut, can alter the extent to which the seed coat is modified (e.g. Murray *et al.*, 1994; Traveset, 1998; Traveset *et al.*, 2001b; Figuerola *et al.*, 2002; Santamaría *et al.*, 2002; Charalambidou *et al.*, 2003, 2005; Pollux *et al.*, 2005). Important morphological traits include the length of the digestive tract and the presence of a gizzard; physiological traits include digestive fluids in the gut (which can vary depending upon the type of food ingested along with the fruits), water content, pH and proportion of plant material. Likewise, retention time can depend upon a variety of factors. Furthermore, plant traits such as pulp composition, seed age, seed size, coat thickness and texture also influence the degree of scarification. For instance, the pulp may have laxative compounds which promote a faster seed deposition (Murray *et al.*, 1994; Wahaj *et al.*, 1998) or constipative substances that slow it down (Cipollini and Levey, 1997).

A long retention time can be disadvantageous because the digestive fluids to which seeds are exposed may damage the embryo and thus decrease seed viability (Janzen *et al.*, 1985; Murray *et al.*, 1994; Santamaría *et al.*, 2002; Charalambidou *et al.*, 2005; Pollux *et al.*, 2005; but see Meyer and Witmer, 1998; Wahaj *et al.*, 1998). Because large seeds generally have shorter retention times in the gut (Levey and Grajal, 1991; Stanley and Lill, 2002a) we might expect their germination to be less affected by digestive fluids than that of small seeds. However, an interspecific comparison in a meta-analysis found that species with large seeds showed more effects of ingestion on germination (compared with hand-cleaned seed) than species with small seeds (Traveset and Verdú, 2002), although this difference was no longer significant when the meta-analysis was phylogenetically controlled (Verdú and Traveset, 2005). Therefore, the implications of seed size for germination after passage through an animal's gut remain unclear and are probably species-dependent. Moreover, seed size is not always correlated with retention time in the gut, as recently found in the duck-dispersed pondweed, *Potamogeton pectinatus* (Potamogetonaceae; L. Santamaría *et al.*, unpublished results). More data are certainly needed both on the association between seed size and gut passage time and on the effect of such time on germination performance, especially for frugivores other than birds. Given the different food retention times that occur among different frugivore taxa (for instance, between small birds and non-flying mammals or reptiles), we might expect strong differences in the effect of seed size on retention time in the gut among these frugivore groups.

In a recent study that compared the effect of different guilds of frugivores on the germination of *Rubia fruticosa* (Rubiaceae), Nogales *et al.* (2005) found that seeds ingested by squirrels (*Atlantoxerus getulus*; Sciuridae) or rabbits (*Oryctolagus cuniculus*; Leporidae) were less viable and germinated in lower proportions than seeds ingested by either birds or lizards. They attributed this difference to the much longer retention time of seeds in the digestive tracts of the mammals. Seeds following frugivore gut passage had a much thinner coat than uningested seeds, presumably due to differences in the rate of seed-coat erosion (Fig. 4.1). In the Mediterranean species *Myrtus communis* (Myrtaceae), germination responses also differed according to the species of frugivore (Traveset *et al.*, 2001b). Again, these differences were attributed to the longer retention times of seeds in mammals compared with birds. However, the fruits of other species of plant ingested by different frugivores with very different retention times showed no differences in seed germination response (Traveset and Willson, 1997; Nogales *et al.*, 2005; Rodríguez-Pérez *et al.*, 2005).

The meta-analysis performed by Traveset and Verdú (2002) further revealed that frugivore taxonomy has an important effect on seed germination responses. This analysis showed that bats and birds tend to have highly positive effects on seed germination (i.e. increasing germination percentages, compared with uningested seeds) whereas non-flying mammals have positive but small effects. Reptiles showed a non-significant effect, although the number of experiments with these frugivores is still small (Traveset and Verdú, 2002). Fishes could not be included in the analyses due to the scarcity of data on this group despite their importance as seed dispersers in many riverine systems. Such different effects among frugivore taxa are again attributed, at least partly, to the different food retention times, which is generally longer in non-flying mammals and reptiles than in birds and bats. Birds that include large amounts of fruits in their diet have a proportionally shorter intestine and a smaller and less muscular gizzard than non-frugivorous birds, which may result in a more gentle treatment of the coats of seeds ingested by specialist frugivores (Pulliainen *et al.*, 1981; Jordano, 2000). Morphological

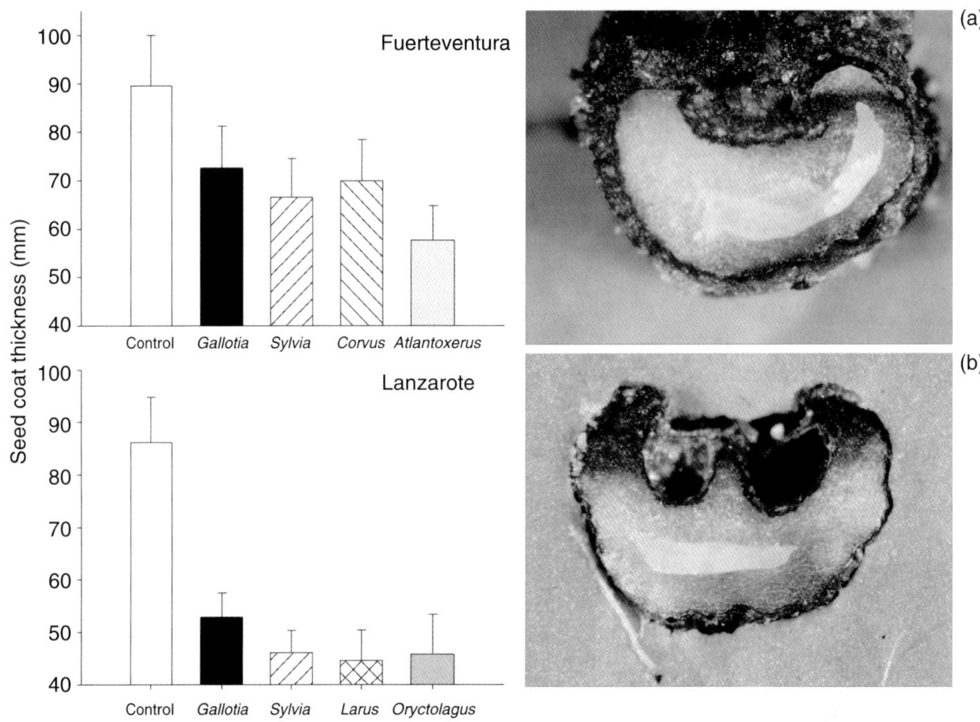

Fig. 4.1. Change in seed coat thickness found in *Rubia fruticosa* seeds after being ingested by different seed dispersers: lizards (*Gallotia atlantica*), warblers (*Sylvia* spp.), ravens (*Corvus corax*), squirrels (*Atlantoxerus getulus*), gulls (*Larus cachinnans*) or rabbits (*Oryctolagus cuniculus*) in (a) Fuerteventura and (b) Lanzarote (Canarian Archipelago). Cross-sections of a seed directly gathered from a fruit (a) and of a seed defecated by the squirrel (b) are shown. More details of the study can be found in Nogales *et al.* (2005).

modifications are also found in frugivorous reptiles and bats; unlike birds, in these animals, frugivores generally have longer intestines than insectivorous species (Kunz and Ingalls, 1994; King, 1996).

Comparative tests of the scarification effect produced by different frugivores feeding on the same plant species often reveal large differences – some imparting a germination enhancement, others an inhibitory effect, and others a neutral effect (see review in Traveset, 1998). Likewise, the same species of frugivore may have different effects on germination depending upon the species of plant studied (Traveset, 1998; Traveset *et al.*, 2001c; Rodríguez-Pérez *et al.*, 2005). To further examine these patterns, we compiled two databases: one with a total of 29 studies in which percentage seed germination of a given species of plant was compared between at least two dispersers, and another (n = 40 studies) in which the effect of a disperser was compared, in the same study, with at least two different species of plant. For each database, a consistency index (dependent variable) was defined as the frequency of the most common effect (positive, negative or neutral) divided by the total number of species tested (either disperser or plant, respectively). The independent factors were the species of disperser in the first database, and the species of plant in the second. The effect of different dispersers on the same plant was found to be more consistent than the effect of a single disperser over different plants. Consistent effects on germination (positive, negative or neutral) were observed in 68.8% of the cases comparing the effect of different dispersers on the same plant, and no significant differences in the consistency index when comparing frugivore groups; $\chi^2 = 3.23$, $df = 3$, $P = 0.357$). In contrast, consistency was much lower (33.8%) when a single disperser was tested with different plants ($\chi^2 = 22.8$, $df = 1$, $P < 0.0001$). These results imply that the particular seed characteristics of each plant (e.g. seed size, pulp composition, seed coat thickness, texture) promote more consistent responses than the specific morphological and/or physiological traits of frugivores.

Despite the fact that seed traits, rather than frugivore traits, dominate the effects of gut passage on germination response, frugivores also contribute to the heterogeneity of germination responses both within plant populations and within plant communities. This heterogeneity, with each species of frugivore having a particular effect on the seeds of each plant consumed, might be especially meaningful in unpredictable climates, such as those found in areas with unpredictable rainfall patterns. For instance, different studies in the eastern and western Mediterranean basin (Izhaki and Safriel, 1990; Barnea *et al.*, 1991; Traveset *et al.*, 2001b,c) have found that the seeds of a particular species of plant are differentially affected by passage through the guts of different bird species. Heterogeneity of seed responses occurs also at an intraspecific level. Meyer and Witmer (1998) reported individual differences within *Prunus virginiana* (Rosaceae) in germination responses, attributing them to differences in fruit nutrient composition as well as in the amount of fruit pulp accompanying the seed. A recent study performed with *Phillyrea angustifolia* (Oleaceae) and *Myrtus communis* with the goal of identifying mechanisms that can explain the different germination responses usually observed, showed that much inter- and intraspecific variation exists in different seed-coat traits. Such traits are: permeability, coat thickness, texture (measured by computing an index from a digital image obtained in the scanning electron microscope) and seed hardness (obtained by means of a testing machine (compression mode) that measures the load needed to break open the seed (A. Traveset, J. Rodríguez-Pérez and B. Pías, unpublished results); variation in all these traits has been found to be associated with variation in germination responses. This implies that, when carrying out studies to test the effects of endozoochory on seed germination, it is important to include in the tests seeds from a representative sample of individual plants. Less variation in germination has been found among individuals within a species of frugivore (A. Traveset, J. Rodríguez-Pérez and B. Pías, unpublished results), although it remains important to test germination using as many individual animals as possible.

Effect of manure/guano on germination and seedling growth: comparison among frugivores

The faecal material accompanying vertebrate-dispersed seeds at deposition sites can play an important role in enhancing seed germination and seedling survival (e.g. Dinerstein and Wemmer, 1988; Traveset *et al.*, 2001a). Although the fertilizing effect of such material has been acknowledged among ecologists concerned with seed dispersal by vertebrates (Willson and Traveset, 2000), few data exist that test the size of this effect. Most frugivores have a very varied diet, which results in a highly variable manure composition; this is especially evident in 'carnivorous' mammals that seasonally consume great quantities of fruits (see, e.g. Herrera, 1989; Willson, 1993). The diet of bears, for instance, consists of vertebrates and invertebrates, as well as plant material such as fruits, stems, leaves or roots (Traveset *et al.*, 2001a). Bear manure consisting of animal material showed a consistently greater effect on seedling growth of *Vaccinium* spp. (Ericaceae) and *Rubus spectabilis* (Rosaceae) than bear manure consisting of vegetation fibre or fleshy pulp (Fig. 4.2; Traveset *et al.*, 2001a). While protein in the diet promotes a supply of nitrogen in the faeces, bones provide a source of calcium

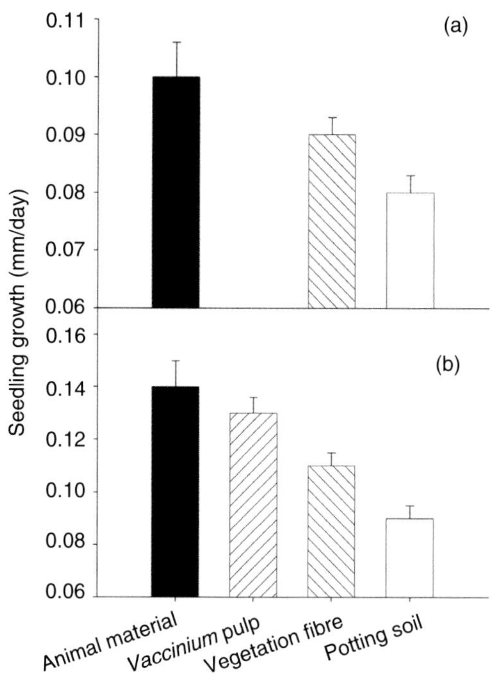

Fig. 4.2. Mean (± SE) seedling length of (a) *Vaccinium ovalifolium/alaskaense* and (b) *Rubus spectabilis* seedlings in different manure treatments. Potting soil was used as the control. Figures drawn from data extracted from Traveset *et al.* (2001a).

phosphate, which may limit growth in poor soils such as those in the temperate rainforest of Alaska (Willson *et al.*, 1998). Other components of the manure may supply some other nutrients or improve the moisture regime for seedlings, thus explaining the greater seedling growth in dung compared with the control (potting soil). Such results are consistent with those reported in other studies. Dinerstein and Wemmer (1988) found a significant increase in seedling growth of *Trewia nudiflora* (Euphorbiaceae) when seeds were embedded in rhinoceros dung compared with those growing on potting soil. Similarly, seedling growth of several *Acacia* (Mimosaceae) species in dung varied significantly between the species of consuming herbivore, probably owing to differences in the nutrient content and in the water-holding capacity of the dung (Miller, 1995).

As well as promoting seedling growth through a fertilizer effect, frugivore manure may also contain toxic compounds that can have negative effects on seed and/or seedling survival (Welch, 1985; Marambe *et al.*, 1993; Malo and Suárez, 1995; Cosyns *et al.*, 2005). Animal wastes are known to contain phenolic compounds and fatty acids that act as germination inhibitors for some plants. These compounds may alter the activity of enzymes that regulate the rate of germination (Marambe *et al.*, 1993, and references therein). Moreover, fruit pulp can also provide a substrate for the growth of fungi or bacteria, which can be very detrimental for germination or for seedling survival (Meyer and Witmer, 1998); thus, despite consuming large quantities of fruits, cedar waxwings were found to be ineffective dispersers of plants such as *Lindera benzoin* (Lauraceae) and *Prunus virginiana* (Rosaceae) because of faecal accumulation around seeds. In contrast, robins did not have these negative effects on the germination of these species as they usually regurgitate the seeds clean (Meyer and Witmer, 1998).

The difference in the influence of manure from frugivores belonging to different vertebrate taxonomic groups on seedling establishment success is likely to be great. The number and density of seeds in faeces, which is known to affect seed and seedling survival, is much higher in frugivorous mammals than in most birds (excepting cassowaries, emus and colonial nesting birds; A.J. Dennis, Queensland, 2005, personal communication) or reptiles (Barnea *et al.*, 1992; Willson, 1993; Andresen and Levey, 2004, and references therein). The large number of seeds in the dung of animals like rhinoceros, elephants or bears can attract seed predators, mainly rodents (Janzen, 1986; Traveset, 1990; Willson, 1993; Bermejo *et al.*, 1998; Andresen and Levey, 2004), and after germination, seedlings are likely to encounter intense competition (Lewis, 1987; Loiselle, 1990). On the other hand, the rich and copious droppings of a large mammal provide more nutrients to a seedling, over a longer period of time, than the small scats of most birds or lizards. In addition, birds excrete relatively high amounts of white urates when eating invertebrates, but watery material often coloured with fruit pigments when eating fruits. While it seems clear that the variation in faecal composition between dispersal agents might affect seed and seedling performance, little is currently known. Thus it would be useful to investigate this in future studies.

The effect of seed aggregations and seed mixing on germination (allelopathic and density-dependent effects)

Depending on a frugivore's size and behaviour, seeds that have been regurgitated or defecated will be deposited over a wide range of densities and with a varying degree of species-mixing (Stiles and White, 1986; Jordano, 2000; Fig. 4.3). Regurgitators tend to process fruits quickly and drop them one at a time beneath the maternal tree, with little species mixing; and the dropped seeds may accumulate to high densities (Stiles and White, 1986; Stiles, 2000). In contrast, fruit-ingesting frugivores deposit seeds in faecal clumps after gut processing. The degree of clustering of these seeds depends on the size of the faecal clump and the density of seeds in each clump, as well as on whether there is an aggregation of clumps beneath roosting or nesting trees, latrines or other regularly used locations (Debussche *et al*., 1982; Debussche and Isenmann, 1994; Dean and Milton, 2000; Stiles, 2000; Takahashi and Kamitani, 2003). The degree of seed-mixing and the chance of deposition with heterospecific neighbours also vary with the size of the frugivore and the gut-retention time (Stiles and White, 1986; Jordano, 2000). Large animals tend to deposit clumps with more seeds and more seed-mixing than do small frugivores.

Frugivorous birds typically deposit two or more species in droppings. Loiselle (1990) recorded an average of 132 seeds of 2.3 plant species per dropping in five species of small to medium-sized frugivorous passerines in Costa Rica. Stanley and Lill (2002a) found that in a temperate woodland in Australia, the white-eye (*Zosterops lateralis*; Zosteropidae) deposited on average 20–25 seeds of 1.1–1.3 species per dropping, and the superb fairy-wren (*Malurus cyaneu*; Maluridae) about 4–10 seeds of just one species. In contrast, one Cassowary dropping (*Casuarius casuarius*; Casuariidae) in North

(a) (b)

Fig. 4.3. Faecal clumps vary greatly in size from (a) bird droppings (from a New Zealand pigeon) with few seeds and a small amount of guano with one or two plant species, to (b) large mixed-species communal dung piles of large vertebrates like these white rhino latrines in South Africa with seedlings of the rhino-dispersed *Datura* sp. The environmental opportunities and challenges are obviously very different in their two extremes and are likely to have influenced the evolution of fruit and seed traits.

Queensland was found to contain 13 'more or less intact fruits' of *Beilschmiedia* sp. (Lauraceae), each with a diameter of about 6 cm and weighing around 52 g (Stocker and Irvine, 1983). The Asian rhino (*Rhinoceros unicornis*; Rhinocerotidae) disperses seeds in huge defecations that can weigh up to 23 kg and presumably potentially contain thousands of seeds (Dinerstein and Wemmer, 1988). Amazonian woolly monkey droppings were found to contain an average of 70 seeds of 2.53 species (Stevenson, Chapter 15, this volume). Pine and stone marten (*Martes martes* and *M. foina*, respectively; Mustelidae) scats collected from a woodland in Germany contained on average about 800 seeds per scat of a variety of fleshy-fruited species (Schaumann and Heinken, 2002). Thus, depending on the frugivore, deposited seeds may find themselves at low densities in the faeces or as part of a large assemblage of conspecific or heterospecific seeds competing for space, light and nutrients.

These very different environments pose different ecological opportunities and challenges, and if there is a consistent pattern to the deposition conditions, there may be selection pressures for the evolution of appropriate germination strategies (Linhart, 1976; Loiselle, 1990; Murray, 1998). In most cases, density-dependent factors will tend to impact negatively on the success of seedlings as they compete for resources and space (Lewis, 1987; Loiselle, 1990), so selection will tend to be for tactics that give an advantage to seedlings in this competition. Potentially these tactics may include early germination to beat the rush, or induced dormancy to wait out the crowd (Loiselle, 1990; Murray, 1998). Alternatively, seeds may produce allelopathic chemicals – or use those contained in fruit pulp – that are active against conspecific or heterospecific neighbouring seeds.

There is an extensive literature on the involvement of allelochemicals found in the seeds and fruits of dry-fruited species, particularly in agriculture. Plants that have been shown to produce water-soluble germination inhibitors that are potent against other pasture plants include the thistles *Carduus nutans* and *Onopordum acanthium* (Asteraceae: Wardle *et al.*, 1991; Qaderi *et al.*, 2003), Bishop's weed *Ammi majus* (Apiaceae; Friedman *et al.*, 1982), *Vigna mungo* (Fabaceae; Suman *et al.*, 2002) and *Lotus tenuis* (Fabaceae; Laterra and Bazzalo, 1999).

Apart from the evidence of pulp inhibitors mentioned earlier (which have all been tested on conspecifics), we are aware of only one study of a fleshy-fruited species that has demonstrated active allelochemicals that are effective on heterospecific seeds. Hruska *et al.* (1982) used a cucumber-seed bioassay to identify germination inhibitors from the lily *Liriope muscari* (Convallariaceae), although they did not test the effect of these inhibitors on seeds of species that are likely to be consumed by the same frugivores that disperse *L. muscari*, and the authors themselves interpret the data as evidence of auto-inhibition to prevent premature germination in-fruit rather than as evidence of allelopathy. Further tests of fruit constituents as sources of allelochemicals effective against potential competitors that are likely to co-occur within droppings would be welcome.

In some situations, there may be advantages to aggregated deposition patterns. Kelly *et al.* (in press) deposited seeds of a New Zealand mistletoe,

Peraxilla tetrapetala (Loranthaceae), on to host branches as single seeds and in clumps of five. Germination and establishment success per seed was significantly higher in the clumps than in single seeds. This appeared to be due to an increased tendency to adhere to the host branch in clumps. However, over the subsequent 57 months, there was a tendency for slightly lower rates of survival in clumps, negating the initial advantage.

The consequences of the aggregation of seeds into clumps as the result of the feeding and deposition patterns of frugivores remains one of the least understood and studied aspects of seed dispersal. Characterizing the effectiveness of different frugivores by their seed dispersal kernels or by their ability to deposit seeds in safe sites ignores the potentially important effects that may occur at the scale of the individual dropping. Differential allelopathic and competitive interspecific effects between seeds within these clumps made by different frugivores have the potential to impose an additional layer of complexity that may distort the view established from simple movement patterns and gut passage times. However, until we have further studies that test the magnitude of these effects, it is difficult to assess their likely importance for the fate of seed.

How do we Experimentally Investigate the Factors that Influence Seed Fate?

Several key questions have emerged from the previous section concerning the interactions between frugivores and fruits and the way that these processes influence the success of seed germination. These questions are both ecological and evolutionary, have implications at the community level as well as for the demography of individual species and, we believe, should be of equal interest to theoretical and applied ecologists as well as to evolutionary biologists. Of central importance are the factors that influence the eventual fates of seeds, which can essentially be boiled down to these: (i) successful germination and establishment to the seedling stage; (ii) death by pre- or post-dispersal predation or disease; and (iii) the induction of dormancy and incorporation into the soil seed bank awaiting future opportunities. The likelihood of each of these is determined, at least in part, by the inherent biological properties of the plants' strategies shaped over millennia by natural selection and also by the changing biotic and abiotic environment and dispersal services provided by frugivores.

1. What happens to fleshy fruits if they are not ingested – are they doomed to fall from the plant and be prevented from germination by autotoxic inhibitors, or are they capable of germinating and maintaining the local population?
2. Does it matter what type of frugivores feed on the fruits? Do frugivores that regurgitate seeds produce different changes in seed traits relevant to germination than frugivores that defecate seeds?
3. How important is the composition of the droppings in which seeds are

deposited? Is there interference between seeds in mixed-species droppings, and to what extent do the other constituents of the dropping facilitate or retard seed germination? We have come some way in determining the consequences of the different deposition patterns that are produced by different frugivores. For example, seeds in dense clumps are frequently worse off than those deposited in small groups, depending on the species. We now need to answer questions such as: (i) are early germination and allelopathy effective strategies for competing with neighbouring seeds; (ii) how important is the fertilization effect and does it alter the relative effectiveness of different dispersers?

Much progress has been made already and we have at least partial answers to some of these questions. However, we believe that further progress and clarification of inconsistent or uncertain results will come from careful attention to the design of experiments, as well as from attention to some factors that hitherto have largely been ignored. An important point that has emerged from the work done to date is the context in which experiments are performed. As we show below, conducting experiments in the field rather than in the laboratory is crucial for answering these questions if we wish to obtain answers that are relevant to the systems we study. In the following sections we outline seven recommendations for the design of future experimental studies that seek to answer these questions.

1. Conduct these studies in field conditions

The micro-ecological conditions prevailing at the site where a seed is deposited can represent another source of variation in its future germination behaviour. Most reported information comes from studies carried out under controlled circumstances, usually in the laboratory (see review of the scarification effect in Traveset and Verdú, 2002), yet the results might not reflect what really happens in nature. Of the 83 studies considered in that review, 56% tested seed germination in laboratory conditions, another 22% performed germination experiments in the field, and 13% used glasshouses. Only a small fraction (7%) used more than one condition. The meta-analysis revealed that laboratory and field conditions were more likely than glasshouse experiments to detect differences between treatments (ingested versus non-ingested seeds), although the lower sample size for glasshouse cases might well have caused such differences. A more recent analysis of the scarification and deinhibition effects, which considered only bird-dispersal systems and included some additional studies that were not in the Traveset and Verdú (2002) dataset, suggested that, in general, the deinhibition effect is larger than the scarification effect, and that both effects are strongly influenced by the testing environment (Robertson *et al.*, 2006; Fig. 4.4). In this analysis, laboratory studies tended to indicate stronger effects than either

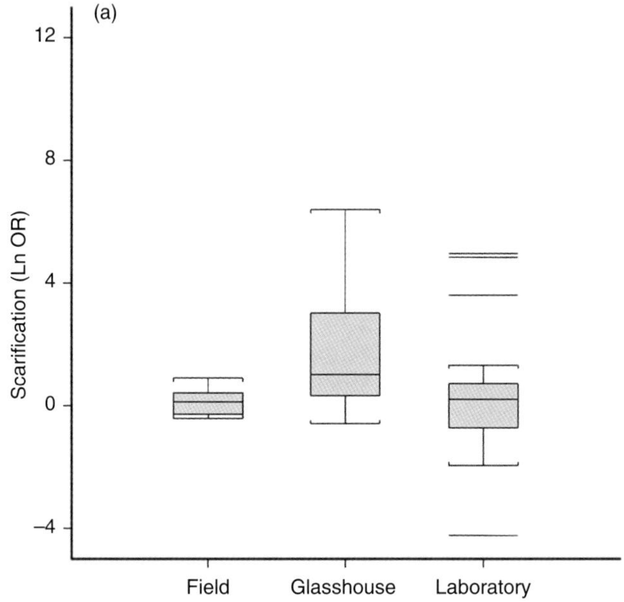

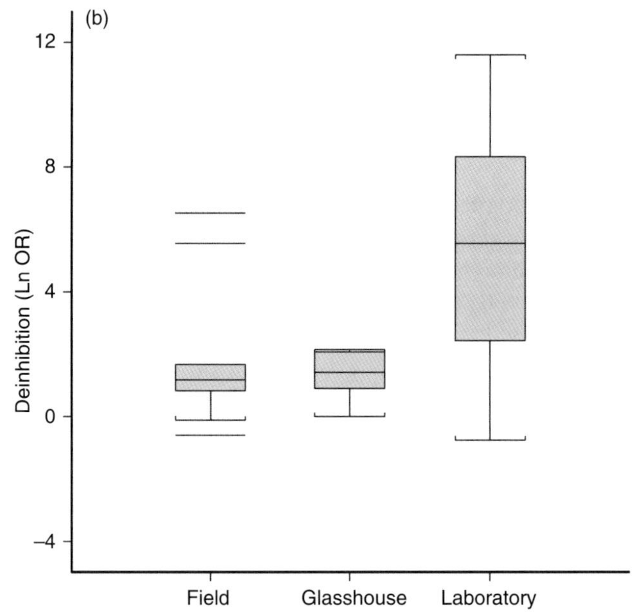

Fig. 4.4. Box plots of the relative size of (a) scarification and (b) deinhibition effects of fruit processing and the effects of the testing environment in fleshy fruited species. In (a) we calculated the size of the scarification effect by calculating the natural log (Ln) of the odds ratios (OR) comparing the final germination percentage of seeds from fruits passed through a gut versus those from hand-cleaned fruits in 56 species, following Traveset and Verdú (2002). In (b) we compared the effect of hand-cleaned versus intact fruits as a measure of the deinhibition effect in 51 species.

glasshouse or field trials, particularly for deinhibition. This might be due to the greater physico-chemical and biological processing of fruits and seeds that occur in the field after they have been deposited. This post-dispersal processing may replicate or replace the processes that happen in the animal gut to some extent and hence reduce the dependence on gut transport for successful germination (Morpeth *et al.*, 1997; Baskin and Baskin, 1998; Morpeth and Hall, 2000; Robertson *et al.*, 2006).

Studies that have tested the scarification effect in more than one experimental condition have often revealed rather inconsistent results (Bustamante *et al.*, 1992; De Figueiredo and Perin, 1995; De Figueiredo and Longatti, 1997; Yagihashi *et al.*, 1998; Traveset *et al.*, 2001c; Figueroa and Castro, 2002; Rodríguez-Pérez *et al.*, 2005; Robertson *et al.*, 2006). Although the constant conditions in the laboratory might seem ideal for detecting differences in germination between ingested and non-ingested seeds, this varies depending upon the plant tested (Table 4.2). Some studies have found a greater scarification effect on germination in the laboratory compared with field experiments (Table 4.2a; e.g. Bustamante *et al.*, 1992, 1993; De Figueiredo and Perin, 1995; Yagihashi *et al.*, 1998), while others have detected an effect of germination only in outdoor conditions and not in a growth chamber or glasshouse (e.g. Traveset *et al.*, 2001c). A few studies have found similar results between laboratory and field conditions (Rust and Roth, 1981; Braun and Brooks, 1987; Izhaki and Safriel, 1990; Barnea *et al.*, 1991). In a recent study, a total of eight species of plants were tested both in an experimental garden and in the field, and only half showed consistent results (Rodríguez-Pérez *et al.*, 2005). Overall, laboratory studies tend to show more benefit of scarification on germination compared with glasshouse or field tests, which on average show that scarification makes little difference to final germination percentage. There are very few cases where the deinhibition effect has been measured in more than one environment (Table 4.2b), but again these few data suggest that the results depend very much on the testing environment used.

The particular characteristics of the habitat where the germination tests are performed in the field may also affect the results. For instance, the salinity of the soil where seeds were sown was found to influence germination responses in two species of halophytes (Espinar *et al.*, 2004). Therefore, if our goal is to evaluate whether a change in germination outcomes is adaptive or not, it is crucial that seed responses to dispersers' gut treatment are examined in the field, in conditions that are similar to those encountered by the subject plant. The laboratory is useful for examining the underlying mechanisms once a field experiment has determined that the outcomes are important. Espinar *et al.*'s (2004) findings also highlight the need to consider the heterogeneity of the habitat in the experimental design, as this may be an additional source of variation in seed germination responses. Habitat effects have already been found in several studies on post-dispersal seed predation, germination success and seedling establishment (e.g. Herrera *et al.*, 1994; Schupp, 1995; Rey and

Table 4.2. Cases where the germination benefit of gut passage has been tested in more than one environment on replicate batches of seeds. The scarification effect compares hand-cleaned to defecated seeds, while the deinhibition effect is the comparison of hand-cleaned seeds to seeds in intact fruits. In both cases, the effect size is measured as the natural log of the odds ratio. Negative figures indicate a suppression of seed germination by either scarification or deinhibition, positive values indicate a germination promotion from these effects.

(a) The scarification effect

			Scarification Ln OR			
Plant	Family	Frugivore	Glasshouse	Field	Laboratory	Authors
Crataegus monogyna	Rosaceae	*Turdus merula*	–0.04	–0.07		Rodríguez-Pérez *et al*., 2005
Cryptocarya alba	Lauraceae	*Dusicyon culpaeus*		0.87	3.58	Bustamante *et al*., 1992, 1993
Ficus luschnathiana	Moraceae	*Plathyrhinus lineatus*		0.32	0.79	De Figueiredo and Perin, 1995
Ficus luschnathiana	Moraceae	*Turdus leucomelas*		–0.02	0.83	De Figueiredo and Perin, 1995
Gaultheria mucronata	Ericaceae	*Unidentified birds*	0.30		0.18	Figueroa and Castro, 2002
Juniperus phoenicea	Cupressaceae	*Turdus merula*		0.10	0.69	Rodríguez-Pérez *et al*., 2005
Myrtus communis	Myrtaceae	*Martes martes*	–0.24	–0.89		Traveset *et al*., 2001a
Myrtus communis	Myrtaceae	*Unidentified birds*	–0.10	0.90	0.33	Traveset *et al*., 2001a
Osyris alba	Santalaceae	*Turdus merula*	0.06	0.08		Rodríguez-Pérez *et al*., 2005
Podophyllum peltatum	Berberidaceae	*Terrapene carolina*		1.99	1.92	Rust and Roth, 1981; Braun and Brooks, 1987
Rhamnus palaestinus	Rhamnaceae	*Turdus merula*	1.06	1.49		Izhaki and Safriel, 1990; Barnea *et al*., 1991
Rubia peregrina	Rubiaceae	*Podarcis pityusensis*	–0.04	–0.46		Rodríguez-Pérez *et al*., 2005
Rubia peregrina	Rubiaceae	*Turdus merula*	0.36	0.14		Rodríguez-Pérez *et al*., 2005
Rubia tenuifolia	Rubiaceae	*Turdus merula*	–0.48	0.07		Izhaki and Safriel, 1990; Barnea et al., 1991
Rubus ulmifolius	Rosaceae	*Podarcis pityusensis*	–0.41	0.69		Rodríguez-Pérez *et al*., 2005
Rubus ulmifolius	Rosaceae	*Turdus merula*	–0.49	0.12		Rodríguez-Pérez *et al*., 2005
Smilax aspera	Smilacaceae	*Turdus merula*	–0.15	0.08	–0.15	Rodríguez-Pérez *et al*., 2005
Solanum luteum	Solanaceae	*Podarcis pityusensis*	–0.08	0.20		Rodríguez-Pérez *et al*., 2005
Solanum luteum	Solanaceae	*Turdus merula*	0.04	–0.15		Rodríguez-Pérez *et al*., 2005
Solanum nigrum	Solanaceae	*Podarcis pityusensis*	0.59	0.12		Rodríguez-Pérez *et al*., 2005
Solanum nigrum	Solanaceae	*Turdus merula*	0.59	0.27		Rodríguez-Pérez *et al*., 2005
Sorbus commixta	Rosaceae	*Turdus naumanni*		0.11	0.62	Yagihashi *et al*., 1998
Average difference from field conditions			–0.10		0.49	

(b) The deinhibition effect

		Deinhibition Ln OR			
Plant species	Family	Glasshouse	Field	Laboratory	Authors
Corynocarpus laevigatus	Corynocarpaceae	3.58	1.06	6.66	Robertson *et al*., 2005; Burrows 1996a
Beilschmiedia tawa	Lauraceae	1.00	1.00	6.98	Robertson *et al*., 2005; Burrows 1999
Ripogonum scandens	Smilacaceae	–0.04	4.06	8.50	Robertson *et al*., 2005; Burrows 1996b
Average difference from field conditions		0.53		5.34	

Alcántara, 2000; Traveset *et al.*, 2003; Schupp, Chapter 20, this volume).

Placing seeds into the field requires a technique that allows the fate of seed to be easily followed. Robertson *et al.* (2006) used small sections of pipe partly pushed into the litter layer and laid out in a regular array to allow easy checking of germination and easy recovery of seeds by sieving the tube contents at the end of experiment to check for viability. Other options for relocating larger seeds include tiny metal tags or monofilament wire tethers attached to the hard parts of seeds (see, e.g. McGuinness, 1997; Steele *et al.*, 2001; Li and Zhang, 2003). Forget and Wenny (2005) provide a review of techniques used to follow seed fate, some of which may be applicable to germination studies.

2. Recognize and appreciate the distinction between the speed of germination and the final germination percentage and avoid talking about the rate of germination

Germination performance may be measured both by the proportion of seeds that have germinated at the end of the study and by the speed with which seeds germinate. Confusingly, both have often been termed germination 'rates' and this should be avoided in future studies (Robertson *et al.*, 2006). While the former is positively related to fitness, the second is not necessarily, as for instance, early germinated seeds might be more likely to die of desiccation, pathogens, predators, etc. (Traveset, 1998, and references therein). Seeds that do not germinate immediately, or that enter dormancy, may nevertheless contribute significantly to plant fitness by dispersing in time rather than in space and may remain in waiting for suitable conditions (Kelly *et al.*, 2004; Robertson *et al.*, 2006). It is therefore worthwhile describing both the distribution of timing of germination as well as recording the final percentage of seed germination and checking for dormant seeds in the remainder.

3. Measure the deinhibition effect as well as the scarification effect

Several recent papers have noted the confusion in the literature about the processes that occur during gut passage, and have emphasized the need to measure both outcomes of gut passage: the deinhibition effect as well as the scarification effect (Kelly *et al.*, 2004; Samuels and Levey, 2005; Robertson *et al.*, 2006). If we are to answer the first two questions (**1** and **2**) from the previous section, we need to compare the performance of seeds from three types of treatment: (i) manually extracted from the pulp; (ii) intact fruits; and (iii) seeds that have been ingested and regurgitated or defecated by frugivores. Comparing (i) and (iii) is a test of the scarification effect, while comparison of (i) and (ii) is a test of the deinhibition effect. Comparing (ii) and (iii) allows an assessment of the combined effect of both processes. In Samuels and Levey's (2005) review of 99 studies, 77% of these considered only the scarification effect, none measured simply the deinhibition effect, and only 18% included all three treatments and could therefore estimate the importance of both effects. The remaining 4% compared seeds from intact fruits and those defecated by frugivores and so were not able to separate deinhibition and scarification effects. The survey of the relative effect size by Robertson *et al.* (2006) suggests that the deinhibition effect is often larger than the scarification effect, although since most studies were conducted in Petri dishes which do not appear to accurately predict the effects in the field, further comparisons are needed from field conditions.

4. Allow sufficient time for seeds to germinate, and recover ungerminated seeds at the end of the study to check for viability/dormancy

As indicated above, seeds have three fates following dispersal – death, germination or secondary dormancy. Distinguishing between dormant and dead seeds is important, since the former may allow seeds to enter the soil seed bank (Kelly *et al.*, 2004) and so they should not be assumed to be dead (Baskin and Baskin, 1998; Robertson *et al.*, 2006). Some seeds may take several years to germinate. For example, seeds of the New Zealand gymnosperm *Prumnopitys ferruginea* (Podocarpaceae) were still germinating in field conditions more than 4 years after bird ingestion (Clout and Tilley, 1992). Seeds that are recovered may be tested for viability either with tetrazolium chloride (Cottrell, 1947) or placed in ideal laboratory conditions for germination.

5. Consider the influence of other components in the diet of frugivores

The diet of a particular species of disperser is known to alter food retention time. For instance, European starlings (*Sturnus migratorius*; Sturnidae) showed decreased gut passage times when their diet was changed from insects to fruits (Karasov and Levey, 1990). Likewise, a seed-based, high-fibre diet has been found to increase seed digestibility in mallards (*Anas platyrhynchos*; Anatidae) and germination of seeds of *Potamogeton pectinatus* decreased as seed retention time in the ducks' guts increased (Charalambidou *et al.*, 2005). For waterfowl, grit quantity (strongly related to diet) in the gizzard can also have large effects on seed germinability (e.g. Santamaría *et al.*, 2002). Thus, for future investigations on the effects of seed ingestion by dispersers on germination, it is important to maintain animals on the same diet when performing experiments in captivity. When defecated seeds are gathered in the field, the manure composition should be considered as a covariate, particularly in cases where it is variable. Transit time of seeds in the digestive tract needs to be considered in future experiments in captivity, especially if it is highly variable and ranges from several days to weeks.

6. Examine the mechanisms by which ingested seeds have a different germination performance from uningested seeds

Seed traits such as weight, coat permeability, coat thickness, texture and hardness may be related to either germination rate or germinability, or to both. For example, Nogales *et al.* (2005) found that *Rubia fruticosa* seeds ingested by squirrels showed the largest reduction in seed coat thickness and at the same time had the lowest viability. Due to variation in traits shown among individuals within species of dispersers and plants (A. Traveset, J. Rodríguez-Pérez and B. Pías, unpublished results), the use of known seed sources (maternal plants) and known individual dispersers will probably shed light on the mechanisms that can explain the differences in seed responses within a species.

7. Experiment with the deposition matrix

A critical aspect of the influence of the frugivores on seed germinability, which has until now been virtually ignored, relates to the nature of the deposition matrix that seeds find themselves in (Stiles and White, 1986; Jordano, 2000). Future studies should recognize that seeds are typically deposited in faecal clumps or droppings and are embedded in guano or other faecal matter and are often competing with a mixture of seeds from one or more species. Carefully designed studies that experiment with the components of this matrix will allow us to consider effects such as the fertilizer effect, seed mixing, the presence of germination inhibitors and/or allelochemicals, and the advantages of rapid or delayed germination in the clump. We know virtually nothing about the relative importance of these effects; however, for plant species that are dispersed by large frugivores that deposit large, highly mixed clumps, as well as by small animals with small droppings with few, unmixed seeds, these effects may be very important for comparisons of disperser effectiveness. All of them could be examined experimentally by following the fate of planted cleaned seeds in different faecal matrices, trying different combinations of neighbours and seed densities (Loiselle, 1990). Distinguishing between allelopathic and direct competition is not straightforward, but the use of stabilized seed extracts or purified secondary metabolites (Cipollini and Levey, 1997) or activated carbon to absorb leachates (Nilsson, 1994) may help shed light on the active mechanisms.

Concluding Remarks

The study of how frugivores influence the germination patterns of endozoochorous plant species is of great relevance to our understanding of these animal–plant interactions from the ecological and evolutionary perspectives. From the plants' viewpoint, we need to know how the array of frugivores that disperse their seeds determine: (i) the speed of germination and the percentage that germinate, which will depend upon fruit and seed processing in the digestive tract (among other factors); (ii) the seed deposition pattern, influenced by the number of seeds and species combination in the deposition (among other factors); and (iii) the plant establishment success within a community, which is partly a function of the type of material in which seeds are embedded at the time of deposition. Most data demonstrating the different mechanisms by which frugivores can affect seed fate are from studies performed with birds and mammals, which are the numerically dominant dispersers in many systems. In contrast, much less information has been gathered from other taxa, such as reptiles or fish. Likewise, we also need more data from field experiments, as most studies have performed germination tests under controlled conditions.

In this chapter, we have reviewed the most relevant information on this topic. From this review we have made a series of methodological recommendations and suggestions for new research directions. We hope these will be useful for those that continue to work on this line of research as well as for those young investigators who are willing to delve into this fascinating aspect of plant–animal interactions.

Acknowledgements

We are grateful to the organizers of the Symposium for all their efforts and help, and to Andrew Dennis, Eric Cosyns and Jordi Figuerola for useful comments on the manuscript. This chapter is framed within project CGL2004-04884-CO2-01/BOS, financed by the Spanish Ministry of Education and Science awarded to A.T. and by project C09X0204 from New Zealand's Public Good Science Fund awarded to A.R.

References

Alcántara, J.M. and Rey, P.J. (2003) Conflicting selection pressures on seed size: evolutionary ecology of fruit size in a bird-dispersed tree, *Olea europaea*. *Journal of Evolutionary Biology* 16, 1168–1176.

Andresen, E. and Levey, D.J. (2004) Effects of dung and seed size on secondary dispersal, seed predation, and seedling establishment of rain forest trees. *Oecologia* 139, 45–54.

Barnea, A., Yom-Tov, Y. and Friedman, J. (1990) Differential germination of two closely related species of *Solanum* in response to bird ingestion. *Oikos* 57, 222–228.

Barnea, A., Yom-Tov, Y. and Friedman, J. (1991) Does ingestion by birds affect seed germination? *Functional Ecology* 5, 394–402.

Barnea, A., Yom-Tov, Y. and Friedman, J. (1992) Effect of frugivorous birds on seed dispersal and germination of multiseeded fruits. *Acta Oecologica* 13, 209–219.

Baskin, C.C. and Baskin, J.M. (1998) Ecologically meaningful germination studies. In: Baskin, C.C. and Baskin, J.M. (eds) *Seeds: Ecology, Biogeography, and Evolution of Dormancy and Germination.* Academic Press, San Diego, California, pp. 5–26.

Bermejo, T., Traveset, A. and Willson, M.F. (1998) Post-dispersal seed predation in the temperate rainforest of Southeast Alaska. *Canadian Field Naturalist* 112, 510–512.

Braun, J. and Brooks, G.R. (1987) Box turtles (*Terrapene carolina*) as potential agents for seed dispersal. *American Midland Naturalist* 117, 312–318.

Burrows, C.J. (1996a) Germination behaviour of seeds of the New Zealand woody species *Alectryon excelsus*, *Corynocarpus laevigatus*, and *Kunzea ericoides*. *New Zealand Journal of Botany* 34, 489–498.

Burrows, C.J. (1996b) Germination behaviour of the seeds of seven New Zealand vine species. *New Zealand Journal of Botany* 34, 93–102.

Burrows, C.J. (1999) Germination behaviour of seeds of the New Zealand woody species *Alseuosmia macrophylla*, *A. pusilla*, *Cordyline banksii*, *Geniostoma rupestre*, *Myrtus bullata* and *Solanum aviculare*. *New Zealand Journal of Botany* 37, 277–287.

Bustamante, R.O., Grez, A.A., Simonetti, J.A., Vásquez, R.A. and Walkowiak, A.M. (1993) Antagonistic effects of frugivores on seeds of *Cryptocarya alba* (Mol) Looser (Lauraceae): consequences on seedling recruitment. *Acta Oecologica* 14, 739–745.

Bustamante, R.O., Simonetti, J.A. and Mella, J.E. (1992) Are foxes legitimate and efficient seed dispersers? A field test. *Acta Oecologica* 13, 203–208.

Charalambidou, I., Santamaría, L., Jansen, C. and Nolet, B.A. (2005) Digestive plasticity in Mallard ducks modulates dispersal probabilities of aquatic plants and crustaceans. *Functional Ecology* 19, 513–519.

Charalambidou, I., Santamaría, L. and Langevoord, O. (2003) Effect of ingestion by five avian dispersers on the retention time, retrieval and germination of *Ruppia maritima* seeds. *Functional Ecology* 17, 147–753.

Cipollini, M.L. and Levey, D.J. (1997) Secondary metabolites of fleshy vertebrate-dispersed fruits: adaptive hypotheses and implications for seed dispersal. *American Naturalist* 150, 346–372.

Clout, M.N. and Tilley, J.A.V. (1992) Germination of miro (*Prumnopitys ferruginea*) seeds after consumption by New Zealand pigeons (*Hemiphaga novaeseelandiae*). *New Zealand Journal of Botany* 30, 25–28.

Cosyns, E., Delporte, A., Lens, L. and Hoffmann, M. (2005) Germination success of temperate grassland species after passage through ungulate and rabbit guts. *Journal of Ecology* 93, 353–361.

Cottrell, H.J. (1947) Tetrazolium salt as a seed germination indicator. *Nature* 159, 748.

Dean, W.R.J. and Milton, S.J. (2000) Directed dispersal of *Opuntia* species in the Karoo, South Africa: are crows the responsible agents? *Journal of Arid Environments* 45, 305–314.

Debussche, M., Escarre, J. and Lepart, J. (1982) Ornithochory and plant succession in Mediterranean abandoned orchards. *Vegetatio* 48, 255–266.

Debussche, M. and Isenmann, P. (1994) Bird-dispersed seed rain and seedling establishment in patchy Mediterranean vegetation. *Oikos* 69, 414–426.

De Figueiredo, R.A. and Longatti, C.A. (1997) Ecological aspects of the dispersal of a Melastomataceae by marmosets and howler monkeys (Primates: Platyrrhini) in a semideciduous forest of southeastern Brazil. *Revue d'Ecologie-La Terre et la Vie* 52, 3–8.

De Figueiredo, R.A. and Perin, E. (1995) Germination ecology of *Ficus luschnathiana* drupelets after bird and bat ingestion. *Acta Oecologica* 16, 71–75.

Dinerstein, E. and Wemmer, C.M. (1988) Fruits rhinoceros eat: dispersal of *Trewia nudiflora* (Euphorbiaceae) in lowland Nepal. *Ecology* 69, 1768–1774.

Espinar, J.L., García, L.V., Figuerola, J., Green, A.J. and Clemente, L. (2004) Helophyte germination in a Mediterranean salt marsh: gut-passage by ducks changes seed response to salinity. *Journal of Vegetation Science* 15, 315–322.

Evenari, M. (1949) Germination inhibitors. *Botanical Review* 15, 153–194.

Figueroa, J.A. and Castro, S.A. (2002) Effects of bird ingestion on seed germination of four woody species of the temperate rainforest of Chiloé Island, Chile. *Plant Ecology* 160, 17–23.

Figuerola, J., Green, A.J. and Santamaría, L. (2002) Comparative dispersal effectiveness of wigeongrass seeds by waterfowl wintering in south-west Spain: quantitative and qualitative aspects. *Journal of Ecology* 90, 989–1001.

Forget, P.-M. and Wenny, D.G. (2005) How to elucidate seed fate? A review of methods used to study seed removal and secondary seed dispersal. In: Forget, P.-M., Lambert, J.E., Hulme, P.E. and Vander Wall, S.B. (eds) *Seed Fate: Predation, Dispersal and Seedling Establishment.* CAB International, Wallingford, UK, pp. 379–393.

Fragoso, J.M.V., Silvius, K.M. and Correa, J.A. (2003) Long-distance seed dispersal by tapirs increases seed survival and aggregates tropical trees. *Ecology* 84, 1198–2006.

Friedman, J., Rushkin, E. and Waller, G.R. (1982) Highly potent germination inhibitors in aqueous eluate of fruits of Bishops weed *Ammi majus* and avoidance of auto inhibition. *Journal of Chemical Ecology* 8, 55–66.

Herrera, C.M. (1989) Seed dispersal by animals: a role in angiosperm diversification? *American Naturalist* 133, 309–322.

Herrera, C.M., Jordano, P., Lopez-Soria, L. and Amat, J.A. (1994) Recruitment of a mast-fruiting, bird-dispersed tree: bridging frugivore activity and seedling establishment. *Ecological Monographs* 64, 315–344.

Herrera, C.M. and Pellmyr, O. (2002) *Plant–Animal Interactions: An Evolutionary Approach.* Blackwell Publishing, Oxford, UK.

Holl, K.D. (1999) Factors limiting tropical rain forest regeneration in abandoned pasture: seed rain, seed germination, microclimate, and soil. *Biotropica* 31, 229–242.

Hruska, A.F., Dirr, M.A. and Pokorny, F.A. (1982) Investigation of anthocyanic pigments and substances inhibitory to seed germination in the fruit pulp of *Liriope muscari. Journal of the American Society for Horticultural Science* 107, 468–473.

Izhaki, I. and Safriel., U.N. (1990) The effect of some Mediterranean scrubland frugivores upon germination patterns. *Journal of Ecology* 78, 56–65.

Janzen, D.H. (1986) Seeds as products. *Oikos* 46, 1–2.

Janzen, D.H., Demment, M.W. and Robertson, J.B. (1985) How fast and why do germinating guanacaste seeds (*Enterolobium cyclocarpum*) die inside cows and horses? *Biotropica* 17, 322–325.

Jordano, P. (2000) Fruits and frugivory. In: Fenner, M. (ed.) *Seeds: The Ecology of Regeneration in Plant Communities.* CAB International, Wallingford, UK, pp. 125–166.

Karasov, W.H. and Levey, D.J. (1990) Digestive system trade-offs and adaptations of frugivorous passerine birds. *Physiological Zoology* 63, 1248–1270.

Kelly, D., Ladley, J.J. and Robertson, A.W. (2004) Is dispersal easier than pollination? Two tests in New Zealand Loranthaceae. *New Zealand Journal of Botany* 42, 89–103.

Kelly, D., Ladley, J.J. and Robertson, A.W. (2007) Is the pollen-limited mistletoe *Peraxilla tetrapetala* (Loranthaceae) also seed-limited? *Austral Ecology*, (in press).

King, G. (1996) *Reptiles and Herbivory.* Chapman and Hall, London.

Kunz, T.H. and Ingalls, K.A. (1994) Folivory in bats: an adaptation derived from frugivory. *Functional Ecology* 8, 665–668.

Laterra, P. and Bazzalo, M.E. (1999) Seed-to-seed allelopathic effects between two invaders of burned Pampa grasslands. *Weed Research* 39, 297–308.

Leishman, M.R., Masters, G.J., Clarke, I.P. and Brown, V.K. (2000) Seed bank dynamics: the role of fungal pathogens and climate change. *Functional Ecology* 14, 293–299.

Levey, D.J. and Grajal, A. (1991) Evolutionary implications of fruit-processing limitations in cedar waxwings. *American Naturalist* 138, 171–189.

Lewis, D.M. (1987) Fruiting patterns, seed germination, and distribution of *Sclerocarya caffra* in an elephant-inhabited woodland. *Biotropica* 19, 50–56.

Li, H.J. and Zhang, Z.B. (2003) Effect of rodents on acorn dispersal and survival of the Liaodong oak (*Quercus liaotungensis* Koidz.). *Forest Ecology and Management* 176, 387.

Linhart, Y.B. (1976) Density-dependent seed-germination strategies in colonizing *versus* non-colonizing plant species. *Journal of Ecology* 64, 375–380.

Loiselle, B.A. (1990) Seeds in droppings of tropical fruit-eating birds: importance of considering seed composition. *Oecologia* 82, 494–500.

Malo, J.E. and Suárez, F. (1995) Herbivorous mammals as seed dispersers in a Mediterranean dehesa. *Oecologia* 104, 246–255.

Marambe, B., Nagaoka, T. and Ando, T. (1993) Identification and biological activity of germination-inhibiting long-chain fatty-acids in animal-waste composts. *Plant and Cell Physiology* 34, 605–612.

Mayer, A.M. and Poljakoff-Mayber, A. (1982) *The Germination of Seeds*, 3rd edn. Pergamon Press, Oxford and New York.

McGuinness, K.A. (1997) Dispersal, establishment and survival of *Ceriops tagal* propagules in a North Australian mangrove forest. *Oecologia* 109, 80.

Meyer, G.A. and Witmer, M.C. (1998) Influence of seed processing by frugivorous birds on germination success of three North American shrubs. *American Midland Naturalist* 140, 129–139.

Miller, M.F. (1995) Acacia seed survival, seed germination and seedling growth following pod consumption by large herbivores and seed chewing by rodents. *African Journal of Ecology* 33, 194–210.

Morpeth, D.R. and Hall, A.M. (2000) Microbial enhancement of seed germination in *Rosa corymbifera* 'Laxa'. *Seed Science Research* 10, 489–494.

Morpeth, D.R., Hall, A.M. and Cullum, F.J. (1997) The involvement of microbes and enzymes in the pre-treatment of woody seeds to overcome dormancy. In: Ellis, R.H., Black, M., Murdoch, A.J. and Hong, T.D. (eds) *Basic and Applied Aspects of Soil Biology*. Kluwer Academic, Dordrecht, The Netherlands, pp. 261–277.

Murray, B.R. (1998) Density-dependent germination and the role of seed leachate. *Australian Journal of Ecology* 23, 411–418.

Murray, K.G., Russell, S., Picone, C.M., Winett-Murray, K., Sherwood, W. and Kuhlmann, M.L. (1994) Fruit laxatives and seed passage rates in frugivores: consequences for plant reproductive success. *Ecology* 75, 989–994.

Nilsson, M.C. (1994) Separation of allelopathy and resource competition by the boreal dwarf shrub *Empetrum hermaphroditum* Hagerup. *Oecologia* 98, 1–7.

Nogales, M., Nieves, C., Illera, J.C., Padilla, D.P. and Traveset, A. (2005) Effect of native and alien vertebrate frugivores on seed viability and germination patterns of *Rubia fruticosa* (Rubiaceae) in the eastern Canary Islands. *Functional Ecology* 19, 429–436.

Pollux, B.J.A., Santamaría, L. and Ouborg, N.J. (2005) Differences in endozoochorous dispersal between aquatic plant species, with reference to plant population persistence in rivers. *Freshwater Biology* 50, 232–242.

Pulliainen, E., Helle, P. and Tunkkari, P. (1981) Adaptative radiation of the digestive system, heart and wings of *Turdus pilaris*, *Bombycilla garrulus*, *Sturnus vulgaris*, *Pyrrhula pyrrhula*, *Pinicola enucleator* and *Loxia pytyopsittacus*. *Ornis Fennica* 58, 21–28.

Qaderi, M.M., Cavers, P.B. and Bernards, M.A. (2003) Isolation and structural characterization of a water-soluble germination inhibitor from scotch thistle (*Onopordum acanthium*) cypselas. *Journal of Chemical Ecology* 29, 2425–2438.

Rey, P. and Alcántara, J.M. (2000) Recruitment dynamics of a fleshy-fruited plant (*Olea europaea*): connecting patterns of seed dispersal to seedling establishment. *Journal of Ecology* 88, 622–633.

Regan, B.C., Julliot, C., Simmen, B., Vienot, F., Charles-Dominique, P. and Mollon, J.D. (2001) Fruits, foliage and the evolution of primate colour vision. *Philosophical Transactions of the Royal Society London Series B: Biological Sciences* 356, 229–283.

Richmond, G.S. and Ghisalberti, E.L. (1994) Seed dormancy and germination mechanisms in *Eremophila* (Myoporaceae). *Australian Journal of Botany* 42, 705–715.

Robertson, A.W., Trass, A., Ladley, J.J. and Kelly, D. (2006) Assessing the benefits of frugivory for seed germination: the importance of the de-inhibition effect. *Functional Ecology* 20, 58–66.

Rodríguez-Pérez, J., Riera, N. and Traveset, A. (2005) Effect of seed passage through birds and lizards on emergence rate of Mediterranean species: differences between natural and controlled conditions. *Functional Ecology* 19, 699–706.

Rust, R.W. and Roth, R.R. (1981) Seed production and seedling establishment in the mayapple, *Podophyllum peltatum* L. *American Midland Naturalist* 105, 51–60.

Samuels, I.A. and Levey, D.J. (2005) Effects of gut passage on seed germination: do experiments answer the questions they ask? *Functional Ecology* 19, 365–368.

Santamaría, L., Charalambidou, I., Figuerola, J. and Green, A.J. (2002) Effect of passage through duck gut on germination of fennel pondweed seeds. *Archiv für Hydrobiologie* 156, 11–22.

Schaumann, F. and Heinken, T. (2002) Endozoochorous seed dispersal by martens (*Martes foina, M. martes*) in two woodland habitats. *Flora* 197, 370–378.

Schupp, E.W. (1995) Seed seedling conflicts, habitat choice, and patterns of plant recruitment. *American Journal of Botany* 82, 399–409.

Schupp, E.W. (1990) Annual variation in seedfall, post-dispersal predation, and recruitment of a Neotropical tree. *Ecology* 71, 504–515.

Stanley, M.C. and Lill, A. (2002a) Avian fruit consumption and seed dispersal in a temperate Australian woodland. *Austral Ecology* 27, 137–148.

Stanley, M.C. and Lill, A. (2002b) Does seed packaging influence fruit consumption and seed passage in an avian frugivore? *Condor* 104, 136–145.

Steele, M.A., Turner, G., Smallwood, P.D., Wolff, J.O. and Radillo, J. (2001) Cache management by small mammals: experimental evidence for the significance of acorn-embryo excision. *Journal of Mammalogy* 82, 35.

Stiles, E.W. (2000) Animals as seed dispersers. In: Fenner, M. (ed.) *Seeds: The Ecology of Regeneration in Plant Communities.* CAB International, Wallingford, UK, pp. 111–124.

Stiles, E.W. and White, D.W. (1986) Seed deposition patterns: influence of season, nutrients, and vegetation structure. In: Estrada, A. and Fleming, T.H. (eds) *Frugivores and Seed Dispersal.* Dr. W. Junk, Dordrecht, The Netherlands, pp. 45–54.

Stocker, G.C. and Irvine, A.K. (1983) Seed dispersal by cassowaries (*Casuarius casuarius*) in North Queensland's rainforests. *Biotropica* 15, 170–176.

Suman, A., Shahi, H.N., Singh, P. and Gaur, A. (2002) Allelopathic influence of *Vigna mungo* (black gram) seeds on germination and radical growth of some crop plants. *Plant Growth Regulation* 38, 69–74.

Takahashi, K. and Kamitani, T. (2003) Colonization of fleshy-fruited plants beneath perch plant species that bear fleshy fruit. *Journal of Forest Research* 8, 169–177.

Tewksbury, J.J. and Nabhan, G.P. (2001) Seed dispersal: directed deterrence by capsaicin in chillies. *Nature* 412, 403–404.

Traveset, A. (1990) *Ctenosaura humilis* Gray (Iguanidae) as a seed disperser in a Central American deciduous forest. *American Midland Naturalist* 123, 402–404.

Traveset, A. (1998) Effect of seed passage through vertebrate frugivores' guts on germination: a review. *Perspectives in Plant Ecology, Evolution and Systematics* 1/2, 151–190.

Traveset, A. and Willson, M.F. (1997) Effect of birds and bears on seed germination of fleshy-fruited plants in temperate rainforest of southeast Alaska. *Oikos* 80, 89–95.

Traveset, A., Bermejo, T. and Willson, M.F. (2001a) Effect of manure composition on seedling emergence and growth of two common shrub species of Southeast Alaska. *Plant Ecology* 155, 29–34.

Traveset, A., Riera, N. and Mas, R.E. (2001b) Ecology of fruit-colour polymorphism in *Myrtus communis* and differential effects of birds and mammals on seed germination and seedling growth. *Journal of Ecology* 89, 749–760.

Traveset, A., Riera, N. and Mas, R.E. (2001c) Passage through bird guts causes interspecific differences in seed germination characteristics. *Functional Ecology* 15, 669–675.

Traveset, A. and Verdú, M. (2002) A meta-analysis of the effect of gut treatment on seed germination. In: Levey, D.J., Silva, W.R. and Galetti, M. (eds) *Seed Dispersal and Frugivory: Ecology, Evolution and Conservation.* CAB International, Wallingford, UK, pp. 339–350.

Traveset, A., Gulías, J., Riera, N. and Mus, M. (2003) Transition probabilities from pollination to establishment in a rare dioecious shrub species (*Rhamnus ludovici-salvatoris*) in two habitats. *Journal of Ecology* 91, 427–437.

Verdú, M. and Traveset, A. (2005) Early emergence enhances plant fitness: a phylogenetically controlled meta-analysis. *Ecology* 86, 1385–1394.

Wahaj, S.A., Levey, D.J., Sanders, A.K. and Cipollini, M.L. (1998) Control of gut retention time by secondary metabolites in ripe *Solanum* fruits. *Ecology* 79, 2309–2319.

Wardle, D.A., Ahmed, M. and Nicholson, K.S. (1991) Allelopathic influence of nodding thistle (*Carduus nutans* L.) seeds on germination and radicle growth of pasture plants. *New Zealand Journal of Agricultural Research* 34, 185–191.

Welch, D. (1985) Studies in the grazing of heather moorland in north-east Scotland. 4. Seed dispersal and plant establishment in dung. *Journal of Applied Ecology* 22, 461–472.

Wenny, D.G. (2001) Advantages of seed dispersal: a re-evaluation of directed dispersal. *Evolutionary Ecology Research* 3, 51–74.

Willson, M.F. (1993) Mammals as seed dispersal mutualists in North America. *Oikos* 67, 159–176.

Willson, M.F., Gende, S.M. and Marston, B.H. (1998) Fishes and the forest. *BioScience* 48, 455–462.

Willson, M.F. and Traveset, A. (2000) The ecology of seed dispersal. In: Fenner, M. (ed.) *Seeds: the Ecology of Regeneration in Plant Communities.* CAB International, Wallingford, UK, pp 85–110.

Yagihashi, T., Hayashida, M. and Miyamoto, T. (1998) Effects of bird ingestion on seed germination of *Sorbus commixta*. *Oecologia* 114, 209–212.

5 Living in the Land of Ghosts: Fruit Traits and the Importance of Large Mammals as Seed Dispersers in the Pantanal, Brazil

C.I. Donatti, M. Galetti, M.A. Pizo, P.R. Guimarães Jr and P. Jordano

Introduction

Seed dispersers play a key role in the ecology and evolution of fleshy-fruited plants; especially in tropical forests where from 70% to 90% of all woody species are dispersed by vertebrates (Fleming *et al.*, 1987; Jordano, 2000). Local plant communities with a range of fruit types are assembled under a variety of selection pressures and historical effects. Analyses of community-level variation in fruit traits indicate that they also reflect variations in the composition of the main seed dispersers in the animal community (Fleming *et al.*, 1987; Fleming, 1991, 2005; Mack, 1993). Fruit size is one of the main traits selected by frugivorous vertebrates and has multiple potential influences on seed mass, and therefore on germination and seedling survival (Jordano, 1995). Not unexpectedly, large-scale, community-wide comparisons of fruit traits have reported variation in traits related to fruit size paralleling changes in the frugivore community (see Fleming *et al.*, 1987; Jordano, 2000; Herrera, 2002, and references therein).

At a community level, the interactions among plants and frugivores often show high diversity and low specificity (Jordano *et al.*, 2003; Forget *et al.*, Chapter 1, this volume; Silva *et al.*, Chapter 26, this volume). In this context, large-scale comparisons between areas with different faunal assemblages have been widely used to investigate local co-adaptations between plants and frugivores (Snow, 1980; Fischer and Chapman, 1993; Voigt *et al.*, 2004) and are a powerful tool in analyses of ecological convergence at the community level (Schluter, 1988; Corlett and Primack, 2006). However, comparative analyses based on extant communities implicitly ignore the fact that these mutualisms build up on highly generalized interactions, where evolutionary change and substitution of the mutualistic partners can occur.

For example, Mack (1993) proposed that the evolution of large fruits and seeds in the Neotropics has been constrained, relative to palaeotropical communities, by a scarcity of large-bodied frugivores. In a different view, Janzen and Martin (1982) proposed that frugivory by large extinct mammals such as native horses, gomphotheres, ground sloths and other extinct megafauna offers an explanation for the dispersal-related reproductive traits of Central American lowland plants. The so-called megafauna syndrome (related to large-bodied mammals > 44 kg) has been the subject of considerable debate. However, the debate suffers from a lack of specific predictions and precise definitions (Howe, 1985; Hunter, 1989; Owen-Smith, 1989; Lord *et al.*, 2002). Comparative biogeographical studies have concluded that large frugivores did not shape the fruit traits of plant communities (Fischer and Chapman, 1993), while other studies have reported a strong relationship between fruit traits and the local fauna (Herrera, 2002; Bollen *et al.*, 2004; Voigt *et al.*, 2004).

In the Palaeotropics, many large-bodied mammals are a major component of the frugivore communities, while the neotropical ecosystems characteristically lack large frugivores. The largest frugivore in the Neotropics is the 300 kg tapir, *Tapirus terrestris* (Tapiridae), while in the Palaeotropics, elephants can weigh at least ten times more (Owen-Smith, 1988). However, hypotheses of co-evolution between fruits and frugivores need to account for both extant and past mutualists. Less than 10,000 years BP, South America was a land of large-bodied mammals (> 44 kg), which experienced relatively recent extinction after millions of years of persistence (Martin and Klein, 1984; Owen-Smith, 1988, 1989). Megafauna taxa include primarily large terrestrial mammals (e.g. large carnivores, xenarthrans, rodents and extinct orders of ungulates), many of them considered mixed grazer-browsers and frugivores (Fariña, 1996). Only 13 mammal genera survive in contemporary neotropical communities, out of 60 whose species had > 44 kg body mass that were present in the Pleistocene fauna. The South American fauna had at least seven genera that included species with body masses ⩾ 1000 kg, yet none is present now. African faunas, in contrast, have 40 extant genera with > 44 kg body mass, including herbivorous and omnivorous species (Martin and Klein, 1984). Five genera ⩾1000 kg are still living in Africa and two genera in Asia (Owen-Smith, 1988).

The recent extinction of a large component of the potential frugivore community undoubtedly has a lasting signal in extant plant–frugivore interactions (Janzen and Martin, 1982), yet we still know very little about its consequences. In fact, it is vital to understand the role of the extinct megafauna on plant–animal relationships because of the ongoing defaunation throughout tropical ecosystems (Fa *et al.*, 2002).

The Pantanal, located in central Brazil and part of Bolivia and Paraguay, is the world's largest freshwater wetland, with 140,000 km^2 of lowland floodplain of the upper Rio Paraguai basin (Swarts, 2000). This area is subjected to seasonal flooding, creating a diverse mosaic of habitats resulting in an extraordinarily rich terrestrial and aquatic biota (Silva *et al.*, 2000). The Pantanal has a high biomass of large-bodied mammals, such as tapirs (*T. terrestris*) and peccaries (*Tayassu pecari* and *Pecari tajacu*; Tayassuidae), and frugivorous birds such as piping-guans, *Pipile cujubi* (Cracidae) and toucans, *Ramphastos toco* (Ramphastidae; Harris *et al.*, 2005). Other conspicuous fauna in the area are cows, *Bos taurus* (Bovidae), feral pigs, *Sus scrofa* (Suidae) and horses, *Equus caballus* (Equidae). Large frugivores are vanishing from most areas in the world, due to selective hunting or fragmentation (Peres, 2000, 2001), but are still abundant in the Pantanal (Lourival, 1997; Trolle, 2003; Harris *et al.*, 2005), mainly due to the low human population density and low hunting pressure (Alho and Lacher, 1991; but see Harris *et al.*, 2005). Therefore, the Pantanal holds the highest concentration of wildlife in South America (Swarts, 2000; Mittermeier *et al.*, 2005) and represents an excellent opportunity to study plant–animal interactions in a pristine habitat.

In this study we present the characteristics of fleshy-fruited plants in the Pantanal and describe the contributions of different animal guilds to seed dispersal. We go on to compare fruit traits in the Pantanal with those in the Atlantic rainforest (Brazil), in a mosaic of forest and savannah in Africa (Ivory Coast), and in an African wetland (Okavango, Botswana), to test similarities in fruit size, colour and shape. Given that open savannahs hold a large diversity and biomass of large-bodied herbivores which can supplement their diet with fruits (Owen-Smith, 1988; Fariña, 1996; Cristoffer and Peres, 2003), we predict that plant communities in the Pantanal should exhibit a distribution of fruit traits across species that is similar to those found in savannahs and savannah-like habitats where megafauna still exist, such as Ivory Coast and Okavango. We expect a different distribution of fruit traits in the Atlantic rainforest, due to the low biomass of large mammals in forest ecosystems compared with the savannahs and savannah-like habitats (e.g. Prins and Reitma, 1989). A comparative test of this hypothesis will help in understanding the historical process of co-evolution with the Pleistocene megafauna and will supplement historical approximations based solely on the study of extant interactions. We also discuss some potential ecological mechanisms that contribute to plant population persistence after the extinction of major seed dispersers, and we argue, based on numerical simulations, that for some long-lived plants there is a possibility that minimal recruitment events allow populations to persist.

Methods

Pantanal fruits and frugivores

Study site

Fazenda Rio Negro is a 7500 ha private area in the Nhecolândia region in the Brazilian Pantanal. Average annual rainfall is 1192.5 mm and mean monthly temperature is 26°C, ranging from 19°C to 33°C (n = 4 years;

D. Eaton, unpublished results). The main vegetation types of the Pantanal include gallery forests, cerrado and semideciduous forests (Prance and Schaller, 1982); all are represented at Fazenda Rio Negro (Silva *et al.*, 2000), where the study was conducted.

Traits of the Pantanal fleshy fruits

We recorded colour, smell and size for 5–40 fruits of each vertebrate-dispersed, fleshy-fruited species on a monthly basis ($n = 30$). Length was measured from the peduncle insertion to the most distal part, and width as the maximum diameter at 90° to length, using digital callipers to the nearest 0.1 mm. Mass was measured with digital scales to the nearest 0.1 g. Colour of ripe fruits was recorded according to human vision. Fruits were collected from different individuals among a sample of 620 marked trees, depending on availability, or from randomly sampled individuals in the field.

For each of the species measured, we recorded life-form and assigned a rank value of smell intensity of the fruits that varied from 0 (without smell) to 2 (very strong, sweet smell), referring to human sensitivity. We used logistic regression to relate fruit smell to both fruit length and width. We also recorded the ability of all measured species for resprouting after disturbance (e.g. fire or logging), human use (according to Pott and Pott, 1994) and the persistence of ripe fruit.

Fruit–frugivore interactions in the Pantanal

Observations of frugivore foraging behaviour were carried out using four techniques:

- *First*, we conducted focal observations of individual fruiting shrubs and trees (Galetti *et al.*, 2002). To detect frugivore activity at selected fruiting plants, those bearing ripe fruits were observed over periods of 4 h, mainly from 06:00 to 10:00 hours. Fruit handling behaviour of animals visiting the trees (i.e. whether they eat the whole fruit, only the seeds, or spit them out) were recorded to classify species as legitimate seed dispersers, fruit-pulp consumers, and/or seed predators (Moermond and Denslow, 1985; Levey, 1987; Jordano and Schupp, 2000).
- *Second*, we monitored frugivore visits to fruiting plants using camera traps in order to record terrestrial and nocturnal consumers of fruits. Camera traps were placed beneath four individuals of each species of plant, focusing on fallen fruits (Miura *et al.*, 1997; Galetti, 2002).
- *Third*, we collected gut contents (fish only – *Piaractus mesopotamicus*; Serrasalmidae) and scats of frugivorous animals (rheas, *Rhea americana*, Rheidae; tapirs, *T. terrestris*; feral pigs, *S. scrofa*; and white-lipped peccaries, *T. pecari*) and the seeds recovered from them were identified to species based on a reference collection.
- *Finally*, we also included personal observations of some fruit–animal interactions.

We combined these data into a plant–frugivore interactions database. We then examined the pattern of interaction between Pantanal fruits and frugivores by means of a canonical correspondence analysis on the matrix of family × frugivore type frequencies. Plant and fruit voucher specimens for this data have been deposited in the EMBRAPA Herbarium in Corumbá, MS, and at Universidade Estadual Paulista (UNESP) at Herbarium Rio Clarense, SP, Brazil.

Megafauna fruit traits and intercontinental patterns

We compared the data set of fruit morphology of the Pantanal with three other plant communities, including one site in the Atlantic rainforest of Brazil (Galetti, 1996; Campassi, 2006; M. Galetti, M.A. Pizo, L.P. Morellato and P. Jordano, unpublished results) and two in Africa: Ivory Coast and Okavango Delta. The Atlantic rainforest does not have a recent history of occupation by large mammals, as we found in savannah and savannah-like habitats in Okavango, Ivory Coast and Pantanal.

Intervales State Park is 49,000 ha of Atlantic rainforest along the south-east coast of Brazil. The average annual rainfall is 4000 mm and the average temperature varies from 21.1°C to 26.8°C. Tropical rainforest dominates the area, including both lowland and highland vegetation (Galetti, 1996).

Camoé National Park is located in the north-eastern part of the Ivory Coast and is 1,150,000 ha in area (Hovestadt *et al.*, 1999). Long-term mean annual rainfall varies from 800 to 1100 mm. The annual mean temperature is 26.5–27°C (Hovestadt *et al.*, 1999). The main vegetation types include a mosaic of shrubby savannah, forest islands and gallery forests along the main rivers.

Okavango Delta in north-western Botswana is a flooded habitat similar to the Pantanal and includes a similar range of habitat types. The area of the wetland is in excess of 1,200,000 ha (McCarthy *et al.*, 1998).

In addition to the data on fruit morphology presented here for the Pantanal, extensive data sets of fruit traits and plant–frugivore interactions are available for the Atlantic rainforest (Galetti, 1996; n = 138) and both African sites: data on fruit morphology from Ivory Coast was based on T. Hovestadt *et al.* (unpublished data, n = 128) and from Okavango on fruit guides (van Wyk, 1997; Thomas and Grant, 2002; n = 44). We used restricted paired comparisons of fruit length, fruit width and fruit colour between confamilial taxa to contrast fruit traits in different communities. Average within-family values for these variables were obtained for each community. Trends were examined by comparing each family-level value across communities. The consistency of a given trend, for example whether fruit diameter is larger for Pantanal versus Atlantic rainforest, was tested at the within-family level. The number of within-family contrasts showing a trend (e.g. increase in fruit size) was tallied and compared with a binomial expectation. The significance of a hypothesized trend was tested by a binomial test on the proportion of paired comparisons that are consistent with the hypothesis when compared with a random expectation of 50% of the within-family trends in each direction. Family-level contrasts control phylogenetic effects when comparing species samples and account for the different representation of genera (see Mack, 1993; Jordano, 1995; Forget *et al.*, Chapter 1, this volume).

Numerical simulations and the persistence of megafauna-dependent plants

Dispersal failure is certainly one of the main potential causes for the decline of plant populations (Cordeiro and Howe, 2001; McConkey and Drake, 2002; Traveset and Riera, 2005; Galetti *et al.*, 2006). Therefore, the long-term persistence of plant populations that produce large, high-cost fruits without efficient seed dispersal remains paradoxical. In this chapter we examine some alternative mechanisms that may allow megafauna-dependent plant populations to persist in the absence of major seed dispersers. However, a more basic question is: How efficient does seed dispersal have to be for a plant species to persist in ecological time?

Simulations were undertaken using BANGU 1.0® (developed by P. Guimarães and P.R. Guimarães Jr), an individual-based, spatially explicit model (see Durrett and Levin, 1994) that simulates individuals of a single species of plant as points occurring over a regular lattice. At each time-step, all plants reproduce (i.e. generate new individuals) following user-defined probabilities for short-distance dispersal events leading to recruitment. Short-distance recruitment was simulated assuming Moore nearest neighbourhoods (Durrett and Levin, 1994); that is, each plant may colonize the eight nearest cells. Individual plants die after a set number of time-steps, opening the possibility for the cell to be colonized by other plants. We analysed how lifespan and short-distance recruitment affect plant population persistence, described by the time until the population becomes extinct. Simulations were performed assuming lattice size = 9.0×10^4 cells, initial plant population = 1000, default lifespan = 100 reproductive events, and default probability of short-distance recruitment = 0.1% per nearest cell.

Results

Traits of the Pantanal fleshy fruits

Approximately 74% of the 620 plants observed in the phenological study produce fleshy vertebrate-dispersed fruits (C.I. Donatti, 2005, unpublished results). We collected information on fruit morphology of 75 fleshy-fruited species found in a 4-year intensive study on fruit–frugivore interactions (Appendix 3). Growth forms included 54 species of trees (72%), nine shrubs (12%), six palms (8.0%), two lianas (2.7%), one bromeliad (1.3%), one cactus (1.3%), one herb (1.3%) and one mistletoe (1.3%).

The fleshy fruits of the Pantanal tend to be large (length = 30.54 ± 23.75 mm and width = 23.00 ± 16.60 mm; n = 75; mean ± SD; Fig. 5.1) varying from the small fruits of *Cissus erosa* (Vitaceae; length = 5.1 ± 0.5 mm, width = 5.8 ± 0.6 mm) to the huge fruits of *Attalea speciosa* (Arecaceae; length = 87.7 ± 7.8 mm, width = 50.7 ± 4.1 mm). The shape of the fruits varies with fruit size: fruits tend to be more elongated with increasing size (Log fruit width (mm) = 0.3228 + 0.8176 Log fruit length (mm); r^2 = 0.7623, n = 75; i.e. with slope < 1.0). Most species in the Pantanal produce dull-coloured fruits, among which the colour is predominantly green (in 20% of species), yellow (16%) or brown (14.6%; see Fig. 5.2).

According to human sensitivity, 32% of the 75 species of plants have a strong sweet smell, 16% have a weak smell, and 52% are odourless. There was a significant relationship between smell and both fruit length (logistic regression, r^2 = 0.22, χ^2 = 33.23, P < 0.0001) and width (r^2 = 0.23,

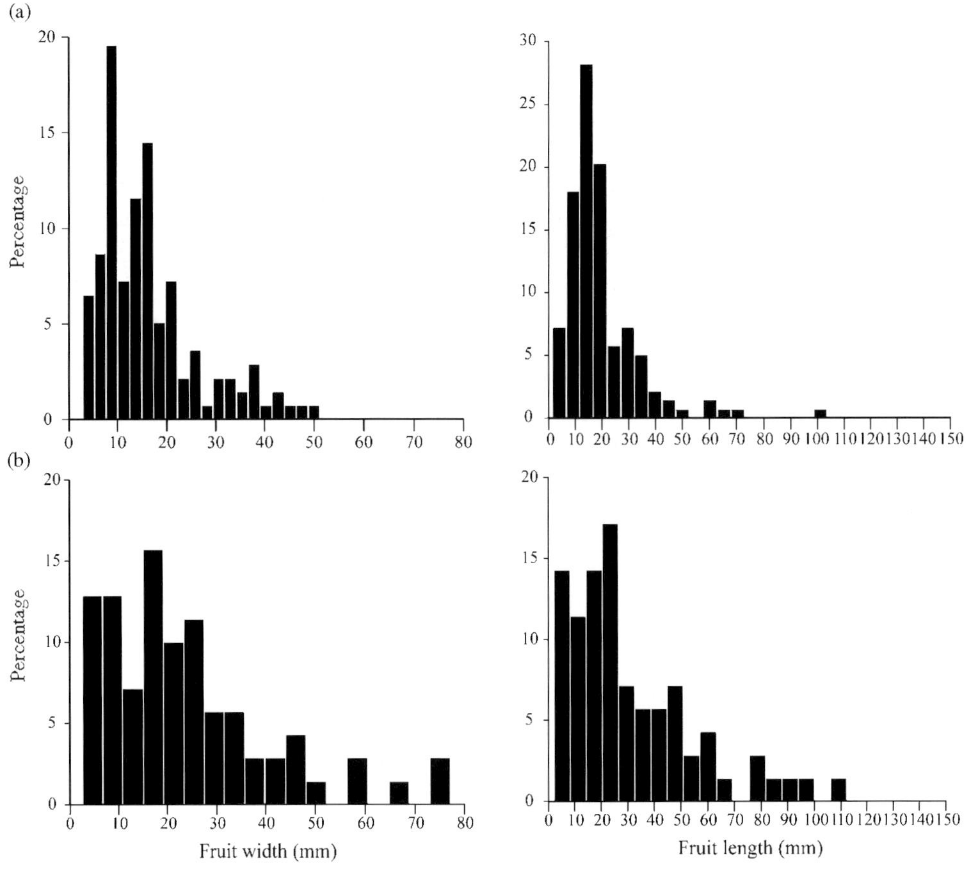

Fig. 5.1. Frequency distributions of fruit width and fruit length (mm) for species in local sites in the (a) Atlantic rainforest (Saibadela) and (b) Pantanal (Fazenda Rio Negro) plant communities, Brazil.

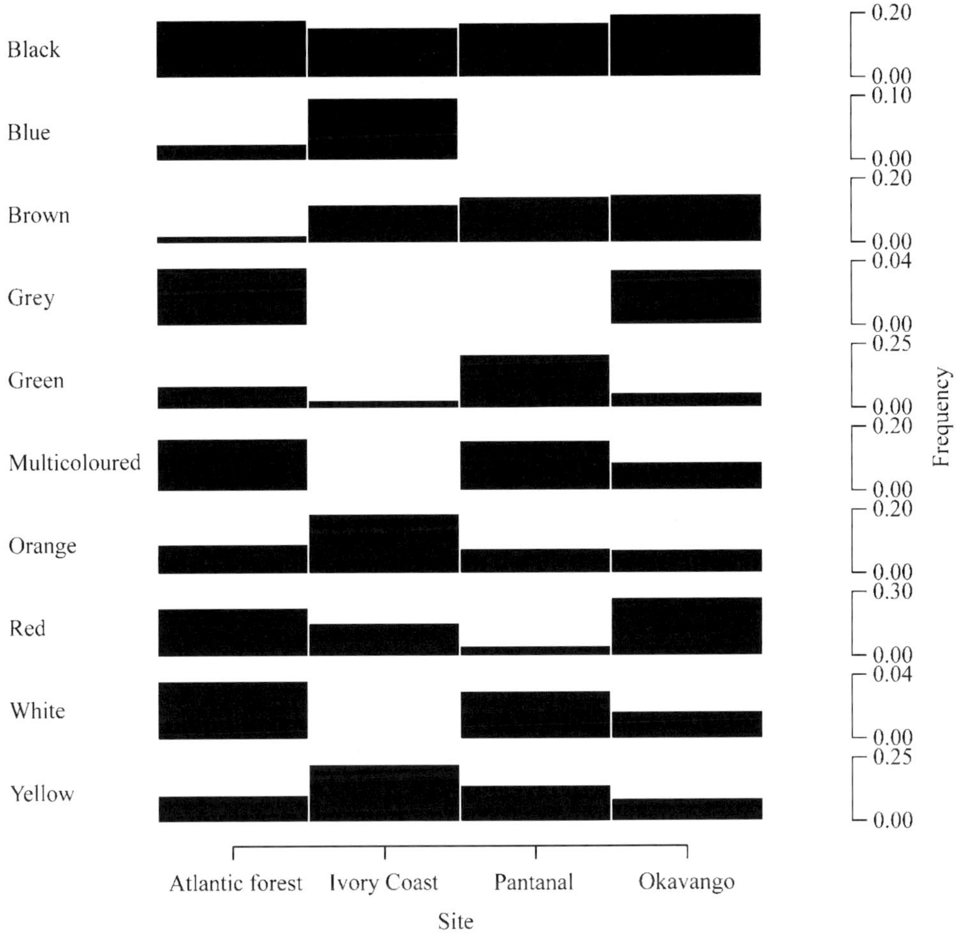

Fig. 5.2. Relative frequencies of fruit colour classes among fleshy-fruited species in the Atlantic rainforest and Pantanal (Brazil) and Ivory Coast and Okavango (Africa). Yellow includes yellow-green, and orange includes yellow-orange.

$\chi^2 = 34.15$, $P < 0.0001$), with larger fruits showing a significant trend to have a strong odour when ripe. Fruits of typical mammal-dispersed species in the Pantanal, such as Sapotaceae and Annonaceae, usually smell stronger than bird-dispersed fruits. Resprouting after fire or clear-cutting was recorded for 32% of these species. Due to their extremely large fruit, some species in the Pantanal lack efficient long-distance seed dispersal (e.g. *Caryocar brasiliense*, Carycaraceae; *A. speciosa*) and in 6.7% of species the ripe fruits persist on the tree (e.g. *Alibertia sessilis*, *Tocoyena formosa*, Rubiaceae; and *Simarouba versicolor*, Simaroubaceae). Usage by humans (47% of species sampled) varied from fruits used locally from wild trees in the neighbourhood of human settlements (e.g. *Annona* spp.) to regional plantation of species with economic value (e.g. *C. brasiliense*, *A. speciosa*).

Fruit–frugivore interactions in the Pantanal

We observed the fruits of 23 species during 690 h of focal observations in the Pantanal and set up camera traps in 27 fleshy-fruited species during 7040 h. We analysed 196 fish guts (*P. mesopotamicus*), 68 scats from *R. americana*, 45 from *T. pecari*, 136 from *S. scrofa* and 213 from *T. terrestris*.

Our observations of non-predatory plant–frugivore interactions indicate that mammals are responsible for the dispersal of 56% of all fleshy-fruited species, while birds disperse 48% and both share 21% of the fruits in the Pantanal. Fish and reptiles disperse 18.6% of the species, but none exclusively (see Appendix 3). This contrasts with other plant communities. In the Atlantic rainforest of Brazil, for instance, 59% of the fleshy-fruited species are dispersed by birds, 28% by mammals and 12% by both groups (Galetti, 1996). In Ivory Coast, Hovestadt *et al.* (1999) found that birds disperse 42% of the fruits and mammals disperse 41%. We do not have information on the seed dispersal syndromes for plants in Okavango.

Colour combinations differed significantly among major disperser categories ($\chi^2 = 96.64$, $df = 7$, $P = 0.02$). Bicoloured, black, and white displays were over-represented in bird- and bird+fish-dispersed species; dull colours (brown and green) were over-represented among species dispersed by mammals in combination with other groups, while yellow was over-represented in fruits consumed by mammals+tortoise (*Geochelone carbonaria*; Testudinidae). Fish, in combination with other frugivore taxa, consumed a variety of colours and showed no specific association with a colour type.

Our data indicate that feral pigs dispersed not only large-sized fruits (e.g. *A. phalerata*: fruit length = 62.7 ± 3.7 mm and fruit width = 34.9 ± 2.9 mm), but also more species than the native fauna: feral pig scats contained 15 species, compared with 11 in tapirs (the largest fruit species dispersed by both was *Dipteryx alata*, Fabaceae: fruit length = 48.8 ± 3.6 mm and fruit width = 39.9 ± 2.4 mm), seven in *R. americana* and five in *T. pecari* (the largest fruit species dispersed by both was *Bactris glaucescens*, Arecaceae: fruit length = 18.8 ± 3.7 mm and fruit width =17.7 ± 1.5 mm). Among the native animals, only the tapir dispersed *A. phalerata* and *D. alata* seeds, species that were also dispersed by feral pigs. In addition, 47% of fruits from all species of plants sampled in the Pantanal, can be consumed by humans.

Megafauna fruit traits and intercontinental patterns

The frequency distributions of fruit length and width differed significantly among the four areas ($F_{(3, 363)} = 10.79$, $P < 0.0001$ and $F_{(3, 371)} = 7.06$, $P < 0.01$, respectively). Pantanal, Okavango and Ivory Coast had more species with fruits > 20 mm than the Atlantic rainforest. Species with fruit width > 55 mm were not represented in the Atlantic rainforest (Fig. 5.3). The

Pantanal showed a frequency distribution of fruit widths much closer to the African sites but still lacked some species with fruits > 100 mm wide, which comprise approximately 8% of the species in the African sites (Fig. 5.3).

Controlling for the phylogenetic effects at family level, the average size of fruits in the Pantanal was bigger (length = 30.54 ± 23.75 mm and width = 23.00 ± 16.60 mm; n = 74) than the Atlantic rainforest (length = 19.51 ± 13.99 mm and width = 16.34 ± 10.05 mm; n = 138; binomial test (one-tailed) for 24 within-family contrasts, $P < 0.001$; see Fig. 5.4). Nineteen families have larger mean fruit size in the Pantanal (e.g. Annonaceae, Anacardiaceae, Clusiaceae and Ebenaceae), while only five families have larger fruits in the Atlantic rainforest than in the Pantanal (Sapotaceae, Myrtaceae, Moraceae, Meliaceae and Lauraceae). In addition, the magnitude of the differences was greater for families where the Pantanal representatives were larger. In contrast, the mean fruit width in each family for the Pantanal did not differ from either the Ivory Coast or Okavango samples. In 26 family contrasts, 15 families had bigger fruits in Ivory Coast and 11 in the Pantanal ($P = 0.577$ for the binomial test; see Fig. 5.4; Ivory Coast mean fruit length = 20.79 ± 17.33 mm and width = 32.71 ± 57.51 mm; n = 128). In 13 contrasts, seven families had bigger fruits in the Pantanal and six in Okavango ($P = 0.538$ for the binomial test; see Fig. 5.4; Okavango mean fruit length = 75.27 ± 154.56 mm and fruit width = 32.51 ± 45.21 mm; n = 44).

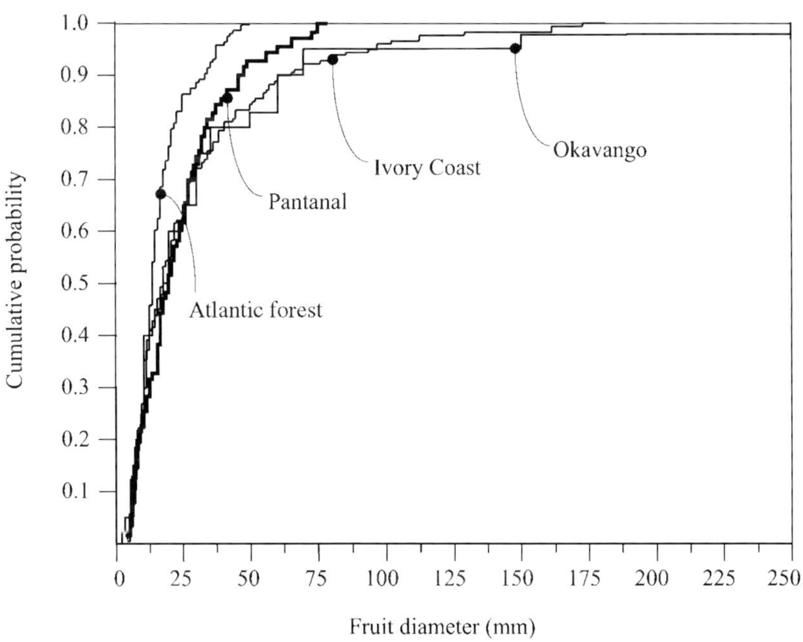

Fig. 5.3. Cumulative frequency distributions of fruit width in different study areas, two African sites (Okavango and Ivory Coast) and two Brazilian sites (Pantanal and Atlantic rainforest).

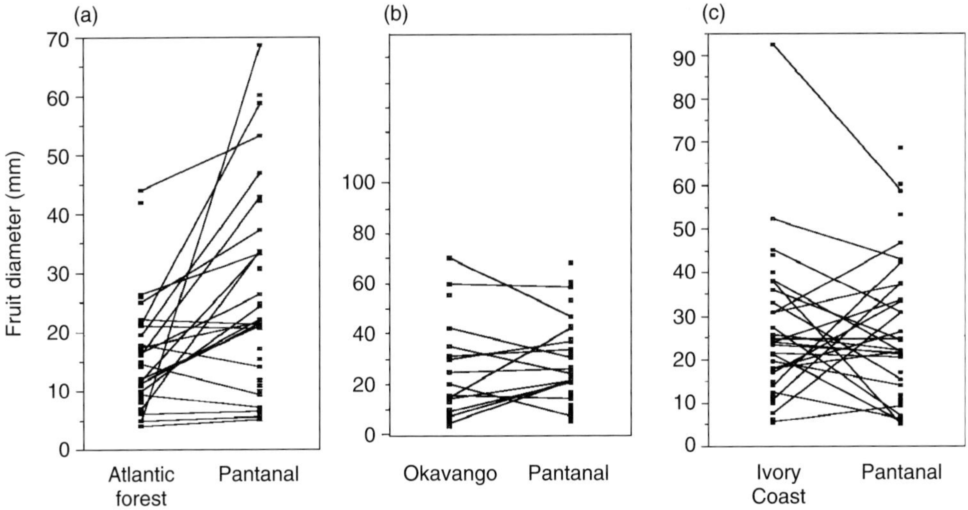

Fig. 5.4. Within-family paired contrasts for fruit width in plant communities around the world: (a) Atlantic rainforest and Pantanal, Brazil; (b) Okavango (Africa) and Pantanal; and (c) Ivory Coast (Africa) and Pantanal. Binomial tests for the comparisons among areas: (a) Atlantic rainforest × Pantanal, $P < 0.001$; (b) Okavango × Pantanal, NS; and (c) Ivory Coast × Pantanal, NS.

The flora of Pantanal has a similar composition of families to both African habitats, especially when considering genera with fruits > 20 mm wide and the predominance of species dispersed by mammals. Within the Pantanal, the variation in the composition of frugivore assemblages feeding on plant families was due to changes in the importance of mammals versuss birds or fish (first canonical variable; Fig. 5.5) or to changes in the importance of rheas and mammals versus other frugivores (second canonical variable, Fig. 5.5).

The distribution of fruit colours varied across locations. While all locations had a similar proportion of black fruits; the species of the Pantanal exhibit relatively high proportions of green or brown fruits; those of Atlantic rainforest exhibit high proportions of white, multicoloured or grey fruits; those of Ivory Coast produce high proportions of blue, orange or yellow fruits; and those of the Okavango produce high proportions of brown and red fruits (Fig. 5.2). There is a significant difference in the frequencies of fruit colours between the Atlantic rainforest and Pantanal ($\chi^2 = 28.99$, $P < 0.0001$), between Ivory Coast and Pantanal ($\chi^2 = 27.56$, $P < 0.0001$) but not between Okavango and Pantanal.

Numerical simulations and the persistence of megafauna-dependent plants

Preliminary simulations showed that lattice size and initial plant population do not affect the qualitative behaviour of simulations. In contrast, lifespan and recruitment probability in the vicinity of the plant had a marked effect

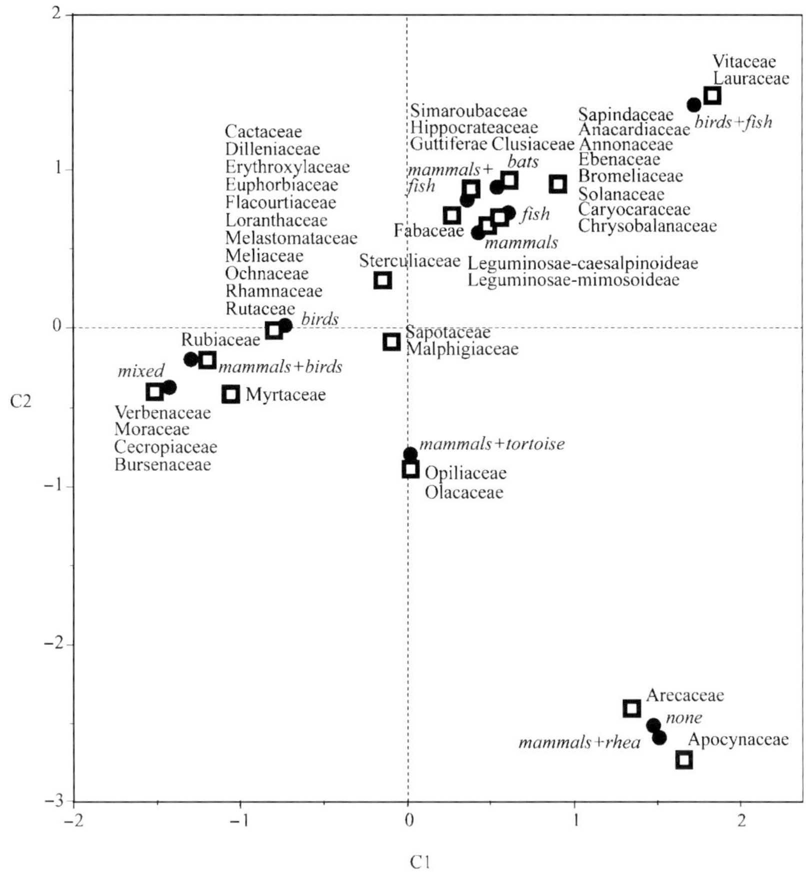

Fig. 5.5. Correspondence analysis of the interaction matrix between species of fruits and frugivores at the family level. The plot of the first two canonical variates groups different plant families (squares) with their major types of seed dispersers (dots). Overlapping family labels have been displaced for clarity.

on population persistence (Fig. 5.6). For short lifespan or small recruitment probabilities the population goes quickly to extinction. However, a very small increase in lifespan or in the probability of recruitment in vicinity of the plant generates a completely new dynamic: plant density increases and eventually stabilizes (Fig. 5.6).

Discussion

Many of the fruits present in the Pantanal fit the classical mammal dispersal syndrome, involving large, green or brown, often sweet-smelling fruits (Janson, 1983; Gautier-Hion *et al.*, 1985; Howe, 1986; Herrera, 1989). The high percentage of mammal-dispersed fruits was also supported by our

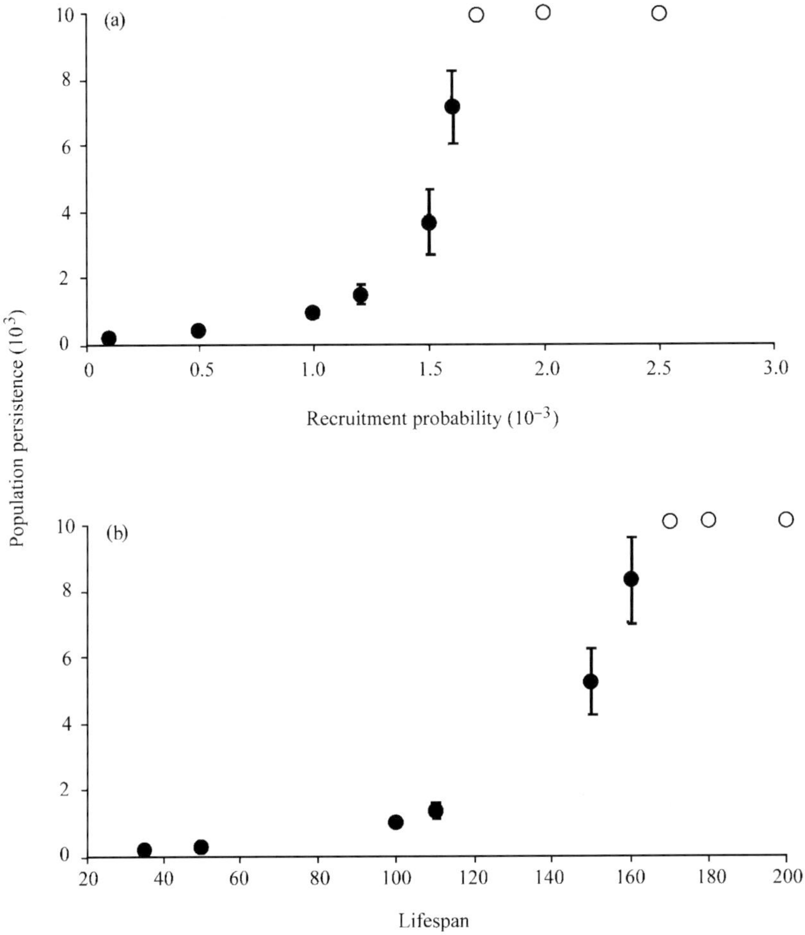

Fig. 5.6. Population persistence after seed disperser extinction under different scenarios of recruitment probability (a) and plant lifespan (b). Closed circles represent average population persistence (n = 10 simulation runs) and open circles represent situations in which the population survived until at least 10^4 reproductive events. Error bars indicate ± 1 SD.

observations of fruit–frugivore interactions. The similarity in fruit morphology between the Pantanal and African communities of savannah type suggests that similar evolutionary processes driven by seed dispersal have been a major force in these communities. The paired contrasts in fruit traits show consistent trends independent of family, indicating the evolutionary relevance of this pattern. Pantanal fruits were significantly more similar to African communities than to the geographically closer Atlantic rainforest assemblages, in relation to fruit size and colour. By the end of the Pleistocene, both South American and African savannahs had a large biomass and diversity of large-bodied mammals (Fariña *et al.*, 1998; Cartelle, 1999) which, combined with the similarities in fruit traits (e.g. large fruit size, dull-coloured fruits predominantly green and brown), suggests that megafauna were an important component of selective pressures on fruits. On the other hand, tropical rain forests have a lower biomass of large mammals than savannah-like ecosystems (Prins and Reitma, 1989). Nowadays, the biomass of native mammals in the Pantanal reaches only 1000 kg/km^2, while that of the exotic fauna (horses, cattle and feral pigs) reaches > 5000 kg/km^2 (Mourão *et al.*, 2002) and it has been estimated that the Pantanal could support about 10,000 kg/km^2 of large-bodied herbivores; a biomass similar to many African savannas (Galetti, 2004). In fact, anyone familiar with both ecosystems will not be surprised at such convergences (Madinah, 1982; Fariña, 1996).

A fundamental question related to megafauna-dispersed fruits in the Pantanal and other neotropical plant communities is: How have these plant species survived for about 10,000 years without the seed dispersers with which they originally co-evolved (Howe, 1985)? Several studies suggest that local extinction of large-bodied frugivores reduces seedling recruitment in vertebrate-dispersed plants. This pattern has been found on intensively studied oceanic islands where birds, bats or lizards were recently extinct (McConkey and Drake, 2002; Traveset and Riera, 2005). The idea that a population of fleshy, vertebrate-dispersed fruits can collapse under the absence of their main seed dispersers started with a study by Temple (1977), who suggested that the tambalacoque tree (*Sideroxylon grandiflorum*; Sapotaceae) relied on the extinct dodo (*Raphus cuculatus*; Raphidae) for successful establishment. Although the dodo–tambalacoque system has been an iconic case in plant–animal interaction studies in the past, it is now established that Temple's analysis was unrealistic and it is an erroneous example of an obligate plant–animal relationship. There is no solid evidence that the dodo was absolutely required for seed germination or that the tambalacoque tree was driven to extinction due to the absence of the dodo (Witmer, 1991). Temple ignored earlier reports of Hill (1941) proving the ability of seeds to germinate without abrasion (Herhey, 2004). More recently, in-depth studies have shown that local extinction of some frugivores can drastically reduce seedling recruitment in vertebrate-dispersed plants (Cordeiro and Howe, 2001; McConkey and Drake, 2002; Traveset and Riera, 2005; Galetti *et al.*, 2006) but rarely drives populations to extinction, especially in long-lived taxa (Herhey, 2004).

The loss of the main seed dispersers, leading to severely limited dispersal and recruitment, is certainly one of the main potential causes for the demise of plant populations (see Traveset and Riera, 2005). Therefore, the survival of plants that produce large, high-cost fruits without efficient seed dispersal remains a paradox. Understanding the ecological persistence of these plants will have implications for understanding the organization of plant communities in the Neotropics after the Pleistocene mass extinctions (Janzen and Martin, 1982) and will highlight the ecological mechanisms that allow plant populations to survive after the extinction of their main seed dispersers (Chapman and Chapman, 1995).

The presence of exotic species in the Pantanal, such as feral pigs and cattle, has contributed to the seed dispersal of many plants, mainly the ones with large fruits and seeds. Fruit consumption and seed dispersal by cattle and horses are described for many species as a way to replace the extinct megafauna (Janzen, 1982; Janzen and Martin, 1982). Our data indicate that feral pigs can disperse more species than the native fauna, and also large-sized fruits (e.g. *A. phalerata*, *Acrocomia aculeate*, Arecaceae), because they are able to swallow and defecate their seeds, which peccaries (*T. pecari* and *P. tajacu*) cannot do. Among the native animals, only the tapir disperses fruits of *A. phalerata*, but the tapirs frequently defecate in salty lakes, a site unsuitable for successful seedling recruitment.

The ability of many species to resprout can also contribute to their survivorship, as can the ability to be dispersed by water (e.g. *Calophyllum brasiliense*, Clusiaceae; *B. glaucescens*; *Garcinia brasiliensis*, Guttiferae). In fact, in a seasonally flooded ecosystem like the Pantanal, dispersal by water may be the most common recruitment mechanism for many fleshy, vertebrate-dispersed fruits. Human use is likely to regularly contribute to the dispersal of seeds from those species used (47% – see Results). Rare events of seed dispersal, mainly done by animals with a low probability of fruit consumption (such as raptors), can also contribute to the dispersal of some species of plants (Galetti and Guimarães Jr, 2004).

However, the survival of some plants, such as *Attalea speciosa*, is surprising when considering the few ecological mechanisms known to explain plant population persistence. They do not display vegetative reproduction, are not known to be dispersed by water, secondary dispersal, humans and/or by rare events. This palm produces large, heavy, well-protected fruits and occurs in areas away from water (*terra firme* or *cordilheira*; Silva *et al.*, 2000), so seed dispersal by water is unlikely. Moreover, we did not find any evidence of scatter-hoarding rodents or large mammals (even the exotic species) dispersing the fruits. In this context, the simple numerical simulations reported here may provide a useful insight: some plant populations may persist simply by a combination of long lifespan and some likelihood of recruitment close to the parental trees (Hubbell, 1980). Indeed, this is apparently the case for *A. speciosa*, where seeds below the mother plant have no bruchid or vertebrate predation. Fallen fruits can germinate and establish below the mother plant, creating large monodominant forests known locally as *babaçuais*.

Conclusions

In this chapter we have shown that fruit traits of the Pantanal species are more similar to those from savannahs in Africa than to those from the neighbouring Atlantic rainforest. Our data indicate that the high diversity of large, fleshy, vertebrate-dispersed fruits in the Pantanal is associated with a prevalence of dispersal by mammals, which is supplemented by other taxa such as birds, fish and reptiles. Among potentially mammal-dispersed fruits, several species demonstrate anachronistic dispersal in the Pantanal and include taxa that probably interacted with the extinct, highly diverse, Pleistocene megafauna. Some of these fruits lack efficient contemporary long-distance seed dispersal and others have extremely low seed removal. A combination of life-history traits has probably allowed these plants to persist after the extinction of their major megafauna dispersers: resprouting, being dispersed by water, extensive secondary seed dispersal by vertebrates (e.g. scatter-hoarding rodents), replacing seed dispersers with exotic fauna or humans, or rare events of seed dispersal (e.g. raptors). Moreover, numerical simulations suggest that, for some species in scenarios of impoverished dispersal, no alternative mechanisms need to be invoked to explain population persistence, other than a combination of long lifespan and an ability to recruit close to parental trees.

Acknowledgements

We would like to thank Earthwatch Institute, FAPESP (04/00810-3), Conservação Internacional do Brasil, and Idea Wild and Programa Iberoamericano CYTED for financial support. For permission to work at Fazenda Rio Negro and logistic support, we thank Conservação Internacional do Brasil, especially R. Lourival and M. Harris. We thank IBAMA for permission to collect fruit and fish samples, S. Salis (EMBRAPA) for identifying the plant samples, and A. Keroughlian, D. Eaton and E. Wang for sharing unpublished data. Also, we thank T. Hovestadt for kindly sharing his data on fruit morphology from Ivory Coast. We thank Andrew Dennis, Pierre-Michel Forget and an anonymous reviewer for their valuable comments. C.I.D. received a CAPES fellowship, M.A.P. are researchers from CnPq and P.R.G. from FAPESP. The Spanish Ministerio de Ciencia y Tecnología (REN2003-00273) and a bilateral agreement between CNPq-CSIC (2004BR00008) and Junta de Andalucía (RNM-305) sponsored P.J. This project had the important help and inspiration of many volunteers from Earthwatch Institute and researchers, for which we are very much indebted.

References

Alho, C.J.R. and Lacher, T.E. Jr (1991) Mammalian conservation in the Pantanal of Brazil. In: Mares, M. and Schmidly, D.J. (eds) *Latin American Mammalogy: History, Biodiversity and Conservation.* University of Oklahoma Press, Norman, Oklahoma.

Bollen, A., Van Elsacker, L. and Ganzhorn, J.U. (2004) Relations between fruits and disperser assemblages in a Malagasy littoral forest: a community-level approach. *Journal of Tropical Ecology* 20, 599–612.

Campassi, F. (2006) PadrÚCes geográficos das síndromes de dispersão e características dos frutos de espécies arbustivo-arbóreas em comunidades vegetais da Mata Atlântica. MSc thesis, Universidade de São Paulo, Brazil.

Cartelle, C. (1999) Pleistocene mammals of the cerrado and caatinga of Brazil. In: Eisenberg, J.F. and Redford, K.H. (eds) *Mammals of the Neotropics: The Central Neotropics.* University of Chicago Press, Chicago, Illinois, pp. 27–46

Chapman, C.A. and Chapman, L.J. (1995) Survival without dispersers: seedling recruitment under parents. *Conservation Biology* 9, 675–678.

Cristoffer, C. and Peres, C.A. (2003) Elephants *versus* butterflies: the ecological role of large herbivores in the evolutionary history of two tropical worlds. *Journal of Biogeography* 30, 1357–1380.

Cordeiro, N.J. and Howe, H.F. (2001) Low recruitment of trees dispersed by animals in African forest fragments. *Conservation Biology* 15, 1733–1741.

Corlett, R.T. and Primack, R.B. (2006) Tropical rainforests and the need for cross-continental comparisons. *Trends in Ecology and Evolution* 21(2), 104–110.

Durrett, R. and Levin, S. (1994) Stochastic spatial models: a user's guide to ecological applications. *Philosophical Transactions of the Royal Society* 343, 329–350.

Fa, J.E., Peres, C.A. and Meeuwig, J. (2002) Bushmeat exploitation in tropical forest: an intercontinental comparison. *Conservation Biology* 16(1), 232–237.

Fariña, R.A. (1996) Trophic relationships among Lujanian mammals. *Evolutionary Theory* 11, 125–134.

Fariña, R., Vizcaíno, S.F. and Bargo, M.R. (1998) Body mass estimations in Lujanian (Late Pleistocene–early Holocene of South America) mammal megafauna. *Mastozoologia Neotropical* 5, 87–108.

Fischer, K.E. and Chapman, C.A. (1993) Frugivores and fruit syndromes: difference in patterns at the genus and species level. *Oikos* 66, 472–482.

Fleming, T.H. (1991) Fruiting plant–frugivore mutualism: the evolutionary theater and the ecological play. In: Price, P.W., Lewinsohn, T.W., Fernández, G.W. and Benson, W.W. (eds) *Plant–Animal Interactions: Evolutionary Ecology in Tropical and Temperate Regions.* John Wiley and Sons, New York, pp. 119–144.

Fleming, T.H. (2005) The relationship between species richness of vertebrate mutualists and their food plants in tropical and subtropical communities differs among hemispheres. *Oikos* 111, 556–562.

Fleming, T.H., Breitwisch, R. and Whitesides, G. (1987) Patterns of tropical vertebrate frugivore diversity. *Annual Review of Ecology and Systematics* 18, 91–109.

Galetti, M. (1996) Fruits and frugivores in a Brazilian Atlantic forest. PhD thesis, University of Cambridge, UK.

Galetti, M. (2002) Seed dispersal of mimetic seeds: parasitism, mutualism, aposematism or exaptation? In: Levey, D.J., Silva, W.R. and Galetti, M. (eds) *Seed Dispersal and Frugivory: Ecology, Evolution and Conservation.* CAB International Publishing, Wallingford, UK, pp. 177–192.

Galetti, M. (2004) Parques do Pleistoceno: recriando o cerrado e o pantanal com a megafauna. *Natureza & Conservação* 2, 18–24.

Galetti, M. and Guimarães, P.R. Jr (2004) Seed dispersal of *Attalea phalerata* (Palmae) by crested caracaras (*Caracara plancus*) in the pantanal and a review of frugivory by raptors. *Ararajuba* 12(2), 59–61.

Galetti, M., Pizo, M.A. and Morellato, P.C. (2002) Métodos para o estudo da fenologia, frugivoria e dispersão de sementes. In: Cüllen, L., Rudran, R. and Pádua, C.V. (eds) *Manual para o Manejo da Vida Silvestre no Brasil.* Smithsonian Institution Press and Instituto de Pesquisas Ecológicas, Nazaré Paulista, Brazil, pp. 395–422.

Galetti, M., Donatti, C.I., Pires, A.S., Guimarães, P.R. Jr and Jordano, P. (2006) Seed survival and dispersal of an endemic Atlantic forest palm: the combined effects of defaunation and forest fragmentation. *Botanical Journal of the Linnean Society* 151, 141–149

Gautier-Hion, A., Duplantier, J.M., Quris, R., Feer, F., Sourd, C., Decoux, J.P., Dubost, G., Emmons, L., Erard, C. and Hecketsweiler, P. (1985) Fruit characters as a basis of fruit choice and seed dispersal in a tropical forest vertebrate community. *Oecologia* 65, 324–337.

Harris, M.B., Tomas, W., Da Silva, C.J., Guimarães, E., Sonada, F. and Fachim, E. (2005) Safeguarding the Pantanal wetlands: threats and conservation initiatives. *Conservation Biology* 19, 714–720.

Herhey, D.R. (2004) The widespread misconception that the tambalacoque or *Calvaria* tree absolutely required the dodo bird for its seeds to germinate. *Plant Science Bulletin* 50, 105–109.

Herrera, C.M. (1989) Seed dispersal by animals: a role in angiosperm diversification? *American Naturalist* 133, 309–322.

Herrera, C.M. (2002) Seed dispersal by vertebrates. In: Herrera, C.M. and O. Pellmyr (eds) *Plant–Animal Interactions: An Evolutionary Approach*. Blackwell Science, Oxford, UK, pp. 185–208.

Hill, A.W. (1941) The genus *Calvaria*, with an account of the stony endocarp and germination of the seed, and description of the new species. *Annals of Botany* 5, 587–606.

Hovestadt, T., Yao, P. and Linsenmair, E. (1999) Seed dispersal mechanisms and the vegetation of forest islands in a West African forest-savanna mosaic (Comoé National Park, Ivory Coast). *Plant Ecology* 144, 1–25.

Howe, H.F. (1985) Gomphothere fruits: a critique. *American Naturalist* 125, 853–865.

Howe, H.F. (1986) Seed dispersal by fruit-eating birds and mammals. In: Murray, D.R. (ed.) *Seed Dispersal*. Academic Press, Sydney, Australia, pp. 123–189.

Hubbell, S. (1980) Seed predation and the coexistence of tree species in tropical forests. *Oikos* 35, 214–229.

Hunter, J.R. (1989) Seed dispersal and germination of *Enterolobium cyclocarpum* (Jacq.) Griseb. (Leguminosae: Mimosoideae): are megafauna necessary? *Journal of Biogeography* 16, 369–378.

Janzen, D.H. (1982) Fruit traits, and seed consumption by rodents, of *Crescentia alata* (Bignoniaceae) in Santa Rosa National Park, Costa Rica. *American Journal of Botany* 69, 1258–1268.

Janzen, D.H. and Martin, P.S. (1982) Neotropical anachronisms: the fruits the Gomphoteres ate. *Science* 215, 19–27.

Janson, C.H. (1983) Adaptation of fruit morphology to dispersal agents in a neotropical forest. *Science* 219, 187–189.

Jordano, P. (1995) Angiosperm fleshy fruits and seed dispersers: a comparative analysis of adaptation and constraints in plant–animal interactions. *American Naturalist* 145, 163–191.

Jordano, P. (2000) Fruits and frugivory. In: Fenner, M. (ed.) *Seeds: The Ecology of Regeneration in Plant Communities*. CAB International, Wallingford, UK, pp. 125–166.

Jordano, P. and Schupp, E.W. (2000) Seed disperser effectiveness: the quantity component and patterns of seed rain for *Prunus mahaleb*. *Ecological Monographs* 70, 591–615.

Jordano, P., Bascompte, J. and Olesen, J.M. (2003) Invariant properties in coevolutionary networks of plant–animal interactions. *Ecology Letters* 6, 69–81.

Levey, D.J. (1987) Seed size and fruit-handling techniques of avian frugivores. *American Naturalist* 129, 471–485.

Lord, J.M., Markey, A.S. and Marshall, J. (2002) Have frugivores influenced the evolution of fruit traits in New Zealand? In: Levey, D.J., Silva, W.R. and Galetti, M. (eds) *Seed Dispersal and Frugivory: Ecology, Evolution and Conservation*. CAB International, Wallingford, UK, pp. 55–68.

Lourival, R.F.F. (1997) Análise de sustentabilidade do modelo de caça tradicional, no Pantanal da Nhecolandia, Corumbá, MS. In: Valladares-Padua, C. and Bodmer, R. (eds) *Manejo e conservação de vida silvestre no Brasil*. MCT-CNPq, Sociedade Civil Mamiraua, Brazil, pp. 123–172.

Mack, A.L. (1993) The sizes of vertebrate-dispersed fruits: a Neotropical-Paleotropical comparison. *American Naturalist* 142, 840–856.

Martin, P.S. and Klein, R. (1984) *Quaternary Extinctions: A Prehistoric Revolution.* University of Arizona Press, Tucson, Arizona.
McCarthy, T.S., Ellery, W.N. and Dangerfield, J.M. (1998) The role of biota in the initiation and growth of islands on the floodplain of the Okavango alluvial fan, Botswana. *Earth Surface Processes and Landforms* 23, 291–316.
McConkey, K.R. and Drake, D.R. (2002) Extinct pigeons and declining bat populations: are large seeds still being dispersed in the tropical pacific? In: Levey, D.J., Silva, W.R. and Galetti, M. (eds) *Seed Dispersal and Frugivory: Ecology, Evolution and Conservation.* CAB International, Wallingford, UK, pp. 381–395.
Madinah, E. (1982) Physiological ecology of neotropical savannas. In: Huntley, B.J. and Walker, B.H. (eds) *Ecology of Tropical Savannas.* Ecological Studies Vol. 42. Springer, Berlin, Germany.
Mittermeier, R.A., Harris, M.B., Mittermeier, C.G., da Silva, J.M.C, Lourival, R., Fonseca, G.A.B. and Seligmann, P.A. (2005) *Pantanal: South America's Wetland Jewel.* Conservation International, Washington, DC.
Miura, S., Yasuda, M. and Ratnam, L. (1997) Who steals the fruits? Monitoring frugivory of mammals in a tropical rain forest. *Malayan Nature Journal* 50, 183–193.
Moermond, T.C. and Denslow, J.S. (1985) Neotropical avian frugivores: patterns of behaviour, morphology, and nutrition, with consequences for fruit selection. In: Buckley, P.A., Foster, M.S., Morton, E.S., Ridgely, R.S. and Buckley, F.G. (eds) *Neotropical Ornithology.* Ornithological Monographs No. 36. American Ornithologists Union, McLean, Virginia, pp. 865–889.
Mourão, G.M., Coutinho, M.E., Mauro, R., Tomás, W.M. and Magnusson, W.E. (2002) Levantamentos aéreos de espécies introduzidas no Pantanal: porcos ferais (porco monteiro), gado bovino e búfalos. EMPRABA, Corumbá, MS, Brazil.
Owen-Smith, R.N. (1988) *Megaherbivores: The Influence of Very Large Body Size on Ecology.* Cambridge University Press, Cambridge, UK.
Owen-Smith, N. (1989) Megafaunal extinctions: the conservation message from 11,000 years. *Conservation Biology* 3, 405–412.
Peres, C.A. (2000) Effects of subsistence hunting on vertebrate community structure in Amazonian forests. *Conservation Biology* 14, 240–253.
Peres, C.A. (2001) Synergistic effects of subsistence hunting and habitat fragmentation on Amazonian forest vertebrates. *Conservation Biology* 15, 1490–1505.
Pott, A. and Pott, V.J. (1994) *Plantas do Pantanal.* EMBRAPA, Corumbá, MS, Brazil.
Prance, G.T. and Schaller, G.B. (1982) Preliminary studies of some vegetation types of the Pantanal, Mato Grosso, Brazil. *Brittonia* 34, 228–251.
Prins, H.H. and Reitma, J.M. (1989) Mammalian biomass in an African equatorial rain forest. *Journal of Animal Ecology* 58, 851–861.
Schluter, D. (1988) The evolution of finch communities on islands and continents: Kenya vs. Galápagos. *Ecological Monographs* 58, 229–249.
Silva, M.P., Mauro, R., Mourão, G. and Couttinho, M. (2000) Distribuição e quantificação de classes de vegetação do Pantanal através de levantamento aéreo. *Revista Brasileira de Botânica* 23, 143–152.
Snow, D.W. (1980) Regional differences between tropical floras and the evolution of frugivory. In: Nohring, R. (ed.) *Acta XVII Congressus Internationalis Ornithologici.* Deuschen Ornithologen-Gesellschaft, Berlin, Germany, pp. 1192–1198.
Swarts, F.A. (2000) The Pantanal in the 21st century. In: Swarts, F.A. (ed.) *The Pantanal: Understanding and Preserving the World's Largest Wetland.* Paragon House, St Paul, Minnesota, pp. 1–22.
Temple, S.A. (1977) Plant–animal mutualism: coevolution with dodo leads to near extinction of plant. *Science* 197, 885–886.

Thomas, V. and Grant, R. (2002) *Sappi Tree Spotting*. Jacana Publishing, Johannesburg, South Africa.

Traveset, A. and Riera, N. (2005) Disruption of a plant–lizard seed dispersal system and its ecological effects on a threatened endemic plant in the Balearic Islands. *Conservation Biology* 19, 421–431.

Trolle, M. (2003) Mammal survey in the southeastern Pantanal, Brazil. *Biodiversity and Conservation* 12, 823–836.

van Wyk, P. (1997) *Southern African Trees: A Photographic Guide*. Struik Publishers, Cape Town, South Africa.

Voigt, F.A., Bleher, B., Fietz, J., Ganzhorn, J.U., Schwab, D. and Bohning-Gaese, K. (2004) A comparison of morphological and chemical fruit traits between two sites with different frugivore assemblages. *Oecologia* 141, 94–104.

Witmer, M.C. (1991) The dodo and the tambalacoque tree: an obligate mutualism reconsidered. *Oikos* 61, 133–137.

6 The Importance of Lizards as Frugivores and Seed Dispersers

A. Valido and J.M. Olesen

Introduction

Despite the accumulation of a considerable body of literature about frugivory and seed dispersal of fleshy-fruited plants by vertebrates during the last decades (e.g. reviews in Howe and Smallwood, 1982; van der Pijl, 1982; Janzen, 1983; Howe, 1986; Howe and Westley, 1988; Stiles, 1989; Jordano, 2000; Herrera, 2002), and also previous editions of the *Frugivores and Seed Dispersal: International Symposium/Workshop* series (namely Estrada and Fleming, 1986; Fleming and Estrada, 1993; Levey *et al.*, 2002), the habits of reptiles as frugivorous animals and their role as seed dispersers have rarely been considered. At most, they are regarded as agents of the past (Labandeira, 2002; Tiffney, 2004).

Understandably, the literature has been dominated by studies on frugivorous birds, primates, ungulates and bats, which behave as important seed dispersers of vascular plants in many habitats. However, the pulp of fleshy fruits is also an important food resource for tortoises and turtles (MacDonald and Mushinsky, 1988; Milton, 1992; Jerozolimski, 2003) and for herbivorous and omnivorous lizards (Schoener *et al.*, 1982; Dearing and Schall, 1992; Valido *et al.*, 2003). Thus, one might expect that species included in both reptile groups would act as potential seed dispersers, but this has hardly ever been seriously considered. The exceptions are classic studies demonstrating the importance of herbivorous chelonians as seed dispersers (see Beccari, 1877; Borzí, 1911; Ridley, 1930; Rick and Bowman, 1961) and also some more recent publications (e.g. Racine and Downhower, 1974; Hnatiuk, 1978; Rust and Roth, 1981; Cobo and Andreu, 1988; Milton, 1992; Moll and Jansen, 1995; Varela and Bucher, 2002; Liu *et al.* 2004). In contrast to these animals, the reptile group upon which we focus in this chapter – the lizards (order Squamata: suborder Sauria), including around 4800 species worldwide – have received little attention as modern plant mutualists although there is a growing interest in their role as seed dispersers, including both the herbivorous species (e.g. Klein, 1977; Janzen, 1982; Iverson, 1985; Traveset, 1990; Hartley *et al.*, 2000) and the omnivorous ones (e.g. Whitaker, 1987; Fialho, 1990; Valido and Nogales, 1994; Willson *et al.*, 1996; Lord and Marshall, 2001; Olesen and Valido, 2003; Valido *et al.*, 2003; Rodríguez-Pérez *et al.*, 2005).

Based on a large number of observations made by naturalists, many tortoises and turtles are regarded as herbivorous (25%; King, 1996) and most small, diurnal lizards as carnivorous. This commonly accepted wisdom is possibly related to the fact that most lizards primarily feed on small invertebrates (e.g. Greene, 1982; van Damme, 1999; Cooper and Vitt, 2002; Vitt and Pianka, 2005). However, a herbivorous diet is observed among lizards too. A strict herbivorous habit (70–100% plant volume in the diet; see Cooper and Vitt, 2002; Espinoza *et al.*, 2004) is generally associated with folivorous species, which are characterized by large body size, live in warm climates, and maintain high body temperatures (but see Espinoza *et al.*, 2004, for exceptions to this rule in small, cold-climate-living lizards). For example, all Iguanidae lizards (36 species) and also the 16 *Uromastyx* and three *Hydrosaurus* species (Agamidae), and some species and populations of skinks (e.g. *Corucia zebrata*, *Egernia* spp., *Tiliqua* spp., Scincidae) among others, are all specialized herbivores and possess the digestive specializations necessary for efficient foliage and stem processing (e.g. Iverson, 1982; Troyer, 1984). Thus, the conclusion is that an almost pure plant diet is uncommon in this widespread and large group of vertebrates. Estimates of the proportion of true herbivorous species are in the region of 1–3% (Iverson, 1982; Cooper and Vitt, 2002), although these values are slowly increasing as more data are becoming available on lizard diets (e.g. Espinoza *et al.*, 2004; O'Grady *et al.*, 2005). However, a number of small/medium-sized lizards regularly or occasionally eat some vegetable matter as well (i.e. they are omnivorous). For example, many insular species and also some mainland lizards from areas (or seasons) poor in arthropods are known to consume nectar, pollen and fruit (e.g. Pérez-Mellado and Corti, 1993; Valido and Nogales, 1994; Traveset and Sáez, 1997; van Damme, 1999; Cooper and Vitt, 2002; Olesen and Valido, 2003, 2004; Valido *et al.*, 2003; Espinoza *et al.*, 2004; O'Grady *et al.*, 2005). Fleshy fruits are, in general, low in fibre and protein, but high in soluble carbohydrates and water content (Jordano, 2000), and are thus easier to digest than foliage and stems. Therefore, these lizards can use such plant resources when other food resources, such as arthropods, are in short supply without any adaptive changes being required in the digestive tract, which is also typical of more folivorous species (Pearson, 1954; Rand, 1978; Búrquez *et al.*, 1986; Cooper and Vitt, 2002; Olesen and Valido, 2003; Valido and Nogales, 2003; O'Grady *et al.*, 2005).

In the rich herpetological literature, fruits are generally included within a category called 'vegetative or plant matter' and are rarely scored separately. Consequently, their importance in the diet of omnivorous lizard species has usually been underestimated. However, in the 'literature forest', the understorey is rich in natural history reports of lizards feeding on fruits. The first testimonies we have at hand are painted motifs on pottery from northern Peru. These ceramics date back to the Mochica culture, which lasted from 200 to 100 BC to AD 700 (Larco, 2001). The first written evidence appears to relate to lizards inhabiting oceanic islands: the Canary Islands (Fructuoso, 1590, reprinted in 1964; Steindacher, 1891), the Mascarene Archipelago (Leguat, 1708, reprinted in 1981; Liénard, 1842), the Galapagos Islands (Darwin, 1845) and Jamaica (Gosse, 1851, cited in Schwartz and Henderson, 1991). These early references indicate that this phenomenon was already known by some of the naturalists of this time. We also wish to draw attention to a small specific report about the importance of lizards as seed dispersers published almost 100 years ago by Antonino Borzí (Borzí, 1911). In this first true review about saurochory, ten species of lizards were included as fruit eaters and potential seed dispersers. Recently, more than 200 examples of species of lizards feeding on fleshy fruit have been cited (Olesen and Valido, 2003); thus, we conclude that they may have a wide importance as seed dispersers. Unfortunately, most of these reports do not include any data on the quantitative importance of lizards as plant mutualists, and the accounts are rarely synthesized into generalizations of broader value to ecology and evolutionary biology. Based on a search of the literature, personal communications, and our own field research, we found that lizard–plant mutualistic interactions were much more frequent than hitherto realized (A. Valido and J.M. Olesen, unpublished results). However, quantitative information about fruit removal and seeds in scats, or quality data on dispersal of seeds to safe sites and subsequent germination are extremely scarce (see Table 2 in Godínez-Alvarez, 2004). Thus, in only a few cases do we have sufficient data to make valid comparisons with the more traditional modes of seed dispersal mediated by birds or mammals (Valido, 1999; Wotton, 2002).

Our main aim in this chapter is to revisit and synthesize the available evidence for fruit-eating lizards around the world in order to call attention to the fact that lizards need to be considered as frugivores and potential seed dispersal agents, especially in certain habitats and situations. In particular in this chapter we address:

1. The geographical distribution of saurochory;
2. The morphological characteristics of fleshy fruits consumed by lizards;
3. The taxonomic affiliation of both lizard and plant species;
4. The characteristics of their role as seed dispersers.

Inclusions and Exclusions from our Review

The available data about lizards using fleshy fruits as an alimentary resource vary in quality. Most reports and personal observations are only brief, qualitative observations of lizards feeding on fleshy fruits (e.g. *Uromastyx ornate*; Agamidae; see Bronstein *et al.*, Chapter 7, this volume). We have excluded all reports that did not explicitly mention fruit as a separate food item. It is known that many lizards readily feed on fruit in captivity (e.g. Rogner, 1997). Such observations were also excluded if the report did not include additional field observations. Thus, only data on lizards observed feeding on fruit in nature were included. In some cases, we confirmed the presence of fruit in the diet of a lizard after personal communication with authors who only reported 'plant material' in their papers. In addition, reports without clear, unambiguous statements about the direct ingestion of fruits have not been included. For example, although infrequent fruit-eating has been reported for the tuatara (*Sphenodon punctatus*; Sphenodontidae) from Stephens and Middle Islands, New Zealand (Walls, 1981; Southey, cited in Whitaker, 1987; Ussher, 1999), based on seeds in droppings, we did not include these observations because the fruits may have originated indirectly from the fruit-eating *Hoplodactylus* gecko, which is commonly consumed by tuataras (Whitaker, 1968). A summary of our data is presented in Table 6.1 (the complete list will be published elsewhere). The systematic arrangement of lizards follows the March 2005 edition of the *European Molecular Biology Laboratory – EMBL – Reptile Database* (Uetz, 2005). We also classified lizards in relation to their geographical distribution: insular, mainland and mixed species (inhabiting islands and mainland) according to the EMBL database. Near-

Table 6.1. Taxonomic and geographical distribution of fruit-eating lizards.

Fruit-eating lizards	Number of species	Fruit-eating lizards	Islands	Mixed	Mainland	P	%
Scincidae	1306	63	36	3	24	***	4.8
Gekkonidae	1081	20	18	0	2	***	1.9
Polychrotidae	393	45	41	0	4	***	11.5
Agamidae	379	11	5	0	6	*	2.9
Tropiduridae	309	38	17	0	21	***	12.3
Lacertidae	280	27	17	2	8	***	9.6
Phrynosomatidae	125	7	1	0	6	ns	5.6
Teiidae	122	24	9	3	12	***	19.7
Anguidae	112	1	1	0	0	ns	0.9
Varanidae	59	2	2	0	0	*	3.4
Cordylidae	54	2	0	0	2	ns	3.7
Iguanidae	36	20	15	3	2	*	55.6
Gerrhosauridae	33	6	4	0	2	ns	18.2
Xantusiidae	24	3	2	0	1	**	12.5
Dibamidae	20	5	5	0	0	**	25.0
Crotaphytidae	10	2	0	1	1	ns	20.0
Corytophanidae	9	3	0	0	3	ns	33.3
Opluridae	7	1	1	0	0	ns	14.3
Total	4359	280	174	12	94	***	6.4
As percentage			62	4	34		

P, the probability that the sample of lizards differs from random (see Results). The last column is the percentage of the total of fruit-eating lizards.

shore islets were regarded as part of the adjacent mainland. For plant genera we followed Mabberley (1997). In the analysis we used MATLAB (Matworks, Natrick, Massachusetts, USA).

Synthesis of Data

Frugivorous lizards

From all the reports and observations included in our review, we found a total of 280 species of lizards from 85 genera and 18 families that consume fleshy fruits at least occasionally (Table 6.1). These observations originate from 45 islands/archipelagos and 30 mainland countries (e.g. Spain, Iran, Argentina). Fruit consumption is more often reported from islands (62%) than from mainland populations (34%). Species observed to eat fruit on both mainland and island regions only constituted 4% (see Fig. 6.1). Overall, there were many more frugivorous species on islands than would be expected by chance (Table 6.1; see below). Island species were significantly over-represented as fruit consumers in 11 out of the 18 lizard families, especially in Polychrotidae (91.1% of the registered cases), Gekkonidae (90.5%), Lacertidae (63%) and Scincidae (56.9%).

To test the null hypothesis stating that the observed distribution of frugivorous lizards is equivalent to the distribution of species between islands, islands + mainland, and mainland, we used a one-tailed Fisher's exact test for N × M contingency tables (Wells and King, 1980). Here we use as our database only the 18 lizard families (including a total of 4359

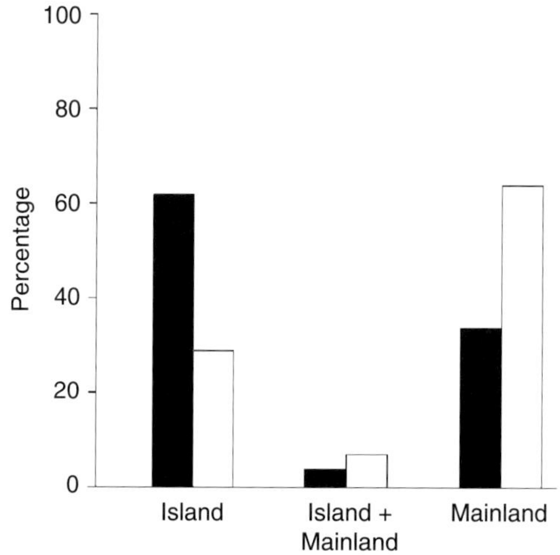

Fig. 6.1. The proportion of fruit-eating lizards (solid bars) compared with all other species (open bars) of lizard (Scincomorpha: Iguania, Gekkota, Scincomorpha) inhabiting islands, mainland or mixed. Data on geographical distributions of lizards from Uetz (2005).

species) in which frugivory is known. According to the EMBL Reptile Database, 30%, 7% and 63% of all species in these families are from islands (I), islands + mainland (IM) and mainland (M), respectively (Uetz, 2005). If we then draw a species randomly from the total pool of 4359 species, the probability of drawing an I-species must be $p_I = 0.30$, an IM-species is $p_{IM} = 0.07$, and an M-species is $p_M = 0.63$. The probability P of drawing two island species in a draw of two species is $P = p_I^2 = 0.30^2$. The probability P of drawing, for example, one I-species, two IM-species and two M-species from a trinomial random draw of five species is:

$$P = \left(\frac{5!}{1!2!2!}\right) p_I^1 p_{IM}^2 p_M^2 = 0.017$$

The general likelihood is

$$P = \left(\frac{n!}{r_I! r_{IM}! r_M!}\right) p_I^{r_I} p_{IM}^{r_{IM}} p_M^{r_M}$$

where n is sample size, r_I is number of I-species, r_{IM} is the number of IM-species and r_M is the number of M-species.

We have compiled a list of 280 species of lizards consuming fleshy fruits. Our review documented that 174 were island species, 12 occurred both on islands and mainland, and 94 were from mainland locations only (Table 6.1). We now want to know whether this is a random draw from the total species pool. To find out the exact probability, we computed the sum of P values by generating all tables that are more extreme than the table given, i.e. more island species. Exact p-values tend to be more conservative, since we are including all cell frequencies with the same fixed marginal totals as the observed one. The probability of getting our sample or one that is even more island-skewed was $P = 1.55 \times 10^{-29}$. Thus, many more frugivorous species occur on islands than would be expected by chance.

This pattern also holds true at the family level. We repeated the same calculations as above for each family (Table 6.1). Island species consuming fruit are highly over-represented in 11 of the 18 families. Exceptions are: Anguidae, Crotaphytidae, Cordylidae, Phrynosomatidae, Corytophamidae, Opluridae and Gerrhosauridae. However, in these families, the number of fruit-eating lizards was very small.

Lizard plants

Worldwide, a total of 414 species, 206 genera and 101 families of plants with fleshy fruits have been reported in the diet of these 280 species of lizards. This figure is clearly an underestimate, since many reports lack data about the identity of the species whose fruits are consumed (e.g. Auffenberg and Auffenberg, 1988; Belver and Avila, 2002). In spite of this shortcoming, these results reveal some interesting patterns that need to be considered in future studies. For example:

1. Lizards consume fruits that are very diverse in type, size and colouration.
2. Fruit consumption occurs on island and mainland areas from a wide latitudinal range.
3. Fruit consumption by lizards is recorded from all continents and many oceanic islands.

In short it is a very widespread phenomenon affecting plants with a wide range of fruit structures.

From this lizard–fruit database, five families: Moraceae (with 30 species), Palmae (27), Rubiaceae (25), Solanaceae (25) and Cactaceae (24) account for 34% of all species of plants that lizards feed on. Most species are scattered widely in the phylogenies of the angiosperms and gymnosperms. The fruit types consumed are highly diverse: including syconium (e.g. *Ficus* spp.; Moraceae); arcestid (e.g. *Juniperus phoeniceae*; Cupressaceae); berry (e.g. *Hippomane mancinella*; Euphorbiaceae); drupe (e.g. *Corynocarpus laevigatus*; Corynocarpaceae); polydrupe (e.g. *Rubus inermis*; Rosaceae); and arillate seed (*Annona araracuara;* Anonnaceae).

This diversity in fruit types is also seen in their coloration. We looked for associations between fruit colour and consumption by lizards by comparing the colours of species consumed by lizards to the colour spectrum of the fleshy fruits available. We used data from three distant archipelagos, the Canary Islands (Fig. 6.2), the Balearic Archipelago (Fig. 6.3) and New Zealand

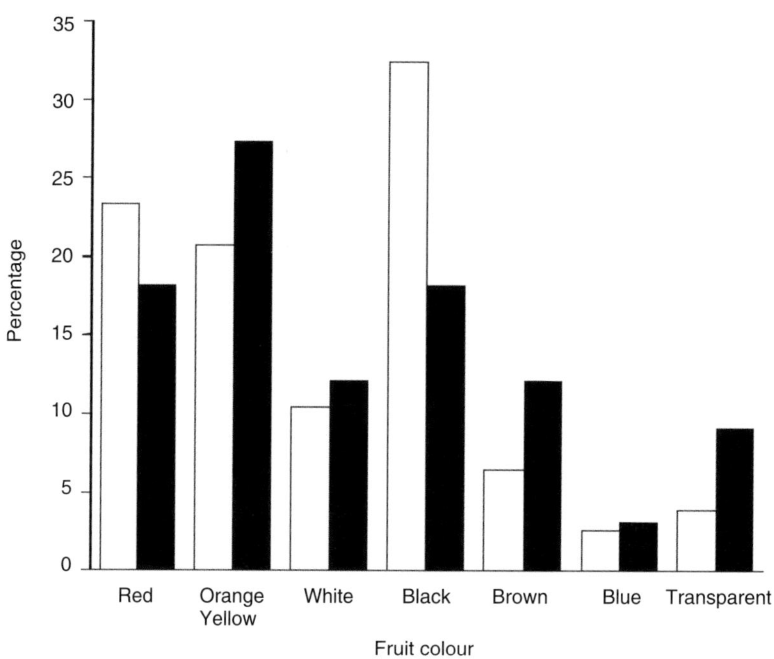

Fig. 6.2. A comparison of the proportion of colours of fleshy-fruited native plants from the Canary Islands (open bars) and those reported in the diet of the endemic lacertid species (*Gallotia* spp.) inhabiting these islands (solid bars). Data from Valido (1999).

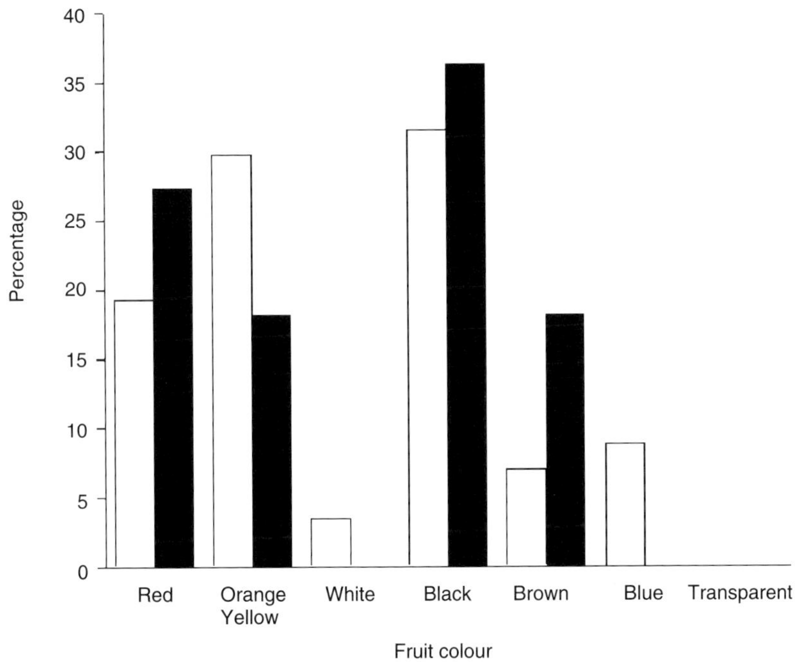

Fig. 6.3. A comparison of the proportion of colours of fleshy-fruited native plants from the Balearic Archipelago (open bars) and those reported in the diet of the lacertid *Podarcis lilfordi* (solid bars). Data from Pérez-Mellado and Traveset (1999) and the Balearic Virtual Herbarium (http:// herbarivirtual.uib.es).

(Fig. 6.4), which are all well represented in our lizard–fruit data set. We looked for any association between colour of fruit consumed by lizards and colour spectrum of fleshy fruits on these three archipelagos. Lord and Marshall (2001) and Lord *et al.* (2002) presented frequency distributions of colours of fleshy fruit for a large sample of New Zealand species. In our analysis, we used their data but lumped several of their colour categories (pink into red, yellow and orange; and purple into black). In addition, we introduced a 'transparent' fruit colour category, and placed the New Zealand *Muehlenbeckia* and the Canarian *Rubia fruticosa*, *Plocama pendula* (Rubiaceae) and *Jasminum odoratissimum* (Oleaceae) into this category. Canarian and Balearic data came from our own database and the Balearic 'virtual herbarium' (http://herbarivirtual.uib.es/), respectively. The numbers of species producing fleshy fruit were 77, 57 and 216, from the Canary Islands, the Balearic Archipelago and New Zealand, respectively. The numbers of fleshy-fruited species consumed by lizards in these three archipelagos were 41, 11 and 14, respectively.

The distributions of fruit colours in the flora from these distant areas were similar. As in most floras, colours such as red, yellow-orange and black constitute a majority of species (> 75%). Less well represented (12–16%) are white, blue and transparent colours. However, in spite of very similar fruit

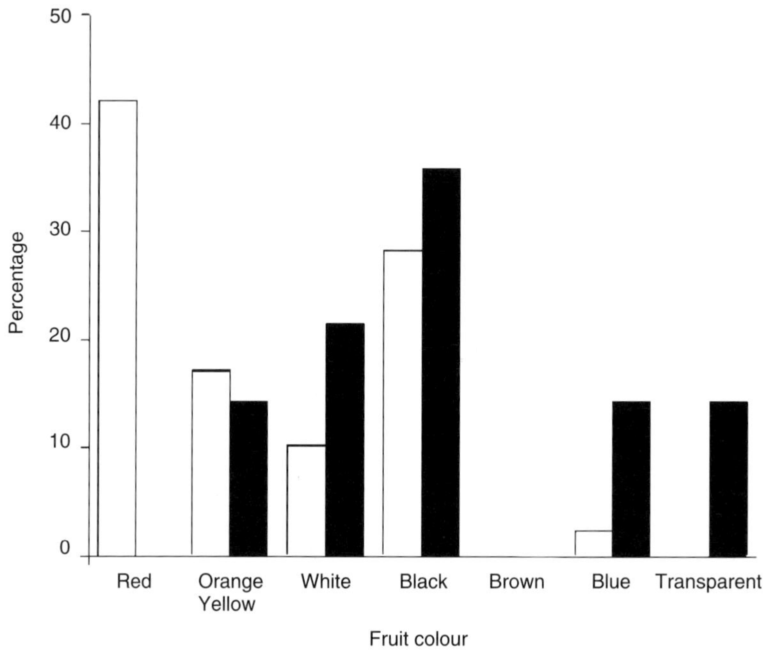

Fig. 6.4. A comparison of the proportion of colours of fleshy-fruited native plants from New Zealand (open bars) and those reported in the diet of lizard species (solid bars). Data from Whitaker (1987), Lord and Marshall (2001) and Lord *et al.* (2002).

colour spectra from the three archipelagos, lizard preferences varied. For example, in contrast with data obtained from the New Zealand flora, there were no fruit colours particularly associated with saurochory in either the Canarian flora (Fig. 6.2) or the Balearic ones (Fig. 6.3). For example, unlike the New Zealand frugivorous lizards (see Lord and Marshall, 2001; Lord *et al.*, 2002), red (*Tamus edulis*; Dioscoreaceae), black (*Rhamnus crenulata*; Rhamnaceae), white (*Bencomia caudata*; Rosaceae), orange (*Canarina canariensis*; Campanulaceae) and translucent (*Plocama pendula*; Rubiaceae) fruits, which can be up to 27 mm in diameter (*Canarina canariensis*) with seeds up to 9 mm in diameter and the number of seeds per fruit up to 1395 (in *Canarina canariensis*) are heavily consumed in different habitats and islands by the Canarian lizards (*Gallotia* spp.; Lacertidae).

Lizards as Frugivorous Vertebrates

Our review and analysis reveal that fleshy fruits may be a more important food resource to some species of lizards than hitherto recognized in the specific literature about frugivores and seed dispersal. A remarkable result of our compilation of 280 fruit-eating species is that it is not only classical herbivorous lizards, such as the Iguanidae, that eat fruits. The omnivorous and, surprisingly, many other species commonly considered to be truly insectivorous and/or carnivorous consume fruits as well. For example, within the Varanidae, which is usually considered a strictly carnivorous family, some species from Malaysia and the Philippines have been reported as feeding extensively (*Varanus olivaceus* – Auffenberg, 1988; Bennett, 2005), mostly (*Varanus mabitang* – Struck *et al.*, 2002) or sporadically on fleshy fruits (*Varanus* spp. – Yasuda *et al.*, 2005).

All major lineages of vertebrates use fruit as an alimentary resource, but birds and mammals are generally considered to be the only significant dispersers of seeds for vertebrate-dispersed plants. Fleming (1991) reported that 35.5% of bird and 19.6% of terrestrial mammal families contain species that, in some instance, use fruits as a food resource. However, among birds these values are lower (15.5%) if we only consider strictly frugivorous species with a nearly total fruit diet (Jordano, 2000). For lizards, we found that species from 69% of the 26 lizard families (at least 6% of all species) consume fleshy fruit in some quantity. However, since more attention is now being paid to this type of plant–animal mutualism, more cases are likely to be added in the near future. For example, there are records of lizards consuming flowers (e.g. *Sceloporus jarrovi* – Ballinguer and Ballinguer, 1979; *Liolaemus donosoi* – Ortiz, 1975), seeds (e.g. *Agama yemenensis* – Al-Johany, 1995), nectar (e.g. *Anolis conspersus* – Echternacht *et al.*, 2000; *Rhacodactylus auriculatus* – Bauer and Sadlier, 2000), sap (e.g. *Gehyra australis* – Greer, 2005; *Phelsuma* spp. from Madagascar – Fölling *et al.*, 2001), pollen (e.g. *Phelsuma sundbergi* – Murphy and Myers, 1996; *Liolaemus platei* – Weiser *et al.*, 1975) and foliage (e.g. *Acanthodactylus erythrurus* – Busack and Jasick, 1982; *Podarcis milensis* – Adamopoulou and Legakis, 2002). At least some of these omnivorous species may be found to eat fleshy fruits as well. However, in order to make a more precise comparison of the relative importance of birds, mammals, reptiles and fish as frugivores, we need to know both the total number of frugivorous species in each group and also the quantitative importance of fruit in their diet. Unfortunately these data are not available for all groups.

Seed dispersal by reptiles was probably the most primitive condition (Tiffney, 2004). As birds and mammals evolved, they became the main dispersal agents in many habitats. However, the herbivory (or in some cases, frugivory) condition was retained in some chelonians and also in the most primitive extant lizard family, the Iguanidae (Estes, 1983). These two reptile groups are the two most commonly used examples of seed dispersers in the frugivory literature when reptiles are cited. However, some modern species of lizards, traditionally considered primarily carnivorous or insectivorous, 'regain' this relict function when other food resources (i.e. arthropods) are no longer available (see below).

Another interesting conclusion to be drawn is that frugivorous lizards are geographically widespread but occur disproportionately on islands. For example, although insular species constitute less than one-third of all recognized species worldwide (Fig. 6.1; Uetz, 2005), they are significantly over-represented in our lizard-frugivory database.

Why Islands?

Island food-webs are relatively more simple than those from the mainland areas. The difference arises because of a lower species diversity (Carlquist, 1974), taxonomic disharmony (some groups are absent or poorly represented; Baur, 1891), and high densities of certain taxa (i.e. density compensation: an increase in abundance in the absence of interspecific competitors; MacArthur *et al.*, 1972) on islands. In this respect, species-poor islands are generally characterized by the presence of an extremely high density of lizards (see Table 2 in Rodda and Dean-Bradley, 2002). For example, the terrestrial vertebrate with the highest recorded density is *Sphaerodactylus macrolepis* (Gekkonidae) from Guana Island (67,600 individuals ha^{-1}; Rodda *et al.*, 2001). Only limited data are available on arthropods. It is known that island areas are likely to have relatively fewer species of insects than nearby mainland areas (Janzen, 1973; Ebenman and Nilsson, 1982). This pattern is evident in flower visitors; a lower number of pollinators have been detected visiting island plants compared with conspecific mainland populations (Barrett, 1998; Olesen and Jordano, 2002; J.M. Olesen and A. Valido, unpublished results). Thus, high densities in island lizards, together with arthropod scarcity and a reduced exposure to predation, may force predominantly arthropod-eating lizards to expand their trophic niche by exploiting other available resources: fruits being an important one (Olesen and Valido, 2003). Pulp from fleshy fruit is less energy-rich than other animal food items (i.e. insects), providing fewer calories and less protein per gram than animals do (see Table 6.3 in Jordano, 2000), but it is a food resource present in most island habitats.

In several families of lizards, frugivory (or omnivory) and insularity are significantly correlated (van Damme, 1999; Cooper and Vitt, 2002; see also Table 6.1). This pattern is also evident when comparing island–mainland populations within species. For example, in 1960 the large (> 500 mm snout-vent length) *Tupinambis merianae* (Teiidae) was introduced from the South American mainland to Fernando de Noronha Archipelago 345 km off the north-eastern Brazilian coastline (Péres, 2003). On the Archipelago it includes a higher proportion of plant material, especially fruits of both native and introduced species, into its diet than in its original mainland habitat (Mercolly and Yanosky, 1994; Kiefer and Sazima, 2002). Other known examples from omnivorous species are populations of *Podarcis hispanica* (Lacertidae) from the Iberian Peninsula (Mellado *et al.*, 1975) compared with populations from Columbretes Islands (Castilla and Bawens, 1991), and *Cnemidophorus lemniscatus* and *Ameiva ameiva* (Teiidae) from Providence Island (Janzen, 1973) which are more omnivorous than populations in several localities on the mainland (e.g. Vega *et al.*, 1988; Magnusson and da Silva, 1993; Vitt and Carvalho, 1995; Colli *et al.*, 1997). Although clearly more data are necessary, these intraspecific comparisons suggest that the increase in the use of fleshy fruits by island lizards is in part due to some common ecological factors occurring in these insular habitats. The use of phylogenetic contrast analysis or independent pair-wise comparisons (in the absence of a fully resolved phylogeny) and controlling for lizard size variation would be the next step in testing this ecological pattern.

The arthropod-scarcity hypothesis can also be extended to other habitats characterized by a constant or seasonal scarcity of animal prey. For example, high mountains, arid regions and caves are lacking in arthropods compared with most other habitats (Mautz and Lopez-Forment, 1978; Pietruszka *et al.*, 1986). Several skinks (Scincidae) from arid regions consume plant material during dry periods when arthropods are scarce (Kuchling, 1970; Symon, 1979; Schleich *et al.*, 1996). In general, insects are scarcer in dry seasons than in the wet seasons (Churchill, 1994; Griffiths and Christian, 1996; Christian *et al.*, 2003), and such a seasonal scarcity of arthropod food may select for omnivory in lizards.

Lizards as Seed Dispersers?

In 1985, the herpetologist John B. Iverson used this query to title a short note calling attention to the role of herbivorous lizards as seed dispersers. In the note, he analysed the percentage of germination for seeds removed from the droppings of the iguanas *Cyclura rileyi* and *C. carinata* from the Bahamas and the Turks and Caicos Islands. An increase (although not statistically significant) in seed germination in *Coccoloba uvifera* (Polygonaceae) seeds passing through the digestive tract (43.8%), compared with control seeds (27.5%), was observed. Although without control seeds for comparison, high levels of germination (85%) were obtained for *Casasia clusiaefolia* (Rubiaceae) seeds collected from *Cyclura* faecal pellets. In this pioneer contribution, the author pointed out that lizards can be considered as legitimate seed dispersers, since seeds were not damaged and germination was enhanced in some cases, even though the average passage time through a lizard gut is around 4 days (Iverson, 1985). As far as we know, this is the first published experimental evidence showing that lizards are legitimate seed dispersers. However, more examples are necessary in order to generalize Iverson's (1985) results and to compare them with results obtained from other frugivore groups (birds and mammals).

Fortunately, in recent decades there have been some studies undertaken, and about 25 papers dealing with the effects of seed passage through lizard guts have been published (reviewed in Traveset, 1998; Traveset and Verdú, 2002; A. Valido, unpublished results). They cover 40 plant species belonging to 17 families, and 17 lizards belonging to seven families. Although the number of experiments with lizards were much fewer than similar bird (153) and mammal (119) experiments, seeds consumed by lizards germinated just as well as seeds consumed by birds and mammals. The proportion of examples where germination of the seeds passing through the digestive tract of lizards was enhanced (25%), unaffected (57%) or inhibited (18%) was similar (χ^2 test; P = 0.28) to those observed when birds (36%, 48%, 16%, respectively), non-flying mammals (39%, 42%, 19%) and bats (25%, 67%, 8%) were the frugivorous species (Traveset, 1998).

A more interesting comparison would be to know the outcome of passage through a lizard digestive tract relative to other sympatric native frugivores (birds and mammals) for the seeds of a given species of plant. Greeff and Whiting (1999) observed no significant differences in germination success of *Ficus cordata* (Moraceae) seeds collected from lizard (48%; *Platysaurus broadleyi*; Cordylidae), bird (34%; mostly *Onychognathus nabouroup* and *Pycnonotus nigricans*) and rock hyrax (29%; *Procavia capensis*) droppings and controls (48%). In other tri-species comparisons, Morales-Mávil *et al.* (2005) showed that seeds being dispersed by *Iguana iguana* had a higher percentage of germination (86.7%) than those from spider monkey, *Ateles geoffroyii* (45%), toucan, *Ramphastos sulfutarus* (33.3%) or controls (20%). In contrast, Mandujano *et al.* (1994) showed that endocarps consumed by *Ctenosaura pectinata* (Iguanidae) germinated less successfully (35%) than those consumed by white-tailed deer (*Odocoileus virginianus*; 72%) or controls (63%).

In studies comparing the germination of seeds passed through lizards and birds, similar results were obtained. For example, seeds from *Plocama pendula* and *Rubia fruticosa* (Rubiaceae) dispersed by sylviid warblers (*Sylvia* spp.), lacertid lizards (*Gallotia* spp.) and control (Valido, 1999; Nogales *et al.*, 2005) all showed a similar percentage germination. An analogous result was also found by Rodríguez-Pérez *et al.* (2005) for several Mediterranean plant species being dispersed by lizards and birds. Reanalysing their data, we found that seeds obtained from blackbird (*Turdus merula*) and *Podarcis pityusensis* (Lacertidae) droppings and controls resulted in a similar germination success for *Rhamnus alaternus* (Rhamnaceae), *Rubia peregrina* (Rubiaceae), *Rubus ulmifolius* (Rosaceae), *Solanum luteum* and *Solanum nigrum* (Solanaceae). In addition, an increase in the germination success of seeds passed through lizards was obtained for *Rhamnus ludovici-salvatoris* (Rhamnaceae) compared with seeds that had passed through birds.

Thus, according to these comparisons, a cautious preliminary answer to Iverson's question would be 'yes': lizards can be regarded legitimate seed dispersers in the same way as frugivorous birds and mammals with regard to the effects on germination of the passage of seeds through the digestive tract.

Whether an animal is a legitimate disperser is just one of the aspects to take into account when we want to know the contribution of seed dispersers to plant fitness (seed dispersal effectiveness; *sensu* Schupp, 1993). To be effective as a seed disperser, lizards must transport viable seeds beyond the parent plant, and deposit them in a site where their chances of survival to adult plants are good. Accordingly, effectiveness can be analysed quantitatively (related to number of seeds being dispersed) and qualitatively (related to the probability that a seed will recruit a new adult individual) (Schupp, 1993; Jordano and Schupp, 2000; Schupp, Chapter 20, this volume). Only two studies (both from islands: Valido, 1999; Wotton, 2002) simultaneously included several aspects of the qualitative and quantitative components of lizard disperser effectiveness. The species studied were *Gallotia galloti* (Lacertidae) and *Plocama pendula*

(Rubiaceae) from the Canary Islands (Valido, 1999), and *Hoplodactylus maculatus* (Gekkonidae) and *Croprosma propinqua* (Rubiaceae) from New Zealand (Wotton, 2002). The studies showed that these omnivorous lizards appeared to be effective seed dispersal agents because: (i) the lizards removed a considerable proportion of fruits from plants; (ii) the ingestion of seeds had no detrimental effect on germination success; and (iii) the lizards tended to deposit seeds in microhabitats where the probability of germination and seedling establishment was improved more frequently than in hostile microhabitats. These results, although only based on two examples, demonstrate that lizards may be considered to be effective seed dispersers.

Although more the study of cases is necessary before we can conclude that lizards are generally effective dispersers, additional evidence, including some of the aspects of the quantitative component of effectiveness, have been found with different species of lizards which may support this conclusion. Extensive use of fleshy fruits (high frequency of occurrence or volume; > 40%) in the diet has been found in: *Ameiva exsul* (Patterson, 1928), *Ameiva ameiva* and *Cnemidophorus lemniscatus* (Janzen, 1973), *Anolis equestris* (Brach, 1976), *Liolaemus nigromaculatus* (Ortiz and Riveros, 1976), *Podarcis* spp. (Salvador, 1986; Pérez-Mellado *et al.*, 2005), *Leciocephalus carinatus* (de Armas, 1987), *Darevskia rudis* (Franzen, 1991), *Klentropyx striatus* (Magnusson and da Silva, 1993), *Tropidurus torquatus* (Côrtes-Figueira *et al.*, 1994), *Lacerta lepida* (Hódar *et al.*, 1996), *Cnemidophorus lemniscatus* (Vitt *et al.*, 1997), *Gallotia* spp. (Valido *et al.*, 2003; Valido and Nogales, 2003), *Anolis garmani* (Herrel *et al.*, 2004) and *Psammodromus algirus* (Calviño-Cancela, 2005), among others.

Although most lizards are not as mobile as many birds and mammals, and long-distance seed dispersal is uncommon, lizards can be effective in transporting seeds sufficiently far away to avoid density-dependent mortality, which may occur below the parent plants. For example, Greeff and Whiting (1999) observed dispersal distances up 187 m in the case of seeds of *Ficus cordata* (Moraceae) being dispersed by *Platysaurus broadleyi* (Cordylidae) from South Africa. Whitaker (1968) recorded 73 m as the maximum distance movement by *Hoplodactylus duvaucelli* (Gekkonidae) in New Zealand. This author also found dispersed seeds 12 m away from the nearest fruiting plants (Whitaker, 1987). Barquín and Wildpret (1975) cited up to 50 m for seeds of *Plocama pendula* (Rubiaceae) being dispersed by *Gallotia galloti* from the nearest potential parent plant. Côrtes-Figueira *et al.* (1994) reported 6.9 m as maximum dispersal distance of *Melocactus violaceus* (Cactaceae) seeds by *Tropidurus torquatus* (Tropiduridae). Using spatial distribution of seedlings of *Daphne rodriguezii* (Thymelaeaceae) from a population where seeds of this species are only being dispersed by lizards (*Podarcis lilfordi*), Traveset and Riera (2005) estimated dispersal distances up to 4 m away from any reproductive adult. However, longer seed dispersal distances would be possible because a gut passage time of 4 days has been reported for this species (Traveset and Riera, 2005). These different seed dispersal distances may be due to differences in lizard body size, gut passage time, alternative food sources, environmental temperature, foraging patterns and social behaviour, among others.

Another important aspect to consider in relation to quality of seed dispersal is that lizards may also deposit considerable quantities of seeds in places suitable for germination and subsequent plant recruitment for the plants they are dispersing, including cracks and crevices in rocks, or sheltered places adjacent to stones (e.g. Whitaker, 1987; Valido and Nogales, 1994; Valido, 1999; Wotton, 2002; Benítez-Malvido *et al.* 2003; but also see Fialho, 1990). Generally, these areas show some common characteristics, for example they have relatively high humidity, low aridity and low predation rates. Lizards may potentially also be responsible for vegetation composition and distribution in such habitats (Valido and Nogales, 1994).

A Lizard Fruit Syndrome?

Fruit syndromes (correlated fruit traits associated with some particular group of frugivores) have been a classical topic in the frugivory literature. For example, small brightly coloured fruits persisting on plants after ripening have generally been associated with birds, while large green/brown-coloured fruits that fall to the ground after ripening are associated with mammals (e.g. Herrera, 2002). A fruit syndrome for reptiles was first proposed by Borzí (1911), possibly based on earlier observations by Beccari (1877). He proposed three tentative types of saurochorous fruits based on what was known at that time: (i) the Scioide (which included *Ficus*, *Capparis* and *Opuntia* fruits); (ii) the *Arum*; and (iii) the *Viola* types. In addition, Borzí formulated the idea that the presence of fruits on the trunk near the ground (caulicarpy) could be an adaptation to seed dispersal by reptiles. Later, van der Pijl (1969, re-edited in 1982) proposed that reptile-fruits are aromatic, coloured, and may be borne near the ground on a trunk or dropped to the ground when ripe. This description is probably based on the fact that the classical reptile-fruit known in van der Pijl's time, *Gingko biloba* (Ginkgoaceae), clearly fits this description: 'greenish drupe-like ovule-seeds with a faint smell which fall to the ground' (p. 131; van der Pijl, 1982). Nowadays, many examples are known that do not fit this syndrome; many lack odour and colours vary widely. For example, Varela and Bucher (2002) found that the odourless fruit of *Celtis pallida* (Ulmaceae) and *Capparis atamisquea* (Capparaceae) had a high frequency of occurrence (59%) in *Teious teyou* (Teiidae) droppings. Several odourless lizard-fruits are known from the Balearic Archipelago (Pérez-Mellado and Traveset, 1999) and the Canary Islands (Valido *et al.*, 2003) as well.

Recently, and using a comprehensive database of fruit characters from the New Zealand flora (n = 246 plant species), Lord and Marshall (2001) and Lord *et al.* (2002) found some support for a lizard-fruit syndrome. They suggest that lizards may be of importance in the evolution of fruit from shrub species with divaricated branches, growing in open, dry habitats. They also proposed that fruit-eating lizards favour small, elliptically shaped, white or pale-blue to sky-blue fruits. Captive *Oligosoma* lizards prefer blue and white *Coprosma* fruit to white and red ones. However, as Lord *et al.* (2002) also point out, this lizard dispersal syndrome cannot be applied to other floras, and differences can be related to the taxonomic composition of the lizard fauna. The guild of frugivorous lizards from New Zealand consists of six species of Gekkonidae and 13 species of Scincidae (Whitaker, 1987, and references therein; Patterson, 1992; Cogger *et al.*, 1993; Spencer *et al.*, 1998). Many of these lizards (e.g. the geckos *Hoplodactylus* spp.) are nocturnal, whereas practically all the *Oligosoma* skinks are diurnal. It is likely that this lizard-fruit syndrome described for New Zealand is location-specific and cannot be applied to other areas. For example, a comparison of the frequency of fruit colours in lizard diets from the Canary Islands (Fig. 6.2) and the Balearic Archipelago (Fig. 6.3) floras with the New Zealand flora (Fig. 6.4) show distinct differences. This is likely to be caused (in part) by practically all the frugivorous lizards from the Canaries and the Balearic Archipelago being diurnal lacertids.

In addition, we need to incorporate information about the evolution of Squamata feeding strategies (Vitt, 2004) and plant phylogeny (Jordano, 1995), in order to test for any association of fruit traits with frugivory in lizards. For example, the Iguania lizards use visual signals to detect prey and their tongue to capture it. Gekkota species, on the other hand, use visual and olfactory prey discrimination, and use their mouth for capture and tongue for cleaning face and eyes. Scincomorpha use visual cues and their vomeronasal organs in chemical discrimination. Visual prey detection may be the primitive state (see Fig. 1 in Vitt, 2004). This means that plants whose fruits are consumed and seeds dispersed by different lizard clades are under very different selection pressures according to the predominant cues used by different lizard taxonomic groups for food selection. With respect to plants, Jordano (1995) found that the evolutionary history of lineages accounts for a great proportion of the variation in fruit traits, and only fruit size (diameter) showed an important interspecific variability independent of phylogeny. Thus, if lizards are important selective agents on fruit traits, we may one day discuss a gecko fruit syndrome and an iguanid fruit syndrome, among others, similar to the distinction between different mammal groups, such as bat-fruit and primate-fruit syndromes.

Avenues for Future Research

Although lizard frugivory and seed dispersal are of particular relevance in some specific geographical situations (principally islands, but also xeric habitats), this topic is a slowly emerging research focus within mutualism studies in general, and seed dispersal research in particular. However, since little has been published on the topic of seed dispersal by lizards, any current interest on ecological and evolutionary plant–frugivore interactions (e.g. Herrera, 2002), including some aspects already published in this volume (e.g. digestive treatment (Traveset *et al.*, Chapter 4), spatial patterns of seed shadows (Dennis and Westcott, Chapter 9), demographic consequences (Godínez-Alvarez and Jordano, Chapter 17) and consequences of frugivore extinction (Silva *et al.*, Chapter 26)) are also relevant for studies focusing on lizards. We have shown that a frugivorous diet is more common in lizards than previously appreciated, but at the same time, many information gaps (e.g. lizard diet) have hampered our efforts to make solid generalizations, especially in relation to island species (see a compilation of the herpetological literature at http://www.herplit.com/).

Apart from this necessary basic natural history information, the next step is to compare patterns of frugivory and the process of seed dispersal by lizards to that by birds and mammals in an attempt to establish the importance of lizards for plant regeneration. Earlier we highlighted several future research directions and several key questions that have emerged from our analyses, which focus on the interactions between frugivorous lizards and plants. When more data become available from more sites, a clearer understanding of the importance of lizards as seed dispersers will emerge. Particularly important is their relevance to conservation, since lizards may be the only seed dispersers of some species of plants, for example *Withania aristata* (Solanaceae), *Neochamaelea pulverulenta* (Cneoraceae), and *Kunkeliella subsuculenta* (Santalaceae) from the Canary Islands (A. Valido, personal observation), and *Daphne rodriguezii* (Thymelaeaceae) from the Balearic Islands (Traveset and Riera, 2005), among others.

Acknowledgements

The work reported here was supported by grants from the Marie Curie European programme (grants MCFI-2000-1995 and MERG-CT-2004-510260) and postdoctoral I3P (Ministerio de Ciencia y Tecnología, Spain) to A.V., and the Danish Science Research Council (J.M.O.). Many other researchers contributed with personal observations about frugivorous lizards, including L.J. Avila, A.M. Bauer, D. Bennett, D.G. Broadley, T.A. Carlo, C.L. Celedon-Neghme, E.J. Censky, J. Davenport, A. Fong, G. García, M. Gaulke, G. Gottsberger, D.H. Hansen, J.D. Lazell, L.V. Moreno, V. Parra-Tabla, R. Powell, L. Rodríguez-Schettino, G. Sorci, R. Swain, L.J. Vitt and A.H. Whitaker. Andrew Dennis, Hector Godínez-Alvarez, Pedro Jordano and Pierre-Michael Forget offered valuable comments on a previous draft. Finally, we would like to thank the organizers of the Symposium/Workshop, Andrew Dennis, Ronda Green, Eugene Schupp and David Westcott, for the invitation to contribute to this volume.

References

Adamopoulou, C. and Legakis, A. (2002) Diet of a lacertid lizard (*Podarcis milensis*) in an insular dune ecosystem. *Israel Journal of Zoology* 48, 207–219.

Al-Johany, A. (1995) The ecology of *Agama yemenensis* Klausewitz (Lacertilia: Agamidae) in south-western Arabia. *Journal of Arid Environments* 29, 495–503.

Auffenberg, W. (1988) *Gray's Monitor Lizard.* University of Florida Press, Gainesville, Florida.

Auffenberg, W. and Auffenberg, T. (1988) Resource partitioning in a community of Philippine skinks (Sauria: Scincidae). *Bulletin of the Florida State Museum, Biological Science* 32, 151–219.

Ballinguer, R.E. and Ballinguer, R.A. (1979) Food resource utilization during periods of low and high food availability in *Sceloporus jarrovi* (Sauria: Iguanidae). *Southwestern Naturalist* 24, 347–363.

Barquín, E. and Wildpret, W. (1975) Diseminación de plantas canarias: datos iniciales. *Vieraea* 5, 38–60.

Baur, G. (1891) On the origin of the Galapagos Islands. *American Naturalist* 25, 217–229.

Bauer, A.M. and Sadlier, R.A. (2000) *The Herpetofauna of New Caledonia.* SSAR Publications, Ithaca, New York.

Barrett, S.C.H. (1998) The reproductive biology and genetics of island plants. In: Grant, P.R. (ed.) *Evolution on Islands.* Oxford University Press, Oxford, UK, pp. 18–34.

Beccari, O. (1877) Della disseminazione delle Palmae. *Bolletino della Real Societá Toscana di Orticoltura Malesia* 1, 1–16.

Belver, L.C. and Avila, L.J. (2002) Diet composition of *Liolaemus bibronii* (Iguania: Liogaemidae) in southern Rio Negro province, Argentine. *Herpetological Journal* 12, 39–42.

Benítez-Malvido, J., Tapia, E., Suazo, I., Villaseñor, E. and Alvarado, J. (2003) Germination and seed damage in tropical dry forest plants ingested by iguanas. *Journal of Herpetology* 37, 301–308.

Bennett, D. (2005) The ecology of *Varanus olivaceus* on Polillo Island and implications for other giant frugivorous lizards in the Philippines. Polillo Butaan Project: 2002–2005 [available at http://www.mampam.com/gfl].

Borzí, A. (1911) Ricerche sulla disseminazione delle piante per mezzo di Sauri. *Memorie della Societe italiana delle scienze, detta dei XL Rome ser III* 17, 97–115.

Brach, V. (1976) Habits and food of *Anolis equestris* in Florida. *Copeia* 1976, 187–189.

Búrquez, A., Flores-Villela, O. and Hernández, A. (1986) Herbivory in a small iguanid lizard, *Sceloporus torquatus torquatus. Journal of Herpetology* 20, 262–264.

Busack, S.D. and Jasick, F.M. (1982) Autoecological observations of *Acanthodactylus erythrurus* (Sauria: Lacertidae) in southern Spain. *Amphibia-Reptilia* 3, 237–255.

Calviño-Cancela, M. (2005) Fruit consumers and seed dispersers on the rare shrub *Corema album*, Empetraceae in coastal sand dunes. *Revue d'Écologie La Terre et la Vie* 60, 97–106.

Carlquist, S. (1974) *Island Biology.* Columbia University Press, New York.

Castilla, A.M. and Bawens, D. (1991) Observations on the natural history, present status, and conservation of the insular lizard *Podarcis hispanica atrata* on the Columbretes archipelago, Spain. *Biological Conservation* 58, 69–84.

Christian, K.A., Webb, J.L. and Schultz, T.J. (2003) Energetics of bluetongue lizards (*Tiliqua scincoides*). *Oecologia* 136, 515–523.

Churchill, S.K. (1994) Diet, prey selection and foraging behaviour of the orange horseshoe-bat. *Rhinonycteris aurantius. Wildlife Research* 21, 115–130.

Cobo, M. and Andreu, A.C. (1988) Seed consumption and dispersal by the spur-thighed tortoise *Testudo graeca. Oikos* 51, 267–273.

Cogger, H.G., Cameron, E.E., Sadlier, R.A. and Eggler, P. (1993) *The Action Plan for Australian Reptiles.* Australian Nature Conservation Agency. Project 124 [available at http://www.deh.gov.au/biodiversity/threatened/action/reptiles/index.html].

Colli, G.R., Péres, A.K. and Zatz, M.G. (1997) Foraging mode and reproductive seasonality in tropical lizards. *Journal of Herpetology* 31, 490–499.

Cooper, W.E. and Vitt, L.J. (2002) Distribution, extent, and evolution of plant consumption by lizards. *Journal of Zoology* 257, 487–517.

Côrtes-Figueira, J.E., Vasconcellos-Neto, J., García, M.A. and Teixeira, A.L. (1994) Saurochory in *Melocactus violaceus* (Cactaceae). *Biotropica* 26, 295–301.

Darwin, C. (1845) *The Voyage of the Beagle.* Murray, London.

Dearing, M.D. and Schall, J.J. (1992) Testing models of optimal diet assembly by the generalist herbivorous lizard *Cnemidophorus murinus. Ecology* 73, 845–858.

de Armas, L.F. (1987) Notas sobre la alimentación de *Leiocephalus carinatus cayensis* (Sauria: Iguanidae). *Poeyana* 350, 1–7.

Ebenman, B. and Nilsson, S.G. (1982) Components of niche width in a territorial bird species: habitat utilization in males and females of the chaffinch (*Fringilla coelebs*) on islands and mainland. *American Naturalist* 119, 331–344.

Echternacht, A.C., Gerber, G.P. and Burton, F.J. (2000) *Anolis conspersus* (Grand Cayman blue-throated anole). Nectivory. *Herpetological Review* 31, 173.

Espinoza, R.E., Wiens, J.J. and Tracy, C.R. (2004) Recurrent evolution of herbivory in small, cold-climate lizards: breaking the ecophysiological rules of reptilian herbivory. *Proceedings of the National Academy of Sciences* 101, 16819–16824.

Estes, R. (1983) The fossil record and early distribution of lizards. In: Rhodin, A.G.J. and Miyata, K. (eds) *Advances in Herpetology and Evolutionary Biology.* Harvard University Press, Cambridge, Massachusetts, pp. 365–398.

Estrada, A. and Fleming, T.H. (1986) *Frugivores and Seed Dispersal.* Dr. W. Junk, Dordrecht, The Netherlands.

Fialho, R.F. (1990) Seed dispersal by a lizard and a treefrog: effect of dispersal site on seed survivorship. *Biotropica* 22, 423–424.

Fleming, T.H. (1991) Fruiting plant–frugivore mutualism: the evolutionary theater and the ecological play. In: Price, P.W., Lewinsohn, T.M., Fernandes, G.W. and Benson, W.W. (eds) *Plant-Animal Interactions: Evolutionary Ecology in Tropical and Temperate Regions.* Wiley, New York, pp. 119–144.

Fleming, T.H. and Estrada, A. (1993) *Frugivory and Seed Dispersal: Ecological and Evolutionary Aspects.* Kluwer, Dordrecht, The Netherlands.

Fölling, M., Knogge, C. and Böhmes, W. (2001) Geckos are milking honeydew-producing planthoppers in Madagascar. *Journal of Natural History* 35, 279–284.

Franzen, M. (1991) Beobachtungen zur phytophagen ernährung von *Lacerta rudis* und *Lacerta clarkorum. Die Eidechse* 2, 22–23.

Fructuoso, G. (1964) *Las Islas Canarias de "Saudades de Terra".* Instituto de Estudios Canarios, S/C Tenerife.

Godínez-Alvarez, H. (2004) Pollination and seed dispersal by lizards: a review. *Revista Chilena de Historia Natural* 77, 569–577.

Greeff, J.M. and Whiting, M.J. (1999) Dispersal of Namaqua fig (*Ficus cordata cordata*) seeds by the Augrabies flat lizard (*Platysaurus broadleyi*). *Journal of Herpetology* 33, 328–330.

Greene, H.W. (1982) Dietary and phenotypic diversity in lizards: why are some organisms specialized? In: Mossakowski, D. and Roth, G. (eds) *Environmental Adaptation and Evolution.* Gustav Fischer, Stuttgart, Germany, pp. 107–128.

Greer, A. (2005) *Encyclopedia of Australian Reptiles.* Australian Museum online [http://www.amonline.net.au/herpetology/research/encyclopedia.pdf, accessed 19 March 2005].

Griffiths, A.D. and Christian, K.A. (1996) Diet and habitat use of frillneck lizards in a seasonal tropical environment. *Oecologia* 106, 39–48.

Hartley, L.M., Glor, R.E., Sproston, A.L., Powell, R. and Parmer-Lee, J.S. (2000) Germination rates of seeds consumed by two species of rock iguanas (*Cyclura* spp.) in the Dominican Republic. *Caribbean Journal of Science* 36, 149–151.

Herrel, A., Vanhooydonck, B. and Joachim, R. (2004) Frugivory in polychrotid lizards: effects of body size. *Oecologia* 140, 160–168.

Herrera, C.M. (2002) Seed dispersal by vertebrates. In: Herrera, C.M. and Pellmyr, O. (eds) *Plant–Animal Interactions: An Evolutionary Approach.* Blackwell Publishing, Oxford, UK, pp. 185–208.

Hnatiuk, S.H. (1978) Plant dispersal by the Aldabran giant tortoise *Geochelone gigantea* (Schweigger). *Oecologia* 36, 345–350.

Hódar, J.A., Campos, F. and Rosales, B.A. (1996) Trophic ecology of the ocellated lizard *Lacerta lepida* in an arid zone of southern Spain: relationships with availability and daily activity of prey. *Journal of Arid Environments* 33, 95–107.

Howe, H.F. (1986) Seed dispersal by fruit-eating birds and mammals. In: Murray, D.R. (ed.) *Seed Dispersal.* Academic Press, New York, pp. 123–189.

Howe, H.F. and Smallwood, J. (1982) Ecology and seed dispersal. *Annual Review of Ecology and Systematics* 13, 201–228.

Howe, H.F. and Westley, L.C. (1988) *Ecological Relationships of Plants and Animals.* Oxford University Press, New York and Oxford.

Iverson, J.B. (1982) Adaptations to herbivory in iguanine lizards. In: Burghardt, G.M. and Rand, A.S. (eds) *Iguanas of the World*, Noyes Publishers, Park Ridge, New Jersey, pp. 60–76.

Iverson, J.B. (1985) Lizards as seed dispersers? *Journal of Herpetology* 19, 292–293.

Janzen, D.H. (1973) Sweep samples of tropical foliage insects: effects of seasons, vegetation types, elevation, time of day, and insularity. *Ecology* 54, 687–708.

Janzen, D.H. (1982) Natural history of guacimo fruits (Sterculiaceae: *Guazuma ulmifolia*) with respect to consumption by large mammals. *American Journal of Botany* 69, 1240–1250.

Janzen, D.H. (1983) Dispersal of seeds by vertebrates guts. In: Futuyma, D.J. and Slatkin, M. (eds) *Coevolution.* Sinauer, Sunderland, Massachusetts, pp. 232–262.

Jerozolimski, A. (2003) Dispersao de sementes por jabutis, *Geochelone denticulata* e *G. carbonaria*, na Amazònia oriental. *VI Congresso de Ecologia do Brasil*, Fortaleza, pp. 29–30.

Jordano, P. (1995) Angiosperm fleshy fruits and seed dispersers: a comparative analysis of adaptation and constraints in plant–animal interactions. *American Naturalist* 145, 163–191.

Jordano, P. (2000) Fruits and frugivory. In: Fenner, M. (ed.) *Seeds: The Ecology of Regeneration in Plant Communities.* CAB International, Wallingford, UK, pp. 125–165.

Jordano, P. and Schupp, E.W. (2000) Determinants of seed disperser effectiveness: the quantity component and patterns of seed rain for *Prunus mahaleb. Ecological Monographs* 70, 591–615.

Kiefer, M.C. and Sazima, I. (2002) Diet of juvenile tegu lizard *Tupinambis merianae* (Teiidae) in southeastern Brazil. *Amphibia-Reptilia* 23, 105–108.

King, G. (1996) *Reptiles and Herbivory.* Chapman and Hall, London.

Klein, E.H. (1977) Una contribución a los hábitos alimenticios del garrobo gris (*Ctenosaura similis*) en el sur de Honduras. *Ceiba* 21, 41–50.

Kuchling, G. (1970) Beobachtungen an *Eumeces schneideri. Die Aquarien- und Terrarienzeitschrift* 23, 216–217.

Labandeira, C.C. (2002) The history of associations between plants and animals. In: Herrera, C.M. and Pellmyr, O. (eds) *Plant–Animal Interactions: An Evolutionary Approach.* Blackwell Publishing, Oxford, UK, pp. 26–74.

Larco, R. (2001) *Los Mochicas II.* Museo Arqueológico Rafael Larco Herrera y Fundación Telefónica, Lima.

Leguat (1981) *The Voyage of Francois Leguat of Bresse to Rodriguez, Mauritius, Java, and the Cape of Good Hope. Vols I and II.* Burt Franklin Publisher, New York.

Levey, D.J., Silva, W.R. and Galetti, M. (2002) *Seed Dispersal and Frugivory: Ecology, Evolution and Conservation.* CAB International, Wallingford, UK.

Liénard, F. (1842) Description de deux, espèces de Geckos de l'ile Rodrigues. *Rapport annuel sur les travaux de la Société d'Histoire Naturelle de l'île Maurice* 13, 55–57.

Liu, H., Platt, S.G. and Borg, C.K. (2004) Seed dispersal by the Florida box turtle (*Terrapene carolina bauri*) in pine rockland forests of the lower Florida Keys, USA. *Oecologia* 138, 539–546.

Lord, J.M. and Marshall, J. (2001) Correlations between growth form, habitat, and fruit colour in the New Zealand flora, with reference to frugivory by lizards. *New Zealand Journal of Botany* 39, 567–576.

Lord, J.M., Markey, A.S. and Marshall, J. (2002) Have frugivores influenced the evolution of fruits traits in New Zealand? In: Levey, D., Silva, W.R. and Galetti, M. (eds) *Seed Dispersal and Frugivory: Ecology, Evolution and Conservation.* CAB International, Wallingford, UK.

Mabberley, D.J. (1997) *The Plant-Book: A Portable Dictionary of the Vascular Plants.* Cambridge University Press, Cambridge, UK.

MacArthur, R.H., Diamond, J.M. and Karr, J.R. (1972) Density compensation in island faunas. *Ecology* 53, 330–342.

MacDonald, L.A. and Mushinsky, H.R. (1988) Foraging ecology of the gopher tortoise, *Gopherus polyphemus*, in a sandhill habitat. *Herpetologica* 44, 345–353.

Magnusson, W.E. and da Silva, E.V. (1993) Relative effects of size, season and species on the diets of some Amazonian savanna lizards. *Journal of Herpetology* 27, 380–385.

Mandujano, S., Gallina, S. and Bullock, S.H. (1994) Frugivory and dispersal of *Spondias purpurea* (Anacardiaceae) in a tropical deciduous forest in México. *Revista de Biología Tropical* 42, 107–114.

Mautz, W.J. and Lopez-Forment, W. (1978) Observations on the activity and diet of the cavernicolous lizard *Lepidophyma smithii* (Sauria: Xantusiidae). *Herpetologica* 34, 311–313.

Mellado, J., Amores, F., Parreño, F.F. and Hiraldo, F. (1975) The structure of a Mediterranean lizard community. *Doñana, Acta Vertebrata* 2, 145–160.

Mercolly, C. and Yanosky, A. (1994) The diet of adult *Tupinambis teguixin* (Sauria, Teiidae) in the eastern Chaco of Argentina. *Herpetological Journal* 4, 15–19.

Milton, S.J. (1992) Plants eaten and dispersed by adult leopard tortoises *Geochelone pardalis* (Reptilia: Chelonii) in the southern Karoo. *South African Journal of Zoology* 27, 45–49.

Moll, D. and Jansen, K.P. (1995) Evidence for a role in seed dispersal by two tropical herbivorous turtles. *Biotropica* 27, 121–127.

Morales-Mávil, J.E., Sánchez-Marín, M.L. and Domínguez-Domínguez, L.E. (2005) Comparación de la germinación de semilllas de *Spondias mombin* ingeridas por la Iguana verde (*Iguana iguana*), el Tucán (*Ramphastos sulfuratus*) y el Mono araña (*Ateles geoffroyii*). In: Reynoso, V.H. and Madinah, W. (eds) *VII Reunión Nacional sobre Iguanas en México.* IBUNAM, Michoacán, Mexico, pp. 93–99.

Murphy, T.J. and Myers, A.A. (1996) The behavioural ecology of *Phelsuma astriata semicarinata* on Aridae Island Nature Reserve, Seychelles. *Journal of Herpetology* 30, 117–123.

Nogales, M., Nieves, C., Illera, J.C., Padilla, D.P. and Traveset, A. (2005) Effect of native and alien vertebrate frugivores on seed viability and germination patterns of *Rubia fruticosa* (Rubicaceae) in the eastern Canary Islands. *Funtional Ecology* 19, 429–436.

O'Grady, S.P., Morando, M., Avila, L. and Dearing, M.D. (2005) Correlating diet and digestive tract specialization: examples from the lizard family Liolaemidae. *Zoology* 108, 201–210.
Olesen, J.M. and Jordano, P. (2002) Geographic patterns in plant–pollinator mutualistic networks. *Ecology* 83, 2416–2424.
Olesen, J.M. and Valido, A. (2003) Lizards as pollinators and seed dispersers: an island phenomenon. *Trends in Ecology and Evolution* 18, 177–181.
Olesen, J.M. and Valido, A. (2004) Lizards and birds as generalized pollinators and seed dispersers of island plants. In: Fernández-Palacios, J.M. and Morici, C. (eds) *Island Ecology.* AEET/ Cabildo Insular de La Palma, pp. 229–249.
Ortiz, J.C. (1975) Una nueva especie de lagartija chilena, *Liolaemus donosoi* sp. nov. (Squamata-Iguanidae). *Anales del Museo de Historia Natural del Valparaíso* 8, 62–65.
Ortiz, J.C. and Riveros, J.M. (1976) Hábitos alimenticios de *Liolaemus nigromaculatus kuhlmani*, Müller y Hellmich. *Anales del Museo de Historia Natural Valparaíso* 9, 131–140.
Patterson, G.B. (1992) The ecology of a New Zealand grassland lizard guild. *Journal of the Royal Society of New Zealand* 22, 91–106.
Patterson, K. (1928) *Amphibians and Land Reptiles of Porto Rico, with a list of those reported from the Virgin Islands: Scientific Survey of Porto Rico and the Virgin Islands.* Volume X, Part I. New York Academy of Sciences, New York.
Pearson, O. (1954) Habits of the lizard *Liolaemus multiformis multiformis* at high altitudes in southern Peru. *Copeia* 2, 111–116.
Péres, A.K. (2003) Sistemática e Conservação de lagartos do gênero *Tupinambis* (Squamata, Teiidae). PhD thesis, University of Brasília, Brazil.
Pérez-Mellado, V. and Corti, C. (1993) Dietary adaptations and herbivory in lacertid lizards on the genus *Podarcis* from western Mediterranean islands (Reptilia: Sauria). *Bonner Zoologische Beiträge* 44, 193–220.
Pérez-Mellado, V. and Traveset, A. (1999) Relationships between plants and Mediterranean lizards. *Natura Croatica* 8, 275–285.
Pérez-Mellado, V., Riera, N., Perera, A. and Martín-García, S. (2005) The lizard, *Podarcis lilfordi* (Squamata: Lacertidae) as a seed disperser of the Mediterranean plant, *Phillyrea media* (Oleaceae). *Amphibia-Reptilia* 26, 105–108.
Pietruszka, R.D., Hanrahan, S.A., Mitchell, D. and Seely, M.K. (1986) Lizard herbivory in a sand dune environment: the diet of *Angolosaurus skoogi. Oecologia* 70, 587–591.
Racine, C.H. and Downhower, J.F. (1974) Vegetative and reproductive strategies of *Opuntia* (Cactaceae) in the Galapagos Islands. *Biotropica* 6, 175–186.
Rand, A.S. (1978) Reptilian arboreal folivores. In: Montgomery, G.G. (ed.) *The Ecology of Arboreal Folivores.* Smithsonian Institution Press, Washington, DC.
Rick, C.M. and Bowman, R.I. (1961) Galápagos tomatoes and tortoises. *Evolution* 15, 407–417.
Ridley, H.N. (1930) *The Dispersal of Plants Throughout the World.* Lovell Reeve & Co., Ashford, UK.
Rodda, G.H. and Dean-Bradley, K. (2002) Excess density compensation of island herpetofauna assemblages. *Journal of Biogeography* 29, 623–632.
Rodda, G.H., Perry, G., Rondeau, R.J. and Lazell, J. (2001) The densest terrestrial vertebrate. *Journal of Tropical Ecology* 17, 331–338.
Rodríguez-Pérez, J., Riera, N. and Traveset, A. (2005) Effect of seed passage through birds and lizards on emergence rate of Mediterranean species: differences between natural and controlled conditions. *Functional Ecology* 19, 699–706.
Rogner, M. (1997) *Lizards.* Vols I and II. Krieger Publishing Company, Malabar, Florida.
Rust, R.W. and Roth, R.R. (1981) Seed production and seedling establishment in the mayapple, *Podophyllum peltatum. American Midland Naturalist* 105, 51–60.
Salvador, A. (1986) *Podarcis pityusensis* (Boscá, 1883) – Pityusen-Eidechse. In: Böhme, W. (ed.) *Handbuch der Reptilien und Amphibien Europas: Echsen III (Podarcis).* Aula-Verlag Wiesbaden, Wiesbaden, Germany, pp. 231–253.

Schleich, H.H., Kästle, W. and Kabisch, K. (1996) *Amphibians and Reptiles of North Africa.* Koeltz Scientific, Koenigstein, Germany.

Schoener, T.W., Slade, J.B. and Stinson, C.H. (1982) Diet and sexual dimorphism in the very catholic lizard genus *Leiocephalus* of the Bahamas. *Oecologia* 53, 160–169.

Schupp, E.W. (1993) Quantity, quality and the effectiveness of seed dispersal by animals. *Vegetatio* 107/108, 15–29.

Schwartz, A. and Henderson, R.W. (1991) *Amphibians and Reptiles of the West Indies: Descriptions, Distributions, and Natural History.* University of Florida Press, Gainesville, Florida.

Spencer, N.J., Thomas, B.W., Mason, R.F. and Dugdale, J.S. (1998) Diet and life history variation in the sympatric lizards *Oligosoma nigriplantare polychronma* and *Oligosoma lineoocellatum. New Zealand Journal of Zoology* 25, 457–463.

Steindacher, F. (1891) Üeber die Reptilien und batrachier der westl. und östl. Gruppe der Canarischen Inseln. *Annalen des k. k. naturhistorischen Hofmuseums* 6, 287–306.

Stiles, E.W. (1989) Fruits, seeds, and dispersal agents. In: Abrahamson, W.G. (ed.) *Plant–Animal Interactions.* McGraw-Hill, New York, pp. 87–122.

Struck, U., Altenbach, A.V., Gaulke, M. and Glaw, F. (2002) Tracing the diet of the monitor lizard *Varanus mabitang* by stable isotope analyses ($\delta^{15}N$, $\delta^{13}C$). *Naturwissenschaften* 89, 470–473.

Symon, D.E. (1979) Fruit diversity and dispersal in *Solanum* in Australia. *Journal of the Adelaide Botanical Gardens* 16, 321–331.

Tiffney, B.H. (2004) Vertebrate dispersal of seed plants through time. *Annual Review of Ecology and Systematic* 35, 1–28.

Traveset, A. (1990) *Ctenosaura similis* Gray (Iguanidae) as a seed disperser in a Central American deciduous forest. *American Midland Naturalist* 123, 402–404.

Traveset, A. (1998) Effect of seed passage through vertebrate frugivores' guts on germination: a review. *Perspectives in Plant Ecology, Evolution and Systematics* 1/2, 151–190.

Traveset, A. and Riera, N. (2005) Disruption of a plant–lizard seed dispersal system and its ecological effects on a threatened endemic plant in the Balearic Islands. *Conservation Biology* 19, 421–431.

Traveset, A. and Sáez, E. (1997) Pollination of *Euphorbia dendroides* by lizards and insects: spatio-temporal variation in patterns of flower visitation. *Oecologia* 111, 241–248.

Traveset, A. and Verdú, M. (2002) A meta-analysis of the effect of gut treatment on seed germination. In: Levey, D., Silva, W.R. and Galetti, M. (eds) *Seed Dispersal and Frugivory: Ecology, Evolution and Conservation.* CAB International, Wallingford, UK, pp. 339–350.

Troyer, K. (1984) Diet selection and digestion in *Iguana iguana*: the importance of age and nutrient requirements. *Oecologia* 61, 201–207.

Uetz, P. (2005) *The EMBL Reptile Database* (CD, March 2005).

Ussher, G.T. (1999) Tuatara (*Sphenodon punctatus*) feeding ecology in the presence of kiore (*Rattus exulans*). *New Zealand Journal of Zoology* 26, 117–125.

Valido, A. (1999) Ecología de la dispersión de semillas por los lagartos endémicos Canarios (g. *Gallotia,* Lacertidae). PhD thesis, University of La Laguna, Tenerife, Canary Islands.

Valido, A. and Nogales, M. (1994) Frugivory and seed dispersal by the lizard *Gallotia galloti* (Lacertidae) in a xeric habitat of the Canary Islands. *Oikos* 70, 403–411.

Valido, A. and Nogales, M. (2003) Digestive ecology of two omnivorous Canarian lizard species (*Gallotia*, Lacertidae). *Amphibia-Reptilia* 24, 331–344.

Valido, A., Nogales, M. and Medina, F.M. (2003) Fleshy fruits in the diet of Canarian lizard *Gallotia galloti* (Lacertidae) in a xeric habitat of the island of Tenerife. *Journal of Herpetology* 37, 741–747.

van Damme, R. (1999) Evolution of herbivory in lacertid lizards: effects of insularity and body size. *Journal of Herpetology* 33, 663–647.

van der Pijl, L. (1982) *Principles of Dispersal of Higher Plants.* Springer, Berlin, Heidelberg, and New York.

Varela, R.O. and Bucher, E.H. (2002) The lizard *Teius teyou* (Squamata: Teiidae) as a legitimate seed disperser in the dry Chaco Forest of Argentina. *Studies on Neotropical Fauna and Environment* 37, 115–117.

Vega, L.E., Chani, J.M. and Trivi, M. (1988) Observations on the feeding habits of *Ameiva ameiva. Herpetological Review* 19, 53–54.

Vitt, L.J. (2004) Shifting paradigms: herbivory and body size in lizards. *Proceedings of the National Academy of Sciences* 101, 16713–16714.

Vitt, L.J. and Carvalho, C.M. (1995) Niche partitioning in a tropical wet season: lizards in the Lavrado area of Northern Brazil. *Copeia* 1995, 305–329.

Vitt, L.J. and Pianka, E.R. (2005) Deep history impacts present-day ecology and biodiversity. *Proceedings of the National Academy of Sciences* 102, 7877–7881.

Vitt, L.J., Zani, P.A., Caldwell, J.P., Araújo, M.C. and Magnusson, W.E. (1997) Ecology of whiptail lizards (*Cnemidophorus*) in the Amazon region of Brazil. *Copeia* 1997, 745–757.

Walls, G.Y. (1981) Feeding ecology of the tuatara (*Sphenodon punctatus*) on Stephens Island, Cook Strait. *New Zealand Journal of Ecology* 4, 89–97.

Wells, H. and King, J.L. (1980) A general 'exact test' for N $\times$ M contingency tables. *Bulletin of the Southern California Academy of Science* 79, 65–77.

Weiser, P.J., Weiser, J.N. and Robres, L. (1975) Notes on cactus flower predation by lizards in the Atacama desert. *Aloe* 13, 117–118.

Whitaker, A.H. (1968) The lizards of the Poor Knights Islands, New Zealand. *New Zealand Journal of Science* 11, 623–651.

Whitaker, A.H. (1987) The roles of lizards in New Zealand plant reproductive strategies. *New Zealand Journal of Botany* 25, 315–328.

Willson, M.F., Sabag, C., Figueroa, J., Armesto, J.J., and Caviedes, M. (1996) Seed dispersal by lizards in Chilean rainforest. *Revista Chilena de Historia Natural* 69, 339–342.

Wotton, D.M. (2002) Effectiveness of the common gecko (*Hoplodactylus maculatus*) as a seed disperser on Mana Island, New Zealand. *New Zealand Journal of Botany* 40, 639–647.

Yasuda, M., Miura, S., Ishii, N., Okuda, T. and Hussein, N.A. (2005) Fallen fruits and terrestrial vertebrate frugivores: a case study in a lowland tropical rainforest in peninsular Malaysia. In: Forget, P.-M., Lambert, J.E, Hulme, P.E. and Vander Wall, S.B. (eds) *Seed Fate: Predation, Dispersal and Seedling Establishment.* CAB International, Wallingford, UK, pp. 151–174.

7 Fleshy-fruited Plants and Frugivores in Desert Ecosystems

J.L. Bronstein, I. Izhaki, R. Nathan, J.J. Tewksbury, O. Spiegel, A. Lotan and O. Altstein

Introduction

Fleshy-fruited species dominate the flora in many ecosystems. However, they make up a remarkably small proportion of taxa in desert ecosystems (Willson *et al.*, 1989; van Rheede van Oudtshoorn and van Rooyen, 1999). Among woody plants, for example, species adapted for seed dispersal by frugivorous vertebrates represent 35–42% of the species in temperate forests, 23–56% in Mediterranean scrublands, 69–75% in humid forests, and 90% in tropical rain forests (Jordano, 2000). In contrast, in desert ecosystems, they constitute 15% of the flora in South Africa (van Rooyen *et al.*, 1990), 7% in central Australia (Jurado *et al.*, 1991), and perhaps only 2% in Israel (Ellner and Shmida, 1981).

A lack of dispersal adaptations (and the presence of adaptations to reduce or prevent dispersal) in desert plants has been documented and discussed extensively (Ellner and Shmida, 1981; van Rheede van Oudtshoorn and van Rooyen, 1999). But there is surprisingly little ecological information about the fleshy-fruited plants that do, in fact, occur in arid environments. One can easily imagine that the ecological importance of these species will belie their taxonomic rarity, because they serve as one of the few reliable sources of water for desert animals. This observation also suggests that fleshy-fruited desert plants should be particularly important targets for conservation efforts.

In this chapter, we provide a global perspective on fleshy-fruited plants, their consumers, and their dispersal biology in desert environments. To our knowledge, this is the first review of this topic (but see van Rheede van Oudtshoorn and van Rooyen, 1999). The chapter is organized as follows:

- *First* we review the current state of knowledge on desert fruit–frugivore interactions from both the plant and animal perspective. We discuss why frugivory might be rare in deserts, and point to plant and animal taxa that are commonly associated in these habitats. Note that we largely exclude ant dispersal from this review, as ant-dispersed propagules are not fleshy-fruited. Myrmecochory is both common and of great ecological significance in desert ecosystems (see Beattie, 1985; Westoby *et al.*, 1991).

- *Second*, we consider whether fruit consumption results in seed dispersal, a question that has been considerably less thoroughly investigated in deserts than elsewhere.
- *Third*, we explore the extent to which frugivory can influence spatial patterning of vegetation in desert systems. We review evidence from both observational and experimental studies in an array of habitats; we also provide new data on the distribution of perennial and annual vegetation in the Sonoran Desert as a function of fruit type.
- *Two detailed case studies* of interactions between desert shrubs and fruit-consuming birds are presented in the next section, one involving the Middle Eastern species *Ochradenus baccatus* (Resedaceae), and the other the North American chilli, *Capsicum annuum* (Solanaceae). In both systems, patterns of frugivory and frugivore behaviour appear to be tightly linked to the spatial structure of the fruiting plants, picking up on themes raised in the previous section.
- *The final section* suggests avenues for further research in desert fruit–frugivore systems. We argue that the relative simplicity of desert ecosystems provides unique opportunities to gain insight into major unanswered questions about seed dispersal ecology.

Desert Fruit–Frugivore Interactions: a Review

How common are fleshy-fruited plants in desert ecosystems?

Fleshy-fruited species clearly make up a very small percentage of taxa in deserts (see Introduction) but there has been surprisingly little discussion of why this is the case (but see Willson *et al.*, 1989, 1990; French and Westoby, 1992; Hughes *et al.*, 1994; van Rheede van Oudtshoorn and van Rooyen, 1999). Nor is it known whether individuals of fleshy-fruited species are rare within arid habitats; it is certainly the case in some systems that their diversity is low, but their absolute abundance is not (see Case studies: *Capsicum annuum*).

Willson *et al.* (1990) found that biotic dispersal generally decreases with decreasing vertical diversity of vegetation structure and with decreasing soil fertility. In addition, Willson *et al.* (1989) presented data showing a negative correlation between the proportion of fleshy-fruited species and moisture availability across Australian and New Zealand plant communities, and provided some evidence that the pattern is global. They considered an array of factors that might explain this pattern. First, precipitation may affect the capacity of plants to construct fleshy fruits, by limiting either soil moisture or soil nutrients (Milewski, 1986). The largest fleshy fruits in deserts are borne by succulents, but even these fruits have been noted to be larger and have a higher water content in relatively wetter environments (de la Barrera and Nobel, 2004). Certain non-succulent shrubs do bear large crops of small fleshy fruits; intriguingly, the ripening periods of these species are often greatly prolonged. Second, frugivorous vertebrates may be less effective dispersers in some habitats. Ellner and Shmida (1981) have argued that selection to protect seeds from destruction by desert seed predators may have restricted the evolution of biotic dispersal mechanisms in desert plants. Finally, phylogenetic, historical, and geographical constraints may be responsible for between-habitat differences (e.g. Moles *et al.*, 2005). Some habitats, for example, may be particularly low in species of vertebrate frugivores (see, e.g. Janzen, 1980; French and Westoby, 1992). Arid habitats, however, are not among these (Janzen, 1986).

The dispersal of cacti by frugivorous vertebrates has attracted the greatest attention (e.g. Janzen, 1986; Bregman, 1988; Silvius, 1995; Montiel and Montana, 2000; Godínez-Alvarez *et al.*, 2002). Other relatively well-studied desert plant–frugivore interactions involve acacias (Davidson and Morton, 1984) and mistletoes (Larson, 1996; Aukema and Martinez del Rio, 2002), as well as *Capsicum* and *Ochradenus*, the genera that we highlight later in this chapter (see also Janzen, 1986; Tester *et al.*, 1987; Dean *et al.*, 1999). Although few in number, some of these species can occur in large populations; they can even dominate large proportions of the desert ecosystems in which they occur. For example, *O. baccatus* (see below) occupies many parts of the deserts of Israel. It is often the most abundant perennial species, with the largest ground cover. *Opuntia* (prickly pear; Cactaceae) cacti form large, dense stands in parts of arid Mexico (Janzen, 1986), and *C. annuum* (see below) can be found under virtually every shrub in some areas in Mexico (Perramond, 2005). Such abundance is one reason why the ecological importance of these few plant species with fleshy fruits is potentially very high in desert communities.

Who consumes fleshy fruits in desert ecosystems?

The ecological importance of fleshy fruits in desert habitats is also suggested by the great diversity of the animals that consume them, as well as by the importance of fruit in some of their diets.

Birds are probably the best-studied desert frugivores. Mistletoe–bird interactions, which are rather specialized among seed dispersal mutualisms, have been studied in both Australian and North-American deserts (Reid, 1989; Murphy *et al.*, 1993; Larson, 1996; Aukema and Martinez del Rio, 2002). Birds are also well-documented consumers of many North-American cactus fruits. Examples of cacti (Cactaceae) whose fruit are fed upon by birds include *Opuntia rastrera* in Mexico (Montiel and Montana, 2000), *Stenocereus griseus* (cardon) in Venezuela (Silvius, 1995) and *Carnegiea gigantea* (saguaro) in Arizona (Wolf and Martinez del Rio, 2000, 2003).

Mammalian frugivores common in desert habitats include ungulates, canids, rodents and bats (Janzen, 1986; Bregman, 1988; Fleming and Sosa, 1994; Campos and Ojeda, 1997; Montiel and Montana, 2000; Milton and Dean, 2001; Geffen *et al.*, 2004), but only the latter two groups have yet been studied in any depth. Fleming and Sosa (1994) report that both volant and terrestrial mammals are effective dispersers of three columnar cacti in the Sonoran Desert. *Leptonycteris* (Phyllostomidae) bats in these habitats remove 10–80% of the seeds available in open fruits at night, subsequently defecating many of the seeds under their day roosts. More recently, a comparison between bats and birds in the desert of Tehuacan Valley, Mexico, revealed that bats, particularly *Leptonycteris curasoae*, had significantly higher effectiveness than birds in dispersing the cactus *Neobuxbaumia tetetzo* (Godínez-Alvarez *et al.*, 2002). In other cases, mammals and birds can be equally effective (see Naranjo *et al.*, 2003).

Lizards and tortoises have also been reported to consume certain fruits in arid habitats (Racine and Downhower, 1974; Janzen, 1986; Montiel and Montana, 2000; Godínez-Alvarez, 2004; Valido and Olesen, Chapter 6, this volume). The importance of fruits in their diets is poorly known. However, Milton (1992) reported that, in the southern Karoo of South Africa, leopard tortoises (*Geochelone pardalis*; Testudinidae) consumed at least 75 species of grasses, succulents, and fruits belonging to 26 plant families; flowers, fruits and seeds made up 67% of identified plant fragments in their faeces, and defecated seeds of seven plant families were viable. The Augrabies flat lizard (*Platysaurus broadleyi*; Cordylidae) in the xeric Northern Cape Province in South Africa, occurs sympatrically with the Namaqua fig (*Ficus cordata cordata*; Moraceae). The lizards form large congregations under fruiting trees, where they consume the energy-rich fruits (Whiting and Greeff, 1997).

The disperser assemblage of a given desert plant species commonly cuts across taxonomic groups. Despite growing awareness of the need to look beyond the effects of a single standard dispersal agent (Higgins *et al.*, 2003; Forget *et al.*, 2004; Vander Wall and Longland, 2004; Levey *et al.*, Chapter 2, this volume; Dennis and Westcott, Chapter 9, this volume; Nathan, Chapter 11, this volume), there is still a paucity of studies on fleshy-fruited plants dispersed by multiple vectors in general, and in desert ecosystems in particular. Multiple vectors may be common during primary dispersal (Chambers and MacMahon, 1994), as seeds are often removed and transported by herbivores, insectivores or abiotic dispersal vectors such as winds or flash-floods, as well as by legitimate frugivores (Howe and Smallwood, 1982; Janzen, 1984; Milton and Dean, 2001; Chang *et al.*, 2005). Multiple vectors are also likely to be common during secondary dispersal (Levey and Byrne, 1993; Chambers and MacMahon, 1994; Vulinec, 2002). One of the best examples focuses on the tree *Prosopis flexuosa* (Fabaceae) of the central Monte Desert of Argentina. This species is dispersed mostly by frugivorous mammals such as maras (*Dolichotis patagonum*; Caviidae), foxes (*Vulpes* spp., Canidae) and armadillos (Dasypodidae) (Villagra *et al.*, 2002), yet three species of leafcutter ants quickly remove seeds after the primary dispersal season and occasionally may place seeds in higher quality sites near their nest or along their foraging paths (Milesi and De Casenave, 2004).

What is the significance of fruits to desert animals?

Animals inhabiting deserts are often faced with dry environmental conditions, as well as scanty and unpredictable food supplies. This suggests that natural selection may have sculpted phenotypes within desert environments to reduce their evaporative water losses (Schmidt-Nielsen, 1964). Indeed, desert birds and mammals have a significantly lower basal metabolic rate (BMR) and total evaporative water loss (TEWL) than species from more mesic environments (Williams and Tieleman, 2002). Yet, because animals living in deserts often do not have access to drinking water, they must rely on preformed water in the diet and metabolic water to supply their needs (Williams and Tieleman, 2002). Thus, succulent fruits are important water sources for many desert animals. Furthermore, fruits probably represent a crucial source of energy and nutrients for maintenance, breeding and migration in many desert animals.

The saguaro cactus (*Carnegiea gigantea*) is a resource of unparalleled importance to the bird community during the summer in the Sonoran Desert of North America, leading Steenbergh and Lowe (1977) to assert that saguaros are ‘keystone’ elements of the Sonoran Desert flora. This is because saguaros represent a crucial source of energy and water for many desert animals, since they produce fruit during the hottest and driest periods and their reproductive output is predictable from year to year (Wolf and Martinez del Rio, 2000). Wolf and Martinez del Rio (2003) reviewed the importance of saguaros for the community of birds that inhabit the Sonoran Desert during the summer, concluding that the resources of the saguaro span dietary guilds and account for a large proportion of the diet of both granivorous and frugivorous birds. Concurrent measurements of ^{13}C and deuterium have revealed that saguaro fruit provide nutrients to many species, and nutrients and water to a subset of these species (Wolf and Martinez del Rio, 2003). Wolf *et al.* (2002) report that, over a given summer, saguaro fruit comprised 50% and 14% of the diet of white-winged doves (*Zenaida asiatica*; Columbidae) and mourning doves (*Zenaida macroura*; Columbidae), respectively; saguaro fruit are an important source of both nutrients and water to the former species, but a major source of nutrients only in the latter. A similar phenomenon has been observed with fruits of the desert mistletoe (*Phoradendron californicum*; Viscaceae), which are often the only succulent fruits available in certain parts of the Sonoran Desert during the winter. These fruits are eaten by many specialists and opportunists. The breeding season of the specialist bird *Phainopepla nitens* (Bombycillidae) coincides with mistletoe's fruiting season; if the fruit crop is insufficient, the birds fail to breed (Walsberg, 1977; Anderson and Ohmart, 1978).

The reliance upon fruits shifts seasonally with the annual cycles of both animals and plants. For example, the Arabian babbler (*Turdoides squamiceps*; Turdoideae) in the Negev Desert in Israel consumes 79% insects and 21% fruit (calculated on dry matter) in the summer, but 35% insects and 65% fruit in winter (Anava *et al.*, 2000). In the Sahel Desert in Africa, fleshy fruits are very important for migrants prior to the spring migration, but they do not appear to be important in midwinter (Jones *et al.*, 1996). In contrast, in the Neotropics, fruits seem to be critical food items for wintering birds. Grey vireos (*Vireo vicinior*; Vireonidae) undergo a dietary shift from being primarily insectivorous on the breeding grounds to relying heavily on elephant tree (*Bursera microphylla*; Burseraceae) fruits during winter. Furthermore, the wintering range of grey vireos in Sonora, Mexico is strongly associated with the distribution of elephant trees in lowland desert scrub areas (Bates, 1992a; Russell and Monson, 1998). In winter, both males and females of several vireo species vigorously defend territories dominated by fruiting trees (Bates, 1992b; Greenberg *et al.*, 1993, 1995).

For many migrants, wide deserts and seas are hostile environments, and only pre-migratory deposition of large fuel stores enables them to succeed during long flights across deserts or seas (Moreau, 1972). In both seasons, fat accumulation is associated with seasonal shifts in food selection, food intake and the efficiency of food use (Bairlein and Simons, 1995). Seasonal frugivory is the most striking shift in diet composition for many migrants (Berthold, 1976; Greenberg, 1981; Herrera, 1984; Izhaki and Safriel, 1985). Due to their relatively high content of some fatty acids and carbohydrates, fruits provide high nutritional quality to promote fat accumulation (Bairlein and Simons, 1995). Furthermore, switching to a diet dominated by fruit is also associated with decreased catabolism of endogenous body protein (Gannes, 2001). The importance of fruit for fat deposition in pre-migratory birds, particularly on desert and sea edges just before long, non-stop flights, has been demonstrated in both autumn and spring. Many autumn migrants stop over in habitats that are rich in fruit, such as *Pistacia atlantica* (Anacardiaceae) groves (Sapir *et al.*, 2004) and scrublands (e.g. Izhaki and Safriel, 1985), to take advantage of the high food availability at the desert edge just before the demanding desert crossing. This phenomenon has also been observed in spring in the xeric Sahel in Africa, where warblers (*Sylvia* spp.; Sylviidae) consume large quantities of berries from the shrub *Salvadora persica* (Salvadoraceae) (71–95% of their diet; Fry *et al.*, 1970; Moreau, 1972; Stoate and Moreby, 1995). Their fat stores just before trans-Saharan migration are associated with their fruit consumption (Stoate and Moreby, 1995).

Does fruit consumption benefit or harm desert plants?

We certainly know more about who consumes desert fruits and seeds than about the costs and benefits that the plants experience from this consumption. Essentially, any consumer can function either as a seed predator or as a dispersal mutualist of a given plant, or play any intermediate role along the continuum defined by these two extremes. Seed consumers can be divided into two groups: (i) frugivores, carnivores and small/medium-sized browsing/grazing herbivores, all of which swallow seeds as food contaminants; and (ii) those that actively harvest seeds as food. The first group of consumers may kill seeds they ingest. However, some or all of these seeds typically survive consumption and are defecated or regurgitated in a germinable state, often in a beneficial location (see Traveset *et al.*, Chapter 4, this volume). The second category, which includes an array of rodents, ants and granivorous birds, comprises animals best known as seed predators or granivores, particularly of plants lacking fleshy fruit. However, they can also function as seed dispersers when they lose or discard seeds during transport, storage, cleaning and consumption, and they may open or consume fleshy fruits to obtain those seeds. A given species of fleshy-fruited plant may well encounter all of these types of consumers, which vary from the beneficial to the harmful (e.g. Silvius, 1995; Larson, 1996; Godínez-Alvarez *et al.*, 2002). In the case of *Opuntia* cactus in North American deserts, for example, Janzen (1986) suggests that mammals such as coyotes, rabbits, ground squirrels, deer, antelope and peccaries are fruit consumers that incidentally disperse or destroy seeds, whereas harvester ants and small/medium-sized rodents are seed harvesters that may confer dispersal benefits at times.

Only a handful of studies have so far been conducted to quantify the relative costs and benefits conferred upon desert plants by fruit and seed consumers. The majority of this work has focused on granivorous rodents, birds and ants rather than on frugivores (e.g. Longland *et al.*, 2001), although some work has examined vertebrate-dispersed cacti (Silvius, 1995; Montiel and Montana, 2000; Godínez-Alvarez *et al.*, 2002), mistletoes (Larson, 1996; Aukema and Martinez del Rio, 2002) and the two species we highlight in this chapter. Given the likely cost of water investment into fleshy fruits, one can imagine that desert plants must experience an exceptionally large payoff from consumption by at least a subset of their consumers. This idea has not, to our knowledge, been tested.

Spatial aspects of desert frugivory and seed dispersal

Movement patterns of animals depend strongly on landscape heterogeneity (Ims, 1995). This is certainly the case for frugivores that inhabit diverse arid environments (Valido and Nogales, 1994; Tewksbury *et al.*, 1999; Dean and Milton, 2000; Aukema and Martinez del Rio, 2002). Less often discussed is the extent to which animal dispersers can create this heterogeneity (Carlo, 2005). A few examples are worth noting. Potential consequences of this non-random movement were demonstrated for the fleshy-fruited shrub *Grewia flava* (Tiliaceae) of the semiarid Kalahari savannah of South Africa, a region characterized by *Acacia* (Fabaceae) trees scattered within grassland vegetation (Tews *et al.*, 2004). There, the deposition of seeds into the grassland matrix is very low and due mostly to carnivorous mammals. Most seeds are deposited beneath *Acacia* trees by small frugivorous birds. This differential arrival explains why the shrubs are usually found beneath trees, as emergence rates beneath *Acacia* were similar to those in the grassland matrix. Spatially explicit simulations demonstrated complex interactions between dispersal distance, seed removal rate and environmental variability in mediating the species' long-term population persistence (Tews *et al.*, 2004).

A dispersal curve or 'kernel' plots the change in the density or the numbers of offspring (seeds or recruits) against distance from the source. It provides a useful summary of the spatial patterns of dispersal or recruitment (Nathan and Muller-Landau, 2000; Nathan, Chapter 11, this volume). At the relatively small scales that characterize the movement range of the major dispersers, we expect to find multi-peaked dispersal curves of both seeds and recruits (Schupp *et al.*, 2002), reflecting the small-scale patchiness of desert ecosystems. The gradually decreasing dispersal curves commonly used as general descriptors of dispersal and recruitment patterns (Nathan and Muller-Landau, 2000; Nathan and Casagrandi, 2004) are not expected to be found in arid and semiarid environments at these relatively small scales. Presumably, such gradually declining dispersal curves adequately describe dispersal patterns at spatial scales that are sufficiently large to distinguish between the frequent occurrence of short-distance dispersal events and the rarity of long-distance dispersal events. Below we illustrate this scale-dependent feature of dispersal kernels, as quantified in one desert system (*O. baccatus*).

In cases where fleshy-fruited plants are extremely attractive to frugivores, positive feedbacks may occur. These are likely to intensify spatial aggregation (Morales and Carlo, 2006). For example, a specialized avian consumer of mistletoe fruits, *Phainopepla nitens*, preferentially uses heavily infected host trees, resulting in a highly aggregated distribution of mistletoe (Aukema and Martinez del Rio, 2002). In this special case of 'directed dispersal', seed deposition sites are not necessarily the most favourable ones, because intraspecific competition among mistletoes is high in heavily infested trees. Directed dispersal in its classic interpretation (disproportionate seed arrival to favourable establishment sites) has been documented in fleshy-fruited plants of several arid ecosystems (Tester *et al.*, 1987; Tewksbury *et al.*, 1999; Tewksbury and Nabhan, 2001).

The high patchiness of desert systems is typically expressed across various spatial and temporal scales (Levin, 1992) representing a range of processes that overlap in their spatial extent. For example, at large spatial scales of a few kilometres and more, patchiness is commonly attributed to water relations (Evenari, 1985) or to elevation and topography (Aukema, 2004). At smaller scales and down to a few metres, desert ecosystems are often characterized by a mosaic of vegetated patches within a matrix of bare ground away from ephemeral watercourses, often intersected by more heavily vegetated regions along watercourses. In sparsely vegetated areas, patches vegetated by perennial woody species tend to amplify aggregation by catching nutrients and generating 'islands of fertility' where nutrient concentrations are increased (Schlesinger *et al.*, 1996) and where water relations are more favourable due to reduced evaporation (Breshears *et al.*, 1998). In addition, seed rain is often higher in these areas, because vegetated patches act as traps for seeds tumbling across a non-vegetated smooth matrix and because of the tendency of many dispersers to concentrate their behaviour in these shaded environments (Bullock and Moy, 2004).

The increased patchiness often observed in arid ecosystems (Noy-Meir, 1985) is often attributed to an increase in the importance in facilitative interactions (Bertness and Callaway, 1994; Callaway, 1995, 1997, 1998; Callaway and Walker, 1997; Hacker and Gaines, 1997; Hacker and Bertness, 1999; Hector *et al.*, 1999), which become more common when abiotic stress is higher (Bertness and Callaway, 1994). In arid systems, patchiness often increases with increasing water stress (Tewksbury and Lloyd, 2001), as vegetation clumps around larger, longer-lived 'nurse plants' (Shreve, 1931, 1951). The facilitative effects of these long-lived perennials have typically been considered in terms of direct habitat-amelioration impacts, in which long-lived perennials create microclimates favourable to seed germination and seedling establishment (Shreve, 1931, 1951; Turner *et al.*, 1966; McAuliffe, 1984; Parker, 1988; Valiente-Banuet and Ezcurra, 1991; Schlesinger *et al.*, 1996; Holmgren *et al.*, 1997; Breshears *et al.*, 1998; Tewksbury and Lloyd, 2001). Comparatively little research has examined the role of non-random or directed dispersal by frugivores in establishing and maintaining the highly aggregated plant associations seen in arid systems (but see Martinez del Rio *et al.*, 1996; Tewksbury *et al.*, 1999; Bullock and Moy, 2004; Morales and Carlo, 2006). Realistically, an array of biotic processes may together strengthen the degree of aggregation. These include pre-dispersal and post-dispersal seed predation (Hulme and Benkman, 2002; Kwit *et al.*, Chapter 19, this volume), rate and quality of fruit removal (Jordano and Schupp, 2000; Carlo, 2005; Carlo *et al.*, Chapter 16, this volume; Schupp, Chapter 20, this volume), and seedling establishment and competition (Aguiar and Sala, 1999). However, few studies have attempted to separate the relative importance of frugivores from abiotic and other biotic factors in creating patterns in desert systems. Because fleshy-fruited plants are typically perennial woody species (Willson, 1986; Willson *et al.*, 1989), the patchiness in their distribution created by directed dispersal often creates 'hotspots', or concentrations of fruiting resources that may in turn attract greater visitation by frugivorous birds (Carlo *et al.*, Chapter 16, this volume). Whether this synergistic effect has positive or negative effects for the plants involved may depend on the relative strength of a series of biological processes, from fruit removal to seed predation, and how each responds to density (Janzen, 1970; Schupp, 1992; Kwit *et al.*, Chapter 19, this volume). The aggregations of long-lived fleshy-fruited perennials caused by directed dispersal may further change microclimates and shift abiotic conditions under nurse plants, but whether this increases or decreases the likelihood of other species colonizing these areas remains an open question (Tewksbury and Lloyd, 2001).

We illustrate the extent to which seed dispersal by frugivores and granivores can influence spatial patterning of vegetation in desert systems by examining the distribution of perennial and annual vegetation in Sonoran Desert study sites as a function of fruit type, separating species that bear fleshy fruits from those that do not. The data for this analysis come from work by Tewksbury and others (Tewksbury and Petrovich, 1994; Tewksbury and Lloyd, 2001), collected to examine the importance of the longest-lived perennial tree in the Sonoran Desert, *Olneya tesota* (Fabaceae), as a nurse plant. Data were collected on ten 1-ha study sites, five in xeric plains and five along ephemeral watercourses (see Tewksbury and Petrovich, 1994; Tewksbury and Lloyd, 2001, for detailed site descriptions). All vegetation was sampled at each site under all *O. tesota* canopies and under an equal number of random sites equal in size to the average canopy size on that site. Vegetation was categorized based on growth-form (trees, shrub species generally growing to more than 1 m in height, shrubs generally less than 1 m but greater than 50 cm, small perennials, annuals, and cacti). Each species was also categorized by fruit type (fleshy-fruited or dry-fruited). To explore the importance of fleshy-fruited species in creating spatial structure and, by extension, the potential importance of frugivores in creating and maintaining spatial patterns in arid systems, the abundance and richness of each growth-form were examined as a function of fruit type, both in random plots and under the canopies of *O. tesota*.

In random plots (n = 144), fleshy-fruited species made up 21% (± 1.9% SE) of perennial species abundance, and 28% (± 2.0% SE) of perennial species richness. This is a significantly greater percentage than reported in other arid systems (Ellner and Shmida, 1981; van Rooyen *et al.*, 1990; Jurado *et al.*, 1991) but still far lower than found in more humid systems. Under the canopies of *O. tesota*, fleshy-fruited perennials increased to 29% (± 1.9% SE) of perennial abundance and 37% (± 1.8% SE) of species richness (n = 144). Fleshy-fruited species dominated large shrubs, and were more common under canopies of *O. tesota* (Fig. 7.1; Table 7.1). Three aspects of these findings are worth noting. First, fleshy-fruited plants dominate abundance and species richness in large shrubs and cacti (Fig. 7.1; Table 7.1), making up 68–77% of all large shrubs and virtually all cacti. The dominance of fleshy-fruited large shrubs (measured as the percentage of all species and individuals that bear fleshy fruits) also varied by habitat type. When random points are picked to represent the characteristics of the habitat in general, fleshy-fruited species and individuals were more common in mesic ephemeral waterways than in open xeric habitat. In contrast, under the long-lived shade of *O. tesota*, where shrubs producing fleshy fruit make up > 75% of the large shrubs, the dominance of fleshy-fruited shrubs was lower along mesic ephemeral waterways than in open xeric areas (Table 7.1). Thus the importance of long-lived perennial canopies in concentrating fruiting species appears to be greater in more xeric areas. Secondly, shrubs bearing fleshy fruit are more abundant under

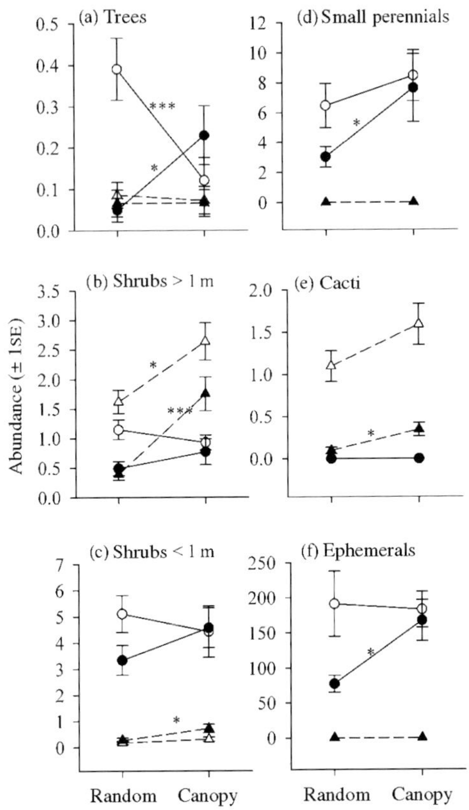

Fig. 7.1. Abundance and distribution of fleshy-fruited (triangles, dashed lines) and dry-fruited (circles, solid lines) species in ten study sites in Sonora, Mexico. Five sites were along ephemeral washes (mesic habitat; white symbols, $n = 83$ canopies and random sites) and five were away from any ephemeral water sources (xeric habitat; black symbols, $n = 61$ canopies and random sites). Differences in abundance between random and canopy plots were determined by Mann–Whitney U tests (*< 0.05, ***< 0.001).

the canopies of *O. tesota* when compared to random points, whereas shrubs bearing dry fruits do not increase in abundance under *O. tesota* canopies compared with random points. This pattern is particularly strong in large shrubs (Fig. 7.1b) but is also seen for smaller shrubs (Fig. 7.1c). This community pattern strongly implicates frugivore behaviour as a driving factor creating fruiting 'hotspots' in desert systems. Finally, increases in the abundance of fleshy-fruited species under *O. tesota* canopies are generally greater in xeric habitats compared with more mesic habitats (Fig. 7.1; see also Tewksbury and Lloyd, 2001). This is true both for fleshy-fruited species in the growth-forms they dominate (large shrubs and cacti), and for dry-fruited species of other growth-forms, such as smaller perennials and trees (Fig. 7.1). This pattern suggests that opportunities for establishment and growth become more limited in increasingly arid systems, increasing the positive impacts of directed dispersal to shaded microhabitats.

Table 7.1. Proportion (± 1 SD) of vegetation bearing fleshy fruit, based on growth-form, in random plots and under the canopies of Olneya tesota (O.t.). Data are from 144 random plots and canopies from ten sites in the Sonoran Desert of northern Mexico, five located along ephemeral watercourses (mesic) and five located on xeric plains. Proportion data are calculated on a per-plot basis, only using plots where the growth-form was present. Cacti are all fleshy-fruited. Small perennials and annuals all lacked fleshy fruit.

			Random	O. tesota	n (random, O.t.)
Richness	Trees*	Mesic	0.21 (0.072)	0.417 (0.149)	31,12
		Xeric	0.6 (0.187)	0.269 (0.122)	5,13
	Shrubs > 1 m	Mesic	0.584 (0.047)	0.677 (0.039)	73,77
		Xeric	0.525 (0.087)	0.774 (0.047)	27,43
	Shrubs < 1 m	Mesic	0.12 (0.038)	0.148 (0.038)	63,66
		Xeric	0.141 (0.035)	0.257 (0.049)	36,51
Abundance	Trees*	Mesic	0.21 (0.072)	0.417 (0.149)	31,12
		Xeric	0.6 (0.187)	0.269 (0.122)	5,13
	Shrubs > 1 m	Mesic	0.58 (0.049)	0.686 (0.04)	73,77
		Xeric	0.501 (0.088)	0.749 (0.051)	27,43
	Shrubs < 1 m	Mesic	0.11 (0.038)	0.144 (0.038)	63,66
		Xeric	0.084 (0.027)	0.244 (0.051)	36,51

* Values for proportionate richness and abundance of fleshy-fruited trees are identical due to plot size – in a given plot, we found either one or zero individuals of a given species.

Case Studies

In this section, we provide examples of two particularly well-studied desert plant–frugivore systems, one from the desert regions in the Middle East, and a second from temperate deserts in the south-western USA and northern Mexico. Our aim is to illustrate the fruit–frugivore interactions surrounding each of these plant species, focusing particularly on aspects of these interactions that are intimately linked to their desert setting. In many ways, the two systems outlined here provide models for the dispersal of fleshy fruits in deserts. In both systems, patterns of frugivory and frugivore behaviour appear to be tightly linked to the spatial structure of the fruiting plants, a phenomenon that may characterize fruit–frugivore relationships in desert systems.

Ochradenus baccatus

Autoecology

Ochradenus (Resedaceae) is a genus of six species found in desert and semi-desert steppe and mountains from Pakistan (east) to Libya (west) and from Israel and Jordan (north) to Somalia (south) (Miller, 1984). *Ochradenus baccatus* is a widespread Saharo-Sindian species. It is probably the dominant fleshy-fruited shrub throughout many hot desert regions in the Middle East. In Israel, it is particularly common in *wadis* (dry river beds) throughout the Judean Desert and the Arava Valley (Zohary, 1966).

Ochradenus baccatus is subdioecious (male, female and hermaphroditic plants all present) in some populations and gynodioecious (plants with bisexual flowers and plants with female flowers present) in others (Wolfe and Shmida, 1995, 1997). *Ochradenus baccatus* has a rare phenological strategy, with prolonged flowering and fruiting periods occurring year-round, with peaks during spring (February to May) and autumn (September to December) (Wolfe and Burns, 2001; Altstein, 2005; Lotan, 2005). Wolfe and Shmida (1997) reported that fruit yield in hermaphrodites was six times lower than in females (25.2 ± 3.62 and 141.4 ± 5.24 fruit per plant during the 18-month study, respectively, n_{total} = 132 plants). A more recent study (Lotan, 2005) quantified fruit yields much higher than this, with considerable variation among sites, seasons and years (in spring 12,000 ± 1850 (SE) fruit, n = 40 plants; in autumn 10,100 ± 1880 (SE) fruit, n = 33 plants), yet retaining the same pattern of females being more fecund than hermaphrodites. The fruit is a white berry, 3–6 mm in diameter, with an average of 3–8 tiny seeds per fruit (1–1.5 mm, 0.6–1.0 mg; Altstein, 2005; Lotan, 2005). The number of fruits and seeds per plant, individual seed mass and germination success are all highest in autumn and early winter (Wolfe and Burns, 2001). The nutritional value of female *O. baccatus* fruit (Lotan, 2005) is similar to the general nutritional profile for fruit from 27 species of the Mediterranean mesic areas (Izhaki, 2002). For example, it has low protein (2.5% dry mass basis) and lipid (1.4%) contents but high non-soluble carbohydrate (85%) content. However, its water content is comparatively high (86%). Thus, it is probably an important water source for its many consumers in this arid environment. Fruit attributes such as nutritional content and morphology are related to abiotic conditions, especially water stress and soil mineral content. As water condition improves, along with some enhancement of nitrogen and other soil minerals, *O. baccatus* fruits are bigger, although they are lower in dry weight, protein and fat (Lotan, 2005).

Frugivory

Year-round and throughout its distributional range, *O. baccatus* fruits are probably the single most important food for many desert frugivores. Birds are the most diverse group that consume *O. baccatus* fruits. The most common resident consumers are yellow-vented bulbuls (*Pycnonotus xanthopygos*; Pycnonotidae), Tristram's grackles (*Onychognathus tristramii*; Sturnidae), Spanish sparrows (*Passer hispaniolensis*; Passeridae) and blackstarts (*Cercomela melanura*; Muscicapidae) (Altstein, 2005). Migrant birds such as blackcaps (*Sylvia atricapilla*; Sylviidae), Orphean warblers (*S. hortensis*; Sylviidae) and lesser whitethroats (*S. curruca*; Sylviidae) also consume its fruits during spring migration (I. Izhaki, O. Spiegel, A. Lotan and O. Altstein, Judean Desert, personal observation; R. Yosef, Eilat, 2002, personal communication). Frugivory on *O. baccatus* fruits has also been observed in several mammals, including Blanford's fox (*Vulpes cana*; Canidae) (B. Shalmon, Eilat, 2002, personal communication), and even in reptiles such as the ornate spiny-tailed lizard (*Uromastyx ornatus*; Agamidae). The latter spends 53% of its foraging time on *Ochradenus* shrubs, which supply its most significant food items (mainly fruits but also flowers) across the year (D. Molcho and O. Ben-David, 2000, unpublished manuscript). *Ochradenus baccatus* also represents a common food resource for herbivores; it is browsed by mammals such as camels, gazelles and ibex (Ward and Saltz, 1994).

Bulbuls deposit most *O. baccatus* seeds within 30 min of intake. Seeds may remain in the gut up to 145 min (Altstein, 2005). Seed passage through the digestive tract of bulbuls increases the percentage of seeds that germinate by 29% (± 3.6% SE) and shortens the time to germination by 153 h (± 3.5 SE) compared with seeds in untreated intact fruits (control 1), which face chemical and physical germination inhibitors in the surrounding pulp (Altstein, 2005). Grackles have considerably longer gut retention time, with a mean of 135.1 ± 14.1 min and maximum of 328.3 ± 12.6 min. Passage through their gut similarly increases the percentage of seeds that germinate by 31% (± 3.3% SE) and shortens the time to germination by 358 h (± 31 SE; Spiegel, 2006). However, germination of digested seeds by either bulbuls or grackles is 6–9% lower when compared to germination of seeds from another type of control in which seeds were manually extracted from the fruit pulp (Altstein, 2005; Spiegel, 2006). We emphasize that, in nature, fruits not treated by frugivores remain intact; thus, seeds within intact fruits are more appropriate controls for treatment by frugivores than are manually extracted seeds (see also Samuels and Levey, 2005; Traveset *et al.*, Chapter 4 this volume). These data thus indicate that the two bird species enhance *O. baccatus* germination by removing germination inhibitors within the pulp rather than by eroding the seed coat (see Traveset *et al.*, Chapter 4 this volume). Furthermore, seeds retained in the gut for different lengths of time did not differ in their germination success in either bulbuls (Altstein, 2005) or grackles (Spiegel, 2006).

Spatial aspects of frugivory and seed dispersal

In the Judean Desert of Israel, bulbuls and grackles spend most of their time on *Acacia* (Fabaceae) and *Tamarix* (Tamaricaceae) trees and on *O. baccatus* and other shrubs, largely avoiding open spaces (Altstein, 2005; Spiegel, 2006). Most seed rain should therefore be generated under these woody species. Recruitment of *O. baccatus* seedlings is, in fact, dense under acacia and tamarix trees (up to 200 seedlings under a single acacia tree), and higher than expected from the predicted bird-mediated seed rain (O. Spiegel, 2005 unpublished results). Seedling survival is also higher under those trees and shrubs compared with open habitats; probably due to reduced insolation and enhanced shielding against soil washout by floods.

These disproportionate rates of arrival and survival support the hypothesis of directed dispersal in this system (see above). For example, the yellow-vented bulbul, a major disperser of *O. baccatus*, occurs only in wadis and oases. The non-random movement of bulbuls within a wadi generates a clumped seed dispersion pattern, with most seeds deposited beneath trees and shrubs; these occupy only a small fraction of their home range, which typically spans more than a few hundred metres (Fig. 7.2). At larger spatial scales, bulbul-generated dispersal curves exhibit rapid declines with increasing distance from the source plant. For a desert plant such as *O. baccatus* facing very unfavourable conditions in most sites, the microhabitat of the deposited seed may be much more important than the distance from the parent plant (Altstein, 2005).

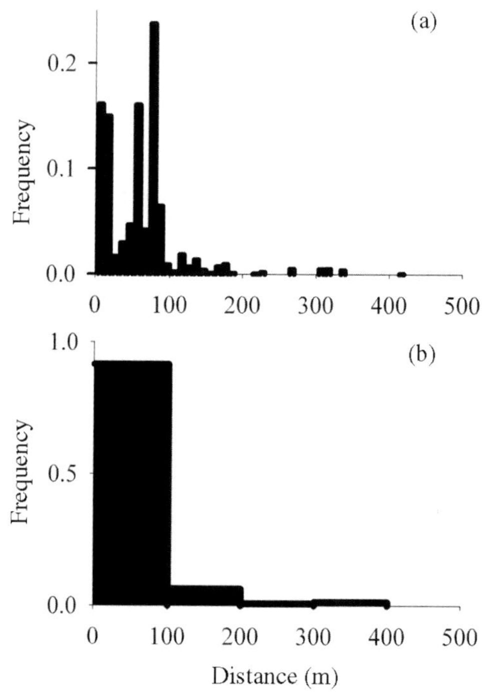

Fig. 7.2. Dispersal curves for *Ochradenus baccatus* seeds dispersed by yellow-vented bulbuls (*Pycnonotus xanthopygos*) in the Judean Desert of Israel. The curves were generated by a simple mechanistic model as the product of movement rate (data from laser rangefinder field measurements) and gut retention time (data from laboratory experiments). When model predictions are summarized in relatively fine resolution (10 m in this case), dispersal curves show multiple peaks especially at relatively short distances (a), resulting from the non-random movements of the bulbuls in the highly patchy wadi habitat. When model predictions are summarized at a coarser resolution (100 m in this case), dispersal curves exhibit rapid decline with increasing distance (b), reflecting the abundance of short-range movements of bulbuls within the wadi, and the scarcity of their long-range movements outside this habitat (Spiegel, 2006; O. Altstein and R. Nathan, 2005, unpublished results).

Capsicum annuum

Autoecology

The genus *Capsicum* is best known for the > 2000 cultivars that populate grocery stores and spice markets around the world (Bosland, 1994). The native range of this genus, which contains 23–27 wild species, extends from northern Argentina to southern Arizona (Pickersgill and Heiser, 1977). The most derived wild chilli species, *Capsicum annuum* L. var. *glabriusculum* (Dunal) Heiser & Pickersgill) (Walsh and Hoot, 2001), known as 'bird pepper' or 'chiltepín', is also the most northerly, with populations as far north as Texas, New Mexico and Arizona.

Despite its specific name, *C. annuum* is a long-lived perennial shrub, and can reach 1.5 m in height. It grows in a wide range of habitats (Pickersgill *et al.*, 1979), but reaches its highest density in arid mid-elevation regions in the state of Sonora, Mexico (Perramond, 2005; J.J. Tewksbury, 2006, unpublished results). Here we focus on the most thoroughly studied populations, those occurring from northern Mexico to southern Arizona, and all information below pertains to these populations. *Capsicum annuum* flowers in July and August, with fruits ripening into a red berry 4–8 mm in diameter in August through September. Fruit removal continues for 2 to 6 months, as the fruits persist until the following summer. Typical fruits contain 8–10 seeds in southern Arizona (range = 2–15 seeds); each seed weighs 4–10 mg (J.J. Tewksbury and T.A. Carlo, 2006, unpublished results). In southern Arizona, fruit production can vary over three orders of magnitude depending on the quality of summer rains. In particularly dry years, plants produce almost no fruit and die back almost to the ground, only to resprout during winter rains.

Frugivory

Capsicum annuum fruit is high in lipids and protein (24% and 14% dry mass respectively) and low in carbohydrates (16% dry mass) (Levey *et al.*, 2006); in contrast, *Ochradenus* is low in protein and lipids but high in non-soluble carbohydrates. These differences in nutrient composition are reflected in differences in disperser assemblages. Mammals are thought to generally avoid lipid-rich fruits (such as *Capsicum*) and are more likely to eat high-carbohydrate fruits such as *O. baccatus* (Martin *et al.*, 1951; Debussche and Isenmann, 1989; Jordano, 1995; Corlett, 1996; Bollen *et al.*, 2004). In contrast, it is thought that at least some birds specialize on lipid-rich fruits (McKey, 1975; Moermond and Denslow, 1985; Snow, 1985; Place and Stiles, 1992; Fuentes, 1994; Witmer, 1996; Witmer and Van Soest, 1998). The species we highlight in the case studies presented here support this separation. *Capsicum annuum* fruits are consumed almost entirely by birds, and in particular by Mimidae, a pattern of consumption similar to that observed for lipid-rich fruits in other neotropical systems (Loiselle and Blake, 1990; Stiles, 1993; Witmer, 1996; Witmer and Van Soest, 1998). In contrast, *O. baccatus* is consumed by a wide range of birds, mammals and lizards (see above), as might be expected given its high water content and low lipid levels.

An obvious next step will be to determine the relative prevalence of different fruit types in arid environments (see following section).

There is a second reason why *C. annuum* is not consumed by mammals: secondary metabolites. Capsicums are known for the presence of capsaicinoids, chemicals responsible for the painful 'heat' or 'pungency' associated with chilli fruit (Iwai *et al.*, 1979; Kawada *et al.*, 1985; Cordell and Araujo, 1993). In wild *C. annuum*, capsaicin and dihydrocapsaicin are the primary secondary metabolites responsible for the pungency of the fruit, together making up between 0.2% and 0.4% dry mass (2200 to 4400 ppm) in a typical year (J.J. Tewksbury, D.J. Levey, A. Traveset and M. Huzinga, 2006, unpublished results), and like many secondary metabolites, they increase in concentration when water stress is more severe (Sung *et al.*, 2005). The capsaicinoid concentrations found in wild chillies are more than enough to completely stop feeding by rodents (Tewksbury and Nabhan, 2001), but birds do not sense these compounds as painful (Mason *et al.*, 1991; Jordt and Julius, 2002), and thus dispersal of *C. annuum* is accomplished almost exclusively by birds (Levey *et al.*, 2006).

The fruits of *C. annuum* are occasionally removed by birds directly after they ripen, but fruit removal is variable and many fruits remain on plants for 6 weeks or more before they are consumed (J.J. Tewksbury and T.A. Carlo, 2006, unpublished results). Variation in removal rates is a function of fruit density, but the effects appear strongly scale-dependent. At large spatial scales, such as across populations, variation in the rate of fruit removal may be a function of intraspecific fruit density: when crops are very large, consumers are effectively saturated, and are unable to remove all the fruit rapidly (cf. Howe and Estabrook, 1977; Davidar and Morton, 1986; Sallabanks, 1993). At smaller spatial scales, heterospecific fruiting may cause resource clumping, creating fruiting neighbourhoods (Sargent, 1990; Carlo, 2005). For example, *C. annuum* under *Celtis palida* (Cannabaceae) trees and other large species bearing fleshy fruit experience higher rates of fruit removal (Tewksbury and Nabhan, 2001).

The most common consumers of *C. annuum* in Arizona and northern Mexico are curve-billed thrashers (*Toxistoma curvirostre*; Mimidae) (Levey *et al.*, 2006). However, we have observed a wide range of birds consuming *C. annuum* in these locations. Gila woodpeckers (*Melanerpes uropygialis*; Picidae) and white-crowned sparrows (*Zonotrichia leucophrys*; Emberizidae) use the fruits primarily for water: they actively squeeze the fruits to remove the water content before dropping the unconsumed component near the parent plant (J.J. Tewksbury, 2000, unpublished results). Where *C. annuum* is most abundant, in northern Mexico, it is not uncommon for it to make up more than 99% of the fruit biomass available to birds for 3–4 months, from September to December (J.J. Tewksbury, 2000, unpublished results). The extent to which birds rely on this resource has not been studied. However, as in *O. baccatus* and certain other arid systems, consumption of fruits can be as much a function of maintaining a positive water balance as it is about nutrition, and the water value of *Capsicum* fruits should not be underestimated (Herrera, 1982).

From the plant's perspective, avian consumption appears very important, illustrating the links between consumer foraging behaviour and community-wide patterns of fruiting plants (Carlo, 2005). *Capsicum annuum* seeds are not harmed by gut passage and occasionally remain in the digestive tracts of curve-billed thrashers for longer than 8 h without any loss of viability (Tewksbury and Nabhan, 2001; J.J. Tewksbury, D.J. Levey, A. Traveset and M. Huzinga, 2006, unpublished results).

Spatial aspects of frugivory and seed dispersal

Similar to bulbuls and grackles on *O. baccatus* in the Middle East, curve-billed thrashers and other primary consumers of *C. annuum* in Arizona spend most of their time moving from one large fruiting shrub to another. They favour *Celtis* spp. (Cannabaceae), which have consistently large fruit crops (Tewksbury *et al.*, 1999). In the process, they consume *C. annuum* fruit and create a highly clumped pattern of seed rain. This produces a strongly non-random association of *C. annuum* plants with fleshy-fruited species in the genera *Celtis*, *Lycium* (Solanaceae), *Condalia* and *Zizyphus* (Rhamnaceae) (Tewksbury *et al.*, 1999). *Capsicum annuum* seedlings, like seedlings of *O. baccatus*, do not survive in open sunlight (Tewksbury *et al.*, 1999). Therefore, the directed pattern of seed rain enhances survival. Importantly, this close association with fleshy-fruit-bearing trees appears to be relaxed somewhat in locations where *C. annuum* is the dominant fruiting species. In these locations, *C. annuum* is again found almost exclusively in shaded environments, but the association with fleshy-fruit-bearing shrubs may be less consistent (Perramond, 2005; J.J. Tewksbury, 2000, unpublished results). *Capsicum annuum* may provide an ideal system in which to examine the roles of dominant and subordinate fruiting resources in the creation of interspecific clumping of fruit–frugivore interactions.

Avenues for further research

Nearly all questions of current interest in the study of plant–frugivore interactions are also relevant in desert ecosystems. Despite this, they have been less well investigated in deserts than in more mesic habitats. However, there are also questions about these interactions that are either particularly relevant to arid systems, or that can be studied with particular ease in these systems. We highlight several of the most interesting of these in this section. Our aim is to point to directions in which further work would be especially promising.

Is there such a thing as a 'desert fruit'?

Are there any general patterns in fruit nutritional characters and secondary chemistry that are more abundant in arid systems? Careful phylogenetic studies and within-species comparisons across wide aridity gradients will help us to understand the degree to which desert systems impose convergent selection on fruit traits.

Do desert systems favour directed dispersal?

Our case studies have highlighted the presence of directed dispersal in desert fruit–frugivore interactions. Does directed dispersal increase in importance as moisture availability decreases, or are patterns of non-random establishment simply easier to observe in these relatively uncomplicated systems? Answering this question will require careful experimental work linking a series of processes taking place from fruit development though to seedling establishment.

Do fruiting 'hotspots' determine seed fate?

To what extent do fruiting neighbourhoods, or multispecies fruiting 'hotspots', change the fate of plants? Addressing this question will require integration across a series of processes that include pre-dispersal seed predation, dispersal patterns, and post-dispersal seed and seedling survival, as well as explicit attention to the effects of plant neighbourhoods on the quantity and quality of dispersal.

Are deserts ideal systems for studying disperser movement and contagion?

Disperser movements are considerably easier to observe in the desert compared with the tropical and temperate regions in which fleshy-fruited plants and frugivores are more common. These features make desert ecosystems attractive for developing and testing advanced models of seed dispersal by animals. This is important because very little progress has been achieved thus far in modelling seed dispersal by animals compared with the rapid progress seen in recent years in modelling wind-mediated seed dispersal (but see Dennis and Westcott, Chapter 9, this volume; Nathan, Chapter 11, this volume). Subsequent refinements of these models can make them appropriate for more complicated and speciose systems, such as those in the tropics.

Are the super-abundant, fleshy-fruited species in deserts 'keystone species'?

We have emphasized that, although relatively few fleshy-fruited species exist in deserts, individuals of some of these species can be very common. For example, *O. baccatus* is locally abundant in nearly every wadi in the Israeli deserts and is consumed by a wide spectrum of animals, including species that are typically insectivorous. In northern Mexico, *C. annuum* can make up more than 99% of the fruit biomass available to birds from September to December each year. Are species like these representative of desert ecosystems worldwide? If so, they should be considered to be urgent targets for conservation. This observation also begs comparison with tropical forests:

are super-abundant, fleshy-fruited species with exceptionally wide spectra of dispersers common there as well?

Are desert systems ideal for estimating total dispersal kernels?

The wide spectrum of dispersers seen for *Ochradenus* and for numerous other desert fleshy-fruited plants, in combination with the relative ease of observing and quantifying disperser movements in open desert conditions, provide a unique opportunity for studying 'total dispersal kernels' in these habitats. The emerging concept of total dispersal kernels (see Nathan, Chapter 11, this volume) recognizes the need to go beyond quantifying dispersal kernels generated by a single dispersal agent (e.g. one frugivore species), because dispersal is almost always mediated by multiple vectors, each of which can generate a different pattern. Current work on *O. baccatus* is proceeding in this direction. A remaining challenge, however, is to quantify dispersal kernels for multiple vectors simultaneously, dissecting what each one of them does separately and what they do jointly.

Is the temporal dimension of seed dispersal more important in desert systems?

We have emphasized the important role of frugivores in the *spatial* dynamics of fleshy-fruited plants in deserts. Do they play a role also in the *temporal* dimension of seed dispersal (e.g. in dormancy, which is extremely important in the unpredictable climatic conditions of desert ecosystems; see Gutterman, 1993)? Virtually nothing is known in this regard. Researchers have placed most of their emphasis on the spatial aspect of 'directed dispersal' (placing seeds at the appropriate site), but it could well be, in deserts in particular, that adding the temporal aspect of this phenomenon (placing the seeds at the appropriate site *at the appropriate time*) can be critically important. Do we see signs of selection for frugivore-increased variance in the timing of seed germination, as was previously detected in Mediterranean scrublands (Izhaki and Safriel, 1990)? What are the possible mechanisms for such a selective force in desert plant–frugivore systems?

Are fleshy-fruited species more persistent in deserts than in other habitats?

Certain fleshy-fruited desert species fruit year-round, or nearly so. Although this is a much rarer reproductive strategy in relatively dry environments than in humid ones (Schütz and Milberg, 1997), many examples can be identified. As we discussed previously, *O. baccatus* flowers year-round (see also Wolfe and Burns, 2001). In the Sonoran Desert, the fishhook barrel cactus (*Ferocactus wislizeni*; Cactaceae) flowers in July–August and produces fruits in August–September, but these fruits may remain in a ripened state on the plant for many months until they are removed by animals (J.L. Bronstein, 2005, unpublished results). Other examples of prolonged fruiting include the xerophilous perennial shrub *Launaea arborescens* (Asteraceae) studied in Spain by Schütz and Milberg (1997), and the hemiparasitic mistletoe *Tristerix corymbosus* (Lauranthaceae) in north-western Patagonia (Aizen, 2003).

Given the harsh abiotic conditions in arid environments, where high radiation and low water availability intensify the costs of maintaining flowers and fleshy fruits, prolonged fruiting of this nature raises both proximate and ultimate questions. What physiological processes allow fruits to be maintained for such long periods, and under what conditions is prolonged fruiting adaptive? The latter is a particularly intriguing question; in most arid habitats, establishment and germination can occur only in the wetter season. Hence, fruit production at other times of year would appear to be extremely costly.

Does the patchy nature of the desert vegetation and the contagious dispersal by frugivores result in a more readily invaded and perturbed system, or is it more resilient?

Because plant–frugivore associations are typically diffusive (Levey and Benkman 1999; Richardson *et al.*, 2000), frugivores (both native and non-native) commonly provide equivalent dispersal service to both native and invasive fleshy-fruited plants (for recent reviews of this subject, see Richardson *et al.*, 2000; Gosper *et al.*, 2005; Buckley *et al.*, 2006). Plant invasion is a major threat to many ecosystems around the world, and deserts are no exception. Yet, the question of whether deserts are more or less susceptible to invasion by fleshy-fruited plants has received little attention. Similarly, assessment of the relative impact of invasive fleshy-fruited plants on invaded desert ecosystems has rarely been investigated.

We hypothesize that the relative impact of invasive fleshy-fruited plants on invaded desert ecosystems depends on human intervention. In deserts with relatively low human intervention, fleshy-fruited plants are not expected to induce drastic changes; however, high levels of human intervention are likely to change the foraging patterns of frugivores and thus could induce a strong impact. Drastic impacts by invasive plants have been observed, for example, in North-American deserts, where invasive annual grasses, which rarely bear fleshy fruits and are not usually dispersed by frugivores, have converted native shrublands with extremely low fire frequency into alien annual grasslands dominated by fire-prone species (Brooks and Pyke, 2001). The major impact was the transition from sparsely distributed shrubs, which provide low fuel for fires and impede their spread, to continuous grasslands, which fuel fires and facilitate their spread. While invasive grasses altered the typical patchiness of desert vegetation, invasive fleshy-fruited plants are not expected to have a similar impact, because of the typical contagious dispersal by frugivores. Yet, if humans alter frugivore foraging patterns, fleshy-fruited species may spread rapidly and generate an emergent problem. For example, densities of invasive prickly pears (*Opuntia ficus-indica*) in the Karoo desert of South Africa are much greater below telegraph and transmission poles and next to wire fences along roads than in open rangelands, mainly due to dispersal by the Cape (*Corvus capensis*) and the pied (*Corvus albus*) crows (Corvideae).

It is worth noting, however, that desert systems are likely to be much more 'closed' to invasion by other species than are more mesic systems, due to the very thing that makes them deserts: a lack of water. Such a small percentage of the earth's biodiversity is well adapted to high heat and low water that invasion may be rare. Furthermore, the plants that humans select for use are typically modified to grow in moist environments (crops are irrigated), and humans rarely traffic in the types of desert plants that could invade. Unfortunately, it seems possible that the 'right' invader could have devastating impact on desert ecosystems, because directed dispersal would move it to the most diverse areas.

Conclusions

A traditional emphasis on the rarity of fleshy fruits in deserts has contributed to a paucity of information about the importance of frugivory in desert systems. In both of the systems we examined as case studies, we found ample evidence that frugivory plays an important role in the ecology of deserts, generating a particularly high degree of spatial structure in fleshy-fruited species. This spatial structure in fruiting species may itself be a general emergent property of xeric systems that exerts a great deal of influence on frugivore behaviour, creating positive feedbacks and a yet higher degree of resource aggregation. Thus, fleshy-fruited species found in arid environments may be disproportionately important ecologically, since they provide precious resources for a wide diversity of animals, and the increased activity of animals in the vicinity of fleshy-fruited desert plants can generate islands of productivity, structuring both plant and animal communities. The patterns of aggregation and the potential effects on animals are not unique to xeric systems, but the relative simplicity of desert ecosystems makes these patterns easier to study, thus providing unique opportunities to gain insight into major unanswered questions in the study of seed dispersal ecology.

Acknowledgements

Much of the work on *Ochradenus* reported here was supported by a grant from the International Arid Lands Consortium (No. 03R-25) to J.L.B., I.I. and R.N. R.N. also gratefully acknowledges support from the Israel Science Foundation (ISF-474/02 and ISF-FIRST 1316/05), and from the Simon and Ethel Flegg Fellowship. R.N., O.S. and O.A. are grateful to S. Ben-Yosef, M. Avraham, S. Friedman, members of the Movement Ecology Laboratory at the Hebrew University of Jerusalem, and many others who have assisted our *Ochradenus* studies in the Israeli desert. Research on *Capsicum* was supported by a grant to J.J.T. by the National Science Foundation (DEB 0129168).

References

Aguiar, M.R. and Sala, O.E. (1999) Patch structure, dynamics and implications for the functioning of arid ecosystems. *Trends in Ecology and Evolution* 14, 273–277.

Aizen, M. (2003) Influences of animal pollination and seed dispersal on winter flowering in a temperate mistletoe. *Ecology* 84, 2613–2627.

Altstein, O. (2005) Seed dispersal of a fleshy fruited desert shrub by birds: effects of gut passage, disperser movement and local spatial structure on seed dispersion patterns. MSc thesis, Ben-Gurion University of the Negev, Beer Sheva, Israel.

Anava, A., Kam, M., Shkolnik, A. and Degen, A.A. (2000) Seasonal field metabolic rate and dietary intake in Arabian babblers (*Turdoides squamiceps*) inhabiting extreme deserts. *Functional Ecology* 14, 607–613.

Anderson, B.W. and Ohmart, R.D. (1978) *Phainopepla* utilization of honey mesquite forests in the Colorado River Valley. *Condor* 80, 334–338.

Aukema, J.E. (2004) Distribution and dispersal of desert mistletoe is scale-dependent, hierarchically nested. *Ecography* 27, 137–144.

Aukema, J.E. and Martinez del Rio, C. (2002) Where does a fruit-eating bird deposit mistletoe seeds? Seed deposition patterns and an experiment. *Ecology* 83, 3489–3496.

Bairlein, F. and Simons, D. (1995) Nutritional adaptations in migrating birds. *Israel Journal of Zoology* 41, 357–367.

Bates, J.M. (1992a) Frugivory on *Bursera microphylla* (Burseraceae) by wintering gray vireos (*Vireo vicinior*, Vireonidae) in the coastal deserts of Sonora, Mexico. *Southwestern Naturalist* 37, 252–258.

Bates, J.M. (1992b) Winter territorial behaviour of gray vireos. *Wilson Bulletin* 104, 425–433.

Beattie, A.J. (1985) *The Evolutionary Ecology of Ant–Plant Mutualisms.* Cambridge University Press, Cambridge, UK.

Berthold, P. (1976) The control and significance of animal and vegetable nutrition in omnivorous songbirds. *Ardea* 64, 140–154.

Bertness, M.D. and Callaway, R. (1994) Positive interactions in communities. *Trends in Ecology and Evolution* 9, 191–193.

Bosland, P.W. (1994) Chiles: history, cultivation, and uses. In: Charalambous, G. (ed.) *Spices, Herbs and Edible Fungi.* Elsevier, Amsterdam, pp. 347–366.

Bollen, A., Van Elsacker, L., and Ganzhorn, J.U. (2004) Relations between fruits and disperser assemblages in a Malagasy littoral forest: a community-level approach. *Journal of Tropical Ecology* 20, 599–612.

Bregman, R. (1988) Forms of seed dispersal in Cactaceae. *Acta Botanica Neerlandica* 37, 395–402.

Breshears, D.D., Nyhan, J.W., Heil, C.E. and Wilcox, B.P. (1998) Effects of woody plants on microclimate in a semiarid woodland: soil temperature and evaporation in canopy and intercanopy patches. *International Journal of Plant Sciences* 159, 1010–1017.

Brooks, M.L. and Pyke, D.A. (2001) Invasive plants and fire in the deserts of North America. In: Galley, K.E.M. and Wilson, T.P. (eds) *Proceedings of the Invasive Species Workshop: The Role of Fire in the Control and Spread of Invasive Species.* Tall Timbers Research Station, Tallahassee, Florida, pp. 1–14.

Buckley, Y.M., Anderson, S., Catterall, C.P., Corlett, R.T., Engel, T., Gosper, C.R., Nathan, R., Richardson, D.M., Setter, M., Spiegel, O., Vivian-Smith, G., Voigt, F.A., Weir, J.E.S. and Westcott, D.A. (2006) Management of plant invasions mediated by frugivore interactions. *Journal of Applied Ecology* 43, 848–857.

Bullock, J.A. and Moy, I.L. (2004) Plants as seed traps: inter-specific interference with dispersal. *Acta Oecologica/International Journal of Ecology* 25, 35–41.

Callaway, R.M. (1995) Positive interactions among plants. *Botanical Review* 61, 306–349.

Callaway, R.M. (1997) Positive interactions in plant communities and the individualistic-continuum concept. *Oecologia* 112, 143–149.

Callaway, R.M. (1998) Competition and facilitation on elevation gradients in subalpine forests of the northern Rocky Mountains, USA. *Oikos* 82, 561–573.

Callaway, R.M. and Walker, L.R. (1997) Competition and facilitation: a synthetic approach to interactions in plant communities. *Ecology* 78, 1958–1965.

Campos, C.M. and Ojeda, R.A. (1997) Dispersal and germination of *Prosopis flexuosa* (Fabaceae) seeds by desert mammals in Argentina. *Journal of Arid Environments* 35, 707–714.

Carlo, T.A. (2005) Interspecific neighbours change seed dispersal pattern of an avian-dispersed plant. *Ecology* 86, 2440–2449.

Chambers, J.C. and MacMahon, J.A. (1994) A day in the life of a seed: movement and fates of seeds and their implications for natural and managed systems. *Annual Review of Ecology and Systematics* 25, 263–292.

Chang, E.R, Zozaya, E.L., Kuijper, D.P.J. and Bakker, J.P. (2005) Seed dispersal by small herbivores and tidal water: are they important filters in the assembly of salt-marsh communities? *Functional Ecology* 19, 665–673.

Cordell, G.A. and Araujo, O.E. (1993) Capsaicin: identification, nomenclature, and pharmacotherapy. *Annals of Pharmacotherapy* 27, 330–336.

Corlett, R.T. (1996) Characteristics of vertebrate-dispersed fruits in Hong Kong. *Journal of Tropical Ecology* 12, 819–833.

Davidar, P. and Morton, E.S. (1986) The relationship between fruit crop sizes and fruit removal rates by birds. *Ecology* 67, 262–265.

Davidson, D.W. and Morton, S.R. (1984) Dispersal adaptations of some *Acacia* species in the Australian arid zone. *Ecology* 65, 1038–1051.

de la Barrera, E. and Nobel, P.S. (2004) Carbon and water relations for developing fruits of *Opuntia ficus-indica* (L.) Miller, including effects of drought and gibberellic acid. *Journal of Experimental Botany* 55, 719–729.

Dean, W.R.J. and Milton, S.J. (2000) Directed dispersal of *Opuntia* species in the Karoo, South Africa: are crows the responsible agents? *Journal of Arid Environments* 45, 305–314.

Dean, W.R.J., Milton, S.J. and Jeltsch, F. (1999) Large trees, fertile islands, and birds in arid savanna. *Journal of Arid Environments* 41, 61–78.

Debussche, M. and Isenmann, P. (1989) Fleshy fruit characters and the choices of bird and mammal seed dispersers in a Mediterranean region. *Oikos* 56, 327–338.

Ellner, S. and Shmida, A. (1981) Why are adaptations for long-range seed dispersal rare in desert plants? *Oecologia* 51, 133–144.

Evenari, M. (1985) The desert environment. In: Evenari, M., Noy-Meir, I. and Goodall, D.W. (eds) *Hot Deserts and Arid Shrublands.* Elsevier, Amsterdam, pp. 1–22.

Fleming, T.H. and Sosa, V.J. (1994) Effects of nectarivorous and frugivorous mammals on reproductive success of plants. *Journal of Mammalogy* 75, 845–851.

Forget, P.-M., Lambert, J.E., Hulme, P.E. and Vander Wall, S.B. (eds) (2004) *Seed Fate: Predation, Dispersal and Seedling Establishment.* CAB International, Wallingford, UK.

French, K. and Westoby, M. (1992) Removal of vertebrate-dispersed fruit in vegetation on fertile and infertile soils. *Oecologia* 91, 447–454.

Fry, C.H., Ash, J.S. and Ferguson-Lees, I.J. (1970) Spring weights of some Palaearctic migrants at Lake Chad. *Ibis* 112, 58–82.

Fuentes, M. (1994) Diets of fruit-eating birds: what are the causes of interspecific differences? *Oecologica* 97, 134–142.

Gannes, L.Z. (2001) Comparative fuel use of migrating passerines: effects of fat stores, migration distance, and diet. *Auk* 118, 665–677.

Geffen, E., Hefner, R. and Wright, P. (2004) North Africa and the Middle East (Ethiopian). In: Sillero-Zubiri, C., Hoffmann, H. and Macdonald, D.W. (eds) *Canids: Foxes, Wolves, Jackals and Dogs*. IUCN/SSC Canid Specialist Group. Gland, Switzerland and Cambridge, UK, pp. 194–228.

Godínez-Alvarez, H. (2004) Pollination and seed dispersal by lizards: a review. *Revista Chilena de Historia Natural* 77, 569–577.

Godínez-Alvarez, H., Valiente-Banuet, A. and Rojas-Martinez, A. (2002) The role of seed dispersers in the population dynamics of the columnar cactus *Neobuxbaumia tetetzo*. *Ecology* 83, 2617–2629.

Gosper, C.R., Stansbury, C.D. and Vivian-Smith, G. (2005) Seed dispersal of fleshy-fruited invasive plants by birds: contributing factors and management options. *Diversity and Distributions* 11, 549–558.

Greenberg, R. (1981) Frugivory in some migrant tropical forest wood warblers. *Biotropica* 13, 215–223.

Greenberg, R., Foster, M.S. and Marquez-Valdelamar, L. (1995) The role of the white-eyed vireo in the dispersal of *Bursera* fruit on the Yucatan Peninsula. *Journal of Tropical Ecology* 11, 619–639.

Greenberg, R., Niven, D.K., Hopp, S. and Boone, C. (1993) Frugivory and coexistence in a resident and a migratory vireo on the Yucatan Peninsula. *Condor* 95, 990–999.

Gutterman, Y. (1993) *Seed Germination in Desert Plants*. Springer, Berlin

Hacker, S.D. and Gaines, S.D. (1997) Some implications of direct positive interactions for community species diversity. *Ecology* 78, 1990–2003.

Hacker, S.D. and Bertness, M.D. (1999) Experimental evidence for factors maintaining plant species diversity in a New England salt marsh. *Ecology* 80, 2064–2073.

Hector, A., Schmid, B., Beierkuhnlein, C., Caldeira, M.C., Diemer, M., Dimitrakopoulos, P.G., Finn, J.A., Freitas, H., Giller, P.S., Good, J., Harris, R., Hogberg, P., Huss, D.K., Joshi, J., Jumpponen, A., Korner, K., Leadley, P.W., Loreau, M., Minns, A., Mulder, C.P.H., O'Donovan, G., Otway, S.J., Pereira, J.S., Prinz, A., Read, D.J., M. Scherer, E.D., Schulze, A.S.D., Siamantziouras, E.M., Spehn, A.C., Terry, A.Y., Troumbis, L., Woodward, F.I., Yachi, S. and Lawton, J.H. (1999) Plant diversity and productivity experiments in European grasslands. *Science* 286, 1123–1127.

Herrera, C.M. (1982) Seasonal variation in the quality of fruits and diffuse coevolution between plants and avian dispersers. *Ecology* 63, 773–85.

Herrera, C.M. (1984) A study of avian frugivores, bird-dispersed plants, and their interaction in Mediterranean scrublands. *Ecological Monographs* 54, 1–23.

Higgins, S.I., Nathan, R. and Cain, M.L. (2003) Are long-distance dispersal events in plants usually caused by nonstandard means of dispersal? *Ecology* 84, 1945–1956.

Holmgren, M., Scheffer, M. and Huston, M.A. (1997) The interplay of facilitation and competition in plant communities. *Ecology* 78, 1966–1975.

Howe, H.F. and Estabrook, G.F. (1977) On intraspecific competition for avian dispersers in tropical trees. *American Naturalist* 111, 817–832.

Howe, H.F. and Smallwood, J. (1982) Ecology of seed dispersal. *Annual Review of Ecology and Systematics* 13, 201–28.

Hughes, L., Dunlop, M., French, K., Leishman, M.R., Rice, B., Rodgerson, L. and Westoby, M. (1994) Predicting dispersal spectra: a minimal set of hypotheses based on plant attributes. *Journal of Ecology* 92, 933–950.

Hulme, P.E. and Benkman, C.W. (2002) Granivory. In: Herrera, C. and Pellmyr, O. (eds) *Plant–Animal Interactions: An Evolutionary Approach*. Blackwell, Oxford, UK, pp. 132–156.

Ims, R.A. (1995) Movement patterns related to spatial structures. In: Hansson, L., Fahrig, L. and Merriam, G. (eds) *Mosaic Landscapes and Ecological Processes*. Chapman and Hall, London, pp. 85–109.

Iwai, K., Suzuki, T. and Fujiwake, H. (1979) Formation and accumulation of pungent principle of hot pepper fruits, capsaicin and its analogues, in *Capsicum annuum* var. *annuum* cv. Karayatsubusa at different growth stages after flowering. *Agricultural and Biological Chemistry* 43, 2493–2498.

Izhaki, I. (2002) The role of fruit traits in determining fruit removal in East Mediterranean ecosystems. In: Levey, D.J., Silva, W.R. and Galetti. M. (eds) *Dispersal and Frugivory: Ecology, Evolution and Conservation.* CAB International, Wallingford, UK, pp. 161–175.

Izhaki, I. and Safriel, U.N. (1985) Why do fleshy-fruit plants of the Mediterranean scrub intercept fall- but not spring-passage of seed-dispersing migratory birds? *Oecologia* 67, 40–43.

Izhaki, I. and Safriel, U.N. (1990) The effect of some Mediterranean scrubland frugivores upon germination patterns. *Journal of Ecology* 98, 56–65.

Janzen, D.H. (1970) Herbivores and the number of tree species in tropical forests. *American Naturalist* 104, 501–528.

Janzen, D.H. (1980) When is it coevolution? *Evolution* 34, 611–612.

Janzen, D.H. (1984) Dispersal of small seeds by big herbivores: foliage is the fruit. *American Naturalist* 123, 338–53.

Janzen, D.H. (1986) Chihuahuan desert nopaleras: defaunated big mammal vegetation. *Annual Review of Ecology and Systematics* 17, 595–636.

Jones, P., Vickery, J., Holt, S. and Cresswell, W. (1996) A preliminary assessment of some factors influencing the density and distribution of palearctic passerine migrants wintering in the Sahel zone of West Africa. *Bird Study* 43, 73–84.

Jordano, P. (1995) Angiosperm fleshy fruits and seed dispersers: a comparative analysis of adaptation and constraints in plant–animal interactions. *American Naturalist* 145, 163–191.

Jordano, P. (2000) Fruits and frugivory. In: Fenner, M. (ed.) *Seeds: The Ecology of Regeneration in Natural Plant Communities.* CAB International, Wallingford, UK, pp. 105–151.

Jordano, P. and Schupp, E.W. (2000) Seed disperser effectiveness: the quantity component and patterns of seed rain for *Prunus mahaleb. Ecological Monographs* 70, 591–615.

Jordt, S.E. and Julius, D. (2002) Molecular basis for species-specific sensitivity to ‘hot’ chili peppers. *Cell* 108, 421–430.

Jurado, E., Westoby, M. and Nelson, D. (1991) Diaspore weight, dispersal, growth form and perenniality of central Australian plants. *Journal of Ecology* 79, 811–830.

Kawada, T., Watanabe, T., Katsura, K., Takami, H. and Iwai, K. (1985) Formation and metabolism of pungent principle of *Capsicum* fruits. 15. Microdetermination of capsaicin by high-performance liquid-chromatography with electrochemical detection. *Journal of Chromatography* 329, 99–105.

Larson, D.L. (1996) Seed dispersal by specialist versus generalist foragers: the plant’s perspective. *Oikos* 76, 113–120.

Levey, D.J. and Benkman, C.W. (1999) Fruit-seed disperser interactions: timely insights from a long-term perspective. *Trends in Ecology and Evolution* 14, 41–43.

Levey, D.J. and Byrne, M.M. (1993) Complex ant–plant interactions: rain forest ants as secondary dispersers and postdispersal seed predators. *Ecology* 74, 1802–1812.

Levey, D.J., Tewksbury, J.J., Cipollini, M. and Carlo, T. (2006) A field test of the directed deterrence hypothesis in two species of wild chile. *Oocologia* 150, 61–68.

Levin, S.A. (1992) The problem of pattern and scale in ecology. *Ecology* 73, 1943–1967.

Loiselle, B.A. and Blake, J.G. (1990) Diets of understory fruit-eating birds in Costa Rica: seasonality and resource abundance. *Studies in Avian Biology* 13, 91–103.

Longland, W.S., Jenkins, S.H., Vander Wall, S.B., Veech, J.A. and Pyare, S. (2001) Seedling recruitment in *Oryzopsis hymenoides*: are desert granivores mutualists or predators? *Ecology* 82, 3131–3148.

Lotan, A. (2005) The impact of environmental conditions on the morphology, nutritional value and secondary metabolites in *Ochradenus baccatus* fruits. MSc thesis, University of Haifa, Israel.

Martin, A.C., Zim, H.S. and Nelson, A.L. (1951) *American Wildlife and Plants: A Guide to Wildlife Food Habitats.* McGraw-Hill, New York.

Martinez Del Rio, C., Silva, A., Medel, R. and Hourdequin, M. (1996) Seed dispersers as disease vectors: bird transmission of mistletoe seeds to plant hosts. *Ecology* 77, 912–921.

Mason, J.R., Bean, N.J., Shah, P.S. and Clark, L. (1991) Taxon-specific differences in responsiveness to capsaicin and several analogs: correlates between chemical structure and behavioural aversiveness. *Journal of Chemical Ecology* 17, 2539–2551.

McAuliffe, J.R. (1984) Sahuaro–nurse tree associations in the Sonoran Desert: competitive effects of sahuaros. *Oecologia* 64, 319–321.

McKey, D. (1975) The ecology of coevolved seed dispersal systems. In: Gilbert, L.E. and Raven, P.H. (eds) *Coevolution of Animals and Plants.* University of Texas Press, Austin, Texas, pp. 159–191.

Milesi, F.A. and De Casenave, J.L. (2004) Unexpected relationships and valuable mistakes: non-myrmecochorous *Prosopis* dispersed by messy leafcutting ants in harvesting their seeds. *Austral Ecology* 29, 558–567.

Milewski, A.V. (1986) A comparison of bird–plant relationships in southern Australia and southern Africa. In: Ford, H.A. and Paton, D.C. (eds) *The Dynamic Partnership: Birds and Plants in Southern Africa.* Flora and Fauna of South Australia Handbooks Committee, pp. 111–118.

Miller, A.G. (1984) A revision of *Ochradenus. Notes from the Royal Botanic Garden of Edinburgh* 41, 491–504.

Milton, S.J. (1992) Plants eaten and dispersed by adult leopard tortoises *Geochelone pardalis* (Reptilia, Chelonii) in the southern Karoo. *South African Journal of Zoology* 27, 45–49.

Milton, S.J. and Dean, W.R.J. (2001) Seeds dispersed in dung of insectivores and herbivores in semi-arid southern Africa. *Journal of Arid Environments* 47, 465–483.

Moermond, T.C. and Denslow, J.S. (1985) Neotropical avian frugivores: patterns of behaviour, morphology, and nutrition, with consequences for fruit selection. *Ornithological Monographs* 36, 865–897.

Moles, A.T., Ackerly, D.D., Webb, C.O., Tweddle, J.C., Dickie, J.B., Pitman, A.J. and Westoby, M. (2005) Factors that shape seed mass evolution. *Proceedings of the National Academy of Sciences USA* 102, 10540–10544.

Montiel, S. and Montana, C. (2000) Vertebrate frugivory and seed dispersal of a Chihuahuan Desert cactus. *Plant Ecology* 146, 221–229.

Morales, J.M. and Carlo, T.A. (2006) The effects of plant distribution and frugivore density on the scale and shape of dispersal kernels. *Ecology* 87, 1489–1496.

Moreau, R.E. (1972) *The Palaearctic–African Bird Migration Systems.* Academic Press, London.

Murphy, S.R., Reid, N., Zhaogui, Y. and Venables, W.N. (1993) Differential passage time of mistletoe fruits through the gut of honeyeaters and flowerpeckers: effects on seedling establishment. *Oecologia* 93, 171–176.

Naranjo, M.E., Rengifo, C. and Soriano, P.J. (2003) Effect of ingestion by bats and birds on seed germination of *Stenocereus griseus* and *Subpilocereus repandus* (Cactaceae). *Journal of Tropical Ecology* 19, 9–15.

Nathan, R. and Muller-Landau, H.C. (2000) Spatial patterns of seed dispersal, their determinants and consequences for recruitment. *Trends in Ecology and Evolution* 15, 278–285.

Nathan, R. and Casagrandi, R. (2004) A simple mechanistic model of seed dispersal, predation and plant establishment: Janzen–Connell and beyond. *Journal of Ecology* 92, 733–746.

Noy-Meir, I. (1985) Desert ecosystem structure and function. In: Evenari, M., Noy-Meir, I. and Goodall, D.W. (eds) *Hot Deserts and Arid Shrublands*. Elsevier, Amsterdam, pp. 93–103.

Parker, C. (1988) Environmental relationships and vegetation associations of columnar cacti in the northern Sonoran Desert. *Vegetatio* 78, 125–140.

Perramond, E. (2005) The politics of ecology: local knowledge and wild chili collection in Sonora, Mexico. *Journal of Latin American Geography* 4, 59–75.

Pickersgill, B. and Heiser, C.B. (1977) Origins and distribution of plants domesticated in the New World tropics. In: Reed, C.A. (ed.) *Origins of Agriculture*. Mouton, The Hague, The Netherlands, pp. 803–886.

Pickersgill, B., Heiser, C. and McNeill, J. (1979) Numerical taxonomic studies on variation and domestication in some species of *Capsicum*. In: Hawkes, J.G., Lester, R.N. and Skelding, A.D. (eds) *The Biology and Taxonomy of the Solanaceae*. Academic Press, London, pp. 679–700.

Place, A.R. and Stiles, E.W. (1992) Living off the wax of the land: bayberries and yellow-rumped warblers. *Auk* 109, 334–345.

Racine, C.H. and Downhower, J.F. (1974) Vegetative and reproductive strategies of *Opuntia* (Cactaceae) in the Galapagos Islands. *Biotropica* 6, 175–186.

Reid, N. (1989) Dispersal of mistletoes by honeyeaters and flowerpeckers: components of seed dispersal quality. *Ecology* 70, 137–145.

Richardson, D.M., Allsopp, N., D'Antonio, C.M., Milton, S.J. and Rejmanek, M. (2000) Plant invasions: the role of mutualisms. *Biological Reviews of the Cambridge Philosophical Society* 75, 65–93.

Russell, S.M. and Monson, G. (1998) *The Birds of Sonora*. University of Arizona Press, Tucson, Arizona.

Sallabanks, R. (1993) Hierarchical mechanisms of fruit selection by an avian frugivore. *Ecology* 74, 1326–1336.

Samuels, I.A. and Levey, D.J. (2005) Effects of gut passage on seed germination: do experiments answer the questions they ask? *Functional Ecology* 19, 365–368.

Sapir, N., Abramsky, Z., Shochat, E. and Izhaki, I. (2004) Scale-dependent habitat selection in migratory frugivorous passerines. *Naturwissenschaften* 91, 544–547.

Sargent, S. (1990) Neighborhood effects on fruit removal by birds: a field experiment with *Viburnum dentatum* (Caprifoliaceae). *Ecology* 71, 1289–1298.

Schlesinger, W.H., Raikes, J.A., Hartley, A.E. and Cross, A.E. (1996) On the spatial pattern of soil nutrients in desert ecosystems. *Ecology* 77, 364–374.

Schmidt-Nielsen, K. (1964) *Desert Animals: Physiological Problems of Heat and Water*. Clarendon Press, Oxford, UK.

Schupp, E.W. (1992) The Janzen–Connell model for tropical tree diversity: population implications and the importance of spatial scale. *American Naturalist* 140, 526–530.

Schupp, E.W., Milleron, T. and Russo, S.E. (2002) Dissemination limitation and the origin and maintenance of species-rich tropical forests. In: Levey, D.J., Silva, W.R. and Galetti, M. (eds) *Seed Dispersal and Frugivory: Ecology, Evolution and Conservation*. CAB International, Wallingford, UK, pp. 19–33.

Schütz, W. and Milberg, P. (1997) Seed germination in *Launaea arborescens*: a continuously flowering semi-desert shrub. *Journal of Arid Environments* 36, 113–122.

Shreve, F. (1931) Physical conditions in sun and shade. *Ecology* 12, 96–104.

Shreve, F. (1951) *Vegetation of the Sonoran Desert*. Carnegie Institution Publication 591, Washington, DC.

Silvius, K.M. (1995) Avian consumers of cardon fruits (*Stenocereus griseus*, Cactaceae) on Margarita Island, Venezuela. *Biotropica* 27, 96–105.

Snow, D.W. (1985) *The Web of Adaptation: Bird Studies in the American Tropics*. Cornell University Press, Ithaca, New York.

Spiegel, O. (2006) Frugivory and seed dispersal in desert ecosystems: scale-dependent dispersal effectiveness. MSc thesis, The Hebrew University of Jerusalem, Jerusalem, Israel.

Steenbergh, W.F. and Lowe, C.H. (1977) *Ecology of the Saguaro. II. Reproduction, Germination, Establishment, Growth, and Survival of the Young Plant.* US Government Printing Office, Washington, DC.

Stiles, E.W. (1993) The influence of pulp lipids on fruit preference by birds. *Vegetatio* 108, 227–235.

Stoate, C. and Moreby, S.J. (1995) Premigratory diet of trans-Saharan migrant passerines in the western Sahel. *Bird Study* 42, 101–106.

Sung, Y., Chang, Y.Y. and Ting, N.L. (2005) Capsaicin biosynthesis in water-stressed hot pepper fruits. *Botanical Bulletin of Academia Sinica* 46, 35–42.

Tester, M.D., Paton, C., Reid, N. and Lange, R.T. (1987) Seed dispersal by birds and densities of shrubs under trees in arid South Australia. *Transactions of the Royal Society of South Australia* 111, 1–5.

Tewksbury, J.J. and Lloyd, J.D. (2001) Positive interactions under nurse-plants: spatial scale, stress gradients and benefactor size. *Oecologia* 127, 425–434.

Tewksbury, J.J. and Nabhan, G.P. (2001) Seed dispersal: directed deterrence by capsaicin in chillies. *Nature* 412, 403–404.

Tewksbury, J.J. and Petrovich, C.A. (1994) The influence of ironwood as a habitat modifier species. In: Nabhan, J.P. and Carr, J.L. (eds) *Ironwood: An Ecological and Cultural Keystone of the Sonoran Desert.* University of Chicago Press, Chicago, Illinois, pp. 29–54.

Tewksbury, J.J., Nabhan, G.P., Norman, D., Suzan, H., Tuxill, J. and Donovan, J. (1999) In situ conservation of wild chiles and their biotic associates. *Conservation Biology* 13, 98–107.

Tews, J., Moloney, K. and Jeltsch, F. (2004) Modeling seed dispersal in a variable environment: a case study of the fleshy-fruited savanna shrub *Grewia flava*. *Ecological Modelling* 175, 65–76.

Turner, R.M., Alcorn, S.M., Olin, G. and Booth, J.A. (1966) The influence of shade, soil and water on saguaro seedling establishment. *Botanical Gazette* 127, 95–102.

Valido, A. and Nogales, M. (1994) Frugivory and seed dispersal by the lizard *Gallotia galloti* (Lacertidae) in a xeric habitat of the Canary Islands. *Oikos* 70, 403–411.

Valiente-Banuet, A. and Ezcurra, E. (1991) Shade as a cause of the association between the cactus *Neobuxbaumia tetetzo* and the nurse plant *Mimosa luisana* in the Tehuacan Valley, Mexico. *Journal of Ecology* 79, 961–971.

van Rheede van Oudtshoorn, K. and van Rooyen, M.W. (1999) *Dispersal Biology of Desert Plants.* Springer, Berlin.

van Rooyen, M.W., Theron, G.K. and Grobbelaar, N. (1990) Life form and dispersal spectra of the flora of Namaqualand, South Africa. *Journal of Arid Environments* 19, 133–145.

Vander Wall, S.B. and Longland, W.S. (2004) Diplochory: are two seed dispersers better than one? *Trends in Ecology and Evolution* 19, 155–161.

Villagra, P.E., Marone, L. and Cony, M.A. (2002) Mechanisms affecting the fate of *Prosopis flexuosa* (Fabaceae, Mimosoideae) seeds during early secondary dispersal in the Monte Desert, Argentina. *Austral Ecology* 27, 416–421.

Vulinec, K. (2002) Dung beetle communities and seed dispersal in primary forest and disturbed land in Amazonia. *Biotropica* 34, 297–309.

Walsberg, G.E. (1977) Ecology and energetics of contrasting social systems in *Phainopepla nitens* (Aves: Ptilogonatidae). *University of California Publications in Zoology* 108, 1–63.

Walsh, B.M. and Hoot, S.B. (2001) Phylogenetic relationships of *Capsicum* (Solanaceae) using DNA sequences from two noncoding regions: the chloroplast atpB-rbcL spacer region and nuclear waxy introns. *International Journal of Plant Science* 162, 1409–1418.

Ward, D. and Saltz, D. (1994) Foraging at different spatial scales: Dorcas gazelles foraging for lilies in the Negev desert. *Ecology* 75, 48–58.

Westoby, M., French, K., Hughes, L., Rice, B. and Rodgerson, L. (1991) Why do more plant species use ants for dispersal on infertile compared with fertile soils? *Australian Journal of Ecology* 16, 445–455.

Whiting, M.J. and Greeff, J.M. (1997) Facultative frugivory in the Cape flat lizard, *Platysaurus capensis* (Sauria: Cordylidae). *Copeia*, 811–818.

Williams, J.B. and Tieleman, B.I. (2002) Ecological and evolutionary physiology of desert birds: a progress report. *Integrative and Comparative Biology* 42, 68–75.

Willson, M.F. (1986) Avian frugivory and seed dispersal in eastern North America. *Current Ornithology* 3, 223–279.

Willson, M.F., Irvine, A.K. and Walsh, N.G. (1989) Vertebrate dispersal syndromes in some Australian and New Zealand plant communities, with geographic comparisons. *Biotropica* 21, 133–147.

Willson, M.F., Rice, B.L. and Westoby, M. (1990) Seed dispersal spectra: a comparison of temperate plant communities. *Journal of Vegetation Science* 1, 547–562.

Witmer, M.C. (1996) Do some bird-dispersed fruits contain natural laxatives? A comment. *Ecology* 77, 1947–1948.

Witmer, M.C. and Van Soest, P.J. (1998) Contrasting digestive strategies of fruit-eating birds. *Functional Ecology* 12, 728–741.

Wolf, B.O. and Martinez del Rio, C. (2000) Use of saguaro fruit by white-winged doves: isotopic evidence of a tight ecological association. *Oecologia* 124, 536–543.

Wolf, B.O., Martinez del Rio, C. and Babson, J. (2002) Stable isotopes reveal that saguaro fruit provides different resources to two desert dove species. *Ecology* 83, 1286–1293.

Wolf, B.O. and Martinez del Rio, C. (2003) How important are columnar cacti as source of water and nutrients for desert consumers? A review. *Isotopes in Environmental and Health Studies* 39, 53–67.

Wolfe, L.M. and Shmida, A. (1995) Regulation of gender and flowering behaviour in a sexually dimorphic desert shrub (*Ochradenus baccatus* Delile [Resedaceae]). *Israel Journal of Plant Sciences* 43, 325–337.

Wolfe, L.M. and Shmida, A. (1997) The ecology of sex expression in a gynodioecious Israeli desert shrub (*Ochradenus baccatus*). *Ecology* 78, 101–110.

Wolfe, L.M. and Burns, J.L. (2001) A rare continual flowering strategy and its influence on offspring quality in a gynodioecious plant. *American Journal of Botany* 88, 1419–1423.

Zohary, M. (1966) *Flora Palaestina. I.* Israel Academy of Sciences and Humanities, Jerusalem, Israel.

8 Ecological Redundancy in Seed Dispersal Systems: A Comparison Between Manakins (Aves: Pipridae) in Two Tropical Forests

B.A. LOISELLE, P.G. BLENDINGER, J.G. BLAKE AND T.B. RYDER

Introduction

Redundancy in ecological function may enhance ecosystem resilience in the face of disturbance, and thus serve to conserve biological diversity (Naeem *et al.*, 1994; Walker, 1995). For example, high-diversity grassland ecosystems are more resistant to disturbance and recover more fully than lower-diversity grasslands (Tilman and Downing, 1994). When more species within an ecosystem offer equivalent ecological services, such as dispersal of plant seeds, then decline or local extinction of one species is offset by other species because of their ecological equivalency. Such redundancy in ecological services, in essence, buffers ecosystems from disturbance. Given present-day concerns about how diversity and stability of ecosystems are affected by disturbance, understanding the degree to which species are redundant in ecological function has received considerable attention in recent years (Ehrlich and Walker, 1998; Naeem, 1998; Chong *et al.*, 2000; Wellnitz and Poff, 2001; Slater and du Toit, 2002).

Seed dispersal is an important ecological process that provides a number of ecosystem services and economic benefits, such as removal of seeds from parent plants where seed and seedling mortality is high, escape from seed predators, enhanced germination, increased gene flow, and regeneration in and restoration of natural and disturbed habitats (Şekercioğlu *et al.*, 2004). Unless compensated, loss of seed dispersers can be expected to result in a reduction in numbers of seeds removed, clumping of seeds below the parent plant, increased predation of seeds and seedlings, reduction in recruitment and gene flow, and declines in (or extinction of) dependent species of plants (see, e.g. Cardoso da Silva and Tabarelli, 2000; Cordeiro and Howe, 2003). From a plant perspective, ecologically redundant dispersers are those that, at the least, remove a proportionally equivalent number of seeds from the plant and offer similar treatment to seeds. Given the importance of establishment limitation in recruitment, it can also be argued that ecologically equivalent species also deposit seeds in similar environments. Here, we address the question of whether seed dispersers in species-rich forests overlap more in ecological function than do seed dispersers in less diverse forests. If so, then increased biological diversity, at least in terms of seed dispersal processes, may increase ecosystem resilience.

Interactions between fruiting plants and their seed dispersers are now largely recognized as diffuse mutualisms with seeds dispersed by an array of fruit consumers, whose behaviour and composition may vary both geographically and temporally (Wheelwright and Orians, 1982; Herrera, 1985, 1986; Gautier-Hion *et al.*, 1993; Chapman and Chapman, 2002; Dennis and Westcott, Chapter 9, this volume). Tight co-evolutionary relationships between plants and seed dispersers are considered rare and limited to cases where seed size constrains a potential disperser assemblage, or to areas where the disperser community is impoverished (Rick and Bowman, 1961; Chapman *et al.*, 1992; Chapman and Chapman, 2002). If interactions between fruiting plants and seed dispersers are diffuse, and if they vary over space and time, then the relative importance of individual dispersers as selective agents on plant traits and their influence on plant community structure and composition are expected to be weak. Diffuse relationships, however, do not preclude dispersers from potentially having selective impacts on plant fitness. If dispersers largely overlap in ecological function, then these species as a group directly influence quantity of seeds removed and early patterns of seed and seedling establishment and, thus, act as selective forces on plant traits.

In neotropical wet forests, a vast majority of woody plant species rely on animals to disperse their seeds (Gentry, 1982) and a rich assemblage of fruit-eating vertebrates can be found (e.g. Karr *et al.*, 1990; Terborgh *et al.*, 1990; Robinson *et al.*, 2000; Blake and Loiselle, 2000, 2001). The high diversity of fleshy-fruited plants and vertebrate frugivores has resulted in complex webs of interactions among plants and their consumers (e.g. McDiarmid *et al.*, 1977; Lambert, 1989; Silva *et al.*, 2002; see Silva *et al.*, Chapter 26, this volume). Yet, from a plant's perspective, the ecological roles of dispersers are not necessarily equivalent and, in some cases, an individual species may be irreplaceable (e.g. Stocker and Irvine, 1983; Wenny and Levey, 1998; see Dennis and Westcott, Chapter 9, this volume). In understorey plants of lowland wet forest in Costa Rica, we found that plant–disperser interactions can be largely asymmetric, with a relatively small suite of species dominating the interactions (Loiselle and Blake, 1999, 2002; see also Silva *et al.*, 2002). Dominance by a few dispersers may have ecological consequences for plants and, potentially, evolutionary consequences if gene flow is limited and interaction strengths remain relatively constant over time. Moreover, if the interactions are dominated by species that are at risk of extinction, whose roles are largely irreplaceable, then plants may be subject to dissemination limitation and population reductions following the decline or loss of strong interactors (e.g. Rainey *et al.*, 1995; Hamann and Curio, 1999; Tabarelli *et al.*, 1999; McConkey and Drake, 2002; Cordeiro and Howe, 2001, 2003; see Silva *et al.*, Chapter 26, this volume).

Here we are interested in whether plant populations in species-rich areas (i.e. areas with many dispersal agents) are potentially buffered from disturbance of disperser communities as a result of increased overlap in ecological function (i.e. redundancy) of seed dispersers. We measure ecological redundancy in seed dispersers based on overlap in species consumed (cf. Jaksic *et al.*, 1996; Slater and du Toit, 2002) and overlap in environments that dispersers occupy and, presumably, disseminate seeds into. We compare our results from species-rich forests in eastern Ecuador to those from Costa Rica, which although biologically diverse, are less so than Amazonian forests. We focus on manakins (Pipridae), which are the predominant arboreal frugivores in the understorey of these two forests (Levey, 1988; Loiselle and Blake, 1991; English, 1998; J.G. Blake and B.A. Loiselle, 2006, unpublished results). We restrict our analyses to manakins, which are the numerically dominant frugivores captured in mist-nets at these sites, to control, in part, for potential effects of phylogeny on fruit selection. We later discuss how manakins fit within the entire frugivore community.

Manakins, which feed primarily on fruit, exhibit lek-mating systems where the males spend much of their day at display arenas (Höglund and Alatalo, 1995). Females visit the leks to select males for mating and are solely responsible for nest-building and all parental duties. These large sexual differences in the behavioural ecology of manakins may influence their respective roles as seed dispersers, such that differences between sexes may rival, or even excel, differences among species. At La Selva Biological Station in Costa Rica, two species of manakins are regularly found in the understorey of old-growth forest, *Pipra mentalis* and *Corapipo altera*, although the latter species breeds at higher elevations and is only a seasonal visitor (4–6 months) at this site (Levey, 1988; Loiselle and Blake, 1991; Rosselli, 1994). In eastern Ecuador, six species of manakins are regularly observed and breed in the forest understorey.

To examine ecological redundancy in seed disperser roles of manakins in these two forests, which differ in biological diversity, we specifically ask:

1. Whether species significantly overlap in fruit diet;
2. Whether species overlap in use of forest environments;
3. Whether the diet and environmental overlap among manakins is greater in the more species-rich forests of eastern Ecuador when compared with Costa Rica.

In addition, we examine whether seed-dispersal roles of manakins in Ecuador differ between sexes, which would imply that studies focusing on ecological function should treat male and female manakins as separate experimental units. Our prediction is that higher species diversity in eastern Ecuador forests will lead to greater redundancy in the ecological functions of species, when compared to Costa Rican forests where local extinction of even one species would theoretically have significant impacts on the spatial dissemination of seeds (Loiselle and Blake, 2002).

Methods

Data on diets of manakins from Ecuador were collected during a 3-year period (2001–2004) at Tiputini Biodiversity Station (TBS), Orellana Province, Ecuador (~0°38′ S, 76°08′ W). TBS was established in 1994 by the Universidad San Francisco de Quito (USFQ) on a tract of undisturbed lowland rainforest within the 1.5-million-ha Yasuní Biosphere Reserve, a region noted for its biological diversity. There are approximately 300+ species of trees (> 10 cm diameter at breast height) per hectare (Kreft *et al.*, 2004), over 525 species of birds, including 12 within the family Pipridae, and more than 150 species of mammals. The station and nearby areas contain a variety of habitats that include *terra firme* and *várzea* forest, palm swamps and other wetlands, as well as numerous areas of natural regrowth (i.e. blow-down areas in various stages of succession). Within TBS, two 100 ha plots (~1 km × 1 km each) were established in *terra firme* forest during 2001 (see Ryder *et al.*, 2006; Loiselle *et al.*, 2007, for a description of Harpia and Puma plots).

Between 2001 and 2004 (February–March 2001; May–August 2002; December 2001–March 2002; May–August 2003; December 2002–March 2003; December 2003–March 2004), the distribution and abundance of manakins was studied on the two plots, using mist-nets, spot-mapping and intensive observations. When possible, faecal samples were collected from manakins captured in mist-nets (12 m × 2.8 m; 36 mm mesh); mist-nets were distributed throughout the study plots. Here, we report on the diets of four species of manakins based on a total of 428 faecal samples that contained plant seeds (Table 8.1). We recognize, however, that faecal samples are biased towards small-seeded plants, as larger seeds are often regurgitated. Therefore our study describes the component of manakin diets that achieves dispersal through the gut. We also examine overlap in habitat use by comparing the spatial distribution of captures at 96 net-sites on each plot. We restrict our comparisons to these 96 net-sites because

Table 8.1. Number of faecal samples containing plant seeds and number of plant species found in faecal samples for four manakin (Pipridae) species in eastern Ecuador based on data collected in 2001–2004.

Species	Number of faecal samples	Number of plant species
Lepidothrix coronata	206	49
Pipra erythrocephala	43	27
Pipra filicauda	56	33
Pipra pipra	123	44
Total	428	85

these sites were systematically sampled during January and March of all four years. We assume that capture sites of birds are likely to represent environments where plant seeds are deposited (cf. Levey, 1988; Loiselle and Blake, 1993, 2002; Blake and Loiselle, 2001).

Diets of manakins from Costa Rica were studied in tropical lowland wet forest at La Selva Biological Station between 1985 and 1993; except for 1985, diets were sampled in conjunction with systematic mist-net studies between December and April (for a complete description of study methods from Costa Rica see Loiselle and Blake, 1990, 1991, 1993, 2002). These months correspond to the time when both *Pipra mentalis* and *Corapipo altera* co-occur in old-growth forest at this site. Comparisons of overlap in diet and habitat-use follow methods used for manakins in Ecuador. In Costa Rica, nets were placed at similar distance intervals, but covered a smaller area.

We evaluated the degree to which diets of manakins were sampled by plotting species accumulation curves and examined differences in the diversity of plant species in manakin diets using rarefaction. Rarefaction uses probability theory to derive an expected number of species and its variance for a sample of a given size (Hurlbert, 1971). We used EcoSim version 6.18 (Gotelli and Entsminger, 2001) to randomly draw a set of diet samples of a specified abundance. We repeated the random draws for each abundance level 1000 times to generate a mean and variance of plant species found in the diet. We repeated this process for each species of manakin separately, as well as for all species combined, at each study site.

To determine whether or not species of manakin overlapped significantly in fruit diet, we constructed a bird-species × plant-species matrix, where the frequency of faecal samples that contained seeds of a given plant filled the cells. We then compared interspecific niche overlap against a random model using EcoSim version 6.18 (Gotelli and Entsminger, 2001). We used Pianka's (1973) pairwise index to calculate niche overlap where

$$O_{12} = O_{21} = \frac{\sum_{i\_l}^{n} p_{2i} p_{1i}}{\sqrt{\sum_{i\_l}^{n} \left(p_{2i}^{2}\right)\left(p_{1i}^{2}\right)}}$$

O_{12}, which equals O_{21}, is the overlap of species 1 with species 2, p_{2i} and p_{1i} is the proportion of plant species i in the diet of species 2 and species 1, respectively. We then examined whether species overlapped more or less than expected by chance using the mean of observed overlaps between pairs of species compared to that produced by an algorithm that retains the niche breadth of each species of manakin, but randomizes which particular species of plant is included in the diet (Winemiller and Pianka, 1990; Gotelli and Entsminger, 2001). This randomization algorithm simply reshuffles cells within each row of the matrix, thus retaining the same niche breadth of the species while assuming that each species of manakin can potentially feed on any given species of plant (i.e. it assumes that birds are physiologically and morphologically capable of feeding on all food items in the matrix). Given the similarities in morphology, body size and foraging manoeuvres among these four species of manakins, we felt the latter assumption was justified. We used 1000 simulations to derive the expected values of niche overlap for species pairs.

To determine whether the spatial distributions of manakin captures were similar on the two study plots, we used GEODISTN.BAS, a QuickBasic 4.5 program that tests for differences between two spatial distributions (Syrjala, 1996). This non-parametric test examines the null hypothesis that two populations have the same spatial distribution. We used the Cramér–von Mises test statistic because it is relatively insensitive to a small number of extreme observations (Syrjala, 1996). We ran 1000 permutations to determine the test statistic. For each species pair on each plot, we provided the *x*–*y* location of the net-site and the number of captures for each species of manakin. A significant test statistic indicates that the spatial distribution of captures for species 1 is different from the spatial distribution of captures for species 2. In Costa Rica, we compared the spatial distribution of manakin captures over a total of 60 net-sites. In Ecuador, manakin captures were compared over 96 net-sites in each of the two study plots.

To compare the forest environments where the manakins were captured and assumed to disseminate seeds (cf. Levey, 1988; Loiselle and Blake, 1993, 2002) in Costa Rica, we assigned net-sites to one of five habitats (ridge, upper-slope, mid-slope, lower-slope, and valley bottoms and other low-lying habitats) (see Loiselle and Blake, 1993, 2002). We then compared the distribution of captures of manakin species among habitats using a χ^2 heterogeneity test. In Ecuador, where plots were georeferenced and GIS databases existed for topography and streams, we established a sample area of 12.5 m radius around the centre of each mist-net location where a manakin was captured. This area was divided into 1 × 1 m cells, within which we measured elevation (m), slope (degrees), aspect and distance from stream (m). We then used zonal statistics in GIS (Spatial Analyst; ESRI), with each net-site defining a zone, to calculate means and standard deviations for each environmental variable within each zone. To visualize the variation in environmental characteristics of sample areas around net-sites within each plot, we used principal components analysis where zone/net-sites are the rows and environmental variables (means and standard deviations) are the columns. Net-sites were ordinated in environmental space using a correlation matrix and Euclidean distance measure, using factor analysis in SPSS version 10 (SPSS, 1999). To determine whether species of manakins in Ecuador were captured in different forest environments and, thus, expected to disseminate seeds to qualitatively different sites, we used the net-site × environment data matrix for analysis of similarity (ANOSIM, described by Clarke and Warwick, 2001). ANOSIM is a non-parametric permutation procedure that is combined with a Monte Carlo test to determine the level of similarity of captures among net-sites across species. The significance of the ANOSIM test statistics was determined by comparison with the values obtained by the randomization procedure.

Results

Diet overlap

We found 81 species of plants in the diets of *Pipra mentalis* and *Corapipo altera* in old-growth forests of Costa Rica (see also Loiselle and Blake, 1990, 1999, 2002). *Pipra mentalis* contained 70 species in 441 faecal samples, while *C. altera* had 39 species in 152 faecal samples; the seeds of 35 species were recorded only once in the diets. Genera of plants commonly encountered in the diets of Costa Rican manakins included *Henriettea*, *Clidemia*, *Ossaea*, *Miconia* and *Leandra* (Melastomataceae), *Psychotria* (Rubiaceae), *Hampea* (Malvaceae) and *Pinzona* (Dilleniaceae). Using all species of plant in the diets, we found that niche overlap between these two manakins was 0.874; using only the 46 species that were recorded at least twice, niche overlap was 0.876. In either case, manakin diets overlapped significantly more than expected by chance ($P < 0.001$). Species accumulation curves for plants revealed that species were still being added to manakin diets and, thus, diet breadth was underestimated for both species in old-growth forests at La Selva (Fig. 8.1). Using rarefaction analysis (Gotelli and Entsminger, 2001), we found that the two species did not differ significantly in diet diversity when samples were corrected for effort. Costa Rican manakins, therefore, generally feed on a similar suite of plants in similar proportions and diversity.

We found 85 species of plants in the diets of the four species of manakins in Ecuador (Table 8.1). Genera of plants commonly encountered in Ecuadorian manakin diets included *Anthurium* and *Philodendron* (Araceae), *Coussapoa* and *Cecropia* (Cecropiaceae), *Miconia* and *Henriettea* (Melastomataceae), *Besleria* (Gesneriaceae) and *Ficus* (Moraceae). Nearly half of all species (n = 42 species) were recorded only once in the diets. Thirty species of plants were represented in at least four faecal samples, and thus could potentially be found in the diets of all four species of manakins. Thirteen of these were included in the diets of all four manakins, and an additional eight were found in the diets of at least three species of manakins. Using only those 30 species of plants that were recorded four or more times in the diets of all manakins combined, we found significantly greater niche overlap (mean pairwise niche overlap = 0.874; $P < 0.001$) than expected by chance (Table 8.2). When all 85 plant species were used to calculate

Table 8.2. Pairwise niche overlap values based on plant species found in at least four faecal samples (n = 30 plant species) for species of manakin found in Ecuador. When all 85 plant species are included, pairwise values differed by less than 0.02 in all cases.

Species	*L. coronata*	*P. erythrocephala*	*P. filicauda*	*P. pipra*
L. coronata	–	0.962	0.834	0.936
P. erythrocephala	–	–	0.854	0.919
P. filicauda	–	–	–	0.741

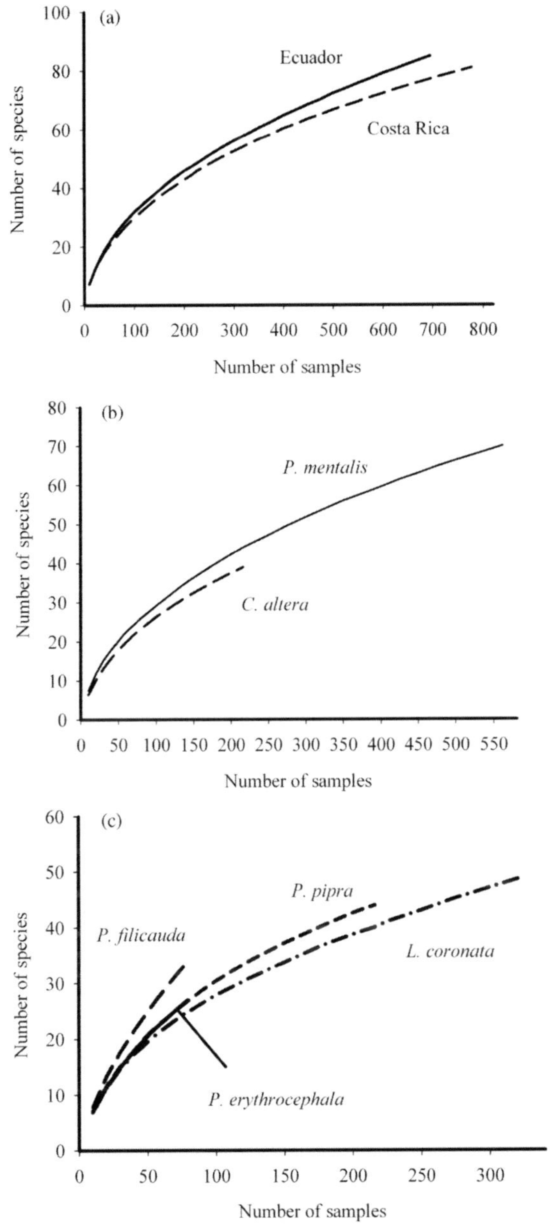

Fig. 8.1. Species accumulation curves for number of species of plants in the diet of manakins in lowland wet forests of Costa Rica and Ecuador: (a) all species combined in Costa Rica (2 species; dashed line) and Ecuador (4 species; solid line), which includes a total of 81 species of plant in 778 samples in Costa Rica and 85 species of plants in 696 samples in Ecuador (number of samples exceeds number of faecal samples reported in the text as > 1 species of plant can be found in a single faecal sample); (b) the two Costa Rican manakins separately (*P. mentalis*, solid line; *C. altera*, dashed line); (c) the four Ecuadorian manakins (*P. filicauda*, *P. pipra*, *P. erythrocephala* and *L. coronata*) separately.

pairwise niche overlap values, we found that values were nearly identical (<0.02 difference in all cases) to this reduced set of 30 species; overall the mean value for pairwise niche overlap was 0.862). Species accumulation curves for plants revealed that more effort is needed to characterize the diets of these four species of manakins in Ecuador; species are still being added even after all samples are combined (Fig. 8.1). Using rarefaction analysis (Gotelli and Entsminger, 2001), we found that the diet of *P. filicauda*, with 33 species, was significantly more diverse ($P < 0.05$) when sample effort was standardized at 76 samples (the cumulative number of samples for *P. filicauda*) than the other three species of manakins; the latter three species did not differ significantly in number of plant species consumed. These results suggest that these four species of manakins are, for the most part, ecologically equivalent in the quantitative and diversity components of seed-disperser effectiveness because they feed on a similar suite of plant species in similar proportions. However, *P. filicauda* was the most different in diet, with relatively lower pairwise values for niche overlap (0.741−0.854) and more plant species consumed after correcting for sampling effort. When *P. filicauda* is excluded, mean pairwise overlap among the remaining three species increased to 0.939 (with the 30 most common species; or 0.934 with all 85 plant species).

Habitat overlap

In old-growth forest of La Selva, 181 *Corapipo altera* and 470 *Pipra mentalis* were captured between 1985 and 1993. The spatial distribution of captures between these two species differed significantly among the 60 mist-net sites (Cramér–von Mises test, $P = 0.043$). Differences in the spatial dissemination of seeds do not necessarily mean that seeds are deposited into different forest environments. When we compared the distribution of captures as a function of forest environments, we found that *P. mentalis* was captured more frequently in ridge habitats, while *C. altera* was captured more frequently in valley bottoms and other low-lying habitats within the sample area ($\chi^2 = 20.3$, $df = 4$, $P < 0.005$) (see also Loiselle and Blake, 1999, 2002). Consequently, these two manakins apparently differ in the local spatial dissemination of seeds, as well as the proportional deposition of seeds into different forest environments. Consequently, differential habitat use by Costa Rican manakins results in a divergence of ecological roles (i.e. lower ecological redundancy) for this component of the seed dispersal process.

In Ecuador, the relative abundance of manakins captured on the two 100 ha study plots differed, with *P. filicauda* captured more frequently on Puma plot, and *P. pipra* more frequently on Harpia plot; *Lepidothrix coronata* was the most frequently captured manakin on both plots (Table 8.3). The relative ranking of each species on each plot based on captures is supported by census data (J.G. Blake, eastern Ecuador, 2006, unpublished results). Differences in these large-scale patterns of

Table 8.3. Number of manakin captures on the two 100 ha study plots in lowland wet forest of eastern Ecuador. Sampling effort is equivalent on the two plots (see Methods).

Species	Harpia	Puma
L. coronata	182	183
P. erythrocephala	29	18
P. filicauda	28	143
P. pipra	85	30

captures were also reflected locally within plots. We found that the spatial distribution of captures between species of manakin differed significantly in nine of 12 pairwise comparisons (Table 8.4). However, patterns were not necessarily consistent between the two study plots. These results suggest that manakins deposit seeds in different locations, except for *P. filicauda* and *L. coronata* on Puma plot, and *L. coronata* and *P. erythrocephala*, or *P. erythrocephala* and *P. pipra* on Harpia plot (Table 8.4).

When we examined the forest environment at net-sites where manakins were captured, as defined largely by topographic features and their heterogeneity (elevation, slope, aspect, and distance to stream), we found few significant differences between sites (Harpia ANOSIM: Global $R = 0.026$, $P = 0.062$; Puma ANOSIM: Global $R = 0.024$, $P = 0.143$). Only on Harpia plot did we find that *P. filicauda* generally frequented different forest environments than *P. erythrocephala* ($P = 0.001$) or *P. pipra* ($P = 0.008$) did (Fig. 8.2). Consequently, these results suggest that although manakins may differ in the spatial dissemination of seeds (results of Cramér–von Mises test above), seeds are expected to arrive into similar forest environments, except as noted above. Thus, there appears to be considerable ecological redundancy among species of Ecuadorian manakins in the forest environments where seeds are disseminated; a qualitative component of disperser effectiveness. However, as we observed in diet overlap, *P. filicauda* was the most distinct of the four species in terms of forest environments frequented.

Table 8.4. Results of the Cramér–Von Mises test examining differences in the spatial distribution of captures between species of manakin in Ecuador. *P* values for Harpia plot are shown above the diagonal, while values for Puma plot are shown below the diagonal.

Species	*L. coronata*	*P. erythrocephala*	*P. filicauda*	*P. pipra*
L. coronata	–	0.538	0.001	0.003
P. erythrocephala	0.029	–	0.004	0.340
P. filicauda	0.288	0.028	–	0.001
P. pipra	0.001	0.020	0.001	–

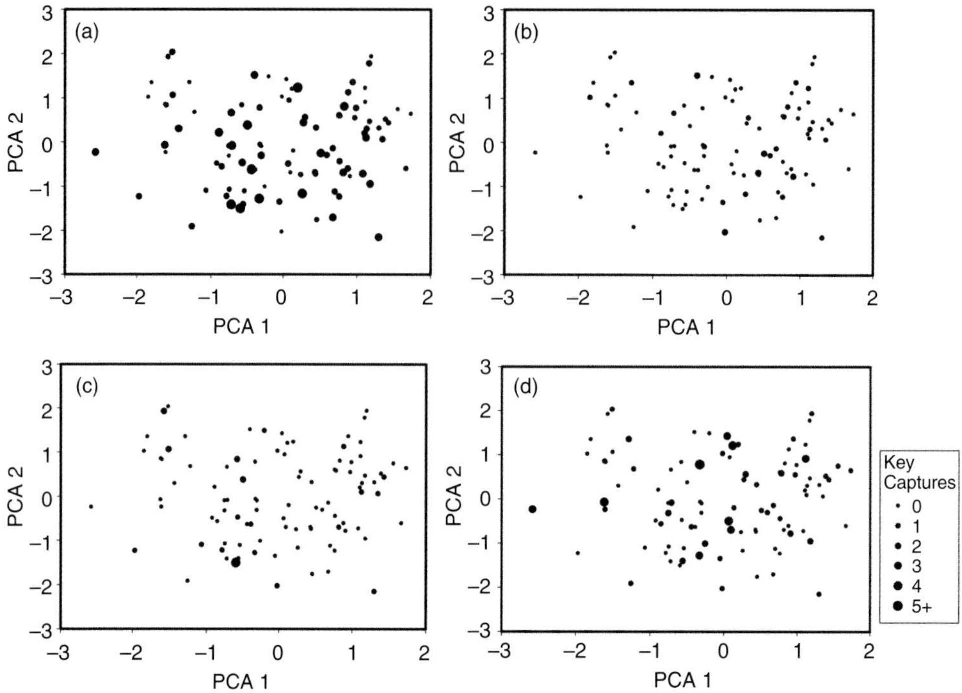

Fig. 8.2. Overlay of manakin captures on net-sites plotted in environmental space based on results from a principal components analysis (PCA) using data from Harpia study plot in Ecuador. The first two principal components accounted for 55.6% of the total variance. Species are displayed as follows: (a) *Lepidothrix coronata*, (b) *Pipra erythrocephala*, (c) *Pipra filicauda* and (d) *Pipra pipra*.

Discussion

Local extinction of strong interactors can result in significant changes in ecosystem structure and function (e.g. Paine, 1969; Flecker, 1996). Consequently, the existence of ecologically redundant species in ecosystems has been hypothesized to buffer ecosystems from disturbance or loss of component species (Naeem *et al.*, 1995; Loreau, 2000). Evidence that redundancy in ecological function exists has been shown experimentally in studies that examined aquatic microbial decomposition (Naeem and Li, 1997; Scarff and Bradley, 2002) and plant productivity (Walker *et al.*, 1999). Moreover, some studies have demonstrated that increased biological diversity buffers ecological function (Naeem *et al.*, 1994; Tilman and Downing, 1994). Increased biological diversity theoretically leads to increased ecological overlap among species. Consequently, the presence of more species performing similar ecological roles offers greater resilience in the face of disturbance, as ecologically equivalent species can compensate for the loss of one or more other species within the same functional group

(Walker, 1995). However, even in biologically diverse communities, the ecological function of some species may be unique. For example, the experimental exclusion of *Prochilodus mariae*, an Andean detritivorous fish, led to large changes in sediment deposition and algal and invertebrate communities (Flecker, 1996). Similarly, the ecological roles of some seed dispersers in species-rich forests appear to be irreplaceable (Stocker and Irvine, 1983; Wenny and Levey, 1998). For example, bellbirds in Costa Rica uniquely disseminate seeds of Lauraceae to favourable sites for germination (Wenny and Levey, 1998). The local extinction of this species would hypothetically cause declines of these trees over time due to recruitment limitation. In a number of cases, local extinction or declines of dominant seed dispersers have resulted in declines in dependent plant species (e.g. Tabarelli *et al.*, 1999; Cordeiro and Howe, 2001, 2003).

Our prediction that manakins in species-rich forests of Ecuador are more ecologically redundant in their roles as seed dispersers than manakins in relatively less diverse forests of Costa Rica was largely upheld. In both forests, pairwise niche overlap values were nearly identical (0.874 versus 0.862 using all plant species in Costa Rica and Ecuador, respectively) suggesting that different species fed on a similar suite of plant species in similar proportions. Yet when *Pipra filicauda* is excluded, pairwise niche overlap for the three remaining species in Ecuador increased to 0.934 (range 0.912–0.957 using all plant species; see also Table 8.2). In addition, manakins in Ecuador were much more likely to disseminate seeds into similar forest environments when compared with Costa Rican manakins, which differed significantly in their use of forest habitats. We also expect that species of manakin overlap considerably in other qualitative components of disperser effectiveness. Given the similar body size and digestive morphologies, we expect that seeds receive similar treatment as they pass through the digestive tract of these manakins (see Traveset *et al.*, Chapter 4, this volume). Fruit-eating birds directly affect plant recruitment through patterns of fruit removal, seed treatment, gut-mixing of seeds, and habitat selection (Schupp, 1993; Jordano and Herrera, 1995). In species-rich forests in Ecuador, the seed-dispersal roles of manakins appear to be quite redundant. Consequently, plants in Ecuador are expected to be less vulnerable to declines or variation in population abundance of one or more of their seed dispersers when compared with Costa Rica or other relatively species-poor forests, where fewer species dominate disperser–plant interactions (Loiselle and Blake, 2002; McConkey and Drake, 2002; Silva *et al.*, Chapter 26, this volume). In addition, ecological equivalency may also mean that different species of dispersers probably exert similar selective pressures on plant recruitment and, thus, may influence the evolution of plant traits.

Our conclusions that manakins from species-rich forests overlap more in ecological function than do manakins in less diverse forests, and therefore that these forests may be more resilient to disturbance, should be treated with caution for several reasons. First, species accumulation curves revealed that we have yet to adequately sample the diets of these manakins. In all cases, species accumulation curves have not levelled out, even when samples are combined across species (Fig. 8.1). Second, many species of plants are only recorded rarely in the diet and, thus, for many species the ecological overlap in seed dispersal roles remains unknown. Furthermore, rare events may be particularly significant in seed dispersal (e.g. rare long-distance dispersal events), and if such rare dispersal events are restricted to one species of disperser, then loss or decline of this species could significantly impact plant populations.

Our comparisons among manakins may also mask important differences in disperser effectiveness between sexes within a species. As lek-breeding birds, male manakins spend considerable time at aggregated display sites during an extended breeding season (Snow, 1962; Sick, 1967; Lill, 1976; Schwartz and Snow, 1978; Tello, 2001). Consequently, movement patterns and habitat use of males may differ from those of females, and such differences are likely to affect qualitative components of disperser effectiveness. None the less, when we separately examined the forest environments where males and females were captured, our results paralleled those reported for the species as a whole, such that few or no differences were found between species in forest environments where manakins were captured. However, we have found that the environmental conditions at lek sites where males aggregate does differ significantly among species of manakins (Loiselle *et al.*, 2007). Thus, even though males overlap substantially in diet, they are more likely to differ in the forest environments into which they disseminate seeds, given the substantial amount of their time spent at lek sites (see also Théry, 1992). If male manakins bring plant seeds to different forest environments, then we might expect that seed survival and seedling establishment in certain lek environments of some plant species may be limited, whereas in others the environmental conditions might favour establishment. Thus, if the original template of seeds dispersed at leks by manakins is similar (as indicated by large diet overlap) and plant recruitment probabilities do not change markedly over time at a site (but see Schupp, Chapter 20, this volume), then we would expect seed and seedling communities to diverge due to post-dispersal processes acting on plant recruitment. The consequences would be that species composition of manakin food plants around leks should differ among species. Indeed, Ryder *et al.* (2006) found significant differences in the species composition of fruiting plants in the vicinity of leks at our site in a study involving three of the four species of manakins. Other studies involving lek-mating birds have also demonstrated that plants found within leks differ from those found in surrounding environments, and that they are dominated by fruiting species consumed by lek inhabitants (e.g. Théry and Larpin, 1993; Krijger *et al.*, 1997). Moreover, Krijger and Opdam (1995) found that seed germination of Melastomataceae was lower in manakin leks when compared with treefall gaps, which probably results in lower effectiveness of seed dispersal by male manakins for Melastomataceae. These results suggest that the mating system of manakins has probably resulted in intersexual differences in some components of seed-disperser effectiveness and, thus, in the case of male manakins, our estimates of ecological redundancy may be overestimated.

Our analysis of ecological redundancy in seed-dispersal function in these two forests which differ in biological diversity was restricted to manakins. Although manakins are the predominant understorey frugivores in these two forests, many other species also feed on fruits at these sites (see, e.g. Karr *et al.*, 1990; English, 1998; B.A. Loiselle, P.G. Blendinger and J.G. Blake, 2006, unpublished results). In Costa Rica, we found that a few species tended to dominate plant–disperser interactions in the forest understorey (Loiselle and Blake, 1999, 2002). Although we have yet to quantify the diets of other frugivores in the forest understorey in Ecuador, including two other species of manakins regularly found in the understorey (*Chiroxiphia pareola*, *Machaeropterus regulus*), our preliminary results point to considerable diet overlap between these other frugivorous species and manakins, as well as some differences (B.A. Loiselle *et al.*, 2006, unpublished results). For example, plants from Araceae (*Anthurium*, *Philodendron*) were important components of manakin diets in Ecuador and Costa Rica. Araceae have also been recorded regularly in diets of at least six other understorey frugivores in Ecuador (e.g. *Chiroxiphia pareola*, *Chloropipo holochlora*, Pipridae; *Turdus albicollis*, *Catharus minimus*, Turdidae; *Euphonia xanthogaster*, *E. rufiventris*, Emberizidae), but not others (e.g. *Pteroglossus azarae*, Ramphastidae; *Momotus momota*, *Barypthengus martii*, Momotidae). In Costa Rica we only recorded Araceae (i.e. *Anthurium* spp.) in the diets of two other frugivores, *Euphonia gouldi* and *Mionectes oleaginous* (Tyrannidae). Thus, we expect that the trend of increased ecological redundancy in Ecuadorian forests will continue when more seed dispersers are included.

Despite this apparent increased redundancy in ecological roles, the presence of ecologically equivalent species does not guarantee that ecosystem function and structure will be conserved under scenarios of disturbance. Seed dispersers have been shown to be more extinction-prone than species from other functional groups (e.g. Terborgh and Winter, 1980; Renjifo, 1999; Şekercioğlu *et al.*, 2004), and even in species-rich forests, plants whose seeds are dispersed by a suite of apparently ecologically equivalent species have experienced significant declines in recruitment following forest fragmentation (Cordeiro and Howe, 2001, 2003). Consequently, conservation efforts should be targeted to conserve key ecological processes such as seed dispersal. To do so requires an understanding of the complex interactions between plants and their dispersers, which include identifying suites of functionally equivalent species. Although ecological redundancy may offer greater ecosystem resilience under conditions of disturbance, natural resource managers should recognize that it is no guarantee.

Acknowledgements

We are very grateful to the following individuals who helped establish the 100 ha study plots or assisted in collecting field data: Jendry Narvaez, Franklin Narvaez, Alvaro Garcia, Juan Carlos Rodriguez, Javier Andy, Kimberly Holbrook, Tina Sommers, Wendy Tori, Renata Durães, José Hidalgo, Ursula Valdez and Jeanne Fair. We also thank the staff of the Tiputini Biodiversity Station, especially Jaime Guerra, Kelly Swing, David Romo, Consuelo de Romo, Ruben Bustamante, Ovideo Godoy and Jose Fabara, who have made our visits to the site so rewarding. We thank Terry Erwin and Tom Kunz, who initially encouraged our investigations in this megadiverse region. We thank Daniel Wenny, Andrew Dennis, David Westcott and an anonymous reviewer for comments and improvements to the manuscript. We sincerely thank the organizers of the *Frugivores and Seed Dispersal Symposium*, especially Andrew Dennis, David Westcott, Ronda Green and Eugene Schupp, for holding a spectacular meeting, and the National Science Foundation OISE programme which supported B.A.L.'s travel to Australia. Finally, we are grateful to the Ministry of the Environment of Ecuador for granting us permission to work at Tiputini Biodiversity Station (No. 13-IC-FAU-DFN) and the National Science Foundation (IBN 0235141 and DEB 0304909), the National Geographic Society (7113-01), the Fulbright US Scholars Program, and the University of Missouri–St Louis for providing research funds.

References

Blake, J.G. and Loiselle, B.A. (2000) Diversity of birds along an elevational gradient in the Cordillera Central, Costa Rica. *Auk* 117, 663–686.

Blake, J.G. and Loiselle, B.A. (2001) Bird assemblages in second-growth and old-growth forests, Costa Rica: perspectives from mist nets and point counts. *Auk* 118, 304–326.

Cardoso da Silva, J.M. and Tabarelli, M. (2000) Tree species impoverishment and the future flora of the Atlantic forest of northeast Brazil. *Nature* 404, 72–74.

Chapman, C.A. and Chapman, L.J. (2002) Plant–animal coevolution: is it thwarted by spatial and temporal variation in animal foraging? In: Levey, D.J., Silva, W.R. and Galetti, M. (eds) *Seed Dispersal and Frugivory: Ecology, Evolution and Conservation.* CAB International, Wallingford, UK, pp. 275–290.

Chapman, L.J., Chapman, C.A. and Wrangham, R. (1992) *Balanites wilsoniana*: elephant dependent dispersal? *Journal of Tropical Ecology* 8, 275–283.

Chong, C.T., Larned, S.T., Covich, A.P. and Kinzie, R.A. III (2000) Species interactions between estuarine detritivores: inhibition or facilitation? *Hydrobiologia* 434, 11–16.

Clarke, K.R. and Warwick, R.M. (2001) *Change in Marine Communities: An Approach to Statistical Analysis and Interpretation*, 2nd edn. PRIMER-E, Plymouth, UK.

Cordeiro, N.J. and Howe, H.F. (2001) Low recruitment of trees dispersed by animals in African forest fragments. *Conservation Biology* 15, 1733–1741.

Cordeiro, N.J. and Howe, H.F. (2003) Forest fragmentation severs mutualism between seed dispersers and an endemic African tree. *Proceedings of the National Academy of Sciences* 100, 14052–14056.

Ehrlich, P.R. and Walker, B. (1998) Rivets and redundancy. *BioScience* 48, 387.

English, P.H. (1998) Ecology of mixed-species understory flocks in Amazonian Ecuador. PhD dissertation, University of Texas, Austin, Texas.

Flecker, A.S. (1996) Ecosystem engineering by a dominant detritivore in a diverse tropical stream. *Ecology* 77, 1845–1854.

Gautier-Hion, A., Emmons, L.H. and Dubost, G. (1993) Seed dispersal versus seed predation: an inter-site comparison of two related African monkeys. *Vegetatio* 107/108, 237–244.

Gentry, A.H. (1982) Patterns of neotropical plant species diversity. *Evolutionary Biology* 15, 1–84.

Gotelli, N.J. and Entsminger, G.L. (2001) EcoSim: Null models software for ecology. Version 7.0. Acquired Intelligence Inc. & Kesey-Bear, Jericho, Vermont [available at http://homepages.together.net/~gentsmin/ecosim.htm].

Hamann, A. and Curio, E. (1999) Interactions among frugivores and fleshy fruit trees in a Philippine submontane forest. *Conservation Biology* 13, 766–773.

Herrera, C.M. (1985) Habitat–consumer interactions in frugivorous birds. In: Cody, M.L. (ed.) *Habitat Selection in Birds*. Academic Press, Orlando, Florida, pp. 341–365.

Herrera, C.M. (1986) Vertebrate-dispersed plants: why they don't behave the way they should. In: Estrada, A. and Fleming, T.H. (eds) *Frugivores and Seed Dispersal*. Dr. W. Junk, Dordrecht, The Netherlands, pp. 5–20.

Hőglund, J. and Alatalo, R.V. (1995) *Leks*. Princeton University Press, Princeton, NJ.

Hurlbert, S.H. (1971) The nonconcept of species diversity: a critique and alternative parameters. *Ecology* 52, 577–585.

Jaksic, F.M., Feinsinger, P. and Jiménez, J.E. (1996) Ecological redundancy and long-term dynamics of vertebrate predators in semiarid Chile. *Conservation Biology* 10, 252–262.

Jordano, P. and Herrera, C.M. (1995) Shuffling the offspring: uncoupling and spatial discordante of multiple stages in vertebrate seed dispersal. *EcoScience* 2, 230–237.

Karr, J.R., Robinson, S.K., Blake, J.G. and Bierregaard, R.O. Jr (1990) Birds of four neotropical forests. In: Gentry, A.H. (ed.) *Four Neotropical Forests*. Yale University Press, New Haven, Connecticut, pp. 237–269.

Kreft, H., Köster, N., Küper, W., Nieder, J. and Barthlott, W. (2004) Diversity and biogeography of vascular epiphytes in Western Amazonia, Yasuní, Ecuador. *Journal of Biogeography* 31, 1463–1476.

Krijger, C.L. and Opdam, M. (1995) *Consequences of Dispersal of Melastomataceae Seeds towards Manakin Leks*. PhD thesis No. 95-11, Wageningen Agricultural University, Wageningen, The Netherlands.

Krijger, C.L., Opdam, M., Théry, M. and Bongers, F. (1997) Courtship behaviour and seed bank composition in a French Guianan rain forest. *Journal of Tropical Ecology* 13, 631–636.

Lambert, F. (1989) Fig-eating by birds in a Malaysian lowland rain forest. *Journal of Tropical Ecology* 5, 401–412.

Levey, D.J. (1988) Tropical wet forest treefall gaps and distributions of understory birds and plants. *Ecology* 69, 1076–1089.

Lill, A. (1976) Lek behaviour in the golden-headed manakin, *Pipra erythrocephala*, in Trinidad (West Indies). *Zeitschrift für Tierpsychologie*, Suppl. 18, 1–84.

Loiselle, B.A. and Blake, J.G. (1991) Temporal variation in birds and fruits along an elevational gradient in Costa Rica. *Ecology* 72, 180–193.

Loiselle, B.A. and Blake, J.G. (1993) Spatial dynamics of understory avian frugivores and fruiting plants in lowland wet tropical forest. *Vegetatio* 107/108, 177–189.

Loiselle, B.A. and Blake, J.G. (1990) Diets of understory fruit-eating birds in Costa Rica. *Studies in Avian Biology* 13, 91–103.

Loiselle, B.A. and Blake, J.G. (1999) Dispersal of melastome seeds by fruit-eating birds of tropical forest understory. *Ecology* 80, 330–336.

Loiselle, B.A. and Blake, J.G. (2002) Potential consequences of extinction of frugivorous birds for shrubs of a tropical wet forest. In: Levey, D.J., Silva, W.R. and Galetti, M. (eds) *Seed Dispersal and Frugivory: Ecology, Evolution and Conservation.* CAB International, Wallingford, UK, pp. 397–406.

Loiselle, B.A., Blake, J.G., Durães, R., Ryder, T.B. and Tori, W. (2007) Environmental segregation in lek sites among six co-occurring species of manakins (Aves: Pipridae) in eastern Ecuador. *Auk* 124, 420–431.

Loreau, M. (2000) Biodiversity and ecosystem functioning: recent theoretical advances. *Oikos* 91, 3–17.

McConkey, K.R. and Drake, D.R. (2002) Extinct pigeons and declining bat populations: are large seeds still being dispersed in the tropical Pacific? In: Levey, D.J., Silva, W.R. and Galetti, M. (eds) *Seed Dispersal and Frugivory: Ecology, Evolution and Conservation.* CAB International, Wallingford, UK, pp. 381–395.

McDiarmid, R.W., Ricklefs, R.E. and Foster, M.S. (1977) Dispersal of *Stemmadenia donnell-smithii* (Apocynaceae) by birds. *Biotropica* 9, 9–25.

Naeem, S. (1998) Species redundancy and ecosystem reliability. *Conservation Biology* 12, 39–45.

Naeem, S. and Li, S. (1997) Biodiversity enhances ecosystem reliability. *Nature* 390, 507–509.

Naeem, S., Thompson, J., Lawler, S.P., Lawton, J.H. and Woodfin, R.M. (1994) Declining biodiversity can alter the performance of ecosystems. *Nature* 368, 734–737.

Naeem, S., Thompson, J., Lawler, S.P., Lawton, J.H. and Woodfin, R.M. (1995) Empirical evidence that declining species diversity may alter the performance of terrestrial ecosystems. *Philosophical Transactions of the Royal Society London B* 347, 249–262.

Paine, R.T. (1969) A note on trophic complexity and community stability. *American Naturalist* 103, 91–93.

Pianka, E.R. (1973) The structure of lizard communities. *Annual Review of Ecology and Systematics* 4, 53–74.

Rainey, W.E., Pierson, E.D., Elmqvist, T. and Cox, P.A. (1995) The role of flying foxes (Pteropodidae) in oceanic island ecosystems of the Pacific. *Symposium of the Zoological Society of London* 67, 79–96.

Renjifo, L.M. (1999) Composition changes in subAndean avifauna after long-term forest fragmentation. *Conservation Biology* 13, 1124–1139.

Rick, C.M. and Bowman, R.I. (1961) Galapagos tomatoes and tortoises. *Evolution* 15, 407–417.

Robinson, W.D., Brawn, J.D. and Robinson, S.K. (2000) Forest bird community structure in central Panama: influence of spatial scale and biogeography. *Ecological Monographs* 70, 209–235.

Rosselli, L. (1994) The annual cycle of the white-ruffed manakin *Corapipo leucorrhoa*, a tropical frugivorous altitudinal migrant, and its food plants. *Bird Conservation International* 4, 143–160.

Ryder, T.B., Blake, J.G. and Loiselle, B.A. (2006) A test of the environmental hotspot hypothesis for lek placement in three species of manakins (Pipridae) in Ecuador. *Auk* 123, 247–258.

Scarff, F.R. and Bradley, J.S. (2002) Ecosystem function and species loss: a microcosm study. *Journal of Biogeography* 29, 641–651.

Schupp, E.W. (1993) Quantity, quality and the effectiveness of seed dispersal by animals. *Vegetatio* 107/108, 15–29.

Schwartz, P. and Snow, D.W. (1978) Display and related behaviour of the wire-tailed manakin *Pipra filicauda. Living Bird* 17, 51–78.

Şekercioğlu, C.H., Daily, G.C. and Ehrlich, P.R. (2004) Ecosystem consequences of bird declines. *Proceedings of the National Academy of Sciences* 101, 18042–18047.

Sick, H. (1967) Courtship behaviour in the manakins (Pipridae): a review. *Living Bird* 6, 5–22.

Silva, W.R., De Marco, P. Jr, Hasui, E. and Gomes, V.S.M. (2002) Patterns of fruit–frugivore interactions in two Atlantic forest bird communities of south-eastern Brazil: implications for conservation. In: Levey, D.J., Silva, W.R. and Galetti, M. (eds) *Seed Dispersal and Frugivory: Ecology, Evolution and Conservation.* CAB International, Wallingford, UK, pp. 423–436.

Slater, K. and du Toit, J.T. (2002) Seed dispersal by chacma baboons and syntopic ungulates in southern African savannas. *South African Journal of Wildlife Research* 32, 75–79.

Snow, D.W. (1962) A field study of the golden headed manakin, *Pipra erythrocephala*, in Trinidad. *Zoologica* 47, 183–198.

Stocker, G.C. and Irvine, A.K. (1983) Seed dispersal by cassowaries *Casuarius casuarius* in North Queensland Australia rain forests. *Biotropica* 15, 170–176.

SPSS (1999) *SPSS for Windows.* Release 10.0.5. SPSS Inc, Chicago, Illinois.

Syrjala, S.E. (1996) A statistical test for a difference between the spatial distributions of two populations. *Ecology* 77, 75–80.

Tabarelli, M., Mantovani, W. and Peres, C.A. (1999) Effects of habitat fragmentation on plant guild structure in the montane Atlantic forest of southeastern Brazil. *Biological Conservation* 91, 119–127.

Tello, J.G. (2001) Lekking behaviour of the round-tailed manakin. *Condor* 103, 298–321.

Terborgh, J. and Winter, B. (1980) Some causes of extinction. In: Soulé, M.E. and Wilcox, B. (eds) *Conservation Biology: An Ecological-Evolutionary Perspective.* Sinauer Associates, Sunderland, Massachusetts, pp. 119–133.

Terborgh, J., Robinson, S.K., Parker, T.A. III, Munn, C.A. and Pierpont, N. (1990) Structure and organization of an Amazonian forest bird community. *Ecological Monographs* 60, 213–238.

Théry, M. (1992) The evolution of leks through female choice: differential clustering and space utilization in six sympatric manakins. *Behavioral Ecology and Sociobiology* 30, 227–237.

Théry, M. and Larpin, D. (1993) Seed dispersal and vegetation dynamics at a Cock-of-the-Rock's lek in the tropical forest of French Guiana. *Journal of Tropical Ecology* 9, 109–116.

Tilman, D. and Downing, J.A. (1994) Biodiversity and stability in grasslands. *Nature* 367, 363–365.

Walker, B. (1995) Conserving biological diversity through ecosystem resilience. *Conservation Biology* 9, 747–752.

Walker, B., Kinzig, A. and Langridge, J. (1999) Plant attribute diversity, resilience, and ecosystem function: the nature and significance of dominant and minor species. *Ecosystems* 2, 95–113.

Wellnitz, T. and Poff, N.L. (2001) Functional redundancy in heterogeneous environments: implications for conservation. *Ecology Letters* 4, 177–179.

Wenny, D.G. and Levey, D.J. (1998) Directed seed dispersal by bellbirds in a tropical cloud forest. *Proceedings of the National Academy of Sciences* 95, 6204–6207.

Wheelwright, N.T. and Orians, G.H. (1982) Seed dispersal by animals: contrasts with pollen dispersal, problems of terminology, and constraints on coevolution. *American Naturalist* 119, 402–413.

Winemiller, K.O. and Pianka, E.R. (1990) Organization in natural assemblages of desert lizards and tropical fishes. *Ecological Monographs* 60, 27–55.

11 Seed and Seedling Shadows

David A. Westcott

Introduction

For animal-dispersed plants, fruit selection and consumption by frugivores marks the beginning of an active phase of dispersal. This is the phase in which the long trajectory of fruit and frugivore evolution, the cycle of fruiting phenology in a forest and fruit selection by a community of frugivores are all translated, through the movement of dispersers, into the spatial arrangement of potential recruits and, ultimately, that of the future adult population. In the relatively simple act of transporting seeds away from the parent plant and depositing them elsewhere, dispersal vectors of all kinds precipitate the myriad spatial 'outcomes' that are contingent on the dispersal process. This act sets the scene for the survival or demise of individual seeds and for changes in population dynamics, population ranges and community structure (Nathan and Muller-Landau, 2000; Levine and Murrell, 2003).

Understanding and predicting the outcomes of the evolutionary, biogeographical and ecological processes that are influenced by dispersal, at a minimum, requires a quantitative description of how many seeds are dispersed and how far from the parent plant they are taken. The frequency distribution of these dispersal distances within a crop or a population has been termed the dispersal curve or kernel (Levin *et al.*, 2003). Variation in the distances that different proportions of seed are dispersed and the distances over which dispersal occurs (shape and scale of the kernel), can have a profound influence on subsequent dynamics. For example, rapid spread of a population or a genotype is most likely when a large proportion of seeds are dispersed long distances (Kot *et al.*, 1996). In contrast, a greater investment in local processes is made and a greater proportion of seeds are likely to be exposed to density-dependent effects when most seeds are dispersed only short distances (Bolker and Pacala,

1999). Thus it is only through an understanding of the shape and the scale of dispersal kernels, and the patterns of variation in these, that we can develop a predictive understanding of how dispersal impacts population, species and community level patterns and processes. This same understanding will also contribute to our ability to predict and manage species responses to some of the major environmental challenges currently facing natural ecosystems around the world (Trakhtenbrot *et al.*, 2005; see Silva *et al.*, Chapter 26; Wright, Chapter 28; Pizo, Chapter 29).

Documenting dispersal distances has rarely proved to be simple. The earliest success came in those circumstances where dispersal distances were short and individual seeds easily tracked. However, such cases are relatively rare, and most dispersal occurs in complex circumstances involving a diverse range of dispersers or vector types. Recent inroads into the description of complex dispersal kernels have been made with wind-dispersed species. Here, the basic physics of dispersal was relatively well understood. The challenge was to incorporate the enormous complexity introduced by local conditions, such as updraughts and turbulence, that seeds encounter in nature and which can so dramatically alter their dispersal outcomes (Nathan *et al.*, 2002; Tackenberg, 2003).

Descriptions of the dispersal kernels produced by vertebrates have proven more difficult. They are the product of complex disperser behaviours that are generally conducted on spatial scales and in environments that make their documentation difficult. Despite the difficulties, our understanding of vertebrates as seed dispersers is growing steadily. For example, Stevenson (Chapter 15) demonstrates that even in ecosystems with high disperser diversity, individual species can still be responsible for the movement of huge numbers of seeds; potentially providing as much as one-third of all dispersal in a plant community. Even where a species' contribution in terms of the quantity of seed removed is more modest, it might still provide quite specific and critical dispersal services, either due to the scale over which it disperses seeds, through the manner of its seed treatment, or the pattern of seed rain it produces (see Dennis and Westcott, Chapter 9; Holbrook and Loiselle, Chapter 13).

While a focus on the contribution of particular dispersers allows an exploration of their role within a community, in most ecosystems plants are dispersed by multiple dispersers, and it is the combined contribution of these species that ultimately produces the dispersal service the plant receives. Integrating across the diverse contributions of these dispersers, from their fruit selection and removal to their physical transportation of seeds to deposition sites, is necessary in order to be able to describe the total dispersal kernel: the kernel that incorporates the contributions of all dispersers (see Dennis and Westcott, Chapter 9; Nathan, Chapter 11). Dennis and Westcott (Chapter 9) provide such a description of a total dispersal kernel and do so in a manner that preserves the identity of the contributions of the different dispersers at the spatial scale at which these contributions are made. Such detailed approaches allow for the exploration of a variety of questions about the role of diversity and similarity among dispersers and plants in structuring the outcomes of the dispersal process, and Nathan (Chapter 11) develops the analytical tools required to do this.

While these developments in ecological approaches represent a dramatic step forward in our ability to describe and understand dispersal, they are simultaneously being matched by advances in the application of molecular genetic techniques to dispersal questions. These approaches allow the documentation of actual and effective dispersal distances within study plots (see Jordano, Chapter 10; Hardesty, Chapter 12; Steele *et al.*, Chapter 14) and the estimation of these distances at larger spatial scales (see Jordano, Chapter 10). Importantly, these studies confirm that the regular, long-distance dispersal estimated from ecological studies (e.g. Dennis and Westcott, Chapter 9; Holbrook and Loiselle, Chapter 13) does actually occur (see Jordano, Chapter 10; Hardesty, Chapter 12), and that in some cases actual dispersal distances greatly exceed those previously predicted (see Hardesty, Chapter 12; Steele *et al.*, Chapter 14).

The chapters in this section clearly demonstrate that research that seeks to describe dispersal kernels and to apply that knowledge to our understanding of the ecological processes on which dispersal impinges is entering a new and exciting phase. Central to this has been both the application of new technologies and the refinement and recombination of already existing approaches. These developments signal not only more accurate descriptions of dispersal but also the development of new questions and applications that will contribute to our ability to conserve and manage natural systems in an increasingly threatened world.

References

Bolker, B.M., Pacala, S.W. (1999) Spatial moment equations for plant competition: understanding spatial strategies and the advantages of short dispersal. *American Naturalist* 153, 575–602.

Kot, M., Lewis, M.A. and van den Driessche, P. (1996) Dispersal data and the spread of invading organisms. *Ecology* 77, 2027–2042.

Levin, S.A., Muller-Landau, H.C., Nathan, R. and Chave, J. (2003) The ecology and evolution of seed dispersal: a theoretical perspective. *Annual Review of Ecology, Evolution and Systematics* 34, 575–604.

Levine, J.M. and Murrell, D.J. (2003) The community-level consequences of seed dispersal patterns. *Annual Review of Ecology, Evolution and Systematics* 34, 549–574.

Nathan, R. and Muller-Landau, H.C. (2000) Spatial patterns of seed dispersal, their determinants and consequences for recruitment. *Trends in Ecology and Evolution* 15, 278–285.

Nathan, R., Katul, G.G., Horn, H.S., Thomas, S.M., Oren, R., Avissar, R., Pacala, S.W. and Levin, S.A. (2002) Mechanisms of long-distance dispersal of seeds by wind. *Nature* 418, 409–413.

Tackenberg, O. (2003) Modelling long-distance dispersal of plant diaspores by wind. *Ecological Monographs* 73, 173–189.

Trakhtenbrot, A., Nathan, R., Perry, G. and Richardson, D.M. (2005) The importance of long-distance dispersal in biodiversity conservation. *Diversity and Distributions* 11, 173–181.

9 Estimating Dispersal Kernels Produced by a Diverse Community of Vertebrates

A.J. Dennis and D.A. Westcott

Introduction

The dispersal of seeds from parent trees is a critical step in the process of plant regeneration and in determining community structure and diversity (Terborgh *et al.*, 2002). In tropical forests, seed dispersal processes are dominated by vertebrates (Willson *et al.*, 1989; Jordano, 1992) and can directly involve individuals belonging to hundreds of species (e.g. Snow, 1981; Terborgh, 1986). These plants and animals interact in a complex and shifting network of relationships (Herrera, 1986); plants may be dispersed by a spatially and temporally varying number of animals, and each disperser may visit a varying number of species of plants (Snow, 1981; Witmer and Cheke, 1991; see Carlo *et al.*, Chapter 16, this volume; Silva *et al.*, Chapter 26, this volume). The outcome of these biotic interactions, combined with abiotic processes that disperse seeds (e.g. wind, water and gravity), ultimately result in the spatial distribution of dispersed seeds from a plant, the seed shadow (Nathan and Muller-Landau, 2000).

An important descriptor of dispersal processes and their influences on ecological and evolutionary processes is the dispersal curve or kernel, which is the frequency distribution of dispersed seeds relative to distance from a parent plant (Levin *et al.*, 2003). The shape and scale of dispersal kernels has implications for the ability of seeds to arrive at favourable sites (Wenny and Levey, 1998) as well as the potential for individuals to recruit in the face of conspecific effects (Janzen, 1970; Connell, 1971; Wilkinson, 1997; Harms *et al.*, 2000; Tewksbury and Lloyd, 2001). Within populations, the form of dispersal kernels can influence the spatial dispersion of individuals (Hamill and Wright, 1986; Bleher *et al.*, 2002), genetic structuring and the spread of advantageous alleles (Fisher, 1937; Loiselle *et al.*, 1995; Petit and Grivet, 2002). Conversely, the spatial characteristics of a population of plants, habitat and landscape characteristics can also affect the shape and scale of dispersal kernels and their variance within the population (Morales and Carlo, 2006). At a species level, the form of dispersal kernels can influence rates of spread (Reid, 1899; Skellam, 1951; Mollison, 1977; Turchin, 1998) and the ability to exploit available habitat within a species' range (Bolker and Pacala, 1999) and can have large impacts on metapopulation dynamics by influencing subpopulation size, overall abundance and persistence (Hanski, 2001). Empirical and theoretical work also suggest that dispersal influences community structure (Harms *et al.*, 2000; Chave *et al.*, 2002; Levine, 2003; Tuomisto *et al.*, 2003), although empirical tests of these hypotheses suggest that dispersal is just one of a number of factors influencing community structure and that its importance cannot yet be assumed (Levine and Murrell, 2003).

While researchers are well aware of the relevance of describing dispersal kernels, our ability to do so empirically has been limited, especially in the case of vertebrate seed dispersal. In part, this has been a function of the difficulty in describing the distances seeds are dispersed by vertebrates. Despite these difficulties, a variety of methods have been used more or less successfully. These range from following animals from ingestion to egestion of seeds (see Stevenson, Chapter 15, this volume), to marking and following seeds (Mack and Druliner, 2003), to estimating distances based on seed retention times and the displacement rates of dispersers (Westcott *et al.*, 2005a; see Holbrook and Loiselle, Chapter 13, this volume). Such studies have been limited to considering the interaction between, at most, a few species of dispersers and a few species of plants. Important though such results are, the dispersal kernels produced represent only a fraction of the total dispersal received by any given plant, since only a fraction of the disperser suite that visits a plant is considered.

Still elusive, then, is the estimation of complete or total dispersal kernels for vertebrate-dispersed plants; dispersal kernels that incorporate the relative contributions of all dispersal vectors, biotic and abiotic, at the scale at which the process occurs (see also Nathan, Chapter 11, this volume). Indeed, we still lack a framework for even guiding an attempt at such a description (but see Carlo *et al.*, Chapter 16, this volume). Achieving an empirical description of total dispersal kernels would allow a realistic assessment of the contribution of dispersal to the full range of population-, metapopulation- and species-level processes that are influenced by dispersal (Levine and Murrell, 2003). It would also contribute to the assessment of a variety of current threats to natural ecosystems, including climate change, invasion, disperser loss and fragmentation.

Our goal in this chapter is to outline an approach to describing vertebrate seed dispersal processes in tropical rain forests. The framework we describe allows for the empirical estimation of total dispersal kernels for individual plants or for species of plant. It can in fact be extended to the plant community as a whole, but here we focus on an example of estimating a total dispersal kernel for a particular species in order to illustrate the process. The species we use is *Elaeocarpus grandis* (Elaeocarpaceae), a rainforest canopy tree with large (25.6 mm mean diameter), blue fruits, whose three to five seeds are encased in a highly sculpted, thick, bony endocarp.

A conceptual model of process interactions

Figure 9.1 provides a conceptual model of the process of seed dispersal and its interactions. This model identifies each of the sub-processes that we describe in order to estimate a total dispersal kernel for any individual plant, species or plant functional group. ***Step 1*** in the model is to identify the participants and reduce the complexity of this interaction network by creating functional classifications of both fruits and seed dispersal vectors (Fig. 9.1a). As our example is for a species of plant, we will focus our discussion of functional classification on dispersers and provide a plant functional classification elsewhere (A.J. Dennis and D.A. Westcott, unpublished results). ***Step 2*** is to document what proportion of a crop is dispersed by each dispersal vector, biotic and abiotic (Fig. 9.1b,c). Next we need to determine what proportion of seeds are dispersed over what distances by each of the relevant disperser functional groups and vectors. A variety of techniques are used to do this, but for most vertebrate dispersers this requires identification of: (i) the length of time they retain seeds; and (ii) how far they move from the seed source during that time. Consequently, in ***Step 3*** we need to document the factors influencing seed retention, such as gut passage rates (Fig. 9.1d) and in ***Step 4*** disperser movement patterns (Fig. 9.1e). Combining the data from Steps 3 and 4 for a disperser functional group gives us the dispersal kernel that it produces for the species or fruit functional group in question (Fig. 9.1f). However, this is just a fraction of the complete dispersal received by that plant, and in ***Step 5*** we combine the dispersal kernels of all relevant disperser functional groups, weighted by the percentage of fruit crop removed by each (Fig. 9.1c again), and also abiotic dispersal mechanisms, to produce the total dispersal kernel (Fig. 9.1g). While variation is inherent in each sub-process, some of this variation is consistent and pervasive for particular questions, such as differences between habitat fragments and continuous forest. Where such effects exist, sampling over the whole range of conditions is necessary (Fig. 9.1b). In what follows, we consider the task and the issues that each step or sub-process raises and our approach to completing it using examples from Australian tropical rainforest. Our framework allows consideration of post-dispersal mortality factors (Fig. 9.1h), effectively estimating a recruitment kernel (Fig. 9.1i). However, in this chapter we only consider dispersal processes up to the point of seed arrival. A methodology for incorporating post-primary dispersal processes into total dispersal kernels is briefly described in the discussion (Fig. 9.1h).

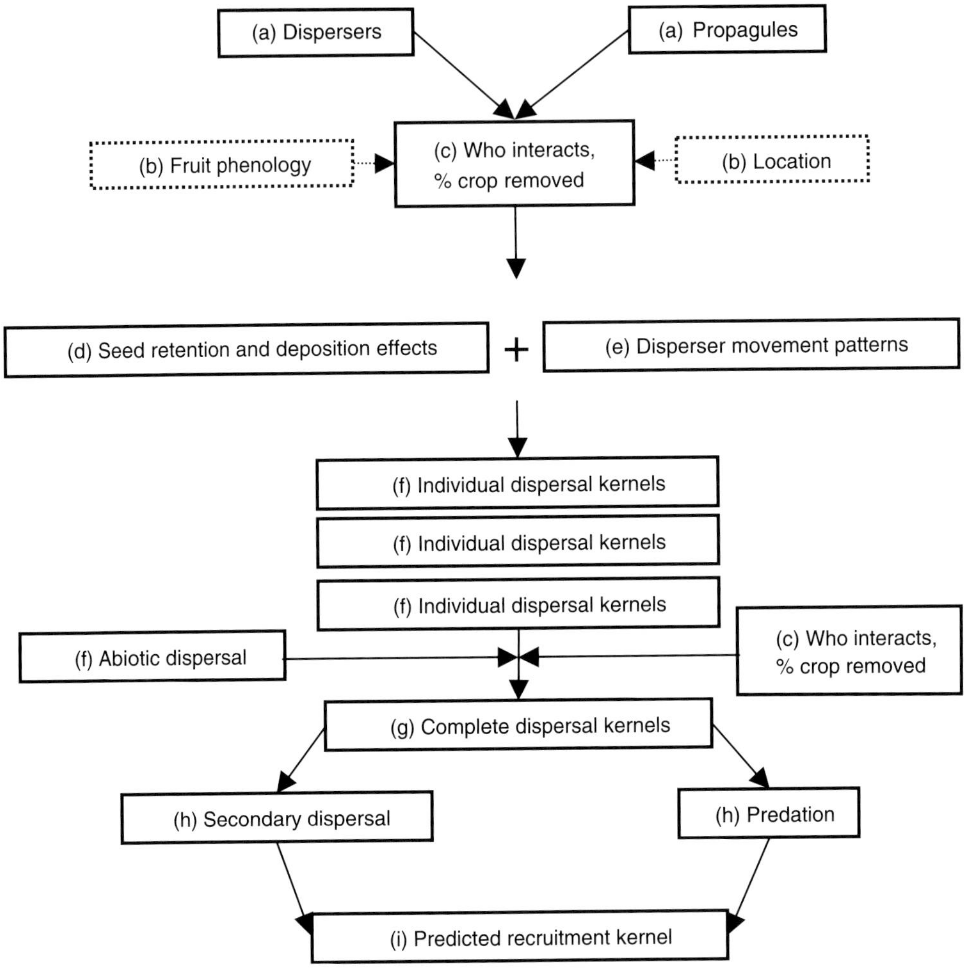

Fig. 9.1. A conceptual model of components of the process of seed dispersal contributing to the formation of a total dispersal kernel. Letters in boxes refer to sections in the text.

Reducing Complexity in Studying Seed Dispersal at Community Scales (Fig. 9.1a)

The diversity of species and functional roles in tropical forests mean that seed shadows are the cumulative product of many contributions. Many authors have recognized that both plants and dispersers interact with multiple other species (e.g. Murray, 1988; Holbrook *et al.*, 2002; see Carlo *et al.*, Chapter 16, this volume; Silva *et al.*, Chapter 26, this volume), creating a complex network. The daunting number of interactions that must be included in a complete description of dispersal in these systems means that few workers have attempted to take this understanding forward to an integrated description that includes all interactions and their relative contributions to a total dispersal kernel (see Nathan, Chapter 11, this volume). How to move beyond the bewildering complexity of interactions becomes the first challenge in attempting to describe total dispersal kernels.

Our approach to reducing complexity

We use functional classifications of dispersers and fruits as a means of reducing complexity while simultaneously preserving the important mechanistic components in a description of complex dispersal processes (Dennis and Westcott, 2006; A.J. Dennis and D.A. Westcott, 2006, unpublished data). Grouping species with similar functional roles incorporates ecological redundancy by treating species providing similar services as a single class. A consequence of this is a loss of species-level detail. However, the description of total dispersal kernels in complex ecosystems is rendered feasible and interpretable. If the selection of traits to define functional classes is performed appropriately, sufficient detail to realistically describe the entire process is retained and the patterns produced are maintained in the description.

For Australia's tropical rain forests, we classified 65 vertebrate dispersers into 15 functional classes; and 441 plants, whose fruit were known to be consumed by those dispersers, into 9 functional types. In doing so we reduced 28,665 potential species-by-species interactions to 135 potential interactions. Here we provide a brief overview of the classification of dispersers; as it is this classification that is used to estimate a total dispersal kernel for *E. grandis*.

Disperser functional groups

We classify dispersers by summarizing the mechanistic elements of the service they provide to plants, and combine species that provide similar services (Dennis and Westcott, 2006). Measures of the quantities of fruit handled, the quality of handling (see Schupp, 1993), and the diversity of plants to which a service is provided, were the three themes under which we refined a list of 26 variables in 10 trait categories to define disperser functional groups (Table 9.1). We then used the combinations of characters to define functional classes either by logical decisions, where species or groups of species clearly provided unique services, or using cluster analysis where divisions between services were less obvious. This process resulted in 15 functional groups of disperser, each providing a distinct service in terms of the outcomes for a plant (Table 9.2).

Table 9.1. Functional traits of seed dispersers used in our classification process.

Trait measured	Justification	Variable or classes
Quantity of fruits removed		
Number swallowed per visit	Affects number of fruit able to be carried for a given fruit size	*Body weight (C)
Frequency of visitation	A surrogate measure of the relative abundance or frequency of visitation for dispersers, it affects quantity of fruit likely to be moved from a crop of fruit on a plant	*Occasional (N) *Regular individuals or pairs (N) *Regular groups (N)
Reliability of visitation	Describes seasonality of visitation and species that are not frugivorous but occasionally eat fruit; affects probability of removing fruit from a crop	*Nomadic/migratory (N) Rarely frugivorous (N)
Quality of fruit handling		
Gentle	Affects number dispersed together, subsequent germination and subsequent likelihood of predation	Through gut (N) Carrying (N) Caching (N)
Predatory	Affects likelihood of a handled seed being killed; e.g. crushing predators more likely to pass small hard seeds intact than chewing or grinding predators	Chewing (N) Grinding/digesting (N) Crushing (N)
Deposition	Affects subsequent seedling competition and likelihood of predation or secondary dispersal	*Clump (N) Scattered (N) Buried (N)
Distance	Each variable has a strong influence on dispersal distance for swallowed seeds	*Movement range (O) *Movement rate (O) *Gut passage rate (O)
Diversity		
Taxonomic	Affects the taxonomic breadth of service provision in the ecosystem	*No. families eaten (C)
Size	Affects the range of plants serviced	*Fruit size range dispersed (C)
Foraging locations	Affects the range of plants serviced and the presentation of fruit taken plus the locations seeds may be deposited	Terrestrial (N) Scansorial (N) *Understorey to canopy (N) *Predominantly canopy (N) Within forest (N) Across landscape (N)

* Traits used in the cluster analysis for the volant birds. The type of variable is indicated by a letter in parentheses: continuous (C); nominal (N); or ordinal (O) (adapted from Dennis and Westcott, 2006).

Table 9.2. A general functional classification of granivores and frugivores describing attributes that differentiate each group. Division 1 is between granivores and frugivores (adapted from Dennis and Westcott, 2006).

Division 2	Division 3	Division 4	Characteristics
Granivores			
Poor dispersers		Gristmill predators	Kill most seeds; rarely pass some intact. Sometimes secondary dispersers
		Chewing predators	Kill most seeds; rarely carry and drop some over short distances[a]
Significant for some species		Mega predators	Kill most seeds; small hard seeds from multiseeded fruit may pass intact, potentially long distances[a]; mega body size[b]
		Predatory rodents	Kill most seeds; also scatter-hoard. Sometimes secondary dispersers.
Frugivores			
Short distance	Within forest	Small to medium size[b]	Short distance dispersal; highly frugivorous; rapid gut[a]
		Medium to large size[b]	Short to moderate distance dispersal[a], highly frugivorous; wide range of seed sizes; moderate to slow gut[a]
		Terrestrial	Short distance[a] dispersers and/or scatter-hoarders; kill few seeds
	Throughout landscape[c]	Frugivores	Short to moderate distance dispersal[a]; partially to highly frugivorous; small seeds; occur throughout landscape; small to medium size
		Facultative	Short dispersal distances[a]; small fruit only; small body size; regular visitors to few species
		Opportunists	Occasionally incorporate fruit in diet, generally short distance dispersal[a]
Long distance	Cross landscape[d]	Wide-ranging slow gut	Long dispersal distances[a], birds fly above canopy, partially or highly frugivorous
		Wide-ranging rapid gut	Long dispersal distances[a], fly above canopy, highly frugivorous; rapid gut[a]
		Wide-ranging large fruit	Long dispersal distances[a], nomadic or migratory, forage throughout strata; eat widest range of fruit sizes
		Wide-ranging small seeds	Long dispersal distances for seeds < 5 mm, Short distances[a] for larger seeds, high levels of seed wastage
	Terrestrial	Mega terrestrial frugivores	Long dispersal distances[a], highly frugivorous, clump dispersers; mega body size[b]

[a] (i) Short, <100 m; (ii) moderate, 100–200 m; (iii) wide, 200–800 m; and (iv) very wide, > 800 m. [b] Small body size, < 50 g; medium body size, 50–200 g; large body size, 201–5000 g; mega body size, > 5000 g. At division 4 'Small' and 'Large' both include medium. [c] Species occurs throughout the landscape matrix. [d] Individuals cross the landscape matrix.

Spatial and Temporal Variation (Fig. 9.1b)

Schupp (Chapter 20, this volume) provides a convincing argument that the arrival of seeds at 'sites' and the suitability of those sites depends on a range of factors. Chief among these are spatial and temporal variations in a raft of conditions, for example, changes in herbivore distribution and abundance, climatic variation across years and locations, and spatial variation in biotic and abiotic conditions. Likewise, the size and timing of fruit crops and their availability relative to other resources (Westcott *et al.*, 2005b), the abundance and distribution of dispersers (Warburton, 1997; Dennis *et al.*, 2005), rates of caching (Forget *et al.*, 2002), and animal behaviour such as movement (Westcott and Graham, 2000) and food choice (Gautier-Hion *et al.*, 1985; Chapman and Chapman, 2002; see Schaefer and Schaefer, Chapter 3, this volume) all vary in time and space. This variability leads to constantly shifting links between dispersers and fruits, both in space and over time (Terborgh, 1986; Herrera, 1998; also see Silva *et al.*, Chapter 26, this volume); as a consequence, what happens in one year and at one location is not always repeated in another year or location. This suggests that generalizing about patterns of seed dispersal is not simple and cannot be done on the basis of short-term studies in particular locations, an issue that is increasingly well recognized (see Schupp, Chapter 20, this volume).

Our approach to dealing with spatial and temporal variation

We recognize two levels of variation in dispersal processes and deal with them differently. 'Local' or small-scale variation we consider to be variation that is inherent to the process and, as such, should be incorporated into estimates of dispersal kernels. Therefore, our approach has been to measure most aspects of seed dispersal (in Fig. 9.1) in different places and at different times. For example, interactions between plants and animals have been measured in multiple locations for each species of plant and fruit functional type; and where relevant and possible, we have also done so at different times of year and in different years. Similar patterns of measurement apply to animal movement, fruit phenology and seed predation. The variation inherent in these data thus becomes incorporated into our description of the process.

There are situations where incorporating variation requires a different approach; namely, when differences are consistent under given circumstances. For example, both between-individual and seasonal differences in movement contribute to a reasonably representative overall estimate of movement for a disperser species. However, if a species shows consistently different movement patterns in different contexts – for example, in fragments of habitat compared with in continuous forest – then combining movement in these distinct contexts will obscure the actual patterns. For example, pied currawongs, *Strepera graculina* (Artamidae), which are known to disperse *E. grandis* seeds, consistently exhibit greater displacement rates in fragmented landscapes than in continuous forest (Fig. 9.2), presumably due to the increased spacing of forest resources resulting from fragmentation. Similar landscape context effects also occur for other stages in the process, for example different suites of dispersers in fragments result in different patterns of fruit removal (Dennis *et al.*, 2004). Thus, landscape context has a pervasive and consistent effect on dispersal outcomes and, if treated as inherent variation, will have the effect of obscuring actual predictive patterns of dispersal in each context. Consequently, we have considered such consistent effects to be the equivalent of separate experimental treatments and have conducted complete sampling under each treatment condition, estimating kernels for fragments and for continuous forest separately.

Relative Contributions to Dispersal (Fig. 9.1c)

In order to realistically reflect the actual contributions of different dispersers to total dispersal kernels it is necessary to estimate which dispersers are relevant to which plants, what proportion of the crop is removed by different vectors, and the manner in which seeds are handled. This requires

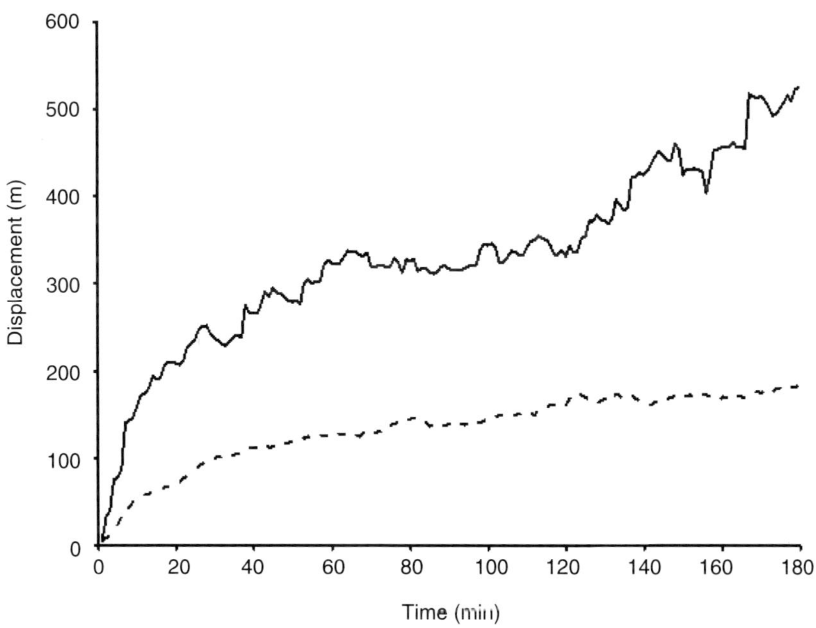

Fig. 9.2. Mean displacement of pied currawongs, *Strepera graculina* (Artamidae), in habitat fragments (solid line) compared to continuous forest (dashed line). Pied currawongs contribute to the dispersal of *Elaeocarpus grandis* (Elaeocarpaceae) and fall in the wide-ranging large fruit group. Data are based on 12 days of tracking in fragmented habitat and 13 in continuous forest for two individuals in each. For details on methods for calculating displacement curves, see Westcott *et al.* (2005a).

documentation of patterns of fruit removal. Two points need to be made. First, it is common for studies of fruit removal to focus on a particular fraction of the process, for example, that done by diurnal frugivores in the plant canopy or a particular taxonomic group. However, most plants are visited by a wide variety of dispersers that together visit all strata from the ground to the canopy around the clock (Jayasekara *et al.*, 2003; Yasuda *et al.*, 2005). Second, not all seeds in a crop are dispersed by animals; some abscise and fall or are dropped beneath the parent, others are dispersed by abiotic means such as wind, rain or gravity. This means that any sampling strategy to determine the fate of seeds needs to be sufficiently flexible to incorporate the range of potential dispersal agents and their times and modes of activity.

Our approach to measuring relative contributions to dispersal

We use a range of techniques to determine the different proportions of fruit dispersed by each dispersal agent. These include:

1. Observations at fruiting trees (night and day);
2. Measuring removal rates of fruits and seeds placed on the forest floor;
3. Measuring fruit production in the canopy and fruit fallen to the ground.

All these data are then used to create a table of the contribution of each disperser functional group and mechanism to total fruit removal (Table 9.3). Below, we outline an example of one such table and briefly describe the methods used to collect the data for it.

Observations at fruiting trees were conducted by pairs of observers. One person scanned the tree at 5 min intervals, recording all activity within and below the tree and recording the animal species, number of individuals and their behaviour (scans). The second person recorded the behaviour of individual animals from their point of entry to the tree or first sighting until their departure or when lost to sight (focal individuals). Data recorded were the number of fruits swallowed, chewed at (where only part of a fruit is swallowed) or dropped, the number of defecations, the duration of the observation, and whether it was an observation of the entire visit or only part of the animal's visit.

Data on the removal of fruits from the forest floor were collected during focal tree observations and by placing fruits and simulated fruits on the forest floor and tracking their fates (Collins, 2002; Dennis, 2003; Dennis *et al.*, 2005; A.J. Dennis, 1999, unpublished results). In some instances (for large fruit and simulated fruit) we used spool and line tracking; smaller fruits were tagged using thread and flagging tape markers (see Forget and Wenny, 2005, for a review of techniques).

The phenology of fruit production and fruit-fall to the ground was measured monthly along twelve 250 m × 5 m transects spread across our study region. We recorded the species, age of fruit (green, ripe and senescent), and abundance of fruit encountered on transects (see Westcott *et al.*, 2005b, for details). The abundance of canopy fruit was estimated

Table 9.3. Summary table describing where the fruits from a crop of *Elaeocarpus grandis* (Elaeocarpaceae) go. The figures are based on data described in the text and represent the average for 24 trees that do not overhang rivers or streams. The disperser groups correspond to those described in Table 9.2 (some names combine words from different divisions in Table 9.2). For a list of species in each group refer to Dennis and Westcott (2006). The No. column is calculated from the mean percentages for the 24 sampled trees and rounded to the nearest whole number.

	%	No.
Fruit in tree	100	6500
Falls to ground	2.8	182
Abiotic dispersal	0	0
Available for birds	97.2	6318

	Dropped		**Volant dispersal**				
Disperser group	%	No.	%	No.	% Surv.	No. Surv.	No. Dead
Unaccounted	0.00	0	0	0			
Small within forest	3.86	251	17.00	1105	100	1105	0
Chewing predators	0.00	0	0	0	0	0	0
Facultative frugivores	0.00	0	0	0	100	0	0
Digestive predators	0.00	0	0	0	0	0	0
Large within forest	1.31	85	45.58	2963	100	2963	0
Occasional frugivores	0.00	0	0	0	100	0	0
Throughout landscape frugivores	0.00	0	0	0	100	0	0
Wide-ranging slow-gut	3.46	225	19.94	1296	100	1296	0
Wide-ranging large fruit	0.23	15	4.86	316	100	316	0
Wide-ranging rapid-gut	0.00	0	0.97	63	100	63	0
Wide-ranging small seeds	0.00	0	0	0	100	0	0
Subtotal	**8.85**	**575**	**88.36**	**5743**		**5743**	**0**
Fallen + dropped	11.65	757					
Eaten on ground	11.18	757	Terrestrial dispersal				
Mega terrestrial frugivores			3.88	252	100	252	0
Terrestrial within forest			3.88	252	100	252	0
Predatory rodents			3.89	253	2	5	248
Subtotal			**11.18**	**727**		**509**	**248**
Total dispersed	96.18	6252					
Predated upon	3.82	248					
Left beneath parent	0.00	0					

using Connell and Green's (2000) approximately logarithmic scale, whereas fruits on the forest floor were counted directly.

Table 9.3 provides a summary of 'where' fruit goes for *E. grandis*, based on 148 h of observation at 24 individual trees that were visited by 665 individuals of 16 species of frugivores. The crop size quoted is the mean for all trees observed. We used the proportion of fruits counted on the ground compared with the total crop in the canopy as an estimate of the number of fruits falling to the ground. The proportions dropped and dispersed by birds were calculated from the mean numbers eaten and dropped per visit for each disperser functional group (focal individual data), multiplied by the number of individuals from each disperser functional group occurring in scans during focal tree observations. Where a species of disperser had a mean visit length of less than 5 min, occurrence of that species in a scan was taken as equivalent to a visit. Where a species of disperser had a mean visit length greater than 5 min, the number of visits was recalculated using the mean visit length, so that a single visit was equal to > 1 scan; the number of scans depended on the mean visit length.

The relative contributions to dispersal by terrestrial animals were estimated on the basis of their relative biomass (Dennis, 1997). Predatory rodents consume most of the seeds that they handle, but *E. grandis* has four seeds enclosed in a woody endocarp and some individual seeds remain uneaten in a small proportion of endocarps. Our observations suggested a roughly 2% survival for endocarps handled by rodents. During phenology surveys we recorded green, ripe and senescent fruit on the forest floor. A complete lack of senescent fruit for *E. grandis* beneath parent crowns indicated that all fruit were consumed or removed.

A high proportion of *E. grandis* were dispersed, primarily by just a few groups of volant dispersers. This is due largely to the fruits being large and therefore swallowed and dispersed almost entirely by large dispersers, a small proportion of the community. A few small species disperse them by carrying them away for consumption or caching. In addition, their seeds are encased in a highly sculpted, bony endocarp that deters most invertebrate and vertebrate seed consumers. Only two murid rodents (*Uromys caudimaculatus* and *Melomys cervinipes*, Muridae) are known to consume the seeds, the latter less frequently than the former. Despite this the seeds are sought out by *U. caudimaculatus*, resulting in all (or most) being removed or eaten from beneath parent crowns.

Seed Retention Times (Fig. 9.1d)

A standard method of estimating dispersal kernels produced by vertebrates is to measure how long animals retain seeds and how far they move during these times. This is a useful technique for species that swallow seeds and pass them through the gut, carry them in cheek pouches, beak or crop, or where seeds adhere externally on the animal. Retention times have been measured in several ways. Generally, the time from ingestion to egestion is recorded during observation of wild (e.g. Yumoto, 1999) or captive animals (e.g. Sun *et al.*, 1997) or, when both seed and disperser are large, with marked seeds or telemetry (Mack, 1995; Mack and Druliner, 2003). Because many fruits contain compounds that are either costive or laxative (see, e.g. see Levey *et al.*, Chapter 2, this volume) and both seed size and flesh texture and adherence influence gut passage rates (Traveset and Verdú, 2002) the use of artificial markers such as dyes, beads or commercially available seeds, is less desirable than using the actual fruits in question.

The difficulty of recovering all seeds and in identifying the ingestion and egestion times of particular seeds in the wild has meant that most seed retention studies have been conducted in captivity. This approach is not without problems. The question of whether changes in activity, diet and behaviour of captive animals influence gut passage rates needs to be considered. There is little evidence with which this can be directly examined. However, the similarity between gut passage and dispersal distances estimated using captive *Casuarius casuarius* (Casuariidae) gut passage data (Westcott *et al.*, 2005a) and telemetry of seeds consumed by wild *C. bennetti* (Mack and Druliner, 2003) suggest that these effects are negligible, in some cases at least. Regardless of this, the captive situation needs to be carefully considered; the space available to animals should be large enough to allow movement but small enough that passed seeds can be reliably retrieved. The period of captivity and potential effects of diet changes during this time should also be considered.

Our approach to estimating seed retention

We estimated retention times for species swallowing and defecating or regurgitating seeds by bringing wild-caught individuals (range 1–10, mean 4.1 individuals per species; 13 species) into captivity for short periods (range 3–10 days), with the exception of cassowaries, C. *casuarius*, for which we used long-term captive animals (Westcott *et al.*, 2005a). These animals represented 10 disperser functional groups (1–2 species per group, mean 1.4). Animals were housed individually and fed an *ad libitum* maintenance diet (without seeds) between experimental feeding bouts. To determine transit times we recorded the ingestion time (measured as the mid-point of each feeding bout) and the time of egestion for each seed. We continued to collect regurgitations or droppings until all seeds were accounted for. We did this using the fruits of as many species of plants representing different plant functional types as were available or would be consumed by the disperser. For the current example, we fed fruits to ten species of dispersers representing the six disperser function groups that feed on *E. grandis* and ingest the seed. These fruits came from 17 species whose fruits and seeds are in the same fruit functional group as *E. grandis* (medium-sized, few-seeded) and included *E. grandis*. In total 1707 medium-sized few-seeded fruits were passed during 90 feeding trials. Retention times for animals that carry and cache seeds were not measured; instead we measured dispersal distances for these species directly (see below)

The species sampled showed a range of retention times (see Fig. 9.3). Seeds could be egested after as little as 1 min or after as much as 28 h, with significant variation both within and among species of dispersers. This wide range of potential passage times for seeds suggests that the effects of gut passage on germinability could vary widely (see Traveset *et al.*, Chapter 4, this volume).

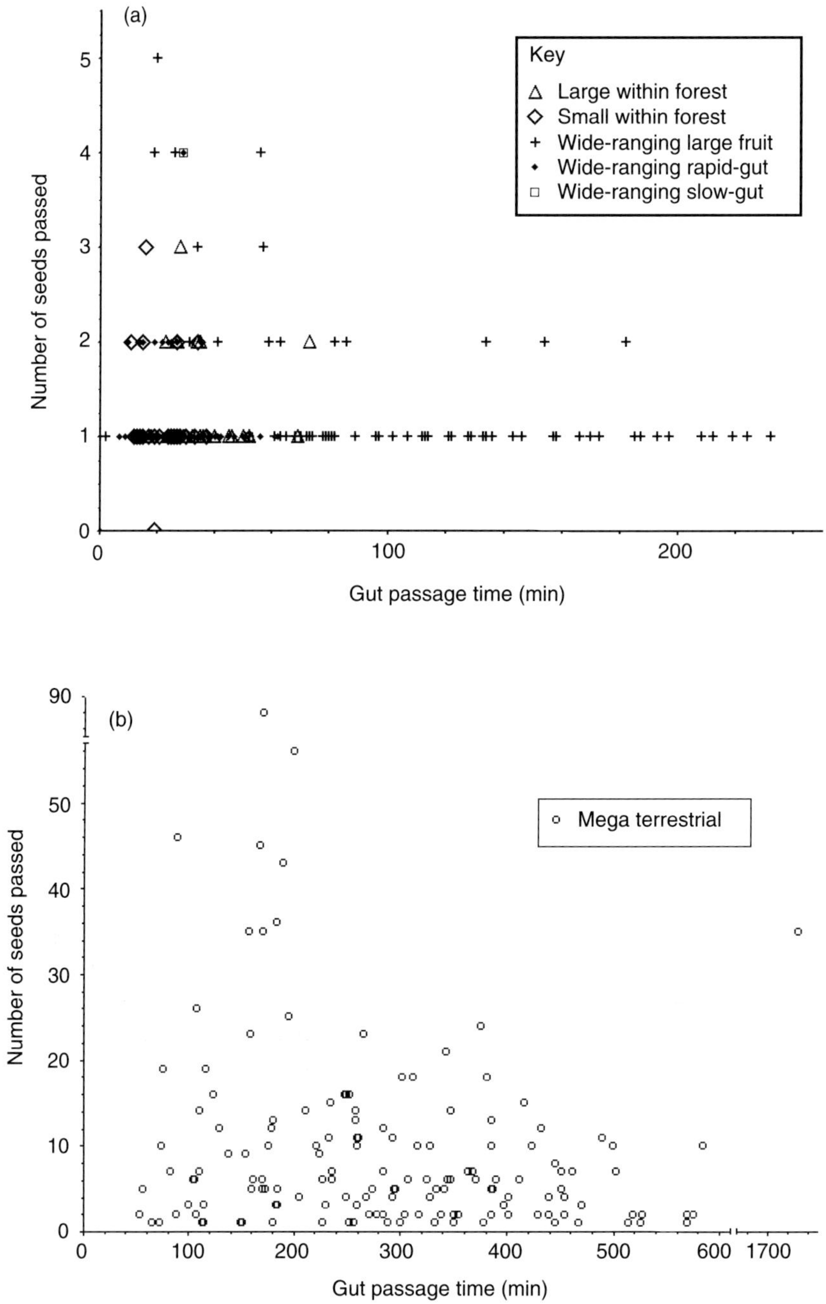

Fig. 9.3. Retention times for the seeds of medium-sized, few-seeded fruits through species of the different disperser functional groups that ingest *Elaeocarpus grandis* (Elaeocarpaceae). Each point on the graph represents the time from ingestion to egestion for the seed(s) contained in a single dropping. Note scale breaks on both axes in graph (b).

Disperser Movement Patterns (Fig. 9.1e)

If dispersal distance is estimated as the product of the time that seeds are retained by dispersers and their displacement rate, then having documented the distribution of seed retention time by a disperser functional group, one must estimate the distance the disperser is likely to have moved in any given time interval. Describing disperser displacement rates requires the documentation of movement in space as a function of time. This can be done by direct observation of the animal (Yumoto, 1999) or through the use of telemetry (Westcott and Graham, 2000; Westcott *et al.*, 2005a; see Holbrook and Loiselle, Chapter 13, this volume). Attempts have been made to use proxies for animal movement such as mark–recapture data (Stansbury, 2001), but our lack of knowledge about short-term patterns of individual movement and the inability to document where and how fast individuals have travelled between captures make this an approach of limited value.

Our approach to measuring movement

We use continuous radio-telemetry, i.e. long sequences of telemetry 'fixes' with high temporal resolution, to describe frugivore movement patterns. A radio-tagged disperser's location at any given point in time, a 'fix', is determined by triangulating bearings from GPS-mapped stations. Three to six simultaneous bearings are used to estimate each fix and we strove to keep the interval between fixes at ≤5 min. To avoid observer effects, no attempt is made to approach or sight an animal. We track animals for as long as possible but use the data from any session longer than the maximum gut passage time. We collect data from a minimum of 25 sessions for each disperser functional group, being composed of data from five sessions from each of five individuals drawn from as many species as possible in that functional group. The data produced during these sessions is a series of fixes represented as x and y coordinates against time.

Estimating Dispersal Kernels for a Disperser (Fig. 9.1f)

Estimating dispersal kernels for species that ingest or carry seeds has been done in a number of ways. These include:

1. Direct observation of dispersers foraging and subsequently depositing seeds (Sun *et al.*, 1997; see Stevenson, Chapter 15, this volume);
2. Marking seeds at parent trees and monitoring their removal by dispersers (Mack, 1995; Vander Wall, 2001; Mack and Druliner, 2003; Forget and Wenny, 2005; Levey *et al.*, 2005);
3. Using biological markers, such as the DNA of an individual seed and its parents (Godoy and Jordano, 2001; Grivet *et al.*, 2005; Pairon *et al.*, 2006);

4. Monitoring patterns of seed arrival in seed traps (Clark *et al.*, 2001);
5. Estimating dispersal distances as a function of the time an animal retains a seed and its displacement distance during that time (Fleming and Heithaus, 1981; Murray, 1988).

Estimating dispersal kernels using a combination of seed retention and disperser movement has both disadvantages and advantages relative to other methods. The major disadvantage is that the results are estimates of dispersal distances and no actual dispersal is documented. Also, because foraging locations are usually not observed, referencing dispersal to specific examples in the field is usually done in terms of expected outcomes. On the other hand, because the approach relies on average behaviours, the results can be readily generalized to a variety of situations. This generality allows for application in a variety of modelling and applied contexts and is more likely to include a full range of dispersal outcomes, including long-distance dispersal events, than other methods.

Our approach for estimating dispersal kernels

Our approach to estimating dispersal kernels has evolved rapidly from estimating average dispersal rates based on average displacement rates derived from a single starting point in each telemetry session (Westcott and Graham, 2000) to incorporating daily patterns of disperser behaviour into the estimation of dispersal kernels using average displacements derived from multiple starting points in each telemetry session (Westcott *et al.*, 2005a). Our current approach has moved away from the use of average displacement models since the process of averaging tends to underestimate the frequency of short-distance dispersal events. Instead, for each telemetry session we choose 15 random starting times such that their frequency distribution matches the distribution of foraging by the disperser, as determined during focal tree watches. From each starting point we then calculate the displacement distance for each gut passage time recorded for a fruit type through a disperser. This process is repeated for each telemetry session and for each individual disperser, tracked to give the dispersal kernel for a plant or fruit functional type that is produced by an individual, a species or a disperser functional group.

For species that carry seeds, such as rodents and musky-rat kangaroos (*Hypsiprymnodon moschatus*; Hypsiprymnodontidae), dispersal distances, while affected by retention time, are determined by factors affecting how far the seed is taken before being deposited, for example, risk of predation, social interactions, or the distribution of appropriate caching sites. For such species, the most effective means of estimating dispersal distances is to tag the fruits or seeds and measure dispersal distances directly. For this purpose we used either spool-lines (Dennis, 2003) or nylon thread and flagging tape markers (Theimer, 2001; Dennis *et al.*, 2005) attached directly to seeds or fruits. Fruits or seeds were placed out under fruiting trees and their removal distances recorded until such time as they were either killed or no longer handled. Final dispersal distances were then combined to produce the dispersal kernel (see Dennis, 2003, for an example).

Estimating Total Dispersal Kernels (Fig. 9.1g)

Total dispersal kernels incorporate the contributions of all dispersers relevant to the focal plant individual, species or functional type. Most descriptions of dispersal kernels deal with those produced by a single disperser for a single species of plant (Forget, 1991) or the kernels produced by a single disperser for several species of plants (Murray, 1988; Westcott and Graham, 2000; Holbrook *et al.*, 2002; Westcott *et al.*, 2005a; see Nathan, Chapter 11, this volume). To date there are astonishingly few studies that compare the dispersal provided by multiple dispersers to a single species of plant, and those that do simply calculate the set of dispersal kernels without attempting to describe their cumulative product (e.g. Mouissie *et al.*, 2005). However, estimating total dispersal kernels is not simply a matter of combining all the relevant disperser's kernels. Different dispersers disperse seeds different distances but also contribute differentially to fruit removal, so simply combining the kernels risks over- or underestimating the proportions of seeds dispersed any particular distance. The net result might provide a correct estimate of the scale of the dispersal kernel but, because of the differing contributions of dispersers to fruit removal, an incorrect estimate of its shape. This is important, since models of population processes dependent on dispersal have shown that they are highly sensitive to the shape of the dispersal kernel (Kot *et al.*, 1996).

Our approach to estimating total dispersal kernels

To ensure that the shape of the total dispersal kernel reflects the relative contributions of each disperser functional group we scale the *y* axis of each disperser's kernel relative to that of other dispersers according to the relative proportion of the total crop they remove. The kernels of each disperser are then combined and should add up to 100% of the crop (see above; Fig. 9.4). In the example given for *E. grandis*, a majority of seeds are dispersed within hundreds of metres of the parent plant, with a small proportion dispersed beyond 500 m, resulting in an extended tail reaching well over 2 km from the source (Fig. 9.4). Disperser functional groups contribute differentially to the distribution of seeds at different distances, some dominating dispersal close to the plant, others at distances, and some contributing at the greatest distances (Fig. 9.4b).

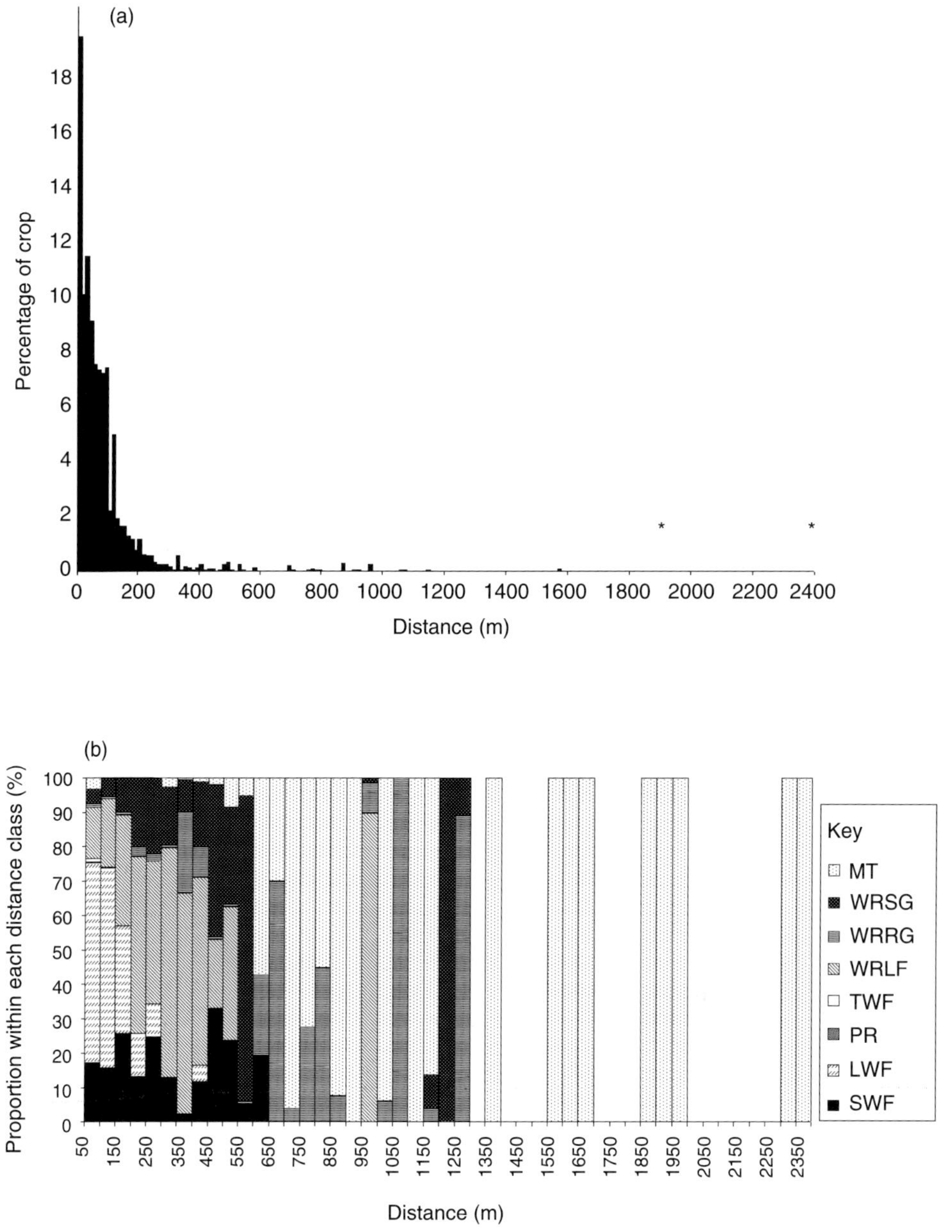

Fig. 9.4. Total dispersal kernel for *Elaeocarpus grandis* (Elaeocarpaceae) incorporating all dispersers and their relative contributions: (a) total kernel, (b) contributions of each disperser functional group. Disperser functional groups include: MT – mega-terrestrial; WRSG – wide-ranging slow-gut; WRRG – wide-ranging rapid-gut; WRLF – wide-ranging large fruit; TWF – terrestrial within-forest; PR – predatory rodents; LWF – large within-forest; SWF – small within-forest.

Discussion

Total dispersal kernels of *E. grandis*

Our estimated total dispersal kernel for *E. grandis* (Fig. 9.4) shows a strongly leptokurtic distribution, which roughly approximates (but does not fit) a negative exponential distribution in the first 300–400 m. We found that 77% of seeds are dispersed less than 100 m away, while just 3% are dispersed beyond 400 m in a long and thin tail that extends out to 2309 m. Disperser functional groups contribute differentially across this kernel. Some provide dispersal only in the first metres or tens of metres (e.g. terrestrial within-forest frugivores and predatory rodents), others across only the first hundreds of metres (e.g. small and large within-forest frugivores), while others contribute across the entire range (e.g. mega-terrestrial and the wide-ranging groups). The different disperser functional groups vary not only in dispersal distances provided, but also in their contribution to crop removal. Most fruits are removed from the canopy, predominantly by birds, with terrestrial groups providing only a small proportion of the dispersal. Even within these two broad categories, different disperser functional groups provided very different contributions. Not surprisingly, groups that comprise large-bodied species or that contain abundant large species (e.g. large within-forest and wide-ranging slow-gut) provided the greatest proportion of dispersal to this relatively large-fruited and large-seeded plant. These patterns of variation in the contribution of dispersers to dispersal distances and crop removal are central to understanding dispersal outcomes and their consequences. Such data make diversity analyses such as those outlined by Nathan (Chapter 11, this volume) possible and provide a powerful approach to predicting the consequences of dispersal.

Our estimates of a total dispersal kernel for *E. grandis* indicate that a small, yet reasonable, proportion of the crop receives long-distance dispersal, despite these events frequently being considered rare (Harper, 1977; Willson, 1993; Clark *et al.*, 1999). Criteria for long-distance dispersal are based either on a threshold distance (absolute) or proportion of dispersal events (proportional) beyond which dispersal is identified as long-distance (Nathan *et al.*, 2005). Both can entail an arbitrary decision (e.g. Cain *et al.*, 2000; Tackenberg, 2003; Soons and Ozinga, 2005) but one that should be underpinned by sound reasoning. For example, some absolute definitions assume that long-distance dispersal crosses a system boundary, such as the distance over which habitats are connected by physical processes (Green and Figuerola, 2005) or the scale of inter-patch distances (Trakhtenbrot *et al.*, 2005). Proportional definitions assume that long-distance dispersal is rare, and a boundary is set accordingly (Green and Figuerola, 2005).

In practice, one's definition of long-distance dispersal is determined by the questions being addressed, the relevant distances or proportions associated with these, and the data on hand. Given that we have set out to describe a total dispersal kernel, rather than consider it in the context of a particular process or question, we adopt a proportional definition based on a statistical definition of extreme values based on the statistical moments of the kernel (i.e. the 75th percentile + 3× the quartile range; StatSoft, 2005). For *E. grandis*, this corresponds to 330 m; only 3.9% of the crop is dispersed further than this distance, though seeds may travel as much as 2309 m. These distances suggest that *E. grandis* seeds will have access to most parts of the landscape and, given that some of its long-distance dispersers use both continuous forest and habitat fragments, that it will be capable of recolonizing areas within the human-dominated matrix.

A framework for describing total dispersal kernels

We have presented a framework for ecological studies of seed dispersal processes and, in particular, for studies of how patterns of interaction between fruits and frugivores produce total dispersal kernels. This framework is built on a mechanistic understanding of the dispersal process and incorporates two key assumptions.

- Dispersal processes, while complex, comprise sets of interactions between fruits and frugivores, many of which are similar in terms of their functional traits relevant to dispersal. These patterns of similarity allow fruits and frugivores to be assigned to groups on the basis of their shared functional traits, greatly reducing the complexity that must be dealt with in order to describe dispersal.
- While dispersal outcomes for plant propagules are ultimately the focus of our interest, understanding the spatial aspects of this requires a focus on the vectors of seed movement, which are primarily vertebrates in tropical forests.

While we feel that our approach and these assumptions are reasonable, the focus of the dispersal literature on individual species interactions, on seeds, and on hypothetical kernels makes their reiteration worthwhile.

In the format presented here, our approach focuses primarily on seed arrival after biotic dispersal. Our consideration of abiotic dispersal is limited to fruit-fall below the parent plant, as the large endocarps of *E. grandis* are rarely dispersed by abiotic means. Neither do we include the effects of post-dispersal mortality. However, these processes can be fully incorporated into our framework with relative ease (Fig. 9.1). To incorporate abiotic dispersal, kernels for water, wind or gravity must be measured or estimated along with the relative proportion of seeds dispersed by the vector. These dispersal kernels can then be incorporated, with the kernels of biotic vectors, into the total dispersal kernel. We have assumed that abiotic dispersal due to short-distance, over-ground movement by gravity and water would result in only a slight rearrangement of the dispersal kernel for *E. grandis*. In general, this is a reasonable assumption for this large fruit; however, we acknowledge that when *E. grandis* canopies overhang streams or rivers some seeds may be carried long distances, potentially resulting in novel dispersal outcomes (Higgins *et al.*, 2003).

Incorporating post-primary dispersal processes, such as secondary dispersal, seed predation and mortality through disease, into predictions of a 'recruitment' or 'effective' dispersal kernel is more difficult, although ultimately an important goal. The difficulty stems from needing to describe several additional processes. Essentially, the relationship between distance from the source and mortality needs to be described to estimate a 'mortality kernel' that can be subtracted from the total dispersal kernel. Where there is no relationship between mortality and dispersal distance, the scale and shape of the recruitment kernel will remain similar to the total dispersal kernel but with a reduced proportion of the crop surviving – perhaps dramatically. Where a relationship does exist, the subtraction of the mortality kernel will determine the shape and scale of the recruitment kernel.

The approach we outline here has both advantages and disadvantages. A disadvantage is the investment required. While this is large, the use of functional classifications reduces the investment that would otherwise be required. To date, ecologists have invested heavily in partial descriptions of dispersal processes. That descriptions of total dispersal kernels require serious effort merely validates past experience and should be considered in the context of the benefits of being able to understand the process as a whole.

Though our approach provides a means of describing some spatial aspects of dispersal, it does not preserve the hereditary linkages between individuals, a shortcoming shared by all ecological approaches. For many ecological questions this is not a concern; for example, if we are simply interested in the shape and scale of the dispersal kernel. However, the spatial pattern of relatedness in dispersal becomes important in considering the effects of dispersal on genetic structuring within and between populations. The use of simulation modelling based on total dispersal kernels and defined patterns of mating would allow estimation of these hereditary linkages (Bohrer *et al.*, 2005) but their empirical documentation is reliant upon the use of genetic approaches.

One key advantage of our approach is the detailing and incorporation of the contributions of all dispersers in a system, be that the dispersal for a plant individual, species, functional class or community. This is a very different approach to that where the contributions of one or a few dispersers are considered or, at the other extreme, where all disperser contributions are pooled and indistinguishable. Our results suggest that care must be taken in extrapolating studies of the contributions of single dispersers to the complete dispersal received by a plant (see Nathan, Chapter 11, this volume). For example, if we had measured only the 'wide-ranging slow-gut' functional group, we would estimate a kernel with a peak between 25 and 75 m from a tree and a tail that extended to 525 m. In contrast, the total dispersal kernel has its peak at the source and a tail that extends to 2309 m (Fig. 9.4). Conclusions based on just 'wide-ranging slow-gut' dispersers would underestimate the frequency of both short- and long-distance dispersal and thus potential recruitment outcomes at both local and larger scales. For natural resource managers, such under- (or over-) estimates of the scale of dispersal could be catastrophic, for example, in the management of weeds or in assessments of a species' ability to persist in a fragmented landscape or to reach new or habitats or refuges.

Being able to describe the combined service provided by all dispersers is an important step in seed dispersal research. But, equally, the description of the relative contributions of different dispersers matches it in importance. This has significant scientific and applied advantages as it allows for exploration of the consequences of changes in disperser communities. For example, our classification of dispersers can be used as a key part of a toolkit for predicting the functional consequences of changes in the distribution and/or abundance of dispersers, be that through decline, loss or increase in dispersers. This may be used for comparing different species of plants; communities in different locations; disturbed or fragmented and undisturbed forests; or for predicting the spread of weeds. For example, some dispersers occur in lower abundance, or not at all, in fragments of habitat compared to continuous forest (Warburton, 1997; Moran *et al.*, 2004). If those species belong to a functional group with other members that remain common in habitat fragments, the loss of one species would not result in a decline in services to the plants in habitat fragments. Where all the members of a functional group occur in low densities or not at all in habitat fragments, a clear loss of services to the plants in fragments is predicted. If a number of disperser functional groups decline, then dispersal services to those plants may be severely reduced (Dennis *et al.*, 2005)

Another important advantage of the approach is that it allows for estimation of the tail of the dispersal kernel. This has been a notoriously difficult task because only a tiny proportion of seeds occur in this part of the kernel and these may be spread over a vast area, making documentation extraordinarily difficult (Cain *et al.*, 2000; Higgins *et al.*, 2003). Using genetic methods may provide a solution (Cain *et al.*, 2000). However, in the context of contemporary dispersal, these methods are generally applied at a plot-scale using techniques that can monitor dispersal from known sources (Jones *et al.*, 2005; Pairon *et al.*, 2006). At larger spatial and temporal scales, the difficulty of sampling populations exhaustively means that most genetic approaches estimate dispersal through measures of relatedness between populations and rely on a variety of assumptions that may be violated in nature; for example, selectively neutral markers, discrete or equal-sized populations, no linkage disequilibrium among loci, or Hardy–Weinberg equilibrium. In contrast, our approach is not reliant on finding those few individuals within a population that received long-distance dispersal. By focusing on the process rather than the outcome, it allows direct estimation of the scale and frequency of long-distance dispersal events. Such estimates can subsequently be used to estimate long-term dispersal outcomes.

This approach is just one of a suite of approaches that are part of our toolkit for exploring questions relating to dispersal. It is a powerful method, but one that should be used only for certain questions and applied in concert with other approaches when appropriate. For example, in many instances genetic approaches would provide a viable alternative to the method we have outlined. However, these are complementary rather than competing approaches. Genetic methods provide a pattern-level description of effective dispersal kernels (i.e. dispersal to the recruitment stage; see Hardesty, Chapter 12, this volume). The ecological approach we have described provides mechanistic insights into the processes leading up to that pattern. Both approaches provide singular insights. For example, the genetic approach identifies the interaction of individual plants and the spatial component of relatedness at the individual and population level, while the ecological approach identifies the contributions of particular dispersers and is better at describing the tail. Combined, the ecological and genetic approaches will prove powerful not only in describing dispersal but also in exploring the factors that influence the success of genotypes. An important contribution that the different methods can make when applied together is in informing experimental design and providing independent tests of each other's predictions.

Conclusions

Measuring dispersal kernels is an important component in attempts to predict and model a variety of ecological processes in tropical forests, including population structuring and expansion, invasions and community dynamics. To date, much effort has focused on partial or indirect methods for describing dispersal curves, for example, documenting patterns of seed rain (Clark *et al.*, 1998) or describing the contributions of individual dispersers (Dennis, 2003; Westcott *et al.*, 2005a). The assumption in the literature appears to be that documenting total dispersal kernels is too difficult or too complex to be achievable. The framework and the data we have presented shows that, though the difficulties are not to be underestimated, the data can be obtained and estimates of dispersal distances based on a mechanistic understanding of the process can be made in a manner that allows the identification of the contributions of different dispersers.

Acknowledgements

We thank Matt Bradford and Adam McKeown for excellent technical assistance. Earthwatch Institute volunteers contributed to various aspects of data collection and management. The Earthwatch Institute provided additional financial support. We thank T. Carlo, R. Nathan and E.W. Schupp for helpful comments on the manuscript. This research was conducted under permits WITK02096404 and WISP02096504 and animal ethics approval OB15/12.

References

Bleher, B., Oberrath, R. and Bohning-Gaese, K. (2002) Seed dispersal, breeding system, tree density and the spatial pattern of trees: a simulation approach. *Basic and Applied Ecology* 3, 115–123.

Bohrer, G., Nathan, R. and Volis, S. (2005) Effects of long-distance dispersal for metapopulation survival and genetic structure at ecological and spatial scales. *Journal of Ecology* 93, 1029–1040.

Bolker, B.M. and Pacala, S.W. (1999) Spatial moment equations for plant competition: understanding spatial strategies and the advantages of short dispersal. *American Naturalist* 153, 575–602.

Cain, M.L., Milligan, B.G. and Strand, A.E. (2000) Long distance seed dispersal in plant populations. *American Journal of Botany* 87, 1217–1227.

Chapman, C.A. and Chapman, L.J. (2002) Plant–animal coevolution: is it thwarted by spatial and temporal variation in animal foraging. In: Levey, D.J., Silva, W.R. and Galetti, M. (eds) *Seed Dispersal and Frugivory: Ecology, Evolution and Conservation.* CAB International, Wallingford, UK, pp. 275–290.

Chave, J., Muller-Landau, H.C. and Levin, S.A. (2002) Comparing classical community models: theoretical consequences for patterns of diversity. *American Naturalist* 159, 1–23.

Clark, J.S., Macklin, E. and Wood, L. (1998) Stages and spatial scales of recruitment limitation in southern Appalachian forests. *Ecological Monographs* 68, 213–235.

Clark, J.S., Silman, M., Kern, R., Macklin, E. and Hille Ris Lambers, J. (1999) Seed dispersal near and far: patterns across temperate and tropical forests. *Ecology* 80, 1475–1494.

Clark, C.J., Poulsen, J.R. and Parker, V.T. (2001) The role of arboreal seed dispersal groups on the seed rain of a lowland tropical forest. *Biotropica* 33, 606–620.

Collins, E.M. (2002) Seed predation and dispersal by rodents in rain forest fragments and continuous forests. Honours thesis, James Cook University, Townsville, Australia.

Connell, J.H. (1971) On the roles of natural enemies in preventing competitive exclusion in some marine animals and in rain forest trees. In: den Boer, P.J. and Gradwell, G.R. (eds) *Dynamics of Numbers in Populations.* Centre for Agricultural Publication and Documentation, Proceedings of the Advanced Studies Institute, Oosterbeek, Wageningen, The Netherlands, pp. 298–312.

Connell, J.H. and Green, P.T. (2000) Seedling dynamics over thirty-two years in a tropical rain forest tree. *Ecology* 81, 568–584.

Dennis, A.J. (1997) Musky rat-kangaroos, *Hypsiprymnodon moschatus*: cursorial frugivores in Australia's wet tropical rain forests. PhD thesis, James Cook University of North Queensland, Townsville, Australia.

Dennis, A.J. (2003) Scatter-hoarding by musky rat-kangaroos, *Hypsiprymnodon moschatus*, a tropical rain-forest marsupial from Australia: implications for seed dispersal. *Journal of Tropical Ecology* 19, 619–627.

Dennis, A.J. and Westcott, D.A. (2006) Reducing complexity in studying seed dispersal at community scales: a functional classification of vertebrate seed dispersers in tropical forests. *Oecologia* 149, 620–634.

Dennis, A.J., Westcott, D.A., McKeown, A., Bradford, M.G. and Harrington, G.N. (2004) Fruit removal rates across a landscape. In: Kanowski, J., Catteral, C.P., Dennis, A.J. and Westcott, D.A. (eds) *Animal–Plant Interactions in Rainforest Conservation and Restoration.* Cooperative Research Centre for Tropical Rainforest Ecology and Management, Rainforest CRC, Cairns, Australia, pp. 9–12.

Dennis, A.J., Lipsett-Moore, G.J., Harrington, G.N., Collins, E. and Westcott, D.A. (2005) Seed predation, seed dispersal and habitat fragmentation: does context make a difference in tropical Australia. In: Forget, P.-M., Lambert, J.E., Hulme, P.E. and Vander

Wall, S.B. (eds) *Seed Fate: Predation, Dispersal and Seedling Establishment.* CAB International, Wallingford, UK, pp. 117–135.
Fisher, R.A. (1937) The wave of advance of advantageous genes. *Annual Eugenics* 7, 355–369.
Fleming, T.H. and Heithaus, E.R. (1981) Frugivorous bats, seed shadows and the structure of tropical forests. *Biotropica* 13, 45–53.
Forget, P.-M. (1991) Scatterhoarding of *Astrocaryum paramaca* by *Proechimys* in French Guiana: comparison with *Myoprocta exilis. Tropical Ecology* 32, 155–167.
Forget, P.-M. and Wenny, D. (2005) How to elucidate seed fate: a review of methods used to study seed removal and secondary seed dispersal. In: Forget, P.-M., Lambert, J.E., Hulme, P.E. and Vander Wall, S.B. (eds) *Seed Fate: Predation, Dispersal and Seedling Establishment.* CAB International, Wallingford, UK, pp. 379–393.
Forget, P.-M., Hammond, D.S., Milleron, T. and Thomas, R. (2002) Seasonality of fruiting and food hoarding by rodents in neotropical forests: consequences for seed dispersal and seedling recruitment. In: Levey, D.J., Silva, W.R. and Galetti, M. (eds) *Seed Dispersal and Frugivory: Ecology, Evolution and Conservation.* CAB International, Wallingford, UK, pp. 241–256.
Gautier-Hion, A., Duplantier, J.M., Feer, Q.R., Sourd, C., Decoux, J.P., Dubost, G., Emmons, L., Erard, C., Hecketsweiler, P., Moungazi, A., Roussihon, C. and Thiollay, J.M. (1985) Fruit characters as a basis of fruit choice and seed dispersal in a tropical forest vertebrate community. *Oecologia* 65, 324–337.
Godoy, J.A. and Jordano, P. (2001) Seed dispersal by animals: exact identification of source trees with endocarp DNA microsatellites. *Molecular Ecology* 10, 2275–2283.
Green, A.J. and Figuerola, J. (2005) Recent advances in the study of long-distance dispersal of aquatic invertebrates via birds. *Diversity and Distributions* 11, 149–156.
Grivet, D., Smouse, P.E. and Sork, V.L. (2005) A novel approach to an old problem: tracking dispersed seeds. *Molecular Ecology* 14, 3585–3595.
Hamill, D.N. and Wright, S.J. (1986) Testing the dispersion of juveniles relative to adults: a new analytic model. *Ecology* 67, 952–957.
Hanski, I. (2001) Population dynamic consequences of dispersal in local populations and in metapopulations. In: Clobert, J., Danchin, E., Dhondt, A.A. and Nichols, J.D. (eds) *Dispersal.* Oxford University Press, Oxford, UK, pp. 283–298.
Harms, K.E., Wright, S.J., Calderón, O., Hernández, A. and Herre, E.A. (2000) Pervasive density-dependent recruitment enhances seedling diversity in a tropical forest. *Nature* 404, 493–495.
Harper, J.L. (1977) *Population Biology of Plants.* Academic Press, London, UK.
Herrera, C.M. (1986) Vertebrate dispersed plants: why they don't behave the way they should. In: Estrada, A. and Fleming, T.H. (eds) *Frugivores and Seed Dispersal.* Dr. W. Junk, Dordrecht, The Netherlands, pp. 5–18.
Herrera, C.M. (1998) Long-term dynamics of Mediterranean frugivorous birds and fleshy fruits: a 12-year study. *Ecological Monographs* 68, 511–538.
Higgins, S.I., Nathan, R. and Cain, M.L. (2003) Are long distance dispersal events in plants usually caused by non-standard means of dispersal? *Ecology* 84, 1945–1956.
Holbrook, K.M., Smith, T.B. and Hardesty, B.D. (2002) Implications of long-distance movements of frugivorous rain forest hornbills. *Ecography* 25, 745–749.
Janzen, D.H. (1970) Herbivores and the number of tree species in tropical forests. *American Naturalist* 104, 501–528.
Jayasekara, P., Takatsuki, S., Weerasinghe, U.R. and Wijesundara, S. (2003) Arboreal fruit visitors in a tropical forest in Sri Lanka. *Mammal Study* 28, 161–165.
Jones, F.A., Chen, J., Weng, G.-J. and Hubbell, S.P. (2005) A genetic evaluation of seed dispersal in the Neotropical tree *Jacaranda copaia* (Bignoniaceae). *American Naturalist* 166, 543–555.

Jordano, P. (1992) Fruits and frugivory. In: Fenner, M. (ed.) *Seeds: The Ecology of Regeneration in Plant Communities*. CAB International, Wallingford, UK, pp. 105–156.
Kot, M., Lewis, M.A. and van den Driessche, P. (1996) Dispersal data and the spread of invading organisms. *Ecology* 77, 2027–2042.
Levey, D.J., Bolker, B.M., Tewksbury, J.J., Sargent, S. and Haddad, N.M. (2005) Effects of landscape corridors on seed dispersal by birds. *Science* 309, 146–148.
Levin, S.A., Muller-Landau, H.C., Nathan, R. and Chave, J. (2003) The ecology and evolution of seed dispersal: a theoretical perspective. *Annual Review of Ecology, Evolution, and Systematics* 34, 575–604.
Levine, J.M. (2003) A patch modelling approach to the community-level consequences of directional dispersal. *Ecology* 84, 1215–1224.
Levine, J.M. and Murrell, D.J. (2003) The community-level consequences of seed dispersal patterns. *Annual Review of Ecology, Evolution, and Systematics* 34, 549–574.
Loiselle, B.A., Sork, V.L. and Graham, C. (1995) Comparison of genetic variation in bird-dispersed shrubs of a tropical wet forest. *Biotropica* 27, 487–494.
Mack, A.L. (1995) Distance and non-randomness of seed dispersal by the dwarf cassowary, *Casuarius bennetti*. *Ecography* 18, 286–295.
Mack, A.L. and Druliner, G. (2003) A non-intrusive method for measuring movements and seed dispersal in cassowaries. *Journal of Field Ornithology* 74, 193–196.
Mollison, D. (1977) Spatial contact models for ecological and epidemic spread. *Journal of the Royal Statistical Society, Series B* 39, 283–326.
Morales, J.M. and Carlo, T.A. (2006) The effects of plant distribution and frugivore density on the scale and shape of dispersal kernels. *Ecology* 86, 1489–1496.
Moran, C., Catterall, C.P., Green, R.J. and Olsen, M.F. (2004) Functional variation among frugivorous birds: implications for rainforest seed dispersal in a fragmented subtropical landscape. *Oecologia* 141, 584–595.
Mouissie, A.M., Lengkeek, W. and van Diggelen, R. (2005) Estimating adhesive seed-dispersal distances: field experiments and correlated random walks. *Functional Ecology* 19, 478–486.
Murray, K.G. (1988) Avian seed dispersal of three neotropical gap-dependent plants. *Ecological Monographs* 58, 271–298.
Nathan, R. and Muller-Landau, H.C. (2000) Spatial patterns of seed dispersal: their determinants and consequences for recruitment. *Trends in Ecology and Evolution* 15, 278–285.
Nathan, R., Sapir, N. and Trakhtenbrot, A. (2005) Long-distance biological transport processes through the air: can nature's complexity be unfolded in silico? *Diversity and Distributions* 11, 131–137.
Pairon, M., Jonard, M. and Jacquemart, A.L. (2006) Modelling seed dispersal of black cherry, an invasive forest tree: how microsatellites may help? *Canadian Journal of Forest Research* 36, 1385–1394.
Petit, R.J. and Grivet, D. (2002) Optimal randomization strategies when testing the existence of a phylogeographic structure. *Genetics* 161, 469–471.
Reid, C. (1899) *The Origin of the British Flora*. Dulua, London.
Schupp, E.W. (1993) Quantity, quality and the effectiveness of seed dispersal by animals. *Vegetatio* 107/108, 15–29.
Skellam, J.G. (1951) Random dispersal in theoretical populations. *Biometrika* 38, 196–218.
Snow, D.W. (1981) Tropical frugivorous birds and their food plants: a world survey. *Biotropica* 13, 1–14.
Soons, M.B. and Ozinga, W.A. (2005) How important is long-distance seed dispersal for the regional survival of plant species? *Diversity and Distributions* 11, 165–172.
Stansbury, C.D. (2001) Dispersal of the environmental weed bridal creeper, *Asparagus asparagoides*, by silvereyes, *Zosterops lateralis*, in south-western Australia. *Emu* 101, 39–45.

StatSoft (2005) STATISTICA (Data analysis software system), version 7.1 [www.statsoft.com].

Sun, C., Ives, A.R., Kraeuter, H.J. and Moermond, T.C. (1997) Effectiveness of three turacos as seed dispersers in a tropical montane forest. *Oecologia* 112, 94–103.

Tackenberg, O. (2003) Modeling long-distance dispersal of plant diaspores by wind. *Ecological Monographs* 73, 173–189.

Trakhtenbrot, A., Nathan, R., Perry, G. and Richardson, D.M. (2005) The importance of long-distance dispersal in biodiversity conservation. *Diversity and Distributions* 11, 173–181.

Terborgh, J. (1986) Community aspects of frugivory in tropical forests. In: Estrada, A. and Fleming, T.H. (eds) *Frugivores and Seed Dispersal.* Dr. W. Junk, Dordrecht, The Netherlands, pp. 371–384.

Terborgh, J., Pitman, N., Silman, M., Schichter, H. and Núñez V.P. (2002) Maintenance of tree diversity in tropical forests. In: Levey, D.J., Silva, W.R. and Galetti, M. (eds) *Seed Dispersal and Frugivory: Ecology, Evolution and Conservation.* CAB International, Wallingford, UK, pp. 1–17.

Tewksbury, J.J. and Lloyd, J.D. (2001) Positive interactions under nurse-plants: spatial scale, stress gradients and benefactor size. *Oecologia* 127, 425–434.

Theimer, T.C. (2001) Seed scatterhoarding by white-tailed rats: consequences for seedling recruitment by an Australian rain forest tree. *Journal of Tropical Ecology* 17, 177–189.

Traveset, A. and Verdú, M. (2002) A meta-analysis of the effect of gut treatment on seed germination. In: Levey, D.J., Silva, W.R. and Galetti, M. (eds) *Seed Dispersal and Frugivory: Ecology, Evolution and Conservation.* CAB International, Wallingford, UK, pp. 339–350.

Tuomisto, H., Ruokolainen, K. and Yli-Halla, M. (2003) Dispersal, environment and floristic variation of western Amazonian forests. *Science* 299, 241–244.

Turchin, P. (1998) *Quantitative Analysis of Movement: Measuring and Modelling Population Redistribution in Animals and Plants.* Sinauer, Sunderland, Massachusetts.

Vander Wall, S.B. (2001) The evolutionary ecology of nut dispersal. *Botanical Review* 67, 74–117.

Warburton, N.H. (1997) Structure and conservation of forest avifauna in isolated rainforest remnants in tropical Australia. In: Laurance, W.F. and Bierregaard, R.O. (eds) *Tropical Forest Remnants: Ecology, Management and Conservation of Fragmented Communities.* University of Chicago Press, Chicago, Illinois, pp. 190–206.

Wenny, D.G. and Levey, D.G. (1998) Directed seed dispersal by bellbirds in a tropical cloud forest. *Proceedings of the National Academy of Sciences USA* 95, 6204–6207.

Westcott, D.A. and Graham, D.L. (2000) Patterns of movement and seed dispersal of a tropical frugivore. *Oecologia* 122, 249–257.

Westcott, D.A., Bentrupperbaumer, J., Bradford, M.G. and McKeown, A. (2005a) Incorporating disperser movement and behaviour patterns into models of seed dispersal. *Oecologia* 146, 57–67.

Westcott, D.A., Bradford, M.G., Dennis, A.J. and Lipsett-Moore, G.J. (2005b) Key fruit resources in Australia's tropical rain forests. In: Dew, J.L. and Boubli, J.P. (eds) *Fruits and Frugivores: The Search for Strong Interactors.* Kluwer Academic, New York, pp. 236–260.

Wilkinson, D.M. (1997) The role of seed dispersal in the evolution of mycorrhizae. *Oikos* 78, 394–396.

Willson, M.F. (1993) Dispersal mode, seed shadows, and colonization patterns. *Vegetatio* 107/108, 261–280.

Willson, M.F., Irvine, A.K. and Walsh, N.G. (1989) Vertebrate dispersal syndromes in some Australian and New Zealand plant communities, with geographic comparisons. *Biotropica* 21, 133–147.

Witmer, M.C. and Cheke, A.S. (1991) The dodo and the tamalacoque tree: an obligate mutualism reconsidered. *Oikos* 61, 133–137.

Yasuda, M., Miura, S., Ishii, N., Okuda, T. and Hussein, N.A. (2005) Fallen fruits and terrestrial vertebrate frugivores: a case study in a lowland tropical rain forest in Peninsular Malaysia. In: Forget, P.-M., Lambert, J.E., Hulme, P.E. and Vander Wall, S.B. (eds) *Seed Fate: Predation, Dispersal and Seedling Establishment.* CAB International, Wallingford, UK, pp. 151–174.

Yumoto, T. (1999) Seed dispersal by Salvin's curassow, *Mitu salvini* (Cracidae), in a tropical forest of Colombia: direct measurements of dispersal distance. *Biotropica* 31, 654–666.

10 Frugivores, Seeds and Genes: Analysing the Key Elements of Seed Shadows

P. Jordano

Introduction

A seed shadow is the spatial distribution of all seeds dispersed from an individual plant (Janzen, 1970; Nathan and Muller-Landau, 2000; Schupp *et al.*, 2002). The sum of seed shadows across all individuals in a population plus those seeds dispersed from other populations makes up seed rain. For vertebrate-dispersed species, seed shadows are the primary outcome of plant–frugivore interactions, yet we know very little about how different frugivore species contribute to them. For instance, a few frugivore species might contribute disproportionately to long-distance dispersal (LDD) events, while another subset of dispersers might contribute to the local, short-distance dispersal (see Dennis and Westcott, Chapter 9, this volume). In addition, seed delivery to specific microhabitats might be provided by only a few species out of the whole frugivore assemblage (Reid, 1989; Wenny and Levey, 1998; Jordano and Schupp, 2000; Jordano *et al.*, 2007).

For years, studies of seed dispersal have been severely limited in their potential to characterize seed shadows and to assess the contribution different species of frugivores make to their extent and reach in distance and space (i.e. the dispersal kernels; see Portnoy and Willson, 1993; Clark *et al.*, 1999; Bullock and Clarke, 2000; Nathan, Chapter 11, this volume). The main difficulty has been to succeed in assigning a tree source for a dispersed seed and then link that information with both the species of frugivore that delivered the seed and the subsequent establishment success for the seedling. These are the three basic pieces of information that we need to account for seed shadows: source of seeds, dispersal vector, and establishment outcome.

Variations in seed shadow pattern and extent between individual plants in a population can be large, and the summed seed shadows can generate very complex patterns of seed rain. Marked peaks and troughs in 2-D seed density patterns (i.e. the two-dimensional spatial distribution of seeds on the ground), including extensive areas with no seed rain at all, are all possible, resulting in an enormous range of recruitment patterns (see Kwit *et al.*, Chapter 19, this volume). It is the interaction between frugivore foraging and the structure of complex landscapes that generates these peaks and troughs in the seed shadows, marking 'hot-spots' and empty spots of seed rain and, consequently, potential plant recruitment (see Carlo *et al.*, Chapter 16, this volume). Our present understanding of these complex dynamics is extremely poor, despite unprecedented advances in recent years, especially when we refer to the influences on genetic structure or the relevance of LDD events (see Hardesty, Chapter 12, this volume). The seed shadow is the immediate outcome of interactions with frugivores and represents the first stage of recruitment; it is therefore of paramount importance to understand whether this 'signal' of the interaction with frugivores lasts to the final stage of adult recruitment.

Recent advances in field techniques (e.g. use of radioactive markers, radio-tracking; Primack and Levy, 1988; see Dennis and Westcott, Chapter 9, this volume), molecular genetics tools (e.g. hypervariable simple-sequence DNA repeats – SSRs or microsatellites) (Godoy and Jordano, 2001; see Hardesty, Chapter 12, this volume), and GIS-based techniques (Loiselle and Blake, 1993) allow a thorough analysis of seed shadows (Nathan and Muller-Landau, 2000). Thus, several recent studies have marked an extraordinary advance in our ability to overcome long-standing methodological limitations in seed dispersal research (Nathan 2006). These advances fall into two distinct approaches to analysing seed shadows: (i) a *'source tree' perspective*; and (ii) a *'target or seed delivery site' perspective*. Future advances in the field will most probably stem from the combination of the two approaches, as they are complementary rather than exclusive.

The deconstruction of a seed shadow involves two steps:

1. Working from the source tree and proceeding away from it by determining dispersal distances and inferring which frugivore species contribute the dispersal events (the 'seed shadow' analysis in its strict sense);
2. Working from the microhabitat patches where seeds arrive and inferring which frugivore species contribute the seed rain from where.

Focal observations at fruiting trees have been routine in seed dispersal studies since the late 1970s (Howe and Kerckhove, 1980; Herrera and Jordano, 1981; Snow and Snow, 1988). The simple approach of watching a tree and attempting to infer where seeds go on the basis of the animals removing them implicitly emphasized a 'from the source tree' perspective and led to the tracking of the movement of seeds away from maternal trees. Therefore, tracking methods, either direct or indirect, have been widely used to assess seed shadows and understand the processes generating them: direct mapping of frugivore locations (Julliot, 1997; Whitney *et al.*, 1998; Holbrook *et al.*, 2002; Mack and Druliner, 2003); direct observations of frugivore foraging and seed delivery (Reid, 1989;

Gómez, 2003; see Stevenson, Chapter 15, this volume); combining frugivore tracking with gut retention times (Fleming, 1988; Murray, 1988; Sun *et al.*, 1997; Westcott and Graham, 2000; Westcott *et al.*, 2005); and using inert markers to track seed movement (Mack, 1995; Vander Wall, 2000; Levey *et al.*, 2005), among others.

On the other hand, a number of studies have used the 'target or seed delivery site' perspective by emphasizing seed rain analysis (Izhaki *et al.*, 1991; Houle, 1992; Nakashizuka *et al.*, 1995; Kollmann and Goetze, 1997; Clark *et al.*, 1999; Alcántara *et al.*, 2000; Harms *et al.*, 2000; Jordano and Schupp, 2000; Rey and Alcántara, 2000; Muller-Landau *et al.*, 2002; Clark *et al.*, 2004; McEuen and Curran, 2004). These studies grew from the analysis of seed banks, but extended this work to encompass the arrival of seeds (Schupp *et al.*, 1989). Work with seed-trap designs has significantly increased our understanding of seed rain patterns and of the initial template from which regeneration processes and recruitment unfold. Additionally, the specific design of the seed-trap monitoring might infl uence the robustness and precision of seed shadow and dispersal kernel estimation (Kollmann and Goetze, 1997; Skarpaas *et al.*, 2005). However, inferences about the causal processes for the observed seed rain patterns remain elusive unless a direct connection between process and pattern can be constructed by combining direct watches at the fruiting trees and seed-trap monitoring (Jordano and Schupp, 2000).

Both approaches to deconstruct seed shadows are methodologically limited by their ability to determine the source tree for dispersed seeds, especially for medium- and long-distance dispersal events and when these involve seed delivery from other populations. Frugivore tracking methods invariably fail to characterize the frequency and extent of extreme LDD events; and even for shorter dispersal, the precise location of the animals when delivering seeds cannot be established (Westcott *et al.*, 2005). Interpretation of seed-trap methods can be limited when the number of potential source trees is large or when seed shadow overlap is extensive (Godoy and Jordano, 2001); additional limitations relate to the specific spacing patterns of the seed traps relative to source trees and the spatial design (Jones *et al.*, 2005; Skarpaas *et al.*, 2005).

In this chapter, I analyse the main components of seed shadows and discuss how hypervariable molecular markers can be used to assess them and the advantages and disadvantages involved in doing so. I present a brief discussion of seed shadow analysis, some illustrative data with *Prunus mahaleb*, and an analysis of perspectives to link frugivore foraging and seed shadow patterns. Despite enormous recent advances in our understanding of seed shadows and dispersal, many exciting aspects remain under-investigated. Some of these include the lasting consequences of frugivore foraging on genetic structure in the plant populations, the frequency and extent of LDD events, and the survival prospects of propagules from LDD events.

Analysing Seed Shadows

Fruiting trees are key elements in plant–frugivore interactions because they are the source of dispersed seeds, but they also shape the foraging movements of frugivores, are sinks for dispersed seeds and hot-spots for pathogens and post-dispersal seed predators (Schupp *et al.*, 2002; Clark *et al.*, 2004; see Carlo *et al.*, Chapter 16, this volume; Gallery *et al.*, Chapter 22, this volume). Their role as seed sources has traditionally been assessed from a tree's perspective; attempting to track the direction and distance the seeds are dispersed. When we consider the opposite view, from the target patch receiving dispersed seeds, different frugivores contribute in different ways to the seed rain arriving at given patches, in terms of dispersal distance, proportion of long-distance dispersal events, and diversity of contributing mother trees (Westcott *et al.*, 2005; see Dennis and Westcott, Chapter 9, this volume; Nathan, Chapter 11, this volume). This may generate extensive variation in the makeup of the seed shadows, both in terms of local seed density and in its fine-scale genetic composition (García *et al.*, 2007a,b). Even in highly diversified plant–frugivore interactions, involving many species with dozens of interactions, only a few specific frugivores may contribute a major fraction of the long-distance dispersal events, or may contribute disproportionately to the seed rain in hot-spots for successful recruitment (Jordano and Schupp, 2000; Fragoso *et al.*, 2003). Long-distance dispersal events, both within and between populations, are probably more frequent than previously thought for endozoochorous species, but it is not clear how they contribute to patterns of colonization and gene flow. Therefore, their pervasive implications for the maintenance of fragmented populations and for the demographic and genetic make-up of plant populations in complex landscapes remain poorly understood.

There are two key questions in seed dispersal studies:

1. Which frugivore species contribute dispersed seed where?
2. Which source trees contribute dispersed seed where?

These are very complex issues, especially for dispersal systems where the diversity of frugivore species is high. Answering the first question requires precise frugivore tracking; answering the second question requires precise seed tracking. The main problem is that tracking methods lose precision when the tracking time or the tracking distance increase, thus severely compromising our precision in assessing LDD events (Nathan *et al.*, 2003). These are inherent problems for any of the methods used to date to assess dispersal distances.

The main components of seed shadow analysis are outlined in Table 10.1. From the perspective of the demographic and genetic effects of the frugivores, there are three main components of seed shadows: distance, density and location relative to the immediate neighbourhood (e.g. of conspecific trees). Let us consider the perspective of a dispersed seed just after successful dispersal to a microsite (target site) on the forest floor.

Table 10.1. The main components of a seed shadow in relation to distance, density and neighbourhood aspects and from the perspectives of the source tree for dispersed seeds and from the target microsite where seeds are delivered. Consider a dispersed seed in the seed shadow and its density, distance and neighbourhood characteristics.

Component	Source tree perspective	Target site perspective
Distance	Distance from source tree	Distance from source tree and to nearest conspecific
Density	Density of dispersed seeds from the source tree at target site	Density of dispersed seeds in target site Density of trees in target site neighbourhood
Neighbourhood	Overlap of seed shadow with conspecifics Genetic relatedness of source trees contributing seed	Genetic relatedness of dispersed seeds in target site Diversity of source trees contributing seed

Traditional seed shadow analysis has considered the relevance of both distance and density effects relative to the source trees (Janzen, 1970): the distance the seed has moved from the source tree, and the density of seeds from that source tree at the target point. By taking a 'target site perspective', it is also relevant to consider the distance to other conspecific trees, and the density of both dispersed seeds (from other conspecific trees) and adult trees in the neighbourhood of the target site (Table 10.1). The 'target site perspective' emphasizes that survival conditions for dispersed seeds might vary enormously in a context-dependent way (see Schupp, Chapter 20, this volume) because the 'safeness' of a site depends not only on distance to the seed source, but also on the environmental context of the target site.

Direct observation

Direct observations of frugivores while foraging for fruits reveal patterns of potential seed delivery, although it is extremely difficult to estimate actual dispersal from direct observation (Wenny and Levey, 1998; Jordano and Schupp, 2000; Gómez, 2003). In addition, reliable estimation of long dispersal distances, even within a local population, is generally not feasible. Direct observation of beween-population seed dispersal via LDD events, involving actual seed immigration, is almost impossible. Direct observation methods have been successfully used in acorn-caching jays (*Garrulus glandarius*) (Gómez, 2003), enabling the reliable measurement of long-distance flights ending in successful caching. However, the limitation persists for direct observation of acorn transfer between populations, especially between distant patches (J.M. Gómez, 2005, personal communication). For legitimate frugivores ingesting whole fruits there is the added complication of ingestion, retention time, and delayed delivery after gut mixing of fruits consumed during various feeding bouts (Holbrook and Smith, 2000; see

Traveset *et al.*, Chapter 4, this volume). The best approximations are obtained when watching habituated animals directly (Yumoto, 1999), using direct observations combined with non-intrusive methods (Mack and Druliner, 2003; Forget and Wenny, 2005; Levey *et al.*, 2005) or with domestic species (Manzano and Malo, 2006).

Almost the same limitations arise when using radio-tracking techniques to infer seed-source tree distances and spatial relations. Despite a high reliability of the frugivore location at a given time interval, there is considerable uncertainty about the actual seed delivery event. One has to resort to functional digestion models and gut passage time estimates to infer the seed delivery events and estimate dispersal distances relative to potential source trees. This leads to a potential bias in the estimation of true seed-source tree distance patterns. Despite these potential limitations, radio-tracking studies have shown very detailed patterns of seed delivery by frugivores involving LDD events (Fleming, 1988; Sun *et al.*, 1997; Holbrook and Smith, 2000; Westcott and Graham, 2000; Holbrook *et al.*, 2002; Westcott *et al.*, 2005). Direct observation or monitoring methods, if combined with seed rain monitoring (Wenny and Levey, 1998; Jordano and Schupp, 2000) will allow very robust estimates of who contributes seeds where, whereas the origin and distance components will be difficult to assess.

Hypervariable DNA markers

DNA-based methods to assess the parentage (maternity) of dispersed seeds have recently been developed, largely as an extension of seedling parentage analysis and pollen paternity analyses (Ennos, 2001). Briefly, parentage analysis uses genetic marker data from adults and established seedlings in a population to infer the male and female parents of focal offspring. Maternity analysis represents a subset of parentage techniques by assuming that established offspring (e.g. seedlings) derive from random mating of maternal trees with a homogeneous outcross pollen pool (Ennos, 2001; Jones and Ardren, 2003). Therefore, these techniques allow an inference based on exclusion probabilities and likelihood estimation. However, by taking advantage of tissue of maternal origin still attached to the dispersed seed in some way (e.g. endocarp tissue, ancillary structures of winged seeds like pappus, wings, acorn pericarp, etc), assignment of the most likely source tree can be inferred directly and robustly (Godoy and Jordano, 2001; Jordano and Godoy, 2002; Ziegenhagen *et al.*, 2003; Grivet *et al.*, 2005; Jones *et al.*, 2005; García *et al.*, 2007a). In contrast to direct observation methods, parentage and maternity analysis of dispersed seeds works from the dispersed seed backwards to infer the origin.

Methods not relying on the genotyping of seed maternal tissue use established seedlings and indirect inference methods to assign maternity and, eventually, paternity (Alvarez-Buylla *et al.*, 1996; Schnabel *et al.*, 1998; Bacles *et al.*, 2006; Hardesty *et al.*, 2006; Sato *et al.*, 2006). While these methods provide direct evidence of LDD events (see Hardesty, Chapter 12, this volume) they might underestimate the actual frequency of LDD events mediated by frugivore movements, since they measure established seedlings, i.e. realized, or effective, dispersal (Cain *et al.*, 2000) after the effect of post-delivery mortality factors (failure to germinate, consumption by seed predators). Therefore, actual immigration rates of seeds can be underestimated and differentiation between the seed recruitment curve and the seedling recruitment curve cannot be accomplished (Jones *et al.*, 2005). However, these methods can be very informative about the patterns of realized dispersal – a key ingredient of seed shadow analysis.

Genotyping methods based on maternal tissue can provide a useful tool when combined with direct watches of frugivores and seed rain analysis with seed traps. Based on direct assignment, they can overcome some of the limitations of likelihood-based assignment techniques (Jones and Ardren, 2003). By relying on sampling of dispersed seed, their main limitation would be the identification of the dispersal agent. Below I summarize a case study with *Prunus mahaleb* (Rosaceae) and discuss potential shortcomings of direct maternity estimation based on hypervariable DNA markers and endocarp tissues.

Model fitting to dispersal data

Valuable efforts in the last decade have been directed to model dispersal events, given the serious difficulties involved in empirically estimating dispersal distances. The two alternative approaches to dispersal modelling have been mechanistic and phenomenological (Nathan and Muller-Landau, 2000). *Mechanistic models* adopt a basic 'source tree' perspective and attempt to infer seed dispersal patterns from characteristics of the dispersers. They have been extensively used for anemochorous and abiotically dispersed species (Nathan *et al.*, 2003). *Phenomenological models* implicitly adopt a 'target site' perspective, aiming at reconstructing the seed dispersal curve given seed-trap data and locations of potential source trees (Sato and Hiura, 1998; Clark *et al.*, 1999). Given a pattern of dispersed seeds (e.g. derived from seed-trap sampling), inverse modelling is used to find out the best fit to a dispersal function that adequately reproduces the pattern, given the characteristics of the trees in the population (location, fecundity, etc). Additionally, these models can also take a source-tree perspective by assuming hypothetical distributions to represent the shape and scale of dispersal curves. Examples would be studies which estimate spread using different dispersal functions in the absence of any arrival data.

Invariably, the main difficulty with modelling attempts has been fitting the 'fat tail' of the dispersal curve: obtaining robust fits for the LDD events (Cain *et al.*, 2000; Nathan, 2006), even for abiotically dispersed species for which the models can be adequately parametrized. The situation with mechanistic models for animal-dispersed species is similar, although there are very few examples available (Westcott and Graham, 2000; Westcott *et al.*,

2005; see Dennis and Westcott, Chapter 9, this volume). Recent approaches to modelling frugivore-mediated dispersal from a mechanistic perspective have successfully incorporated movement patterns that depend on frugivore foraging behaviour and habitat preferences, combined with gut passage time dynamics (Morales and Carlo, 2006). Basically, they model the seed dispersal kernels as a function of frugivore movement and seed passage times after regurgitation or defecation (Murray, 1988). Assuming that, after consuming fruit, animals perform a random walk, it is possible to approximate movement with a diffusion equation (Morales and Carlo, 2006). The solution of the diffusion equation is then combined with a probability density function for gut-passage time for seeds in order to solve for the distances at which they would be deposited. However, simple diffusion may be a poor approximation for frugivore movements (Holbrook *et al.*, 2002; Westcott *et al.*, 2005) that frequently entail more complex patterns than a pure random walk (Boyer *et al.*, 2004). Furthermore, as stated by Morales and Carlo (2006), the spatial pattern of plant distribution will feed back into the characteristics of seed dispersal kernels via its effects on frugivore movements.

A Case Study

An account of Prunus mahaleb dispersal

Our protocols for the genetic evaluation of seed dispersal in *Prunus mahaleb* combine seed sampling in the field by means of seed traps that passively capture seeds dispersed (regurgitated or defecated) by frugivores, genetic analysis of adult trees and seed endocarps, and direct observations of frugivores at feeding trees. This enables us to identify uniquely the source tree for a dispersed seed and to infer the contribution of different species of frugivores to the seed rain in different microhabitats (Godoy and Jordano, 2001; Jordano and Godoy, 2002). Other applications of this approach can be found in recent work by Ziegenhagen *et al.* (2003), Grivet *et al.* (2005) and Jones *et al.* (2005).

Prunus mahaleb is a rosaceous tree that in south-eastern Spanish populations is gynodioecious, with some individuals producing hermaphrodite flowers and others with androsterile flowers, which behave as functional females. In the southern Iberian Peninsula, this species flowers between mid-May and mid-June at high elevations (over 1300 m), and insects, mainly bees (Hymenoptera: Andrenidae, Apidae) and flies (Diptera: Calliphoridae, Syrphidae), act as pollen vectors. *Prunus mahaleb* produces fleshy fruits (drupaceous) with one seed per fruit. In late July fleshy fruits are produced and consumed by frugivorous animals that disperse the seeds until late August or early September.

Our nine local *P. mahaleb* populations are located in Parque Natural de las Sierras de Cazorla, Segura y Las Villas (Jaén province, south-east Spain). In this area *P. mahaleb* naturally occurs as isolated small (ten trees) to medium-sized (150 trees) distinct populations. A few populations might reach approx. 3000 trees. These populations are naturally isolated from each other and occupy approximately 150 km^2. The main study population was located in Nava de las Correhuelas (NCH, hereafter), at 1615 m elevation. Detailed descriptions of the area and general methods can be found in Jordano and Schupp (2000) and references therein. The site is dominated by grasslands with scattered patches of deciduous vegetation, gravelly soil or rock outcrops covered by shrubs or small isolated trees. The rocky slopes are dominated by open pine forest (*Pinus nigra* subsp. *salzmannii*; Pinaceae) and juniper (*Juniperus communis*; Cupressaceae).

Frugivorous birds and mammals visiting *P. mahaleb* trees in Spanish populations usually behave as legitimate seed dispersers, swallowing the fruits whole and defecating and/or regurgitating the seeds, usually after leaving the tree. Most seed rain of *P. mahaleb* in the study areas is contributed by frugivorous birds. Seed rain and the resulting recruitment pattern of seedlings and saplings are highly patchy, and largely restricted to microhabitats beneath woody cover in the vicinity of fruiting trees (Jordano and Schupp, 2000).

Our general approach to assess seed dispersal distances and spatial patterns is to compare the multilocus genotype of the endocarp of dispersed seeds (either defecated or regurgitated by frugivorous animals) with those of potential maternal source trees in the population. Both hermaphrodite and female trees can act as seed sources in the study population, while only hermaphrodites act as pollen donors. The sex ratio is ~1:1 for the two gender types in this population. The endocarp is a maternal tissue (2n) with an identical genotype to the mother tree and in *Prunus* is derived from the carpellar wall. Thus, a full matching of the multilocus genotypes of a dispersed seed and a maternal tree unequivocally identifies the tree as the source for the dispersed seed, enabling a direct estimation of the dispersal distance (Godoy and Jordano, 2001).

Source tree identification

For this study of seed dispersal, a total of 180 adult trees and 95 dispersed seed endocarps were initially genotyped by Godoy and Jordano (2001). We later expanded this sample to include 263 trees from NCH (of which we used only the 196 reproductives in 2003; García *et al.*, 2005, 2007a,b) and 557 endocarps, and this is currently our main analysis data set. We have also genotyped trees from another eight populations in addition to NCH, totalling 472 trees.

Each adult tree in the NCH population showed a unique multilocus genotype. The source tree for individual dispersed seeds was identified by comparing the endocarp multilocus genotype with the complete set of genotypes of reproductive trees in the population. To assign the source tree for each dispersed seed we carried out an identity check by matching the multilocus genotype of the endocarp at nine microsatellite loci with those of the adult trees; for the adult trees we assessed 11 SSR loci. We used the full seed sample from the 1996 (n = 95) and 1997 (n = 462) cohorts (Table 10.2). We dropped $n = 27$ endocarps due to amplification failure for > 2 loci. We used CERVUS (Marshall *et al.*, 1998) and GIMLET (Valière, 2002) to identify the mismatches and multiple-matches among endocarps and putative source trees. For each sampled seed, the adult individual having a genotype matching the seed endocarp genotype was assigned as the mother tree. In a few cases (n = 27 endocarps), we failed to find evidence that an NCH tree was the mother source, but they had ≤ 2 loci missing and thus we were unable to assign them as immigrant seeds in NCH and so we considered them to be reassigned NCH seeds. Of the endocarps which could be assigned to NCH trees, all but two were assigned to a single tree; 97 endocarps were not assignable to any tree in NCH and we inferred that they were immigrant seeds dispersed from fruiting trees in other populations. We thus had 437 endocarps assigned to the NCH study population. The two seeds with double matching had failed amplifications for two loci and resulted in ambiguous matching (i.e. with two putative maternal trees). We assigned the seeds to the tree nearest to the sampling location, due to the fact that this procedure would minimize the estimation errors of dispersal distances when these are distributed with high skew. Incorrect assignment of a rare long-distance event would seriously bias the estimate of the dispersal function. Significant matches between endocarp and adult genotypes were found by testing a hypothesis of identity ($r_l = 1$, $r_m = 1$) in all possible pairwise comparisons between endocarps and adult trees and obtaining significance estimates by a jackknife resampling method (Queller and Goodnight, 1989).

Problems with SSR genotyping

There can be some problems at the genotyping and assignment stages when working with seed endocarps, and several of them are frequently encountered in maternity analyses.

Table 10.2. Estimates of the frequency (%) of within- and between-population dispersal events of pollen and seeds in *Prunus mahaleb*. Pollen dispersal was inferred from paternity analysis of progeny samples taken from maternal tree canopies before consumption by frugivores. Seed dispersal events were inferred from seed progeny sampled in seed traps after dispersal by frugivores. n = 200 seeds for pollen dispersal analysis (1997 cohort); n = 557 seeds for seed dispersal analysis (n = 95 seeds for the 1996 cohort; n = 462, 1997 cohort). See Godoy and Jordano (2001), García *et al.* (2005, 2007a,b), Jordano *et al.* (2007).

Stage	Within population	Between populations
Pollen	90.50	9.50
Seeds	79.66	20.34

- **Stutter bands** are fairly common when analysing microstellites, especially with dinucleotide repeats. We have observed them in amplifications from endocarp extracts but also, to the same extent, with leaf extracts and with all the other species we have worked on. Alleles were assigned to the largest, and most abundant, fragment. Heterozygotes for two close alleles were recognized by the shorter allele showing a higher intensity than the larger (the intensity of the band being the sum of the short allele and the stutter bands of the long allele). This results in a characteristic pattern clearly distinguishable from the homozygote pattern, in which the intensity of the bands decreases progressively with size. This is described by Hoelzel (1998).
- **Null alleles** are alleles that do not amplify, probably because of a mismatch in one of the primers used. They can be suspected if a heterozygote deficit is detected only in some of the loci; they are frequent in parentage analysis (Björklund, 2005). If null alleles are present, they should appear both in leaf and endocarp extracts. Therefore, we think it unlikely that they would affect the identity checks between seeds and trees: a seed showing a false homozygote will match a tree showing the same pattern. However, this will affect and limit the resolution power and bias the significance values for assignments (Jones and Ardren, 2003) (see below). We did not discount the possibility that null alleles were present in our samples, but we did not detect them. We observed a heterozygote deficit in most loci resulting from high inbreeding in the study population due to frequent selfing. We have not detected null alleles in paternity analyses (embryo genotyping), where we used progeny with known paternal and maternal trees, these being obtained from hand pollinations.
- **Allelic dropout** can also be a serious problem for the assignment of seeds to mother trees. It occurs when one of the alleles in a heterozygote is not amplified stochastically when using limited template DNA, as can be the case when endocarp extracts are used. In the controlled comparisons we made between leaf and endocarp genotypes, allelic dropout was not observed. However, this can be a potential problem if the DNA yield or quality is limiting. Repeating the amplification of homozygote loci several times (Taberlet, 1996) and accurately determining the concentration of DNA and excluding those samples in the limiting range where allelic dropout can occur (Morin *et al.*, 2001) are two strategies to deal with the problem at the genotyping stage. At the analysis level, any exclusion of identity between a seed and a potential mother tree based on only 1 or 2 loci mismatching (i.e. < 10% of the seeds in our sample) was rechecked.

Seed dispersal distances

Seed dispersal distances can readily be determined with the direct assignment method based on endocarp genotypes. For *P. mahaleb* we have already reported distances with a reduced sample of seeds (Godoy and Jordano, 2001; Jordano and Godoy, 2002). The pattern with a more extensive sample is quite similar, revealing highly limited dispersal in terms of distance, with most seeds dispersed ≤50 m from the source tree (Fig. 10.1) (Jordano and Godoy, 2002; García *et al.*, 2007a). Median dispersal distance (including only within-population events) was 32.0 m with quartiles Q_{25} = 10.5 m and Q_{75} = 100.3 m (n = 437 seeds). It is worth noting that $Q_{99.5}$ = 1025.0 m, indicating that LDD events within a population can occur with low frequency; thus, 3.5% of the dispersal events had a mean dispersal distance of 793.8 m. The resulting seed dispersal curve agrees well with the intra-population distances estimated from direct watches of birds while foraging at feeding trees (Jordano and Schupp, 2000) but, as expected, gives a much more precise and robust estimation of the LDD events. The dispersal curve was markedly leptokurtic (kurtosis = 16.68), with mode at ≤ 50 m and maximum at 1025 m. Note that in Fig. 10.1, dispersal events within 1200 m distance indicate within-population dispersal: seeds dispersed from trees growing in the NCH population. Only a small fraction of the dispersal events recorded correspond to immigrant seeds dispersed from other populations, and these LDD events are represented in Fig. 10.1 by the bars beyond 1200 m.

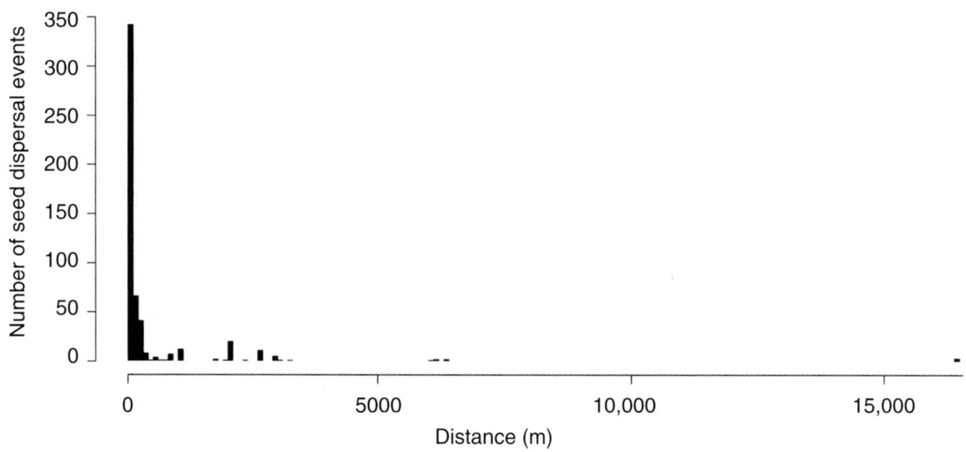

Fig. 10.1. Distance patterns in the seed rain of *Prunus mahaleb* (n = 558 seeds): the frequency distribution of dispersal distances in the study population including the long-distance between-population dispersal estimates. The dispersal curve combines results of seed dispersal distances from trees growing in the population (direct assignment) with distance estimates for immigrant seeds (assignment to the most likely source population). The abcissa shows the actual distance values between the location of a dispersed seed (defecated or regurgitated by a frugivore) and its maternal tree (log-scale, actual distance +1 m; distance 0 m includes seeds falling beneath the canopy of the mother tree) or source population. The dispersal distances within the population (i.e. not including seeds originating from other populations), reach 1200 m; those corresponding to immigrant seeds are indicated by the bars beyond this distance (n = 97 seeds with provenance assigned to other populations).

We were recently able to estimate the relative contribution of different frugivore types to the dispersal curve, estimating the total dispersal kernel (TDK; see Nathan, Chapter 11, this volume) (Jordano *et al.*, 2007). We analysed endocarps from faecal samples, pellets or remains that we identified as having been dispersed by different frugivore groups including small- and medium-sized birds (warblers, redstarts, thrushes, crows), and carnivorous mammals (red fox, stone marten and badger). This way, were able to assign dispersal distances to seeds that we know had been dispersed by these frugivores. Whereas small-sized birds were by far the main seed dispersers up to 250 m, larger frugivores dispersed most of the seeds to distances up to 1100 m, indicating that seed dispersal within the population can be extensive but highly structured spatially. Medium-sized birds (*Turdus viscivorus*, Turdidae; *Corvus corone*, Corvidae; and, most likely, *Columba palumbus*, Columbidae) contributed most of the LDD events within the population by dispersing seeds mostly beyond 100 m; but they also contributed to short dispersal distances (especially *T. viscivorus*) (Jordano *et al.*, 2007). Seed dispersal distances by carnivorous mammals ranged from 0 m (i.e. under the source tree canopy) up to 990 m, with a peak at 650–700 m (Jordano *et al.*, 2007). These distance intervals obviously correspond to within-population dispersal events (seeds consumed in trees growing in the study population); but seeds can be moved longer distances when being dispersed from other populations.

We tracked the origin of immigrant seeds dispersed from other populations by using assignment methods based on Bayesian inference techniques (Wilson and Rannala, 2003). Briefly, endocarps not assigned to any tree within the focal NCH population can be assigned to one of the eight sampled populations based on the allelic frequencies of the trees growing in them. Given that an endocarp sampled in a seed trap in the focal population (NCH) does not match any source tree in the population, we basically ask what is the likelihood that the seed came from any of the eight populations (located between 1.5 and 17 km away from NCH), given its multilocus genotype and the allelic frequencies of each population. We performed this test with the non-NCH seeds and included the distance interval between NCH and each of these populations in Figure 10.1, thus obtaining the 'extended' seed dispersal curve that includes all the LDD events recorded in the seed sample studied. This reveals an unexpectedly high frequency, up to 20.34% (Table 10.1), of immigrant seeds (Godoy and Jordano, 2001; Jordano and Godoy, 2002; P. Jordano and J. Godoy, 2006, unpublished results; García *et al.*, 2007a; Jordano *et al.*, 2007).

When including the distance estimates for immigrant seeds, the dispersal curve extends to median = 62.0 m with quartiles Q_{25} = 13.5 m and Q_{75} = 274.0 m (n = 557 seeds, the whole seed sample), with a maximum of 16,400 m for a LDD event recorded for the Torcal Llano population, ≥ 17 km from NCH. We have evidence that most immigrant seeds in the study population are dispersed by carnivorous mammals, including badger (*Meles meles*; Mustelidae), stone marten (*Martes foina*; Mustelidae) and red fox (*Vulpes vulpes*; Canidae) (Jordano *et al.*, 2007).

Their weighted contribution to the immigrant seed pool (considering both the fraction of immigrant seeds in their samples and their quantitative seed removal) was 66.9%, while frugivorous birds accounted for the remaining 33.1%. When considering each dispersal vector separately, we observed that 74.2% of the seeds dispersed by mammals were imported from outside the population, whereas 21.9% of the seeds dispersed by birds came from other populations. While these are relative figures, it must be noted that, quantitatively, frugivorous birds disperse a considerably larger amount of *P. mahaleb* seeds than mammals do (Herrera and Jordano, 1981; Jordano and Schupp, 2000). Whereas small-sized birds dispersed seeds mainly beneath the canopies of *P. mahaleb* and other fleshy-fruited trees or shrubs, mammals deposited dispersed seeds preferentially in open sites (rocky soils and open ground with little woody vegetation or grass cover). Medium-sized birds dispersed seeds mainly to open areas (*C. corone*) and beneath pine trees (*T. viscivorus*).

A number of studies have reported very long dispersal events mediated by frugivores, ranging between hundreds of metres (Wenny and Levey, 1998; Gómez, 2003; Forget and Wenny, 2005) to thousands of metres (Fleming, 1988; Mack, 1995; Yumoto, 1999; Levey *et al.*, 2005), to kilometres (Holbrook and Smith, 2000; Westcott and Graham, 2000; Oddou-Muratorio *et al.*, 2001; Holbrook *et al.*, 2002; Fragoso *et al.*, 2003; Westcott *et al.*, 2005; Jordano *et al.*, 2007). In addition, reported dispersal distances for established seedlings from animal-dispersed seeds range up to kilometres (Hardesty *et al.*, 2006). Despite this increase in our understanding of LDD events, we still lack a clear picture of the actual frequency associated with events in different parts of the dispersal curve and, most especially, those at the tail of the distribution (Cain *et al.*, 2000; Nathan, 2006). We still lack detailed data on the frequency of immigrant seeds dispersed from other populations into a given population or fragment and on how different frugivore species contribute to this. We especially need a tighter integration of ecological and molecular DNA-based methods to be able to assess dispersal patterns in high-density situations (Grivet *et al.*, 2005; Jones *et al.*, 2005; Hardesty *et al.*, 2006) and in complex systems where the total dispersal kernel (see Nathan, Chapter 11, this volume) can have very distinct components. However, it is becoming more and more evident that frugivores potentially have a dramatic influence on LDD events and patterns of gene flow via seed dispersal (Godoy and Jordano, 2001; Oddou-Muratorio *et al.*, 2001; Webb and Peart, 2001; García *et al.*, 2005; Bacles *et al.*, 2006; Hardesty *et al.*, 2006; Jordano *et al.*, 2007) compared with pollinators; a view not traditionally supported in the literature (Ennos, 2001). Our data with *P. mahaleb* suggest that seed-mediated gene flow probably doubles pollen-mediated gene flow in the studied populations, and this scenario might well be generalizable to other species of trees and shrubs with animal pollinators and dispersers (Oddou-Muratorio *et al.*, 2001; García *et al.*, 2007a,b). Moreover, these data are compatible with recent interpretations of old paradoxes about LDD in biogeographic scenarios, like the postglacial colonization of Northern continental areas in Europe and North America (Nathan, 2006). The data on LDD of *P. mahaleb* also emphasize a central role of frugivores in the connectivity of fragmented populations via LDD events (see also Levey *et al.*, 2005), a central issue in conservation of altered habitats. Given that plant–frugivore interactions are typically of low specificity and high diversity, involving many species of dispersers for a given species of plant, this evidence shows that probably only a small subset of the frugivore coterie has a disproportionately high contribution to LDD events (Jordano *et al.*, 2007; see Dennis and Westcott, Chapter 9, this volume; Nathan, Chapter 11, this volume).

Spatial heterogeneity: 2-D patterns and the seed shadow

Seed dispersal curves emphasize a distance component of seed dispersal. The frequency distribution of dispersal distances is usually represented as a function of directional distance (Nathan and Muller-Landau, 2000). But it is clear that seed dispersal is a 2-D process (sometimes a 3-D process, when, e.g. epiphytes and mistletoes are involved), as emphasized by Janzen (1970) in his seminal paper. Two-dimensional patterns in seed shadows originate chiefly from the interaction of three components; namely, the location of fruiting trees, frugivores' foraging behaviour and habitat preferences, and the major elements of complex landscapes (e.g. forest edges, rock outcrops, ravines, gaps, etc.). Recent findings from animal-movement analysis have revealed that characteristic 2-D patterns in frugivore movement may markedly deviate from purely random walks (Fig. 10.2) (Viswanathan *et al.*, 2002) and can potentially influence the structure of seed shadows (Morales and Carlo, 2006).

While purely Brownian random walks are built up from relatively similar displacements or steps, with a characteristic step length ℓ and a random turning angle, movements and search patterns of biological organisms typically follow a particular type of random walk termed a Lévy walk or, more generally, a Lévy flight pattern (Fig. 10.2) (Viswanathan *et al.*, 2002; Bartumeus, 2005). In a random walk there is a characteristic scale of displacement, defined by a mean step length $\bar{\ell}$; most movements imply a similar distance, the average displacement $\bar{\ell}$, and there are no extremely long steps (i.e. LDD events do not exist). The probability of observing a movement of a given step length ℓ_j is given by a Gaussian distribution. In contrast, actual movement patterns of foraging animals have been found to be better represented by Lévy walks (Viswanathan *et al.*, 2002), where the probability of observing a path of length ℓ_j is given by:

$$P(\ell_j) \sim \ell_j^{-\mu} \qquad (10.1)$$

Thus, the probability of finding a given step length is best described as a function that follows a power-law distribution in which the frequency of LDD events decreases as a negative power function of their length. We find many small dispersal distances but we can also find extremely long

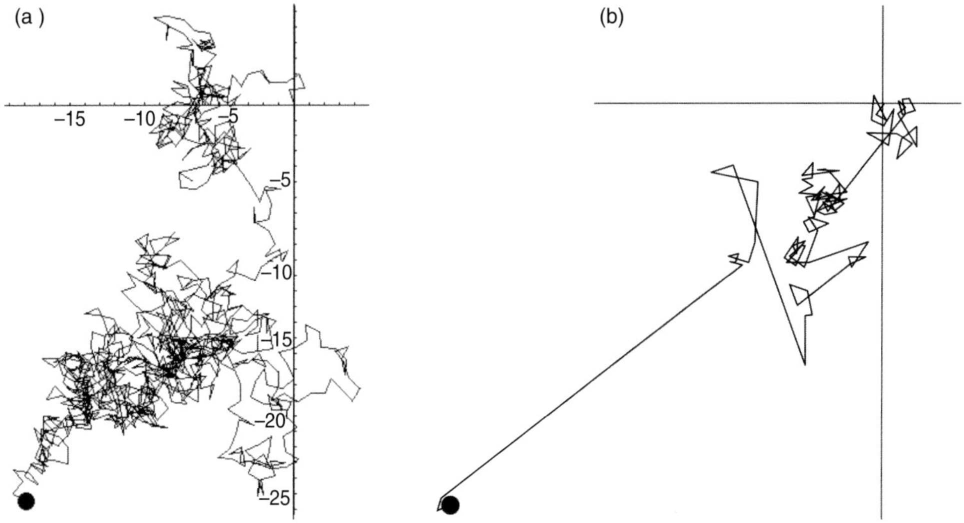

Fig. 10.2. Two simulated trajectories of movement patterns by animal frugivores: (a) a random walk pattern (*n* = 5000 trajectories); (b) a Lévy walk pattern (*n* = 190 trajectories). Axes are reference lines.

dispersal events, which occur with very low probability. In a purely Brownian motion model, these long steps simply do not exist. The parameter μ is the exponent that fully characterizes such a distribution and can be determined from empirical data. In contrast with random walks, Lévy walks have spatial scale invariance in step lengths ℓ_j and in the duration of intervals between steps (Viswanathan *et al.*, 2002). As a consequence, Lévy walks do not have a characteristic step scale and steps of extremely long length can be recorded (Fig. 10.2) (Ramos-Fernandez *et al.*, 2004; Bartumeus, 2005). Ramos-Fernandez *et al.* (2004) have recently reported foraging patterns of *Ateles geoffroyi* (Cebidae) in Yucatán best described as Lévy walks. In addition, some of the results reported by frugivore radio-tracking studies (Westcott and Graham, 2000; Holbrook *et al.*, 2002; Fragoso *et al.*, 2003; Westcott *et al.*, 2005; see Dennis and Westcott, Chapter 9, this volume) suggest similar patterns.

The seed dispersal curve of *P. mahaleb* is extremely leptokurtic and has a characteristic Lévy walk pattern. That is, if we represent all the steps moved by the seeds dispersed from a given tree, or from the tree population as a whole (derived from Fig. 10.1), we should obtain a pattern of step lengths markedly deviating from a Brownian random walk and approaching a Lévy walk (Fig. 10.3a). The dispersal curve for *P. mahaleb* seeds deviates markedly from what would be expected if the frugivores were simply following the location of the fruiting trees while foraging: it has a fatter tail, with a higher probability of steps (dispersal distances) beyond >100 m. Moreover, the associated exponent $\mu = 1.67$ is very close to that expected for a Lévy walk. In comparison, the tail of the distribution of inter-tree distances (Fig. 10.3a) drops suddenly at shorter distances

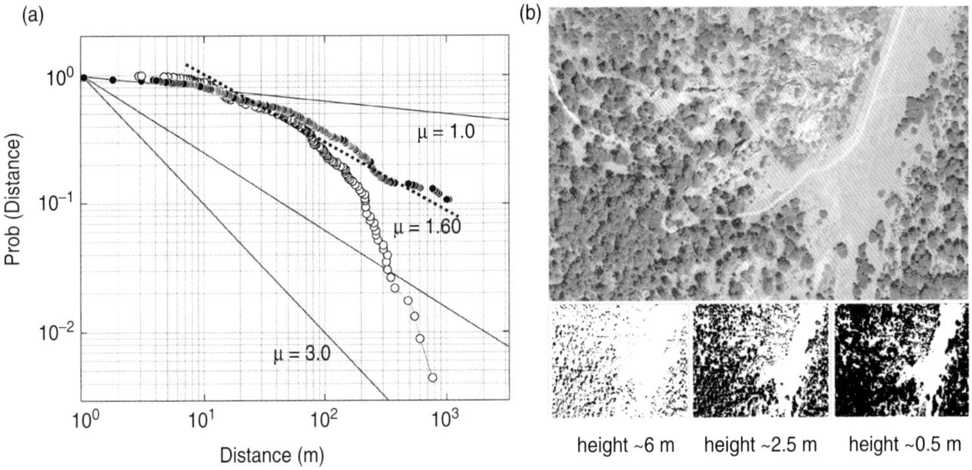

Fig. 10.3. The distribution of seed dispersal distances in relation to habitat characteristics and landscape heterogeneity. (a) Cumulative frequency distributions for dispersal distances. Filled dots, the observed distribution of seed dispersal distances; open dots, the distance distribution that would be expected under a pattern of frugivore foraging that follows exactly the location of fruiting trees, based on tree nearest-neighbour distances. The slopes for μ = 1.0–3.0, drawn for reference, indicate the expected slopes for three different movement patterns, with μ = 1.60 approaching a Lévy walk. Note that the slope for seeds (dashed line) closely approaches the expected pattern with Lévy walk movement (μ = 1.67) and markedly deviates from the slope of a random walk with purely Brownian motion (μ = 3.00). (b) Aerial photograph of the study area and three slices of plant cover at three different heights (0.5, 2.5 and 6 m), illustrating the different patchiness and heterogeneity of plant cover defining major landscape elements such as the pine forest edge, rocky slopes (light grey areas) and open grassland with no shrubs (large grey area).

(100 m), reflecting the aggregation pattern of the trees in the area. This implies that the dispersal distances contributed by frugivores closely map the spacing patterns of fruiting trees, but only up to a certain distance (≤ 100 m; Fig. 10.3a). Beyond this, frugivores are probably responding to other major landscape elements (e.g. rock outcrops, forest edges, large patches of open grassland, etc.) that cause the fat tail of the seed dispersal distribution, adding more frequent LDD events than expected from a Brownian random walk pattern generated by a tracking of the crops of the fruiting trees. For instance, the long flights performed by *T. viscivorus* (Jordano and Schupp, 2000) frequently face the pine forest edge, at distances ≥ 100 m of most *P. mahaleb* fruiting trees (Fig. 10.3b). If these medium sized birds are selecting habitat with tall woody vegetation (e.g. pines > 6 m height), then they should be 'perceiving' a much more patchy landscape, and thus requiring longer flights, than for example, small warblers seeking vegetation cover < 0.5 m (Fig. 10.3b). In addition, these major landscape elements certainly have a lasting effect on non-random directionality patterns in the population-level seed dispersal, as reported recently for *P. mahaleb* (García *et al.*, 2007b).

Two-dimensional patterns in the *P. mahaleb* seed rain and the individual seed shadows, accurately tracked with DNA-based genotyping methods, thus reflect the complex effects of frugivore foraging, habitat preferences and heterogeneous landscapes. This situation is probably generalizable to other plant–frugivore interactions (Morales and Carlo, 2006) where the combined spatial dynamics of habitat use and digestion processes determine complex seed shadows (Jordano *et al.*, 2007; see Dennis and Westcott, Chapter 9 this volume; Nathan, Chapter 11, this volume). Much of this complexity can be adequately handled by mechanistic models (Nathan *et al.*, 2002; Morales and Carlo, 2006) incorporating very simple rules (Viswanathan *et al.*, 2002) that successfully recover the patterns revealed by molecular genetic techniques and direct watches of foraging animals (Jordano *et al.*, 2007). However, a better integration of these different approaches is needed in future research.

Perspectives: Extending the Seed Shadow

Recent advances in the study of seed dispersal by frugivores have dramatically increased our ability to assess seed shadows and to understand dissemination patterns mediated by animals. The most promising avenues for research in this area are clearly transdisciplinary in nature, requiring at least three main ingredients:

1. Solid field data on how plant distribution influences disperser behaviour and movements;
2. Careful use of molecular tools and assignment/parentage analysis based on hypervariable DNA markers;
3. Robust modelling techniques that explicitly incorporate spatial heterogeneity.

This point has been repeatedly emphasized in recent reviews specific to each of these three components (Cain *et al.*, 2000; Ennos, 2001; Jordano and Godoy, 2002; Wang and Smith, 2002; Jones and Ardren, 2003; Nathan *et al.*, 2003; Forget and Wenny, 2005; Nathan, 2006; García, 2007). Adequate and viable means of synthesizing from among these approaches have been identified and are starting to be applied in current research.

Extending the seed shadow means increasing our understanding of the extremely rare, yet recurrent, LDD events that occur in nature (Nathan, 2006) and a better understanding of their reach, extent and frequency. Dispersal events have a pervasive influence on gene-flow patterns, genetic structuring, colonization ability, connectivity between fragmented areas, and cohesiveness in metapopulation scenarios. Despite the high intrinsic diversity of plant–frugivore interactions, only a few species, independently of their quantitative importance in fruit removal, might account for these rare events of far-reaching biological significance. Conversely, short-distance events are a major influence on local population recruitment and in the successful build-up of a large propagule source able to realize successful LDD.

Acknowledgements

I am indebted to Manolo Carrión, Juan Luis García-Castaño, Jesús Rodríguez and, especially, Cristina García, Juan Miguel Arroyo and José A. Godoy for generous help with field and laboratory work and for making this study possible. Discussions with José A. Godoy, Cristina García, Juan Miguel Arroyo, Eugene W. Schupp, Janis Boettinger, Juan Luis García-Castaño, Miguel Angel Fortuna and Jordi Bascompte were very helpful during all the stages of this manuscript and, over the years, have helped enormously to focus my research on *Prunus* genetics. I would like to thank the organizers of the *Fourth International Symposium/Workshop on Frugivores and Seed Dispersal* – Andrew Dennis, Ronda Green, Eugene W. Schupp and David Westcott – for their invitation to contribute to this volume and for their help and advice during the editing process. Comments by Cristina García, Alfredo Valido, Arndt Hampe and two anonymous reviewers helped to improve drafts of the manuscript. The study was supported by the Spanish Ministerio de Ciencia y Tecnología (BOS2000-1366-C02-01, REN2003-0273 and CGL2006-00373 to P.J.). Over the years, the Consejería de Medio Ambiente, Junta de Andalucía, provided generous facilities that made possible this study in the Sierra de Cazorla and authorized my work there.

References

Alcántara, J., Rey, P., Valera, F. and Sánchez-Lafuente, A. (2000) Factors shaping the seedfall pattern of a bird-dispersed plant. *Ecology* 81, 1937–1950.

Alvarez-Buylla, E., Chaos, A., Piñero, D. and Garay, A. (1996) Demographic genetics of a pioneer tropical tree species: patch dynamics, seed dispersal, and seed banks. *Evolution* 50, 1155–1166.

Bacles, C.F.E., Lowe, A.J. and Ennos, R.A. (2006) Effective seed dispersal across a fragmented landscape. *Science* 311, 628.

Bartumeus, F. (2005) Lévy processes in animal movement and dispersal. PhD thesis, University of Barcelona, Spain.

Björklund, M. (2005) A method for adjusting allele frequencies in the case of microsatellite allele drop-out. *Molecular Ecology Notes* 5, 676–679.

Boyer, D., Miramontes, O., Ramos-Fernandez, G., Mateos, J.L. and Cocho, G. (2004) Modeling the searching behaviour of social monkeys. *Physica A* 342, 329–335.

Bullock, J.M. and Clarke, R.T. (2000) Long-distance seed dispersal by wind: measuring and modelling the tail of the curve. *Oecologia* 124, 506–521.

Cain, M., Milligan, B. and Strand, A. (2000) Long-distance seed dispersal in plant populations. *American Journal of Botany* 87, 1217–1227.

Clark, C.J., Poulsen, J.R., Connor, E.F. and Parker, V.T. (2004) Fruiting trees as dispersal foci in a semi-deciduous tropical forest. *Oecologia* 139, 66–75.

Clark, J., Silman, M., Kern, R., Macklin, E. and Hilleris-Lambers, J. (1999) Seed dispersal near and far: patterns across temperate and tropical forests. *Ecology* 80, 1475–1494.

Ennos, R. (2001) Inferences about spatial processes in plant populations from the analysis of molecular markers. In: Silvertown, J. and Antonovics, J. (eds) *Integrating Ecology and Evolution in a Spatial Context.* Blackwell Science, London, pp. 45–72.

Fleming, T. (1988) *The Short-Tailed Fruit Bat: A Study in Plant–Animal Interactions.* Chicago University Press, Chicago, Illinois.

Forget, P.-M. and Wenny, D. (2005) How to elucidate seed fate? A review of methods used to study seed removal and secondary seed dispersal. In: Forget, P.-M., Lambert, J.E., Hulme, P.E. and Vander Wall, S.B. (eds) *Seed Fate: Predation, Dispersal and Seedling Establishment.* CAB International, Wallingford, UK, pp. 379–393.

Fragoso, J.M.V., Silvius, K.M. and Correa, J.A. (2003) Long-distance seed dispersal by tapirs increases seed survival and aggregates tropical trees. *Ecology* 84, 1998–2006.

García, C. (2007) A unified framework for the study of dispersal in plants. *Evolutionary Ecology* (in press).

García, C., Arroyo, J.M., Godoy, J.A. and Jordano, P. (2005) Mating patterns, pollen dispersal, and the ecological maternal neighbourhood in a *Prunus mahaleb* population. *Molecular Ecology* 14, 1821–1830.

García, C., Jordano, P. and Godoy, J.A. (2007a) Spatial and maternal source tree correlations in a frugivore-generated seed rain. *Journal of Ecology* (in press).

García, C., Jordano, P. and Godoy, J.A. (2007b) Contemporary pollen and seed dispersal in a *Prunus mahaleb* population: patterns in distance and direction. *Molecular Ecology* 16, 1947–1955.

Godoy, J.A. and Jordano, P. (2001) Seed dispersal by animals: exact identification of source trees with endocarp DNA microsatellites. *Molecular Ecology* 10, 2275–2283.

Gómez, J.M. (2003) Spatial patterns in long-distance dispersal of *Quercus ilex* acorns by jays in a heterogeneous landscape. *Ecography* 26, 573–584.

Grivet, D., Smouse, P.E. and Sork, V.L. (2005) A novel approach to an old problem: tracking dispersed seeds. *Molecular Ecology* 14, 3585–3595.

Hardesty, B.D., Hubbell, S.P. and Bermingham, E. (2006) Genetic evidence of frequent long-distance recruitment in a vertebrate-dispersed tree. *Ecology Letters* 9, 516–525.

Harms, K.E., Wright, S.J., Calderón, O., Hernández, A. and Herre, E.A. (2000) Pervasive density-dependent recruitment enhances seedling diversity in a tropical forest. *Nature* 404, 493–495.

Herrera, C. and Jordano, P. (1981) *Prunus mahaleb* and birds: the high efficiency seed dispersal system of a temperate fruiting tree. *Ecological Monographs* 51, 203–221.

Hoelzel, A.R. (1998) A method for adjusting allele frequencies in the case of microsatellite allele drop-out. In: Hoelzel, A.R. (ed.) *Molecular Genetic Analysis of Populations: A Practical Approach.* Oxford University Press, Oxford, UK, pp. 234–245.

Holbrook, K.M. and Smith, T.B. (2000) Seed dispersal and movement patterns in two species of ceratogymna hornbills in a west African tropical lowland forest. *Oecologia* 125, 249–257.

Holbrook, K.M., Smith, T.B. and Hardesty, B.D. (2002) Implications of long-distance movements of frugivorous rain forest hornbills. *Ecography* 25, 745–749.

Houle, G. (1992) Spatial relationship between seed and seedling abundance and mortality in a deciduous forest of north-eastern North America. *Journal of Ecology* 80, 99–108.

Howe, H.F. and Kerckhove, G.A.V. (1980) Nutmeg dispersal by tropical birds. *Science* 210, 925–927.

Izhaki, I., Walton, P. and Safriel, U. (1991) Seed shadows generated by frugivorous birds in an eastern Mediterranean scrub. *Journal of Ecology* 79, 575–590.

Janzen, D. (1970) Herbivores and the number of tree species in tropical forests. *American Naturalist* 104, 501–528.

Jones, A.G. and Ardren, W.R. (2003) Methods of parentage analysis in natural populations. *Molecular Ecology* 12, 2511–2523.

Jones, A., Chen, J., Weng, G. and Hubbell, S. (2005) A genetic evaluation of seed dispersal in the neotropical tree, *Jacaranda copaia* (Bignoniaceae). *American Naturalist* 166, 543–555.

Jordano, P. and Godoy, J. (2002) Frugivore-generated seed shadows: a landscape view of demographic and genetic effects. In: Levey, D., Silva, W. and Galetti, M. (eds) *Frugivores and Seed Dispersal: Ecology, Evolution, and Conservation.* CAB International, Wallingford, UK, pp. 305–321.

Jordano, P. and Schupp, E. (2000) Seed disperser effectiveness: the quantity component and patterns of seed rain for *Prunus mahaleb. Ecological Monographs* 70, 591–615.

Jordano, P., García, C., Godoy, J.A. and García-Castaño, J.L. (2007) Differential contribution of frugivores to complex seed dispersal patterns. *Proceedings of the National Academy of Sciences USA* 104, 3278–3282.

Julliot, C. (1997) Impact of seed dispersal of red howler monkeys *Alouatta seniculus* on the seedling population in the understorey of tropical rain forest. *Journal of Ecology* 85, 431–440.

Kollmann, J. and Goetze, D. (1997) Notes on seed traps in terrestrial plant communities. *Flora* 192, 1–10.

Levey, D.J., Bolker, B.M., Tewksbury, J.J., Sargent, S. and Haddad, N.M. (2005) Effects of landscape corridors on seed dispersal by birds. *Science* 309, 146–148.

Loiselle, B. and Blake, J. (1993) Spatial distribution of understory fruit-eating birds and fruiting plants in a neotropical lowland wet forest. In: Fleming, T. and Estrada, A. (eds) *Frugivory and Seed Dispersal: Ecological and Evolutionary Aspects.* Kluwer Academic, Dordrecht, The Netherlands, pp. 177–189.

Mack, A. (1995) Distance and non-randomness of seed dispersal by the dwarf cassowary *Casuarius bennetti. Ecography* 18, 286–295.

Mack, A.L. and Druliner, G. (2003) A non-intrusive method for measuring movements and seed dispersal in cassowaries. *Journal of Field Ornithology* 74, 193–196.

Manzano, P. and Malo, J.E. (2006) Extreme long-distance seed dispersal via sheep. *Frontiers in Ecology and the Environment* 5, 244–248.

Marshall, T.C., Slate, J., Kruuk, L.E.B. and Pemberton, J.M. (1998) Statistical confidence for likelihood-based paternity inference in natural populations. *Molecular Ecology* 7, 639–655.

McEuen, A.B. and Curran, L.M. (2004) Seed dispersal and recruitment limitation across spatial scales in temperate forest fragments. *Ecology* 85, 507–518.

Morales, J.M. and Carlo, T.A. (2006) The effects of plant distribution and frugivore density on the scale and shape of dispersal kernels. *Ecology* 87, 1489–1496.

Morin, P.A., Chambers, K.E., Boesch, C. and Vigilant, L. (2001) Quantitative polymerase chain reaction analysis of DNA from non-invasive samples for accurate microsatellite genotyping of wild chimpanzees (*Pan troglodytes verus*). *Molecular Ecology* 10, 1835–1844.

Muller-Landau, H., Wright, S., Calderón, O., Hubbell, S. and Foster, R. (2002) Assessing recruitment limitation: concepts, methods and case-studies from a tropical forest. In: Levey, D., Silva, W. and Galetti, M. (eds) *Seed Dispersal and Frugivory: Ecology, Evolution and Conservation.* CAB International, Wallingford, UK, pp. 35–53.

Murray, K. (1988) Avian seed dispersal of three neotropical gap-dependent plants. *Ecological Monographs* 58, 271–298.

Nakashizuka, T., Iida, S., Masaki, T., Shibata, M. and Tanaka, H. (1995) Evaluating increased fitness through dispersal: a comparative study on tree populations in a temperate forest, Japan. *Écoscience* 2, 245–251.

Nathan, R. (2006) Long-distance dispersal of plants. *Science* 313, 786–788.

Nathan, R. and Muller-Landau, H.C. (2000) Spatial patterns of seed dispersal, their determinants and consequences for recruitment. *Trends in Ecology and Evolution* 15, 278–285.

Nathan, R., Horn, H., Chave, J. and Levin, S. (2002) Mechanistic models for tree seed dispersal by wind in dense forests and open landscapes. In: Levey, D., Silva, W. and

Galetti, M. (eds) *Seed Dispersal and Frugivory: Ecology, Evolution and Conservation.* CAB International, Wallingford, UK, pp. 69–82.

Nathan, R., Perry, G., Cronin, J.T., Strand, A.E. and Cain, M.L. (2003) Methods for estimating long-distance dispersal. *Oikos* 103, 261–273.

Oddou-Muratorio, S., Petit, R., Le Guerroue, B., Guesnet, D. and Demesure, B. (2001) Pollen- versus seed-mediated gene flow in a scattered forest tree species. *Evolution* 55, 1123–1135.

Portnoy, S. and Willson, M. (1993) Seed dispersal curves: behaviour of the tail of the distribution. *Evolutionary Ecology* 7, 25–44.

Primack, R. and Levy, C. (1988) A method to label seeds and seedlings using gamma-emitting radionuclides. *Ecology* 69, 796–800.

Queller, D. and Goodnight, K. (1989) Estimating relatedness using genetic markers. *Evolution* 43, 258–275.

Ramos-Fernandez, G., Mateos, J.L., Miramontes, O., Cocho, G., Larralde, H. and Ayala-Orozco, B. (2004) Lévy walk patterns in the foraging movements of spider monkeys (*Ateles geoffroyi*). *Behavioral Ecology and Sociobiology* 55, 223–230.

Reid, N. (1989) Dispersal of mistletoes by honeyeaters and flowerpeckers: components of seed dispersal quality. *Ecology* 70, 137–145.

Rey, P. and Alcántara, J. (2000) Recruitment dynamics of a fleshy-fruited plant (*Olea europaea*): connecting patterns of seed dispersal to seedling establishment. *Journal of Ecology* 88, 622–633.

Sato, H. and Hiura, T. (1998) Estimation of overlapping seed shadows in a northern mixed forest. *Forest Ecology and Management* 104, 69–76.

Sato, T., Isagi, Y., Sakio, H., Osumi, K. and Goto, S. (2006) Effect of gene flow on spatial genetic structure in the riparian canopy tree *Cercidiphyllum japonicum* revealed by microsatellite analysis. *Heredity* 96, 79–84.

Schnabel, A., Nason, J. and Hamrick, J. (1998) Understanding the population genetic structure of *Gleditsia triacanthos* L.: seed dispersal and variation in female reproductive success. *Molecular Ecology* 7, 819–832.

Schupp, E.W., Howe, H.F., Augspurger, C.K. and Levey, D.J. (1989) Arrival and survival in tropical treefall gaps. *Ecology* 70, 562–564.

Schupp, E., Milleron, T. and Russo, S. (2002) Dissemination limitation and the origin and maintenance of species-rich tropical forests. In: Levey, D., Silva, W. and Galetti, M. (eds) *Seed Dispersal and Frugivory: Ecology, Evolution and Conservation.* CAB International, Wallingford, UK, pp. 19–33.

Skarpaas, O., Shea, K. and Bullock, J.M. (2005) Optimizing dispersal study design by Monte Carlo simulation. *Journal of Applied Ecology* 42, 731–739.

Snow, B. and Snow, D. (1988) *Birds and berries.* T. & A.D. Poyser, Calton, UK.

Sun, C., Ives, A., Kraeuter, H. and Moermond, T. (1997) Effectiveness of three turacos as seed dispersers in a tropical montane forest. *Oecologia* 112, 94–103.

Taberlet, P. (1996) Reliable genotyping of samples with very low DNA quantities using PCR. *Nucleic Acids Research* 24, 3189–3194.

Valière, N. (2002) GIMLET: a computer program for analysing genetic individual identification data. *Molecular Ecology Notes* 2, 377–379.

Vander Wall, S.B. (2000) Secondary dispersal of jeffrey pine seeds by rodent scatter-hoarders: the roles of pilfering, recaching and a variable environment. In: Levey, D., Silva, W. and Galetti, M. (eds) *Seed Dispersal and Frugivory: Ecology, Evolution and Conservation.* CAB International, Wallingford, UK, pp. 193–208.

Viswanathan, G.M., Bartumeus, F., Buldyrev, S.V., Catalan, J., Fulco, U.L., Havlin, S., da Luz, M.G.E., Lyra, M.L., Raposo, E.P. and Stanley, H.E. (2002) Lévy flight random searches in biological phenomena. *Physica A* 314, 208–213.

Wang, B. and Smith, T. (2002) Closing the seed dispersal loop. *Trends in Ecology and Evolution* 17, 379–385.

Webb, C.O. and Peart, D.R. (2001) High seed dispersal rates in faunally intact tropical rain forest: theoretical and conservation implications. *Ecology Letters* 4, 491–499.

Wenny, D. and Levey, D. (1998) Directed seed dispersal by bellbirds in a tropical cloud forest. *Proceedings of the National Academy of Sciences USA* 95, 6204–6207.

Westcott, D.A. and Graham, D.L. (2000) Patterns of movement and seed dispersal of a tropical frugivore. *Oecologia* 122, 249–257.

Westcott, D.A., Bentrupperbaumer, J., Bradford, M.G. and McKeown, A. (2005) Incorporating patterns of disperser behaviour into models of seed dispersal and its effects on estimated dispersal curves. *Oecologia* 146, 57–67.

Whitney, K., Fogiel, M., Lamperti, A., Holbrook, K., Stauffer, D., Hardesty, B., Parker, V. and Smith, T. (1998) Seed dispersal by ceratogymna hornbills in the Dja Reserve, Cameroon. *Journal of Tropical Ecology* 14, 351–371.

Wilson, G.A. and Rannala, B. (2003) Bayesian inference of recent migration rates using multilocus genotypes. *Genetics* 163, 1177–1191.

Yumoto, T. (1999) Seed dispersal by Salvin's curassow, *Mitu salvini* (Cracidae), in a tropical forest of Colombia: direct measurements of dispersal distance. *Biotropica* 31, 654–660.

Ziegenhagen, B., Liepelt, S., Kuhlenkamp, V. and Fladung, M. (2003) Molecular identification of individual oak and fir trees from maternal tissues of their fruits or seeds. *Trees – Structure and Function* 17, 345–350.

11 Total Dispersal Kernels and the Evaluation of Diversity and Similarity in Complex Dispersal Systems

R. Nathan

Introduction

Interest in dispersal has risen rapidly over the last 15 years (Clobert *et al.*, 2001; Bullock *et al.*, 2002), particularly interest in seed dispersal (Levey *et al.*, 2002; Forget *et al.*, 2004). Progress in this subject has been motivated by the importance of dispersal processes for the structure and dynamics of populations, metapopulations and communities (Nathan and Muller-Landau, 2000; Levin *et al.*, 2003; Levine and Murrell, 2003), impacting on conservation concerns such as biological invasions, transgene escape, connectivity in fragmented landscapes, and range shifts following climate changes (Neilson *et al.*, 2005; Trakhtenbrot *et al.*, 2005). Over this period, plant dispersal research has shown a marked tendency towards a more quantitative approach, employing more advanced methods and closer integration between theory and fieldwork, resulting in greater realism of dispersal models and better understanding of dispersal mechanisms than previously (Nathan, 2006). These advances have set the stage for addressing challenging questions about the high complexity characterizing dispersal systems.

Two of the most prominent features of complex dispersal systems have been recognized for a long time (Darwin, 1859; Ridley, 1930), yet have remained largely under-investigated: dispersal of nearly any given species of plant is mediated by multiple dispersal vectors, and, nearly any dispersal vector disperses multiple species of plants and diaspore types. These aspects have rarely been incorporated within the basic quantitative description of dispersal systems, known as the dispersal curve or dispersal kernel, which is the frequency distribution of distances travelled by the plant dispersal units. Dispersal kernels should reliably and accurately account for the dispersal systems because their statistical properties (shape and scale, statistical moments), especially those related to the distribution tail, are critically important determinants of key processes such as spread rate, connectivity and gene flow (Clark, 1998; Cain *et al.*, 2000; Levin *et al.*, 2003). Dispersal kernels accounting for the existence of multiple vectors dispersing a single species of plant, or of multiple species dispersed by a single vector – termed here total dispersal kernels (TDKs) – are likely to describe dispersal of this particular plant or dispersal vector more realistically and accurately than dispersal kernels assuming a single dispersal vector or species of plant. The few recent examples of TDKs, reviewed below, show that different dispersal vectors can generate very different dispersal kernels for the same plant, and the same dispersal vector can generate very different dispersal kernels for different species of plants. This illustrates the complexity associated with dispersal systems, and motivates the allocation of research efforts towards quantifying and modelling TDKs.

In this chapter, I first overview the empirical evidence for complex dispersal systems in which a single species of plant is dispersed by multiple vectors, and a single dispersal vector disperses multiple species of plants. In particular, I highlight recent studies that have estimated dispersal kernels in such complex systems. Then I define the TDK concept, classify different types of TDKs and explain their importance. Next, I illustrate how this concept provides a quantitative basis for addressing questions about diversity and similarity in complex dispersal systems. I conclude by highlighting some central issues that should be considered in the design of research approaches and of specific methods for estimating TDKs.

Complex Dispersal Systems

Studies quantifying dispersal kernels differ in the level of complexity they incorporate, ranging from the simplest system – a single species of plant dispersed by a single vector – to the most complex system in which each species of plant is dispersed by multiple vectors, and each vector disperses multiple species of plants (Table 11.1). These differences reflect either genuine differences in the complexity of the system being studied or, more likely, differences in the level of simplification employed by researchers, who tend to focus on specific elements of the studied system presumed to be of primary importance.

Single species of plant dispersed by multiple vectors

The monumental monographs of Ridley (1930) and van der Pijl (1982) on plant dispersal reflect the common conception held during most of the previous century among scholars of plant dispersal, that the identity of the presumed principal dispersal agent is the primary criterion for classifying dispersal systems. Although both Ridley (1930, p. xiv) and van der Pijl (1982, pp. 96–101) have explicitly acknowledged that many species of plants are dispersed by multiple agents, their use of vector-specific terminology and

Table 11.1. Examples of studies estimating total dispersal kernels (TDKs) in dispersal systems of different complexity.

Dispersal vector	One	Few	Many
Dispersed plant			
One	*Quercus*/jay (Gómez, 2003) Jacaranda/wind (Jones *et al.*, 2005) (and *many* other examples)	*Spergularia*/wind, water (Redbo-Torstensson and Telenius, 1995) *Trillium*/ants*, deer (Vellend *et al.*, 2003) *Ochradenus*/2 bird species (Spiegel, 2006)	*Elaeocarpus*/9 vertebrate disperser *functional types* (Dennis and Westcott, Chapter 9, this volume) *Prunus*/2 bird species, 1 functional group of birds, 1 functional group of mammals (Jordano *et al.*, 2007)
Few	4 forb species/wind (Soons *et al.*, 2004) 3 wind-dispersed tree species/wind 3 bird-dispersed tree species/birds (pooled) 3 bat-dispersed tree species/bats (pooled) (Clark *et al.*, 2005)	4 diaspore types/4 mammalian species (Mouissie *et al.*, 2005)	No known examples
Many	6 plant species/ochre-bellied flycatcher (Westcott and Graham, 2000) 11 plant species/cassowary (Westcott *et al.*, 2005)	No known examples	No known examples

*Ant-generated dispersal kernels not estimated, but the maximum recorded dispersal distance is given for comparison.

their detailed descriptions of vector-specific dispersal modes have inevitably led researchers to associate a certain species of plant with a single dispersal vector. Ridley (1930) and van der Pijl (1982) termed dispersal of plants by multiple dispersal vectors as polychory, 'in contrast to the *normal* case of haplochory' (van der Pijl, 1982, p. 96; *my italics*). Haplochory, dispersal by a single vector, has been the primary criterion for classifying dispersal systems comprising many specific types such as anemochory (dispersal by wind) and myrmecochory (dispersal by ants). This classification is regularly inferred from the diaspore morphology (e.g. wing and hairs versus elaiosome, respectively). Despite numerous cautionary notes emphasizing the pitfalls inherent in haplochory-based classification, and, consequently, in single-vector research (e.g. Howe and Smallwood, 1982; McEvoy and Cox, 1987; Bakker *et al.*, 1996; Nathan and Muller-Landau, 2000; Tackenberg *et al.*, 2003), this classification still dominates the literature on plant dispersal. Only a few studies have estimated the prevalence of polychorous systems thus far. A notable exception is the study of Ozinga *et al.* (2004), who classified five long-distance dispersal (LDD) vectors of 900 species of plants and concluded that for 123 plant communities in the Netherlands polychory is the rule rather than the exception, even though the level of polychory was underestimated in this study, which focused only on five presumed LDD vectors. Even fewer studies have quantified dispersal kernels generated by two or more vectors (Table 11.1).

Polychory is a general term for plant dispersal by multiple dispersal vectors. A more specific term used by van der Pijl (1982) in the same context, *diplochory*, accounts for the sequential dispersal by different dispersal vectors. This phenomenon has received much attention recently (Chambers and MacMahon, 1994; Forget *et al.*, 2004; Vander Wall and Longland, 2004; Schurr *et al.*, 2005). Another specific phenomenon, termed *heterodiaspory* (van der Pijl, 1982), accounts for the occurrence of two or more different dispersal morphologies in the same plant. This phenomenon has attracted much attention because it facilitates the examination of evolutionary forces underlying selection for dispersal (Sorensen, 1978; Lloyd, 1984; Venable, 1985; Imbert, 2002). Several studies found differences in dispersal distances measured for seeds of different morphs dispersed by different dispersal vectors (e.g. McEvoy and Cox, 1987; Redbo-Torstensson and Telenius, 1995), while others found that the distances travelled by the two morphs are usually similar, and that differences occur only under specific conditions (e.g. Telenius and Torstensson, 1989; Ronsheim, 1994). However, heterodiaspory is rather uncommon in the plant world (van der Pijl, 1982), and polychory occurs much more frequently when plants are dispersed by different dispersal agents, often irrespective of their dispersal morphology (Darwin, 1859; Ridley, 1930; van der Pijl, 1982; Higgins *et al.*, 2003; Ozinga *et al.*, 2004). For example, the plant, *Mimulus guttatus* (Phrymaceae), has monomorphic seeds which are dispersed effectively by water, wind, birds and deer (Vickery *et al.*, 1986). The examples provided below show that the dispersal kernels of exactly the same diaspore can differ markedly in their shape, scale and other statistical properties. Therefore, the TDK concept should not be restricted to a specific kind of polychory but should accommodate this widespread phenomenon *sensu lato*.

Dennis and Westcott (Chapter 9, this volume) provide the most comprehensive TDK study to date. Their work incorporates no less than 65 vertebrate dispersers classified into 15 functional types which consume, in total, fruits/seeds of 441 species of plants classified into nine functional types. From this remarkably rich data set, they illustrate their practical approach for estimating TDKs for one species of plant, *Elaeocarpus grandis* (Elaeocarpaceae), dispersed by multiple vectors. The resulting TDK (Figs 9.4a,b, this volume) clearly illustrates the differential scale-dependent contribution of different vector functional types. I consider this work to be the single most important attempt thus far to estimate TDK in real-life systems, not only by proving that the daunting task of estimating TDKs is not impossible, but also by providing a detailed account of how this can be accomplished. Furthermore, inspired by this chapter, I adopt here the use of functional types, rather than species, in the formal definition of the TDK concept described below.

Spiegel (2006) provides a sound example for the potential differences in dispersal kernels generated by two vectors simultaneously dispersing the same species of plant. He compared dispersal kernels of seeds of the fleshy-fruited desert shrub, *Ochradenus baccatus* (Resedaceae), generated by two avian dispersal vectors, yellow-vented bulbuls (*Pycnonotus xanthopygos*; Pycnonotidae) and Tristram's grackles (*Onychognathus tristramii*; Sturnidae) (see also Bronstein *et al.*, Chapter 7, this volume). These species do not differ in the amount of seeds they disperse, nor in their gut treatment of the seeds. Yet, due to the difference in their body mass (~40 g versus 120 g, respectively), they differ markedly in their gut retention times (mean ~30 min versus 2 h, respectively) and scale of movements (tens to a few hundred metres versus hundreds to a few thousand metres, respectively). A simple mechanistic model coupling the measured components of this dispersal system shows notable differences between the dispersal kernels predicted for seeds dispersing by these two avian species. Bulbuls generate dispersal kernels with much higher concentrations at short distances and with thinner tails that extend over considerably shorter distances compared with the dispersal kernels generated by the grackles. Thus, the two species switch roles as a function of spatial scale; bulbuls are responsible for most within-habitat dispersal, while grackles are almost exclusively responsible for LDD between habitat patches (dry river beds separated by a hostile matrix).

Mouissie *et al.* (2005) estimated dispersal kernels for adhesive dispersal of seeds of four different morphologies by four different mammalian dispersal vectors. The four seed types were distinguished by the mean time they remained attached to the dispersal vectors. The four vectors were distinguished by their movement patterns, which were indicated by three parameters of a correlated random walk (CRW) model: the mean velocity of the animal, the typical size of one step and the standard deviation of the turning angle. The CRW simulations predicted considerable differences among dispersal vectors, with 99% of the seeds dispersed up to ~12 m by wood mouse, 400–800 m by fallow deer, 800 m by cattle, and 2800 m by sheep. Large herbivores can transport seeds over long distances not only externally (epizoochory; see also Couvreur *et al.*, 2004; Manzano and Malo, 2006), but also internally (endozoochory; see Pakeman, 2001). For example, *Trillium grandiflorum* (Trilliaceae) is a forest herb of eastern North America, with morphological adaptations for dispersal by ants (van der Pijl, 1982). Because ants disperse seeds over scales of tens of metres at the most, dispersal by ants cannot account for the fast (hundreds of metres per year) postglacial expansion rates estimated for ant-dispersed forest herbs (Cain *et al.*, 1998). This Reid's paradox (Clark *et al.*, 1998a) can be resolved when considering dispersal by white-tailed deer, which consume large amounts of *Trillium* berries and can transport viable seeds, a quarter of them to distances further than 1 km (Vellend *et al.*, 2003). In summary, these three examples show that dispersal vectors can differ by several orders of magnitude in the scale of distances to which they disperse seeds of the same species of plant, or diaspores of the same dispersal morphology.

Single dispersal vector dispersing multiple diaspore types

Buoyancy of the diaspore and its viability during transport are considered the main adaptations for plant dispersal by water (hydrochory), either by sea currents or streams. Seed buoyancy duration, measured for many species (e.g. Ridley, 1930), has been used as a primary criterion for classifying hydrochorous species. Yet, seeds with long wings and other clear adaptations for wind dispersal can also float for long periods in the water. Furthermore, plant vegetative parts can float and propagate as well (hence serving as dispersal units), regardless of the morphological adaptations of seeds of the same plant. For example, Boedeltje *et al.* (2003) reported that several species classified according to their seed morphology as wind-dispersed, such as *Phragmites australis* (Poaceae) and *Typha latifolia* (Typhaceae), were trapped in a stream in the Netherlands. Minimum dispersal distances (distance to the nearest stand) for these two species were much longer than the minimum distances known for typical wind-dispersed seeds (estimated as 700 and 800 m, respectively), with the former species being the seventh most abundant among the 65 species in the samples; a surprisingly high rank for a wind-dispersed species in a hydrochorous system. Thus water, rather than wind, may be the key dispersal vector determining the abundance and distribution of these two species bearing morphological adaptations for wind dispersal. Higgins *et al.* (2003) analysed data collected in the volcanic island Surtsey that emerged from the sea 33 km south of Iceland. Of the 38 species of plants that were observed to be carried to the island by sea currents between 1963 and 1972, only nine (24%) are known to bear morphological adaptations for hydrochory, while 12 others (32%) were classified as wind-dispersed, and six (16%) as bird-dispersed. The remaining 11 species (28%) had no known specialized adaptations for dispersal.

Wind is another important abiotic dispersal vector (Ridley, 1930; van der Pijl, 1982). Tackenberg *et al.* (2003) used a mechanistic model of wind dispersal to calculate the potential for LDD among species of different dispersal morphologies. They found that species with unspecialized dispersal morphologies can also be dispersed by wind over relatively long distances. It is important to note, however, that this group of species was found to have the lowest potential for LDD, compared with groups of species with standard wind-dispersal morphologies. Higgins *et al.* (2003) also used a (different) mechanistic model of wind dispersal to calculate the expected dispersal distance of maple (*Acer rubrum*; Aceraceae) seeds and hickory (*Carya glabra*; Juglandaceae) nuts in different wind conditions. Maple seeds are 14 mg, asymmetric samaras (winged seeds) which autorotate during fall and reach an average terminal velocity (rate of fall in still air) of 0.66 m/s. The 8 g hickory nuts, in comparison, have a terminal velocity of 7.84 m/s. Under extreme wind conditions occurring, for example, during severe thunderstorms, 1% of the maple seeds were predicted to travel distances greater than 11.3 km. These extreme wind conditions are also predicted to carry 1% of the hickory nuts over distances greater than ~650 m. It is clear that the heavy hickory nuts have no morphological adaptations for wind dispersal, yet thunderstorms with such strong winds typically occur 1–3 times each year in eastern North American forests. Hence, wind dispersal of hickory nuts for distances of hundreds of metres may not be an extraordinarily rare event. It should be noted that wind-generated dispersal kernels can also vary markedly among species with known adaptations for wind dispersal (Soons *et al.*, 2004; Clark *et al.*, 2005).

Biotic dispersal systems are characterized by unspecialized relationships between animal dispersers and dispersed plants and do not exhibit the specific mutualism typical of some pollination systems (Herrera, 1982, 1985; Howe, 1984; Jordano, 1987; Levey and Benkman, 1999; Lord *et al.*, 2002; see Corlett, Chapter 24, this volume; Dennis and Westcott, Chapter 9, this volume). Medium/large-sized mammalian herbivores can effectively disperse propagules of a large number of plant species of diverse dispersal morphologies (including species with no apparent adaptations) either internally or externally (Janzen, 1984; Pakeman, 2001; Bruun and Fritzbøger, 2002; Pakeman *et al.*, 2002; Couvreur *et al.*, 2004; Myers *et al.*, 2004). For example, Pakeman *et al.* (2002) showed that a wide range of species germinated from rabbit and sheep dung, comprising 37% of the species of plants in the grazed ecosystem they studied. The species dispersed internally by these herbivores generally have small round seeds, but there were also species with large seeds and seeds with known adaptations for epizoochory and anemochory.

Although the above-mentioned studies suggest a potential for considerable variation among dispersal kernels of different diaspore types dispersed by the same biotic vector, only a small number of studies have actually estimated such seed dispersal kernels (Table 11.1). Mouissie *et al.* (2005) estimated dispersal kernels of seeds of different dispersal morphologies, generated by four epizoochorous dispersal vectors (see above). Each of the four vectors generated similar kernels for hooked and bristly seeds, but dispersal kernels of smooth seeds were drastically shorter than those of the other morphologies for all vectors, except sheep. Clark *et al.* (2005) used inverse modelling techniques to estimate dispersal kernels for seeds of three species of plants dispersed by birds and three others dispersed by monkeys. The best-fitting dispersal function for all bird- and monkey-dispersed species was the inverse power function; but significant variation has been found in parameters fitted for different species dispersed by the same vector. Monkey-dispersed species differ in their fecundity parameter but not in the dispersal (scale) parameter, whereas bird-dispersed species differ in both fecundity and dispersal parameters. It should be noted that the identity of the dispersal vectors has been inferred in this study from intensive observations on the predominant dispersers for each species of tree, and the comparisons assume that each is dispersed exclusively by a single (type of) dispersal vector. Thus, this study does not account for the potential role of other vectors in dispersing seeds of each tree species.

Westcott *et al.* (2005) reported that in Queensland, Australia, cassowaries disperse seeds of 238 species of plants of different sizes and shapes. They used a mechanistic model based on radio-tracking of cassowary movements and data on their gut retention times to estimate the dispersal kernels they generate during different times of the day, for 11 different species of plants. This unique study illustrates many of the key points I wish to emphasize in this chapter.

- *First*, it demonstrates the considerable variation among dispersal kernels generated by the same vector for the 11 different species of plants. This motivates the quantitative examination of dispersal kernels generated by other dispersal vectors for seeds of different species of plants.
- *Second*, despite the notable variation in the frequency of seeds expected to be deposited in short versus long distances, many of the estimated kernels exhibited a mode at about 330 m. Thus, an important take-home message is to be careful about inferring dispersal processes from information about the mode (or median or mean) alone, which could lead to the erroneous conclusion that cassowaries disperse seeds of different plant species to similar distances. This emphasizes the importance of estimating the entire dispersal kernels to reveal differences that are obscured when using statistics of central tendency alone.
- *Third*, the clear finding that cassowaries generate very different dispersal kernels at different times of the day (e.g. much extended in the morning as compared with the afternoon) and potentially with different behaviours, emphasizes the complexity of the dispersal

system, and the need to deeply understand the biology and life history of both the dispersal vectors and the dispersed plants in order to properly evaluate the potential outcomes of dispersal processes.

- *Finally*, and perhaps most important, this study and the related study of Dennis and Westcott (Chapter 9, this volume) prove that it is possible to address the daunting task of estimating TDKs.

Total Dispersal Kernel (TDK): the Concept

Overview and importance

The TDK concept accounts for the overall effect of all active dispersal vectors and dispersed organisms in the system of interest. The 'system' can be an ecosystem or a particular area of interest; it could also be a particular plant or animal community, but the role of abiotic dispersal vectors might be overlooked under such a definition. The concept is developed here specifically for dispersal of plant seeds, but directly applies to any passively dispersed organism and essentially can be applied to any actively dispersed organism as well. The vector can be either biotic or abiotic, and the diaspore can be of any form (van der Pijl, 1982). For example, a single biotic dispersal vector can be a particular species of bird or bat that disperses viable seeds, and abiotic dispersal vectors may include wind, water or floating platforms. Following Dennis and Westcott (Chapter 9, this volume) I adopt here the use of functional types of both dispersed plants (*plant functional type*, hereafter PFT) and dispersal vectors (*vector functional type*, hereafter VFT). Indeed, classification into functional types is not a simple task (see Dennis and Westcott, 2006; Dennis and Westcott, Chapter 9, this volume), and success in addressing this task is not always guaranteed. Yet, such classification is likely to be highly instrumental both to reduce redundancy among plants or vectors (Dennis and Westcott, 2006; Dennis and Westcott, Chapter 9, this volume), and to facilitate the practical implementation of the TDK concept.

In its broadest sense, TDK_{system} is defined as the frequency of dispersal distances travelled by seeds of all PFTs dispersed by all active VFTs in this system, such that dispersal in this system is described by a single dispersal kernel. Thus, TDK_{system} is a characteristic of the system that can be used to compare different systems (see below) in a manner similar to a diversity index, which characterizes species diversity in a community. An interesting example for the potential use of TDK_{system} data, suggested by O. Steinitz (Jerusalem, 2005, personal communication), is to couple TDK_{system} and plant fecundity data to calculate the sum of all dispersal distances in the system as a coarse measure for the level of kinetic energy allocated to maintain dispersal in the system. This (or a more elaborate) measure can provide the means to examine how sensitive dispersal processes are to, for example, spatial heterogeneity and environmental changes and disturbances, by comparing TDK_{system} for different landscape structures along environmental gradients before and after disturbance. Because dispersal processes tend to stabilize ecosystem and community dynamics (e.g. Loreau *et al.*, 2003), TDK_{system} may even serve as a measure of ecosystem stability.

More specific TDKs emphasize the perspective of either the dispersed plants or the dispersal vectors. TDK_{plant} would summarize the frequency of dispersal distances of seeds of a certain PFT dispersed by all its active VFTs in the system. TDK_{vector} would summarize the frequency of dispersal distances of seeds of all PFTs dispersed by a particular VFT. More specific cases, $TDK_{plant,\ vector}$ would summarize the frequency of dispersal distances of seeds of one (or several) PFTs dispersed by one (or several) VFTs. These types of TDK provide a framework for assessing the relative importance of different plants and different vectors in shaping the dispersal processes of the system under study. This assessment can contribute to our understanding of the ecology and evolution of plant–animal interactions, and is also of practical importance for evaluating how removing a particular PFT, or altering the operation of a certain VFT, may affect the dynamics of the system. It also provides the means to guide conservation plans, for example by identifying VFTs that provide a unique (and hence irreplaceable) dispersal service for particular (or all) PFTs and, hence, whose protection should be given high priority. The methodology for addressing questions of this kind is illustrated below.

To investigate dispersal processes in the system under study, the TDK concept must be applied to dispersal kernels of seeds (or other types of diaspores) per se. To investigate post-dispersal (establishment) processes, one should quantify the distance-dependent patterns of seed and seedling survival in the system (Janzen, 1970; Nathan and Casagrandi, 2004). The TDK concept can also be extended to total effective dispersal kernels (TEDKs), which are evaluated from data on seedlings, saplings or new reproductive individuals, thereby reflecting the joint effects of dispersal and establishment (Nathan *et al.*, 2003; see Hardesty, Chapter 12, this volume). A major caveat relates to our ability to interpret the relative contribution of different vectors in structuring TEDKs, not only because vectors can rarely, if ever, be identified by observing newly established plants, but also because effects of dispersal can differ significantly from those of post-dispersal processes (Jordano and Herrera, 1995; Schupp and Fuentes, 1995; Nathan and Muller-Landau, 2000; Nathan and Casagrandi, 2004). Furthermore, TEDKs reflect the net effects of early recruitment processes, which are indeed more important to the structure and dynamics of a system than the effects of either dispersal or post-dispersal processes separately. An ability to link corresponding TDKs and TEDKs may still be feasible, for example by long-term inspection of seed fate trajectories (Forget *et al.*, 2004). Success on this front promises profound insights into the relative roles of the different players (plants and vectors) of complex dispersal systems, in determining the structure and dynamics of the communities and ecosystems.

Terminological clarifications

The choice of the term 'total dispersal kernel' requires several terminological clarifications. First, two basic varieties of the probability density function (pdf) can be used to summarize dispersal distances in a population: the probability of seed arrival to distance r, and the same probability per unit area. These two varieties were termed, respectively, *distance distribution* and *dispersal kernel* by Nathan and Muller-Landau (2000), and *1-D pdf* and *2-D pdf* by Cousens and Rawlinson (2001). Assuming isotropic dispersal, the two terms differ only by a factor of $2\pi r$. The concept of TDK does not require a distinction between the two varieties, as it can be applied to pdf of any kind. The term *dispersal kernel* was chosen chiefly because it has become increasingly popular over recent years, and is now widely recognized by dispersal ecologists. Second, the common related term, *seed shadow,* defined as 'the spatial distribution of all seeds dispersed from a single plant' (Nathan and Muller-Landau, 2000) should be preserved to describe spatially realistic seed dispersion patterns in two or three dimensions (e.g. Schupp *et al.*, 2002) and thus is too wide for the pdf-based concept of TDKs. Third, the term 'total' has been preferred over 'community-wide' to allow for the incorporation of abiotic dispersal vectors as well, which are not included in the ecological interpretation of communities. Finally, 'total' does not dictate perfect knowledge about the system in question, but in the ordinary statistical sense we are interested in identifying the most significant plants (or PFTs) and vectors (or VFTs) in the system under study, and attempting to estimate their contribution to the true TDK of this system.

A formal definition of TDK

Consider a system of interest with m PFTs and n potential VFTs. Denote $C_{i,j,x}$ as the number of seeds of PFT i dispersed by VFT j to distance category x, and $S_{i,j}$ as the total number of seeds dispersed from PFT i by VFT j. Assuming complete information about this system, $\mathrm{TDK}_{a,b}$ represents the TDK of a PFTs ($a = 1,\ldots, m$) dispersed by b VFTs ($b = 1,\ldots, n$). If a particular PFT (e.g. $i = 1$) is exclusively dispersed by a single VFT (e.g. $j = 1$) and the same VFT exclusively disperses this PFT, the TDK ($\mathrm{TDK}_{1,1}$) is quantified by plotting $C_{1,1,x}/S_{1,1}$ as a function of x. Focusing on the plant perspective, the $\mathrm{TDK}_{\mathrm{plant}}$ of the ith PFT ($\mathrm{TDK}_{i,b}$) is quantified by plotting $\sum_{j=1}^{b} C_{i,j,x} / \sum_{j=1}^{b} S_{i,j}$ as a function of x. Similarly, focusing on the vector perspective, the $\mathrm{TDK}_{\mathrm{vector}}$ of the jth VFT ($\mathrm{TDK}_{a,j}$) is quantified by plotting $\sum_{i=1}^{a} C_{i,j,x} / \sum_{i=1}^{a} S_{i,j}$ as a function of x. Seed dispersal in this system can also be summarized by a single overall TDK_{system} (= $TDK_{a,b}$); the frequency of dispersal events at each distance category x is equal to $\sum_{i=1}^{a}\sum_{j=1}^{b} C_{i,j,x} / \sum_{i=1}^{a}\sum_{j=1}^{b} S_{i,j}$ in this case. Figure 11.1 illustrates the different types of TDK for a hypothetical system.

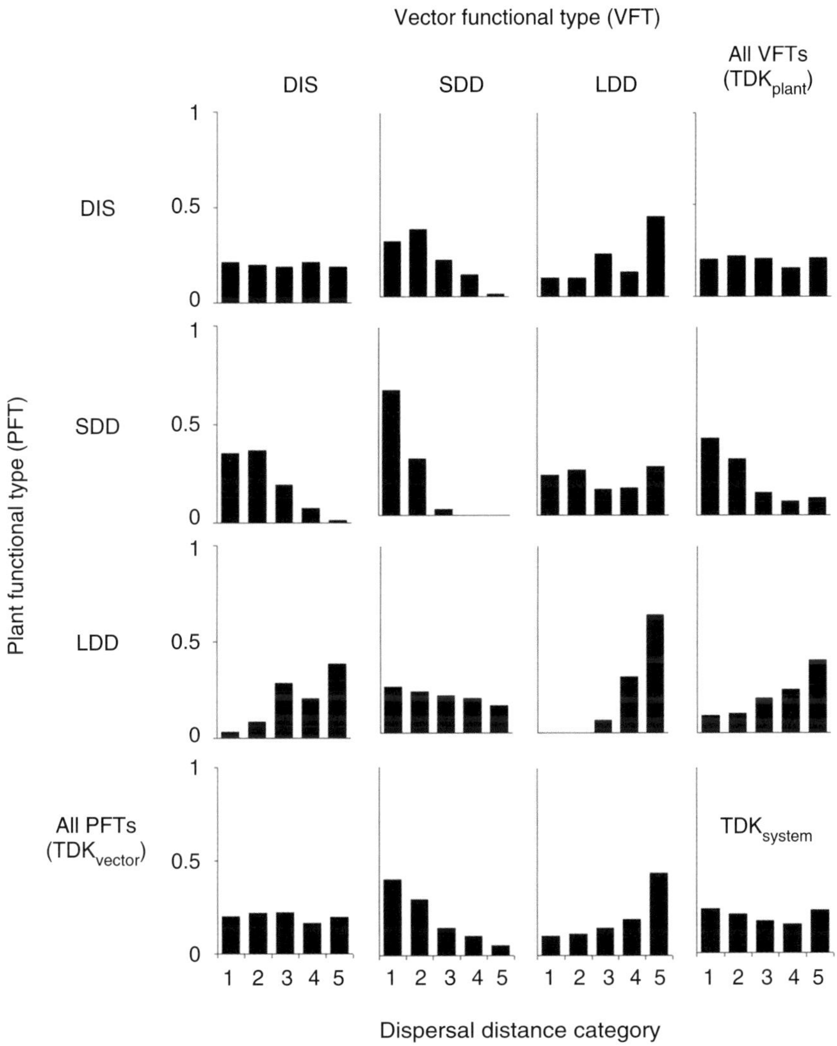

Fig. 11.1. Hypothetical example of a complex dispersal system in which dispersal kernels of seeds of three different plant functional types (PFTs) dispersed by three different vector functional types (VFTs) were estimated. Both VFTs and PFTs represent cases of distance-insensitive (DIS), short-distance (SDD) and long-distance (LDD) dispersal (see text). The concept of total dispersal kernels (TDK) is illustrated: TDK_{plant} is the TDK of seeds of a certain PFT dispersed by all its VFTs; TDK_{vector} is the TDK of seeds of all PFTs dispersed by a certain VFT; TDK_{system} is the TDK of seeds of all PFTs dispersed by all VFTs in the system under study.

A Quantitative Framework for Analysing Diversity and Similarity in Dispersal Systems

General considerations

By accounting for the complex associations between multiple plants and multiple dispersal vectors typical of nearly any community or ecosystem, the TDK concept provides the means to explore many questions about the processes and patterns of dispersal in any system under study. For example, data collected to estimate TDK provide the means to address questions about the diversity and similarity among different VFTs dispersing seeds of the same PFT or about different PFTs dispersed by the same VFT. As emphasized above, addressing these questions is important both for basic research (e.g. for understanding the interactions that structure complex dispersal systems) and for practical reasons (e.g. which VFTs are irreplaceable, and hence should be protected).

The analysis suggested here is not intended to replace the common practice of fitting functional forms to the dispersal kernel data (e.g. Clark *et al.*, 1998b, 2005; Nathan and Muller-Landau, 2000). This fitting can provide important information about the properties of the observed TDK and its implications, for example for population spread. However, the suggested analysis does provide more straightforward and less ambiguous answers to questions on diversity and similarity in dispersal systems. Furthermore, this method can accommodate dispersal kernels of any shape, including multi-peaked and disorganized patterns that are likely to be fairly common in nature (Schupp *et al.*, 2002), whereas all the standard functional forms commonly used in inverse modelling can only describe single-peaked gradually decreasing dispersal patterns.

To illustrate how such analysis can be performed, I have selected commonly used indices of diversity and similarity. These indices are simple and widely known and therefore were selected to simplify the first introduction of this kind of analysis. It should be emphasized that a similar analysis could (and perhaps should) be performed by using other indices and metrics as well.

Simpson's index, D, in its standard use for measuring species diversity (dominance), gives the probability that any two individuals drawn at random from a large community belong to the same species. Its complement $(1-D)$ thus gives the probability that two randomly chosen individuals are different species. Although heavily weighted toward the most abundant species in the sample, Simpson's index is one of the most meaningful and robust diversity measures available (Lande, 1996; Magurran, 2004). The index takes into account both richness (number of species) and evenness (in the relative abundance of species): diversity increases both with higher richness and higher evenness. In the context of species diversity, the Simpson's complement $(1-D)$ is defined as

$$1 - D = 1 - \sum \frac{n_x(n_x - 1)}{N(N - 1)} \tag{11.1}$$

where n_x and N are the number of individuals of the xth and of all species, respectively. The value of this index ranges between 0 and 1: the greater the value, the greater the sample diversity.

Among quantitative indices of similarity in species composition, the Morisita–Horn index (C_{MH}) is less sensitive to species richness and sample size (Wolda, 1981; Magurran, 2004) in comparison with many alternative indices. The index is defined as

$$C_{MH} = \frac{2\sum(p_x q_x)}{\left(\frac{\sum p_x^2}{N_p^2} + \frac{\sum q_x^2}{N_q^2}\right) N_p N_q} \tag{11.2}$$

where p_x and q_x are the number of individuals of the xth species at site P and Q, respectively, and N_p and N_q are the total number of individuals at site P and Q, respectively. The index measures the extent to which two samples are similar in the composition of species and their abundances, and its values can range from 0 (no similarity) to 1 (complete similarity).

In the following two sections, the procedures for analysing TDKs are explained and illustrated. It should be emphasized that the TDK concept applies for general polychorous systems, but the examples given here do not account for cases of diplochory and other forms of sequential dispersal. To illustrate the approach, I use a hypothetical system with three VFTs and three PFTs, for which dispersal distance data are classified into five distance categories (Fig. 11.1). The three VFTs represent 'distance-insensitive' (DIS), short-distance (SDD) and long-distance (LDD) dispersers. A similar division has been applied to the three PFTs. While the three patterns are fairly straightforward in relation to dispersal vectors, their application for plants can be confusing, as it requires the plant's control over dispersal distances. Indeed, plants can control dispersal distances to a certain degree; for example, all else being equal, seeds detached en masse from a plant in weak winds may reach only short distances, while seeds detached en masse from a plant only in very strong winds may reach only long distances. Similarly, the chemical effects of fruits in a frugivore gut can control, at least to some extent, the distances of dispersal.

Several other simplifying assumptions in this example should be clarified. First, I assume no variance in plant fecundity or the total number of seeds dispersed by each VFT (i.e. each VFT disperses exactly the same number of seeds of each PFT). Indeed, this variance is an important feature of complex dispersal systems which could certainly be accounted for in such analyses, but would make the examples too complicated. Similarly, the within-kernel variance around the mean number of seeds dispersed to each distance category is also of primary importance and can also be accounted for, but is excluded here for simplicity. Finally, it should be clarified that the statistical comparison between the values of the indices of diversity and similarity should be based on methods for constructing confidence intervals (see Magurran, 2004), which are not calculated here because questions of statistical inference are of little value in the context of this ad hoc hypothetical dataset.

Analysis of diversity in dispersal distance

To examine patterns of diversity in the dispersal system of interest, we first use TDK data to examine the diversity of seed dispersal distances for a certain PFT among its VFTs. To this end, we calculate the Simpson's complement ($1-D$) for each kernel $\mathrm{TDK}_{i,j}$ where i is the focal PFT, and j is the VFT ($j = 1,\ldots, b$). Here, the parameter n_x in Eqn (11.1) is equal to $C_{i,j,x}$: the number of seeds of the focal PFT i dispersed by VFT j to distance category x. The parameter N in the same equation is equal to $S_{i,j}$: the total number of seeds of the focal PFT dispersed by the same VFT over all distances. We then repeat the calculations for each of the b VFTs. Comparisons of $1-D$ values estimated for the focal PFT among its VFTs (i.e. among values of the same row in Table 11.2) would help evaluate which VFT spreads the seeds of this PFT in a more diverse manner, with higher values reflecting VFTs dispersing seeds more evenly among a greater number of dispersal distance categories. For example, the lowest $1-D$ value for the SDD PFT is found for the SDD VFT, because this combination strongly limits dispersal and hence seeds are unevenly distributed and are concentrated in a small number of (short) distance categories. For the same SDD PFT, the highest value is found for the LDD VFT, because the opposing tendencies for SDD (plant) and LDD (vector) result in seeds being dispersed more evenly to a greater number of (both short and long) distance categories. It should be emphasized that this

Table 11.2. Spatial (dispersal distance) diversity matrices calculated for the hypothetical dispersal data shown in Fig. 11.1. Higher values indicate higher diversity in the number of seeds among different distance categories. The functional types of both dispersal vectors (VFTs) and plant species (PFT) represent cases of distance-insensitive (DIS), short-distance (SDD) and long-distance (LDD) dispersal (see text).

VFT PFT	DIS	SDD	LDD	All
DIS	0.80	0.73	0.72	0.80
SDD	0.70	0.47	0.79	0.71
LDD	0.72	0.80	0.51	0.74
All	0.80	0.70	0.72	0.80

metric alone cannot help distinguish between VFTs operating at different scales; see, for example, the almost identical $1-D$ values of the least similar VFTs (SDD versus LDD) dispersing the DIS PFT (Table 11.2).

We can repeat a similar exercise to examine the diversity of dispersal distances of seeds dispersed by a certain VFT, among the different PFTs it disperses. We now calculate $1-D$ for each dispersal kernel $TDK_{i,j}$ where i is the dispersed PFT ($i = 1,..., a$), and j is the focal VFT. The parameter n_x is equal to $C_{i,j,x}$, i.e. the number of seeds of PFT i dispersed by the focal VFT j to distance category x, and N is equal to $S_{i,j}$, as above. Comparisons of $1-D$ values estimated for the focal VFT among different PFTs (i.e. among values of the same column in Table 11.2) would help evaluate for which PFT the same VFT spreads the seeds in a more diverse manner, with higher values reflecting PFTs whose seeds are being dispersed more evenly to a greater number of dispersal distance categories. For example, the highest $1-D$ value for the DIS VFT is found for the DIS PFT, because seeds are being dispersed in fairly equal proportions to many distance categories in this combination. When the same VFT disperses either the SDD or the LDD PFTs, the dispersal properties of the plant impose dispersal of its seeds to a lower number of distance categories, with marked uneven distribution, resulting in lower $1-D$ values.

Two more general questions about diversity of dispersal systems can be addressed by using TDK data:

- Do VFTs in the system differ in the diversity of seed dispersal distances they generate for all PFTs they disperse?
- Do PFTs in the system differ in the diversity of seed dispersal distances generated collectively by all their VFTs?

To address these questions, the same procedures are repeated but the parameter n_x now equals either $\sum_{i=1}^{a} C_{i,j,x}$ (for the first question) or $\sum_{j=1}^{b} C_{i,j,x}$ (for the second question). To address the first question, $1-D$ values estimated from TDK_{vector} data for the b different VFTs should be compared. Addressing the second question requires comparison of $1-D$ values estimated from the TDK_{plant} data for the a different PFTs. In the example, the answer to the first question is given in the bottom row of Table 11.2. The highest value is found for the DIS VFT as expected from a disperser defined as having no preference for any particular distance category. The values obtained for the SDD and the LDD VFT are lower because these VFTs deposit seeds in a lower number of distance categories. The answer to the second question is given in the right-hand column of Table 11.2. The highest value is found, similarly, for the DIS PFT, because its seeds are dispersed to many distance categories. Seeds of the SDD and the LDD PFTs are dispersed to particular distance categories, hence their lower $1-D$ values.

A coarser analysis of diversity in dispersal systems can be obtained from TDK data, though it does not require information about dispersal distances. This analysis addresses two general questions:

- Do VFTs in the system differ in the diversity of the PFTs they disperse?
- Do PFTs in the system differ in the diversity of their VFTs?

To address these questions, the same procedures shown above are repeated, but the parameter n_x is equal in this case to either $n_j = \sum_x \sum_{i=1}^{a} C_{i,j,x}$ (for the first question) or to $n_i = \sum_x \sum_{j=1}^{b} C_{i,j,x}$ (for the second question), and N is, again, either $\sum_{i=1}^{a} S_{i,j}$ or $\sum_{j=1}^{b} S_{i,j}$, respectively. To address the first question, $1-D$ values estimated for the a different PFTs should be compared; addressing the second question requires comparison of $1-D$ values estimated for the b different VFTs. Because of the equal fecundity assumption, these analyses would reveal no differences for the hypothetical system, illustrated in Figure 11.1.

Finally, TDK data may also be used to compare diversity among different systems. Each system can be characterized, in addition to the above-mentioned measures, by three general $1-D$ values: one spatial (dispersal distance) and two non-spatial measures. The spatial measure quantifies the total diversity among distance categories of seeds of all PFTs dispersed by all VFTs together (i.e. the $\text{TDK}_{\text{system}}$). Here, $n_x = \sum_{i=1}^{a} \sum_{j=1}^{b} C_{i,j,x}$ and $N = \sum_{i=1}^{a} \sum_{j=1}^{b} S_{i,j}$, the total of all seeds of all PFTs dispersed by all VFTs. The value obtained for the hypothetical system is 0.80 (Table 11.2). The two non-spatial measures (not illustrated) express the total diversity among VFTs in the total number of seeds (of all PFTs) they disperse, and among PFTs in the total number of their seeds, dispersed by all VFTs. The parameter n_x equals $n_j = \sum_x \sum_{i=1}^{a} C_{i,j,x}$ in the former and $n_i = \sum_x \sum_{j=1}^{b} C_{i,j,x}$ in the latter; in both cases, $N = \sum_{i=1}^{a} \sum_{j=1}^{b} S_{i,j}$.

Analysis of similarity in dispersal distance

To examine patterns of similarity in a dispersal system of interest, we first use TDK data to examine the similarity in dispersal distances among all

$b(b-1)/2$ pairs of VFTs. To this end, we calculate the Morisita–Horn index (C_{MH}) for each pair of dispersal kernels $TDK_{i,j=P}$ and $TDK_{i,j=Q}$ where i is the PFT ($i = 1,..., a$) and j indicates the two focal VFTs P and Q. Here, the parameters p_x and q_x in Eqn (11.2) are equal to $C_{i,j,x}$: the number of seeds of each PFT i dispersed by the two focal VFTs ($j = Q$ and $j = P$) to distance category x. The parameters N_p and N_q of the same equation are equal to $S_{i,j}$: the total number of seeds of the same PFT dispersed by the two focal VFTs over all distances. We then repeat the calculations for each of the a PFTs. Comparisons of C_{MH} values estimated for different PFTs would help evaluate for which PFTs the two focal VFTs disperse the seeds in a more similar manner, with lower values reflecting lower similarity in the way the two VFTs distribute the seeds over the different dispersal distance categories.

Again, we can repeat a similar exercise to examine the similarity in dispersal distances among all $a(a-1)/2$ pairs of PFTs, by calculating C_{MH} for each pair of dispersal kernels $TDK_{i=P,j}$ and $TDK_{i=Q,j}$ where i indicates the two focal PFTs P and Q and j is the VFT ($j = 1,..., b$). Here, the parameters p_x and q_x are equal to $C_{i,j,x}$: the number of seeds of each of the two focal PFTs ($i = Q$ and $i = P$), dispersed by VFT j to distance category x. The parameters N_p and N_q are equal to $S_{i,j}$: the total number of seeds of each of the two focal PFTs dispersed by the same VFT over all distances. We then repeat the calculations for each of the b VFTs. Comparisons of C_{MH} values estimated for the different VFTs would help evaluate which VFTs disperse the two focal PFTs in a more similar manner, with higher values reflecting VFTs that are more similar in the way they distribute the seeds among the different dispersal distance categories.

TDK data can also be used to address two more general questions:

- How similar are the different TDK_{vector}s – the dispersal kernels generated by each VFT collectively for all the PFTs it disperses?
- How similar are the TDK_{plant}s – the dispersal kernels of each PFT generated collectively by all its VFTs?

To address these questions, the same procedures above are repeated with the parameters p_x and q_x now equal to, respectively, $\sum_{i=1}^{a} C_{i,j=P,x}$ and $\sum_{i=1}^{a} C_{i,j=Q,x}$ (for the first question) or $\sum_{j=1}^{b} C_{i=P,j,x}$ and $\sum_{j=1}^{b} C_{i=Q,j,x}$ (for the second question). The parameters N_p and N_q are equal, respectively, to $\sum_{i=1}^{a} S_{i,j=P}$ and $\sum_{i=1}^{a} S_{i,j=Q}$ (for the first question) and $\sum_{j=1}^{b} S_{i=P,j}$ and $\sum_{j=1}^{b} S_{i=Q,j}$ (for the second question). To address the first question, C_{MH} values estimated for the $b(b-1)/2$ different pairs of VFTs should be compared.

Addressing the second question requires comparison of C_{MH} values estimated for the $a(a-1)/2$ different pairs of PFTs. In the hypothetical example, the lowest similarity is found, as expected, between the SDD and the LDD VFTs (Table 11.3a). Kernels generated by each of these two VFTs are much more similar to the one generated by the DIS VFT. Similar results are obtained for the comparisons among PFTs (Table 11.3b).

Finally, two general questions about similarity in dispersal systems can be addressed by using TDK data, although not requiring information about dispersal distances:

- Do two different VFTs in the system disperse a similar proportion of seeds of the different PFTs?
- Are seeds of two different PFTs in the system being dispersed in a similar proportion by the different VFTs?

Again, we repeat the same procedures above, but the parameters p_x and q_x are now equal, respectively, to $\sum_x \sum_{i=1}^{a} C_{i,j=P,x}$ and $\sum_x \sum_{i=1}^{a} C_{i,j=Q,x}$ (for the first question) and to $\sum_x \sum_{j=1}^{b} C_{i=P,j,x}$ and $\sum_x \sum_{j=1}^{b} C_{i=Q,j,x}$ (for the second question).

The parameters N_p and N_q are exactly as in the previous analysis. To address the first question, C_{MH} values estimated for the $b(b-1)/2$ different pairs of VFT should be compared. Addressing the second question requires comparison of C_{MH} values estimated for the $a(a-1)/2$ different pairs of VFTs.

Table 11.3. Spatial (dispersal distance) similarity matrices calculated for the hypothetical dispersal data shown in Fig. 11.1. Higher values indicate higher similarity in the number of seeds among different distance categories. The functional types of (a) dispersal vectors (VFTs) and (b) plant species (PFT) represent cases of distance-insensitive (DIS), short-distance (SDD) and long-distance (LDD) dispersal (see text).

(a) VFT	DIS	SDD
SDD	0.84	
LDD	0.82	0.50
(b) PFT	**DIS**	**SDD**
SDD	0.84	
LDD	0.86	0.53

Future Research

The TDK concept emphasizes several components of complex dispersal systems that require special attention.

- *First*, because the operation of dispersal vectors is affected by spatial and temporal variation at different scales (e.g. McEvoy and Cox, 1987; Redbo-Torstensson and Telenius, 1995; Nathan *et al.*, 2002; Gómez, 2003; Nathan and Katul, 2005), any attempt to estimate TDK must explicitly consider issues of scale and spatiotemporal variation. A high-priority challenge we need to address is how to assess how TDK depends on the spatial heterogeneity and the temporal dynamics typical of real-life landscapes (see Dennis and Westcott, Chapter 9, this volume).
- *Second*, to comprehensively assess the full spectrum of VFTs dispersing seeds of each PFT, one should quantify seed fate trajectories for a sufficiently large sample of individual plants. Similarly, to comprehensively assess the full spectrum of plants dispersed by each VFT, one should quantify seed intake by a sufficiently large sample of individual vectors. Classification of plants and vectors into functional types is therefore a major challenge that requires a thorough investigation and test of traditional assumptions.
- *Third*, attempts to make inferences about dispersal per se from its outcome (seedlings, sapling, next generation individuals), for example by using genetic variation to quantify TEDKs (see Hardesty, Chapter 12, this volume), are likely to fail because this kind of analysis does not allow, in most cases, for the identification of the dispersal vectors. But perhaps the greatest challenge is to identify the role of specific VFTs in determining patterns of *effective* dispersal; to link TDKs to the corresponding TEDKs and vice versa. As stressed above, successful attempts to address this challenge will shed light on the ultimate role of the different players in complex dispersal systems in determining the structure and dynamics of the communities and ecosystems.

Estimating TDKs is a challenging task, but our methodologies for quantifying dispersal distances, including LDD, are advancing rapidly (Cain *et al.*, 2000; Nathan, 2001, 2003, 2005, 2006; Wang and Smith, 2002; Cain *et al.*, 2003; Nathan *et al.*, 2003; Dennis and Westcott, Chapter 9, this volume). This is evident in the brief overview of studies that have successfully estimated TDKs (see above and Table 11.1), nearly all of them products of recent studies. The TDK concept emerged from the need to closely accommodate basic properties observed in nearly all dispersal systems. Yet, the practical difficulties in quantifying TDKs are indeed considerable and should not be underestimated. Mechanistic models are inherently suitable to address these challenges because they simulate the impact of dispersal vectors (Dennis and Westcott, chapter 9, this volume; Nathan, 2006). Genetic methods can directly estimate seed dispersal kernels (e.g. Jones *et al.*, 2005) and the contribution of different animal dispersal vectors can be disentangled by identifying the vector-specific scats and pellets (Jordano *et al.*, 2007). The great complexity characterizing dispersal systems may be reduced by defining functional groups of plants and dispersal vectors (see Dennis and Westcott, Chapter 9, this volume; Dennis and Westcott, 2006). Studies of the evolution of dispersal, and its ecological consequences, must account for multiple vectors and move beyond the simplistic and often unrealistic assumption of dispersal by a single vector (Nathan, 2006). Successful attempts to quantify and analyse TDKs thus promise better understanding, and more efficient management, of complex dispersal systems.

Acknowledgements

The ideas described in this chapter have been inspired by studies supported by the National Science Foundation (NSF IBN 9981620 and DEB 0453665), the German–Israeli Foundation (GIF I-2006-1032.12/2000), the Israeli Science Foundation (ISF 474/02 and ISF-FIRST 1316/05), the International Arid Land Consortium (IALC 03R/25) and the Simon and Ethel Flegg Fellowship. Discussions with A. Dennis, P. Jordano, O. Spiegel and D. Westcott during the *Fourth International Symposium/Workshop on Frugivory and Seed Dispersal* (July 2005, Brisbane, Australia) were very helpful in formulating the TDK concept. I am also grateful to N.M. Ben-Eliahu, A. Dennis, F. Jeltsch, O. Spiegel, O. Steinitz, A. Trakhtenbrot, D. Troupin, A. Tsoar, D. Westcott and two anonymous reviewers for valuable comments on earlier drafts of this chapter.

References

Bakker, J.P., Poschlod, P., Strykstra, R.J., Bekker, R.M. and Thompson, K. (1996) Seed banks and seed dispersal: important topics in restoration ecology. *Acta Botanica Neerlandica* 45, 461–490.

Boedeltje, G., Bakker, J.P., Bekker, R.M., Van Groenendael, J.M. and Soesbergen, M. (2003) Plant dispersal in a lowland stream in relation to occurrence and three specific life-history traits of the species in the species pool. *Journal of Ecology* 91, 855–866.

Bruun, H.H. and Fritzbøger, B. (2002) The past impact of livestock husbandry on dispersal of plant seeds in the landscape of Denmark. *Ambio* 31, 425–431.

Bullock, J.M., Kenward, R.E. and Hails, R. (2002) *Dispersal Ecology.* Blackwell, Malden, Massachusetts.

Cain, M.L., Damman, H. and Muir, A. (1998) Seed dispersal and the Holocene migration of woodland herbs. *Ecological Monographs* 68, 325–347.

Cain, M.L., Milligan, B.G. and Strand, A.E. (2000) Long-distance seed dispersal in plant populations. *American Journal of Botany* 87, 1217–1227.

Cain, M.L., Nathan, R. and Levin, S.A. (2003) Long-distance dispersal. *Ecology* 84, 1943–1944.

Chambers, J.C. and MacMahon, J.A. (1994) A day in the life of a seed: movement and fates of seeds and their implications for natural and managed systems. *Annual Review of Ecology and Systematics* 25, 263–292.

Clark, C.J., Poulsen, J.R., Bolker, B.M., Connor, E.F. and Parker, V.T. (2005) Comparative seed shadows of bird-, monkey-, and wind-dispersed trees. *Ecology* 86, 2684–2694.

Clark, J.S. (1998) Why trees migrate so fast: confronting theory with dispersal biology and the paleorecord. *American Naturalist* 152, 204–224.

Clark, J.S., Fastie, C., Hurtt, G., Jackson, S.T., Johnson, C., King, G.A., Lewis, M., Lynch, J., Pacala, S., Prentice, C., Schupp, E.W., Webb, T. III and Wyckoff, P. (1998a) Reid's paradox of rapid plant migration: dispersal theory and interpretation of paleoecological records. *BioScience* 48, 13–24.

Clark, J.S., Macklin, E. and Wood, L. (1998b) Stages and spatial scales of recruitment limitation in southern Appalachian forests. *Ecological Monographs* 68, 213–235.

Clobert, J., Danchin, E., Dhondt, A.A. and Nichols, J.D. (2001) *Dispersal.* Oxford University Press, Oxford, UK.

Cousens, R.D. and Rawlinson, A.A. (2001) When will plant morphology affect the shape of a seed dispersal 'kernel'? *Journal of Theoretical Biology* 211, 229–238.

Couvreur, M., Christiaen, B., Verheyen, K. and Hermy, M. (2004) Large herbivores as mobile links between isolated nature reserves through adhesive seed dispersal. *Applied Vegetation Science* 7, 229–236.

Darwin, C. (1859) *The Origin of Species by Means of Natural Selection.* John Murray, London.

Dennis, A.J. and Westcott, D.A. (2006) Reducing complexity in studying seed dispersal at community scales: a functional classification of vertebrate seed dispersers in tropical forests. *Oecologia* 149, 620–634.

Forget, P.M., Lambert, J.E., Hulme, P.E. and Vander Wall, S.B. (2004) *Seed Fate: Predation, Dispersal and Seedling Establishment.* CAB International, Wallingford, UK.

Gómez, J.M. (2003) Spatial patterns in long-distance dispersal of *Quercus ilex* acorns by jays in a heterogeneous landscape. *Ecography* 26, 573–584.

Herrera, C.M. (1982) Seasonal variation in the quality of fruits and diffuse coevolution between plants and avian dispersers. *Ecology* 63, 773–785.

Herrera, C.M. (1985) Determinants of plant–animal coevolution: the case of mutualistic dispersal of seeds by vertebrates. *Oikos* 44, 132–141.

Higgins, S.I., Nathan, R. and Cain, M.L. (2003) Are long-distance dispersal events in plants usually caused by nonstandard means of dispersal? *Ecology* 84, 1945–1956.

Howe, H.F. (1984) Constraints on the evolution of mutualisms. *American Naturalist* 123, 764–777.

Howe, H.F. and Smallwood, J. (1982) Ecology of seed dispersal. *Annual Review of Ecology and Systematics* 13, 201–228.

Imbert, E. (2002) Ecological consequences and ontogeny of seed heteromorphism. *Perspectives in Plant Ecology Evolution and Systematics* 5, 13–36.

Janzen, D.H. (1970) Herbivores and the number of tree species in tropical forests. *American Naturalist* 104, 501–528.

Janzen, D.H. (1984) Dispersal of small seeds by big herbivores: foliage is the fruit. *American Naturalist* 123, 338–353.

Jones, F.A., Chen, J., Weng, G.-J. and Hubbell, S.P. (2005) A genetic evaluation of seed dispersal in the neotropical tree *Jacaranda copaia* (Bignoniaceae). *American Naturalist* 166, 543–555.

Jordano, P. (1987) Patterns of mutualistic interactions in pollination and seed dispersal: connectance, dependence asymmetries, and coevolution. *American Naturalist* 129, 657–677.

Jordano, P. and Herrera, C.M. (1995) Shuffling the offspring: uncoupling and spatial discordance of multiple stages in vertebrate seed dispersal. *Ecoscience* 2, 230–237.

Jordano, P., García, C., Godoy, J.A. and García-Castaño, J.L. (2007) Differential contribution of frugivores to complex seed dispersal patterns. *Proceedings of the National Academy of Sciences of the USA of America* 104, 3278–3282.

Lande, R. (1996) Statistics and partitioning of species diversity, and similarity among multiple communities. *Oikos* 76, 5–13.

Levey, D.J. and Benkman, C.W. (1999) Fruit–seed disperser interactions: timely insights from a long-term perspective. *Trends in Ecology and Evolution* 14, 41–43.

Levey, D.J., Silva, W.R. and Galetti, M. (2002) *Seed Dispersal and Frugivory: Ecology, Evolution and Conservation.* CAB International, Wallingford, UK.

Levin, S.A., Muller-Landau, H.C., Nathan, R. and Chave, J. (2003) The ecology and evolution of seed dispersal: a theoretical perspective. *Annual Review of Ecology Evolution and Systematics* 34, 575–604.

Levine, J.M. and Murrell, D.J. (2003) The community-level consequences of seed dispersal patterns. *Annual Review of Ecology Evolution and Systematics* 34, 549–574.

Lloyd, D.G. (1984) Variation strategies of plants in heterogeneous environments. *Biological Journal of the Linnean Society* 21, 357–385.

Lord, J.M., Markey, A.S. and Marshall, J. (2002) Have frugivores influenced the evolution of fruit traits in New Zealand? In: Levey, D.J., Silva, W.R. and Galetti, M. (eds) *Seed Dispersal and Frugivory: Ecology, Evolution and Conservation.* CAB International, Wallingford, UK, pp. 55–68

Loreau, M., Mouquet, N. and Gonzalez, A. (2003) Biodiversity as spatial insurance in heterogeneous landscapes. *Proceedings of the National Academy of Sciences of the USA* 100, 12765–12770.

Magurran, A.E. (2004) *Measuring Biological Diversity.* Blackwell, Malden, Massachusetts.

Manzano, P. and Malo, J.E. (2006) Extreme long-distance seed dispersal via sheep. *Frontiers in Ecology and Environment* 4, 244–248.

McEvoy, P.B. and Cox, C.S. (1987) Wind dispersal distances in dimorphic achenes of ragwort, *Senecio jacobaea. Ecology* 68, 2006–2015.

Mouissie, A.M., Lengkeek, W. and van Diggelen, R. (2005) Estimating adhesive seed dispersal distances: field experiments and correlated random walks. *Functional Ecology* 19, 478–486.

Myers, J.A., Vellend, M., Gardescu, S. and Marks, P.L. (2004) Seed dispersal by white-tailed deer: implications for long-distance dispersal, invasion, and migration of plants in eastern North America. *Oecologia* 139, 35–44.

Nathan, R. (2001) The challenges of studying dispersal. *Trends in Ecology and Evolution* 16, 481–483.

Nathan, R. (2003) Seeking the secrets of dispersal. *Trends in Ecology and Evolution* 18, 275–276.

Nathan, R. (2005) Long-distance dispersal research: building a network of yellow brick roads. *Diversity and Distributions* 11, 125–130.

Nathan, R. (2006) Long-distance dispersal of plants. *Science* 313, 786–788.

Nathan, R. and Casagrandi, R. (2004) A simple mechanistic model of seed dispersal, predation and plant establishment: Janzen–Connell and beyond. *Journal of Ecology* 92, 733–746.

Nathan, R. and Katul, G.G. (2005) Foliage shedding in deciduous forests lifts up long-distance seed dispersal by wind. *Proceedings of the National Academy of Sciences of the USA* 102, 8251–8256.

Nathan, R. and Muller-Landau, H.C. (2000) Spatial patterns of seed dispersal, their determinants and consequences for recruitment. *Trends in Ecology and Evolution* 15, 278–285.

Nathan, R., Horn, H.S., Chave, J. and Levin, S.A. (2002) Mechanistic models for tree seed dispersal by wind in dense forests and open landscapes. In: Levey, D.J., Silva, W.R. and

Galetti, M. (eds) *Seed Dispersal and Frugivory: Ecology, Evolution and Conservation.* CAB International, Wallingford, UK, pp. 69–82.

Nathan, R., Perry, G., Cronin, J.T., Strand, A.E. and Cain, M.L. (2003) Methods for estimating long-distance dispersal. *Oikos* 103, 261–273.

Neilson, R.P., Pitelka, L.F., Solomon, A.M., Nathan, R., Midgley, G.F., Fragoso, J.M.V., Lischke, H. and Thompson, K. (2005) Forecasting regional to global plant migration in response to climate change. *BioScience* 55, 749–759.

Ozinga, W.A., Bekker, R.M., Schaminée, J.H.J. and Van Groenendael, J.M. (2004) Dispersal potential in plant communities depends on environmental conditions. *Journal of Ecology* 92, 767–777.

Pakeman, R.J. (2001) Plant migration rates and seed dispersal mechanisms. *Journal of Biogeography* 28, 795–800.

Pakeman, R.J., Digneffe, G. and Small, J.L. (2002) Ecological correlates of endozoochory by herbivores. *Functional Ecology* 16, 296–304.

Redbo-Torstensson, P. and Telenius, A. (1995) Primary and secondary seed dispersal by wind and water in *Spergularia salina. Ecography* 18, 230–237.

Ridley, H.N. (1930) *The Dispersal of Plants Throughout the World.* Lovell Reeve & Co., Ashford, UK.

Ronsheim, M.L. (1994) Dispersal distances and predation rates of sexual and asexual propagules of *Allium vineale* L. *American Midland Naturalist* 131, 55–64.

Schupp, E.W. and Fuentes, M. (1995) Spatial patterns of seed dispersal and the unification of plant population ecology. *Ecoscience* 2, 267–275.

Schupp, E.W., Milleron, T. and Russo, S.E. (2002) Dissemination limitation and the origin and maintenance of species-rich tropical forests. In: Levey, D.J., Silva, W.R. and Galetti, M. (eds) *Seed Dispersal and Frugivory: Ecology, Evolution and Conservation.* CAB International, Wallingford, UK, pp. 19–33.

Schurr, F.M., Bond, W.J., Midgley, G.F. and Higgins, S.I. (2005) A mechanistic model for secondary seed dispersal by wind and its experimental validation. *Journal of Ecology* 93, 1017–1028.

Soons, M.B., Heil, G.W., Nathan, R. and Katul, G.G. (2004) Determinants of long-distance seed dispersal by wind in grasslands. *Ecology* 85, 3056–3068.

Sorensen, A.E. (1978) Somatic polymorphism and seed dispersal. *Nature* 276, 174–176.

Spiegel, O. (2006) *Frugivory and Seed Dispersal in Desert Ecosystems: Scale-dependent Dispersal Effectiveness.* MSc thesis, The Hebrew University of Jerusalem, Jerusalem.

Tackenberg, O., Poschlod, P. and Bonn, S. (2003) Assessment of wind dispersal potential in plant species. *Ecological Monographs* 73, 191–205.

Telenius, A. and Torstensson, P. (1989) The seed dimorphism of *Spergularia marina* in relation to dispersal by wind and water. *Oecologia* 80, 206–210.

Trakhtenbrot, A., Nathan, R., Perry, G. and Richardson, D.M. (2005) The importance of long-distance dispersal in biodiversity conservation. *Diversity and Distributions* 11, 173–181.

van der Pijl, L. (1982) *Principles of Dispersal in Higher Plants*, 3rd edn. Springer, Berlin.

Vander Wall, S.B. and Longland, W.S. (2004) Diplochory: are two seed dispersers better than one? *Trends in Ecology & Evolution* 19, 155–161.

Vellend, M., Myers, J.A., Gardescu, S. and Marks, P.L. (2003) Dispersal of *Trillium* seeds by deer: implications for long-distance migration of forest herbs. *Ecology* 84, 1067–1072.

Venable, D.L. (1985) The evolutionary ecology of seed heteromorphism. *American Naturalist* 126, 577–595.

Vickery, R.K. Jr, Phillips, D.R. and Wonsavage, P.R. (1986) Seed dispersal in *Mimulus guttatus* by wind and deer. *American Midland Naturalist* 116, 206–208.

Wang, B.C. and Smith, T.B. (2002) Closing the seed dispersal loop. *Trends in Ecology & Evolution* 17, 379–385.

Westcott, D.A. and Graham, D.L. (2000) Patterns of movement and seed dispersal in a tropical frugivore. *Oecologia* 122, 249–257.

Westcott, D.A., Bentrupperbäumer, J., Bradford, M.G. and McKeown, A. (2005) Incorporating patterns of disperser behaviour into models of seed dispersal and its effects on estimated dispersal curves. *Oecologia* 146, 57–67.

Wolda, H. (1981) Similarity indices, sample size and diversity. *Oecologia* 50, 296–302.

12 How Far Do Offspring Recruit from Parent Plants? A Molecular Approach to Understanding Effective Dispersal

B.D. Hardesty

Introduction

Dispersal is a fundamental process because of its ecological and evolutionary consequences. While dispersal underlies many of the hypotheses formulated to explain the maintenance and diversity of tropical forest communities (Janzen, 1970; Connell, 1971; Hubbell *et al.*, 1990; Tilman, 1994; Hurtt and Pacala, 1995; Hubbell, 2001), it is perhaps the least understood plant life-history stage (Anderson, 1991; Eriksson and Jakobsson, 1999). Effective or realized dispersal (dispersal that results in a germinated seedling) determines plant migration rates. This, in turn, defines future plant distributions and patterns of diversity (Pitelka, 1997). Together, dispersal and recruitment reflect the complicated patterns and processes of seed arrival, germination, establishment, survival and growth in both contemporary and historical time. Although effective dispersal plays a key role in determining patterns of tree diversity and distribution; until recently, quantifying the distances that seeds move and germinate from parent plants has remained an unsolved problem in ecology.

Seed dispersal is an extremely important life-history stage for plants, yet we know relatively little about the net effect of dispersal in tropical trees. This is partly because dispersal has been studied in relatively few of the many species of tropical trees and their associated dispersers and partly because the patterns observed have been variable, making generalizations difficult. Because an estimated 60–90% of tropical tree species are adapted for dispersal by vertebrates (Willson *et al.*, 1989; Jordano, 1992), understanding the effectiveness of dispersal by vertebrates (Schupp, 1993; Bustamante and Canals, 1995) is central to our understanding of the processes underlying ecosystem dynamics.

The interaction between vertebrate dispersers and the plants they disperse is diverse. For vertebrate-dispersed tree species, we can generally assume that active dispersal has taken place for seeds that arrive and survive away from the crown of parent trees. Fruit-consuming animals may ingest seeds and subsequently defecate them; or they may spit, drop, regurgitate seeds intact, or crush them. Seeds may be scatter-hoarded or may be transported by attachment to animals' coats or feathers. Furthermore, animals may disproportionately deposit seeds in particular locations such as latrines (Fragoso, 1997) or near gaps (Wenny and Levey, 1998). Seeds may be dispersed underneath hetero- (Clark *et al.*, 2001) or conspecific adults (Jordano and Godoy, 2002). Also, seed handling by animals may result in improved, reduced or negligible impacts on subsequent germination and establishment. Ultimately, whether dispersal to a particular site will be beneficial is context-dependent both in space and time (Schupp, Chapter 20, this volume; also see Schupp, 1993).

Understanding the consequences of dispersal rather than the individual dispersal events themselves may lead to qualitatively new insights. Molecular techniques can help to elucidate patterns of historical and contemporary dispersal and to understand the genetic diversity that underpins plant populations across a broad geographical range. We can also empirically measure distances between parent plants and their offspring and reliably demonstrate the frequency and extent of seed arrival and effective dispersal events. Such knowledge provides new insights into the processes underlying plant population dynamics.

In this chapter I provide a brief overview of the relevance of long-distance dispersal. This is followed by an introduction to molecular marker utility for assessing dispersal-related questions, what genetic diversity estimates can provide, and a brief mention of some user-friendly computer programs for analysing molecular data. I next present some findings of parent–offspring recruitment distances from a case study of a vertebrate-dispersed tree in central Panama. In the final section, I discuss problematic issues and promising areas of development in dispersal biology and some avenues for future research.

Long-distance Dispersal

Long-distance dispersal (LDD) (or a lack thereof) underlies the evolutionary history of and influences the ecological trajectories of plant populations through colonization and gene flow. At a local scale, such events are difficult to study because of their presumably low frequency of occurrence (Harper, 1977; Willson, 1993; Clark *et al.*, 1999; Ouborg *et al.*, 1999). At a landscape scale, LDD events are biologically important for plants because they are fundamental to population and species level persistence and spread (Clark, 1998; Cain *et al.*, 2000; Levin *et al.*, 2003). Although such events may occur infrequently at the level of an individual seed, LDD events are common at a population level, as evidenced by tree migrations over large geographical expanses during postglacial range expansion (Davis, 1976; McLachlan *et al.*, 2005).

Locally, longer dispersal distances not only enable the offspring of a single parent to colonize a larger area more rapidly, but also permit offspring to escape from host-specific pathogens or other specialized natural enemies that may concentrate around parent plants (Howe and Smallwood, 1982; Muller-Landau and Adler, Chapter 18, this volume; Gallery *et al.*, Chapter 22, this volume). LDD also promotes avoidance of competition with related individuals and allows the offspring of individual parents to sample a greater number and variety of microsites suitable for regeneration, serving as a bet-hedging strategy.

Understanding LDD requires clear terminology to ensure that we are communicating effectively regarding this process that is so fundamentally spatially and temporally scale-dependent. Nathan (2005) identifies two general LDD definitions. The first is an absolute method in which LDD is defined as dispersal events which exceed some threshold dispersal distance. Defining LDD in this manner may involve selecting somewhat arbitrary threshold values, so it is important that the order of magnitude appropriately reflects the scale of the process (and system) in question. The second is a proportional method whereby LDD events are those beyond a certain high percentile in the tail of the total dispersal distribution, for example: the 2.5% of seeds receiving the longest distance dispersal. This definition of LDD may be more difficult to practically employ because it requires one to know the entire distribution of dispersal distances. Considering the definitions as complementary and reporting both whenever possible, rather than treating the two as mutually exclusive, is optimal (Nathan, 2005). Explicitly defining LDD, however it is done, and providing an explanation of why a particular threshold or definition was chosen, will inform the reader and avoid possible ambiguity or misinterpretation, while facilitating comparisons between species, sites and systems.

Molecular Marker Utility, Genetic Diversity and Analytic Tools

Molecular markers provide a practical tool to gain insight into a wide array of dispersal-related questions that have long eluded dispersal ecologists using traditional methods. Often, seeds or nearby seedlings are attributed to the nearest reproductive source (Augspurger, 1983; Clark *et al.*, 1999, 2005; Nathan *et al.*, 2002), without confirmation. However, contagious seed dispersal of vertebrate-dispersed tropical trees is common (Clark *et al.*, 2001), immigrant seed rain occurs regularly (Hardesty and Parker, 2002), and establishment of seedlings from species with no nearby adult plants is frequent (Webb and Peart, 2001). In actuality, genetic data demonstrate that seedlings infrequently germinate near their parent plants (Sezen *et al.*, 2005; Hardesty *et al.*, 2006). Furthermore, with molecular markers, we can empirically determine the relative contribution of reproductive individuals to the offspring in a particular cohort or future generation of a population which may be highly variable in space and time.

Insights into the level and pattern of gene flow are fundamental to our understanding of plant population dynamics, and genetic techniques can further our understanding of the complicated relationships between plants, pollinators and dispersers. Dispersal promotes genetic diversity (Hamrick, 2004) and aids in colonization (Petit *et al.*, 1997; Ouborg *et al.*, 1999), while introducing novel alleles into a population and providing a lever to counteract the effects of genetic drift. Dispersal further impacts on the maintenance and spatiotemporal structuring of genetic variation (Young *et al.*, 1996; Jacquemyn *et al.*, 2006) within and between populations.

Genetic approaches to dispersal-related questions will also provide insight into the future impact of habitat loss or degradation. The maintenance and spatial distribution of genetic diversity becomes an increasingly important question as forest populations are fragmented. Landscape fragmentation may result in a loss of genetic heterogeneity, increased inbreeding, genetic drift and isolation (Young *et al.*, 1996; Aldrich and Hamrick, 1998; Sezen *et al.*, 2005), particularly as distances between patch sizes increase. Without sufficient dispersal capabilities, species with restricted range sizes and small populations may be more readily threatened with extinction due to demographic, genetic and microsite stochasticity (Trakhtenbrot *et al.*, 2005). Furthermore, historical geographical barriers may result in genetic homogeneity within individual plant populations even if, across a broad range, the species retains high levels of genetic diversity. It is worth noting that recent loss of genetic diversity may not be detected in a recently fragmented landscapes, particularly if generation times are long and only adult plants are sampled.

As molecular techniques have improved, we have also seen a corresponding improvement and increase in the analytical tools available to address genetic diversity and dispersal-related questions. Many of these tools allow researchers to cut to the heart of questions that have proved to be intractable using traditional ecological approaches, such as describing the spatial extent of gene flow, the patterns of relatedness of cohorts, genders or other targeted individuals within and between populations, and the true spatial extent of (genetic) neighbourhoods. Analytical advances also facilitate our efforts to obtain good empirical estimates of distances between parents and their offspring, or to calculate the effective numbers of individuals required to sustain present levels, or attain desired levels, of genetic diversity.

To understand spatial relatedness, ancestry and kinship coefficients, gene dispersal and other applications, ***SPAGeDi*** (Vekemans and Hardy, 2004) is a simple and effective freeware package. For direct dispersal estimates, parentage analysis can be done using ***CERVUS*** (Marshall *et al.*, 1998) to match progeny with their parents. From this, it is an easy step to calculate the distance between the two. Grivet *et al.*'s (2005) ***probability of maternal identity*** (***PMI***) is useful for estimating the effective number of seed donors and to gauge the degree of structure within the local seed pool. To estimate genetic differentiation among pollen gamete pools and to understand the spatial extent of pollen movement, ***TwoGener*** is appropriate (Smouse *et al.*, 2001; Austerlitz and Smouse, 2002), and Burczyk *et al.*'s ***neighbourhood model*** (1996,

2002) provides another paternity-based estimator of pollen donors. Some of the analytical approaches mentioned above will help researchers to overcome the inherent difficulty of tracking the fate of all seeds and/or genotyping all reproductive adults within a study area – the sheer numbers of which may be prohibitive in terms of time, labour and/or cost.

It is worth noting that, to date, the majority of studies employing molecular markers for ecological plant studies have focused on gene flow in general (whether local spatial genetic structure within and between populations) and/or pollen-mediated gene movement (Ellstrand, 1992; Sork *et al.*, 1999; Hamrick and Nason, 2000; Austerlitz *et al.*, 2004; Smouse and Sork, 2004). Few studies have empirically measured effective dispersal distances from both maternal and paternal sources for vertebrate-dispersed tropical trees across entire populations (but see Aldrich and Hamrick, 1998; Sezen *et al.*, 2005; Hardesty *et al.*, 2006). Even fewer have measured dispersal in wind-dispersed tropical species using genetic techniques (but see Hufford, 2000; Jones *et al.*, 2005).

Case Study

Now let us turn to a genetic-marker-assisted study of dispersal and recruitment of a widespread vertebrate-dispersed neotropical tree, *Simarouba amara* Aubl. (Simaroubaceae). The research was carried out on the *S. amara* population in and around the 50-ha Forest Dynamics Project (FDP) plot on Barro Colorado Island (BCI), Panama (9°10′ N, 79°51′ W; Hubbell and Foster, 1983). We used variable nuclear microsatellite genetic markers previously developed for *S. amara* (Rodriguez *et al.*, 2000).

The main objectives were:

1. To estimate effective dispersal (i.e. seedling recruitment) distances between offspring and parents, with the goal of quantifying the relative frequency of LDD events;
2. To compare gene movement via seed and pollen in a natural population of the dioecious tree, *S. amara*, testing the hypothesis that gene movement is more extensive for pollen than for seed.

The focus on dispersal and recruitment away from parent plants is relevant because seeds arriving and germinating away from the canopy of the parent plant are presumably disperser-assisted. Such dispersal reflects the relative importance of vertebrate dispersers to plant recruitment and, hence, community structure. Here, ‘long-distance’ was defined operationally as > 100 m from the maternal parent, because previous results for *S. amara* dispersal indicate a mean seed movement of < 10 m from presumed parents (median = 21 m), based upon seed-trap data (Muller-Landau, 2001). Moreover, this is a long distance relative to distances of measured density-dependent effects on seedling and sapling growth and survival, which are generally detected within distances of 20 m or less (Hubbell *et al.*, 1990, 2001; Ahumada *et al.*, 2004; Uriarte *et al.*, 2005).

Materials and Methods

Study site

The FDP is located on a plateau near the centre of Barro Colorado Island (BCI), an island formed by the flooding of Lake Gatun following the construction of the Panama Canal in the early 1900s. The lowland moist tropical forest on BCI receives approximately 2600 mm of rainfall annually, and the island experiences a pronounced dry season between December and April (Leigh *et al.*, 1982). The first FDP census was completed in 1982, and since 1985 the FDP has been censused at 5-year intervals. In each census, all woody plants within the FDP with ≥ 1 cm diameter at breast height (dbh) are identified, mapped and measured to assess growth, mortality and recruitment of new individuals (see Condit, 1998, for methodology details).

Study species

Simarouba amara is a widespread neotropical tree found in lowland moist forests from upper Mesoamerica (Honduras, Nicaragua) across the Amazon basin east to French Guiana. *S. amara* is dioecious, insect-pollinated and dispersed primarily by vertebrates. On BCI it grows to approx. 35 m in height with a maximum reported dbh of 70 cm (Croat, 1978). The unisexual flowers occur in terminal panicles comprising numerous, small, pale-yellowish flowers. Flowering persists for several weeks annually on BCI. Pollination takes place by generalist insects such as small bees and moths.

The purplish-black fruits ripen within a few months of pollination and attract numerous vertebrate dispersers including toucans (*Pteroglossus* and *Ramphastos* spp.; Ramphastidae), chachalacas (*Ortalis cinereiceps*; Cracidae), flycatchers (Tyrannidae), motmots (Momotidae), and thrushes (Turdidae) (Croat, 1978), tamarins (*Saguinus geoffroyi*; Cebidae), howler monkeys (*Alouatta palliata*; Atelidae) and spider monkeys (*Ateles geoffroyi*; Atelidae) (Hladik and Hladik, 1969). In addition to birds and primates as the dominant dispersers, leaf-cutter ants move seeds short distances (B.D. Hardesty, personal observation). The fruits are large-seeded and fleshy, approximately 17 × 10 mm, with a seed length of 10–14 mm, and presented in clusters of 3–5 drupes (Croat, 1978). Flowering and fruiting occur annually. *Simarouba amara* seeds do not experience dormancy (Camargo *et al.*, 2002) and seedlings germinate within a few months of dispersal.

Sampling

From 2001 to 2003, leaf tissue was collected from the canopy layer of all individuals ≥ 20 cm dbh within the BCI 50-ha FDP (n = 134) (Fig. 12.1). Although most *S. amara* < 30 cm dbh do not reproduce, in order to ensure

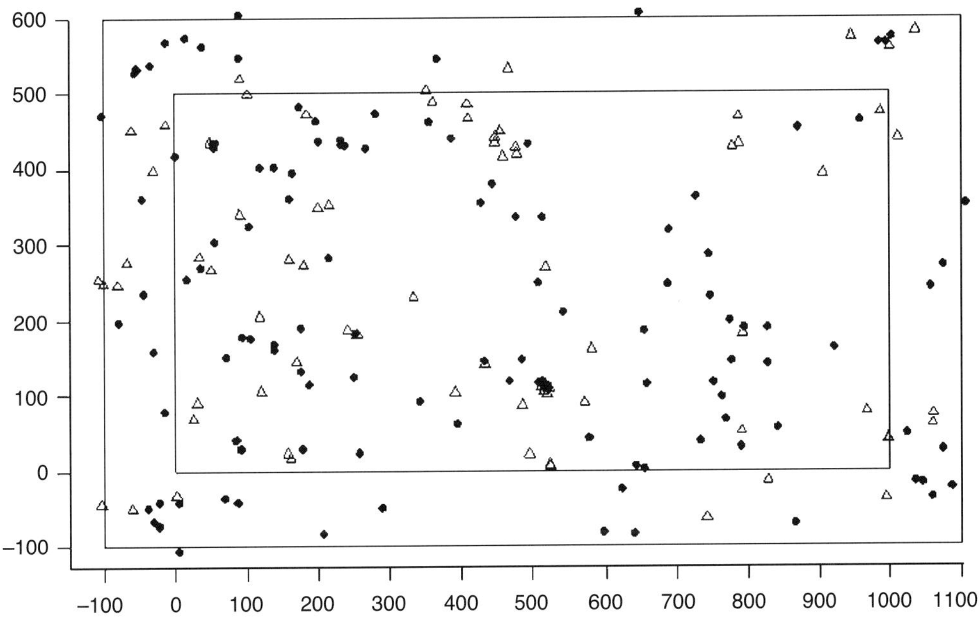

Fig. 12.1. Location of S. amara over the 84-ha sampling area for males or non-fruiting trees (closed circles) and female trees (open triangles).

that all reproductive trees were included and because *S. amara* grows rapidly (based upon BCI FDP demography data), any individual tree ≥ 20 cm dbh was assessed for reproductive output (dbh based upon the 2000 census). Trees were explicitly mapped and tissue was collected from all *S. amara* ⩾20 cm dbh within a 100-m buffer area around the FDP (Fig. 12.1). The buffer area added 84 reproductive-sized individuals and increased the total area sampled for adults by 34 ha. Trees were sexed based upon morphological differences between male and female flowers and the presence of fruit. Non-reproductive trees were not considered for maternal assignment to seedlings germinated in that year. Hence, 49, 56 and 70 female trees from the 84-ha study plot were used for maternal matching in 2001, 2002 and 2003, respectively (total n = 228 trees, 74 female, 61 male, and 83 non-reproductive-status or unknown-gender trees). Trees identified as male were used for testing paternity, and individuals ≥ 20 cm dbh whose reproductive status was 'unknown', but which could have been reproductive, were also considered as potential pollen parents.

From 2001 to 2003, approximately 40 ha of the 50-ha FDP (in 5 × 5 m quadrats) were searched for *S. amara* seedlings, generally avoiding the plot edge (Fig. 12.2). Leaf tissue was collected from each seedling encountered, except in quadrats where ≥ 5 seedlings occurred; in these instances, tissue was sampled from every third seedling. *Simarouba amara* seedlings were assumed to be dispersed in a given year if cotyledons were present (these are

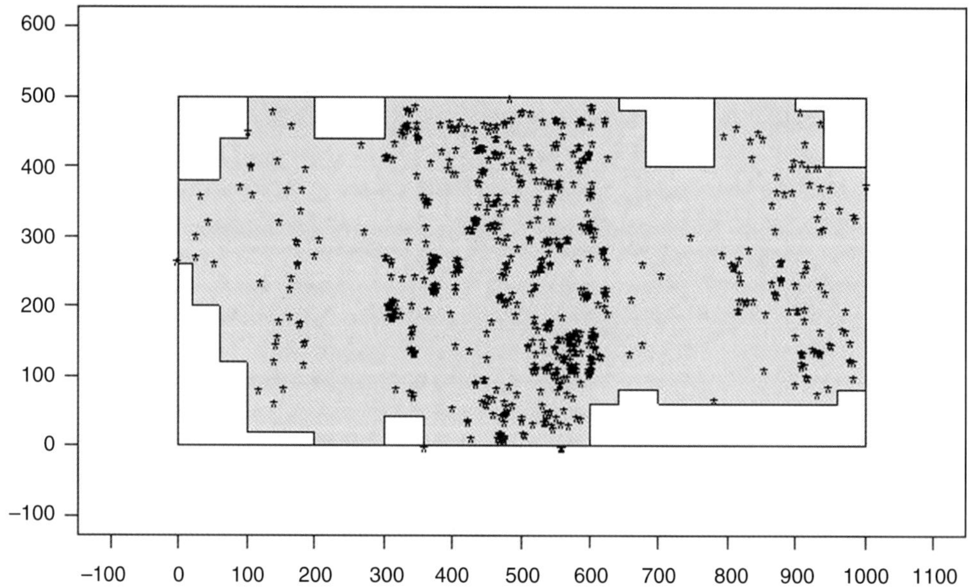

Fig. 12.2. Locations of *S. amara* seedlings (*, $n = 782$) and area searched for seedlings (grey) throughout the BCI 50-ha Forest Dynamics Plot (modified from Hardesty *et al.*, 2006).

retained up to 6 months post-germination), and/or if the stem base was distinctly purplish in colour, which occurs during the first few months of growth (B.D. Hardesty, personal observation). In all, 782 seedlings were genotyped and matched to their maternal source; 540 of these were identified as 'cohort of a given year'.

Tissue collection, DNA extraction and genotyping

Leaf tissue collected from all adult trees and seedlings sampled were either refrigerated in the laboratory and the DNA extracted within 2 days, or they were put into labelled vials and flash-frozen in liquid nitrogen. DNA extractions using 40–50 mg of the leaf tissue were performed using DNEasy kits (Qiagen Corporation). PCR protocols and data analyses followed those described by Hardesty *et al.* (2005, 2006) using five nuclear microsatellite markers developed by Rodriguez *et al.* (2000).

Genetic diversity

Nei's measure of genetic diversity (H_e), calculated as the proportion of individuals that are heterozygous under the assumption of the Hardy–Weinburg equilibrium, and allelic richness (A) were used to assess genetic diversity. The average number of alleles per locus at all five scored loci provides an additional measure of genetic diversity among adults and offspring (Table 12.1).

Maternity and paternity analyses

Maternity assignment was performed using simple exclusion, i.e. comparing multilocus genotypes of seedlings to those of candidate females. Reproductive females were rejected as candidate mothers if alleles at any of the five loci were incompatible with those of the seedling in question, precluding an exact multilocus match. All reproductive female trees within the study area had unique genotypes, with the exception of three trees sharing an exact genotype (located > 100 m apart). For cases in which multilocus matching resulted in > 1 suitable candidate mother for a seedling, two assignments were used: (*option 1*) the nearest reproductive female was assigned as the maternal parent, thereby conservatively estimating the maternal–offspring recruitment distance; or (*option 2*), a randomly assigned genetically possible candidate was assigned as the actual mother. Since the female parent was not definitive for seedlings with > 1 potential mother, parent-pair analyses could not be performed for these seedlings. However, these individuals were tested separately for paternal assignment (as first parent with other parent unknown).

The exact multilocus matching for maternity was coupled with likelihood estimates of parentage for paternity assignment. CERVUS (Marshall *et al.*, 1998) was used to perform paternity analyses to permit comparisons of gene movement via both pollen and seed. Exclusion probabilities differed depending on whether the goal was to identify a single parent (neither parent known) or whether a maternal genotype was already known and paternal exclusion was the goal (Table 12.1). Analyses and discussion describe seedlings for which there is high confidence (80% or higher, as per CERVUS) in assigning the paternal parent (details provided by Hardesty *et al.*, 2006).

Table 12.1. Number of alleles at each locus (A), expected heterozygosity (H_e), and exclusion probabilities for single-parent analyses, and second-parent analysis when one parent is known (paternity analysis) for *S. amara*. Total values represent pooled values for first and second parent exclusion probabilities (modified from Hardesty *et al.*, 2006).

Locus	A	H_e	1st parent exclusion	2nd parent exclusion
SA02	15	0.553	0.167	0.327
SA05	10	0.733	0.327	0.502
SA06	14	0.254	0.034	0.136
SA27	16	0.835	0.509	0.678
SA29	10	0.502	0.145	0.317
Mean	13.0	0.575	–	–
Total	65		0.772	0.936

Results

A total of 65 alleles occurred among the five microsatellite loci sampled across all adults and seedling recruits (range = 10–16 alleles per locus; Table 12.1). All five loci exhibited high levels of variability. Although SA06 was found to have a large number of alleles, it is a relatively uninformative locus because it exhibits low heterozygosity (Table 12.1) and probably contributes little to the probability of excluding potential parents. In contrast, the four other loci exhibited higher levels of heterozygosity, and were therefore more informative for parentage analyses.

On average, pollen movement and dispersal events resulting in germinated seedlings occurred over a similar scale (Fig. 12.3). For the 94 seedlings with a single genetic mother, the mean recruitment distance was 391.6 ± 234.6 m (SD), range 9.3–1000.5 m (Fig. 12.3, solid bars). The average distance that pollen moved between parents was slightly less (334.4 ± 231.0 m (SD), range 8.0–1063.2 m; Fig. 12.3, horizontal striped bars). It is worth noting that seldom were nearby reproductive adults the assigned parents of seedlings: the average distance between nearest reproductive male–female pairs was

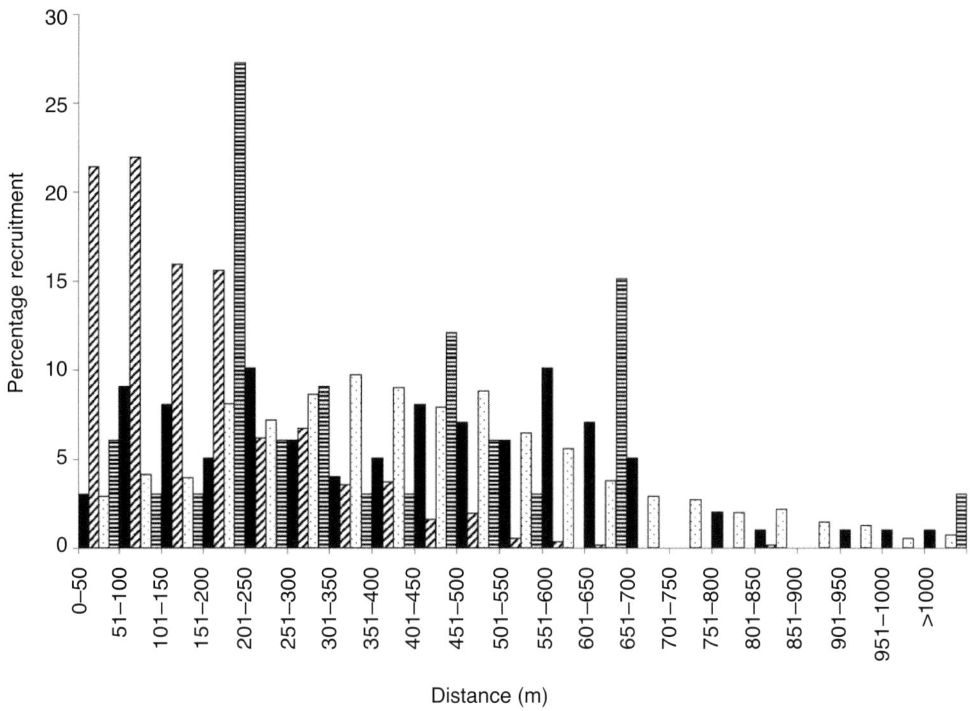

Fig. 12.3. Percentage recruitment distances at 50 m distance classes for unique matched mother (only one possible mother) (solid bars), 2+ parents with nearest assigned as maternal parent (hatched bars), and randomly assigned maternal parent from potential genetic mothers (open dotted bars). Horizontal striped bars depict pollen movement distances (modified from Hardesty et al., 2006).

54.1 ± 35.6 (SD), range 2.3–145.6 m. Similarly, for seedlings with a unique maternal match, seldom was the nearest reproductive female the actual maternal parent (only 4.3% of cases). The average distance between seedlings and the nearest female tree was 50.5 ± 35.5 m (SD), range 0.76–153.1 m.

Because of the infrequency with which the nearest adult was the parent plant, a binary approach to assigning mothers for seedlings with > 1 candidate mothers was applied. First, the nearest of the possible genetic mothers was assigned as the 'maternal parent' to seedlings with 2+ candidate mothers. In this instance the average effective dispersal distance dropped to 148.5 ± 121.2 m (SD), range 3.6–803.6 m (Fig. 12.3, hatched bars), and almost certainly underestimates actual dispersal distances. Alternatively (option 2), a maternal parent was randomly assigned from the pool of genetically possible candidates based upon exact multilocus matching. When using option 2, the assigned dispersal distances paralleled the findings for the unambiguously assigned seedlings and averaged 402.6 ± 219.2 m (SD), range 7.3–1127.4 m (Fig. 12.3, open dotted bars). The minimum, maximum and 10th, 25th, 50th, 75th and 90th percentile dispersal distances for exact matches and those with > 1 candidate maternal parent, based upon each assignment method, are provided in Table 12.2. Furthermore, 16.9% (n = 132) of the 782 genotyped seedlings had no maternal matches within the FDP, indicating that they resulted from long-distance effective dispersal events originating from outside the 84-ha study area.

Finally, the distances between siblings that shared their maternal parent to the average distance between nearest seedlings were compared (Fig. 12.4). By definition, these half and full siblings result from effective seed dispersal events (not pollen dispersal). Pairwise distances between such half and full siblings (mean = 197.7 ± 165.5 m SD) differed significantly ($P < 0.0001$, t-test) from nearest seedling distances (mean = 11.3 ± 15.8 m SD), indicating that germinated siblings did not tend to occur in familial clumps resulting from seed deposition.

Discussion

Case study

This case study highlights the utility of using genetic approaches to further our understanding of effective seed dispersal. Genotyping seedlings and matching them to their maternal and paternal source enables us to empirically resolve long-standing questions in dispersal ecology about the spatial relationships between parent plants and their offspring by providing empirical answers to questions which we previously have been able to address solely through inference.

One advantage of working with a dioecious species such as *S. amara* is that each individual only contributes once to each seed or seedling. This essentially halves the number of potential parent pairs compared with a hermaphroditic species. Working with a dioecious species also permits us to

Table 12.2. Minimum distances, 10th, 25th, 50th, 75th, 90th percentile, and maximum dispersal distances for seedlings with 1–4+ potential maternal parents, and results from all genotyped seedlings (pooled). Effective dispersal distances are based upon exact maternal matches, assignment to nearest of possible genetic mothers (option 1), or random assignment from among possible genetic mothers (option 2).

Assignment	No. possible mothers	Min. distance (m)	10th percentile (m)	25th percentile (m)	50th percentile (m)	75th percentile (m)	90th percentile (m)	Max. distance (m)
Exact	1	9.3	75.3	192.9	402.0	580.3	656.1	1000.5
Nearest	2	11.4	50.4	122.2	274.0	423.1	476.7	803.6
Nearest	3	25.1	65.9	100.9	152.7	249.7	435.6	479.0
Nearest	4+	3.6	27.3	50.9	105.6	176.2	265.1	576.0
Nearest	Pooled	3.6	35.1	64.1	135.4	241.5	587.1	1000.5
Random	2	11.4	57.4	223.8	394.2	595.0	720.2	1079.6
Random	3	44.9	127.2	194.8	340.0	427.2	546.5	667.0
Random	4+	7.3	153.6	241.3	383.9	378.0	719.8	1127.4
Random	Pooled	7.3	137.1	238.1	378.0	536.9	705.2	1127.4

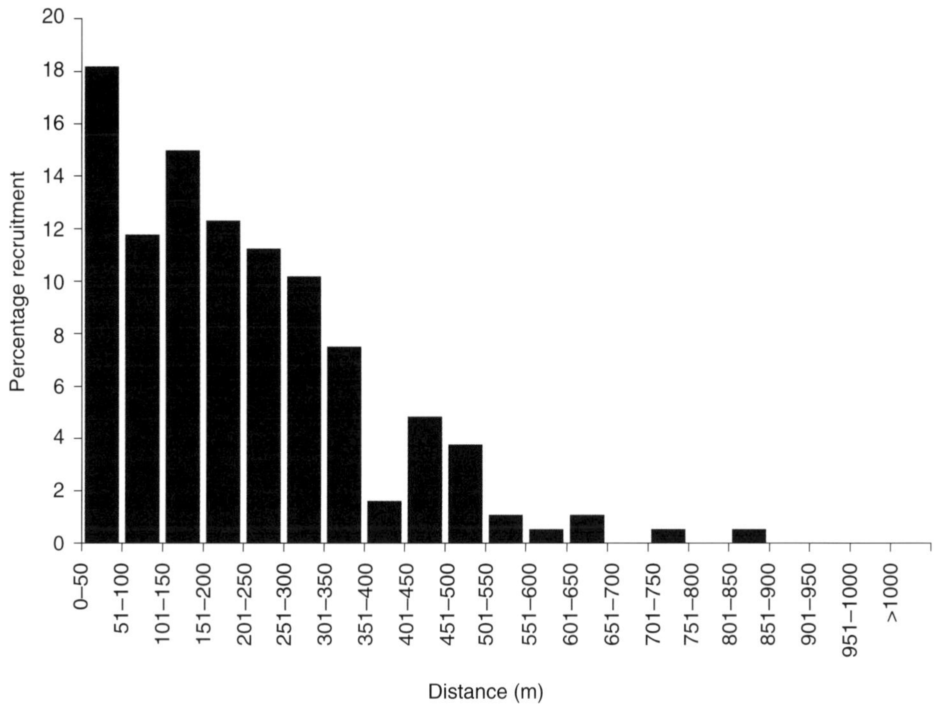

Fig. 12.4. Effective dispersal distances (percentage of events in 50-m distance classes) for siblings sharing their maternal parent.

discern between pollen and seed movement more easily and to estimate the variance in individual seed output among individuals/years; something that may prove more challenging in other systems (see Sezen *et al.*, 2005; Bacles *et al.*, 2006).

Dispersal events were observed 1 km or more from parent plants – essentially at the maximum distance possible for detection within the study system. The high frequency of long-distance recruitment suggests that vertebrates are playing an important role as seed dispersers. Seedling establishment frequently occurred near non-parental trees rather than adjacent to maternal or paternal parents, and distances between siblings that shared maternal parents often exceeded 100 m. These results further support the importance of frugivores in ecosystem function. Given that seed density is highest under the canopy of fruiting trees (B.D. Hardesty, unpublished results) these findings suggest that density dependence near parent plants plays an important role in determining *S. amara* recruitment patterns. Furthermore, it seems that *S. amara* is not strongly dispersal-limited at the scale of 50 (or 84) ha, as one might predict based upon seed rain data, where a mean dispersal distance of 39 m was documented for *S. amara* (Muller-Landau, 2001).

Seeds of *S. amara* that arrive away from parents may experience an elevated probability of recruitment, despite the proximity of nearby non-parent adults. Jordano and Godoy (2002) found parallel results for seed deposition in *Prunus mahaleb*. In their system, microsatellite markers showed that seeds were frequently deposited beneath the canopy of non-maternal conspecific adults. Results from both studies suggest that the mechanisms impacting both seed deposition and seedling recruitment patterns are more complex than non-genetic distance- and density-related Janzen–Connell effects (Janzen, 1970; Connell, 1971) predict. Such findings could not be determined without a molecular-marker-assisted dispersal study.

Assigning the 'actual mother' randomly from among genetically possible reproductive females (option 2) probably provides a more accurate representation of dispersal distances than assigning the nearest genetically possible female as the maternal parent (option 1). With random assignment, the dispersal distribution paralleled results from unambiguously assigned seedlings (Fig. 12.3), whereas proximity assignment (option 1) appears to greatly underestimate dispersal when compared with seedlings of known parentage. Much shorter recruitment distances were found at all percentiles of the dispersal distributions with option 1 than were found using option 2 or with seedlings that had unambiguous maternal assignment (Table 12.2).

In general, we tend to be conservative in estimating dispersal distances. One goal may be to err on the conservative side while reasonably reflecting the scale of the process in question, rather than focusing solely on dispersal distances themselves. It is just as erroneous to seriously underestimate dispersal distances as it would be to presume LDD, and will undoubtedly confound our understanding of the dispersal process within a study system. Overall, the effective seed and pollen movement estimates presented here are inherently conservative, due to the possible false assignment of an offspring to a parent whose true parent may reside outside the 84-ha study site (Devlin and Ellstrand, 1990).

Limitations

Some parentage assignment techniques have been criticized because of the bias of increased likelihood of assignment for homozygous individuals (Devlin and Ellstrand, 1990; Jones and Arden, 2003). While it is a concern that low levels of heterozygosity may lead to reduced ability to discriminate among candidate parents, the use of multilocus exclusion renders the concern a moot point for maternal matching, and the high level of heterozygosity at four of the five loci used in this study (Table 12.1) reduces this problem for paternity assignment. Because of the large number of reproductive-sized adults within the study area, exclusion probabilities were reduced (77% for single parent and 94% for second parent; Table 12.1) and thus it was not possible to discern among all candidate parents in many of the genotyped seedlings. Increased numbers of polymorphic loci would undoubtedly have resulted in unambiguous assignment among candidate parents for additional seedlings.

Even with ambiguous parentage assignment for many individuals, this technique reveals much useful information, and yields insights that non-genetic dispersal studies are unable to provide, such as the important result that seedlings were seldom a genetic match to the nearest reproductive adult. Because many parentage methods are likelihood-based estimates, we can estimate the range of distances for candidate parents and determine whether nearby adults are within the candidate parent pool, or whether germinated seedlings result from dispersal outside a study area (e.g. Bacles *et al.*, 2006). Alternately, fractional parent assignment can be used (Devlin *et al.*, 1988) to determine the relative reproductive output of individuals within a population, although it may be less useful if the goal is to calculate exact dispersal distances.

Recent advances and the present state of the field

Successful dispersal can be inferred from the spatial distribution of genotypes (Loiselle *et al.*, 1995), and can be measured directly using parentage analyses. To date, inferring dispersal distances using spatial autocorrelation (e.g. local relatedness) (Epperson, 2005; Hardy *et al.*, 2005) has been largely unsatisfying, although it may be a useful first step to assessing the extent and relative frequency of historical long-distance gene dispersal (Hardy *et al.*, 2005). We are currently improving estimates of dispersal by incorporating animal movement patterns and gut passage times for estimating seed shadows (Dennis and Westcott, Chapter 9, this volume; Holbrook and Loiselle, Chapter 13, this volume), by modelling dispersal kernels (Nathan, Chapter 11, this volume) and by employing molecular techniques (Jordano, Chapter 10, this volume; Steele *et al.*, Chapter 14, this volume) to match parents and offspring for seeds and seedlings.

As results from additional combined ecological and genetic studies become available, we will probably observe greater dispersal distances, in both frequency and spatial extent. The TwoGener approach (Smouse *et al.*, 2001) for estimating dispersal distances without genotyping all reproductive individuals in an area is an obvious advantage when sampling over a large geographical area. Calculating actual dispersal distances between parents and offspring is not possible with TwoGener, although the extent of genetic neighbourhood sizes can be estimated. Unlike reported pollinator movements of 5–14+ km (Nason *et al.*, 1996, 1998), to date we have been unable to directly measure effective seed dispersal distances at such spatial scales, though dispersal at large scales certainly occurs (Bacles *et al.*, 2006). We have seen evidence that extensive animal movements may occur over short time periods, and dispersers may retain seeds in the gut for several hours (Shilton *et al.*, 1999; Holbrook and Smith, 2000; Holbrook *et al.*, 2002; Westcott *et al.*, 2005), providing strong evidence for the long-distance dispersal capabilities of frugivores.

While it may not be surprising that genetic methods empirically demonstrate longer (and fatter) dispersal distributions than non-genetic model fit dispersal kernel estimates have shown, such findings represent a significant advance in our understanding of the extent and frequency of LDD events. Model fit dispersal estimates still prove frustrating, particularly for animal-dispersed species. Even for wind-dispersed trees, for which model fits generally work best (Nathan *et al.*, 2002), microsatellite matching of seeds to their maternal sources demonstrated that different models fit better in different years (Jones *et al.*, 2005). Furthermore, they describe a problem of non-trivial uncertainty in the tail of their distribution kernels, further highlighting the complexity of dispersal through space and time.

Direct observations of frugivore movements will seldom, if ever, reveal which animals are successfully dispersing the seeds that germinate, grow and ultimately recruit. However, observations will permit evaluation of the efficiency and effectiveness of seed removal by dispersers (Bustamante and Canals, 1995). For vertebrate-dispersed species, the patterns of local and large-scale dispersal are extremely complex. Unpredictable movements of dispersers, social interactions, seed deposition at latrines, caching, roosting, lekking and other behaviours undoubtedly affect patterns of both primary and secondary dispersal (Westcott *et al.*, 2005). These, in turn, impact on the probability of a seed arriving at a site that is suitable for germination and persistence.

With recent decades of research findings and advances in molecular techniques, we are in a position to better measure and model the pattern, process and outcome of seed dispersal and to understand the complex factors involved in plant–frugivore population dynamics. With rapidly changing landscapes, it is more important than ever to understand the multitude of factors, especially the role of effective seed dispersal, which impact ecosystem function.

Long-distance dispersal and changing landscapes

Dispersal, particularly LDD, may result in species-spread into new habitats and may influence the population genetic structure at local and landscape scales. Contemporary long-distance gene dispersal via seed frequently occurs in both temperate (Bacles *et al.* 2006; Steele *et al.*, Chapter 14, this volume) and tropical systems (Aldrich and Hamrick, 1998; Jones *et al.*, 2005; Sezen *et al.*, 2005; Hardesty *et al.*, 2006). Across a 900-ha fragmented stand in Scotland, Bacles *et al.* (2006) have estimated that nearly 50% of seedlings of a wind-dispersed species have established from dispersal events that originated outside their study area. Even within a continuous forest matrix, an estimated 17% of *S. amara* seedlings established from dispersal events that originated outside the study area. This is a minimum estimate – actual values may be much higher if substantial cryptic gene flow occurs (Devlin and Ellstrand, 1990). The regular or frequent occurrence of successful LDD events may help to explain the maintenance and persistence of the diversity observed in tropical forests.

As landscapes are fragmented, the impact of dispersers will become even more critical. For example, forest fragmentation may result in the direct loss of dispersers – the net outcome of which may be far fewer seedling recruits than in continuous forest (Cordeiro and Howe, 2003). Fragmentation may also lead to genetic dominance by a few individuals which are reproductively responsible for the subsequent generations (Aldrich and Hamrick, 1998; Sezen *et al.*, 2005). Such reproductive dominance can result in genetic bottlenecks which may threaten species health and persistence. To date, most genetic studies of effective or realized dispersal (Aldrich and Hamrick, 1998; Sezen *et al.*, 2005; Bacles *et al.*, 2006) have been carried out in fragmented or otherwise disturbed landscapes. We need additional large-scale studies which incorporate seed arrival and effective dispersal, not only in fragmented landscapes, but also within intact forests. Such studies will provide critical baseline data that will allow us to compare patterns and consequences of dispersal in fragmented and intact ecosystems. Simultaneous studies comparing the same species across intact and fragmented landscapes would be ideal. Such research will enable us to develop better predictive models and inform us whether general trends persist across a wide spectrum of species, habitats and landscapes. The application and timeliness of such work cannot be underestimated – such studies will not be feasible once intact landscapes are altered.

Future directions

We need more integrated studies, particularly long-term studies that encompass both the disperser and the plant perspective. By employing similar methods across sites, we can more easily make cross-site and species comparisons. Synthesizing and integrating results from multiple studies over multiple spatial and temporal scales will provide the empirical data necessary to better understand forest dynamics, and will enable us to build more accurate models that provide insight into the local and large-scale frequency of short- and long-distance dispersal.

Combined with information on community structure, local species richness, neighbourhood, and disperser attributes such as movement patterns and seed deposition, we can use genetic data to better predict seed deposition and the likelihood of germination and survival at local and large spatial scales. Improved mechanistic models, in particular, can enable us to predict patterns of dispersal and recruitment and to understand how changes in systems may alter subsequent seed arrival, germination and establishment patterns. This is imperative for making informed habitat and resource management decisions. As we gain further understanding of seed-fate pathways (Vander Wall *et al.*, 2005) and advance mechanistic dispersal models, we will better understand not only the relative frequency but also the outcome of non-standard dispersal events (Higgins *et al.*, 2003).

Ultimately, it is effective dispersal rather than seed movement per se that underlies the present day population and landscape level plant distribution and diversity patterns that we observe. Using molecular techniques, we can gain insights into the historical footprint of dispersal while also measuring current seed and pollen movement. Coupling genetic with ecological field studies will provide a much richer picture of both the evolutionary and ecological impact of dispersal (and dispersers themselves) on the present and future community structure.

Vertebrate-assisted seed dispersal plays a fundamental role in structuring forest communities, particularly in tropical ecosystems. Few studies to date have used molecular markers to document the scale over which seed dispersal is occurring for vertebrate-dispersed tropical trees within intact habitats. Fewer still have addressed the question of long-distance seed dispersal at scales relevant to the home-range sizes encompassed by the main frugivore dispersers within communities. Comprehensive research programmes which combine animal movements (home-range sizes and fine-scale movements), fruit removal rates, and seed passage times (with gut passage effect on germination) with habitat data, fecundity estimates, tree height and crown area, and genetic estimates of dispersal will permit an unprecedented understanding of the net impact that frugivores exert on structuring forest communities.

Making use of the many long-term Forest Dynamics Plots (FDP; 18 or more at present count) with the demography data currently available can greatly aid our research efforts. Generally, within these plots, all woody stems $\geq$1 cm dbh are mapped and identified across permanent large-scale plots (50 ha or more). These data are available through http://ctfs.si.edu/. The automated telemetry system on BCI is available for tracking animal movements, and provides another excellent opportunity to combine frugivore movement patterns with molecular studies of dispersal in a forest where we have >20 years of plant demography data.

Finally, to date we have seen limited research efforts that explicitly target frugivore dispersal of invasive plant species in heavily altered landscapes. Using a combination of field and genetic methods, we can gain insight into the seed dispersal process for invasive species, and we can compare native and invasive species within the landscape. Theory predicts that colonization of new areas will be associated with population bottlenecks that reduce within-population genetic diversity and will increase levels of genetic differentiation among populations. This may be particularly true for weedy or invasive plant species, which often have self-compatible breeding systems and may reproduce vegetatively. Molecular tools may help to clarify the history of introduction and tell us whether single or multiple events are responsible for the spread of particular invasive species. Such knowledge can aid efforts to retard the spread of invasive species. Additional dispersal studies of weedy species and managed landscapes would facilitate management decision-making in order to more effectively combat invasion by non-native species.

While LDD and subsequent establishment are presumed to be rare events, the question of scale is seldom addressed. Bacles *et al.* (2006) have suggested that effective seed dispersal may occur over tens of kilometres for a wind-dispersed tree in a fragmented landscape in Scotland. At what spatial (metres, tens of metres, hundreds of metres, kilometres) and temporal (months, years, decades, centuries) scale is dispersal rare, for vertebrate- or wind-dispersed species? How does this vary among species, populations, forests and landscapes? These are questions to drive future research efforts, paying particular attention to the appropriate scale required for answering the questions we are asking, and for employing successful conservation strategies.

Conclusions

Given the multitude of dispersers, the varied impacts they have on the outcome of dispersal, and the variety of dispersal syndromes, understanding the patterns and consequences of dispersal can be overwhelming. Even in simple systems, spatial and temporal environmental heterogeneity makes it difficult to predict precise patterns and outcomes of dispersal. Combining ecological methods with genetic techniques provides an improved means to address long-standing questions in dispersal biology. Together, the methods complement one another and provide more insight into the dispersal process than either could provide independently (Silvertown, 1991; Ouborg *et al.*, 1999). The case study presented here highlights some striking differences between non-genetic dispersal estimates based upon seed-trap studies and genetically determined effective dispersal distances. The disparity between seed arrival and seedling establishment distributions and the high frequency of recruitment away from parents and siblings demonstrate the significance of utilizing molecular tools to address fundamentally important questions in ecology. Integrating ecological and molecular methods will no doubt increase our understanding of dispersal processes at local and landscape levels. We should expect exciting new insights as more studies combine field and molecular techniques in order to understand the patterns and process of dispersal.

Acknowledgements

I thank the organizers of the *Fourth International Symposium/Workshop on Frugivores and Seed Dispersal* for the opportunity to contribute to this volume: Andrew Dennis, Ronda Green, Eugene Schupp and David Westcott. This research was supported by The Smithsonian Tropical Research Institute, a grant from the Center for Tropical Forest Science, a Sigma Delta Epsilon Graduate Women in Science Fellowship, the Garden Club of America's Tropical Botany Award, and by a final-year fellowship from the Graduate School of the University of Georgia. For assistance in the field, Pablo Ramos, Wendy Miles, David Sabino, Amanda Garman and Kirk Peterson were invaluable. For help in the laboratory I thank Christopher Dick, Maribel Gonzales and Oris Sanjur. Suggestions from Preston Aldrich, Christopher Dick, Kimberly Holbrook, David Westcott and an anonymous reviewer improved this chapter. I am also indebted to Steve Hubbell, Biff Bermingham and Jim Hamrick.

References

Ahumada, J.A., Hubbell, S.P., Condit, R. and Foster, R.B. (2004) Long-term survival in a Neotropical forest: the influence of local biotic neighbourhood. In: Losos, E.C. and Leigh, E.G. Jr (eds) *Tropical Forest Diversity and Dynamism: Findings from a Large-Scale Plot Network*. University of Chicago Press, Chicago, Illinois, pp. 408–432.

Aldrich, P.R. and Hamrick, J.L. (1998) Reproductive dominance of pasture trees in a fragmented tropical forest mosaic. *Science* 281, 103–105.

Anderson, M. (1991) Mechanistic models for the seed shadows of wind-dispersed plants. *American Naturalist* 137, 476–497.

Augspurger, C.K. (1983) Offspring recruitment around tropical trees: changes in cohort distance with time. *Oikos* 40, 189–196.

Austerlitz, F. and Smouse, P.E. (2002) Two-generation analysis of pollen flow across a landscape. IV. Estimating the dispersal parameter. *Genetics* 161, 355–363.

Austerlitz, F., Dick, C.W., Dutech, C., Klein, E.K., Oddou-Muratorio, S., Smouse, P.E. and Sork, V.L. (2004) Using genetic markers to estimate the pollen dispersal curve. *Molecular Ecology* 13, 937–954.

Bacles, C.F.E., Lowe, A.J. and Ennos, R.A. (2006) Effective seed dispersal across a fragmented landscape. *Science* 311, 628.

Burczyk, J., Adams, W.T. and Shimizu, J.Y. (1996) Mating patterns and pollen dispersal in a natural knobcone pine (*Pinus attenuate* Lemmon.) stand. *Heredity* 77, 251–260.

Burczyk, J., Adams, W.T., Moran, G.F. and Griffin, A.R. (2002) Complex patterns of mating revealed in a *Eucalyptus regnans* seed orchard using allozyme markers and the neighbourhood model. *Molecular Ecology* 11, 2379–2391.

Bustamante, R.O. and Canals, L.M. (1995) Dispersal quality in plants: how to measure efficiency and effectiveness of a seed disperser. *Oikos* 73, 133–136.

Cain, M.L., Milligan, B.G. and Strand, A.E. (2000) Long-distance seed dispersal in plant population. *American Journal of Botany* 87, 1217–1227.

Camargo, J.L.C., Ferraz, I.D.K and Imakawa, A.M. (2002) Rehabilitation of degraded areas of Central Amazonia using direct sowing of forest tree seeds. *Restoration Ecology* 10, 636–644.

Clark, C.J., Poulsen, J.R, and Parker, V.T. (2001) The role of arboreal seed dispersal groups on the seed rain of a lowland tropical forest. *Biotropica* 33, 606–620.

Clark, C.J., Poulsen, J.R., Bolker, B.M., Connor, E.F. and Parker, V.T. (2005) Comparative seed shadows of bird-, monkey-, and wind-dispersed trees. *Ecology* 86, 2684–2694.

Clark, J.S. (1998) Why trees migrate so fast: confronting theory with dispersal biology and the paleo record. *American Naturalist* 152, 204–224.

Clark, J.S., Silman, M., Kern, R., Macklin, E. HilleRisLambers, J. (1999) Seed dispersal near and far: generalized patterns across temperate and tropical forests. *Ecology* 80, 1475–1494.

Condit, R. (1998) *Tropical Forest Census Plots*. Springer, New York.

Connell, J.H. (1971) On the role of natural enemies in preventing competitive exclusion in some marine animals and in rainforest trees. In: Den Boer, P.J. and Gradwell, G. (eds) *Dynamics of Populations*. PUDOC, Wageningen, The Netherlands, pp. 298–312.

Cordeiro, N.J. and Howe, H.F. (2003) Forest fragmentation severs mutualism between dispersers and an endemic African tree. *Processing of the National Academy of Sciences* 100, 14052–14056.

Croat, T. (1978) *Flora of Barro Colorado Island*. Stanford University, Port Alto, California.

Davis, M.B. (1976) Pleistocene biogeography of temperate deciduous forests. *Geoscience and Man* 13, 13–26.

Devlin, B. and Ellstrand, N.C. (1990) The development and application of a refined method for estimating gene flow from angiosperm paternity analysis. *Evolution* 44, 248–259.

Devlin, B., Roeder, K. and Ellstrand, N.C. (1988) Fractional paternity assignment: theoretical development and comparison to other methods. *Theoretical and Applied Genetics* 76, 369–380.

Ellstrand, N.C. (1992) Gene flow among seed plant populations. *New Forests* 6, 241–256.

Eriksson, O. and Jakobsson, A. (1999) Recruitment trade-offs and the evolution of dispersal mechanisms in plants. *Evolutionary Ecology* 13, 411–423.

Epperson, B.K. (2005) Estimating dispersal from short distance spatial autocorrelation. *Heredity* 95, 7–15.

Fragoso, J.M.V. (1997) Tapir-generated seed shadows: scale-dependent patchiness in the Amazon rain forest. *Journal of Ecology* 85, 519–529.

Grivet, D., Smouse, P.E. and Sork, V.L. (2005) A novel approach to an old problem: tracking dispersed seeds. *Molecular Ecology* 14, 3585–3595.

Hamrick, J.L. (2004) Response of forest trees to global environmental changes. *Forest Ecology and Management* 197, 323–335.

Hamrick, J.L. and Nason, J.D. (2000) Gene flow in forest trees. In: Young, A., Boshier, D. and Boyle, T. (eds) *Forest Conservation Genetics: Principles and Practice*. CSIRO Publishing, Collingwood, Australia, pp. 81–90.

Hardesty, B.D. and Parker, V.T. (2002) Community seed rain patterns and a comparison to adult community structure in a West African tropical forest. *Plant Ecology* 164, 49–64.

Hardesty, B.D., Dick, C.W., Kremer, A., Hubbell, S.P. and Bermingham, E. (2005) Spatial genetic structure of *Simarouba amara* Aubl. (Simaroubaceae), a dioecious, animal-dispersed Neotropical tree, on Barro Colorado Island, Panama. *Heredity* 95, 290–297.

Hardesty, B.D., Hubbell, S.P. and Bermingham, E. (2006) Genetic evidence of frequent long distance recruitment in a vertebrate-dispersed tree. *Ecology Letters* 9, 516–525.

Hardy, O.J., Maggia, L., Bandou, E., Breyne, P., Caron, H., Chevallier, M.-H., Doligez, A., Dutech, C., Kremer, A., Latouche-Hallé, C., Troispoux, V., Veron, V. and Degen, B. (2006) Fine-scale genetic structure and gene dispersal inferences in 10 Neotropical tree species. *Molecular Ecology* 15, 559–571.

Harper, J.L. (1977) *Population Biology of Plants*. Academic Press, London.

Higgins, S.I., Nathan, R. and Cain, M.L. (2003) Are long-distance dispersal events in plants usually caused by nonstandard means of dispersal? *Ecology* 84, 1945–1956.

Hladik, A. and Hladik, C.M. (1969) Rapports trophiques entre végétation et primates dans le foret de Barro Colorado (Panama). *La Terre et la Vie* 1, 25–117.

Holbrook, K.M. and Smith, T.B. (2000) Seed dispersal and movement patterns in two species of *Ceratogymna* hornbills in a West African tropical lowland forest. *Oecologia* 125, 249–257.

Holbrook, K.M., Smith, T.B. and Hardesty, B.D. (2002) Implications of long-distance movements of frugivorous rain forest hornbills. *Ecography* 25, 745–749.

Howe, H.F. and Smallwood, J. (1982) Ecology of seed dispersal. *Annual Review of Ecology and Systematics* 13, 201–228.

Hubbell, S.P. (2001) *The Unified Neutral Theory of Biodiversity and Biogeography*. Princeton University Press, Princeton, New Jersey.

Hubbell, S.P. and Foster, R.B. (1983) Diversity of canopy trees in a Neotropical forest and implications for conservation. In: Sutten, S.L., Whitmore, T.C. and Chadwick, A.C. (eds) *Tropical Rain Forest: Ecology and Management*. Blackwell Scientific, Oxford, UK.

Hubbell, S.P., Condit, R. and Foster, R.B. (1990) Presence and absence of density dependence in a Neotropical tree community. *Transactions of the Royal Society London* 330, 269–281.

Hubbell, S.P., Ahumada, J.A., Condit, R. and Foster, R.B. (2001) Local neighbourhood effects on long-term survival of individual trees in a Neotropical forest. *Ecology Research* 16, 859–875.

Hufford, K.M. (2000) Viability selection, inbreeding depression and male reproductive success at three early life stages of the tropical tree, *Pladypodium elegans*. PhD thesis University of Georgia. Athens, Georgia.

Hurtt, G.C. and Pacala, S.W. (1995) The consequences of recruitment limitation: reconciling chance, history and competitive differences between plants. *Journal of Theoretical Biology* 176, 1–12.

Jacquemyn, H., Honnay, O., Van Looy, K. and Breyne, P. (2006) Spatiotemporal structure of genetic variation of a spreading plant metapopulation on dynamic riverbanks along the Meuse River. *Heredity* 96, 471–478.

Janzen, D.H. (1970) Herbivores and the number of tree species in tropical forests. *American Naturalist* 104, 501–528.

Jones, A.G. and Arden, W.R. (2003) Methods of parentage analysis in natural populations. *Molecular Ecology* 12, 2511–2523.

Jones, F.A., Chen, J., Weng, G.-J. and Hubbell, S.P. (2005) A genetic evaluation of seed dispersal in the Neotropical tree, *Jacaranda copaia* (Bignonianaceae). *American Naturalist* 166, 543–555.

Jordano, P. (1992) Fruits and frugivory. In: Fenner, M. (ed.) *Seeds: The Ecology of Regeneration in Plant Communities*. CAB International, Wallingford, UK, pp. 105–156.

Jordano, P. and Godoy, J.A. (2002) Frugivore-generated seed shadows: a landscape view of demographic and genetic effects. In: Levey, D.J., Silva, W. and Galetti, M. (eds) *Frugivores and Seed Dispersal: Ecology, Evolution, and Conservation*. CAB International, Wallingford, UK, pp. 305–321.

Leigh, E.G. Jr, Rand, A.S. and Windsor, D.M. (eds) (1982) *The Ecology of a Tropical Forest: Seasonal Rhythms and Long-Term Changes*. Smithsonian Institution Press, Washington, DC.

Levin, S.A., Muller-Landau, H.C., Nathan, R. and Chave, J. (2003) The ecology and evolution of seed dispersal: a theoretical perspective. *Annual Review of Ecology and Systematics* 34, 575–604.

Loiselle, B.A., Sork, V.L., Nason, J. and Graham, C. (1995) Spatial genetic structure of a tropical understory shrub, *Psychotria officinalis* (Rubiaceae). *American Journal of Botany* 82(11), 1420–1425.

Marshall, T.C., Slate, J., Kruuk, L.E.B. and Pemberton, Z.J.M. (1998) Statistical confidence for likelihood-based paternity inference in natural populations. *Molecular Ecology* 7, 639–655.

McLachlan, J.S., Clark, J.S. and Manos, P.S. (2005) Molecular indicators of tree migration capacity under rapid climate change. *Ecology* 86, 2088–2098.

Muller-Landau, H.C. (2001) Seed dispersal in a tropical forest: empirical patterns, their origins, and their consequences for community dynamics. PhD thesis, Princeton University, Princeton, New Jersey.

Nathan, R. (2005) Long-distance dispersal research: building a network of yellow brick roads. *Diversity and Distributions* 11, 125–130.

Nathan, R., Horn, H.S., Chave, J. and Levin, S.A. (2002) Mechanistic models for tree seed dispersal by wind in dense forests and open landscapes. In: Levey, D.J., Silva, W.R. and Galetti, M. (eds) *Seed Dispersal and Frugivory: Ecology, Evolution and Conservation*. CAB International, Wallingford, UK, pp. 69–82.

Nason, J.D., Herre, E.A. and Hamrick, J.L. (1998) The breeding structure of a tropical keystone plant resource. *Nature* 391, 685–687.

Nason, J.D., Herre, E.A. and Hamrick, J.L. (1996) Paternity analysis of the breeding structure of strangler fig populations: evidence for substantial long-distance wasp dispersal. *Journal of Biogeography* 23, 501–512.

Ouborg, N.J., Piquot, Y. and van Groenendael, J.M. (1999) Population genetics, molecular markers and the study of dispersal in plants. *Journal of Ecology* 87, 551–568.

Petit, R.J., Pineau, E., Demesure, B., Bacilieri, R., Ducousso, A. and Kremer, A. (1997) Chloroplast DNA footprints of postglacial recolonization by oaks. *Proceedings of the National Academy of Sciences, USA* 94, 9996–10001.

Pitelka, L.F. (1997) Plant migration and climate change. *American Scientist* 85, 464–473.
Rodriguez von Platen, H., Geistlinger, J., Berlyn, G., Kahl, G. and Weising, K. (2000) Characterization of novel microsatellite loci isolated from the tropical dioecious tree *Simarouba amara. Molecular Ecology Primer Notes* 9, 498–500.
Schupp, E.W. (1993) Quantity, quality and the effectiveness of seed dispersal by animals. *Vegetatio* 107/108, 15–29.
Sezen, U.U., Chazdon, R.L. and Holsinger, K.E. (2005) Genetic consequences of tropical second-growth forest regeneration. *Science* 307, 891.
Shilton, L.A., Altringham, J.D., Compton, S.G. and Whittaker, R.J. (1999) Old World fruit bats can be long-distance seed dispersers through extended retention of viable seeds in the gut. *Proceedings of the Royal Society London, Series B, Biological Sciences* 266, 219–223.
Silvertown, J. (1991) Dorothy's dilemma and the unification of plant population biology. *Trends in Ecology and Evolution* 6, 346–348.
Smouse, P.E. and Sork, V.L. (2004) Measuring pollen flow in forest trees: an exposition of alternative approaches. *Forest Ecology and Management* 197, 21–38.
Smouse, P.E., Dyer, R.J., Westfall, R.D. and Sork, V.L. (2001) Two-generation analysis of pollen flow across a landscape. I. Male gamete heterogeneity among females. *Evolution* 55, 260–271.
Sork, V.L., Nason, J., Campbell, D.R. and Fernandez, J.F. (1999) Landscape approaches to historical and contemporary gene flow in plants. *Trends in Ecology and Evolution* 14, 219–224.
Tilman, D. (1994) Competition and biodiversity in spatially structured habitats. *Ecology* 75, 2–16.
Trakhtenbrot, A., Nathan, R., Perry, G. and Richardson, D.M. (2005) The importance of long-distance dispersal in biodiversity conservation. *Diversity and Distributions* 11, 173–181.
Uriarte, M., Canham, C.D., Thompson, J., Zimmerman, J.K. and Brokaw, N. (2005) Seedling recruitment in a hurricane driven forest: light limitation, density-dependence, and the spatial distribution of parent trees. *Journal of Ecology* 93, 291–304.
Vander Wall, S.B., Forget, P.-M., Lambert, J.E. and Hulme, P.E. (2005) Seed fate pathways: filling the gap between parent and offspring. In: Forget, P.-M., Lambert, J.E., Hulme, P.E. and Vander Wall, S.B. (eds) *Seed Fate: Predation, Dispersal and Seedling Establishment.* CAB International, Wallingford, UK, pp. 1–8.
Vekemans, X. and Hardy, O.J. (2004) New insights from fine-scale spatial genetic structure analyses in plant populations. *Molecular Ecology* 13, 921–935.
Webb, C.O. and Peart, D.R. (2001) High seed dispersal rates in faunally intact tropical rain forests: theoretical and conservation implications. *Ecology Letters* 4, 491–499.
Wenny, D.G. and Levey, D.J. (1998) Directed seed dispersal by bellbirds in a tropical cloud forest. *Proceedings of the National Academy of Sciences USA* 95, 6204–6207.
Westcott, D.A., Bentrupperbaumer, J., Bradford, M.G. and McKeown, A. (2005) Incorporating patterns of disperser behaviour into models of seed dispersal and its effects on estimated dispersal curves. *Oecologia* 146, 57–67.
Willson, M.F. (1993) Dispersal mode, seed shadows, and colonization patterns. *Vegetatio* 107/108, 261–80.
Willson, M.F., Irvine, A.K. and Walsh, N.G. (1989) Vertebrate dispersal syndromes in some Australian and New Zealand plant communities, with geographic comparisons. *Biotropica* 21, 133–147.
Young, A., Boyle, T. and Brown, T. (1996) The population genetic consequences of habitat fragmentation for plants. *Trends in Ecology and Evolution* 11, 10045–10048.

13 Using Toucan-generated Dispersal Models to Estimate Seed Dispersal in Amazonian Ecuador

K.M. Holbrook and B.A. Loiselle

Introduction

Animal-mediated seed dispersal plays a significant role in plant recruitment and thus helps determine tropical forest composition (Gautier-Hion *et al.*, 1985; Willson, 1992; Howe, 1993; Finegan, 1996; Hamann and Eberhard, 1999). Although the majority of tropical trees depend on vertebrates to move their seeds (Gentry, 1982; Howe and Smallwood, 1982; Willson and Crome, 1989; Jordano, 1992; Howe and Westley, 1997), few studies have directly measured the effectiveness of seed dispersers. In a Central American forest, Howe and Vande Kerckhove (1981) examined foraging behaviour and ranked frugivore species by their propensity to move seeds away from maternal *Virola nobilis* (Myristicaceae) trees. They found that chestnut-mandibled toucans (*Ramphastos swainsonii*; Ramphastidae), which swallowed fruits whole and moved away relatively quickly from fruiting trees, were the most effective at removing seeds and were therefore likely to have had the greatest impact on plant fitness from among the assemblage of frugivores feeding on *Virola* fruits. For *V. nobilis* seeds, escaping distance- and density-dependent mortality below the maternal tree was critical to plant recruitment (Howe and Vande Kerckhove, 1981). Certain environments may be particularly favourable for seed and seedling recruitment, and any behaviours of animals that result in seeds being deposited in these more favourable environments clearly benefit plant fitness and affect disperser effectiveness. Wenny and Levey (1998) found that seedling survival was significantly higher for seeds dispersed by male three-wattled bellbirds (*Procnias tricarunculata*; Cotingidae) than for seeds dispersed by four other species of birds, due to the favourable environmental conditions found below bellbird perch sites. Thus, bellbird behaviour and movement following fruit consumption directly influenced plant recruitment patterns. An important step in understanding the role of frugivores in dispersing seeds of tropical trees is studying their seed dispersal ecology and movement patterns on different temporal and spatial scales. Here we combine observations on the foraging and movement behaviour of toucans with fruit crop data to predict spatial patterns of seed dispersal of a neotropical tree.

A seed disperser's effectiveness, measured as its contribution to plant fitness, is defined by qualitative and quantitative components of the dispersal it provides to a particular plant species (Schupp, 1993). For example, the distance a seed is moved from the plant, which is a function of animal movement behaviour, is one of the qualitative components of effectiveness; whereas the average number of seeds removed per visit, a function of animal feeding behaviour, is a quantitative component. One way to measure the effectiveness of a disperser is to estimate its contribution to a particular tree's seed shadow, which is defined as the spatial dispersion of seeds relative to a parent plant (Janzen, 1970). Seed dispersal and resultant seed shadows may influence key processes, such as colonization, population persistence (Ouborg *et al.*, 1999; Cain *et al.*, 2000), and plant population structure (Loiselle *et al.*, 1995). Movement patterns and feeding behaviours of frugivores directly impact on seed shadows (Janzen *et al.*, 1976; Fleming and Heithaus, 1981; Murray, 1988; Fragoso, 1997; Sun *et al.*, 1997; Holbrook and Smith, 2000). Frugivores that remain for long periods in fruiting trees will drop most seeds beneath a parent tree, thereby influencing the seed shadow differently than frugivores that visit trees for short periods and deposit most seeds at sites away from the parent. These differences can result in spatial variability in seed shadows, which can have consequences for seed and seedling survival and population demographics (Kwit *et al.*, Chapter 19, this volume). Further, because some frugivores, such as hornbills and toucans, travel across different habitat types including degraded and fragmented forest areas (Graham, 2001a,b; Holbrook *et al.*, 2002) they are likely to facilitate gene flow and help maintain genetic diversity of plant populations. Jordano and Godoy (2000) found high levels of genetic diversity in *Prunus mahaleb* (Rosaceae), a bird-dispersed species, suggesting that long-distance dispersal by frugivorous birds contributes to within-population genetic diversity.

Theoretically, seed shadows are expected to show a distance–density effect, in which the number of seeds declines with increasing distance away from a parent, regardless of whether the dispersal system involves animals (Janzen, 1970; Fleming and Heithaus, 1981). A dispersal kernel describes the frequency distribution of dispersal distances and is different from the seed shadow, which is the product of the dispersal curve and the total number of seeds dispersed (Levin *et al.*, 2003). Dispersal kernels have typically been described as having a leptokurtic distribution with a peak in seed density at the parent plant (Harper, 1977; Levin, 1979). Such theoretical distributions have been supported by empirical data. Studies that examine seed shadows from the plant perspective traditionally use seed traps to predict dispersal functions for plants dispersed by different vectors (typically wind or animal). These studies have revealed that dispersal curves largely match theoretical expectations and are best fit by distributions that show distance–density effects, such as exponential, lognormal, Gaussian, Student's *t*, Weibull, and inverse power functions (Hoppes, 1988; Willson, 1993; Laman, 1996; Clark *et al.*, 1999, 2005; Jones *et al.*, 2005). Studies that measure seed shadows from an animal perspective, however, have described non-leptokurtic dispersal curves, and in many cases show distributions that are not exponential in shape and thus do not match theoretical expectations (Murray, 1988; Sun *et al.*, 1997; Holbrook and Smith, 2000; Westcott and Graham, 2000; Wehncke *et al.*, 2003; Westcott *et al.*, 2005). These examples suggest a great diversity in seed dispersal patterns and potential difficulty in attributing any one form of dispersal function to a dispersal system.

Why does this apparent conflict in dispersal kernels exist between distributions generated from animal and plant perspectives? Discrepancies might simply result from the species that investigators select for study. Studies that rely on seed traps might tend to select plants that produce large crop sizes with many small seeds, while those that focus on animals might tend to select plants that produce relatively small crop sizes with nutritious, high-quality and often large-seeded fruits (Sun *et al.*, 1997; Holbrook and Smith, 2000; Westcott *et al.*, 2005). Dispersal functions for these two kinds of plants might be expected to differ regardless of the method employed to measure seed shadows. Alternatively, assumptions underlying data strictly dependent on seed traps may not always hold, and seed movement patterns might be underestimated. Recent genetic studies demonstrate that dispersal distances are often underestimated and/or that the assumption that the parent is the nearest reproductive adult to a deposited seed or seedling is frequently incorrect (H. Abe, Brisbane, 2005, personal communication; Jones *et al.*, 2005; Hardesty *et al.*, 2006). Jones *et al.* (2005) found dispersal of *Jacaranda copaia* (Bignoniaceae), a wind-dispersed tropical tree, to be more complex than can be described by a simple, single-distribution dispersal kernel. Furthermore, fitted models potentially underestimated the long-distance dispersal component of *J. copaia* (Jones *et al.*, 2005). Hardesty *et al.* (2006) suggest that even though it may appear that the majority of seeds are dropped beneath a reproductive tree, one must be cautious about assigning parentage; their genetic data clearly showed that germinated seedlings were seldom produced by the nearest, or even nearby, reproductive adults. Model simulations found that movement patterns of frugivorous birds contributed to asymmetry in fruit removal and that, at least in populations where fruiting plants were aggregated, dispersed seeds belonged to the nearest neighbour (Morales and Carlo 2006; Carlo *et al.*, Chapter 16, this volume). Finally, Steele *et al.* (Chapter 14, this volume) demonstrate that dispersal of *Quercus alba* (Fagaceae) is more extensive than behavioural studies would suggest.

Efforts to describe the seed shadows of vertebrate-dispersed plants from a frugivore perspective have focused on measuring dispersal distances and seed passage times of frugivores (Murray, 1988; Sun *et al.*, 1997; Holbrook and Smith, 2000; Westcott and Graham, 2000; Wehncke *et al.*, 2003; Westcott *et al.*, 2005). Dispersal kernels estimated from a frugivore perspective allow a better understanding of the direct contribution of particular seed dispersers to observed seed deposition patterns. Frugivore-generated dispersal kernels also allow estimation of the dispersal tail, which can reflect rare, but important, long-distance dispersal events (Cain *et al.*, 2000; Holbrook and Smith, 2000).

Here we estimate dispersal kernels for *Virola flexuosa* (Myristicaceae) that have been generated by toucans. Although some studies have focused on seed dispersal by toucans (Howe *et al.*, 1985; Wenny, 2000), none has examined dispersal kernels for a particular toucan species. Furthermore, there are no studies that integrate frugivore-generated dispersal kernels estimated from movement data and seed passage times with foraging behaviour of the disperser and fecundity of the plant. We combine our frequency distribution of dispersal distances with information on fruit removal by toucans and crop size of individual trees to present a spatially explicit model that predicts the pattern of seed dispersal for *V. flexuosa* within a large-scale plot at the individual tree and population levels. We refer to the frequency distribution of dispersal distances generated by toucans as dispersal kernels, and the spatially explicit models of seed dispersion patterns as seed shadows, recognizing that the latter are not complete seed shadows. Specific objectives were:

1. To estimate seed dispersal distances for a neotropical tree, *V. flexuosa*, based solely on toucan movements and seed retention times;
2. To present a spatially explicit model, which more realistically 'outlines' the dispersion patterns generated by toucans.

Methods

Study site and species

This study was conducted at the Tiputini Biodiversity Station (Tiputini: 0°38′ S, 76°09′ W) and Yasuní Research Station (Yasuní: 0°40′ S, 76°24′ W) in Orellana Province, Ecuador, from 2001 to 2005. Ecuador is extremely biodiverse, hosting 6% of the world's vascular plant species (>16,000; Jørgensen and Leòn-Yánez, 1999) and 15% of the world's bird species (>1500; Parker *et al.*, 1996). Both stations (approximately 27 km apart), located in the Yasuní Biosphere Reserve (1.5 million ha) in equatorial western Amazonia are floristically similar, with elevations ranging from approx. 200 to 245 m. The reserve represents the largest protected area of mature forest in the Amazon region of Ecuador. The vegetation is evergreen, lowland rainforest and the area receives >3000 mm of rain each year (Jørgensen and Leòn-Yánez, 1999), with the majority of rainfall occurring October to November and May to July (J. Guerra, 2005, unpublished results; Valencia *et al.*, 2004). The presence of canopy towers at Tiputini and Yasuní, access to office and laboratory space, herbarium collections, and existing trail systems at both sites, provided an excellent setting for this study. Additionally, Tiputini has two 100-ha plots (see Loiselle *et al.*, 2007) and Yasuní has a 50-ha Forest Dynamics Plot (FDP) (see Valencia *et al.*, 2004). Research was conducted primarily in the FDP at Yasuní and within 50 ha of one of the 100-ha plots at Tiputini; both study plots are located in mature terra firme forest. All stems >10 cm diameter at breast height (dbh) in the FDP at Yasuní have been mapped and identified (Valencia *et al.*, 2004). In addition, we searched for and mapped all adult *V. flexuosa* in a 100-m strip (34 ha) surrounding the existing FDP. At Tiputini, we located and mapped all adult individuals of *V. flexuosa* within 84 ha, comprising the 50-ha plot mentioned above and the surrounding 34 ha. Data on toucan foraging and movement behaviour were collected at both Yasuní and Tiputini. Data on the fecundity of individual trees were collected only at Tiputini and subsequent dispersal models are presented for Tiputini only. A total of seven female, 14 male and four unknown *V. flexuosa* adults are shown in the plot at Tiputini (Fig. 13.1). Unknown individuals were of reproductive size (>30 cm dbh), but did not produce flowers or fruit during the study period.

This study focuses on toucans (Ramphastidae) because they are important members of seed-disperser communities and are found throughout the Neotropics. Moreover, although toucans are large conspicuous

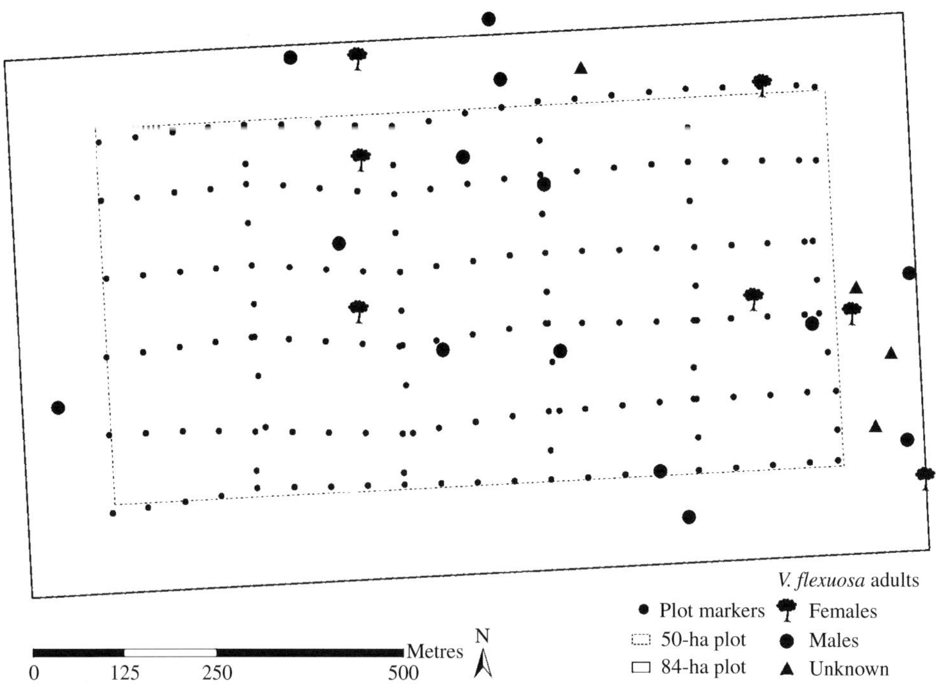

Fig. 13.1. Tiputini plot showing all adult *V. flexuosa* located within the 84-ha study area. The inner 50-ha plot is the site of our original study; the plot was expanded to include all adult *V. flexuosa* within a 100-m strip surrounding the 50-ha plot.

members of the canopy and subcanopy bird community, their ecology has been relatively little studied. The many-banded araçari (*Pteroglossus pluricinctus*), white-throated toucan (*Ramphastos tucanus*) and channel-billed toucan (*R. vitellinus*) are ideal study species because they are abundant in Amazonia and present an opportunity to compare seed dispersal by toucans that differ in size and probably differ in diet, movement patterns, and seed dispersal ecology. *Virola flexuosa* A.C. Sm. (Myristicaceae) is a dioecious, shade-tolerant species widespread throughout South America (Lambright, 1981). In eastern Ecuador, *V. flexuosa* is regularly consumed by toucans (Holbrook, 2006) and is likely to be important in toucan diets, as are other species in the genus (Howe *et al.*, 1985; Galetti, 2000). For example, Galetti (2000) found 11–38% of all foraging observations by *Ramphastos* toucans were concentrated on *Virola*. Howe *et al.* (1985) found that dispersers of *V. nobilis* were restricted to a small, specialized group of frugivores, of which toucans were shown to disperse the majority of seeds away from parent trees. *Virola flexuosa* is relatively easy to identify in the field, in both fertile and sterile condition, and the fruit is distinctive with a bright-red aril enclosed by a capsule, which dehisces during early and mid-morning hours (K.M. Holbrook, 2002, personal observation).

Estimating seed shadows: field methods

Fruit removal

Tree watches were conducted throughout the fruiting season (November–March, 2002–2004) to determine the relative role that toucans play in dispersing the seeds of *V. flexuosa*. A total of 13 focal female trees at Tiputini and Yasuní were observed from 06:00 to 10:00 hours, with a minimum of eight replications per tree. All visiting frugivores were identified during tree-watch observations. We recorded the number of visits, total number of fruits eaten and removed per visit, and time spent foraging for each visit. In addition, for each frugivore species, we recorded the number of seeds ingested, regurgitated and dropped or knocked down from the canopy. Fruit removal was defined as the number of seeds swallowed and taken away from the fruiting tree.

Crop size

Seed traps (1 m^2) made of PVC tube and mosquito screen were placed at randomly selected points between the tree bole and the edge of the crown of each focal tree, in order to estimate crop size. We used enough traps (4–6) to cover approximately 5% of the area below each crown. Each fruit consists of a capsule enclosing an arillate seed, with the arillate seed being the unit of dispersal. Following Howe and Vande Kerckhove (1981), capsules collected in the traps provided an estimate of the total crop matured. Specifically, we counted the number of capsules collected in seed traps and divided by the proportion of the canopy area sampled by traps. To estimate crop size, we used seed-trap data from nine female trees at Tiputini. Seven trees were located within the 84-ha plot area (Fig. 13.1) and two trees were located 1.9 and 3.4 km from the centre of the 50-ha plot.

Seed retention

We determined seed retention times with captive toucans and use these data in combination with movement data to estimate toucan-generated dispersal kernels (see below). Seed passage trials were conducted both in the field in Ecuador and with captive toucans at the Saint Louis Zoo, USA. For field trials, four individuals of *P. pluricinctus* were held captive for up to 2 days in cages constructed of flexible nylon mesh (1 × 1 × 2 m) (Santana *et al.*, 1986). We marked seeds from ripe *V. flexuosa* fruits with short strands of cotton thread for easy discrimination between individual seeds and to differentiate any that may have remained in the gut from feeding in the wild. Seeds were placed in small cubes of papaya (*Papaya carica*; Caricaceae) or a supplemental, gel-like bird food designed for frugivores (PurinaMills; J. Dempsey, St Louis, 2003, personal communication) and were provided from 06:00 to 17:00 hours. In addition to papaya and bird gel with seeds, we provided locally available fruit *ad libitum*, including papaya, watermelon (*Citrullus lanatus*; Cucurbitaceae), and cantaloupe (*Cucumis melo*). Each bird was observed continuously between 06:00 and 18:00 hours from behind a blind near the cage. At the St Louis Zoo, trials were conducted with four *Ramphastos* individuals (one *R. tucanus*, one *R. vitellinus* and two *R. toco*). We placed seeds of *V. flexuosa* (collected and frozen in the field) into fruit pieces (papaya and grapes (*Vitis vinifera*; Vitaceae)) that were provided from 08:00 to 14:30 hours. We observed birds continuously from outside their cage between 08:00 and 17:00 hours. In all trials, seeds were recorded immediately after regurgitation or defecation and retention times calculated. We tested for differences in seed retention times between species using a Mann–Whitney *U* test. Retention data from all three species of *Ramphastos* were used in the estimation of seed shadow models for the genus *Ramphastos*.

Movement patterns

To determine toucan movement patterns, we radio-tracked three species of toucans during two 3-month and two 8-month field seasons (2001–2005) following methods employed in Cameroon by Holbrook and Smith (2000). We captured a total of 25 *P. pluricinctus*, two *R. tucanus*, and two *R. vitellinus* at Tiputini and Yasuní, using canopy nets at fruiting trees and/or nesting sites. We attached radio-transmitters (6 g each; Holohil Systems Ltd, Carp, Ontario, Canada) at the base of the central tail feathers. Body mass of tagged birds ranged from 195 to 684 g, resulting in the transmitter weighing less than 3% of the bird's mass. Transmitter attachment did not appear to affect bird movements. Toucan locations were measured by triangulation using receivers and hand-held two-element Yagi antennas (White and Garrott, 1990; Kenward, 2001). Tracking stations were located on permanent canopy towers, temporary canopy platforms, and ground positions, allowing for minimal error in location data. Station positions were determined using a global positioning system (GPS). Three observers, using two-way radios, collected simultaneous bearings approximately 2 days per week at each site over a period of 3–8 months during each of four field seasons. Tracking periods lasted 4–6 daylight hours, with individual birds located every 15 min. We chose a 15-min interval period to ensure that we could collect bird locations within the time-frame of seed passage or regurgitation, which were expected to be longer than 15 min (Sun *et al.*, 1997; Holbrook and Smith, 2000). In addition, we dedicated several tracking days to following individually tagged birds in order to collect detailed movement and location data, such as tree-to-tree movements and cavity roost locations. These detailed movements complemented the location data collected through triangulation, and were used to supplement data collected for the calculation of toucan-generated dispersal kernels. Bird locations were estimated through triangulation using the program LOAS 2.03 (Ecological Software Solutions). Because of the small sample size for *Ramphastos* we combined movement data from both species to estimate seed shadows for the genus *Ramphastos*; our expectation was that home ranges for the two species would be similar, based on similar territory sizes reported by Terborgh *et al.* (1990).

Estimating seed shadows: dispersal models

Probability of seed deposition

Following Murray (1988) and Holbrook and Smith (2000), we estimated toucan-generated dispersal kernels ('seed shadows' in Murray, 1988; Holbrook and Smith, 2000) using seed retention times and movement data from radio telemetry. Location accuracy using radio telemetry may be reduced in tropical forests, due to wet habitats, thick vegetation and uneven topography (Zimmerman and Powell, 1995; Millspaugh and Marzluff, 2001). Because of this, we used conservative distance categories of 100-m increments to better represent the precision of our telemetry data. We chose time categories (e.g. 15, 30, 60 and 90 min) for the model, based on seed-retention trials. Within each of these time categories, distances (e.g. 100, 200 and 300 m) moved by toucans were grouped and we calculated the probability of movements made within each distance category, within each time category. These are summed across each time category to give a final probability for each distance category. The final calculation is:

$$p_d = \sum_t (a_{dt} \cdot b_t) \tag{13.1}$$

where p = probability of a seed being deposited at a particular distance category (d) from the parent tree, a = probability of a bird being within a particular distance category (d) in time interval (t), and b = probability of a seed being passed in that time interval. We then plotted p against distance to give a probability of seed deposition at various distances. In each case, the first location of the day was considered the 'origin' or parent tree in the model.

Spatially explicit models

Probabilities of seed deposition for each toucan species (Eqn 13.1) were combined with *V. flexuosa* fruit removal data to more realistically estimate seed shadows of *V. flexuosa*. Using a spatially explicit model, all adult female trees (located by x, y coordinates and entered in a GIS database) in the 84-ha plot at Tiputini served as the origin for calculations in the model. Thus, seed dispersal predictions were limited to the scale of the 84-ha plot, which was chosen because we expected this area to encompass the major proportion of a toucan home range, where most seeds would probably be dispersed.

For model estimates, we divided the 84-ha plot into 5 × 5 m cells, each assigned a value for the distance to each adult female tree in the plot. Using *reclassify* in Spatial Analyst (ArcGIS 9.1; ESRI, Inc), a probability of dispersal for each toucan species (p, Eqn 13.1) was then assigned to each cell based on the distance from each female tree that cell falls into (i.e. <100, 100–200, etc.). The estimated dispersal distances that seeds are dispersed were based on the toucan models and were assumed to be the same for each female tree. We then multiplied each set of probabilities by the number of seeds dispersed away from each tree based on observation data (these varied depending on crop size) to provide tree-specific seed shadows. A population-level seed shadow was then generated by adding individual tree shadows on the 84-ha plot. The equation for the final summed model is written as:

$$N_x = \sum_m (p_{xm} \cdot r_m) \tag{13.2}$$

where N = the number of seeds predicted to fall at a particular location (x, individual 5 × 5 m cells within the 84-ha plot), p = the probability of seed deposition at varying distances (x) from each female tree (m), and r = number of fruit removed at each female tree. The numbers of seeds dispersed to each location are reported as the proportion of total *Virola* seeds estimated to be dispersed by toucans in the landscape. This method allows us to draw the 'individual tree seed shadows' that together define the population model. We examined differences in the final summed seed shadows generated by *Ramphastos* and *Pteroglossus* toucans by using GEODIST.BAS, a QuickBASIC 4.5 program that tests for differences between two spatial distributions, based on a modified Cramér–von Mises non-parametric test (Syrjala, 1996). This non-parametric test evaluates the differences between the spatial distributions of two populations. Specifically, we compared the spatial distribution of seed deposition by

Ramphastos with the seed deposition by *P. pluricinctus* by generating 200 random points within the study plot. The minimum distance selected between points was 5 m to account for our 5 × 5 m cell size. Each point was associated with *x*, *y* coordinates and a 'value' corresponding to the relative number of seeds dispersed by *Ramphastos* or *P. pluricinctus*. We tested for differences in the resulting spatial distribution of seeds predicted to be dispersed by *Ramphastos* and *P. pluricinctus* at these 200 random points using 1000 permutations to generate the Cramér–von Mises test statistic. A significant test statistic indicates that the seed deposition pattern produced by *Ramphastos* differs from that produced by *P. pluricinctus*.

Results

Fruit removal and seed retention

More than 400 hours of observations at 13 individual fruiting trees indicate that *Ramphastos* and *P. pluricinctus* represent 64.3% of visits and remove more than 52.0% of dispersed seeds (Table 13.1). To calculate the number of seeds dispersed from individual *Virola* trees, the proportion of seeds removed was multiplied by crop size, which varied widely from 183 to 39,100 fruits (mean ± SD, 14,408 ± 15,526). Multiplying by crop size allowed us to better estimate removal over the entire fruiting period (mean ± SD, 78 ± 25 days), rather than restricting removal to times when foraging behaviour data were collected. The resulting mean (SD; range) number of seeds dispersed per tree was 1689 (2022; 22–3779) for *P. pluricinctus* and 5503 (6589; 73–15,576) for *Ramphastos*. The level of variation in crop size observed here is not unusual for tropical trees (Howe and Vande Kerckhove, 1981; Russo, 2003; Clark *et al*., 2005). Seed retention experiments revealed little difference between *Ramphastos* and *Pteroglossus* toucans (Fig. 13.2), with the average seed retention time approximately 30 min. Most seeds ingested by toucans were regurgitated (97% for *P. pluricinctus* and 91% for *Ramphastos*); some in as little as 5–10 min (n = 19) and others in 60–90 min (n = 14; N_{tot} = 141). A small percentage (5%; n = 7) were retained in the gut for around 100–140 min by *Ramphastos*, suggesting a trend for longer retention times by the larger *Ramphastos* species.

Table 13.1. Visitation and seed removal of *Virola flexuosa* by *Ramphastos* and *Pteroglossus*. *Ramphastos* includes *R. tucanus*, *R. vitellinus*, and *Pteroglossus* includes *P. pluricinctus*.

Bird species	Percentage of total visits (*n*)	Mean ± SD number of seeds removed per visit (*n*)	Percentage of total seeds removed (*n*)
Ramphastos	47.6 (234)	1.6 ± 1.9 (115)	39.8 (115)
Pteroglossus	16.7 (82)	1.4 ± 1.2 (8)	12.2 (8)
Other visitors	35.7 (176)	2.8 ± 5.0 (64)	48.0 (64)

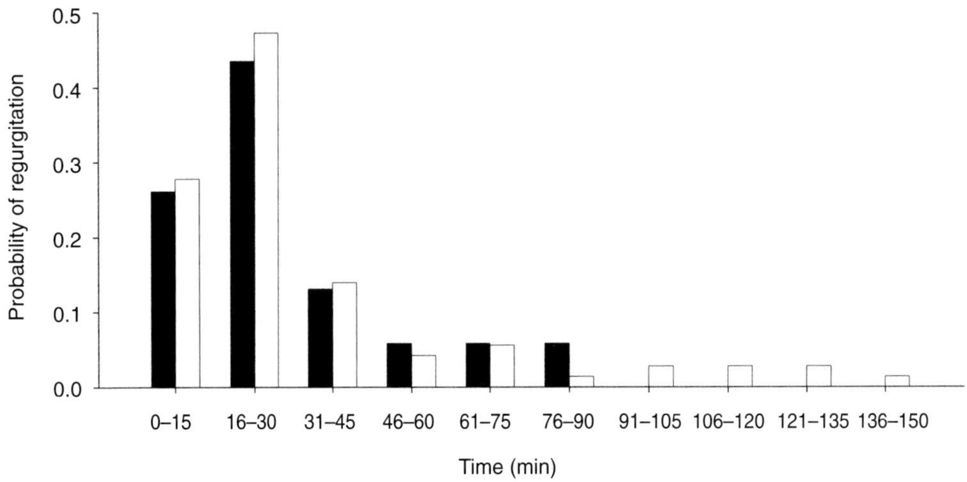

Fig. 13.2. Seed retention times of *V. flexuosa* by *P. pluricinctus* (black bars) and *Ramphastos* spp. (*R. tucanus*, *R. toco*, *R. vitellinus*) (white bars). Trials were conducted with four individuals in each group. Mean (SD) retention times were 28 (20) and 34 (32) min for *P. pluricinctus* and *Ramphastos*, respectively. Comparisons between *P. pluricinctus* and *Ramphastos* (grouped due to equivalent size) showed no significant differences (Mann–Whitney $U = 2544$, $P = 0.486$). Mean body mass for *P. pluricinctus* and *Ramphastos* were 288 and 566 g, respectively.

Movement patterns

We tracked radio-tagged individuals for 3–25 days with the number of locations ranging from 11 to 185. Because birds were radio-tracked using a triangulation method, we were not able to collect continuous movement data on each bird. Rather, data represent locations collected in 15-min intervals. *Ramphastos* and *P. pluricinctus* had significantly different mean (SD) home ranges of 86 (25) ha and 191 (64) ha for *Ramphastos* and *P. pluricinctus*, respectively ($F_{(1, 13)} = 5.36$, $P = 0.049$) (see Holbrook, 2006). These home-range estimates were based on individuals with more than 40 locations recorded. However, movements from all radio-tracked individuals were used for estimating dispersal kernels. Distances travelled per movement bout ranged from 0 to >2000 m (Fig. 13.3). Time calculated between locations varied from 15 to 165 min and from 15 to 135 min for *Ramphastos* and *P. pluricinctus*, respectively. These distributions are strongly leptokurtic, with most movements being <300 m. A greater proportion of *Ramphastos* (28%) versus *P. pluricinctus* (16%) movements were within 100 m, which is reflected in the overall shape of dispersal kernels. The longest recorded movements (both recorded within 30 min) were 3027 and 3665 m for *Ramphastos* and *P. pluricinctus*, respectively.

Supplementary tracking data, in which we directly followed tagged birds for up to 5 h, were collected on more than 20 different days. During

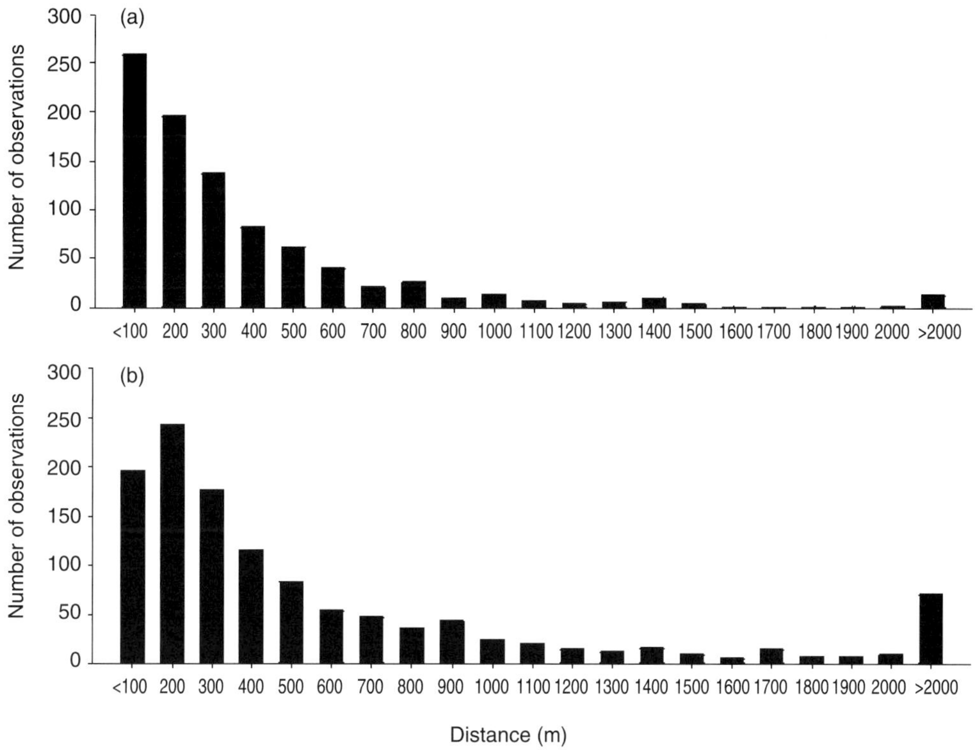

Fig. 13.3. Distances travelled per movement bout for (a) *P. pluricinctus* and (b) *Ramphastos*. These include all recorded movements between temporally consecutive locations. Although most observations were collected within the 15-min time category (*P. pluricinctus*, $n = 401$, $N_{tot} = 1225$; *Ramphastos*, $n = 287$, $N_{tot} = 915$), many locations were collected at longer intervals due to an occasional loss of transmitter signal during telemetry sessions.

these tracking periods, we never observed toucans to travel from one fruiting *V. flexuosa* to another in the same tracking period, suggesting that dispersal to conspecific individuals is rare.

Dispersal models

Dispersal kernels predict that *Ramphastos* and *P. pluricinctus* disperse 72% and 84% of seeds, respectively, >100 m away from parent trees (Fig. 13.4). The shapes of dispersal kernels differed significantly, with the peak of the distribution occurring at <100 m for *Ramphastos*, but between 100 and 200 m for *P. pluricinctus*. The dispersal kernel generated by *Ramphastos* appears to be an exponential distribution, with the peak located within 100 m of the source, while the kernel for *P. pluricinctus* suggests a χ^2 distribution. Both dispersal kernels exhibit very long thin tails, suggesting long-distance dispersal, similar to those reported for the southern cassowary

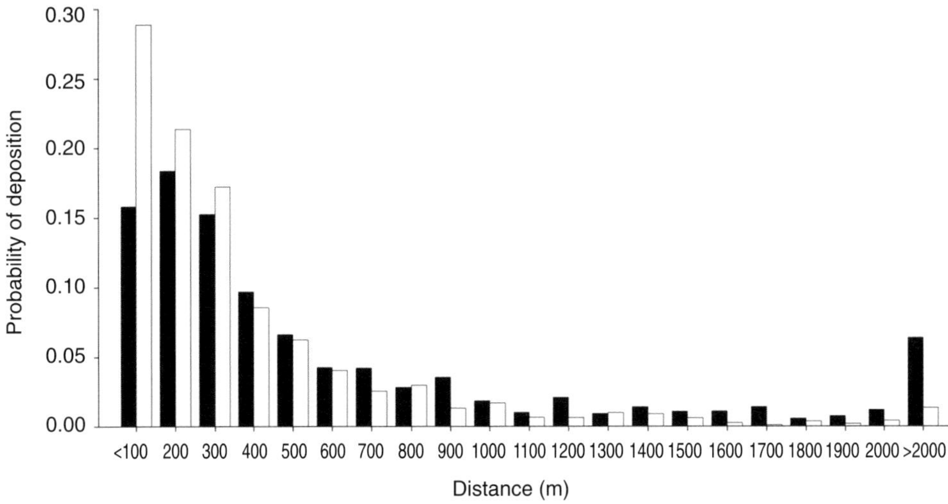

Fig. 13.4. Estimated dispersal kernels for *V. flexuosa* produced by *P. pluricinctus* (black bars) and *Ramphastos* (white bars) toucans. These kernels represent two-dimensional models at the individual tree level. Differences between dispersal kernels were significant (Kolmogorov–Smirnov; $Z = 1.651$; $P = 0.009$).

(*Casuarius casuarius*; Casuariidae) in Australia (Westcott *et al.*, 2005) and African hornbills (*Ceratogymna atrata* and *C. cylindricus*; Bucerotidae) in Cameroon (Holbrook and Smith, 2000).

The spatially explicit models of seed shadows generated from our combined data on plant fecundity, fruit removal rates and dispersal kernels are presented in Fig. 13.5. Using these population-level seed shadows, we predicted that some areas in the landscape were more likely to receive seed-fall than others. The areas with the highest density of seed rain occurred around trees with the highest fecundities and in areas where individual seed shadows overlapped (Fig. 13.5). These regions of high seed deposition also probably represent areas of seed mixing, where dispersed seeds represent a genetic mix of all the adult trees in the study area. Our estimates indicate that seed mixing occurred throughout the study area, with each location receiving different probabilities of seed-fall from different adults, depending on how the toucans moved through space and the fecundity of individual trees. Note that the estimated percentage of seeds dispersed by *Ramphastos* or *P. pluricinctus* represents the proportion of the total seeds dispersed by that species. Most reproductive *V. flexuosa* do not produce fruit every year and there appears to be a general trend towards fruit production every second or third year (K.M. Holbrook, 2006, unpublished results). Consequently, we cannot evaluate whether the differences in fruit production among maternal trees observed during the course of this study will hold over time. Therefore, we restrict the spatial depiction to this time period and do not make any assumption about future years.

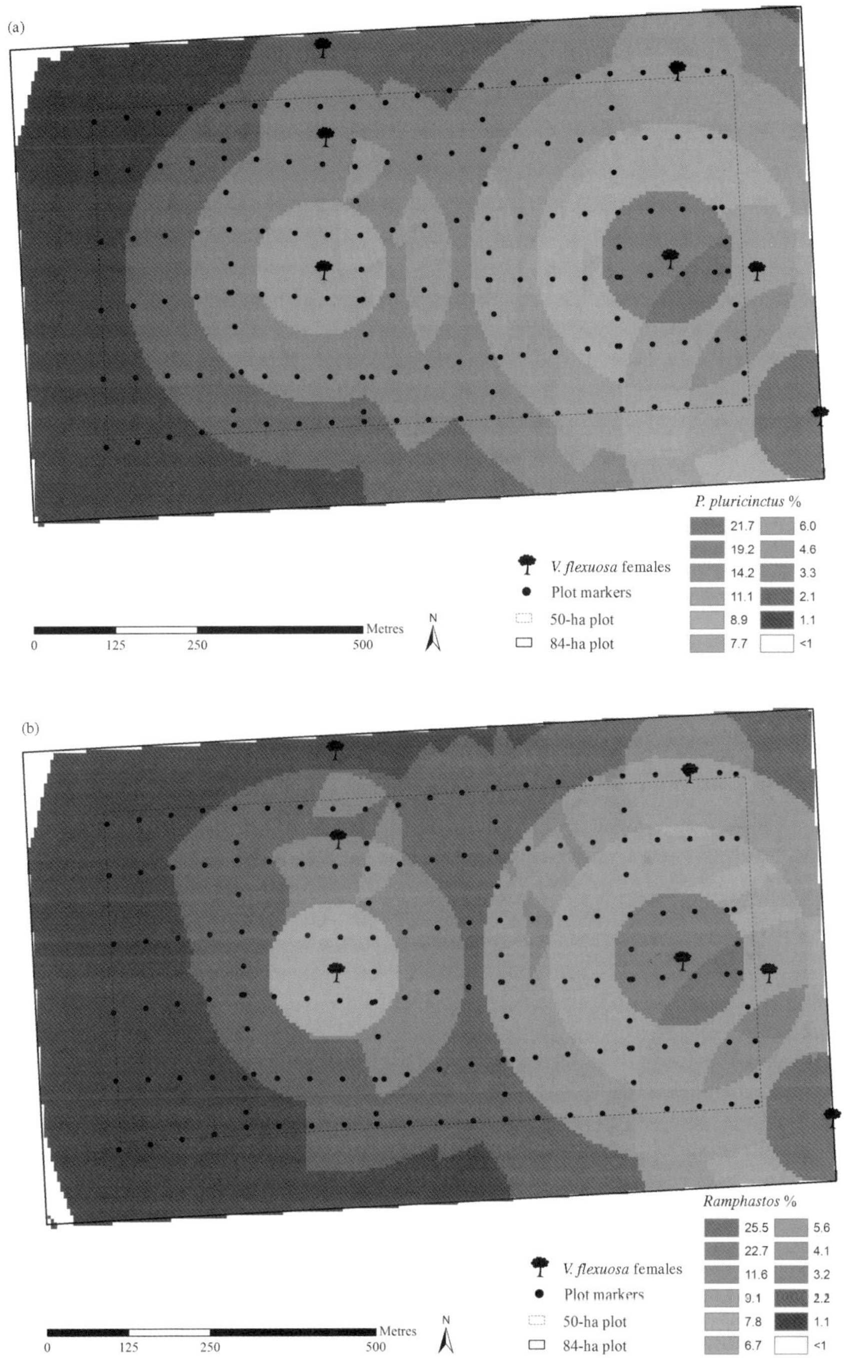

Fig. 13.5. Probability of seed deposition of *V. flexuosa* by (a) *P. pluricinctus* and (b) *Ramphastos*. Seed shadows represent data from toucan-generated model, fruit removal, and crop size of *V. flexuosa*. Differences between distributions are significant (Cramér–von Mises; $\psi = 2.461$; $P = 0.001$). Only female trees are shown.

Discussion

Due in large part to the great diversity of animal dispersers in tropical forests, understanding and predicting seed dispersal patterns have been difficult. In this chapter, we integrate animal behaviour and seed dispersal processes across temporal and spatial scales. In doing so, we have improved upon a two-dimensional dispersal kernel by incorporating aspects of disperser behaviour and crop size of individual female trees (Fig. 13.5). The resulting spatially explicit model of seed dispersion is more realistic in depicting patterns of seed deposition than those using just frugivore-generated dispersal kernels. We predicted that *V. flexuosa* seeds would land in a patchy fashion on our study site, with greater numbers of seeds being deposited in overlapping areas between fruiting trees. At the population level, trees with the highest fecundities clearly drive much of the seed deposition pattern. We observed the highest density of seed-fall in the south-eastern portion of the study plot, where the more fecund trees were located at the time (Fig. 13.5). Our results have demonstrated that patterns of behaviour, including disperser movements, seed retention times, foraging behaviour, in addition to plant parameters such as crop size, can significantly impact the shape and scale of dispersal kernels and the patchy nature of seed shadows.

While we have attempted to incorporate several of the many factors contributing to generating a seed shadow, we recognize that others remain to be incorporated. For example, our models do not detail specific repeated movements of toucans to dispersal foci, such as roost sites, nest cavities or other fruiting trees. A closer examination of movement behaviour may indicate that toucans exhibit measurable non-random movement (see Kwit *et al.*, Chapter 19 this volume). In addition, seed dispersal studies require integrating processes across a wide range of scales. For example, other species contribute to a total seed shadow for *V. flexuosa*, which may change the scale at which a seed shadow is investigated. Although 84 ha represents a large proportion of a toucan home range, it may not be large enough to encompass movement patterns of all dispersers of *V. flexuosa*. In addition, our seed shadow estimates do not incorporate seed rain from adult trees outside the 84-ha area. Furthermore, individual plant recruitment operates at a much finer scale and our study does not address environmental heterogeneity on the study plot at this scale.

Shape and scale of dispersal

The shape and scale of dispersal kernels can be affected by a number of attributes of the process. For example, in simulation models, Morales and Carlo (2006) found the degree of plant aggregation to influence scale, while shape was mostly dominated by frugivore density: overall, mean dispersal distances were reduced as the spatial aggregation of plants increased. In contrast, Westcott and Graham (2000) found that gut passage rate determined the overall shape of the dispersal kernel for an understorey flycatcher (*Mionectes oleaginous*; Tyrannidae), while movements determined scale.

How long seeds are retained by fruit consumers potentially has strong impacts on animal-mediated seed shadows (Westcott and Graham, 2000; Morales and Carlo, 2006; Carlo *et al.*, Chapter 16, this volume). For example, seeds with longer retention times tend to be dispersed farther away from the parent tree and may also be deposited in a greater variety of locations and habitats (Sun *et al.*, 1997). In our retention trials, nearly all *V. flexuosa* seeds were regurgitated within 60 min. However, a small proportion of seeds were defecated and had passage times exceeding 100 min. In addition, there was a trend for the larger *Ramphastos* to have longer retention times than the smaller *P. pluricinctus*. This trend for larger species to have longer seed retention was found in African hornbills (Holbrook and Smith, 2000) and in a range of small neotropical frugivores (Levey, 1986). Avoiding ballast is largely driven by the size of the bird and is likely to shorten the scale of a dispersal kernel. *Pteroglossus pluricinctus* would be more likely to regurgitate large seeds than *Ramphastos* in order to avoid ballast, due to its smaller size (see Levey, 1986). Despite these potential differences, the toucans investigated here demonstrated statistically similar patterns of seed retention.

Due to similar retention times of seeds, differences in dispersal kernels generated by *P. pluricinctus* and *Ramphastos* were largely the result of different movement patterns. Home-range use of *P. pluricinctus*, which forages in family (or single-species) flocks, was nearly twice as large as that of *Ramphastos*. This suggests that *P. pluricinctus* has the ability to move seeds a greater distance than *Ramphastos*. Within 30 min (mean seed retention time), the average distance moved by *P. pluricinctus* and *Ramphastos* was 528 and 348 m, respectively. These dispersal distances are similar to those recently reported for *C. casuarius*, where the average dispersal distance from the parent plant was 337 m (Westcott *et al.*, 2005).

Although many dispersal studies suggest that dispersal kernels peak at the parent plant or fit a negative exponential distribution (Portnoy and Willson, 1993) or some combination of distributions (Clark *et al.*, 1999; Nathan and Muller-Landau, 2000; Clark *et al.*, 2005; Jones *et al.*, 2005), many vertebrate-generated dispersal kernels do not exhibit this pattern (Murray, 1988; Sun *et al.*, 1997; Holbrook and Smith, 2000; Westcott and Graham, 2000; Wehncke *et al.*, 2003; Westcott *et al.*, 2005; but see Dennis and Westcott, Chapter 9, this volume). We found that *P. pluricinctus* dispersal kernels, like many other vertebrate-generated kernels, were not highly leptokurtic and most seeds were not dispersed directly underneath or within metres of a parent tree (Fig. 13.4). Instead, a large proportion of seeds were moved a considerable distance away. *Ramphastos*, on the other hand, did exhibit a roughly negative exponential curve. Our observations at fruiting trees suggest that *P. pluricinctus* arrived, foraged for several minutes, and departed without returning to the tree in that observation block (4 h). *Ramphastos* often arrived to forage for several minutes, departed to adjacent trees and returned. This behaviour was recorded with some frequency and appears to be reflected in radio-tracking data, further supporting the observed differences between species. Both curves, however, had distributions with long thin tails, indicating that some seeds would be dispersed over long distances.

Accurately measuring the frequency and scale of long-distance seed dispersal is well understood to be a difficult process (Clark *et al.*, 1999; Cain *et al.*, 2000; Nathan and Casagrandi, 2004). In Panama, Jones *et al.* (2005) found considerable uncertainty in the tail of the distribution after fitting several dispersal models to the seed dispersal data of a wind-dispersed tropical tree. Nevertheless, seed dispersal studies are challenged to provide measures of long-distance dispersal because of its likely importance in influencing recruitment processes and determining plant population structure. Long-distance dispersal probably allows for greater gene flow and genetic mixing (Schupp, 1993; Hamilton, 1999; Shilton *et al.*, 1999). In addition, longer dispersal distances make it possible for the offspring of a single parent to sample a larger area, thus reducing kin competition and effecting bet-hedging over a greater number and variety of environments (Muller-Landau and Hardesty, 2005). At the population level, longer dispersal distances reduce dispersal limitation (Clark *et al.*, 1998; Nathan and Muller-Landau, 2000). Finally, long-distance dispersal may facilitate the arrival of rare species and colonization of new sites (Holbrook and Smith, 2000; Holbrook *et al.*, 2002).

The effectiveness of toucan dispersal

Toucans appear to be very effective dispersers for *V. flexuosa* at Tiputini. Our results indicate that *Ramphastos* and *P. pluricinctus* toucans effectively disperse seeds beyond the canopy of fruiting trees, with *P. pluricinctus* dispersing seeds farther than *Ramphastos* (Fig. 13.4). Despite fairly short seed retention, which could lead to many seeds being deposited beneath a parent, visits to fruiting trees were short and movement rates were such that seeds were brought well beyond the parent canopy. Dispersal to fruiting conspecifics is unlikely, as the chance of a toucan travelling from one conspecific to another is limited by: (i) *V. flexuosa* being rare, with a mean 3.6 adult female trees per 50 ha; (ii) toucan *core* home ranges ranging from 19 to 28 ha, incorporating only two trees; and (iii) the timing of peak fruiting for individual trees frequently being offset (K.M. Holbrook, 2006, unpublished data). Recent studies on Barro Colorado Island, Panama, found that negative density-dependent recruitment contributed to an increase in diversity when comparing seeds to seedling recruits, confirming the importance of distance and/or density-dependent mortality (Harms *et al.*, 2000). The extensive seed shadows reported here suggest that toucans may decrease density-dependent seed and/or seedling mortality by transporting seeds away from parent plants (Janzen, 1970; Connell, 1971) and creating large areas of low-density seed populations (Fig. 13.5).

Unlike what has been found in many theoretical dispersal kernels, most *V. flexuosa* seeds escape the immediate vicinity of the maternal tree. But what is an effective dispersal distance? Howe *et al.* (1985) found a 44-fold increase in survival of seeds moved greater than 45 m from parent *Virola* trees. Although we do not have seed survival data on *V. flexuosa*, seedling transects conducted at fruiting trees found greater numbers of larger-sized seedlings (>30 cm) beyond 40 m, while smaller-sized seedlings (<30 cm) were found only up to 40 m from trees (K.M. Holbrook, 2006, unpublished data); a pattern that may reflect differential survival in seedlings. Therefore, seeds dispersed beyond 100 or 200 m are likely to be well beyond the distance at which seed and/or seedling survival increases significantly.

The dispersal kernels we estimated for toucans highlight the frequency of long-distance dispersal. Seeds dispersed by toucans may be transported to locations several kilometres away. Although the mean dispersal distance for *P. pluricinctus* was 560 m and 338 m for *Ramphastos*, we recorded individual movements up to 3027 m and 3665 m for *Ramphastos* and *P. pluricinctus*, respectively. A toucan can potentially fly several kilometres within the average retention time of a *V. flexuosa* seed, suggesting that they are capable of dispersing seeds very long distances, particularly individuals that travel outside their core home range.

The importance of toucans as seed dispersers is underscored by their primarily frugivorous diet (Remsen *et al.*, 1993; Galetti, 2000), probable gentle treatment of seeds (Whitney *et al.*, 1998, and references therein), foraging behaviour at fruiting trees, and relatively large home ranges. Furthermore, toucans readily cross open habitat and secondary forest (Graham, 2001a,b), suggesting their importance in facilitating gene flow and forest regeneration. Toucans are also considered to be scatter-dispersers (Howe, 1989), which is likely to be beneficial to plant fitness by lowering mortality caused by density-dependent processes and distributing seeds to more environments, thus enhancing recruitment probabilities.

Conclusions

As reviewed by Harper (1977) and Wang and Smith (2002), many factors interact to determine the density and dispersion patterns of plant populations (see also Russo *et al.*, Chapter 23, this volume). Seed dispersal is only the first step in the process. Ultimately, the spatial and temporal distribution of 'suitable sites' determines seedling establishment patterns (Schupp, Chapter 20, this volume). Although estimating dispersal kernels from a frugivore perspective is limited by the ability to track frugivores throughout a landscape, these models allow us to more accurately quantify seed and seedling shadows and have great potential to be combined with genetic-based models. In the future we intend to address questions of dispersal limitation using both ecologically based and genetic-based models of seed dispersal. This combination will further our understanding of how frugivore behaviour and dynamics influence plant recruitment and population structure.

Acknowledgements

We thank many great field assistants, including C. Cardoni, D. Cotacachi, C. Egan, M. Euaparadorn, C. Fairbanks, J.E. Guevara, E. Guevara, J. Klavins, M. McCaustland, A. Norris, M. Ogburn, A. Van Gunst and A. Wolf, for generous help in the field, and extend a special thank-you to the staff and scientists of Yasuní Research Station and Tiputini Biodiversity Station, especially Jaime Guerra, David Romo, Kelly Swing and Consuelo de Romo. In addition, for support throughout this project, we thank Pontifica Universidad Católica del Ecuador, Universidad San Francisco de Quito and the Smithsonian Tropical Research Institute. For permission to work in Ecuador, we thank the Ministry of Environment (No. 22-IC-FAU-DFN). We gratefully acknowledge funding support from the Chapman Memorial Fund, Cleveland Park Zoo, US Environmental Protection Agency-STAR Fellowship, Fulbright Fellowship, International Center for Tropical Ecology, Lincoln Park Zoo, National Science Foundation Dissertation Enhancement Grant (OISE-0400971), Organization for Tropical Studies, PEO Scholar Award, St Louis Zoo, St Louis Rainforest Alliance and the Wildlife Conservation Society. Finally, we sincerely thank those who organized the 2005 *Frugivory and Seed Dispersal Symposium/Workshop*, especially Andrew Dennis, David Westcott, Ronda Green and Eugene Schupp, and the National Science Foundation OISE programme, which supported K.M.H.'s travel to Australia. David Westcott, Tomás Carlo and Andrew Dennis helped us improve the manuscript with their insightful and detailed comments.

References

Cain, M.L., Milligan, B.G. and Strand, A.E. (2000) Long-distance seed dispersal in plant populations. *American Journal of Botany* 87, 1217–1227.

Clark, C.J., Poulsen, J.R., Bolker, B.M., Connor, E.F. and Parker, V.T. (2005) Comparative seed shadows of bird-, monkey-, and wind-dispersed trees. *Ecology* 86, 2684–2694.

Clark, J.S., Macklin, E. and Wood, L. (1998) Stages and spatial scales of recruitment limitation in southern Appalachian forests. *Ecological Monographs* 68, 213–235.

Clark, J.S., Silman, M., Kern, R., Macklin, E. and HilleRisLambers, J. (1999) Seed dispersal near and far: patterns across temperate and tropical forests. *Ecology* 80, 1475–1494.

Connell, J.H. (1971) On the role of natural enemies in preventing competitive exclusion in some marine animals and in rain forest trees. In: Boer, P.J. and Gradwell, G.R. (eds) *Dynamics of Populations*. Centre for Agricultural Publishing and Documentation, Wageningen, The Netherlands, pp. 298–313.

Finegan, B. (1996) Pattern and process in neotropical secondary rain forests: the first 100 years of succession. *Trends in Ecology and Evolution* 11, 119–124.

Fleming, T.H. and Heithaus, E.R. (1981) Frugivorous bats, seed shadows, and the structure of tropical forests. *Biotropica* 13 (Suppl.), 45–53.

Fragoso, J.M.V. (1997) Tapir-generated seed shadows: scale-dependent patchiness in the Amazon rain forest. *Journal of Ecology* 85, 519–529.

Galetti, M. (2000) Frugivory by toucans (Ramphastidae) at two altitudes in the Atlantic forest of Brazil. *Biotropica* 32, 842–850.

Gautier-Hion, A., Duplantier, J.-M., Quris, R., Feer, F., Sourd, C., DeCoux, J.-P., Dubost, G., Emmons, L., Erard, C., Hecketsweiler, P., Moungazi, A., Roussilhon, C. and Thiollay, J.-M. (1985) Fruit characters as a basis of fruit choice and seed dispersal in a tropical forest vertebrate community. *Oecologia* 65, 324–337.

Gentry, A.H. (1982) Patterns of neotropical plant species diversity. *Evolutionary Biology* 15, 1–84.

Graham, C.H. (2001a) Factors influencing movement patterns of keel-billed toucans in a fragmented tropical landscape in Southern Mexico. *Conservation Biology* 15, 1789–1798.

Graham, C.H. (2001b) Habitat selection and activity budgets of keel-billed toucans at the landscape level. *Condor* 103, 776–784.

Hamann, A. and Eberhard, C. (1999) Interactions among frugivores and fleshy fruit trees in a Philippine submontane rainforest. *Conservation Biology* 13, 766–773.

Hamilton, M.B. (1999) Tropical tree gene flow and seed dispersal. *Nature* 401, 129–130.

Hardesty, B.D., Hubbell, S.P. and Bermingham, E. (2006) Genetic evidence of frequent long-distance recruitment in a vertebrate-dispersed tree. *Ecology Letters* 9, 516–525.

Harms, K.E., Wright, S.J., Calderon, O., Hernandez, A. and Herre, E.A. (2000) Pervasive density-dependent recruitment enhances seedling diversity in a tropical forest. *Nature* 404, 493–495.

Harper, J.L. (1977) *Population Biology of Plants.* Academic Press, London.

Holbrook, K.M. (2006) Seed dispersal limitation in a Neotropical nutmeg, *Virola flexuosa* (Myristicaceae): an ecological and genetic approach. PhD dissertation, University of Missouri–St Louis, Missouri.

Holbrook, K.M. and Smith, T.B. (2000) Seed dispersal and movement patterns in two species of *Ceratogymna* hornbills in a West African tropical lowland forest. *Oecologia* 125, 245–257.

Holbrook, K.M., Smith, T.B. and Hardesty, B.D. (2002) Implications of long-distance movements of frugivorous rain forest hornbills. *Ecography* 25, 745–749.

Hoppes, W.G. (1988) Seedfall patterns of several species of bird-dispersed plants in an Illinois woodland. *Ecology* 69, 320–329.

Howe, H.F. (1989) Scatter- and clump-dispersal and seedling demography: hypothesis and implications. *Oecologia* 79, 417–426.

Howe, H.F. (1993) Specialized and generalized dispersal systems: where does 'the paradigm' stand? *Vegetatio* 107/108, 3–13.

Howe, H.F. and Smallwood, J. (1982) Ecology of seed dispersal. *Annual Reviews in Ecology and Systematics* 13, 201–228.

Howe, H.F. and Vande Kerckhove, G.A. (1981) Removal of wild nutmeg (*Virola surinamensis*) crops by birds. *Ecology* 62, 1093–1106.

Howe, H.F. and Westley, L.C. (1997) Ecology of pollination and seed dispersal. In: Crawley, M.J. (ed.) *Plant Ecology,* 2nd edn. Blackwell Scientific, Oxford, UK, pp. 262–283.

Howe, H.F., Schupp, E.W. and Westley, L.C. (1985) Early consequences of seed dispersal for a neotropical tree (*Virola surinamensis*). *Ecology* 66, 781–791.

Janzen, D.H. (1970) Herbivores and the number of tree species in tropical forests. *American Naturalist* 104, 501–528.

Janzen, D.H., Miller, G.A., Hackforth-Jones, J., Pond, C.M., Hooper, K. and Janos, D.P. (1976) Two Costa Rican bat-generated seed shadows of *Andira inermis* (Leguminosae). *Ecology* 57, 1068–1075.

Jones, F.A., Chen, J., Weng, G.-J. and Hubbell, S.P. (2005) A genetic evaluation of seed dispersal in the neotropical tree *Jacaranda copaia* (Bignoniaceae). *American Naturalist* 166, 543–555.

Jordano, P. (1992) Fruits and frugivory. In: Fenner, M. (ed.) *Seeds: The Ecology of Regeneration in Plant Communities.* CAB International, Wallingford, UK, pp. 105–156.

Jordano, P. and Godoy, J.A. (2000) RAPD variation and population genetic structure in *Prunus mahaleb* (Rosaceae), an animal-dispersed tree. *Molecular Ecology* 9, 1293–1305.

Jørgensen, P.M. and Leòn-Yánez, S. (1999) *Catalogue of the Vascular Plants of Ecuador.* Missouri Botanical Garden Press, St Louis, Missouri.

Kenward, R.E. (2001) *A Manual for Wildlife Radio Tagging.* Academic Press, San Diego, California.

Laman, T.G. (1996) Ficus seed shadows in a Bornean rain forest. *Oecologia* 107, 347–355.

Lambright, D.D. (1981) The comparative anatomy and morphology of *Virola* (Myristicaceae). PhD dissertation, Miami University, Miami, Florida.

Levey, D.J. (1986) Methods of seed processing by birds and seed deposition patterns. In: Estrada, A. and Fleming, T.H. (eds) *Frugivores and Seed Dispersal.* Dr. W. Junk, Dordrecht, The Netherlands, pp. 147–158.

Levin, D.A. (1979) The nature of plant species. *Science* 204, 381–384.

Levin, S.A., Muller-Landau, H.C., Nathan, R. and Chave, J. (2003) The ecology and evolution of seed dispersal: a theoretical perspective. *Annual Review of Ecology and Systematics* 34, 575–604.

Loiselle, B.A., Sork, V.L. and Graham, C. (1995) Comparison of genetic variation in bird-dispersed shrubs of a tropical wet forest. *Biotropica* 27, 487–494.

Loiselle, B.A., Blake, J.G., Durães, R., Ryder, T.B. and Tori, W. (2007) Environmental and spatial segregation of leks among six co-occurring species of manakins (Pipridae) in eastern Ecuador. *Auk* 124, 000–000.

Millspaugh, J.J. and Marzluff, J.M. (2001) *Radio Tracking and Animal Populations.* Academic Press, San Diego, California.

Morales, J.M. and Carlo, T.A (2006) The effects of plant distribution and frugivore density on the scale and shape of dispersal kernels. *Ecology* 86, 1489–1496.

Muller-Landau, H.C. and Hardesty, B.D. (2005) Seed dispersal of woody plants in tropical forests: concept, examples and future directions. In: Burslem, D., Pinard, M. and Hartley, S. (eds) *Biotic Interactions in the Tropics: Their Role in the Maintenance of Species Diversity.* Cambridge University Press, Cambridge, UK, pp. 267–309.

Murray, K.G. (1988) Avian seed dispersal of three neotropical gap-dependent plants. *Ecological Monographs* 58, 271–298.

Nathan, R. and Casagrandi, R. (2004) A simple mechanistic model of seed dispersal, predation and plant establishment: Janzen–Connell and beyond. *Journal of Ecology* 92, 733–746.

Nathan, R. and Muller-Landau, H.C. (2000) Spatial patterns of seed dispersal, their determinants and consequences for recruitment. *Trends in Ecology and Evolution* 15, 278–285.

Ouborg, N.J., Piquot, Y. and Van Groenendael, J.M. (1999) Population genetics, molecular markers and the study of dispersal in plants. *Journal of Ecology* 87, 551–568.

Parker, T.A., Stotz, D.F. and Fitzpatrick, J.W. (1996) Ecological and distributional databases. In: Stotz, D.F., Fitzpatrick, J.W., Parker, T.A. and Moskovits, D.K. (eds) *Neotropical Birds: Ecology and Conservation.* University of Chicago Press, Chicago, IIllinois, pp. 131–436.

Portnoy, S. and Willson, M.F. (1993) Seed dispersal curves: behavior of the tail of the distribution. *Evolutionary Ecology* 7, 25–44.

Remsen, J.V., Hyde, M.A. and Chapman, A. (1993) The diets of Neotropical trogons, motmots, barbets, and toucans. *Condor* 95, 178–192.

Russo, S.E. (2003) Responses of dispersal agents to tree and fruit traits in *Virola calophylla* (Myristicaceae): implications for selection. *Oecologia* 136, 80–87.

Santana, E.C., Moermond, T.C. and Denslow, J.S. (1986) Fruit selection in the collared araçari (*Pteroglossus torquatus*) and the slaty-tailed trogon (*Trogon massena*): two birds with contrasting foraging modes. *Brenesia* 25/26, 279–295.

Schupp, E.W. (1993) Quantity, quality and the effectiveness of seed dispersal by animals. *Vegetatio* 107/108, 15–29.

Shilton, L.A., Altringham, J.D., Compton, S.G. and Whittaker, R.J. (1999) Old World fruit bats can be long-distance seed dispersers through extended retention of viable seeds in the gut. *Proceedings of the Royal Society London, Series B, Biological Sciences* 266, 219–223.

Sun, C., Ives, A.R., Kraeuter, H.J. and Moermond, T.C. (1997) Effectiveness of three turacos as seed dispersers in a tropical montane forest. *Oecologia* 112, 94–103.

Syrjala, S.E. (1996) A statistical test for a difference between the spatial distributions of two populations. *Ecology* 77, 75–80.

Terborgh, J., Robinson, S.K., Parker, T.A., Munn, C.A. and Pierpoint, N. (1990) Structure and organization of an Amazonian forest bird community. *Ecological Monographs* 60, 213–238.

Valencia, R., Condit, R., Foster, R.B., Romoleroux, K., Munoz, G.V., Svenning, J.-C., Magård, E., Bass, M., Losos, E.C. and Balslev, H. (2004) Yasuní Forest Dynamics Plot, Ecuador. In: Losos, E.C. and Leigh, E.G. (eds) *Tropical Forest Diversity and Dynamism.* The University of Chicago Press, Chicago, Illinois, pp. 609–620.

Wang, B.C. and Smith, T.B. (2002) Closing the seed dispersal loop. *Trends in Ecology and Evolution* 17, 379–385.

Wehncke, E.V., Hubbell, S.P., Foster, M.S. and Dalling, J.W. (2003) Seed dispersal patterns produced by white-faced monkeys: implications for the dispersal limitation of neotropical tree species. *Journal of Ecology* 91, 677–685.

Wenny, D.G. (2000) Seed dispersal, seed predation, and seedling recruitment of a Neotropical montane tree. *Ecological Monographs* 70, 331–351.

Wenny, D.G. and Levey, D.J. (1998) Directed seed dispersal by bellbirds in a tropical cloud forest. *Proceedings of the National Academy of Sciences of the USA* 95, 6204–6207.

Westcott, D.A. and Graham, D.L. (2000) Patterns of movement and seed dispersal of a tropical frugivore. *Oecologia* 122, 249–257.

Westcott, D.A., Bentrupperbäumer, J., Bradford, M.G. and McKeown, A. (2005) Incorporating patterns of disperser behaviour into models of seed dispersal and its effects on estimated dispersal curves. *Oecologia* 146, 57–67.

White, G.C. and Garrott, R.A. (1990) *Analysis of Wildlife Radio-tracking Data.* Academic Press, San Diego, California.

Whitney, K.D., Fogiel, M.F., Lamperti, A.M., Holbrook, K.M., Stauffer, D.J., Hardesty, B.D., Parker, V.T. and Smith, T.B. (1998) Seed dispersal by *Ceratogymna* hornbills in the Dja Reserve, Cameroon. *Journal of Tropical Ecology* 14, 351–371.

Willson, M.F. (1992) The ecology of seed dispersal. In: Fenner, M. (ed.) *Seeds: The Ecology of Regeneration in Plant Communities.* CAB International, Wallingford, UK, pp. 61–85.

Willson, M.F. (1993) Dispersal mode, seed shadows, and colonization patterns. *Vegetatio* 107/108, 261–280.

Willson, M.F. and Crome, F.H.C. (1989) Patterns of seed rain at the edge of a tropical Queensland rain forest. *Journal of Tropical Ecology* 5, 301–308.

Zimmerman, J.W. and Powell, R.A. (1995) Radiotelemetry error: location error method compared with error polygons and confidence ellipses. *Canadian Journal of Zoology* 73, 1123–1133.

14 Linking Seed and Seedling Shadows: A Case Study in the Oaks (*Quercus*)

M.A. Steele, J.E. Carlson, P.D. Smallwood, A.B. McEuen, T.A. Contreras and W.B. Terzaghi

Introduction

Seed dispersal is a fundamental process in the life history of plants (van der Pilj, 1972). Dispersal from parental seed sources can allow seeds and seedlings to escape density-dependent predation and local competition near the parent and to establish in sites where the probability of germination, establishment, and recruitment are higher (Janzen, 1970; Connell, 1971; Terborgh *et al.*, 2002; Vander Wall *et al.*, 2005). After successful recruitment, dispersal continues to impact fitness by subsequently influencing reproduction, demography and population genetics (Howe, 1989) and is thus fundamental for understanding a wide range of ecological and evolutionary processes in plants (Howe, 1989; Herrera *et al.*, 1994; Schupp and Fuentes, 1995).

For animal-dispersed plants, movement of seeds away from parental sources depends almost entirely on the activity of seed consumers, but despite numerous studies on animal-mediated dispersal, our understanding of their impact on seed and seedling shadows is still quite poor (Jordano and Schupp, 2000; Jordano and Godoy, 2002; Vander Wall *et al.*, 2005). Here we briefly review some of the key challenges in understanding the impact of seed consumers on plant dispersal and demography (see also Jordano and Herrera, 1995; Wenny and Levey, 1998; Godoy and Jordano, 2001; Vander Wall *et al.*, 2005), and provide an overview of a case study in the oaks that serves to illustrate many of these issues (Smallwood *et al.*, 2001; Steele and Smallwood, 2002; Steele *et al.*, 2005).

Connecting seed dispersal with plant distribution and demography

Although the basic tenets of the original Janzen–Connell model are widely accepted, only one, density dependent mortality near parental sources, has been clearly documented (Wills *et al.*, 1997; Harms *et al.*, 2000; Howe and Miriti, 2000; Wang and Smith, 2002). The link between seed dispersal and plant demography, in particular, remains vague (see Godínez-Alvarez and Jordano, Chapter 17, this volume). Several recent reviews characterize the connection between seed dissemination and plant recruitment as an essential, but difficult, step in showing how the process of seed dispersal influences vegetation structure, plant coexistence, gene flow and the genetic structure of plant populations (Howe and Miriti, 2000, 2004; Nathan and Muller-Landau, 2000; Wang and Smith, 2002). These and other reviews collectively emphasize several non-mutually exclusive reasons for why the consequences of dispersed seeds remain elusive.

- First, the short duration and focus of many studies fails to capture the temporal and spatial variation in seed dispersal. Long-term studies such as those of Herrera (1998; see also Levey and Benkman, 1999), as well as those over larger geographical areas, are absolutely necessary for understanding how variation in seed production, abundance of seed consumers, and the interactions between the two, vary over spatial and temporal scales.
- A second challenge is to better understand the importance of rare, but disproportionately critical, long-distance dispersal events (Clark *et al.*, 1998; Nathan and Muller-Landau, 2000; Wang and Smith, 2002). Growing evidence supports the notion that such rare events, coupled with higher mortality near parental sources, may ultimately drive the patterns of recruitment in some systems. Thus, patterns of dispersal alone may be quite misleading.
- Finally, few studies have attempted to link various life stages in the seed dispersal cycle, possibly owing to the breadth of expertise and approaches necessary to do so (Wang and Smith, 2002). Despite considerable advances in the study of seed movement (e.g. Forget and Wenny, 2005), there have been few attempts to follow seed fates beyond germination. In some systems, this may be accomplished by following a cohort of seeds through the entire seed dispersal cycle (Wenny, 2000). However, in many other systems such an approach is difficult because of the diversity of variables that link one life stage with another. Numerous factors, such as: (i) secondary dispersal (Vander Wall 2002; Vander Wall and Longland, 2005); (ii) effects of microsite on seed deposition (including directed dispersal, see Wenny and Levey, 1998), germination, and seedling growth (Schupp and Fuentes, 1995; Nathan and Muller-Landau, 2000); and (iii) dispersal and establishment limitations, which may be common in many systems, all greatly complicate the path between dispersal and establishment (Hubbell *et al.*, 1999; Nathan and Muller-Landau, 2000).

Together, these factors present significant challenges to connecting patterns of seed dispersal with that of plant distribution. The following case study helps to further illustrate the inherent difficulty in this endeavour.

A Case Study in the Oaks

The oak system

The oaks (Fagaceae), represented by more than 500 species worldwide, are dominant trees in many temperate and subtropical biomes. All members of the genus produce acorns (technically a fruit) that are typically dispersed by scatter-hoarding mammals and birds (Steele and Smallwood, 2002). In eastern deciduous forests, including much of the Central Hardwoods region of the USA, these oaks are limited to two major sections within the subgenus *Quercus* whose characteristics significantly influence the manner in which they are stored by animals: the white oaks (Sect. *Quercus*) and the red oaks (Sect. *Lobatae*; Nixon, 1993), hereafter designated the WO and RO groups, respectively. Although jays (Corvidae) are known to disperse acorns several kilometres, and may be responsible for the northward migration of *Quercus* and other fagaceous tree species following the last glacial retreat in North America, in this chapter, we concentrate on the dispersal of acorns by small mammals whose scatter-hoarding behaviour is thought to be a primary influence within forest stands (Steele and Smallwood, 2002).

RO acorns typically have higher concentrations of both lipids (approx. 20% by dry mass) and tannins (5–15%) than WO acorns (lipids, approx. 10%; tannins, < 2%) especially in the eastern USA (see Table 1 in Smallwood *et al.*, 2001). RO acorns are dormant for the winter, whereas those of WO germinate immediately or soon after seed-fall (Fox, 1982; Young and Young, 1992), and even while still attached to the tree in some species in southern forests of North America (M.A. Steele, Central Mexico, 1997, personal observation).

The dispersal syndromes of oaks

The contrasting characteristics of WO and RO directly influence the behaviour of seed-consuming and scatter-hoarding animals in markedly different ways for these two oak groups (Smallwood and Peters, 1986; Steele and Smallwood, 2002; Steele *et al.*, 2004). The early germination of WO acorns, in particular, poses a significant problem for species that attempt to store them. Germinating WO rapidly convert resources in the cotyledon into a thick, fleshy taproot composed largely of indigestible cellulose (Korstian, 1927; Fox, 1974). This adaptation helps WOs escape seed predation (Barnett, 1977), while the animal is left with a highly perishable food source, unsuitable for long-term storage.

Many small mammals, and possibly jays, respond to precocious germination by selectively consuming WO acorns and dispersing and/or caching RO acorns preferentially over those of WO (Table 14.1). We have now documented this pattern of differential storage and dispersal in a wide range of circumstances by several species of mammals (Smallwood and

Table 14.1. Summary of the response of species of small mammal to acorns of *Q. rubra* red oaks (RO, Sect. *Lobatae*) and *Q. alba* white oaks (WO, Sect. *Quercus*) in deciduous forests of the eastern USA.

Species	Selective hoarding of RO acorns	Selective consumption of WO acorns	Selective dispersal of RO acorns	Embryo excision of WO acorns	Source(s)
Peromyscus leucopus (white-footed mouse)	x	x	x		P.D. Smallwood, personal observation; Steele *et al*., 2001
Glaucomys volans (southern flying squirrel)	x	x			P.D. Smallwood, personal observation
Tamias striatus (eastern chipmunk)	x	x	x		Steele *et al*., 2001
Sciurus carolinensis (eastern grey squirrel)	x	x	x	x	Fox, 1982; Smallwood and Peters, 1986; Steele *et al*., 1996; Hadj-Chikh *et al*., 1996
Sciurus niger (fox squirrel)	x	x	x	x	P.D. Smallwood, personal observation; Moore, 1995
Sciurus aureogaster (Mexican grey squirrel)				x	Steele *et al*., 2001

Peters, 1986; Steele and Smallwood, 2002; Steele *et al.*, 2004; Table 14.1). We also have shown that in at least one scatter-hoarding mammal, the eastern grey squirrel (*Sciurus carolinensis*; Sciuridae), this response is due directly to the differences in germination schedule, rather than any other physical or nutritional differences in the acorn cotyledon (i.e. tannins or lipids; Hadj-Chikh *et al.*, 1996). Moreover, we have found that grey squirrels selectively cache artificial acorns in which the cotyledon has been experimentally modified as long as these artificial acorns are constructed from the shells of RO acorns (Steele *et al.*, 2001a). Acorns made with the shells of WO acorns, in contrast, are consistently eaten. These results led us to hypothesize that a chemical cue in the shell, possibly related to those that control dormancy, may ultimately be used as the proximate cue for determining which acorns are non-perishable and should be stored (Steele *et al.*, 2001a).

For tree squirrels, it also appears that early germination of WO acorns may have selected for another behavioural strategy that allows them to manage this problem (Steele *et al.*, 2006). When presented with WO acorns, grey squirrels and at least two other species of tree squirrels (*Sciurus*) will sometimes cache these acorns but only after they have excised the embryo, thereby killing the seed (Fox, 1982; Steele *et al.*, 2001b). The result is a non-viable seed that stores well for over 6 months (Steele *et al.*, 2001b). Grey squirrels raised in captivity with no previous experience of acorns still perform embryo excision on WO and selective storage of RO acorns, suggesting a strong innate basis for the behaviour (Steele *et al.*, 2006).

Although many of the above studies focused on one species, it appears that an entire guild of granivorous mammals may respond in a similar manner. Table 14.1 summarizes the evidence for selective caching, selective dispersal (and caching), and embryo excision by several species of small mammals, all of which show a strong tendency to store RO acorns preferentially over those of WO. In field experiments in which we selectively provisioned communities of small mammals with tagged acorns, WO acorns consistently experienced high mortality as a result of either consumption or embryo excision (Steele *et al.*, 2001b; Moore, 2005; Fig. 14.1). Most RO acorns, in contrast, were stored intact (Steele *et al.*, 2001b; Moore, 2005; Fig. 14.1) often 20 m or more from their source (Fig. 14.2). In fact, the contrasting fates of RO and WO acorns illustrated in Figs 14.1 and 14.2 have now been documented on numerous occasions whenever ambient food levels are moderate or high (M.A. Steele, north-eastern Pennsylvania, USA, 2004, unpublished data; Moore, 2005). As expected, in low-mast years, survival of both oak types is rare.

Hypotheses

These collective observations have led us to hypothesize that RO acorns are afforded a distinct dispersal advantage and that WO may be dispersal-limited under a wide range of circumstances. We further predicted that

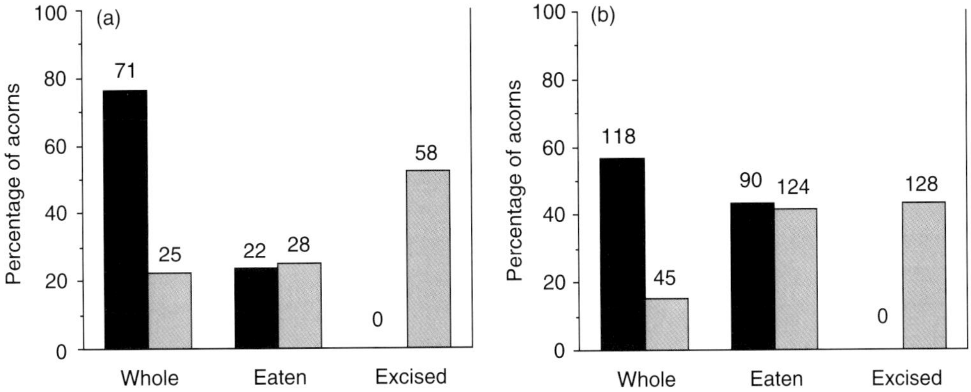

Fig. 14.1. Example of hoarding responses of small mammals at two sites (a, site A3; b, site A2) to acorns of red oak (RO, *Q. rubra*; black bars) and white oak (WO, *Q. alba*; grey bars). Excised acorns are those which had their embryos removed before caching. Note that RO acorns are typically stored and WO acorns are rarely stored. Results modified from Steele *et al.* (2001b).

this differential dispersal of the two oak groups is likely to have a significant impact on the spatial arrangement of WO and RO seedlings and may ultimately influence forest structure and, in turn, a number of other characteristics in the oaks (Steele and Smallwood, 2002; Steele *et al.*, 2004). Although putative seedling shadows generated with the modelling procedures of Ribbens *et al.* (1994) have supported this differential dispersal hypothesis (Smallwood *et al.*, 1998), these results produce only best-fit seedling shadows based on matching seedlings to their closest potential sources.

In the study described in this chapter, we sought to conduct a more definitive test of the differential dispersal hypothesis (DDH) by using DNA fingerprinting to match oak seedlings with their parent trees in order to determine seedling shadows of RO and WO species. Based on our studies of disperser behaviour, we hypothesize that WOs will exhibit short, truncated seedling shadows and that RO species will have longer seedling shadows and will exhibit less clumping closer to parental sources than WOs.

Field studies

To test the DDH, we conducted DNA matching of seedlings and potential parent trees in several mature oak forests (2–12 ha) in Pennsylvania, Maryland, Virginia and southern Indiana. The results shown here are based on data from a site adjacent the Powdermill Biological Station in western Pennsylvania, USA, approximately 20 km south-east of Ligonier,

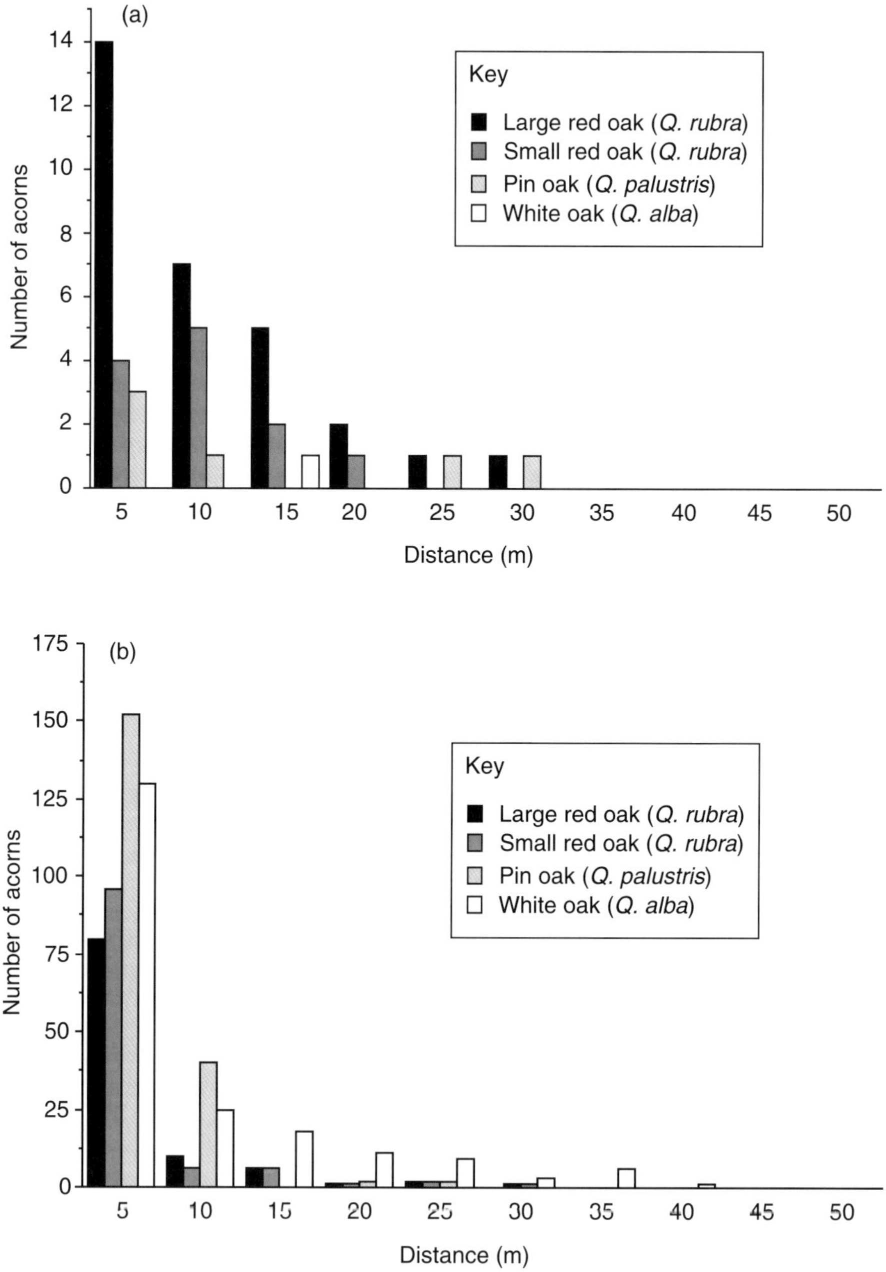

Fig. 14.2. Typical dispersal kernels observed for tagged red oak (RO, *Q. rubra*) and white oak (WO, *Q. alba*) acorns based on consolidated data from five locations at a single study area during a high-mast year (M.A. Steele, north-eastern Pennsylvania, USA, 2004, unpublished data). Results are based on recovery of metal-tagged acorns that were found and (a) stored intact or (b) eaten, as evidenced by recovery of the metal tag only. Note that RO acorns are dispersed and cached, and WO acorns are only moved when eaten (and/or stored, recovered and eaten).

on land owned by the Pennsylvania Bureau of Forestry (40°10′47.94″ N, 79°13′18.59″ W). The site, a second-growth forest, was dominated by mature stands of red, white and black oak (*Q. rubra*, *Q. alba* and *Q. velutina*), hickory (*Carya ovata*; Fagaceae) and red maple (*Acer rubrum*; Aceraceae). Seedlings of all three species of oak were abundant at the site.

To delineate the study area, we first mapped a 100-m transect and from that centre-line determined the position, diameter at breast height (dbh), and species of each adult oak tree (± 0.5 m, with dbh > 11 cm) within 1.6 ha of forest. We mapped the position of all adult oaks within 30 m of one side, 70 m of the other and 30 m from the top and bottom of the transect. Seedlings were mapped (± 1 m) to 4 m on either side of the 100-m transect. Detailed maps of all adult trees and seedlings were then produced with a geographic information system (GIS; Arcview 3.2, ESRI, Redlands, California). A single leaf was removed from each seedling or adult tree, snap-frozen on dry ice, returned to the laboratory and stored at −80°C until processed.

Laboratory analyses

We directly measured seedling–parent relationships using microsatellite DNA markers, previously shown to be useful across a wide range of oak species in both subgenera (Dow *et al.*, 1995; Isagi and Suhandono, 1997; Steinkellner *et al.*, 1997a,b). Dow and Ashley (1998), for example, showed that in burr oak (*Q. macrocarpa*) parentage (and thus seed dispersal and seedling establishment) could be accurately determined by sampling 62 adult trees and 100 saplings in one stand, using only four highly informative microsatellite DNA loci.

DNA was extracted using a modified version of the CTAB procedure of Stewart and Via (1993) and quantified by fluorometry using Hoechst 33258 (Sambrook and Russell, 2001). Additional details of the methodology are posted at http://staffweb.wilkes.edu/william.terzaghi/.

DNA fingerprints were created for seedling and adult tissue samples using 10 fluorescent-labelled primer sets shown to amplify polymorphic simple tandem repeat loci in northern red (*Q. rubra*) and chestnut oaks (*Q. prinus*; W.B. Terzaghi and J.E. Carlson, Wilkes University, Pennsylvania State University, 2002, unpublished data). Six of the primer sets were developed by Steinkellner *et al.* (1997a,b) and four were developed by Isagi and Suhandono (1997). Sizes of the products were determined using an ABI 3700 DNA analyser.

We then used a Java program developed by Damon Terzaghi to match DNA fingerprints. This program compares the fingerprint of each adult with that of all seedlings, recording matches when fragments corresponded within ± 1 base pair (bp). Seedlings that matched adults at a minimum of seven microsatellites (with no match failures) were scored as positives, whereas seedlings with bands failing to match an adult were excluded as offspring from that parent. We estimated seedling distances two ways, by measuring distances between: (i) parent–offspring matches; and (ii) seedlings and the nearest adult of the same species that could not be excluded as a parent (Fig. 14.3). Further details of these methods are available at http://staffweb.wilkes.edu/william.terzaghi/.

Overview of Findings

We generated successful fingerprints for 256 seedlings and 105 parents at the Powdermill Study Area. However, our somewhat conservative estimates of parent–seedling matches, based on a minimum of seven matching loci (and failure to exclude at all loci), resulted in only 83 parent–seedling matches (20 for *Q. alba* and 63 for *Q. rubra*). These procedures for positive matching of seedlings with paternal sources were verified independently

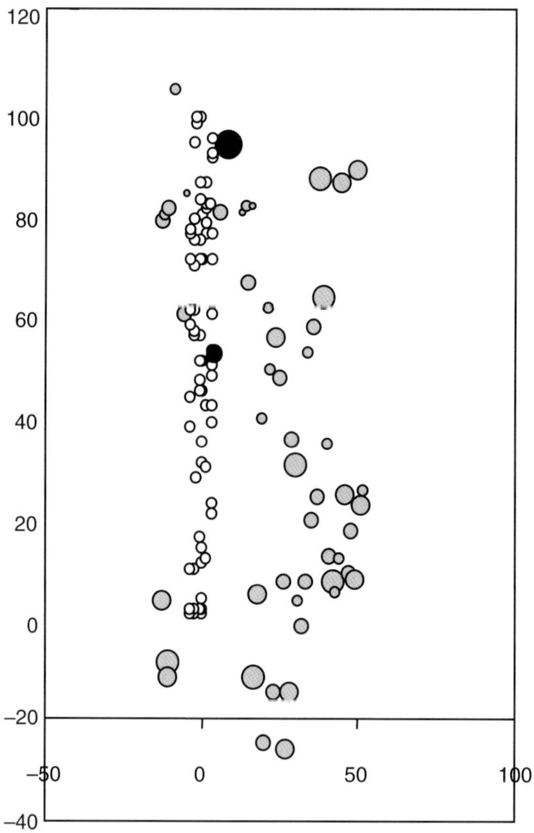

Fig. 14.3. Map of white oak (WO, *Q. alba*) seedlings (small circles) and adults (large circles, proportional to dbh) at the Powdermill Study Area. Note the reference parent (large dark circle), the one matched seedling (small dark circle) and the other seedlings that could be excluded as offspring at a minimum of two loci.

by testing acorns from known maternal sources (W.B. Terzaghi and J.E. Carlson, Wilkes University, 2005, unpublished data). The high percentage of seedlings for which we were unable to match a parental source suggests that a number of parent trees were located outside the study plot or had died after the establishment of the seedling; however, there was no evidence of logging on the site. Distances between these positively matched parents and offspring ranged between 2.4 m and 109 m (mean = 51.3 m ± 8.0 SE) for *Q. alba* and 2.3 m and 112.6 m (mean = 30.8 m ± 4.1 SE) for *Q. rubra*, respectively, and were significantly greater in *Q. alba* than in *Q. rubra* ($df = 81$, $t = -2.40$, $P = 0.019$; Fig. 14.4a).

Although positively matched seedlings confirm a parental source, our procedures do not distinguish between maternal and paternal (pollen) parents. We therefore used the exclusion procedures described above to calculate the minimum distance to an adult of the same species that could not be excluded as a parent. This approach resulted in calculation of minimum dispersal distances for 76, 164 and 16 seedlings of *Q. alba*, *Q. rubra* and *Q. velutina* (black oak, Sect. *Lobatae*), respectively. Mean minimum dispersal distances for *Q. rubra* (17.5 m ± 1.0 SE) were less than those of both *Q. alba* (30.8 m ± 2.05 SE) and *Q. velutina* (30.5 m ± 2.6 SE) and significantly different overall (one-factor ANOVA, $F_{(2, 256)} = 29.30$, $P < 0.0001$; Fig. 14.4b), due primarily to the lower minimum distances for *Q. rubra*. The minimum distances for *Q. rubra* were significantly less than those of both *Q. alba* (Scheffé's $F = 4.78$, $P < 0.0001$) and *Q. velutina* (Scheffé's $F = 8.94$, $P < 0.0004$). By contrast, there was no significant difference between the minimum distances for *Q. alba* and *Q. velutina* (Scheffé's $F = 9.36$, $P = 0.9793$).

Interpretation, Implications and Conclusions

Our results, although based on findings from only one study area, suggest that *Q. alba* (WO) may not exhibit short, truncated seedling shadows, as predicted from previous behavioural studies (Smallwood and Peters, 1986; Steele and Smallwood, 2002; Steele *et al.*, 2004) and inverse modelling of seedling shadows (Smallwood *et al.*, 1998). Instead, these data suggest that recruitment of *Q. alba* occurs at distances far greater than predicted from our determination of seed shadows; also contrary to our predictions, seedling shadows of *Q. alba* may even exceed those of *Q. rubra* (RO). Some of the longer seedling-to-parent distances may be explained as artefacts of our sampling protocol; specifically, there were fewer WO adults within our sampled transect than RO adults, and therefore less opportunity to detect close parents. However, the fact that seedlings that cluster around WO adults can be excluded as seedlings of that adult via DNA fingerprinting (Fig. 14.3) suggests our general conclusion that *Q. alba* is dispersed greater distances (than indicated from behavioural studies) remains valid.

Given previous evidence for dispersal limitation in WO based on the behaviour of an entire granivore community, these results raise important

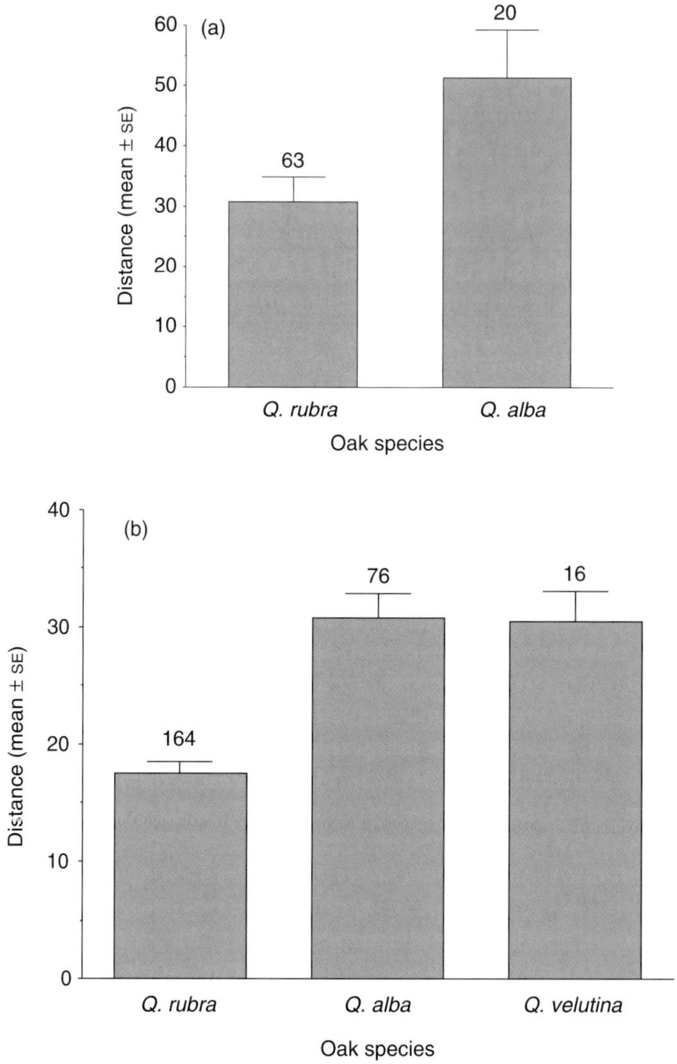

Fig. 14.4. Estimates of seedling shadows based on analysis of microsatellite DNA markers. (a) The known distances (mean ± SE) between positively matched parents and seedlings of red oak (RO, *Q. rubra*) and white oak (WO, *Q. alba*) and (b) the minimum distance (mean ± SE) to the nearest potential tree that could not be excluded based on DNA results for *Q. rubra*, *Q. velutina* and *Q. alba*. (b) Adult trees were excluded if their DNA fingerprint differed from a seedling at a minimum of one locus.

questions regarding the manner in which WO may be dispersed. These findings also have broader implications for understanding the complexity of factors that link seed and seedling shadows, and further call into question the conclusions that can be drawn even from well studied, animal-generated seed shadows.

The seedling shadow of *Q. alba*

Assuming that WO seedlings tend to establish at greater distances from parents, two non-mutually exclusive explanations probably account for the inconsistencies between the seed and seedling shadows of *Q. alba*. First, it is possible that the seedling shadow is the result of a combination of rare establishment events in the tail of the seed shadow, coupled with high mortality nearer to parental sources. Such an outcome would be consistent with the original predictions of Janzen (1970) and Connell (1971). Additionally, repeated caching by several scatter-hoarding individuals or species (i.e. diplochory; see Vander Wall and Longland, 2005) may also contribute to dispersal and establishment farther than expected from the source, although we have found little evidence for this. Our behavioural research, now based on > 15 years of study, indicates that establishment of any *Q. alba* seedlings as a result of the activity of scatter-hoarding mammals is exceedingly rare, even when acorns are abundant. In a 4-year study, in which we followed dispersal and establishment of both RO and WO, we observed infrequent establishment of *Q. alba* acorns (< 0.01%), whereas acorns of two RO species (*Q. rubra* and *Q. palustris*) were regularly cached and germinated in any year of moderate or high acorn abundance (M.A. Steele *et al.*, north-eastern Pennsylvania, USA, 2004, personal observation). Similar results have been reported recently for mammalian granivores in other portions of the Central Hardwoods region of the USA (Moore, 2005). It is plausible, given the lack of synchrony between the acorn crops of these two subgenera (Koenig and Knops, 2002), that the establishment of WO seedlings occurs only in years when WO species mast independently of RO. Although such a pattern would be difficult to document, given the rarity of the events, our behavioural studies suggest that the activity of mammalian granivores might exert strong selective pressure on WOs to mast independently of RO species.

A second explanation for the unexpectedly long seedling shadows of *Q. alba* may be the activity of other seed consumers. Although numerous species depend on oaks for food in eastern deciduous forests (Van Dersal, 1940), most (e.g. deer, bears, racoons, pigeons) masticate acorns in a manner that would not contribute to dispersal, and many others larder-hoard acorns in trees, burrows or other microsites where they are unable to establish (Steele and Smallwood, 2002). However, the movement and accidental loss of seeds during this process is an unexplored mechanism of dispersal (see below).

Several species of birds are also known to contribute to oak dispersal. They include the blue jay (*Cyanocitta cristata* – Darley-Hill and Johnson, 1981; Johnson and Webb, 1989; Johnson *et al.*, 1997), scrub jays (*Aphelocoma* spp. – DeGange *et al.*, 1989; Hubbard and McPherson, 1997), the European jay (*Garrulus glandarius* – Kollmann and Schill, 1996; Monsandl and Kleinert, 1998) and several other corvids (*Corvus* spp. and *Pica pica* – Waite, 1985). However, only one of these species, the blue jay, resides in eastern deciduous forests. However, several aspects of blue jay scatter-hoarding behaviour initially led us to discount them as a major factor in the dispersal of oaks (especially of WOs) within forest patches (see Steele and Smallwood, 2002).

1. Blue jays, because of their gape limitation, are reported to selectively disperse and cache small-seeded red oaks over those species producing larger acorns (e.g. *Q. rubra* and *Q. alba*; Scarlett and Smith, 1991).
2. Jays are known for long-distance dispersal, often moving acorns further than a kilometre (Darley-Hill and Johnson, 1981; Johnson and Webb, 1989; Kollmann and Schill, 1996).
3. Perhaps most importantly, jays are often considered agents of directed dispersal, selectively caching acorns in open successional habitats such as sparsely wooded prairies, forested edges, hedgerows, grasslands and other open habitat patches where conditions are ideal for establishment of some species of oak (Darley-Hill and Johnson, 1981; Johnson and Webb, 1989; Kollmann and Schill, 1996; Monsandl and Kleinert, 1998; Gomez, 2003).
4. In experimental studies of oak dispersal in the Midwestern USA, Moore (2005) failed to see any removal of acorns by blue jays within forest patches when jays were selectively provisioned with acorns by means of semipermeable exclosures.

Although these observations would seem to argue against blue jays as a potential explanation for the dispersion of *Q. alba* in the present study, several additional notes on the bird's behaviour may suggest otherwise. A recent study on acorn selection by captive blue jays confirms the bird's preference for smaller RO acorns, but also indicates that blue jays will consume the larger acorns of *Q. rubra* and *Q. alba* when little else is available (Moore and Swihart, 2006). Moore and Swihart (2006) also noted that, when consuming *Q. alba* acorns, the birds often consumed only a portion before dropping the acorns from their perch, possibly because of their larger size. Elsewhere, the dropping of acorns by birds has been proposed as a mechanism of oak dispersal (Webb, 1986) and it has been shown that jays will eat the top half of acorns, due to tannin gradients in the cotyledon, but then discard the apical end which is still able to germinate (Steele *et al.*, 1993). These observations, coupled with the rather peculiar seedling shadows reported here (Figs 14.3 and 14.4) for *Q. alba* and *Q. velutina*, a species preferred by blue jays (Scarlett and Smith, 1991), lead us to hypothesize that feeding, rather than scatter-hoarding by jays, may account for the dispersal and establishment of *Q. alba*. Although further research is needed to test this hypothesis, it serves to illustrate just how complex the factors contributing to WO establishment may be.

Conclusions

In addition to shedding light on the process of oak establishment, these results also have broader implications for the study of seed dispersal and plant population biology. Schupp and Fuentes (1995) argue that the study of seed dispersal offers a unifying theme in plant population biology by linking the constraints faced by seeds, seedlings and saplings to the specific process of recruitment and to the study of plant demography in general, but also note that this critical link is frequently not made because of the lack of detailed studies in many systems. Our initial findings here, indicating significant disagreement between our previous estimates of the seedling shadows of WO and the dispersion of seedlings in relation to parental sources, illustrate the sheer complexity involved in understanding the establishment process and underscore the argument by Schupp and Fuentes (1995) that 'correspondence, or lack of correspondence, between patterns of seed arrival and adult recruitment, tells us little about the causal relationships between the two'. Clearly, more experimental work is needed in our system, as well as most others, to determine how patterns of seed dispersion translate into patterns of establishment and recruitment. Indeed the process of recruitment limitation, although overestimated in our previous behavioural studies, may be a key factor in determining forest community structure.

Finally these results add to a growing number of studies that demonstrate the importance of genetic research for understanding temporal and spatial patterns of seed dispersal and recruitment (Godoy and Jordano, 2001; Jones *et al.*, 2005; see Hardesty, Chapter 12, this volume). Despite the methodological limitations in our approach (difficulties in identifying known maternal sources of seedlings), these techniques have allowed us to begin to link seed dispersal with recruitment and to identify gaps in our knowledge of seed dispersal in this system.

Acknowledgements

We gratefully acknowledge the financial support of the National Science Foundation (DBI-9978807), Wilkes University, the University of Richmond and the School of Forest Resources of Pennsylvania State University, as well as the logistic support of Dr Joesph Merritt (Powdermill Biological Station) and the Pennsylvania Bureau of Forestry. We thank Judith Mank, Holly Marking, Nadia Terzaghi, Jeffrey Ward and Tiffany Leptuck for their leadership in the field and laboratory, and our small army of undergraduates from Wilkes University and the University of Richmond for their assistance with field collections and DNA extraction. We also thank Damon Terzaghi for writing the Java program that allowed us to match or exclude seedlings and parents. Finally we are grateful to Salvatore Agosta, Jeffery Klemens, Jeffrey Moore and Robert Swihart for helpful comments and discussions.

References

Barnett, R.J. (1977) The effect of burial by squirrels on germination and survival of oak and hickory nuts. *American Midland Naturalist* 98, 19–30.

Clark, J.S., Fastie, C., Hurtt, G., Jackson, S.T., Johnson, C., King, G.A., Lewis, M., Lynch, J., Pacala, S., Prentice, C., Schupp, E.W., Webb, T. III and Wyckoff, P. (1998) Reid's paradox of rapid plant migration: dispersal theory and interpretation of paleoecological records. *BioScience* 48, 13–24.

Connell, J.H. (1971) On the role of natural enemies in preventing competitive exclusion in some marine animals and in forest trees. In: den Boer, P.J. and Gradwell, G.R. (eds) *Dynamics of Populations*. Centre for Agricultural Publishing and Documentation (PUDOC), Wageningen, The Netherlands, pp. 298–312.

Darley-Hill, S. and Johnson, W.C. (1981) Acorn dispersal by the blue jay (*Cyanocitta cristata*). *Oecologia* 50, 231–232.

DeGange, A.R., Fitzpatrick, J.W., Layne, J.N. and Woolfenden, G.E. (1989) Acorn harvesting by Florida scrub jays. *Ecology* 70, 348–356.

Dow, B.D. and Ashley, M.V. (1998) High levels of gene flow in bur oak revealed by paternity analysis using microsatellites. *Journal of Heredity* 89, 62–70.

Dow, B.D., Ashley, M.V. and Howe, H.F. (1995) Characterization of highly variable GA/CT in microsatellites in the burr oak, *Quercus macrocarpa*. *Theoretical and Applied Genetics* 91, 137–141.

Forget, P.-M. and Wenny, D.G. (2005) How to elucidate seed fate? A review of methods used to study seed removal and secondary dispersal. In: Forget, P.-M., Lambert, J.E., Hulme, P.E. and Vander Wall, S.B. (eds) *Seed Fate: Predation, Dispersal and Seedling Establishment*. CAB International, Wallingford, UK, pp. 297–314.

Fox, J.F. (1974) Coevolution of white oak and its seed predators. PhD thesis, The University of Chicago, Chicago, Illinois.

Fox, J.F. (1982) Adaptation of gray squirrel behaviour to autumn germination by white oak acorns. *Evolution* 36, 800–809.

Godoy, J.A. and Jordano, P. (2001) Seed dispersal by animals: exact identification of source trees with endocarp DNA microsatellites. *Molecular Ecology* 81, 1667–1674.

Gomez, J.M. (2003) Spatial patterns of long-distance dispersal of *Quercus ilex* acorns by jays in a heterogeneous landscape. *Ecography* 26, 573–584.

Hadj-Chikh, L.Z., Steele, M.A. and Smallwood, P.D. (1996) Caching decisions by grey squirrels: a test of the handling-time and perishability hypotheses. *Animal Behaviour* 52, 941–948.

Harms, K.E., Wright, S.J., Calderón, O., Hernández, A. and Herre, E.A. (2000) Pervasive density-dependent recruitment enhances seedling diversity in a tropical forest. *Nature* 404, 493–495.

Herrera, C.M. (1998) Long-term dynamics of Mediterranean frugivorous birds and fleshy fruits: a 12-year study. *Ecological Monographs* 68, 511–538.

Herrera, C.M., Jordano, P., Lopez-Soria, L. and Amat, J.A. (1994) Recruitment of a mast-fruiting, bird-dispersed tree: bridging frugivore activity and seedling establishment. *Ecological Monographs* 64, 315–344.

Howe, H.F. (1989) Scatter- and clump-dispersal and seedling demography: hypotheses and implications. *Oecologia* 79, 417–426.

Howe, H.F. and Miriti, M.N. (2000) No question: seed dispersal matters. *Trends in Ecology and Evolution* 15, 434–436.

Howe, H.F. and Miriti, M.N. (2004) When seed dispersal matters. *Bioscience* 54, 651–660.

Hubbard, J.A. and McPherson, G.R. (1997) Acorn selection by Mexican jays: a test of a tri-trophic symbiotic relationship hypothesis. *Oecologia* 110, 143–146.

Hubbell, S.P., Foster, R.B., O'Brien, S.T., Harms, K.E., Condit, R., Wechsler, B., Wright, S.J. and Loo de Lao, S. (1999) Light-gap disturbances, recruitment limitation, and tree diversity in a neotropical forest. *Science* 283, 584–587.

Isagi, Y. and Suhandono, S. (1997) PCR primers amplifying microsatellite loci of *Quercus myrsinifolia* Blume and their conservation between oak species. *Molecular Ecology* 6, 897–899.

Janzen, D.H. (1970) Herbivores and the number of tree species in the tropical forests. *American Naturalist* 194, 501–528.

Johnson, W.C. and Webb, T. (1989) The role of blue jays (*Cyanocitta cristata*) in the postglacial dispersal of fagaceous trees in eastern North America. *Journal of Biogeography* 16, 561–571.

Johnson, W.C., Adkisson, C.S., Crow, T.R. and Dixon, D.M. (1997) Nut caching by blue jays (*Cynaocitta cristata* L.): implications for tree demography. *American Midland Naturalist* 138, 357–370.

Jones, F.A., Chen, J., Weng, G.-J. and Hubbell, S.P. (2005) Genetic evaluation of seed dispersal in the neotropical tree *Jacaranda copaia* (Bignoniaceae). *American Naturalist* 166, 543–555.

Jordano, P. and Godoy, J.A. (2002) Frugivore-generated seed shadows: a landscape view of demographic and genetic effects. In: Levey, D.J., Diva, W.R. and Galetti, M. (eds) *Seed Dispersal and Frugivory: Ecology, Evolution, and Conservation.* CAB International, Wallingford, UK, pp. 305–322.

Jordano, P. and Herrera, C.M. (1995) Shuffling the offspring: uncoupling and spatial discordance of multiple stages in vertebrate seed dispersal. *Ecoscience* 2, 230–237.

Jordano, P. and Schupp, E.W. (2000) Determinants of seed dispersal effectiveness: the quantity component in the *Prunus mahaleb*–frugivorus bird interaction. *Ecological Monographs* 70, 591–615.

Koenig, W.B. and Knops, M.H. (2002) The behavioural ecology of masting in oaks. In: McShea, W.J. and Healy, W.M. (eds) *Oak Forest Ecosystems: Ecology and Management for Wildlife.* Johns Hopkins University Press, Baltimore, Maryland, pp. 182–195.

Kollmann, J. and Schill, H.P. (1996) Spatial patterns of dispersal, seed predation, and germination during colonization of abandoned grassland by *Quercus petrea* and *Corylus avellana*. *Vegetatio* 125, 193–205.

Korstian, C.F. (1927) Factors controlling germination and early survival in oaks. *Yale University School of Forestry Bulletin* 19, 1–122.

Levey, D.J. and Benkman, C. (1999) Fruit–seed disperser interactions: timely insights from a long-term perspective. *Trends in Ecology and Evolution* 14, 41–43.

Monsandl, R. and Kleinert, A. (1998) Development of oaks (*Quercus petrea* (Matt.) Liebl.) emerged from bird-dispersed seeds under old-growth pine (*Pinus silvestris* L.) stands. *Forest Ecology and Management* 106, 35–44.

Moore, J.E. (2005) Ecology of animal-mediated seed dispersal in the fragmented central hardwoods region. PhD thesis, Purdue University, West Lafayette, Indiana.

Moore, J.E. and Swihart, R.K. (2006) Diet choice of captive blue jays: implications for tree dispersal. *Condor* 108, 377–388.

Nathan, R. and Muller-Landau, H.C. (2000) Spatial patterns of seed dispersal, their determinants and consequences for dispersal. *Trends in Ecology and Evolution* 15, 278–285.

Nixon, K.C. (1993) Infrageneric classification of *Quercus* (Fagaceae) and typification of sectional names. *Annual Science of Forestry* 50 (Suppl. 1), 25s–34s,

Ribbens, E., Silander, J.A. and Pacala, S.W. (1994) Seedling recruitment in forests: calibration models to predict patterns of tree seedling dispersion. *Ecology* 75, 1794–1806.

Sambrook, J. and Russell, D.W. (2001) *Molecular Cloning: A Laboratory Manual*, 3rd edn. Cold Spring Harbor Laboratory Press, Cold Spring Harbor, New York.

Scarlett, T.L. and Smith, K.G. (1991) Acorn preferences of urban blue jays (*Cyanocitta crsitata*) during fall and spring in northwestern Arkansas. *Condor* 93, 438–442.

Schupp, E.W. and Fuentes, M. (1995) Spatial patterns of seed dispersal and the unification of plant population ecology. *Ecoscience* 2, 267–275.

Smallwood, P.D. and Peters, W.D. (1986) Grey squirrel food preferences: the effects of tannin and fat concentration. *Ecology* 67, 168–174.

Smallwood, P.D., Steele, M.A., Ribbens, E. and McShea, W.J. (1998) Detecting the effects of seed hoarders on the distribution of tree species: gray squirrels (*Sciurus carolinensis*) and oaks (*Quercus*) as a model system. In: Steele, M.A., Merritt, J.F. and Zegers, D.A. (eds) *Ecology and Evolutionary Biology of Tree Squirrels*. Virginia Museum of Natural History, Special Publication 6, Martinsville, Virginia, pp. 211–222.

Smallwood, P.D., Steele, M.A. and Faeth, S.H. (2001) The ultimate basis of caching preferences of rodents and the oak-dispersal syndrome: tannins, insects, and seed germination. *American Zoologist* 41, 840–851.

Steele, M.A. and Smallwood, P.D. (2002) Acorn dispersal by birds and mammals. In: McShea, W.J. and Healy, W.M. (eds) *Oak Forest Ecosystems: Ecology and Management for Wildlife*. Johns Hopkins University Press, Baltimore, Maryland, pp. 182–195.

Steele, M.A., Knowles, T., Bridle, K. and Simms, E. (1993) Tannins and partial consumption of acorns: implications for dispersal of oaks by seed predators. *American Midland Naturalist* 130, 229–238.

Steele, M.A., Smallwood, P.D., Spunar, A. and Nelsen, E. (2001a) The proximate basis of food-hoarding decisions and the oak dispersal syndrome: detection of seed dormancy by rodents. *American Zoologist* 41, 852–864.

Steele, M.A., Turner, G., Smallwood, P.D., Wolff, J.O. and Radillo, J. (2001b) Cache management by small mammals: experimental evidence for the significance of acorn embryo excision. *Journal of Mammalogy* 82, 35–42.

Steele, M.A., Smallwood, P.D., Terzaghi, W.B., Carlson, J.E., Contreras, T. and McEuen, A. (2004) Oak dispersal syndromes: do red oaks and white oaks exhibit different dispersal strategies? SRS-73, US Department of Agriculture, Forest Service, Southern Research Station, Ashville, North Carolina.

Steele, M.A., Wauters, L.A. and Larsen, K.W. (2005) Selection, predation, and dispersal of seeds by tree squirrels in temperate and boreal forests: are tree squirrels keystone granivores? In: Forget, P.-M., Lambert, J.E., Hulme, P.E. and Vander Wall, S.B. (eds) *Seed Fate: Predation, Dispersal and Seedling Establishment*. CAB International, Wallingford, UK, pp. 205–222.

Steele, M.A., Manierre, S., Genna, T., Contreras, T., Smallwood, P.D. and Pereira, M. (2006) The innate basis of food-hoarding decisions: evidence for coevolution between the tree squirrels and the oaks. *Animal Behaviour* 71, 151–160.

Steinkellner, H., Fluch, S., Turetschek, E., Lexer, C., Streiff, R., Kremer, A., Burg, K. and Glössi, J. (1997a) Identification and characterization of $(GA/CT)_n$-microsatellite loci from *Quercus petraea*. *Plant Molecular Biology* 33, 1093–1096.

Steinkellner, H., Lexer, C., Turetschek, E. and Glössi, J. (1997b) Conservation of $(GA)_n$-microsatellite loci between *Quercus* species. *Molecular Ecology* 6, 1189–1194.

Stewart, C.N. Jr and Via, L.E. (1993) A rapid CTAB DNA isolation technique useful for RAPD fingerprinting and other PCR applications. *BioTechniques* 14, 748–749.

Terborgh, J., Pitman, N., Silman, M., Schicter, H. and Núñez, V.P. (2002) Maintenance of tree diversity in tropical forests. In: Levey, D.J., Diva, W.R. and Galetti, M. (eds) *Seed Dispersal and Frugivory: Ecology, Evolution, and Conservation*. CAB International, Wallingford, UK, pp. 1–18.

Van der Pilj, L. (1972) *Principles of Dispersal in Higher Plants.* Springer, New York.

van Dersal, W.R. (1940) Utilization of oaks by birds and mammals. *Journal of Wildlife Management* 4, 404–428.

Vander Wall, S.B. (2002) Seed dispersal of Jeffrey pine seeds by rodent scatter-hoarders: the roles of pilfering, recaching and a variable environment. In: Levey, D.J., Diva, W.R. and Galetti, M. (eds) *Seed Dispersal and Frugivory: Ecology, Evolution and Conservation.* CAB International, Wallingford, UK, pp. 193–208.

Vander Wall, S.B. and Longland, W.S. (2005) Diplochory and the evolution of seed dispersal. In: Forget, P.-M., Lambert, J.E., Hulme, P.E. and Vander Wall, S.B. (eds) *Seed Fate: Predation, Dispersal and Seedling Establishment.* CAB International, Wallingford, UK, pp. 297–314.

Vander Wall, S.B., Forget, P.-M., Lambert, J.E. and Hulme, P.E. (2005) Seed fate pathways: filling the gap between parent and offspring. In: Forget, P.-M., Lambert, J.E., Hulme, P.E. and Vander Wall, S.B. (eds) *Seed Fate: Predation, Dispersal and Seedling Establishment.* CAB International, Wallingford, UK, pp. 1–8.

Waite, R.K. (1985) Food caching and recovery by farmland corvids. *Bird Study* 32, 45–49.

Wang, B.C. and Smith, T.B. (2002) Closing the seed dispersal loop. *Trends in Ecology and Evolution* 17, 379–385.

Webb, S.L. (1986) Potential role of passenger pigeons and other vertebrates in the rapid Holocene migrations of nut trees. *Quartenary Research* 26, 367–375.

Wenny, D.G. (2000) Seed dispersal, seed predation, and seedling recruitment of a Neotropical montane tree. *Ecological Monographs* 70, 331–351.

Wenny, D.G. and Levey, D.J. (1998) Directed dispersal by bellbirds in a tropical cloud forest. *Proceedings of the National Academy of Sciences USA* 95, 6204–6207.

Wills, C., Condit, R., Foster, R.B. and Hubbell, S.P. (1997) Strong density- and diversity-related effects help to maintain tree species diversity in a neotropical forest. *Proceedings of the National Academy of Sciences USA* 94, 1252–1257.

Young, J.A. and Young, C.G. (1992) *Seeds of Woody Plants in North America.* Dioscorides Press, Portland, Oregon.

15 Estimates of the Number of Seeds Dispersed by a Population of Primates in a Lowland Forest in Western Amazonia

P.R. STEVENSON

Introduction

In order to assess the impact of seed dispersal on plant populations, it is necessary to take into account all components of seed dispersal effectiveness, including the quality and the quantity of seeds dispersed (Schupp, 1993). Seed dispersal quality has two components. First, quality depends on the kind of treatment that the seeds receive when they are manipulated and treated in the mouth and digestive tract of the animal. Second, quality depends also on the conditions of the location where they are deposited and the characteristics of deposition itself. For instance, theoretical models and empirical studies in tropical forests have demonstrated an advantage in recruitment success when seeds fall away from conspecific trees, although this is not always true and there is wide species-specific variation in this pattern (Hamilton and May, 1977; Wills *et al.*, 1997; Wills and Condit, 1999; Terborgh *et al.*, 2002; Wright, 2002 and references therein).

The quantity of seed dispersal depends on the number of visits to a particular plant, the duration of the visits, and the rate at which seeds are removed from the plant by frugivores. Many published studies provide information on the quantity of seeds ingested by frugivorous or granivorous animals (Sallabanks, 1992; Herrera *et al.*, 1994; Graham *et al.*, 1995; Compton *et al.*, 1996; Willson *et al.*, 1996; Lambert, 1999; Nogales *et al.*, 1999; Jordano and Schupp, 2000; Montiel and Montana, 2000; Renne *et al.*, 2000; Godínez-Alvarez *et al.*, 2002; Russo, 2003; Balcomb and Chapman, 2003) and on the number of seeds found in faecal samples (Pigozzi, 1992; Castro *et al.*, 1994; Valido and Nogales, 1994; Nogales *et al.*, 1995, 1998; Caceres *et al.*, 1999; Stevenson, 2000). However, there are fewer studies that integrate observations of both animals eating fruits or seeds and faecal seed analyses with the aim of assessing seed removal from fruiting plants and quantifying the number of seeds defecated by the animals (Guitian *et al.*, 1994; Mandujano *et al.*, 1994; Chapman and Chapman, 1996; Kaplin and Moermond, 1998; Loiselle and Blake, 1999; Andresen, 2002; Figuerola *et al.*, 2002; Nogales *et al.*, 2002; Balcomb and Chapman, 2003). None of these studies has addressed the question of how different the estimates based on behavioural observations are from those based on faecal samples, and how the comparison could be used to characterize the quantitative component of seed dispersal effectiveness. In this study, I followed individual woolly monkeys (*Lagothrix lagothricha*; Atelidae) in order to make such comparisons. The first purpose of this study was to present two estimates of the number of seeds dispersed by woolly monkeys in Tinigua National Park, Colombia, at both the individual and the population levels. The second aim was to compare the number of seeds manipulated in fruiting trees with the number of seeds dispersed by woolly monkeys.

Both approaches to quantifying seed dispersal (observational versus faecal analysis) present biases that may produce dissimilar results. Differences in estimates may occur for four reasons. First, estimating the number of seeds dispersed using direct observation is usually based on visit duration, fruit ingestion rate, and handling behaviour. Feeding time is generally easier to measure than feeding rates (i.e. items manipulated per unit of time), which are usually extrapolated from subsamples made at times of good visibility. Thus, if sample size is insufficient, the estimates of seed dispersal may be biased and dissimilar to the estimates in faecal samples. In this case, a negative correlation would be expected between the difference in the two quantification approaches and sample size for feeding rates. The second potential source of dissimilarity between quantitative estimates is related to seed treatment. It is known that frugivorous primates drop and spit many seeds while feeding (Howe, 1980). Similarly, both bird gizzards and mammal mouths can crush seeds that will not necessarily be recognized in faecal samples, and behavioural observations do not always account for this kind of damage by frugivores. In either case, one expects to find a consistently lower number of seeds in the frugivores' faecal samples. The same effect is predicted for the third reason for differences: it is difficult to find faecal droppings, which is particularly likely when animals travel high in the forest canopy, as primates do. In order to separate the confounding effects of seed predation and the ability to recover seeds, it is important to measure recovery rates of seeds ingested by focal animals during long observation sessions, when the actual number of ingested seeds can be estimated. For this purpose, it is necessary to focus on species of plants with seeds that are unlikely to be dropped or crushed in the mouth. Fourth, if gut passage rate is long in relation to the sampling time of observations, an overall agreement between both estimates would not be expected. However, a better correlation between estimates should result as sample size increases. The predictions derived from all four potential sources of dissimilarity were tested in this investigation.

Recently, researchers have conducted many studies on seed dispersal in order to ascertain the role of dispersal limitation on the maintenance of species richness in tropical forests (Hubbell *et al.*, 1999). Although many predictions and methods have been advanced in order to test all the components of the hypothesis (Muller-Landau *et al.*, 2002), there are few studies quantifying the magnitude of dispersal limitation. In the study reported in this chapter, I include five estimates of the mass of seeds dispersed at the community level, which I hope will promote future research that could clarify the magnitude of dispersal limitation and whether or not it is common in all lowland tropical forests. The final goal of this study was to compare the estimates of seed dispersal by the woolly monkeys with seed-trap data and direct observations on fruiting trees, in order to determine the proportion of the seeds in the community dispersed by woolly monkeys.

Study Site and Methods

The study site is located in a tropical lowland forest on the eastern border of Tinigua National Park (201,875 ha), west of the Macarena mountain range, in the Department of Meta, Colombia (2°40′ N, 74°10′ W, 350–400 m above sea level). The Macarena Center of Ecological Investigations (CIEM), where this study took place, is on the western margin of the Duda River. Rainfall is markedly seasonal in the region, with a 2–3 month dry period occurring between December and March (Stevenson, 2002). Average annual precipitation for the three study years was 2782 mm.

Field protocol

I carried out behavioural observations on focal individuals of two habituated groups of woolly monkeys, CR-12 and CR-D (initially described in Stevenson *et al.*, 1994). I completed 60 h of focal samples per month for 2 years (August 1996–July 1997 and February 2000–January 2001). Observations were evenly distributed across all hours of the day (from 06:00 to 18:00 h). I followed individuals of four different age/sex classes as focal animals for at least 12 h per month for each class (adult male, adult female, female with an infant, and immature animals of both sexes). Sample time was biased toward some individuals of the group that were easier to recognize and to follow (e.g. the largest males, males with particular facial markings, and some females with young infants). In general, I identified individuals according to natural patterns on the genitals (spots on clitoris and penis), body size and facial characteristics. I carefully followed focal monkeys for as long as possible to collect data from the ingestion of fruits until the deposition of seeds.

Fruit diet

When a focal animal was feeding on fruit, I recorded the following information: time, number of minutes spent feeding, the species of plant, and whether the fruit was ripe or unripe. I recorded feeding rates as the number of fruits manipulated by woolly monkeys during periods of 30 s when monkeys were in optimal conditions for observation.

Faecal samples

Whenever possible, I collected all depositions of the focal individual falling to the forest floor or on to low vegetation during the sampling periods. A single deposition was defined as the faecal material dropped by an individual in a particular location within a period of 1 min. It was impossible to collect some of the excrement, because it remained totally or partly in high branches or on leaves. However, a proportion of this material usually dropped later through the action of secondary dispersers, thus an increase in the time over which searching was conducted could lead to a higher recovery rate. To avoid quantification differences due to searching time, I used an upper limit of 5 min to search for each deposition. I stored faeces in marked plastic bags usually for one day before analysis.

I washed the faecal samples and counted and identified all seeds. Large loads of small seeds, less than 3 mm in length, were not counted individually; their numbers in each faecal sample were estimated using four different categories (25–50, 50–200, 200–1000 and 1000–4000 seeds). In preliminary analyses I used conservative quantification estimates based on the lower limit of each category (Stevenson, 2000). For the analyses presented in this chapter, I used the same conservative estimate in addition to a less conservative, but still reasonable, estimate, the mid-range value for the category. I checked for the presence of very small seeds (< 1 mm) and tried to quantify their abundance, but these estimates were not precise because those seeds were not retained in the sieve (= 1 mm).

Woolly monkeys defecate and disperse seeds at night, but I limited my observations to daytime hours and was not able to recover all faecal material dropped by the focal animal during the day. Therefore, the number of seeds actually collected in this study represents an underestimate of the quantity of seeds dispersed. To compensate, my estimates roughly take into account these two limitations. It was very difficult to estimate dropping frequency at night: monkeys defecate at night in response to perturbations or activity on the ground and it is usually unclear whether the faecal material comes just from one individual at a natural rate. Finally, I decided to use a small sample size when a single individual slept directly over my zinc-roofed house (mean = 3, n = 2), to correct for night depositions. Given that focal woolly monkeys defecate on average 13 times in daylight hours, the night estimate represents 19% of the total depositions. Therefore, I calculated the total number of seeds as a fixed correction factor:

Number of seeds (night correction) = Actual number * 100% / (100% – 19%) = Actual number * 1.23.

This correction factor was not used to quantify the number of small seeds because the correction for low recovery rates in the long follows probably included the night factor, given the long retention times for small seeds (see below; Stevenson *et al.*, 2002).

In order to estimate recovery rates of seeds dispersed during the day, I extracted data from some long focal follows and compared the estimated number of seeds ingested with the number of seeds recovered during the focal session. I used only cases when it was possible to estimate the number of seeds ingested from behavioural observations. This value was calculated as the feeding time on one tree of a particular species (visited only once) and multiplied by seed ingestion rates, recorded during the sampling period. These rates were measured in a similar way as fruit feeding rates (number of fruits manipulated in 30 s), with the additional requirement that I only used species that the woolly monkeys ingest whole, without spitting or chewing the seeds. In this way, it was possible to use the number of seeds per fruit (Stevenson *et al.*, 2000; Stevenson, 2002) to estimate the number of seeds actually ingested by the focal animal. I did this analysis separately for small and relatively large seeds (>3 mm) because correcting estimates were very different, depending on seed size. In the case of large seeds, it was possible to recover all the seeds in the same day of observations, as long as the feeding bout was early in the morning and a humped distribution of seeds through time was recognized. I found that recovery rates of dispersed seeds were low: for relatively large seeds (>3 mm) recovery rates varied between 10 and 69% (mean = 32%, n = 23). In contrast, only 9% of the potentially dispersed small seeds were encountered in the faecal samples (range = 1–19%, n = 5). Due to the small sample size, I used two fixed correction factors to account for inefficient recovery:

Number of large seeds (recovery correction) = Number * 100% / 32% = Original number * 3.1

Number of small seeds (recovery correction) = Number * 100% / 9% = Original number * 11.1

Fruit traps

I extracted all fruit and seed contents from 300 fruit traps (800 cm^2 each), located in random places along 12 transects scattered throughout the study site (*c.*450 ha). The fruit traps were plastic containers that were emptied twice a month (March 1990–March 1991) (see details in Stevenson *et al.*, 1994). I estimated the mass of dispersed seeds as the dry weight of all naked diaspores that fell in the traps and that lacked parental plants above each trap. Data from lianas were discarded, as I was not always certain about the presence of an adult plant above the trap. The analysis presented here only includes fleshy animal-dispersed fruits.

Frugivory observations

With the help of several assistants, I carried out a total of 3438 h of diurnal observations on 75 different species of plants present at Tinigua during a period of 2 years (1999–2001). I selected fruiting trees that had good crown visibility, and trees were observed in periods of ripe fruit production, usually between 06:00 and 10:00 hours. I collected additional information on nocturnal animals using three different methods:

1. I used fruit traps that were checked at dawn and dusk for several species (*Gustavia hexapetala*, Lecythidaceae; *Protium crenatum*, Burseraceae; *Garcinia macrophylla*, Clusiaceae; *Virola flexuosa*, Myristicaceae; *Inga alba*, Fabaceae; *Pourouma bicolor* and *Cecropia membranacea*, Urticaceae; *Brosimum lactescens* and *Pseudolmedia laevis*, Moraceae). Traps placed under trees of these species were monitored during a total of 96 nights.
2. I estimated fruit removal from particular branches for three species of *Cecropia* by comparing the size of ripe infructescences at dawn and dusk.
3. I made additional direct observations of all species (e.g. Stevenson *et al.*, 2000).

The total diurnal sampling time varied among species, but all species included were observed for at least 22 h in periods of peak frugivore activity (mean = 46, range = 22–115 h). I avoided sampling during periods of rainfall and I did not use blinds or camouflage. This did not seem to influence the behaviour of the canopy frugivores, which have been subject to observation for more than a decade. However, our presence during observations possibly influenced the approach and feeding by terrestrial mammals such as coatis, tayras and tapirs.

During the fruiting tree observations, I recorded the duration of each visit by every frugivore (in minutes), its species and feeding rates. Feeding rates for primates were taken as the total number of fruits manipulated during periods of 30 s, when a focal animal was clearly visible. Focal sampling for birds was set to periods of 10 s, because it was more difficult to continuously observe birds for longer periods. I noted the behaviour of each species relevant to seed dispersal, such as the type of fruits that were consumed (ripe versus unripe), handling techniques and agonistic interactions.

Results

Dispersal patterns

I collected a total of 1562 depositions during the study, and all but 36 of these contained seeds. In each study year (1996 and 2000), I collected approximately the same number of depositions (754 versus 773). On average a single dropping included 70 seeds, but the number of seeds per deposition was highly variable (range: 0–1409) and dependent on the abundance of small seeds. Each dropping contained seeds from an average of 2.53 different species of plants (range: 0–9). These seeds belonged to at least 147 different species. Over the course of the study I collected a total of 106,869 seeds in depositions (using the lower limit of the range for seeds <3 mm, as in Stevenson, 2000). Given that population densities for woolly monkeys increased from 41 to 50 individuals km^{-2} between 1996 and 2000, I conservatively estimated that the woolly monkeys in the study area were dispersing at least 33,000 seeds km^{-2} day^{-1} in both study years (a = 1996; b = 2000):

(a) (49,118 seeds/individual*[5*12] days)*(41 individuals km^{-2}) = 33,564 seeds km^{-2} day^{-1};

(b) (57,751 seeds/individual*[5*12] days)*(50 individuals km^{-2}) = 48,126 seeds km^{-2} day^{-1}.

However, correcting for night defecations and recovery rates, and using the mid-range value for small seeds (< 3 mm), I estimated that focal animals were dispersing on average 1,390,419 seeds per year during sampling days (5 days and nights per month for each year; see Table 15.1). Based on population densities in the area, woolly monkeys were dispersing around 860,025 and 1,268,553 seeds km^{-2} day^{-1} during the 2 study years.

Seeds smaller than 3 mm, from species such as *Cecropia* spp., *Coussapoa* spp. (Urticaceae), *Ficus* spp. (Moraceae) and *Henriettella* spp. (Melastomataceae) were the most abundant in the faecal samples (83%, Table 15.2). However, species with larger seeds (> 10 mm) were also abundant in the faecal samples (e.g. *Pourouma bicolor* – 1084 seeds collected; *Inga cylindrica* – 496; *Gustavia hexapetala* – 717). About 97% of the seeds dispersed (using the corrected estimates) correspond to small-seeded species. However, the proportional representation of large-seeded species changes considerably when the parameter of quantification is seed mass. In terms of dispersed seed mass, woolly monkeys defecated more seed material from large-seeded species than from small-seeded species (Table 15.2). Only three out of 20 species included in the list of the most important species, ranked by mass, were small-seeded: *Cecropia* spp. and *Ficus trigona*. The opposite pattern was found when species were ranked by the number of seeds dispersed: *Laetia corymbulosa* (Salicaceae), *Apeiba aspera* (Malvaceae) and *Inga alba* being the only listed species with seeds larger than 3 mm.

Comparisons between manipulated and dispersed seeds

Because recovery rates were low, I used the corrected estimates to remove their confounding effects on the quantity of dispersed seeds. In general, there is good agreement between the number of seeds manipulated and

Table 15.1. Estimated number of seeds dispersed by the population of woolly monkeys in Tinigua National Park. This table shows the number of seeds found in the faecal samples (using the mid-range to estimate the number of small seeds) and the number of seeds after correcting for different factors that lowered recovery rates (see Methods). In the last column, the corrected estimate of seeds dispersed by focal animals during the sampling periods has been multiplied by the woolly monkey population density to calculate the number of seeds dispersed by the whole population and the range of variation when calculations are made using the lower and upper limits of the ranks.

			Recovery		Night						
Year	< 3 mm	> 3 mm	< 3 mm (9%)	> 3 mm (32%)	> 3 mm (81%)	< 3 mm	> 3 mm	Total	Ind. km^{-2}	No.	Seeds km^{-2} day^{-1}
1996	110,473	8,060	11.1	3.1	1.2	1,227,478	31,096	1,258,573	41	60	860,025 [332,985–1,378,296]
2000	133,593	9,823	11.1	3.1	1.2	1,484,367	37,897	1,522,264	50	60	1,268,553 [475,359–2,047,118]
Mean								1,390,419			1,064,289

Table 15.2. Species of plant dispersed by woolly monkeys showing both the mass of fresh seeds dispersed and the number of seeds dispersed. Columns show the estimates from the observed recovered seeds (5-day focal follows per month for 2 years) and the corrected estimate adjusted for recovering efficiency and depositions at night.

Dispersed mass (g)				Number of seeds			
Species	Fresh seed mass (g)	Recovered	Corrected	Species	Fresh seed mass (g)	Recovered	Corrected
Gustavia hexapetala	1.09	779	3,007	*Cecropia membranacea*	< 0.01	29,196	939,289
Pourouma bicolor	0.49	526	2,028	*Henriettella fissanthera*	< 0.01	16,529	526,544
Garcinia macrophylla	5.96	393	1,519	*Cecropia sciadophylla*	0.01	12,167	340,467
Spondias mombin	3.13	382	1,474	*Ficus trigona*	< 0.01	6,525	221,000
Inga cylindrica	0.76	374	1,445	*Ficus andicola*	< 0.01	5,450	177,833
Hymenaea oblongifolia	2.42	360	1,390	*Ficus nymphaeifolia*	< 0.01	5,063	171,478
Inga alba	0.23	286	1,104	*Ficus sphenophylla*	< 0.01	3,700	128,278
Cecropia membranacea	< 0.01	29	939	*Bellucia pentamera*	< 0.01	1,650	54,111
Inga edulis	0.50	227	875	*Ficus gomelleira*	< 0.01	1,564	44,600
Protium sagotianum	0.46	215	825	*Ficus obtusifolia*	< 0.01	753	21,867
Cecropia sciadophylla	0.01	24	681	*Ficus membranacea*	< 0.01	625	17,833
Castilla ulei	0.62	162	624	*Laetia corymbulosa*	0.01	4,464	17,222
Spondias venulosa	3.06	153	591	*Cecropia engleriana*	< 0.01	525	13,556
Brosimum lactescens	0.41	130	501	*Coussapoa orthoneura*	< 0.01	550	12,722
Bursera inversa	0.82	122	477	*Ficus insipida*	< 0.01	636	12,289
Inga cf. *acreana*	0.76	115	442	*Apeiba aspera*	0.02	2,035	7,851
Inga stenoptera	0.82	96	361	*Ficus americana*	< 0.01	225	4,889
Pseudomalmea diclina	1.84	88	341	*Inga alba*	0.23	1,255	4,842
Pouteria caimito	1.76	76	292	*Ficus guianensis*	< 0.01	178	4,589
Pseudolmedia laevigata	0.25	72	277	*Pourouma bicolor*	0.49	1,084	4,182
Ficus trigona	< 0.01	8	265	*Ficus usiacurina*	< 0.01	125	2,778
126 remaining species		1,282	5,368	126 remaining species		9,256	35,167

dispersed by woolly monkeys (Fig. 15.1). It was estimated that 30% of the variation in the number of seeds dispersed could be explained by variation in the number of seeds manipulated (slope = 0.71, F = 109.1, P < 0.001, n = 182 species). However, there seemed to be differences related to seed size; the slope of the regression line was higher for the small seeds (<3 mm: slope = 0.99, r^2 = 0.47, F = 19.3, P = 0.0002, n = 24) than for larger seeds (>3 mm: slope = 0.41, r^2 = 0.20, F = 36.3, P < 0.001, n = 158).

I did not find a negative relationship between the residuals from the overall regression model and the sample size of feeding rates (F = 1.2, P = 0.2, n = 182 species). The relationship between the absolute difference in estimates and the number of feeding rate records was positive and low (r^2 = 0.02, F = 4.9.3, P = 0.03, n = 182). Therefore, the difference between the number of seeds manipulated and dispersed was not associated with insufficient sampling of feeding rates.

As expected from long seed retention times, more discrepancies between quantification estimates were found at short timescales (e.g. days) than at an interval of 1 month, when the regression coefficient between manipulated and dispersed seeds was highest (Table 15.3). This suggests that over short time-frames the correlation is poor because of the time lag of

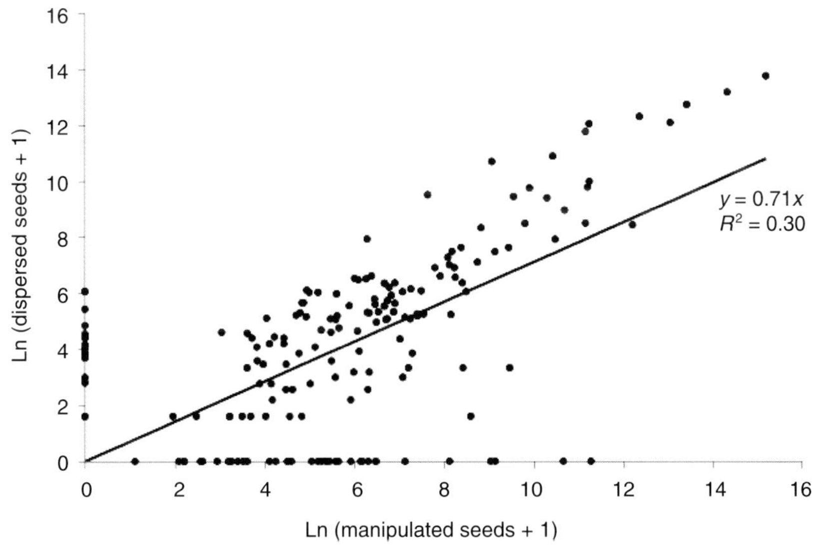

Fig. 15.1. Relationship between the estimated numbers of seeds dispersed and seeds manipulated by woolly monkeys in Tinigua National Park. The line represents the regression model fitted to the empirical data, forcing the intercept to zero. The points on the y axis correspond to seeds of species that were not consumed by the monkeys during sampling periods. The points on the x axis, and away from the origin, represent the species from which the woolly monkeys typically predate or drop seeds.

Table 15.3. Regression statistics for different temporal scales (day, month, year and the whole study) of the relationship between seeds dispersed and the estimated number of seeds manipulated by woolly monkeys. Sample size (*n*) indicates the number of species for which it was possible to have estimates of the number of seeds manipulated and dispersed at each timescale.

	Day	Month	Year	2 years
Slope	0.51	0.64	0.60	0.62
r^2	0.39	0.60	0.34	0.38
F	5.2	26.5	74.9	109.1
P	0.05	<0.001	<0.001	<0.001
n	10	18	149	182

digestion, while at long time-frames (e.g. years) it is possible that seasonal differences in the types of fruit available result in a poor relationship. However, this difference in the number of manipulated versus dispersed seeds persisted, at least for some species, despite increasing sampling time. Long retention also explains why at least 12 species were found in depositions but were never seen consumed during focal follows (located on the y axis in Fig. 15.1). Although none of those species was very abundant in the faeces, their exclusion from the regression analysis led to a higher slope of 0.92.

I found 37 species that were never or only rarely dispersed, suggesting either predation or inefficient dispersal (Fig. 15.1). Eight of those species were rarely consumed and I did not discount the possibility that the woolly monkeys could have dispersed the seeds. The remaining species belong to two main groups (see Table 15.4). First, the species that are not dispersed presumably because seeds are damaged in the mouth when unripe fruits are consumed (13 species, including four with wind-dispersed seeds). Second, 12 species seemed to have seeds too large to be swallowed by the woolly monkeys (Stevenson *et al.*, 2005). Two other species had seeds that could be ingested based on seed size, but that the woolly monkeys usually dropped from the fruiting tree. This pattern is common for many species, but it is not usually as marked as in these cases. Finally, two other species are probably dispersed, but their seeds are so small they might not have been recognized in the faecal samples.

Seed dispersal estimates from fruit traps

I estimated that different species of animals disperse ~56.8 kg ha^{-1} of seeds each year (Table 15.5). This estimate is roughly the same mass as for non-dispersed naked seeds that fall into fruit traps. However, when complete fruits (including seeds) are considered, the proportion of dispersed seeds in the community changes from 50.1% to 19.6% (Table 15.5). Given a population density of 12 individuals km^{-2} in 1991 (Stevenson, 2002) and

Table 15.4. List of plant species that are chewed up or rarely dispersed by woolly monkeys according to the difference between the estimated numbers of manipulated and dispersed seeds (M–D). Species are grouped by the potential cause for the lack of efficient dispersal.

Species	Feeding time	M–D	Species	Feeding time	M–D
Seed predation (mainly consumed unripe)			Large seeds dropped under the tree		
Pseudolmedia obliqua[d]	1,086	12,788	*Socratea exorrhiza*	36	216
Piptadenia pteroclada[w]	108	5,396	*Hymenaea courbaril*	195	195
Brosimum guianensis[d]	193	4,528	*Clarisia racemosa*	11	99
Xylophragma seemannianum[w]	35	3,360	*Antrocaryon amazonicum*	30	90
Batocarpus orinocensis	40	1,248	*Drypetes amazonica*	15	69
Enterolobium schomburgckii[d]	57	1,178	*Syagrus sancona*	29	29
Ceratophytum tetragonolobum[w]	11	660	*Pouteria procera*	14	25
Brosimum utile	104	645	*Iriartea deltoidea*	3	18
Dialium guianensis	55	550	*Pouteria* sp.	7	13
Samanea saman	36	374	*Cayaponia capitata*	4	8
Astrocaryum chambira	59	283	*Pouteria pariry*	2	7
Clarisia biflora	66	264	*Rhodostemonodaphne kunthiana*	1	2
Mansoa kerere[w]	1	60			
Medium seeds dropped under the tree			Unnoticed small seeds		
Pouroma minor[u]	32	179	*Henriettella sylvestris*	35	77,280
Eugenia nesiotica	30	90	*Bellucia grossularioides*	7	8,400

[d] = seeds from ripe fruits usually dropped; [w] = wind-dispersed species; [u] = probably predated unripe.

assuming the same dispersal patterns found for individual woolly monkeys in 1996 and 2000, the woolly monkeys could have been dispersing about 19.4 kg ha^{-1} year^{-1} (roughly one-third of the mass dispersed by the community).

Frugivory observations

I found much variation in the number of animal species consuming ripe fruits for the 75 species of plants included in the analysis (Table 15.6). The species most exclusively used by woolly monkeys were mainly protected fruits or large unprotected fruits, which is in agreement with the expected patterns (Janson, 1983). The highest number of species recorded feeding on a single species of plant was 43 (*Ficus sphenophylla*) and the lowest number was zero. For some, I observed a single species of frugivore during tree observations, but for all of these I found evidence to show that the fruits are consumed by other frugivores (see values in parentheses in Table 15.6 from nocturnal frugivory or sporadic observations – e.g. Stevenson *et*

Table 15.5. Comparison of dry seed mass collected in 300 traps in Tinigua National Park over a year cycle and the dry mass of seeds dispersed by woolly monkeys per year. The weight of seeds dispersed (those without a parent above the fruit trap) were also compared to the total weight of non-dispersed seeds using different methods of estimation.

Seed type	Dry seed mass (kg ha^{-1} $year^{-1}$)	%
Dispersed seeds	56.8	50.1
Naked non-dispersed seeds[a]	56.6	49.9
Dispersed seeds	56.8	19.6
Total non-dispersed seeds[b]	233.7	80.4
Dispersed seeds	56.8	18.5
Total non-dispersed seeds[c]	249.9	81.5
Dispersed seeds	56.8	22.4
Total non-dispersed seeds[d]	196.3	77.6
Dispersed seeds	56.8	21.3
Total non-dispersed seeds[e]	209.9	78.7
Dispersed by woolly monkeys[f]	19.4	34

[a] 'Naked seeds' = seeds without surrounding pulp or other fruit tissues; [b] including naked seeds as well as seeds imbedded in fruit tissues (assuming a constant community seed/fruit ratio of 7/17, from a sample of 120 species); [c] estimating the proportion of seed/fruit on a species-by-species basis; [d] same as 2, assuming that seeds falling under the parents weigh 16% less than intact seeds (an average for 153 cases when the weight of naked seeds was compared to that of intact seeds, a difference due to predation or aberrant development); [e] same as 3, assuming a seed weight loss of 16%; [f] total seed mass estimated from the corrected seed numbers, morphological data, and subtracting 7.2% of the dispersed seeds that fall under parental trees of the same species of plant.

al., 2000). Although woolly monkeys appeared to be the main dispersers for a large number of the species included in the analysis, there was no single species dispersed only by them. In fact, other animals in this community disperse some of the same species that show 100% values of fruit manipulation by woolly monkeys. Thus, the relative importance of each frugivore may change if the sample size is increased. That might be particularly true for species with very low visitation rates, such as *Jacaratia digitata* (Caricaceae), and even for one of the species with a large sample size, *Hymenaea oblongifolia* (Fabaceae).

Nocturnal activity was substantial for only a few species of plants, such as *Cecropia membranacea* and *C. engleriana*. For example, on average 68% of the total fruit removal from four *C. membranacea* trees occurred at night, although there were large variations in this pattern depending on tree size and location (e.g. large monkeys could not feed on small trees near river margins). Some other species were probably dispersed mainly by bats (e.g. *Licania arborea*; Chrysobalanaceae), given that the fruit remains under

Table 15.6. Summary of frugivory observations in the Tinigua community. Sample size is shown in the first column, followed by the percentage of fruits handled by five groups of frugivores. Seed predators included parrots and macaws (Psittacidae). The number of species observed to feed during sampling periods is included, as is the total number of species known to consume the fruits (in parentheses). The last two columns show an index of consumer diversity and the percentage of minutes in which there was feeding activity. The percentage can be more than 100% because it includes the total visit minutes for each individual feeding simultaneously.

Species	*N* (h)	Woolly monkey	Other atelines	Other monkeys	Seed predators	Other birds	Total number of species	Simpson's index (1–D)	% of minutes visited
Anacardiaceae									
Spondias mombin[a]	95	89	11	0	0	0	5	0.2	60
Spondias venulosa[a]	28	92	8	0	0	0	3	0.14	191
Annonaceae									
Guatteria punctata	35	12	25	1	0	62	10	0.75	168
Pseudomalmea diclina	45	30	10	60	0	0	4	0.58	11
Oxandra mediocris	36	64	2	0	0	34	9	0.52	106
Araliaceae									
Dendropanax caucanus	46	0	52	0	0	48	6	0.64	21
Arecaceae									
Oenocarpus bataua	81	0	56	32	12	0	5 (6)	0.57	3
Socratea exorrhiza[b]	30	88	11	0	0	0	3 (>4)	0.21	29
Burseraceae									
Bursera inversa	101	37	37	0	0	26	31	0.72	236
Crepidospermum rhoifolium	35	80	20	0	0	0	2 (4)	0.33	24
Protium glabrescens	36	2	96	2	0	1	6	0.08	67
Protium sagotianum	50	41	41	8	10	0	6	0.65	7
Cannabaceae									
Celtis schippii	38	0	0	91	0	9	14 (15)	0.24	90
Caricaceae									
Jacaratia digitata[c]	40	100	0	0	0	0	1 (2)	0	2
Celastraceae									
Maytenus macrocarpa	23	77	0	3	0	20	13	0.4	140

Continued

Table 15.6. *Continued*

Species	*N* (h)	Woolly monkey	Other atelines	Other monkeys	Seed predators	Other birds	Total number of species	Simpson's index (1–D)	% of minutes visited
Chrysobalanaceae									
Licania arborea[b]	25	0	0	0	0	0		1	0
Clusiaceae									
Clusia nigrolineata	31	0	0	0	0	100	9	0.69	19
Clusia palmicida	32	2	0	0	0	98	21	0.86	142
Garcinia macrophylla	101	63	34	0	3	0	3	0.49	27
Cucurbitaceae									
Cayaponia granatensis	32	49	46	0	0	6	6	0.55	62
Euphorbiaceae									
Hyeronima alchorneoides	30	0	7	13	2	78	40	0.81	1427
Sapium laurifolium	25	0	0	23	0	77	25	0.88	91
Fabaceae									
Erythrina amazonica[d]	35	0	0	0	0	0		1	0
Hymenaea oblongifolia[e]	114	99	0	0	0	1	2 (5)	0.02	7
Inga acreana	50	62	8	30	0	0	5	0.53	45
Inga acrocephala	38	24	10	50	0	17	4	0.66	4
Inga alba	41	48	24	0	8	20	8	0.69	172
Inga edulis[f]	74	54	0	41	0	5	4 (6)	0.61	6
Inga cf. *olivacea*	31	0	33	67	0	0	4	0.6	1
Swartzia leptopetala	30	9	0	82	0	9	5	0.31	61
Lauraceae									
Nectandra membranacea	39	0	12	0	0	88	11	0.54	46
Ocotea longifolia	34	0	74	0	0	26	18	0.43	445
Ocotea oblonga	35	0	1	0	0	99	17	0.79	86
Rhodostemonodaphne synandra	31	2	0	0	0	98	5	0.49	44
Lecythidaceae									
Gustavia hexapetala	94	84	14	0	2	0	4	0.28	5

Loranthaceae									
Psittacanthus cucullaris	34	0	0	0	0	100	7	0.69	36
Malvaceae									
Apeiba aspera	115	63	0	1	37	0	3	0.47	18
Marcgraviaceae									
Souroubea sympetala	31	4	0	0	0	96	23	0.77	441
Melastomataceae									
Clidemia sp.	35	0	0	0	0	100	3	0.52	43
Henriettella fissanthera	98	29	0	3	0	68	31	0.88	371
Meliaceae									
Guarea guidonia	45	0	0	0	0	100	3	0.54	9
Trichilia pallida	33	0	0	0	0	100	7	0.66	26
Trichilia pleeana	46	67	5	4	0	24	6	0.51	33
Trichilia tuberculata	55	22	30	22	0	26	15	0.79	124
Moraceae									
Brosimum alicastrum	36	72	9	0	0	19	7	0.44	21
Brosimum guianensis	31	57	39	0	2	1	11	0.52	106
Brosimum lactescens	37	65	17	18	0	1	10	0.53	185
Castilla ulei	40	4	14	54	0	27	12	0.66	62
Ficus andicola	38	10	41	0	0	49	36	0.83	823
Ficus insipida	34	100	0	0	0	0	1 (> 5)	0	3
Ficus sphenophylla	47	33	15	1	0	51	43	0.79	1322
Helicostylis tomentosa	36	66	33	1	0	0	4 (5)	0.45	8
Pseudolmedia laevigata	32	18	45	4	13	19	32	0.74	942
Pseudolmedia obliqua	39	14	61	13	10	2	15	0.63	750
Myristicaceae									
Iryanthera laevis	23	0	79	0	0	21	3	0.34	8
Virola calophylla	33	6	0	0	0	94	15	0.85	62
Virola flexuosa	69	23	37	0	0	40	15	0.69	107
Virola peruviana	45	8	78	0	0	14	4	0.38	25
Myrtaceae									
Eugenia nesiotica[a]	30	100	0	0	0	0	1 (4)	0	0
Nyctaginaceae									
Neea laxa	33	49	4	24	0	23	11	0.67	190

Continued

Table 15.6. *Continued*

Species	*N* (h)	Woolly monkey	Other atelines	Other monkeys	Seed predators	Other birds	Total number of species	Simpson's index (1–D)	% of minutes visited
Rubiaceae									
Alibertia cf. *hadrantha*	39	36	46	0	0	18	5	0.64	16
Salicaceae									
Casearia aculeata	30	0	0	0	0	100	9	0.72	18
Laetia corymbulosa[c]	36	99	0	0	0	1	2 (3)	0.02	41
Laetia procera	44	6	4	0	0	90	20	0.82	50
Sapindaceae									
Talisia intermedia	38	40	36	24	0	0	3	0.65	136
Sapotaceae									
Pouteria caimito	22	4	0	36	0	60	5	0.57	100
Pouteria procera	33	88	12	0	0	0	3	0.22	68
Solanaceae									
Cestrum racemosum	39	38	44	0	0	17	19	0.72	135
Urticaceae									
Cecropia engleriana[b]	76	16	0	1	0	83	9 (>11)	0.73	3
Cecropia membranacea[h]	109	20	18	27	0	35	28 (>32)	0.85	16
Cecropia sciadophylla	59	29	44	0	0	28	21	0.72	23
Coussapoa orthoneura	32	0	2	0	2	97	58	0.92	907
Pourouma bicolor[i]	70	54	36	5	5	0	7 (8)	0.64	85
Pourouma petiolulata[i]	40	93	7	0	0	0	2 (4)	0.13	36
Violaceae									
Leonia glycycarpa	39	79	17	4	0	0	3	0.35	2
Average	46	35	19	10	1	32	11		

[a] Consumed by tapirs and tortoises; [b] dispersed by bats; [c] consumed by spider monkeys; [d] seed removal was not observed; [e] unripe fruits heavily predated by parrots; [f] consumed by kinkajous and night monkeys; [g] consumed by other monkeys and tapirs; [h] heavily consumed by bats, and also ingested by owl monkeys, tyras and tapirs; [i] consumed by night monkeys.

roosts. Night monkeys (*Aotus brumbacki*; Cebidae) seemed to be responsible for the nocturnal activity noted for species such as *Pourouma bicolor*, *Brosimum lactescens* and *Pseudolmedia leavis* (Moraceae). Nevertheless, nocturnal activity for these plants was very low compared with diurnal activity.

On average, woolly monkeys removed 35% of the fruits in this community-wide analysis, including the most abundant canopy species in the area. Although a preliminary estimate, I believe it is accurate, as at some trees there seemed to be an over-representation of the role of woolly monkeys but in others the opposite situation occurred. Furthermore, this estimate concurs with the one based on the difference between the amount of dispersed seeds falling into traps and the estimate of seeds dispersed by the woolly monkeys (Table 15.5).

Discussion

Woolly monkeys could be considered to be excellent seed dispersers, in terms of the quantity of seeds dispersed, for most species of plants they consume. At least three sources of information lead to the same conclusion: (i) the estimates of seed numbers dispersed; (ii) the relationship between the number of seeds dispersed and manipulated; and (iii) the large proportion of seed mass that they disperse in the study area. Although there have been few studies of this kind, the estimated numbers of seeds dispersed by woolly monkeys in Tinigua National Park were higher than the estimates for other species of primates. For example, considering 'large' seeds (> 2 mm), Chapman (1995) estimated that three frugivorous monkeys (*Ateles geoffroyi* and *Alouatta palliata*, Atelidae; and *Cebus capucinus*, Cebidae) in Santa Rosa National Park (Costa Rica) dispersed approximately 5600 seeds km^{-2} day^{-1}. In comparison, woolly monkeys at Tinigua dispersed 'larger' seeds (> 3 mm) at a higher rate (at least 6781 and possibly 26,160 seeds km^{-2} day^{-1}; the latter estimate uses the corrected values and average population density estimates). Perhaps the only community that approaches the seed-dispersal patterns found at Tinigua corresponds to another west-Amazonian site (Yasuni National Park), where both woolly and spider monkeys are found at high densities (Dew, 2001). Dew (2001) estimated that woolly monkeys in Yasuni disperse roughly 3193 medium and large seeds (> 5 mm) ha^{-1} year^{-1}, but it is not clear whether those values were corrected for recovery rates and dispersal at night. Nevertheless, estimates of long-distance seed dispersal from primates in the Neotropics are higher than those found for palaeotropical monkeys (Chapman, 1995; Poulsen *et al.*, 2001). This pattern is caused by the role of cercopithecine monkeys in the Palaeotropics, which frequently use fruit resources but usually spit up seeds near parental trees (Lambert, 1999; Dominy and Duncan, 2005). Therefore, the large number of seeds dispersed by woolly monkeys in comparison with other sites, and the large proportion of seeds manipulated by them, suggests that the magnitude of dispersal limitation differs between tropical forests.

The linear relationship between the estimated number of seeds manipulated and the corrected number of seeds dispersed indicates that woolly monkeys disperse most of the seeds that they handle. The slope of the regression line for small-seeded species (< 3 mm) was very close to 1, which indicates very efficient dispersal and suggests that it is very difficult for woolly monkeys to consume fruits without swallowing seeds. In fact, for all small-seeded species, woolly monkeys always require a considerable amount of time to separate seeds from pulp, and the time required to separate a standard amount of fruit pulp is a good predictor of the likelihood of dispersal (P. Stevenson, 2006, unpublished results). In only a few cases have I observed woolly monkeys spitting out seeds from small-seeded plants (i.e. figs), and only when the animals have been feeding for a long time in the tree. In other cases, woolly monkeys pull whole fruit spikes from *Cecropia* trees, take a bite from the distal part, and then drop the rest of the fruit, thereby providing an inefficient service as seed dispersers. However, the general removal rates observed for *Cecropia* trees (Stevenson, 2002) indicate that the overall role of frugivores as seed dispersers is positive (mainly due to the efficient role of bats), and that woolly monkeys also disperse large quantities of these seeds. In summary, given how difficult it must be for woolly monkeys to separate the pulp from the seeds, they seem to play a very efficient role in terms of the amount of seeds they disperse.

The slope of the regression between the number of seeds dispersed and the number manipulated, combined with an estimate of the actual number of seeds dispersed, provides a measure of seed dispersal efficiency. However, it is important that both components be used. For example, a slope of 1.0 could be achieved by a frugivore removing only a few seeds, but all that it manipulated. Its ecological effect would be smaller than that of a less efficient disperser (slope < 1) removing more seeds.

In this study, using both behavioural observations and faecal analyses to estimate the quantity of seeds dispersed was important for determining which species were killed or inefficiently dispersed, especially where the frugivores ingest ripe fruits without dropping seeds. For example, species such as *Brosimum utile* and *Clarisia biflora* (Moraceae) were frequently consumed without seeds being dropped under the parental trees. Although these cases appeared to be effective dispersal, this conclusion was rejected because these seeds were never found in faecal samples. Dispersal studies that focus only on faeces are not able to identify species with seeds that are nearly or fully destroyed by the animals during mastication, because seed fragments are not recognizable. This result reinforces the importance of combining behavioural observations of frugivores in fruiting trees with faecal analyses in order to assess the predatory role as well as the dispersal role of primates.

The large proportion of seeds that woolly monkeys seem to be dispersing in this community (34% of the dry seed mass and 35% of the fruits manipulated) suggests that they are playing an important role in plant population dynamics. Combined with evidence that seed dispersal can be an important contributor to plant fitness (Stevenson, 2002), this strongly suggests that woolly monkeys play a substantial ecological role in the Tinigua community. The estimated proportion of seeds manipulated and dispersed by woolly monkeys is similar to that for the entire bird community in Tinigua, composed of at least 156 species. In addition, the role of large ateline primates was disproportionate compared with other frugivores, since the combined contribution of woolly and spider monkeys was 54% of seed handling. Part of this pattern seems to be related to the large contribution that ateline monkeys (*Lagothrix* and *Ateles*) make to the primate community biomass (Table 15.7). However, both species visit more fruiting species and disperse more seeds than other similarly abundant species such as howler monkeys (Stevenson *et al.*, 2002). Therefore, while the role of ateline primates is associated with their abundance, their behaviour as seed dispersal agents is also important. In turn, the extinction of large primates (see Wright, Chapter 28, this volume) may have strong effects on regeneration patterns in this kind of ecosystem.

Table 15.7. Preliminary density estimates (individuals km^{-2}) and biomass for the most common species of primate at the Macarena Center of Ecological Investigations (modified from Stevenson, 2002). Standard errors from monthly averages are also shown, as well as transect length travelled and the estimated width.

Year	Transect length (km)	*Lagothrix lagothricha*	*Ateles belzebuth*	*Cebus apella*	*Saimiri sciureus*	*Alouatta seniculus*
Transect width (m)		22	19	20	20	16
1990–1991	349	42 ± 6	23 ± 3	31 ± 4	13 ± 5	20 ± 5
1996–1997	482	41 ± 4	24 ± 4	24 ± 4	9 ± 3	17 ± 3
2000	534	50 ± 11	25 ± 5	26 ± 9	16 ± 3	30 ± 5
Average		44.7	24.0	27.0	12.7	22.3
Biomass (kg/km^2)		389	180	78	12	145

Acknowledgements

I would like to thank Alicia Madinah, Carolina García, Monica Pineda, Tatiana Samper and Andrés Link for assistance in the field. I thank Charles Janson, Patricia Wright, John G. Fleagle, Anthony DiFiore, Maria Clara Castellanos, Nicole Gibson and the editors of this book for their comments. This study was possible thanks to the logistic support of CIEM (Centro de Investigaciones Ecológicas La Macarena) and permits from Unidad de Parques Nacionales. Financial support came from the following institutions: La Fundación para la Promoción de la Investigación y la Tecnología (Banco de la República), Margot Marsh Foundation, Lincoln Park Zoo, Primate Conservation Inc. and IdeaWild.

References

Andresen, E. (2002) Primary seed dispersal by red howler monkeys and the effect of defecation patterns on the fate of dispersed seeds. *Biotropica* 34, 261–272.

Balcomb, S.R. and Chapman, C.A. (2003) Bridging the gap: influence of seed deposition on seedling recruitment in a primate–tree interaction. *Ecological Monographs* 73, 625–642.

Caceres, N.C., Dittrich, V.A.O. and Monteiro, E.L.A. (1999) Fruit consumption, distance of seed dispersal and germination of Solanaceous plants ingested by common opossum (*Didelphis aurita*) in southern Brazil. *Revue d'Ecologie-La Terre et la Vie* 54, 225–234.

Castro, S.A., Silva, S.I., Meserve, P.L., Gutierrez, J.R., Contreras, L.C. and Jaksic, F.M. (1994) Frugivory and seed dispersal by culpeo fox (*Pseudalopex culpaeus*) in Fray-Jorge-National-Park (IV Region, Chile). *Revista Chilena de Historia Natural* 67, 169–176.

Chapman, C.A. (1995) Primate seed dispersal: coevolution and conservation implications. *Evolutionary Anthropology* 4, 74–82.

Chapman, C.A. and Chapman, L.J. (1996) Frugivory and the fate of dispersed seeds of six African tree species. *Journal of Tropical Ecology* 12, 491–504.

Compton, S.G., Craig, A. and Waters, I.W.R. (1996) Seed dispersal in an African fig tree: birds as high quantity, low quality dispersers? *Journal of Biogeography* 23, 553–563.

Dew, L. (2001) Synecology and seed dispersal in woolly monkeys (*Lagothrix lagotricha poeppigii*) and spider monkeys (*Ateles belzebuth belzebuth*) in Parque Nacional Yasuni, Ecuador. PhD thesis, Department of Anthropology, University of California, Davis, California.

Dominy, N.J. and Duncan, B.W. (2005) Seed-spitting primates and the conservation and dispersion of large-seeded trees. *International Journal of Primatology* 26, 631–649.

Figuerola, J., Green, A.J. and Santamaria, L. (2002) Comparative dispersal effectiveness of wigeongrass seeds by waterfowl wintering in south-west Spain: quantitative and qualitative aspects. *Journal of Ecology* 90, 989–1001.

Godínez-Alvarez, H., Valiente-Banuet, A. and Rojas-Martinez, A. (2002) The role of seed dispersers in the population dynamics of the columnar cactus *Neobuxbaumia tetetzo. Ecology* 83, 2617–2629.

Graham, C.H., Moermond, T.C., Kristensen, K.A. and Mvukiyumwami, J. (1995) Seed dispersal effectiveness by two bulbuls on *Maesa lanceolata*, an African montane forest tree. *Biotropica* 27, 479–486.

Guitian, J., Munilla, I., Guitian, P. and Lopez, B. (1994) Frugivory and seed dispersal by redwings *Turdus iliacus* in Southwest Iceland. *Ecography* 17, 314–320.

Hamilton, W.D. and May, R.M. (1977) Dispersal in stable habitats. *Nature* 269, 578–581.

Herrera, C.M., Jordano, P., Lopez-Soria, L. and Amat, J.A. (1994) Recruitment of a mast-fruiting, bird-dispersed tree: bridging frugivore activity and seedling establishment. *Ecological Monographs* 64, 315–344.

Howe, H.F. (1980) Monkey dispersal and waste of a neotropical fruit. *Ecology* 61, 944–959.

Hubbell, S.P., Foster, R.B., O'Brien, S.T., Harms, K.E., Condit, R., Wechsler, B., Wright, S.J. and de Lao, S.L. (1999) Light-gap disturbances, recruitment limitation, and tree diversity in a neotropical forest. *Science* 283, 554–557.

Janson, C.H. (1983) Adaptation of fruit morphology to dispersal agents in a neotropical forest. *Science* 219, 187–189.

Jordano, P. and Schupp, E.W. (2000) Seed disperser effectiveness: the quantity component and patterns of seed rain for *Prunus mahaleb. Ecological Monographs* 70, 591–615.

Kaplin, B.A. and Moermond, T.C. (1998) Variation in seed handling by two species of forest monkeys in Rwanda. *American Journal of Primatology* 45, 83–101.

Lambert, J.E. (1999) Seed handling in chimpanzees (*Pan troglodytes*) and redtail monkeys (*Cercopithecus ascanius*): implications for understanding hominoid and cercopithecine fruit-processing strategies and seed dispersal. *American Journal of Physical Anthropology* 109, 365–386.

Loiselle, B.A. and Blake, J.G. (1999) Dispersal of melastome seeds by fruit-eating birds of tropical forest understory. *Ecology* 80, 330–336.

Mandujano, S., Gallina, S. and Bullock, S.H. (1994) Frugivory and dispersal of *Spondias purpurea* (Anacardiaceae) in a tropical deciduous forest in Mexico. *Revista de Biologia Tropical* 42, 107–114.

Montiel, S. and Montana, C. (2000) Vertebrate frugivory and seed dispersal of a Chihuahuan desert cactus. *Plant Ecology* 146, 221–229.

Muller-Landau, H.C., Wright, S.J., Calderon, O., Hubbell, S.P. and Foster, R.B. (2002) Assessing recruitment limitation: concepts, methods and case-studies from a tropical forest. In: Levey, D.J., Silva, W.R. and Galetti, M. (eds) *Seed Dispersal and Frugivory: Ecology, Evolution, and Conservation.* CAB International, Wallingford, UK, pp. 35–53.

Nogales, M., Delgado, J.D. and Madinah, F.M. (1998) Shrikes, lizards and *Lycium intricatum* (Solanaceae) fruits, a case of indirect seed dispersal on an oceanic island (Alegranza, Canary Islands). *Journal of Ecology* 86, 866–871.

Nogales, M., Valido, A. and Madinah, F.A. (1995) Frugivory of *Plocama pendula* (Rubiaceae) by the rabbit (*Oryctolagus cuniculus*) in xerophytic zones of Tenerife (Canary Islands). *Acta Oecologica* 16, 585–591.

Nogales, M., Valido, A., Madinah, F.M. and Delgado, J.D. (1999) Frugivory and factors influencing visitation by birds at 'Balo' (*Plocama pendula* Ait., Rubiaceae) plants in the Canary Islands. *Ecoscience* 6, 531–538.

Nogales, M., Quilis, V., Madinah, F.M., Mora, J.L. and Trigo, L.S. (2002) Are predatory birds effective secondary seed dispersers? *Biological Journal of the Linnean Society* 75, 345–352.

Pigozzi, G. (1992) Frugivory and seed dispersal by the European badger in a Mediterranean habitat. *Journal of Mammalogy* 73, 630–639.

Poulsen, J.R., Clark, C.J. and Smith, T.B. (2001) Seed dispersal by a diurnal primate community in the Dja Reserve, Cameroon. *Journal of Tropical Ecology* 17, 787–808.

Renne, I.J., Gauthreaux, S.A. and Gresham, C.A. (2000) Seed dispersal of the Chinese tallow tree (*Sapium sebiferum* (L.) Roxb.) by birds in coastal South Carolina. *American Midland Naturalist* 144, 202–215.

Russo, S.E. (2003) Responses of dispersal agents to tree and fruit traits in *Virola calophylla* (Myristicaceae): implications for selection. *Oecologia* 136, 80–87.

Sallabanks, R. (1992) Fruit fate, frugivory, and fruit characteristics: a study of the hawthorn, *Crataegus monogyna* (Rosaceae). *Oecologia* 91, 296–304.

Schupp, E.W. (1993) Quantity, quality and the effectiveness of seed dispersal by animals. *Vegetatio* 107/108, 15–29.

Stevenson, P.R. (2000) Seed dispersal by woolly monkeys (*Lagothrix lagothricha*) at Tinigua National Park, Colombia: dispersal distance, germination rates, and dispersal quantity. *American Journal of Primatology* 50, 275–289.

Stevenson, P.R. (2002) Frugivory and seed dispersal by woolly monkey (*Lagothrix lagothricha*) at Tinigua National Park, Colombia. PhD thesis, IDPAS–SUNY at Stony Brook, New York.

Stevenson, P.R., Quiñones, M.J. and Ahumada, J.A. (1994) Ecological strategies of woolly monkeys (*Lagothrix lagotricha*) at La Macarena, Colombia. *American Journal of Primatology* 32, 123–140.

Stevenson, P.R., Quiñones, M.J. and Castellanos, M.C. (2000) *Guía de Frutos de los Bosques del Río Duda, La Macarena, Colombia.* Asociación para La Defensa de La Macarena–IUCN, Bogotá, Colombia.

Stevenson, P.R., Castellanos, M.C., Pizarro, J.C. and Garavito, M. (2002) Effects of seed dispersal by three ateline monkey species on seed germination at Tinigua National Park, Colombia. *International Journal of Primatology* 23, 1187–1204.

Stevenson, P.R., Pineda, M. and Samper, T. (2005) Influence of seed size on dispersal patterns of woolly monkeys (*Lagothrix lagothricha*) at Tinigua Park, Colombia. *Oikos* 110, 435–440.

Terborgh, J., Pitman, N., Silman, M., Schichter, H. and Núñez, V.P. (2002) Maintenance of tree diversity in tropical forests. In: Levey, D.J., Silva, W.R. and Galetti, M. (eds) *Seed Dispersal and Frugivory: Ecology, Evolution, and Conservation.* CAB International, Wallingford, UK, pp. 1–17.

Valido, A. and Nogales, M. (1994) Frugivory and seed dispersal by the lizard *Gallotia galloti* (Lacertidae) in a xeric habitat of the Canary Islands. *Oikos* 70, 403–411.

Wills, C., Condit, R., Foster, R.B. and Hubbell, S.P. (1997) Strong density- and diversity-related effects help to maintain tree species diversity in a neotropical forest. *Proceedings of the National Academy of Sciences of the USA* 94, 1252–1257.

Wills, C. and Condit, R. (1999) Similar non-random processes maintain diversity in two tropical rainforests. *Proceedings of the Royal Society London, Series B, Biological Sciences* 266, 1445–1452.

Willson, M.F., Sabag, C., Figueroa, J. and Armesto, J.J. (1996) Frugivory and seed dispersal of *Podocarpus nubigena* in Chiloe, Chile. *Revista Chilena de Historia Natural* 69, 343–349.

Wright, J.S. (2002) Plant diversity in tropical forests: a review of mechanisms of species coexistence. *Oecologia* 130, 1–14.

III Seed Fate and Establishment

EUGENE W. SCHUPP

Introduction

This section spans the transition from spatial patterns of seed-fall to the fates of seeds and seedlings, perhaps the most critical stages determining the success or failure of the seed dispersal process. The field of seed dispersal has been, in a sense, adrift since the 'paradigm' of specialist versus generalist seed dispersal systems (McKey, 1975; Howe and Estabrook, 1977) met the reality that most fruit–frugivore interactions are in fact very diffuse and likely to be non-equilibrial (Herrera, 1998; Levey and Benkman, 1999). While no unifying theoretical framework has yet emerged to to replace the paradigm, exciting progress is none the less being made; although adrift, we are at least adrift in a very fertile sea, with our understanding growing theoretically and empirically.

An exploration of the widespread networks of indirect plant–frugivore and plant–plant interactions is beginning to change our understanding of the process and the consequences of seed dispersal (Carlo, 2005; see Carlo *et al.*, Chapter 16). These interactions among fluid assemblages of unspecialized dispersers foraging on fluid assemblages of co-fruiting, unspecialized plants are by their very nature diffuse and wildly variable both spatially and temporally. A focus on spatially explicit networks of plants as elements connected by frugivores, and on the influence of 'hubs', or plants that capture most dispersal services, suggests that whether co-occurring plants compete for or facilitate each other's dispersal depends on the identity and spatial patterning of neighbouring plants and the identity and availability of frugivores. Furthermore, hubs greatly influence patterns of seed dispersal, creating nested directionality of deposition patterns and very patchy dispersal limitation.

The context of the population and community in which an interaction is imbedded has other important consequences for dispersal. In particular, additional insight into the complexity of patterns of seed-fall and of seed and seedling survival can be gained by expanding on the original Janzen–Connell model (Janzen, 1970; Connell, 1971) to more explicitly include interactions. For example, tree population density can affect the outcome of distance- and density-responsive enemy attack (Schupp, 1992). Fruiting plants do not live in isolation, but interact with other fruiting plants of the same and different species as well as with a host of enemies. Consequently, the overlapping of multiple seed shadows results in much more complex population-level seed shadows – and patterns of recruitment – than predicted by the Janzen–Connell model, even if individual seed shadows monotonically decrease with distance from fruiting trees (see Kwit *et al.*, Chapter 19). In reality, though, seed shadows approaching smooth symmetrical curves are the exception, if they exist at all. The complexity of disperser behaviour generates contagious dispersal, with many seeds arriving in some sites and few to none in others, independent of distance from a fruiting tree (Schupp *et al.*, 2002). The realization that real-world seed shadows created by overlap and by contagious dispersal are extraordinarily complex has major implications for quantitative and spatial patterns of recruitment and species diversity (see Kwit *et al.*, Chapter 19).

Similarly, models suggest that the interaction between patterns of seed dispersal and of natural enemy dispersal can have significant consequences for the impact of enemies on recruitment (Adler and Muller-Landau, 2005; see Muller-Landau and Adler, Chapter 18). Surprisingly, seed dispersal distance might even have contrasting effects on the intensity of mortality and on the density dependence of population growth, with shorter-distance dispersal increasing the former and longer-distance dispersal increasing the latter. Additionally, the consequences of seed dispersal distances for density dependence, population regulation, and species richness appear to be directly dependent on the dispersal distances of natural enemies.

Thus, while there have been extremely clear demonstrations of Janzen–Connell density- and distance-dependent recruitment for decades (e.g. Clark and Clark, 1984; Howe *et al.*, 1985), it is not at all surprising that there are many exceptions to the pattern (Hammond and Brown, 1998). Exceptions may also arise because of variation in natural enemy abundance and behaviour. For example, pathogen-induced mortality varies tremendously beneath different female *Cecropia* sp. crowns within a forest – in fact, the variation can be greater than the difference in mean mortality between forests hundreds of kilometres apart and differing dramatically in climate (see Gallery *et al.*, Chapter 22). In addition, natural enemies may at times be unaware of the model and simply not know exactly how to behave. While dispersal of *Cecropia* seeds away from the parent reduces seed mortality from pathogens, this advantage seems to have nothing to do with distance from a fruiting tree or density of seeds, but simply with the presence of a *Cecropia*, whether male or female (see Gallery *et al.*, Chapter 22). Continued research into the interactions between seed dispersal and pathogens is poised to advance our understanding of the consequences of seed dispersal.

Most of the emphasis on the consequences of dispersal for recruitment, especially in tropical forests, has centred on interactions with the biotic environment as exemplified by the focus on the Janzen–Connell model, or has been related to gap dependence. There are, however, strong edaphic controls of species distributions of at least some species in many forests, driven by slope, moisture, or soils (e.g. Harms *et al.*, 2001; see Russo *et al.*, Chapter 23). While little explored to date, the interactions among dispersal mode, seed size, and edaphic specializations can have major consequences for plant recruitment, population and community structure, and evolution, with the importance of different factors varying with spatial scale (see Russo *et al.*, Chapter 23). Studies incorporating edaphic variables along with biotic interactions in models and experiments of seed dispersal should be especially enlightening. For example, some tropical wind-dispersed trees appear to be gap-dependent largely because of much greater pathogen mortality in the understorey, not because of photosynthetic limitations (Augspurger, 1984); it is the interaction between the abiotic and biotic environments that determines the outcome of dispersal.

While biotic interactions have dominated thinking on the consequences of seed dispersal, not all biotic interactions have been deemed equally important. In particular, post-dispersal mutualistic interactions have received much less attention than have the negative interactions of seed predation, herbivory and pathogen attack, despite growing evidence that in many cases understanding such mutualisms is critical to understanding the overall consequences of dispersal. An excellent example is the secondary dispersal of wind-dispersed pines by seed-caching rodents, which can completely rearrange the initial seed shadow, moving seeds further from the parent and even altering the habitat and microhabitat distributions of seeds (Vander Wall, 1992, 2002). Seed-caching rodents can be involved in even more intricate mutualisms with plants by dispersing not only seeds, but also mycorrhizal inoculum. Rodents can improve plant establishment by directing the dispersal of seeds to sites that have the appropriate inoculum, by directing the dispersal of appropriate inoculum to sites that have or are likely to receive the seeds independently, or by co-dispersing the seeds and the inoculum (see Theimer and Gehring, Chapter 21). Given the abundance of potential post-dispersal mutualisms, including facilitation of recruitment in stressful environments, it is somewhat surprising that they have received minimal attention in studies of the consequences of seed dispersal.

As the field of seed dispersal ecology has advanced, it has become increasingly clear that many of the consequences of dispersal are not fixed, but are in fact context-dependent (see Schupp, Chapter 20). The consequences of any particular 'pattern' of dispersal can vary greatly from place to place and from year to year; even rankings of the suitability of alternative dispersal sites can switch as environmental conditions shift spatially and temporally (Ibáñez and Schupp, 2001). In fact, a major theme running throughout this section, either implicitly or explicitly, is the idea that both the process and the outcome of dispersal are variable and highly dependent on the ecological context. While this complicates our quest for understanding seed dispersal, it is a reality that can and must be incorporated into any comprehensive theory of seed dispersal.

Although our knowledge of the consequences of dispersal has grown steadily over the last decades, our quantitative abilities to address dispersal have lagged in important areas. Modelling approaches have been, and will continue to be, critical for advancing the field (e.g. see Kwit *et al.*, Chapter 19; Muller-Landau and Adler, Chapter 18), but they cannot stand alone without empiricism. What is especially critical is quantitatively linking seed dispersal to its population-level consequences, a need articulated over a decade ago (Howe, 1989; Herrera *et al.*, 1994; Schupp and Fuentes, 1995) and still largely unfulfilled. However, Godínez-Alvarez and Jordano (Chapter 17) present a potential solution: a matrix model approach for integrating studies of the process of seed dispersal with studies of the population-level consequences of dispersal. Such models have the flexibility to incorporate the contributions of distinct species or functional groups of dispersers into the fecundity function of the matrix; thus they can directly link dispersal with the consequences – that is, with the transitions in the matrix. Although projection matrices have limitations, their extreme flexibility can accommodate the wide diversity of both dispersal patterns and delayed consequences of dispersal for recruitment that characterize 'the real world'.

The chapters in this section, whether predominantely empirical, models or syntheses of the literature, explore new territory and expand our view of the consequences of seed dispersal. They reveal new approaches, new ideas, and new enthusiasm. And they point the way for greater progress in the future.

References

Adler, F.R. and Muller-Landau, H.C. (2005) When do localized natural enemies increase species richness? *Ecology Letters* 8, 438–447.

Augspurger, C.K. (1984) Seedling survival of tropical tree species: interactions of dispersal distance, light gaps, and pathogens. *Ecology* 65, 1705–1712.

Carlo, T.A. (2005) Interspecific neighbors change seed dispersal pattern of an avian-dispersed plant. *Ecology* 86, 2440–2449.

Clark, D.A. and Clark, D.B. (1984) Spacing dynamics of a tropical rain-forest tree: evaluation of the Janzen–Connell model. *American Naturalist* 124, 769–788.

Connell, J.H. (1971) On the role of natural enemies in preventing competitive exclusion in some marine animals and in rain forest trees. In: den Boer, P.J. and Gradwell, G.R. (eds) *Dynamics of Populations*. Centre for Agricultural Publishing and Documentation (PUDOC), Wageningen, The Netherlands, pp. 298–313.

Hammond, D.S. and Brown, V.K. (1998) Disturbance, phenology and life-history characteristics: factors influencing distance/density-dependent attack on tropical seeds and seedlings. In: Newbery, D.M., Prins, H.H.T. and Brown, N.D. (eds) *Dynamics of Tropical Communities*. Blackwell, Oxford, UK, pp. 51–78.

Harms, K.E., Condit, R., Hubbell, S.P. and Foster, R.B. (2001) Habitat associations of trees and shrubs in a 50-ha neotropical forest plot. *Journal of Ecology* 89, 947–959.

Herrera, C.M. (1998) Long-term dynamics of Mediterranean frugivorous birds and fleshy fruits: a 12 year study. *Ecological Monographs* 68, 511–538.

Herrera, C.M., Jordano, P., López-Soria, L. and Amat, J.A. (1994) Recruitment of a mast-fruiting, bird-dispersed tree: bridging frugivore activity and seedling establishment. *Ecological Monographs* 64, 315–344.

Howe, H.F. (1989) Scatter- and clump-dispersal and seedling demography: hypothesis and implications. *Oecologia* 79, 417–426.

Howe, H.F. and Estabrook, G.F. (1977) On intraspecific competition for avian dispersers in tropical trees. *American Naturalist* 11, 817–832.

Howe, H.F., Schupp, E.W. and Westley, L.C. (1985) Early consequences of seed dispersal for a neotropical tree (*Virola surinamensis*). *Ecology* 66, 781–791.

Ibáñez, I. and Schupp, E.W. (2001) Positive and negative interactions between environmental conditions affecting *Cercocarpus ledifolius* seedling survival. *Oecologia* 129, 543–550.

Janzen, D.H. (1970) Herbivores and the number of tree species in tropical forests. *American Naturalist* 104, 501–528.

Levey, D.J. and Benkman, C.W. (1999) Fruit–seed disperser interactions: timely insights from a long-term perspective. *Trends in Ecology and Evolution* 14, 41–42.

McKey, D. (1975) The ecology of coevolved seed dispersal systems. In: Gilbert, L.E. and Raven, P. (eds) *Coevolution of Animals and Plants*. University of Texas Press, Austin, Texas, pp. 159–191.

Schupp, E.W. (1992) The Janzen–Connell model for tropical tree diversity: population implications and the importance of spatial scale. *American Naturalist* 140, 525–530.

Schupp, E.W. and Fuentes, M. (1995) Spatial patterns of seed dispersal and the unification of plant population ecology. *Écoscience* 2, 267–275.

Schupp, E.W., Milleron, T. and Russo, S.E. (2002) Dissemination limitation and the origin and maintenance of species-rich tropical forests. In: Levey, D.J., Silva, W.R. and Galetti, M. (eds) *Seed Dispersal and Frugivory: Ecology, Evolution and Conservation*. CAB International, Wallingford, UK, pp. 19–33.

Vander Wall, S.B. (1992) The role of animals in dispersing a "wind-dispersed" pine. *Ecology* 73, 614–621.

Vander Wall, S.B. (2002) Secondary dispersal of Jeffrey pine seeds by rodent scatter hoarders: the roles of pilfering, recaching, and a variable environment. In: Levey, D.J., Silva, W.R. and Galetti, M. (eds) *Seed Dispersal and Frugivory: Ecology, Evolution and Conservation*. CAB International, Wallingford, UK, pp. 193–208.

16 Plant–Frugivore Interactions as Spatially Explicit Networks: Integrating Frugivore Foraging with Fruiting Plant Spatial Patterns

T.A. Carlo, J.E. Aukema and J.M. Morales

Introduction

Dispersal is essential for the persistence of populations in space and time (McArthur and Wilson, 1968; Levins, 1969). Theoretical and empirical studies suggest that seed dispersal is one of the major mechanisms that build and sustain diversity in plant communities (Nathan and Muller-Landau, 2000; Webb and Peart, 2001; Condit *et al.*, 2002). In many plant communities, frugivorous animals are believed to play the major role in structuring and maintaining diversity through their seed dispersal services (Schupp *et al.*, 2002; Terborgh *et al.*, 2002). The relevance of frugivores is implicit in the many adaptations for endozoochory (i.e. seed dispersal through guts) in nearly all plant taxa (van der Pijl, 1972), as well as in the high proportions of woody plant species that are dispersed by frugivores in many plant communities (Herrera, 2003). Therefore, in order to understand the ecology of many plant communities it is necessary to have a mechanistic understanding of frugivore–plant interactions and their relationship to plant population dynamics and diversity. However, we still know very little about how frugivores influence plant distribution, community structure and plant evolution (Wang and Smith, 2002).

Endozoochory is set apart from other dispersal modes, such as wind dispersal, by several important properties. First, frugivores create non-random seed shadows that show strong patterns of directionality towards specific habitats, cover types, species of plants, branches, topographic features and more (Janzen, 1970; Reid, 1991; Wenny, 2001; Schupp *et al.*, 2002; Kwit *et al.*, 2004a; Russo and Augspurger, 2004). Second, fruiting plants usually surpass frugivores in the numbers of individuals and/or species, creating asymmetries that could further limit seed dispersal services among plant species or individuals (Wheelwright, 1985; Jordano,

 Seed Dispersal: Theory and its Application in a Changing World (eds A.J. Dennis *et al.*)

1987; Carlo *et al.*, 2003). This suggests that for frugivore-dispersed plants, frugivory, and thus dispersal, are highly contingent on context and location relative to other co-fruiting individuals or species (Saracco *et al.*, 2004; Carlo, 2005a). Furthermore, frugivores show preferences for particular species of fruiting plant and this can accentuate asymmetries in per capita seed dispersal of plants when frugivores are shared (Carlo *et al.*, 2003). Preferences may then promote more limitation in seed dispersal services among co-fruiting individuals or species. This results in multiple indirect plant–frugivore and plant–plant interactions that are highly variable in space and time because they take place between assemblages of unspecialized frugivores and multiple species of fruiting plants (Janzen, 1983; Jordano, 1987; Fleming *et al.*, 1993; Jordano *et al.*, 2003; Bascompte *et al.*, 2006). In ecological and evolutionary terms, such relationships are diffuse, because interactions between plants and frugivores generally lack strong specificity (Herrera, 1982; Janzen, 1983; Fleming *et al.*, 1993).

In this chapter, we attempt to fit the diffuse characteristics of frugivore–plant interactions into a framework that provides new ideas about frugivory and seed dispersal patterns, plant community associations and evolution. Specifically, we propose that frugivory and seed dispersal patterns should be examined in a spatially explicit manner that will help explain variability at the levels of both plant individuals and species. We argue that the framework can shed light on fruiting plant community structure, diversity and evolution.

A Field Without a Conceptual Framework

The realization that a majority of plant–frugivore relationships are diffuse and apparently non-equilibrial (Herrera, 1998) brought conceptual progress in the field of frugivory and seed dispersal to a halt (Howe, 1993; Levey and Benkman, 1999). This can be attributed to the influence of traditional ecological concepts centred on the individuality of species (Callaway, 1997; Bruno *et al.*, 2003; Lortie *et al.*, 2004), to the notion that omnivory destabilizes food webs (Pimm and Lawton, 1978; but see Agrawal, 2003), and on the principle of competitive exclusion championed in niche theory (Hutchinson, 1957). In accord with these views, the original conceptual framework of the frugivory and seed dispersal field, once known as 'the paradigm', proposed that there are two types of plant frugivore systems, specialized and generalized (McKey, 1975). The paradigm predicted that specific fruit traits (e.g. high nutrient content, seed/pulp ratio, water content), phenology patterns, quantity and quality of dispersal, and specificity of interactions with frugivores would determine whether plants followed generalized or specialized strategies (McKey, 1975; Howe, 1993). For example, species with lipid-rich fruits should show asynchronous fruiting patterns to sustain a handful of highly specialized species of frugivores, thus providing high-quality dispersal, while those with watery and sugary fruits (i.e. cheaper fruits to produce) should produce superabundant and synchronous fruit crops in order to attract multiple unspecialized and opportunistic frugivores and omnivores that provide massive, although low-quality, dispersal on a per-seed basis (McKey, 1975; Howe, 1993).

However, as evidence mounted, the predictions of this paradigm found little support (Howe, 1993). For example, many fig trees show asynchronous phenology patterns and have large groups of frugivore consumers that include specialists and non-specialists as dispersers, while producing small, sugary fruits (Cruz, 1974; Terborgh, 1986; Kannan and James, 1999). Other species, such as *Schefflera morototoni* (Araliaceae), produce small, nutrient-rich fruits in superabundant and synchronous fruit displays that attract multiple non-specialized frugivores (Carlo *et al.*, 2003; Saracco *et al.*, 2005). In some cases, omnivores have been reported to behave, from a plant's perspective, as specialized frugivores (Carlo *et al.*, 2003). For example, *Phainopepla nitens* (Bombycillidae) can be considered a frugivore specializing on *Phoradendron californicum* (Viscaceae) during the winter in the Sonoran Desert. However, *P. nitens* is mostly an insectivore during the summer at high elevations, and *P. californicum* is also effectively dispersed by other species of birds in the desert (Larson, 1996; Chu and Walsberg, 1999). Hence, most (if not all) plant–frugivore systems involve plants sharing multiple frugivores, and frugivores visiting multiple plants and feeding on other organisms at different trophic levels as well.

The impacts of shared consumers on community structure have been poorly explored theoretically and empirically (Agrawal, 2003), particularly in the plant–frugivore literature. This is surprising because:

1. Seed dispersal is central in theories of diversity and community ecology (Hubbell, 2001; Terborgh *et al.*, 2002; Wang and Smith, 2002);
2. Plants that are dispersed by frugivores are prevalent in many communities (Herrera, 2003);
3. Almost all species of plants dispersed by frugivores share multiple frugivore seed dispersers (Jordano, 1987; Carlo *et al.*, 2003; Jordano *et al.*, 2003).

Understanding the distribution and abundance of species of fruiting plants, associations and communities may require a mechanistic approach that explicitly incorporates the importance of multiple fruit resources and the ubiquity of having shared and generalized plant–frugivore interactions.

Competition and Facilitation in Frugivore-dispersed Plants: Neighbourhood Effects

García *et al.* (2001) titled a paper 'Frugivory in *Juniperus communis* depends more on population characteristics than on individual attributes'. Their title reveals not only a common finding in many frugivory and seed dispersal studies, it also represents the type of result that 'the paradigm' could not handle conceptually and that we still lack adequate theory to describe. We have no theory to explain how variability in characteristics that exist beyond an individual plant (e.g. having co-fruiting intra- and heterospecific neighbours) affects frugivory and seed dispersal of plants.

A common source of variation in frugivory services that is beyond the level of the organism is the fruiting neighbourhood of a fruiting plant (Manasse and Howe, 1983; Herrera, 1984; Denslow, 1987; Sargent, 1990; Saracco *et al.*, 2005; Carlo, 2005a; Canham and Uriarte, 2006). Failing to account for intra- and heterospecific neighbourhood interactions may obscure key aspects of the effects of generalized frugivores on seed dispersal and the spatial patterning of plant communities (Carlo and Aukema, 2005). For example, Foster (1990) examined the contribution of 27 plant characteristics (e.g. seed/pulp ratio, nutrients, water content, crop size, height, etc.) to the variation in rates of fruit removal among individuals of *Allophylus edulis* (Sapindaceae) in Paraguay. Of all the variables measured, Foster found only a few significant correlations (fruit crop size being most important; see also Davidar and Morton, 1986). Foster argued that interactions with influential heterospecific neighbours, in some cases, could override intraspecific trait variance in explaining seed removal rates. Explanations for results from both García *et al.* (2001) and Foster (1990) call for spatial and positional information of fruiting plant individuals in relation to others in order to make sense of observed patterns of fruit removal. A simple way to summarize and simplify spatial/positional information is to characterize the co-fruiting neighbourhood of a plant (Manasse and Howe, 1983; Sargent, 1990; Saracco *et al.*, 2005).

Neighbourhoods have long been suspected to be important in explaining frugivory patterns in co-fruiting plants through competition (Manasse and Howe, 1983) or facilitation (Herrera, 1984). However, studies often report conflicting results for the effects of both intra- and interspecific co-fruiting neighbours. Intraspecific neighbours have been reported to decrease fruit removal (Manasse and Howe, 1983) or to increase it (Sargent, 1990); and we know of only one study that reported no effects of intraspecific neighbour density on fruit removal rates (French *et al.*, 1992). Similarly, co-fruiting interspecifics have been noted to increase fruit removal (Herrera, 1984; Whelan *et al.*, 1998; van Ommeren and Whitham, 2002; Saracco *et al.*, 2005) as well as decrease it (Carlo *et al.*, 2003; Saracco *et al.*, 2005). Furthermore, in a field experiment, Carlo (2005a) found that seeds of *Solanum americanum* (Solanaceae) reached more sites and appeared in more faecal samples (i.e. dispersal *units*) when in a neighbourhood with *Cestrum diurnum* (Solanaceae) than in a neighbourhood composed only of other *S. americanum*. This was because mixed-species neighbourhoods were more attractive to frugivores than single-species neighbourhoods. A summary of studies that have looked at neighbourhood effects on frugivory and seed dispersal is presented in Table 16.1.

Competition and facilitation among fruiting plants that share seed-dispersal agents can be accentuated at the scale of plant neighbourhoods

Table 16.1. Studies that have looked at aspects of competition and/or facilitation among neighbouring plants for seed dispersal services by frugivores.

Source	Study species (Family)	Habitat/climate	Effects of intraspecific neighbours	Effects of interspecific neighbours
Herrera and Jordano, 1981	*Prunus mahaleb* (Rosaceae)	Mediterranean scrubland	Competition	N/A
Moore and Wilson, 1982	*Lindera benzoin* (Lauraceae)			
Manasse and Howe, 1983	*Virola surinamensis* (Myristicaceae)	Tropical forest	Competition	N/A
Herrera, 1984	*Rosa canina*, *Crataegus monogyna* (Rosaceae)	Temperate	N/A	Facilitation
Denslow, 1987	*Sambucus pubens* (Caprifoliaceae)	Temperate	Competition	N/A
Sargent, 1990	*Viburnum dentatum* (Caprifoliaceae)	Temperate	Facilitation	N/A
French *et al.*, 1992	*Coprosma quadrifida* (Rubiaceae)	Subtropical montane wet sclerophyll forest	Neutral	N/A
Tewksbury and Nabhan, 2001	*Capsicum annuum* (Solanaceae)	Sonoran Desert/ temperate	N/A	Facilitation
Burns, 2002	Community meta-analysis	Temperate and tropical	N/A	Facilitation
van Ommeren and Whitham, 2002	*Phoradendron juniperinum* (Viscaceae) *Juniperus monosperma* (Cupressaceae)	Temperate evergreen forest	N/A	Facilitation
Aukema and Martínez del Rio, 2002	*Phoradendron californicum* (Viscaceae)	Sonoran Desert	Facilitation	N/A
Saracco *et al.*, 2004	*Schefflera morototoni* (Araliaceae)	Subtropical shaded coffee plantation	Facilitation	Facilitation
Saracco *et al.*, 2005	Bird-dispersed trees	Subtropical montane forest	Competition	Facilitation
Carlo, 2005	*Solanum americanum*, *Cestrum diurnum* (Solanaceae)	Tropical pasture	Neutral	Facilitation
Carlo and Aukema, 2005	*Phoradendron hexastichum* (Viscaceae) *Cecropia schreberiana* (Cecropiaceae)	Tropical moist forest	N/A	Facilitation

because the spatial patterning of fruit resources influences frugivore foraging decisions and movements (Levey *et al.*, 1984; Saracco *et al.*, 2004). Using simulation models developed by Morales and Carlo (2006), we show here that whether plants compete for or facilitate frugivory and seed dispersal services is contingent on two factors (Fig. 16.1):

1. The spatial patterning of the plant population;
2. The availability of frugivores.

Thus, the disparate findings of the relatively few studies that have looked for or found neighbourhood effects (Table 16.1) might be understood when looking at the variance in both the spatial patterning of studied plant populations and the availability of frugivores. What is crucial is that fitness gradients may develop among plant individuals depending on their spatial context regarding co-fruiting intraspecific and interspecific individuals with which they share seed-dispersal agents.

On the other hand, frugivores may help maintain, strengthen, and propagate species associations by the co-dispersal of seeds from different species of plants from within neighbourhoods, and by directional seed dispersal patterns (Wenny, 2001; Clark *et al.*, 2004; Carlo and Aukema, 2005). Understanding interactions involving frugivores that show differential responses to species of plant and to the spatial patterning and abundance of fruit resources requires a conceptual framework that allows all such components (e.g. spatial pattern, frugivore behaviour and preferences) to interact and generate ecological patterns.

Plant–Frugivore Interactions as Spatially Explicit Networks

Plant–frugivore relationships are characterized by indirect interactions, many of which are mediated by density of resources, pathogens and herbivores (Janzen, 1970; Morales and Carlo, 2006), distance between sources (Levey *et al.*, 1984) and the properties of different species of plants, such as fruit nutritional value and size (Moermond and Denslow, 1985) and secondary compounds (Cipollini and Levey, 1997; Tewksbury and Nabhan, 2001). Frugivores physically link plant individuals and also affect how seeds are deposited in space (Herrera and Jordano, 1981), and thus generate the templates for plant recruitment (Schupp and Fuentes, 1995; Nathan and Muller-Landau, 2000). Plant–frugivore interactions can be modelled as a unique type of network: one that integrates plant–plant linkage through frugivore movements in a spatially explicit manner.

In many types of networks, it is common for a few elements to drive most interactions, for example in such diverse systems as the Internet, Bose–Einstein condensates, biochemical pathways, social relationships and trophic webs (Strogatz, 2001; Whitham *et al.*, 2003). Elements of a network that show high connectivity to other network elements are termed 'hubs'. What is most important is that hubs facilitate connection among the elements of the network. In mutualistic networks of ecological interactions, hubs may

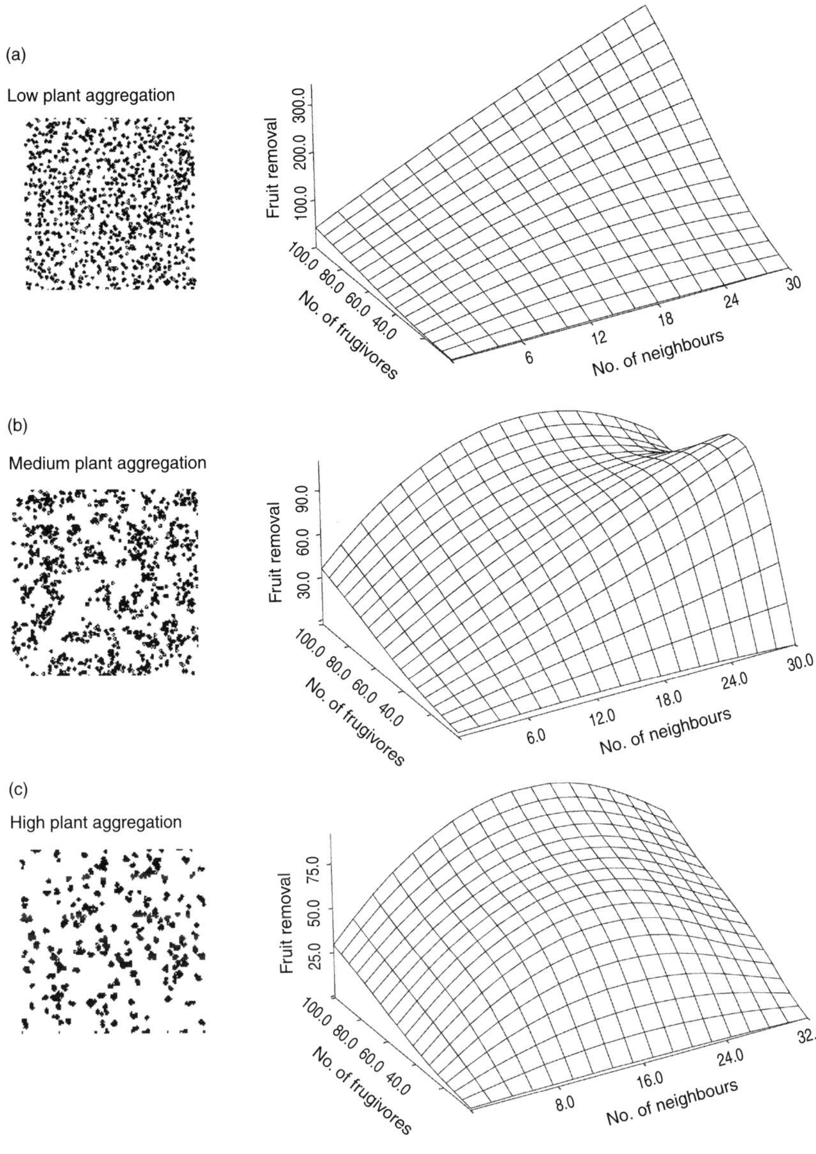

Fig. 16.1. Relationships between fruit removal (vertical axis), neighbourhood density and frugivore abundance. Lines are from a quadratic fit to data smoothed as a response surface (using S-Plus 6.2). (a) Plants with a highly aggregated spatial pattern, (b) plants with medium spatial aggregation, (c) plants distributed at random. Results are from spatially explicit computer simulations developed by Morales and Carlo (2006; see also Carlo, 2005b). Simulated birds moved among plants, ate fruit, and dispersed seeds in a population of 1000 plants in 5 × 5 km landscapes using stochastic movement and foraging rules parameterized with field data (Carlo, 2005b). Three levels of frugivore abundance (1, 10 and 100 birds) were used. Simulated data were averaged across 30 replicate landscapes for each factor combination. The number of neighbours was counted within a 100-m radius around each plant and fruit removal values averaged across the 30 replicate landscapes.

promote the formation of hierarchical relationships among species of fruiting plants in a community (Bascompte *et al.*, 2003, 2006; Jordano *et al.*, 2003; Lázaro *et al.*, 2005). Hierarchical relationships between fruiting plants occur when certain species capture seed dispersal interactions more frequently and from more dispersers than other species.

Hierarchical network properties have already been described in plant–frugivore mutualistic networks (Terborgh, 1986; Jordano, 1987; Sallabanks, 1993; Bascompte *et al.*, 2003, 2006; Carlo *et al.*, 2003; Jordano *et al.*, 2003). Jordano *et al.* (2003) suggested that hierarchical network topologies could quite well describe many community patterns of frugivore–plant interactions worldwide. This idea implies that, in many communities, just a handful of plant and animal species will be responsible for a disproportionate share of the frugivory and seed dispersal interactions (Fig. 16.2). Still, the spatial component of plant–frugivore networks of ecological interactions has not been previously explored. The lack of a spatially explicit approach is remarkable, given that frugivory and seed dispersal patterns depend upon the ways in which frugivores move and interact with the spatial landscape structure of fruiting plant populations (Westcott and Graham, 2000; see Kwit *et al.*, Chapter 19, this volume; Holbrook and Loiselle, Chapter 13, this volume). Furthermore, the last

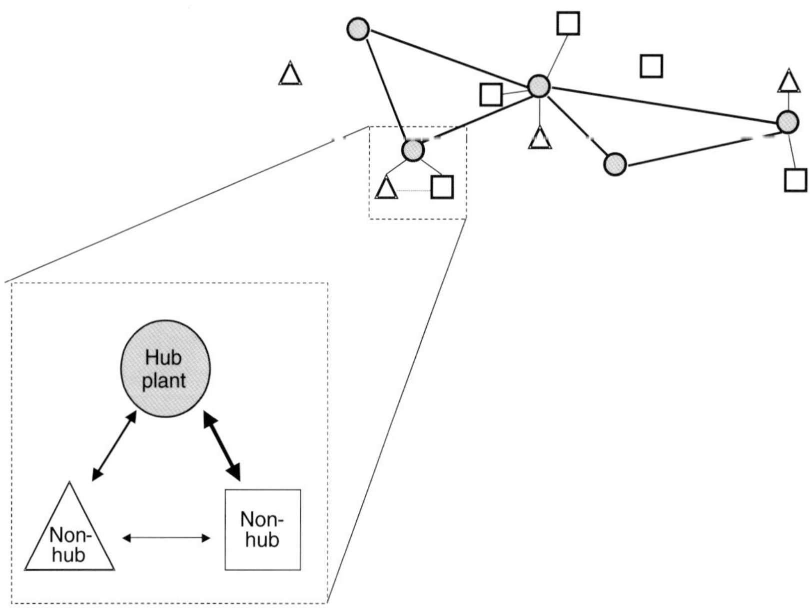

Fig. 16.2. Hypothetical representation of how hub plant species or neighbourhoods affect the flow of frugivore movements within a local plant community. Thickness of a line indicates the frequency of movement. Movement is most common between hubs (grey circles) and least common between non-hub species (triangles and squares). Inset shows that some non-hub species (squares) interact more strongly (heavier arrow) with hubs than others. The probability of movement from hub to non-hub plants decreases with distance.

decade has seen an emphasis on the importance of space in ecological and evolutionary processes (Tilman and Kareiva, 1997). It is time to explicitly incorporate space in the modelling of plant–frugivore interactions.

The spatial structure of landscapes influences movements of animals as different as beetles (Crist *et al.*, 1992; Morales and Ellner, 2002), birds (Levey *et al.*, 2005) and elk (Morales *et al.*, 2004). Frugivores are known to detect and track fruit abundance at small and large spatial and temporal scales (Levey, 1988; Rey, 1995; Kwit *et al.*, 2004b; Márquez *et al.*, 2004). Therefore, it should be expected that fruit removal rates would be affected by both landscape patterns of resource distribution and the local neighbourhoods of plants. But irrespective of the exact mathematical function describing connectivity in plant–frugivore systems (i.e. power-law, truncated power-law, exponential; Jordano *et al.*, 2003), it is crucial that we develop a basic understanding of the role that well-connected (i.e. hub) species of plants play in communities. In this chapter we will explore the case of plant–frugivore networks that have hubs or hub-like elements, and the implications for plant population dynamics, community ecology and diversity.

Hubs in Plant–Frugivore Spatial Networks

Plant–frugivore networks have a dual nature and can be viewed from the plants' perspectives or from the frugivores' perspectives (i.e. plants connecting frugivore consumers or frugivores connecting plants; Jordano *et al.*, 2003). Here we depict the network in the following way. We consider plants as the network elements to be connected by frugivores because plants are sessile elements and frugivores move among them, creating visitation patterns that are influenced by the explicit locations of plants. We define a frugivory hub as a plant individual, species or neighbourhood which has a higher probability of visitation than the rest of the elements forming the network. Thus, frugivory hubs capture a large share of frugivory and dispersal services in a population and/or community. We propose that hubs emerge both among individuals of a species in an intraspecific network, and among different co-fruiting species in an interspecific community network. The network concept has been applied at the interspecific level (Bascompte *et al.*, 2006), although not in a spatially explicit manner, but intraspecific networks have not been previously examined. In the real world, both interspecific and intraspecific networks are simultaneously in action and interacting with each other; such interactions have potentially important effects on community composition and structure.

Frugivory Hubs Within a Single-species Plant Population

In a population of a fruiting plant species, particular individuals can receive disproportionate frugivory and seed dispersal services. We refer to these individuals as intraspecific hubs. An individual may become a hub because of individual attributes (e.g. larger size) and/or spatial location (e.g. neighbourhood). Carlo (2005b) examined distributions of fruit removal (i.e. histograms for average fruit removal per plant individual) in bird-dispersed plant populations, using spatially explicit computer simulations of frugivory and seed dispersal (Fig. 16.3). He found that distributions were highly skewed when frugivore density was in the range of 0.01–0.001 frugivores per plant, regardless of the degree of spatial aggregation of the population (Fig. 16.3a,b). Only at high frugivore densities of 0.1 frugivores per plant were distributions asymmetric, with most individuals having an intermediate number of fruits removed (Fig. 16.3c). Nevertheless, the distributions from the simulation model had much more variance compared with null models where birds chose fruiting plants at random (T.A. Carlo and J.M. Morales, 2006, unpublished results). This means that fruit removal services were monopolized, to a greater degree than expected by chance, by a handful of fruiting plants in the population (Carlo *et al.*, 2003; T.A. Carlo and J.M. Morales, 2006, unpublished results). At the same time, more plants than expected received little or no frugivory services, especially in aggregated landscapes (Fig. 16.3).

Neighbourhood density was generally positively correlated with frugivory (Fig. 16.1). Indeed, positive density dependence is probably the best documented pattern in the frugivory and seed dispersal literature, where positive relationships between crop size and visitation rates have been shown repeatedly (Davidar and Morton, 1986; Foster, 1990; Sargent, 1990; Sallabanks, 1993; Saracco *et al.*, 2005). Positive density dependence in fruit removal rates can be expressed at both individual (Davidar and Morton, 1986; Sallabanks, 1993) and neighbourhood scales (Aukema, 2004; Saracco *et al.*, 2004) because the spatial definition of a fruit 'patch' depends on the spatial perception of resources by frugivores. It is probably largely irrelevant for a frugivore if 500 fruits within a 4-m^2 area are all in one plant or found in five plants each with 100 fruits, or 50 smaller plants each with ten fruits. Spatial autocorrelation in fruit removal services at small spatial scales (relative to the range of animal movements; see Aukema and Martínez del Rio, 2002) is likely to be a general phenomenon (Kwit *et al.*, 2004b; Saracco *et al.*, 2004) and a contributing factor to hub formation. Hence, the emergence of hubs within a fruiting plant population is also dependent on the way in which individuals are distributed in space.

Evidence for the Emergence of Hubs in Single-species Plant Networks

Hubs are not only visible in theoretical models but are also observable in nature. Evidence for hubs can be found by examining the frequency distribution and linkage of frugivore visits to plants or neighbourhoods of plants. For example, mistletoes are aggregated within the host trees they parasitize, so each host tree can be considered as a mistletoe neighbourhood with unique characteristics (location in space, mistletoe fruit crop, physical

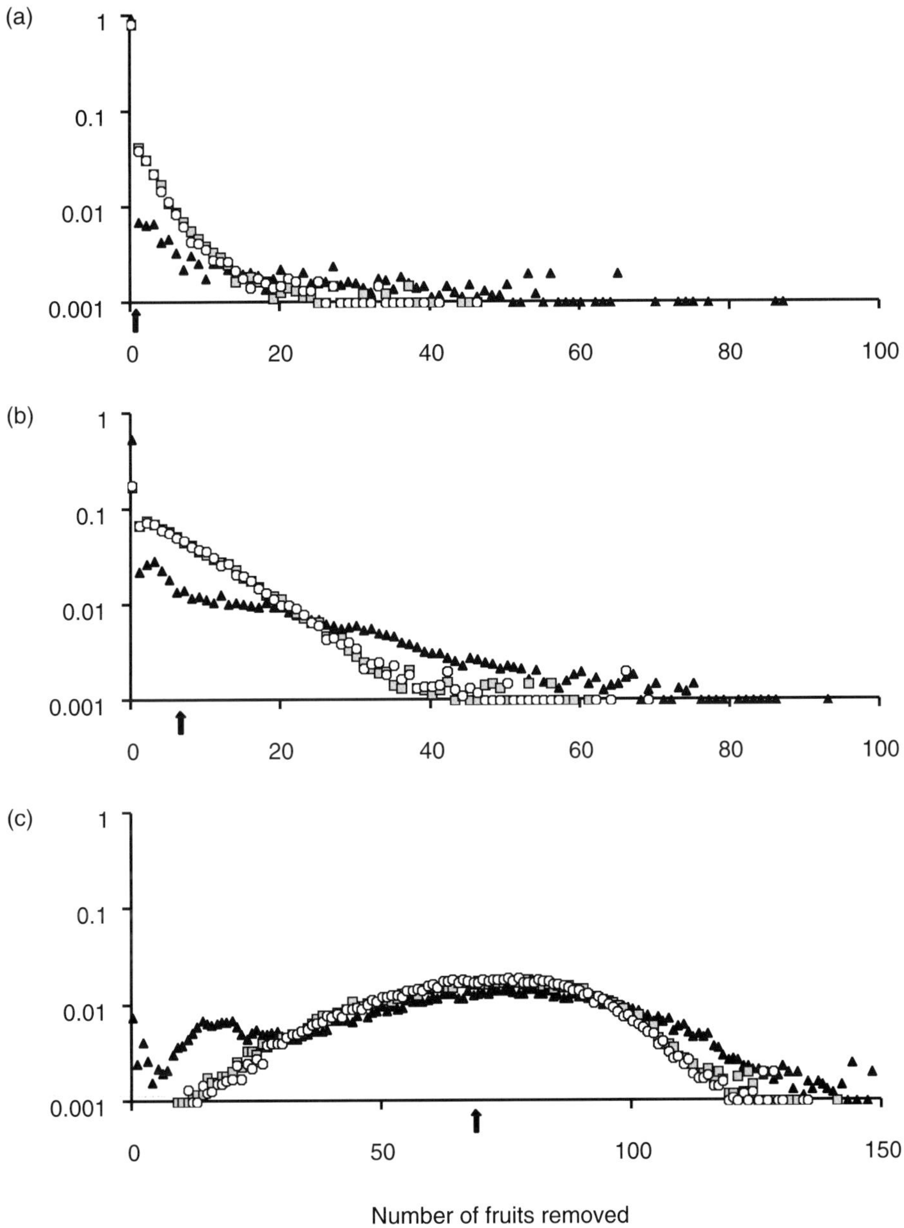

Fig. 16.3. Frequency distributions of fruit removal events by (a) one frugivore, (b) ten frugivores, or (c) 100 frugivores in simulated plant populations with high landscape aggregation (black triangles), medium landscape aggregation (grey squares) or low landscape aggregation (white circles). The *x* axis is the number of fruits removed from a plant. The *y* axis is the average proportion (of 30 replicate runs) of individuals in the population (1000 plants) receiving a given amount of fruit removal. Black arrows indicate the mean values. Data were produced with spatially explicit computer simulations developed by Morales and Carlo (2006; see also Carlo, 2005b and legend of Fig. 16.1).

characteristics of the tree, etc). On a mapped 4-ha plot, J.E. Aukema (2001, unpublished results) observed phainopeplas (*P. nitens*, the primary dispersers of the mistletoe *P. californicum*) visiting a few trees very frequently, while failing to visit most trees (Fig. 16.4a). In addition, she found that phainopepla flights linked a handful of trees to a large number of other trees, while a majority of infected trees were linked to few, if any, other trees (Fig. 16.4b). Nine out of 250 mistletoe-infected trees were responsible for capturing 56% of observed visits, and acted as hubs linking visits to a large number of other mistletoe-infected trees. When looking at the characteristics of the mistletoe-infected trees that acted as frugivory hubs, Aukema (2001) found that the nine most-linked trees, on average, had denser mistletoe populations than other infected trees (t = 2.19, P = 0.029, n = 260). In addition to mistletoe resources, hub trees also had other physical and spatial features that were attractive to phainopeplas. Hub trees were significantly taller than other infected trees (t = 2.85, P = 0.01, n = 260), often had dead branches suitable for perching on (J.E. Aukema, 2006, unpublished results), and were located on drainages more frequently than would be expected by chance (χ^2 = 8.78, P = 0.011, n = 242).

Further evidence for hubs within a species is found in studies such as that of Aldrich and Hamrick (1998) who, using molecular approaches, found that a few trees were producing the most seedlings in a fragmented landscape. Although there were many individuals present in the study area, they hypothesized that the reproductive dominance of the individuals was due to a privileged spatial position in pastures that allowed them to produce more fruit and to attract more frugivores. In the context of plant–frugivore networks, these individuals acted as hubs.

Frugivory Hubs Among Plant Species

Frugivores commonly show preferences for particular species among the available fruiting plants that they use (Levey *et al.*, 1984; Moermond and Denslow, 1985), which often results in a disproportionate use of a subset of fruiting species versus others in the field (Carlo *et al.*, 2003; Bascompte *et al.*, 2006). Therefore, interspecific hubs for frugivory can form when some species serve as magnets (*sensu* Johnson *et al.*, 2003) for mutualists. In other words, in an interspecific context, frugivory hubs are species that receive a disproportionate amount of attention from frugivores and are less dispersal-limited than the rest of the species in a given plant community.

While the emergence of hubs in intraspecific networks seems largely controlled by patterns of local density, relationships among interspecific hubs and other species are likely to be determined largely by species properties. For example, species of plants vary not only in aspects of the quality and quantity of their fruits (Moermond and Denslow, 1985), they also differ in the timing and length of fruiting. For example, Carlo *et al.* (2003) found that the most connected (*sensu* Jordano *et al.*, 2003) and preferred species across forested habitats in Puerto Rico had long fruiting

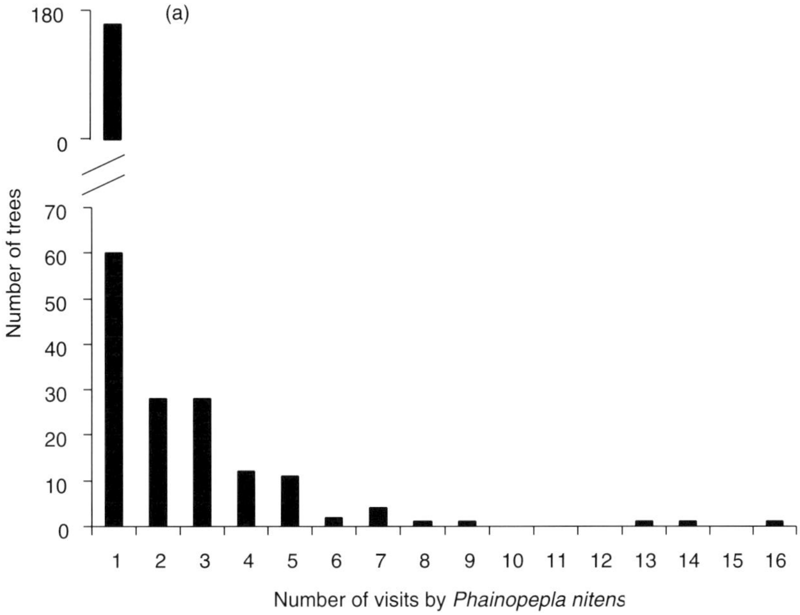

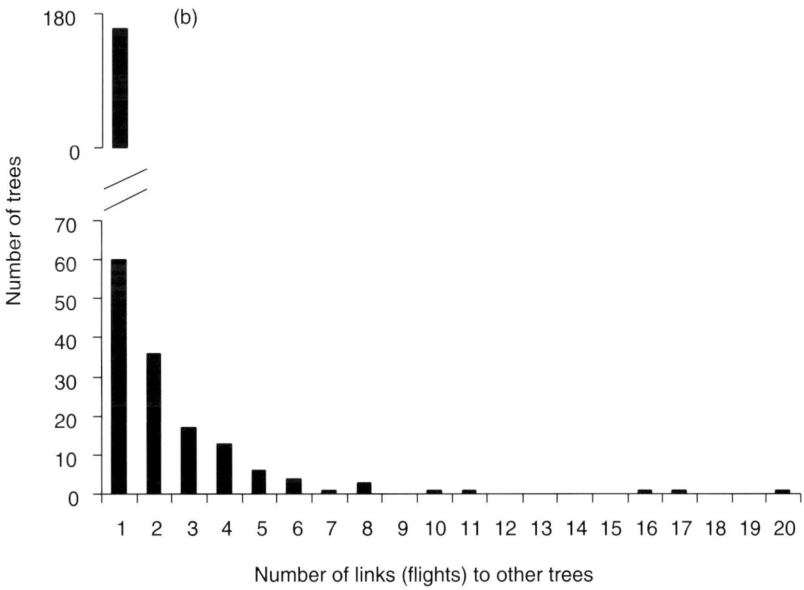

Fig. 16.4. On a 4-ha mapped plot in the Sonoran Desert, J.E. Aukema (2001, unpublished results; see also Aukema and Martínez del Rio, 2002) observed that *Phainopepla nitens*, a bird specializing on mistletoe (*Phoradendron californicum*) berries, visited a few mistletoe-infected mesquite trees (*Prosopis velutina*; Mimosaceae) frequently, but visited most trees very rarely (a). Similarly, only a few trees were linked to many other trees through direct tree-to-tree flights (b). Those trees that received many bird visits and were connected to many other trees through bird flights acted as intraspecific hubs for frugivory.

seasons and could re-fruit faster than less connected species. Therefore, for a species to become an interspecific hub of frugivory and seed dispersal, it appears that it might be necessary to be able to offer fruit for time periods that are longer than those of non-hub species, although the generality of this result still remains to be documented.

The most important effect of interspecific frugivory hubs is that hubs will affect the movement patterns of frugivores and, consequently, the flow of frugivory and seed dispersal services in a locality. Therefore, the proximity of fruiting plants to hub species may determine whether an individual plant experiences facilitation (i.e. by being close to the hub; see Fig. 16.2) or competition (i.e. by being further from hubs). In addition, species serving as hubs could create strong directional patterns of seed dispersal (Wenny, 2001), with seed rain of frugivore-dispersed species higher near hub species.

Evidence for the Existence of Frugivory Hubs Among Plant Species

We were not able to find a single study and know of no data set that could be used to directly examine spatially explicit hubs in interspecific plant–frugivore networks. Co-evolutionary plant–frugivore networks have been explored for interspecific communities, though not in a spatial context (Bascompte *et al.*, 2003, 2006; Jordano *et al.*, 2003). These networks have been found to be asymmetric, with some species of plants interacting more strongly than others with animal dispersers (Bascompte *et al.*, 2006). Similar to the intraspecific case, plants functioning as frugivory hubs in plant communities should show high connectivity and preferential use by the frugivore community. Using data from Carlo *et al.* (2003) we found that five out of 68 endozoochorous species captured the most interactions with frugivores (i.e. connectivity, Fig. 16.5a) in forests of north/central Puerto Rico. The pattern was similar for the distribution of fruit preferences (i.e. fruiting species that were used by frugivores more frequently than expected based on their relative abundance, Fig. 16.5b). *Cecropia schreberiana* (Cecropiaceae) was the species with the greatest number of recorded avian fruit consumers; the same species was preferred by the most species of birds (four) in the community, while most species of plants (75%) were not preferred. *Cecropia schreberiana* also fruits year-round (Carlo *et al.*, 2003), and thus meets the expected conditions for a hub species in Puerto Rican forest communities.

Consequences of Frugivory Hubs for Patterns of Seed Deposition

Because hubs could strongly affect the flow of frugivory events among plant individuals and species, they should play an important role in subsequent patterns of seed deposition. In general, the probability of

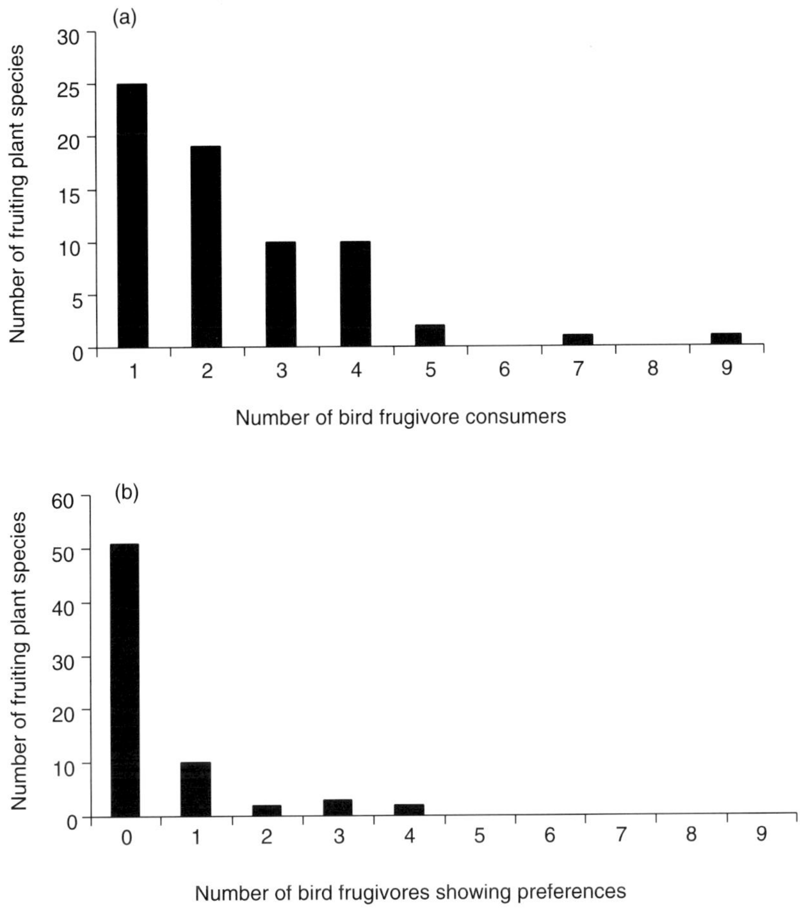

Fig. 16.5. Histograms showing avian fruit use and fruit preferences from Carlo *et al.* (2003) across five moist secondary forests and 8 months of observation in Puerto Rico. (a) The number of plants (*y* axis) in the plant community with the observed number of frugivorous bird consumers (*x* axis, maximum was 9). (b) The number of plant species that were preferred (i.e. observed use was greater than would be expected by the relative abundance of the species' fruits; see Carlo *et al.*, 2003), and how many species of birds preferred them. Note that a few plants captured most of the frugivory services and preferences in the communities.

frugivore movement would be greatest: (i) from one hub to another hub; (ii) from a hub to a non hub plant; and (iii) from a non-hub to a hub; and would be least from one non-hub to another non-hub. The result is a series of 'avenues' of frequent movement accompanied by higher rates of frugivory and particular patterns of seed dispersal (e.g. directional, *sensu* Wenny, 2001) for the plants in or near such avenues (Herrera and Jordano, 1981). Specifically, we believe that at least two effects of hubs on seed fall patterns can be predicted.

1. Hubs create nested and directional patterns of seed deposition

Because frugivores visit hubs much more frequently, hubs then provide dispersal linkage for plants that are visited less frequently. This has at least two obvious implications. First, seeds from non-hub plants will have a disproportionate share of their seeds arriving under or near hubs. By the same token, seeds from hubs will be proportionally more widespread, reaching more locations than those of seeds from non-hubs. This should result in a nested deposition pattern (*sensu* Bascompte *et al.*, 2003) in which seeds (either intra- or interspecific) from most plants will be falling *within* the seed shadow of hubs. In the case of an interspecific hub, nested seed deposition patterns could then translate into nested patterns of plant species distributions like those described by Lázaro *et al.* (2005) in northern Europe. For the case of an intraspecific hub, spatial diversity of genotypes will be nested within the genotypes of hubs, with patches of high diversity (near hubs) and patches of low diversity elsewhere, where the genotypes of the hubs are overrepresented.

2. Hubs create seed dispersal limitation that is patchily distributed in space and time

Hub formation has an inescapable consequence: it alters the spatial distribution of dispersal services. This creates patches where non-hub plants are dispersal-limited and patches where hub plants are dispersal-rich. In the intraspecific case, plants that are far from hubs are expected to experience competition, whereas plants closer to hubs are expected to experience facilitation. The interspecific case is more complex because it involves overlapping spatial networks of several species at once. In this case, proximity to a hub species could result in either competition or facilitation. Hubs can cause competition for non-hub species that are nearby if the frugivores' preference for the fruits of the hub is so strong that they ignore the non-hub fruits nearby. For these non-hub species, the location of intraspecific hubs will not be spatially correlated with the hub species. Conversely, if frugivores are more generalized (which is the rule rather than the exception; see Bascompte *et al.*, 2006), proximity to hub species may facilitate both fruit removal and seed dispersal for non-hub species. In this case, the spatial position of intraspecific hubs is expected to be concordant with the position of the interspecific hubs. Interspecific facilitation will not only cause spatial association of hubs from different species, but will also lead to co-dispersal of seeds from hub and non-hub species.

Frugivore Spatial Networks: Community Ecology, Evolution and Conservation

The spatial network framework we have presented could be used to test diverse mechanistic hypotheses regarding plant community assembly, succession, invasions, range shifts under climatic change, and restoration practices. It also can provide a framework for developing new evolutionary hypotheses and experiments designed to explore in more detail the conditions that drive processes of natural selection in plant–frugivore systems.

The presence of interspecific hubs for frugivory and seed dispersal in plant communities can affect successional dynamics and influence their spatial patterning and diversity. Because hub species within plant communities are the broadest and strongest interactors with frugivore assemblages, they should not suffer from dispersal limitation. This allows them to reach more locations (including more long-distance dispersal) than ordinary frugivore-dispersed plants that (by definition) comprise the majority of the community. The dominance of hubs then serves to bring non-hub species to new locations by directed dispersal. In fact, some hub species may also be early successional and pioneer species, as is the case with *Cecropia schreberiana* and *Schefflera morototoni* in Puerto Rico. Thus, successional dynamics could be driven by the arrival of pioneer hub species (Vieira *et al.*, 1994). Alternatively, the establishment of pioneer species that do not interact with the frugivore community (e.g. *Spathodea campanulata*, see Lugo and Helmer, 2004) may retard the arrival and diversity build-up in, for example, secondary successional forests.

Conservation practices can also benefit from an understanding of frugivory networks and knowledge of the identity of hub species and the strength and frequency of linkages between species in regional communities. For example, exotic species that are animal-dispersed (such as the Russian olive, *Elaeagnus angustifolia* (Eleagnaceae) in western North America) become elements of local frugivore networks. Invasive species can take advantage of the networks in order to spread. In some cases, exotic species could become hubs themselves in their new environments, which could actually have positive effects for native plants (through co-dispersal, directional dispersal and facilitation) and animals (by providing food resources). Knowing the topology of frugivory networks can also help restoration biologists to select which species to use in restoration projects. Planting hub species may accelerate the development of diverse plant and animal communities. In addition, climate change scenarios predict great shifts in the ranges of many species and widespread extinction if they are unable to respond to these changes and move into suitable areas (Thomas *et al.*, 2004). Maintaining plant–frugivore interactions will be critical to the persistence of thousands of plant and animal species. If hub species are able to, or are aided in, moving with the shifts in climate, other species may be able to hitch-hike through the dispersal network. Furthermore, because hub species are by definition important to a diversity of frugivores, if hub species are able to move, they can supply resources for some other frugivores whose ranges are also shifting, as well as for frugivores in their new communities.

Hierarchical network structures can also affect processes of gene flow, spatial genetic structuring and natural selection. In theory, the interaction of frugivores with the spatial patterning of a plant population by itself generates asymmetries in the seed dispersal of individuals based on local neighbourhood density and location (Figs 16.1 and 16.3; Aldrich and Hamrick, 1998; Carlo, 2005b). In the absence of other factors influencing individual fitness, these asymmetries can then produce a type of genetic drift that can influence the genetic make-up of populations. For example, the spread and increase of particular alleles could be increased tremendously if they are found in individuals associated with hubs. Conversely, a potentially beneficial allele can be lost much faster than predicted by models that do not account for spatial interactions (e.g. the model of Levin and Kerster, 1975), given the large dispersal asymmetries produced by spatial positioning. Still, it remains a great challenge to understand the processes of natural selection taking place in the context of a spatial network. Evolution and co-evolution in a network context may depend on how much some species need the presence of others locally in order to receive sufficient seed dispersal services and maintain viable populations over time. Because plant–frugivore networks (including hub species) change regionally, the potential selective pressures of interspecific interactions within local networks can promote patterns of mosaic evolution (*sensu* Thompson, 1994).

In conclusion, theories that consider only competitive interactions fail to explain satisfactorily how multiple species coexist and why some plant communities are so diverse (Terborgh *et al.*, 2002; Lortie *et al.*, 2004). Because the dispersal of endozoochorous plants is contingent on the interaction of frugivores with the spatial distribution and diversity of fruiting plants, interactions in plant–frugivore networks can be important forces in the organization and diversity of communities where plant–frugivore mutualisms are common. Looking at them in spatially explicit ways could help explain patterns of ecological connectance that are just starting to be explored.

Acknowledgements

We thank E.W. Schupp, J. Bascompte, D.G. Wenny, Andrew Dennis and C. Martínez del Rio for helpful comments on earlier versions of this chapter. This work was made possible by an NSF grant (0407826) to T.A.C., a Smith Fellowship to J.E.A., the University of Colorado, E. Santiago, J.A. Collazo, and the Botanical Garden of the University of Puerto Rico. Travel of the senior author to Australia to participate in the Symposium/Workshop was partly supported by the National Science Foundation (OISE No. 0456402 to E.W. Schupp).

References

Agrawal, A.A. (2003) Why omnivory? *Ecology* 84, 2521.

Aldrich, P.R. and Hamrick, J.L. (1998) Reproductive dominance of pasture trees in a fragmented tropical forest mosaic. *Science* 281, 103–105.

Aukema, J.E. (2001) Dispersal and spatial distribution of the desert mistletoe, *Phoradendron californicum*, at multiple scales: patterns, processes and mechanism. PhD dissertation, The University of Arizona, Tucson, Arizona.

Aukema, J.E. (2004) Distribution and dispersal of desert mistletoe is scale-dependent, hierarchically nested. *Ecography* 27, 137–144.

Aukema, J.E. and Martínez del Rio, C. (2002) Where does a fruit-eating bird deposit mistletoe seeds? Seed deposition patterns and an experiment. *Ecology* 83, 3489–3496.

Bascompte, J., Jordano, P. and Olesen, J.M. (2006) Asymmetric coevolutionary networks facilitate biodiversity maintenance. *Science* 312, 431–433.

Bascompte, J., Jordano, P., Melián, C.J. and Olesen, J.M. (2003) The nested assembly of plant–animal mutualistic networks. *Proceedings of the National Academy of Sciences of the USA* 100, 9383–9387.

Bruno, J.F., Stachowicz, J.J. and Bertness, M.D. (2003) Inclusion of facilitation into ecological theory. *Trends in Ecology and Evolution* 18, 119–125.

Burns, K.C. (2002) Seed dispersal facilitation and geographic consistency in bird–fruit abundance patterns. *Global Ecology and Biogeography* 11, 253–259.

Callaway, R.M. (1997) Positive interactions in plant communities and the individualistic-continuum concept. *Oecologia* 112, 143–149.

Canham, C.D. and Uriarte, M. (2006) Analysis of neighbourhood dynamics of forest ecosystems using likelihood methods and modelling. *Ecological Applications* 16, 62–73

Carlo, T.A. (2005a) Interspecific neighbours change seed dispersal pattern in an avian dispersed plant. *Ecology* 86, 2440–2449.

Carlo, T.A. (2005b) Competition and facilitation in bird dispersed plants. PhD dissertation, University of Colorado, Boulder, Colorado.

Carlo, T.A. and Aukema, J.E. (2005) Female-directed dispersal and facilitation between a mistletoe and a dioecious host. *Ecology* 86, 3245–3251.

Carlo, T.A., Collazo, J.A. and Groom, M.J. (2003) Avian fruit preferences across a Puerto Rican forested landscape: pattern consistency and implications for seed removal. *Oecologia* 134, 119–131.

Chu, M. and Walsberg, G.E. (1999) Phainopepla (*Phainopepla nitens*). *Birds of North America* 415.

Cipollini, M.L. and Levey, D.J. (1997) Secondary metabolites of fleshy vertebrate-dispersed fruits: adaptive hypotheses and implications for seed dispersal. *American Naturalist* 150, 346–372.

Clark, C.J., Poulsen, J.R., Connor, E.F. and Parker, V.T. (2004) Fruiting trees as dispersal foci in a semi-deciduous tropical forest. *Oecologia* 139, 66–75.

Condit, R., Pitman, N., Leigh, E.G., Chave, J., Terborgh, J., Foster, R.B., Nunez, P., Aguilar, S., Valencia, R., Villa, G., Muller-Landau, H.C., Losos, E. and Hubbell, S.P. (2002) Beta-diversity in tropical forest trees. *Science* 295, 666–669.

Crist, T.O., Guertin, D.S., Wiens, J.A. and Milne, B.T. (1992) Animal movements in heterogeneous landscapes: an experiment with *Eleodes* beetles in shortgrass prairie. *Functional Ecology* 6, 536–544.

Cruz, A. (1974) Feeding assemblages of Jamaican birds. *Condor* 76, 103–107.

Davidar, P. and Morton, E.S. (1986) The relationship between fruit crop sizes and fruit removal rates by birds. *Ecology* 67, 262–265.

Denslow, J.S. (1987) Fruit removal from aggregated and isolated bushes of red elderberry *Sambucus pubens*. *Canadian Journal of Botany* 65, 1229–1235.

Fleming, T.H., Venable, D.L. and Herrera, L.G. (1993) Opportunism vs. specialization: the evolution of dispersal strategies in fleshy-fruited plants. *Vegetatio* 108, 107–120.

Foster, M. (1990) Factors influencing bird foraging preferences among intraspecific fruit trees. *Condor* 92, 844–854.

French, K., O'Dowd, D.J. and Lill, A. (1992) Fruit removal of *Coprosma quadrifida* (Rubiaceae) by birds in south-eastern Australia. *Australian Journal of Ecology* 17, 35–42.

García, D., Zamora, R., Gómez, J.M. and Hodin, J.H. (2001) Frugivory in *Juniperus communis* depends more on population characteristics than on individual attributes. *Journal of Ecology* 89, 639–647.

Herrera, C.M. and Jordano, P. (1981) *Prunus mahaleb* and birds: the high efficiency seed dispersal system of a temperate tree. *Ecological Monographs* 51, 203–218.

Herrera, C.M. (1982) Seasonal variation in the quality of fruits and diffuse coevolution plants and avian dispersers. *Ecology* 63, 773–785.

Herrera, C.M. (1984) Seed dispersal and fitness determinants in wild rose: combined effects of hawthorn, birds, mice, and browsing ungulates. *Oecologia* 63, 386–393.

Herrera, C.M. (1998) Long-term dynamics of Mediterranean frugivorous birds and fleshy fruits: a 12-year study. *Ecological Monographs* 68, 511–538.

Herrera, C.M. (2003) Seed dispersal by vertebrates. In: Herrera, C.M. and Pellmyr, O. (eds) *Plant–Animal Interactions: An Evolutionary Approach.* Blackwell, Oxford, UK, pp. 185–208.

Howe, H.F. (1993) Specialized and generalized dispersal systems: where does 'the paradigm' stand? *Vegetatio* 107/108, 3–13.

Hubbell, S.P. (2001) *A Unified Theory of Biodiversity and Biogeography.* Princeton University Press, Princeton, New Jersey.

Hutchinson, G.E. (1957) Concluding remarks. *Cold Spring Harbor Symposia on Quantitative Biology* 22, 415–427.

Janzen, D.H. (1970) Herbivores and the number of tree species in tropical forests. *American Naturalist* 104, 501–528.

Janzen, D.H. (1983) Dispersal of seeds by vertebrate guts. In: Futuyma, D.J. and Slatkin, M. (eds) *Coevolution.* Sinauer Associates, Sunderland, Massachusetts, pp. 232–263.

Johnson, S.D., Peter, C.I., Nilsson, L.E. and Agren, J. (2003) Pollination success in a deceptive orchid is enhanced by co-occurring rewarding magnet plants. *Ecology* 84, 2919–2927.

Jordano, P. (1987) Patterns of mutualistic interactions in pollination and seed dispersal: connectance dependence asymmetries, and coevolution. *American Naturalist* 129, 657–677.

Jordano, P., Bascompte, J. and Olesen, J.M. (2003) Invariant properties in coevolutionary networks of plant–animal interactions. *Ecology Letters* 6, 69–81.

Kannan, R. and James, D.A. (1999) Fruiting phenology and the conservation of the great pied hornbill (*Buceros bicornis*) in the western Ghats of southern India. *Biotropica* 31, 167–177.

Kwit, C., Levey, D.J. and Greenberg, C.H. (2004a) Contagious seed dispersal beneath heterospecific fruiting trees and its consequences. *Oikos* 107, 303–308.

Kwit, C., Levey, D.J., Greenberg, C.H., Pearson, S.F., McCarty, J.P., Sargent, S. and Mumme, R.L. (2004b) Fruit abundance and local distribution of wintering hermit thrushes (*Catharus guttatus*) and yellow-rumped warblers (*Dendroica coronata*) in South Carolina. *Auk* 121, 46–57.

Larson, D.L. (1996) Seed dispersal by specialist versus generalist foragers: the plant's perspective. *Oikos* 76, 113–120.

Lázaro, A., Mark, S. and Olesen, J.M. (2005) Bird-made fruit orchards in northern Europe: nestedness and network properties. *Oikos* 110, 321–329.

Levey, D.J. (1988) Spatial and temporal variation in Costa Rican fruit and fruit-eating bird abundance. *Ecological Monographs* 58, 251–269

Levey, D.J. and Benkman, C.W. (1999) Fruit–seed disperser interactions: timely insights from a long-term perspective. *Trends in Ecology and Evolution* 14, 41–42.

Levey, D.J., Moermond, T.C. and Denslow, J.S. (1984) Fruit choice in neotropical birds: the effect of distance between fruits on choice patterns. *Ecology* 65, 844–850.

Levey, D.J., Bolker, B.M., Tewksbury, J.J., Sargent, S. and Haddad, N.M. (2005) Effects of landscape corridors on seed dispersal by birds. *Science* 309, 146–148.

Levin, D.A. and Kerster, H.W. (1975) The effect of gene dispersal on the dynamics and statics of gene substitution in plants. *Heredity* 35, 317–336.

Levins, R. (1969) Some demographic and genetic consequences of environmental heterogeneity for biological control. *Bulletin of the Entomological Society of America* 15, 237–240.

Lortie, C.J., Brooker, R.W., Choler, P., Michalet, R., Pugnaire, F.L. and Callaway, R. (2004) Rethinking plant community theory. *Oikos* 107, 433–438.

Lugo, A.E. and Helmer, E.H. (2004) Emerging forests on abandoned land: Puerto Rico's new forests. *Forest Ecology and Management* 190, 145–161.

Manasse, R.S. and Howe, H.F. (1983) Competition for dispersal agents among tropical trees: influences of neighbors. *Ecology* 59, 185–190.

Márquez, A.L., Real, R. and Vargas, J.M. (2004) Dependence of broad-scale geographical variation in fleshy-fruited plant species richness on disperser bird species richness. *Global Ecology and Biogeography* 13, 295–304.

McArthur, R.H. and Wilson, E.O. (1968) *The Theory of Island Biogeography.* Princeton University Press, Princeton, New Jersey.

McKey, D. (1975) The ecology of coevolved seed dispersal systems. In: Gilbert, L.E. and Raven, P.H. (eds) *Coevolution of Animals and Plants.* University of Texas Press, Austin, Texas, pp. 159–191.

Moermond, T.C. and Denslow, J.S. (1985) Neotropical avian frugivores: patterns of behavior, morphology, and nutrition, with consequences for fruit selection. *Ornithological Monographs* 36, 865–897.

Morales, J.M. and Carlo, T.A. (2006) The effects of plant distribution and frugivore density on the scale and shape of dispersal kernels. *Ecology* 86, 1489–1496.

Morales, J.M. and Ellner, S.P. (2002) Scaling up movement in heterogeneous landscapes: the importance of behavior. *Ecology* 83, 2240–2247.

Morales, J.M., Haydon, D.T., Frair, J., Holsiner, K.E. and Fryxell, J.M. (2004) Extracting more out of relocation data: building movement models as mixtures of random walks. *Ecology* 85, 2436–2445.

Nathan, R. and Muller-Landau, H.C. (2000) Spatial patterns of seed dispersal, their determinants and consequences for recruitment. *Trends in Ecology and Evolution* 15, 278–285.

Pimm, S.L. and Lawton, J.H. (1978) On feeding on more than one trophic level. *Nature* 275, 542–544.

Reid, N. (1991) Coevolution of mistletoes and frugivorous birds. *Australian Journal of Ecology* 16, 457–469.

Rey, P. (1995) Spatio-temporal variation in fruit and frugivorous bird abundance in olive orchards. *Ecology* 76, 1625–1635.

Russo, S.E. and Augspurger, C.K. (2004) Aggregated seed dispersal by spider monkeys limits recruitment to clumped patterns in *Virola calophylla*. *Ecology Letters* 7, 1058–1067.

Sallabanks, R. (1993) Hierarchical mechanisms of fruit selection by an avian frugivore. *Ecology* 74, 1326–1336.

Saracco, J.F., Collazo, J.A. and Groom, M.J. (2004) How do frugivores track resources? Insights from spatial analyses of bird foraging in a tropical forest. *Oecologia* 139, 235–245.

Saracco, J.F., Collazo, J.A., Groom, M.J. and Carlo, T.A. (2005) Crop size and fruit neighborhood effects on visitation to fruiting trees. *Biotropica* 37, 80–86.

Sargent, S. (1990) Neighborhood effects on fruit removal by birds: a field experiment with *Viburnum dentatum* (Caprifoliaceae). *Ecology* 71, 1289–1298.

Schupp, E.W. and Fuentes, M. (1995) Spatial patterns of seed dispersal and the unification of plant population ecology. *Ecoscience* 2, 267–275.

Schupp, E.W., Milleron, T. and Russo, S.E. (2002) Dissemination limitation and the origin of species-rich tropical forests. In: Levey, D.J., Silva, W.R. and Galetti, M. (eds) *Seed Dispersal and Frugivory: Ecology, Evolution and Conservation.* CAB International, Wallingford, UK, pp. 19–43.

Strogatz, S.H. (2001) Exploring complex networks. *Nature* 410, 268–276.

Terborgh, J. (1986) Keystone plant resources in the tropical forest. In: Soulé, M.E. (ed.) *Conservation Biology: The Science of Scarcity and Diversity.* Sinauer Associates, Sunderland, Massachusetts, pp. 33–44.

Terborgh, J., Pitman, N., Silman, M.R., Schichter, H. and Núñez, V.P. (2002) Maintenance of tree diversity in tropical forests. In: Levey, D.J., Silva, W.R. and Galetti, M. (eds) *Seed Dispersal and Frugivory: Ecology, Evolution and Conservation.* CAB International, Wallingford, UK, pp. 1–18.

Tewksbury, J.J. and Nabhan, G.P. (2001) Seed dispersal: directed deterrence by capsaicin in chillies. *Nature* 412, 403–404.

Thomas, C.D., Cameron, A., Green, R.E., Bakkenes, M., Beaumont, L.J., Collingham, Y.C., Erasmus, B.F.N., de Siqueira, M.F., Grainger, A., Hannah, L., Hughes, L., Huntley, B., van Jaarsveld, A.S., Midgley, G.F., Miles, L., Ortega-Huerta, M.A., Peterson, A.T., Phillips, O.L. and Williams, S.E. (2004) Extinction risk from climate change. *Nature* 427, 145–148.

Thompson, J.N. (1994) *The Coevolutionary Process.* University of Chicago Press, Chicago, Illinois.

Tilman, D. and Kareiva, P. (1997) *Spatial Ecology.* Princeton University Press, Princeton, New Jersey.

van der Pijl, L. (1972) *Principles of Dispersal in Higher Plants.* Springer, New York.

van Ommeren, R.J. and Whitham, T.G. (2002) Changes in interactions between juniper and mistletoe mediated by shared avian frugivores: parasitism to potential mutualism. *Oecologia* 130, 281–288.

Vieira, I.G., Uhl, C. and Nepstad, D. (1994) The role of the shrub *Cordia multispicata* Cham. as a succession facilitator in an abandoned pasture in Paragominas, Amazonia. *Vegetatio* 115, 91–99.

Wang, B.C. and Smith, T.B. (2002) Closing the seed dispersal loop. *Trends in Ecology and Evolution* 17, 379–385.

Webb, C.O. and Peart, D.R. (2001) High seed dispersal rates in faunally intact tropical rain forest: theoretical and conservation implications. *Ecology Letters* 4, 491–499.

Wenny, D.G. (2001) Advantages of seed dispersal: a re-evaluation of directed dispersal. *Evolutionary Ecology Research* 3, 51–74.

Westcott, D.A. and Graham, D.L. (2000) Patterns of movement and seed dispersal of a tropical frugivore. *Oecologia* 122, 249–257.

Wheelwright, N.T. (1985) Competition for dispersers and the timing of flowering and fruiting in a guild of tropical trees. *Oikos* 44, 465–477.

Whelan, C.J., Schmidt, K.A., Steele, B.B., Quinn, W.J. and Dilger, S. (1998) Are bird-consumed fruits complementary resources? *Oikos* 83, 195–205.

Whitham, T.G., Young, W.P., Martinsen, G.D., Ghering, C.A., Schweitzer, J.A., Schuster, S.M., Wimp, G.M., Fischer, D.G., Bailey, J.K., Lindroth, R.L., Woolbright, S. and Kuske, C.R. (2003) Community and ecosystem genetics: a consequence of the extended phenotype. *Ecology* 84, 559–573.

17 An Empirical Approach to Analysing the Demographic Consequences of Seed Dispersal by Frugivores

H. GODÍNEZ-ALVAREZ AND P. JORDANO

Introduction

Seed dispersal is a key stage in the life cycle of plants because it contributes to the recruitment of new individuals (Howe and Smallwood, 1982). Many authors have suggested that seed dispersal has important demographic consequences, and as such it is necessary to link comprehensive analyses of seed dispersal with studies of plant demography in order to completely understand the population dynamics of plants (Schupp and Fuentes, 1995; Godínez-Alvarez *et al.*, 2002; Jordano and Godoy, 2002; Wang and Smith, 2002; Howe and Miriti, 2004). In this chapter, we present an approach for integrating these two disparate types of study as a way to better understand the effects of seed dispersal on the population growth rate of plants.

Dispersal of seeds by fruit-eating animals is a complex process whose consequences encompass a series of concatenated stages beyond fruit consumption and seed removal, such as seed deposition, germination, and seedling emergence and establishment (Jordano and Herrera, 1995; Wang and Smith, 2002). Perhaps due to this complexity, no study to date has analysed all these sequential stages and determined their consequences by means of an explicitly demographic approach. Since successful seed dispersal by animals is the outcome of a mutualistic interaction between the animals and the plants, it is expected that it has net positive effects on the populations of each participating species. The effects of seed dispersal should ideally be observed and measured in the population growth rates of both animals and plants (Addicott, 1986), which requires an understanding of frugivory and the behaviour of frugivores beyond fruit removal.

The effects of seed dispersal have traditionally been incorporated into plant demographic studies by means of simulations with population dynamics models that are often built without considering the detailed natural history of the particular plant–frugivore system being examined (Horvitz and Schemske, 1995; Martínez-Ramos and Alvarez-Buylla, 1995; Valverde and Silvertown, 1997). On the other hand, incorporating the consequences of mutualistic interactions into matrix population models has so far had only limited success (Herrera and Jordano, 1981; Horvitz and Schemske, 1994; Godínez-Alvarez *et al.*, 2002). A particular limitation to the viable integration of seed dispersal with its demographic consequences has been an inability to link the effects of frugivores' actions with their delayed consequences at later stages of recruitment.

To accomplish our goal in this chapter, we first review some existing approaches to the study of seed dispersal. We then present a brief discussion of the population projection matrix models which are commonly used in plant demographic studies. Finally, we use empirical data to illustrate our ideas and then discuss the possible benefits and disadvantages of our approach. We argue that matrix population models can incorporate the necessary elements of analysis to clearly integrate the effects of frugivore activity on plant population dynamics. Ultimately, we hope to stimulate more integrated ecological studies, thus contributing to our understanding of the key role of seed dispersal in plant demography.

Approaches to the Study of Seed Dispersal

From a theoretical perspective, in order to understand the demographic consequences of seed dispersal, it is essential to obtain data on all critical stages and ecological processes from the reproduction of the parent plants through to reproductive maturity of the new adults (Schupp and Fuentes, 1995). Relevant data include fruit production, the composition of the frugivore assemblage and the quantity of seeds removed by each species, frugivore effects on seed germination, the microsites into which frugivores deposit seeds, and the effects of those microsites on seed survival, germination and adult recruitment.

Because the evaluation of such broad-ranging data types is methodologically so complex, studies of seed dispersal to date have only analysed certain stages of the overall process. Some studies have concentrated on the pre-dissemination stage, mainly through an analysis of the quantities of seeds dispersed by different frugivore species (Schupp, 1993; Jordano and Schupp, 2000), whereas others have focused on the post-dissemination stage, generally by estimating seedling recruitment probabilities (Jordano and Herrera, 1995; Rey and Alcántara, 2000; García, 2001; Traveset *et al.*, 2003).

The overall consequence of seed dispersal can perhaps be viewed best within the framework of 'disperser effectiveness'. Disperser effectiveness is defined as the relative contribution of a frugivore species to plant reproduction, which depends on the quantity of seeds removed and the quality of treatment given to the seeds during the process of dispersal (Schupp, 1993). At present, no single study has thoroughly analysed both the quantitative and qualitative components for all the frugivores of a particular plant species, but very good beginnings have been made for some systems (Howe, 1977; Herrera and Jordano, 1981; Murray, 1988; Reid, 1989; Jordano and Schupp, 2000). The frequency of visits and the number of seeds removed per visit are among the best-documented quantitative aspects, while fruit handling, post-feeding movement patterns of frugivores, and the effects of handling on seed germination are some of the best-studied qualitative aspects (see Table 17.1 for examples).

Studies addressing recruitment probabilities are generally based on the analysis of seed and seedling survival in different microhabitats where seed rain may or may not have been adequately quantified. Important microhabitat-specific data include the number of seeds dispersed to that site by the frugivore community as a whole, the proportion of seeds suffering predation, the proportion of seeds germinating, and at least some minimal data on survival of the seedlings. Actual data on seedling survival has been extremely variable in duration, ranging from several months up to 4 years post-germination (Table 17.1). Based on such data, recruitment probabilities for particular plant species have been calculated as the product of the transition probabilities among the early life cycle stages (e.g. ovule to seedling, Traveset *et al.*, 2003; seed to seedling, Herrera *et al.*, 1994; Jordano and Herrera, 1995; Rey and Alcántara, 2000). In many cases at least, the processes acting during these stages of the life cycle are independent and the eventual outcomes are site-specific (Jordano and Herrera, 1995; Schupp, 1995) and often context-dependent in other ways (see Schupp, Chapter 20, this volume).

In order to determine the role of frugivores in plant population dynamics, we believe that it is necessary to improve our knowledge of the pre- and post-dissemination stages of seed dispersal and then to identify the methodological limitations for integrating frugivore activity with the success of seeds and seedlings following handling and dissemination. A review of the relevant literature showed that for some species the post-dissemination stage has been intensively studied while data regarding the pre-dissemination stage are scarce or non-existent, and vice versa (Table 17.1). Thus, an imbalance is evident between the amount of data for the pre- and post-dissemination stages of a given species.

In addition, even if we have adequate data we are limited in our ability to effectively link these stages. Effectively linking such data is essential in order to integrate seed dispersal with information on plant demography. We suggest this integration may be achieved through the use of projection matrices, which incorporate data on survival, growth and fecundity of individuals at different stages of the life cycle, and summarize it in relevant population parameters such as the population growth rate, or lambda (λ; Caswell, 2001). In particular, we suggest that one way to better understand the impact of frugivores on the demography of plant populations is to use projection matrices to link the pre- and post-dissemination stages for each individual disperser species.

Table 17.1. A sample of studies that have assessed aspects of the quantitative and qualitative components of disperser effectiveness, sensu Schupp (1993). Numbers indicate distinct aspects of the quantitative or qualitative components of disperser effectiveness: 1 = Relative abundance, expressed mainly as number of individuals per hour of observation; 2 = Frequency of visits to the plant, estimated as the number of visits per hour; 3 = Time spent feeding on the plant; 4 = Number or proportion of seeds and/or fruits removed per dropping, cache or visit; 5 = Post-foraging behaviour, expressed as direction of movements, time between movements, movement distance, and/or number of movements to distinct landscape perches, units or microhabitats; 6 = Number or proportion of seeds defecated per minute; 7 = Number or proportion of seeds germinating after gut passage and/or germination rate; 8 = Number or proportion of seeds and/or seedlings in different microhabitats; 9 = Number or proportion of seedlings surviving for 1–4 years; – = Not evaluated.

Plants/Animals	Quantity	Quality	Reference
Casearia corymbosa (Flacourtiaceae)/Birds (6 spp.)	1, 2, 4	7, 8	Howe, 1977
Prunus mahaleb (Rosaceae)/ Birds (>10 spp.)	1, 2, 3, 4	5, 8	Herrera and Jordano, 1981; Jordano and Schupp, 2000
Plants (3 spp.)/Birds (6 spp.)	1	6, 7	Murray, 1988
Amyema quandang (Loranthaceae)/Birds (2 spp.)	4	7, 8	Reid, 1989
Pinus jeffreyi (Pinaceae)/Rodents (2 spp.)	4	8	Vander Wall, 1993
Juniperus ashei (Cupressaceae)/Birds (2 spp.)	3, 4	5, 8	Chávez-Ramírez and Slack, 1994
Maesa lanceolata (Maecaceae)/Birds (2 spp.)	2, 4	6, 7	Graham *et al.*, 1995
Phillyrea latifolia (Oleaceae)/Birds	4	7, 8, 9	Jordano and Herrera, 1995
Plants (6 spp.)/Frugivores	4	7, 9	Chapman and Chapman, 1996
Phoradendron californicum (Viscaceae)/Birds (3 spp.)	1, 2	–	Larson, 1996
Plants (2 spp.)/Rodents (2 spp.)	–	8, 9	Forget, 1997
Plants (6 spp.)/Birds (3 spp.)	3, 4	5, 6	Sun *et al.*, 1997
Ocotea endresiana (Lauraceae)/Birds (5 spp.)	–	5, 8	Wenny and Levey, 1998
Plants (4 spp.)/Birds (6 spp.)	4	8	Loiselle and Blake, 1999
Plants (9 spp.)/Birds (2 spp.)	–	5, 6	Holbrook and Smith, 2000
Olea europaea (Oleaceae)/Birds	4	7, 8, 9	Rey and Alcántara, 2000
Plants (18 spp.)/*Lagothrix lagothricha* (Atelidae)	4	5, 7	Stevenson, 2000
Beilschmiedia pendula (Lauraceae)/ Birds (4 spp.)	–	7, 8, 9	Wenny, 2000a
Ocotea endresiana/Birds (5 spp.)	–	7, 8, 9	Wenny, 2000b
Juniperus communis (Cupressaceae)/Birds (2 spp.)	4	7, 8, 9	García, 2001
Corema album (Empetraceae)/ Birds (2 spp.), Rodents (1 spp.)	–	7, 8, 9	Calviño-Cancela, 2002, 2004
Ruppia maritima (Ruppiaceae)/Birds (9 spp.)	4	7	Figuerola *et al.*, 2002
Neobuxbaumia tetetzo (Cactaceae) /Bats (1 spp.), Birds (4 spp.)	1, 2	5, 7	Godínez-Alvarez *et al.*, 2002
Plants (12 spp.)/Monkeys (3 spp.)	2, 4	8	Kaplin and Lambert, 2002
Leptonychia usambarensis (Sterculiaceae)/Birds (11 spp.)	4	9	Cordeiro and Howe, 2003
Rhamnus ludovici-salvatoris (Rhamnaceae)/Birds (2 spp.)	4	8, 9	Traveset *et al.*, 2003
Pinus monophylla (Pinaceae)/Rodents (6 spp.)	4	8	Hollander and Vander Wall, 2004

Population Projection Matrices and Seed Dispersal

Projection matrix models are commonly used in the study of plant demography because they provide basic population parameters such as population growth rates, stable size distributions, and reproductive values (Caswell, 2001). Projection matrices used in the study of plants generally use size or life cycle stage for categorizing the population structure (Fig. 17.1a) because these are generally better predictors than age of plant fate. The structure can be depicted with life cycle diagrams in which different stages of the life cycle are represented by n_i, where n is the number of individuals in stage i, along with the probability that an individual in stage i at time t contributes through survival, growth, or reproduction to stage j at time $t+1$. Based on these diagrams, projection matrices are easily derived by considering the links between the different stages as well as their associated transition probabilities (Fig. 17.1b).

The entries in a projection matrix can vary depending on the complexity of the life cycle (Silvertown *et al.*, 1993; Caswell, 2001). For instance, projection matrices for species with relatively simple life cycles in which individuals at one stage can only either grow to the next stage or

(a)

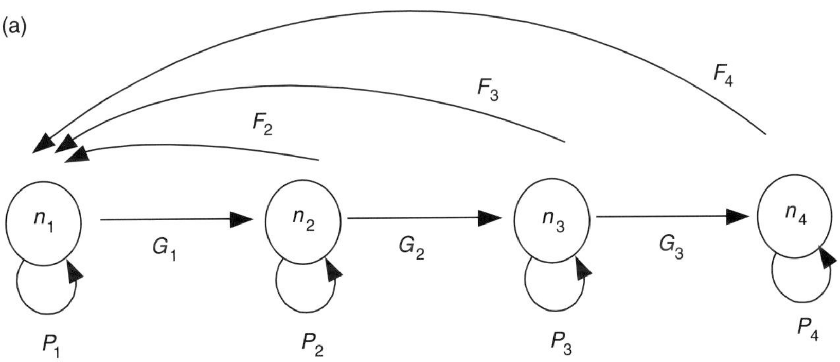

(b)

$$A = \begin{bmatrix} P_1 & F_2 & F_3 & F_4 \\ G_1 & P_2 & 0 & 0 \\ 0 & G_2 & P_3 & 0 \\ 0 & 0 & G_3 & P_4 \end{bmatrix}$$

Fig. 17.1. An example of a life cycle diagram (a) and its corresponding projection matrix (b) for a plant with a relatively simple life cycle categorized in four stages (n_i). Individuals in stage i can survive and grow (G_i) into the next stage, or survive and remain in the same stage (P_1), during a given time interval. Individuals in the last three stages contribute to the first stage through sexual reproduction (F_i).

survive at the same stage during a specified time interval, and in which adults can only reproduce sexually, will only have possible values along the main diagonal, the first lower subdiagonal, and the first row (see Fig. 17.1b). Values along the main diagonal refer to the survival probabilities of individuals that do not grow to the next stage, those in the first subdiagonal represent the transition or growth probabilities, and those in the first row indicate fecundities.

The fecundity value of a given adult stage can be calculated as the product of the probability of reproduction for that stage, the mean number of seeds produced by a reproductive individual of that stage, and the probability of a seed becoming a seedling during the time-step, which is considered constant for all adult stages (Godínez-Alvarez *et al.*, 1999; Box 17.1). Frugivore effects may be incorporated by considering data on quantitative (i.e. frequency of visits to the plants and seed removal) and qualitative (i.e. seed germination after gut passage and number of flights to different landscape patches) aspects of each species of frugivore. The first step is to calculate:

1. The probability of seed removal;
2. The probability of seed germination after gut passage;
3. The probability that seeds would be delivered to each distinct patch type;
4. The probability of making the transition from seed to seedling in each distinct patch type.

Using these probabilities, new fecundity values can be estimated, which in turn may be incorporated in the projection matrix to calculate the population growth rate due to frugivore effects (see Box 17.1 for a detailed description about how to calculate and integrate these four probabilities into fecundity values).

With this approach we can begin to evaluate the effects of frugivores on plant demography. Depending on the system, we can do this for each individual species of frugivore separately or for disperser functional groups composed of species sharing some taxonomical or ecological affinities (see Dennis and Westcott, Chapter 9, this volume). Either way, we can use fairly easily collected data to estimate, respectively, the contribution of particular species or of functional groups to population growth rate (see Box 17.2). In addition, we can use this approach to evaluate, among other things, the consequences for plants of an avian seed predator or of an invasive disperser species, as well as the population collapse or even extinction of a given species of frugivore. By estimating the population growth rate under various scenarios (e.g. with and without a particular frugivore) we can determine the relative importance of given species to the maintenance of plant populations.

Dispersal of *Neobuxbaumia tetetzo* Seeds: an Example

In this section we present a brief overview of seed dispersal of the columnar cactus *Neobuxbaumia tetetzo* (Cactaceae) as an example of the ideas

Box 17.1 Incorporating frugivore effects into the fecundity values of matrix models

Frugivore effects can be incorporated into the fecundity values of matrix models through the estimation of four probabilities:

1. The probability of **seed removal;**
2. The probability of **seed germination after gut passage;**
3. The probability of **delivering seeds to particular patches;**
4. The probability of **making the transition from seed to seedling in each patch**.

To illustrate the estimations of these probabilities, consider an example with three species of frugivores (*A*, *B* and *C*) for which we have data on the frequency of visits to the plants and the number of seeds removed per visit.

To estimate the ***probability of seed removal***, these data are first multiplied together and then this product is summed across all species. From this total, the relative proportion of total seeds removed by each species of frugivore is calculated; this represents the probability of seed removal per species.

Frugivore species	Frequency of visits (1)	No. of seeds removed (2)	Product of 1 and 2	Relative proportion
A	3	50	150	0.22
B	7	5	35	0.05
C	5	100	500	0.73
		Total	685	1.00

Other aspects of seed dispersal such as fruit handling success (e.g. proportion of fruits ingested and successfully removed away from the parent tree relative to the number of fruits handled) can easily be incorporated into this approach.

The ***probability of seed germination after gut passage*** can be estimated with laboratory experiments using seeds defecated by each species of frugivore. The mean proportion of seeds germinating for each species of frugivore may be interpreted as the probability of seed germination after gut passage.

To estimate the ***probability of delivering seeds to particular patches***, one can determine the number of post-feeding flights to different landscape patches for each species of frugivore and then calculate the relative proportion of total flights to a microhabitat type that are made by each species. A hypothetical example with three species of frugivores and two patches is:

	Frugivore species					
	A		B		C	
	No. of flights	Relative proportion	No. of flights	Relative proportion	No. of flights	Relative proportion
Patch 1	12	0.25	27	0.90	13	0.50
Patch 2	36	0.75	3	0.10	13	0.50
Total	48	1.00	30	1.00	26	1.00

Continued

Box 17.1 *Continued*

These probabilities can be combined with the frequency of visits to the plants and the number of seeds removed per visit to estimate the number of seeds in a patch type contributed by a given species.

The ***probability of making the transition from seed to seedling in each patch*** can be estimated through field experiments in those patch types considered relevant. In each of these patches, seeds directly obtained from fruits can be sown in the field to determine the number of emerging seedlings as well as the number of surviving seedlings after 1 year. The mean proportion of surviving seedlings with respect to the total number of seeds sown may be considered as the probability that seeds germinate and survive in each patch.

To incorporate these probabilities into the fecundity values, first consider that the fecundity based on the recruitment of 1-year old seedlings can be calculated as:

$$F_i = R_i S_i C \quad (1)$$

where F is the fecundity for individuals in stage i, R is the probability of reproduction, S is the mean number of seeds produced per individual, and C is the probability of passing from seed to seedling in the field after 1 year, which is constant for all adult stages. Based on this definition, C can be replaced in (1) by the four estimated probabilities above. That is:

$$F_{i,s,m} = R_i S_i P_s G_s \sum_{s=1}^{x} \sum_{m=1}^{y} \left(D_{sm} T_m\right) \quad (2)$$

where, F_i, R_i, and S_i are the same as in (1), P_s is the probability of seed removal by the frugivore s, G_s is the probability of seed germination after gut passage, D_{sm} is the probability that seeds would be delivered in the patch m by s, and T_m is the probability of making the transition from seed to seedling in m. The simplest case is when plants are visited only by one frugivore species (s = 1) and their seeds and seedlings successfully germinate and survive only in one patch type (m = 1). In this case, P_s, G_s, and D_{sm} are the probabilities estimated for the only species of frugivore and T_m is constant. When there is more than one frugivore species (s = x) and patch type (m = y), P_s can be calculated as the sum of the removal probabilities estimated for each species, and G_s as the mean weighted by the seed removal probability.

presented above. Detailed information on this plant and its seed dispersers can be found in Godínez-Alvarez *et al.* (2002).

Neobuxbaumia tetetzo is a common plant in the Tehuacán Valley, a dry zone in south-central México. Fleshy fruits produced on branch tips dehisce at night, exposing a whitish, sugary pulp with hundreds of small black seeds. Fruits are consumed at night by the long-nosed bat *Leptonycteris curasoae* (Phyllostomatidae) and during the day by several species of birds such as cactus wren (*Campylorhynchus brunneicapillus*; Troglodytidae), curved-billed thrasher (*Toxostoma curvirostre*; Mimidae) and grey-breasted woodpecker (*Melanerpes hypopolius*; Picidae). Since the traits of *N. tetetzo* fruits agree with those suggested for fruits consumed by bats (van der Pijl, 1982), it is reasonable to assume that bats have a greater contribution to cactus population growth rate than birds do.

Box 17.2. Disperser functional groups: combining the probabilities of frugivore species with ecological or taxonomical affinities

The probabilities calculated for each species of frugivore can be used individually or in combination with other species that share some ecological or taxonomical affinities in order to evaluate the species-specific or functional group effects of frugivores on population growth rate.

Consider for simplicity that probabilities of seed removal (P_s), seed germination after gut passage (G_s), and seed deposition (D_{sm}) in two particular landscape patches have been calculated for two species of birds (*A* and *B*) and one species of bat (*C*). The probabilities of making the transition from seed to seedling (T_m) in each patch were calculated independent of the frugivore species and remain constant. As an example:

Frugivores			D_{sm}		T_m	
(*s*)	P_s	G_s	$m = 1$	$m = 2$	$m = 1$	$m = 2$
A	0.22	0.90	0.25	0.75	0.1	0.01
B	0.05	0.60	0.90	0.10	0.1	0.01
C	0.73	0.80	0.50	0.50	0.1	0.01

Based on these data and Eqn (2), it is possible to estimate, for example, the effects of birds (*A*+*B*), bats (*C*), and all frugivore species (*A*+*B*+*C*). In the cases of birds and all frugivore species, the probability of seed removal can be calculated as the sum of the probabilities estimated for each species being considered (*A*+*B* or *A*+*B*+*C*), while the probability of seed germination can be estimated as the mean weighted by the seed removal probability.

Frugivores species	P_s	G_s	$\sum_{s=1}^{x}\sum_{m=1}^{y}(D_{sm}T_m)$
A	0.22	0.90	0.0325
B	0.05	0.60	0.091
A+*B*	0.27	0.84	0.1235
C	0.73	0.80	0.055
A+*B*+*C*	1.00	0.81	0.1785

These probabilities may be substituted in the calculation of the fecundities, to obtain new values that integrate the effects of bats, of birds, or of all frugivore species.

To test this assumption, information on seed dispersal for different frugivores was incorporated into a population projection matrix of *N. tetetzo* (Godínez-Alvarez *et al.*, 1999). This projection matrix was built by grouping individuals in size categories based on the height of the principal trunk and then by considering survival, reproduction and growth probabilities of individuals in each category over a 1-year interval. Size categories were: seedling 1 (< 2 cm), seedling 2 (2–8 cm), seedling 3 (8–15 cm), sapling

(15–45 cm), juvenile (45–100 cm), immature (100–150 cm), mature 1 (150–250 cm), mature 2 (250–350 cm), mature 3 (350–450 cm), mature 4 (450–550 cm), mature 5 (550–650 cm) and mature 6 (> 650 cm), yielding a total of 12 categories. The projection matrix only has values in the main diagonal (survival without growth probabilities), the first lower subdiagonal (growth probabilities), and the first row (fecundity values). Fecundity values were estimated as the product of: (i) the probability of reproduction; and (ii) the mean number of seeds produced by a reproducing individual for each reproductive category, as well as (iii) the constant estimated probability of passing from seed to seedling over a 1-year time-step (Fig. 17.2).

The seed dispersal data included quantitative and qualitative aspects of disperser effectiveness (Schupp, 1993). The quantitative aspects were frequency of visits to plants (visits/h) and seed removal by each species (number of seeds removed/fruit). This latter aspect was estimated as number of seeds removed per fruit, instead of per visit, because of the large number of seeds and their small size. The qualitative aspects were the proportion of seed germination after gut passage and the number of post-feeding flights to trees and shrubs for each species of frugivore, and the number of surviving seedlings beneath trees and shrubs after 1 year. The number of post-feeding flights and the number of surviving seedlings were estimated only for trees and shrubs, since these plants provide the only microsites suitable for seed germination and seedling establishment (Godínez-Alvarez *et al.*, 1999). Using these data, an estimate of: (i) the probability of seed removal; (ii) the probability of seed germination after gut passage; (iii) the probability of delivering seeds to tree and shrub microsites; and (iv) the probability of making the transition from a seed to a seedling beneath trees and shrubs, were calculated for each species of

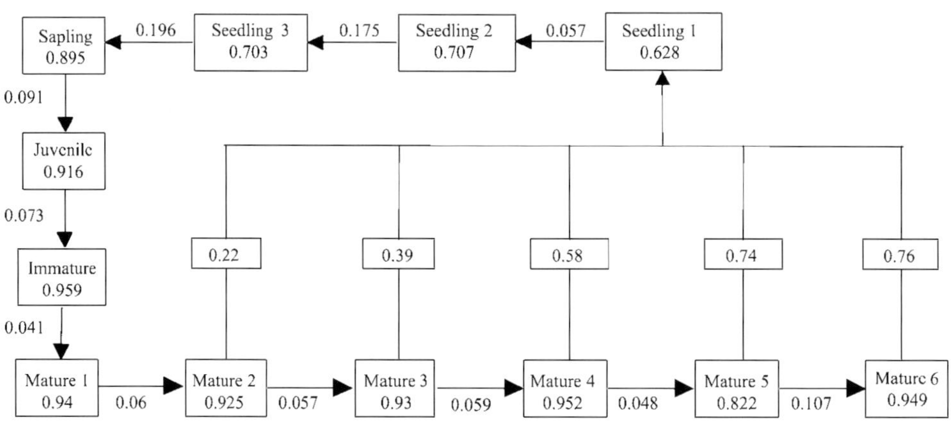

Fig. 17.2. Life cycle diagram of the columnar cactus *Neobuxbaumia tetetzo*. Values inside boxes refer to survival probabilities (P_i values in Fig. 17.1a) and arrows represent transition or growth probabilities (G_i values in Fig. 17.1a). Boxes above mature categories 2–6 indicate fecundity values (F_i values in Fig. 17.1a).

frugivore, and were used to modify the fecundity values of the *N. tetetzo* matrix model. The calculation of these probabilities and the modification of the fecundity values were made following the procedures described in Box 17.1.

The role of different frugivores on the population dynamics of *N. tetetzo* was determined through matrix simulations assessing the effects of individual species as well as the effect of all dispersers acting together. The effect of a species of frugivore was simulated by incorporating the fecundity values calculated for that species into the matrix model. To simulate the effect of all dispersers acting together, the probability of seed removal was calculated as the sum of all the probabilities estimated for each species, whereas the probability of seed germination was estimated as the mean weighted by the seed removal probability (see Box 17.2). The population growth rate (lambda, λ) was calculated for each simulation by multiplying the projection matrix by a vector representing the number of individuals in each size category. This procedure was repeated with each resulting vector until its proportions remained constant, at which point the population grows at a rate equal to lambda.

Results showed that species of frugivore differed in their effects on the population dynamics of *N. tetetzo* (Table 17.2). As expected, the estimated population growth rate due to the long-nosed bat, *L. curasoae*, was higher than those estimated to be due to the birds *C. brunneicapillus*, *M. hypopolius* and *T. curvirostre*. The higher contribution of *L. curasoae* was mostly due to its higher probabilities of seed removal and of delivering seeds to tree and shrub microsites. The matrix simulations conducted for all species of frugivores acting together also showed a high contribution to the population growth rate of this columnar cactus (Table 17.2).

Conclusions and Perspectives

Many authors have pointed out that in order to advance our understanding of seed dispersal by frugivores it is essential to evaluate the demographic consequences of dispersal for plant populations (Schupp and Fuentes, 1995; Godínez-Alvarez *et al.*, 2002; Jordano and Godoy, 2002; Wang and Smith, 2002; Howe and Miriti, 2004). To accomplish this goal, it is necessary to obtain detailed information on the outcome of frugivore activity and their potential effects on critical stages of the life cycle of plants, such as seed germination and seedling, juvenile and adult survival and growth – the delayed consequences of dispersal. The studies conducted up until now have provided some information on the dispersal effectiveness of frugivores and the demographic fate of seeds and seedlings. However, this information has not allowed an evaluation of the demographic consequences of seed dispersal because it is incomplete for most plants and because the methodological approaches needed to integrate this information into plant demography have not been clearly described. The importance of this chapter is the demonstration that an empirical approach based on

Table 17.2. Frugivore effects employed to modify the projection matrix of *Neobuxbaumia tetetzo*. Effects were estimated according to Eqn (2) in Box 17.1, where P_s = probability of seed removal by species *s*, G_s = probability of seed germination after gut passage, D_{sm} = probability of delivering seeds in the patch *m* by *s*, and T_m = probability of making the transition from seed to seedling in *m*.

Frugivore species	P_s	G_s	D_{sm}	T_m	$\sum_{s=1}^{x}\sum_{m=1}^{y}(D_{sm}T_m)$	Estimated rate of increase
Leptonycteris curasoae (bat)	0.987	0.86	0.72	0.00139	0.001	1.003
Campylorhynchus brunneicapillus (bird)	0.002	0.98	0.43	0.00139	6.0×10^{-4}	0.97
Melanerpes hypopolius (bird)	0.008	0.96	0.03	0.00139	4.2×10^{-5}	0.97
Toxostoma curvirostre (bird)	0.003	0.98	0.12	0.00139	1.7×10^{-4}	0.97
All frugivores	1.000	0.86	–	–	1.8×10^{-3}	1.009

P_s is estimated as the product of the frequency of visits and number of seeds removed per fruit for each frugivore; this product was summed across all frugivores, and the relative proportion of the total was calculated for each species. For all frugivores, it was calculated as the sum of the removal probabilities estimated for each species.
G_s is estimated as the mean proportion of germinated seeds in laboratory experiments using seeds defecated by each frugivore species. For all frugivore species, it was estimated as the mean weighted by the seed removal probability.
D_{sm} is estimated as the proportion of post-feeding flights to trees and shrubs out of the total number of flights recorded.
T_m is calculated as the mean proportion of seedlings surviving under trees and shrubs after 1 year.

projection matrices allows us to explore the ecological consequences of seed dispersal in an explicitly demographic context. This approach is based on the incorporation of frugivore effects into the matrix elements concerned with plant reproduction in order to simulate frugivore impacts on the population growth rate of plants. This allows a comprehensive analysis of the effectiveness of different frugivore species or guilds with the best measure possible: their effects on plant population dynamics.

Since this approach may enhance our understanding of seed dispersal, it is essential to conduct more studies using these ideas in order to determine their validity and applicability. In this context, it is important to define those aspects of seed dispersal that provide the minimum information needed to conduct these demographic evaluations. Generally, the basic data needed are:

1. The frequency of visits by a species of frugivore to plants;
2. The number of seeds dispersed per visit;
3. The species-specific effects on seed germination;
4. The species-specific contribution to seed rain in distinct and relevant microhabitats;
5. Data on seed and seedling survival in these microhabitats.

Item 4 requires a simultaneous assessment of seed rain, frugivore foraging and habitat preferences at the landscape level (e.g. Wenny and Levey, 1998; Jordano and Schupp, 2000). With this information, it is possible to estimate the probabilities to be incorporated into the first row of a projection matrix. Other aspects such as the reliability of frugivores, digestive physiology, and post-foraging behaviour, among others, may also be incorporated into the calculation of these probabilities, but much of these data may be most important for dissecting *why* species differ in effectiveness, rather than for actually quantifying differences.

The likelihood of seeds being dispersed to sites where seedling recruitment can successfully occur in the field may be significantly altered by spatio-temporal variability, such as landscape heterogeneity and annual changes in the frugivore community (Jordano and Herrera, 1995; Schupp, 1995; Schupp and Fuentes, 1995; Herrera, 1998; Howe and Miriti, 2004). In addition, the 'quality' of a given type of dispersal site (e.g. microhabitat) may change temporally and spatially (see Schupp, Chapter 20, this volume). Thus, in order to completely understand the demographic consequences of seed dispersal, it would be important to also begin to quantify variability in critical factors and to evaluate the effects of such variability on the population growth rate of plants. These evaluations are especially important, since spatial and temporal variability can affect not only the estimates of the frugivore effects, but also the estimates of the other matrix elements, influencing the rate of population growth and the contribution of different life stages to this rate (i.e. elasticities; see Pfister, 1998). The approach described here may be used as a preliminary basis to incorporate these factors, through monitoring annual changes in plant demography as well as in some selected stages of seed dispersal. Such monitoring also might include different habitats where it is known that some demographic processes such as recruitment or survival probabilities may change throughout the life cycle of plants (Howe and Miriti, 2004). Based on this information, it is possible to build multi-annual matrices for different habitats, which integrate frugivore effects and plant demography. Comparing the population growth rates and elasticities calculated for each of these matrices would permit an evaluation of the extent to which the relative importance of particular species of frugivores and different life stages to plant population dynamics changes among habitats and/or years.

The approach discussed here depends on the availability of projection matrices to estimate the growth rate of populations. These matrix models have assumptions that might limit the use of the proposed approach under some conditions (Caswell, 2001).

- Projection matrices are time-invariant in that survival and growth probabilities and fecundity values of individuals remain constant through time.
- They also assume that populations grow at a constant rate until reaching a stable distribution; thus they do not consider density-dependent effects.

- In addition, there are potential problems associated with multiplying demographic estimates (e.g. fecundity of individuals in a size class) that have variances.

Because of these assumptions, the proposed approach should be used only to generate scenarios of the possible effects of frugivores on the population growth rate, instead of trying to predict the actual population dynamics of plants (Godínez-Alvarez *et al.*, 2002). Such explorations are extremely valuable in and of themselves; indeed, the patterns are more valuable than the actual values.

The relative impacts of survival, growth and fecundity on the growth rate of populations apparently change depending on the life history of plants (Silvertown *et al.*, 1993). In general, fecundity is the most important demographic process for the population persistence of plants with short life cycles, whereas adult survival is the process with the highest contribution for long-lived perennials. Due to these differences, it is possible that frugivore effects on the population growth rate would be more easily detected in plants with short life cycles than in long-lived perennials, since evaluations are based on the modification of the fecundity values. However, field evidence is needed to support these ideas.

An understanding of the demographic consequences of seed dispersal by frugivores depends upon the integration of information on frugivore activities and plant demography. With the approach discussed here, we hope to stimulate more integrative studies that contribute to a better understanding of the ecological consequences of this biological interaction.

Acknowledgements

We thank Eugene W. Schupp and two anonymous reviewers for their critical comments on the manuscript. This chapter was written during H.G.A.'s research stay at the Integrative Ecology Group, Estación Biológica de Doñana. The financial support for this stay was provided by the Dirección General de Asuntos del Personal Académico, UNAM.

References

Addicott, J.F. (1986) On the population consequences of mutualism. In: Diamond, J. and Case, T.J. (eds) *Community Ecology*. Harper and Row, New York, pp. 425–436.

Calviño-Cancela, M. (2002) Spatial patterns of seed dispersal and seedling recruitment in *Corema album* (*Empetraceae*): the importance of unspecialized dispersers for regeneration. *Journal of Ecology* 90, 775–784.

Calviño-Cancela, M. (2004) Ingestion and dispersal: direct and indirect effects of frugivores on seed viability and germination of *Corema album* (*Empetraceae*). *Acta Oecologica* 26, 55–64.

Caswell, H. (2001) *Matrix Population Models: Construction, Analysis, and Interpretation*, 2nd edn. Sinauer Associates, Sunderland, Massachusetts.

Chapman, C.A. and Chapman, L.J. (1996) Frugivory and the fate of dispersed and non-dispersed seeds of six African tree species. *Journal of Tropical Ecology* 12, 491–504.

Chávez-Ramírez, F. and Slack, R.D. (1994) Effects of avian foraging and post-foraging behavior on seed dispersal patterns of Ashe juniper. *Oikos* 71, 40–46.

Cordeiro, N.J. and Howe, H.F. (2003) Forest fragmentation severs mutualism between seed dispersers and an endemic African tree. *Proceedings of the National Academy of Sciences USA* 100, 14052–14056.

Figuerola, J., Green, A.J. and Santamaría, L. (2002) Comparative dispersal effectiveness of wigeongrass seeds by waterfowl wintering in south-west Spain: quantitative and qualitative aspects. *Journal of Ecology* 90, 989–1001.

Forget, P.-M. (1997) Effect of microhabitat on seed fate and seedling performance in two rodent-dispersed tree species in rain forest in French Guiana. *Journal of Ecology* 85, 693–703.

García, D. (2001) Effects of seed dispersal on *Juniperus communis* recruitment on a Mediterranean mountain. *Journal of Vegetation Science* 12, 839–848.

Godínez-Alvarez, H., Valiente-Banuet, A. and Valiente, B.L. (1999) Biotic interactions and the population dynamics of the long-lived columnar cactus *Neobuxbaumia tetetzo* in the Tehuacan Valley, Mexico. *Canadian Journal of Botany* 77, 203–208.

Godínez-Alvarez, H., Valiente-Banuet, A. and Rojas-Martínez, A. (2002) The role of seed dispersers in the population dynamics of the columnar cactus *Neobuxbaumia tetetzo*. *Ecology* 83, 2617–2629.

Graham, C.H., Moermond, T.C., Kristensen, K.A. and Mvukiyumwami, J. (1995) Seed dispersal effectiveness by two bulbuls on *Maesa lanceolata*, an African montane forest tree. *Biotropica* 27, 479–486.

Herrera, C.M. (1998) Long-term dynamics of Mediterranean frugivorous birds and fleshy fruits: a 12-year study. *Ecological Monographs* 68, 511–538.

Herrera, C.M. and Jordano, P. (1981) *Prunus mahaleb* and birds: the high efficiency seed dispersal system of a temperate fruiting tree. *Ecological Monographs* 51, 203–218.

Herrera, C.M., Jordano, P., López-Soria, L. and Amat, J.A. (1994) Recruitment of a mast-fruiting, bird-dispersed tree: bridging frugivore activity and seedling establishment. *Ecological Monographs* 64, 315–344.

Holbrook, K.M. and Smith, T.B. (2000) Seed dispersal and movement patterns in two species of *Ceratogymna* hornbills in a West African tropical lowland forest. *Oecologia* 125, 249–257.

Hollander, J.L. and Vander Wall, S.B. (2004) Effectiveness of six species of rodents as dispersers of singleleaf piñon pine (*Pinus monophylla*). *Oecologia* 138, 57–65.

Horvitz, C.C. and Schemske, D.W. (1994) Effects of dispersers, gaps, and predators on dormancy and seedling emergence in a tropical herb. *Ecology* 75, 1949–1958.

Horvitz, C.C. and Schemske, D.W. (1995) Spatiotemporal variation in demographic transitions of a tropical understory herb: projection matrix analysis. *Ecological Monographs* 65, 155–192.

Howe, H.F. (1977) Bird activity and seed dispersal of a tropical wet forest tree. *Ecology* 58, 539–550.

Howe, H.F. and Miriti, M.N. (2004) When seed dispersal matters. *BioScience* 54, 651–660.

Howe, H.F. and Smallwood, J. (1982) Ecology of seed dispersal. *Annual Review of Ecology and Systematics* 13, 201–228.

Jordano, P. and Godoy, J.A. (2002) Frugivore-generated seed shadows: a landscape view of demographic and genetic effects. In: Levey, D.J., Silva, W.R. and Galetti, M. (eds) *Seed Dispersal and Frugivory: Ecology, Evolution and Conservation.* CAB International, Wallingford, UK, pp. 305–321.

Jordano, P. and Herrera, C.M. (1995) Shuffling the offspring: uncoupling and spatial discordance of multiple stages in vertebrate seed dispersal. *Écoscience* 2, 230–237.

Jordano, P. and Schupp, E.W. (2000) Seed dispersal effectiveness: the quantity component and patterns of seed rain for *Prunus mahaleb*. *Ecological Monographs* 70, 591–615.

Kaplin, B.A. and Lambert, J.E. (2002) Effectiveness of seed dispersal by *Cercopithecus* monkeys: implications for seed input into degraded areas. In: Levey, D.J., Silva, W.R. and Galetti, M. (eds) *Seed Dispersal and Frugivory: Ecology, Evolution and Conservation*. CAB International, Wallingford, UK, pp. 351–364.

Larson, D.L. (1996) Seed dispersal by specialist versus generalist foragers: the plant's perspective. *Oikos* 76, 113–120.

Loiselle, B.A. and Blake, J.G. (1999) Dispersal of melastome seeds by fruit-eating birds of tropical forest understory. *Ecology* 80, 330–336.

Martínez-Ramos, M. and Alvarez-Buylla, E.R. (1995) Seed dispersal and patch dynamics in tropical rainforests: a demographic approach. *Écoscience* 2, 223–229.

Murray, K.G. (1988) Avian seed dispersal of three neotropical gap-dependent plants. *Ecological Monographs* 58, 271–298.

Pfister, C.A. (1998) Patterns of variance in stage-structured populations: evolutionary predictions and ecological implications. *Proceedings of the National Academy of Sciences USA* 95, 213–218.

Reid, N. (1989) Dispersal of mistletoes by honeyeaters and flowerpeckers: components of seed dispersal quality. *Ecology* 70, 137–145.

Rey, P.J. and Alcántara, J.M. (2000) Recruitment dynamics of a fleshy-fruited plant (*Olea europaea*): connecting patterns of seed dispersal to seedling establishment. *Journal of Ecology* 88, 622–633.

Schupp, E.W. (1993) Quantity, quality, and the effectiveness of seed dispersal by animals. In: Fleming, T.H. and Estrada, A. (eds) *Frugivory and Seed Dispersal: Ecological and Evolutionary Aspects*. Kluwer Academic, Dordrecht, The Netherlands, pp. 15–29.

Schupp, E.W. (1995) Seed–seedling conflicts, habitat choice, and patterns of plant recruitment. *American Journal of Botany* 82, 399–409.

Schupp, E.W. and Fuentes, M. (1995) Spatial patterns of seed dispersal and the unification of plant population ecology. *Ecoscience* 2, 267–275.

Silvertown, J., Franco, M., Pisanty, I. and Mendoza, A. (1993) Comparative plant demography: relative importance of life cycle components to the finite rate of increase in woody and herbaceous perennials. *Journal of Ecology* 81, 465–476.

Stevenson, P.R. (2000) Seed dispersal by woolly monkeys (*Lagothrix lagothricha*) at Tinigua National Park, Colombia: dispersal distance, germination, and dispersal quantity. *American Journal of Primatology* 50, 275–289.

Sun, C., Ives, A.R., Kraeuter, H.J. and Moermond, T.C. (1997) Effectiveness of three turacos as seed dispersers in a tropical montane forest. *Oecologia* 112, 94–103.

Traveset, A., Gulias, J., Riera, N. and Mus, M. (2003) Transition probabilities from pollination to establishment in a rare dioecious shrub species (*Rhamnus ludovici-salvatoris*) in two habitats. *Journal of Ecology* 91, 427–437.

Valverde, T. and Silvertown, J. (1997) An integrated model of demography, patch dynamics and seed dispersal in a woodland herb, *Primula vulgaris*. *Oikos* 80, 67–77.

van der Pijl, L. (1982) *Principles of Dispersal in Higher Plants*. Springer, Berlin, Germany.

Vander Wall, S.B. (1993) Cache site selection by chipmunks (*Tamias* spp.) and its influence on the effectiveness of seed dispersal in Jeffrey pine (*Pinus jeffreyi*). *Oecologia* 96, 246–252.

Wang, B.C. and Smith, T.B. (2002) Closing the seed dispersal loop. *Trends in Ecology and Evolution* 17, 379–385.

Wenny, D.G. (2000a) Seed dispersal of a high quality fruit by specialized frugivores: high quality dispersal? *Biotropica* 32, 327–337.

Wenny, D.G. (2000b) Seed dispersal, seed predation, and seedling recruitment of a neotropical montane tree. *Ecological Monographs* 70, 331–351.

Wenny, D.G. and Levey, D.J. (1998) Directed seed dispersal by bellbirds in a tropical cloud forest. *Proceedings of the National Academy of Sciences USA* 95, 6204–6207.

18 How Seed Dispersal Affects Interactions with Specialized Natural Enemies and their Contribution to the Maintenance of Diversity

H.C. Muller-Landau and F.R. Adler

Introduction

Seed dispersal patterns affect the rate at which offspring encounter specialized natural enemies, and thereby the strength of the associated interspecific interaction (Howe and Smallwood, 1982). Relatively specialized natural enemies such as pathogens and insect herbivores are expected to be found in higher densities near plants of their host species (Janzen, 1970). Because dispersing offspring are less likely to land near a conspecific than those remaining near their maternal parent, they are presumed to be less likely to encounter specialized enemies, and thus to have a higher than average survival and growth (Ridley, 1930; Janzen, 1970; Howe and Smallwood, 1982). In line with these predictions, empirical studies have generally found that seed predation by insects declines with increasing distance from conspecific adults, while overall survival increases (Hammond and Brown, 1998; Hubbell *et al.*, 2001). Thus, both offspring and parent gain a fitness advantage from dispersal; an advantage that may be an important factor in the evolution of dispersal (Muller-Landau *et al.*, 2003).

Interactions with specialized natural enemies can contribute powerfully to the maintenance of plant species diversity (Janzen, 1970; Connell, 1971; Armstrong, 1989; Pacala and Crawley, 1992). This contribution depends fundamentally on the population-level density dependence induced by natural enemies: as a species becomes more common, its natural enemies become more common and reduce its reproductive rate; as a species becomes rarer, its natural enemies become rarer and its reproductive rate increases (Ridley, 1930; Gillett, 1962). Such population-level density dependence is the signature of a stabilizing contribution to the maintenance of diversity, *sensu* Chesson (2000). While the tropical forest literature on natural enemy effects has focused mainly on the spatial context of local seed dispersal and local enemy attack since Janzen's seminal paper (but see Schupp, 1992), theoretical studies have demonstrated that specialized natural enemies can contribute to density dependence in populations and the maintenance of diversity in communities, even in completely non-spatial models (Armstrong, 1989; Pacala and Crawley, 1992).

Given the importance of seed dispersal in determining the strength of interactions with specialized natural enemies, and the importance of interactions with specialized natural enemies for the maintenance of diversity, it seems obvious that seed dispersal will be important in determining the contribution of specialized enemies to the maintenance of diversity. However, the form, and even the direction, of the relationship are far from obvious (Adler and Muller-Landau, 2005). On the one hand, because shorter seed dispersal distances lead to higher damage from natural enemies, we might hypothesize that natural enemy interactions will contribute most to the maintenance of diversity when seed dispersal distances are short. On the other hand, longer seed dispersal distances should lead to larger differences in the chance of encountering natural enemies between low and high population densities; thus, we might hypothesize that population-level density dependence and contributions to the maintenance of diversity will be strongest when seed dispersal distances are long. The influences of seed dispersal patterns will, of course, also depend on natural enemy dispersal patterns, which determine how enemy encounter rates change with seed dispersal distances.

Janzen (1970) predicted that longer seed dispersal distances will promote higher seedling survival and higher species diversity. The reasoning is that, as seed dispersal distances increase, the mode of the population recruitment curve shifts farther away from the parents, even as the total area under the curve increases (Fig. 18.1). Janzen further hypothesized that longer parent–offspring distances will allow more species to coexist (Janzen, 1970). Thus, higher seed dispersal distances should lead to higher diversity, facilitated by natural enemy interactions. Nathan and Casagrandi (2004) showed, in their simple spatial model, that the peak of the seedling recruitment curve does indeed move away from the adult as seed dispersal increases. But the strong link drawn by Janzen (1970) and later Hubbell (1980) between spacing dynamics and the maintenance of diversity has been shown to be faulty. Becker *et al.* (1985) showed that distance- or density-dependent predation can maintain diversity even when it does not result in a shift of the peak of seedling recruitment curve away from adults. The question remains whether such a shift in seedling recruitment due to increased seed dispersal would facilitate stronger contributions of natural enemy interactions to the maintenance of diversity.

The need for a better understanding of the consequences of seed dispersal has particular urgency today because anthropogenic influences are changing seed dispersal patterns in many populations. Abundances of vertebrate frugivores and seed dispersers in many tropical forests have been reduced by hunting and habitat fragmentation, potentially reducing seed dispersal of associated plant species (Emmons, 1989; Redford, 1992; Bond,

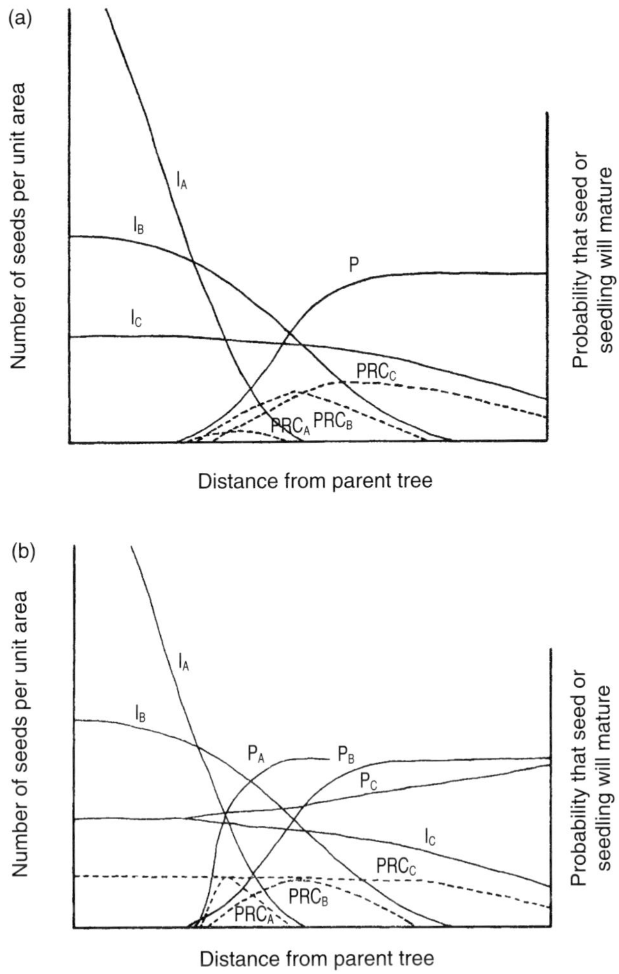

Fig. 18.1. Janzen's (1970) conception of the influences of different seed dispersal functions (I_A, I_B, I_C) on seedling recruitment curves (PRC_A, PRC_B, PRC_C) for (a) distance-responsive seed or seedling predation (P) and (b) density-responsive seed or seedling predation (P_A, P_B, P_C). Modified from Figs 4 and 5 of Janzen (1970).

1994; Wright, 2003). Anthropogenic influences may also increase dispersal of some species of plant through habitat modification, intentional and unintentional transport of propagules, as well as increases in wind velocity associated with global climate change (Soons *et al.*, 2004). Decreased (or increased) seed dispersal rates and distances potentially expose larger (or smaller) numbers of seeds and seedlings to natural enemies concentrated near parent trees, and thereby might alter the strength and influences of natural enemy interactions (Wright and Duber, 2001; Muller-Landau, 2007). Largely for this reason, many authors have warned that the potential consequences of defaunation for tropical plant species diversity could be dire (Emmons, 1989; Redford, 1992; Bond, 1994). However, the theoretical framework for understanding the impacts expected via changes in natural enemy interactions and other aspects of population and community dynamics remains severely underdeveloped (Muller-Landau, 2007).

Our objective here is to investigate how seed dispersal distances interact with natural enemy dispersal distances to affect the strength of natural enemy interactions and the influences of these interactions on plant populations and communities. We use a simulation model that captures key features of tropical tree populations and communities, including implicit local spatial competition for resources among adults, concentration of natural enemies around adults, and enemy attack functions appropriate for pathogens and insects. We specifically examine the impacts on the proportion of offspring killed by natural enemies, the form of population-level density dependence induced by enemies, and the species diversity maintained by natural enemy interactions alone. Finally, we discuss the implications of our results for understanding the consequences of anthropogenic changes in seed dispersal distances.

Methods

Model landscape

The simulation model is discrete-space, spatially explicit and individual-based. The model landscape is a rectangular grid of cells, each of which can contain exactly one adult tree (of one species) as well as many seeds, natural enemies, and juvenile plants potentially of multiple species. Individual plants are tracked through three age or size classes: seeds, juveniles, and adults. Individual spores, eggs or offspring of natural enemies, henceforth referred to simply as 'spores', are also tracked in all simulations. To eliminate edge effects, seeds or spores that go off one side of the landscape wrap back around on the other side (i.e. the model landscape is a torus, having periodic boundary conditions).

Simulations were conducted on a landscape of 900 (30 × 30) adult tree grid cells. To put this in familiar units, we say that each cell (and thus each canopy tree crown) has dimensions of 10 m × 10 m. Thus, the total simulated landscape area is 9 ha.

Model processes

Simulations are in discrete time, and model processes unfold in a set order within each time-step (Fig. 18.2). Seeds first disperse around parent trees, according to specified seed production and seed dispersal functions (Fig. 18.2a). Next, spores disperse from their sources, which are here taken to be all adult trees of the host species (Fig. 18.2b). This scenario is meant to reflect

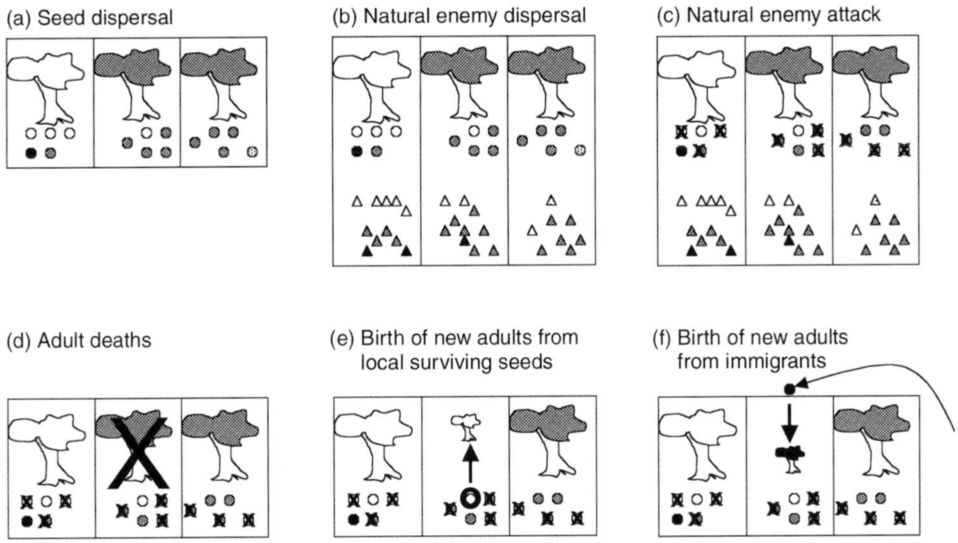

Fig. 18.2. The sequence of events that occurs during each time-step in the simulation model, in which different species are represented by different colours/stippling. (a) Seeds (circles) are dispersed away from adults according to a seed dispersal function. (b) Spores of natural enemies (triangles) are dispersed away from adult hosts according to a spore dispersal function. (c) Natural enemies attack seeds within their local neighbourhood, with the probability of a seed surviving dependent on the local density of host-specific natural enemies according to an infectivity function. (d) A fixed number of adults are randomly chosen to die. (e) A new adult is randomly chosen to replace each dead adult from among the local surviving seeds, or (f) with a small probability from an immigrant seed.

soil-borne pathogens that are often lethal to seedlings and are abundant on the roots of conspecific adult trees, where they have no apparent ill effects. Seeds then either survive to become juveniles or succumb to natural enemies, depending on the density of spores of their specialized natural enemy within the same grid cell (Fig. 18.2c). Next, a set proportion of adult trees is killed at random (Fig. 18.2d). Finally, for each adult killed, a replacement is chosen from among the surviving juveniles in that grid cell or else, with very low probability, the replacement is chosen at random (uniformly) from the species pool, effectively mimicking immigration or speciation (Fig. 18.2e). There is no dormancy or persistence of seeds, spores or juveniles from one time-step to the next: only adults survive between time-steps. The adult mortality rate was set at 10% per time-step, making each time-step equivalent to 5–10 years in a typical tropical forest with an annual canopy tree mortality rate of 1–2%.

The functional forms that specify seed dispersal, spore dispersal, and the probability of seeds surviving to become juveniles match some commonly measured in the field. Seed and spore dispersal distances both follow two-dimensional Gaussian distributions (Clark *et al.*, 1998). Dispersal in all directions is equally likely. Thus, the total probability of a seed or spore going distance r (integrating over the entire circle at this distance) is given by the probability density function

$$p(x) = \frac{2r}{\sigma^2} \exp\left(\frac{-r^2}{\sigma^2}\right) \tag{18.1}$$

where σ is the dispersal distance parameter, henceforth referred to as the dispersal scale, and mean dispersal distance is equal to approximately 0.886σ (Clark *et al.*, 1998). The number of seeds of a given species in a grid cell is drawn from a Poisson distribution with expectation equal to the sum of expected contributions from all adult trees of that species in the model landscape (with the expected contribution for any given tree to that cell calculated by integrating over 100 evenly spaced points within the cell). Similarly, the number of spores of a given species in a grid cell is drawn from a Poisson distribution with expectation equal to the sum of expected contributions from all spore sources (adults of its host species) in the landscape. Following a functional form commonly employed in epidemiological studies (e.g. Baker, 1978), the probability that a seed will become a juvenile is a negative exponential function of the local density (density within its grid cell) of spores of its natural enemies (P):

$$\Pr(survival) = \exp(-\beta P) \tag{18.2}$$

We refer to β as the infectivity. For each grid cell, the number of seeds of a species that survive to become juveniles is drawn from a binomial distribution with the given survival probability.

Simulations

At the start of each simulation, the species identity of each adult tree was assigned at random from among an initial 100 species, out of a total species pool of 200 species. Once in every 9000 births, an immigrant was chosen at random from this larger species pool (here a birth is defined as a new adult chosen to replace a dead adult). Within any simulation, all species were equivalent in every way (mortality rate, seed production, seed dispersal scale), except that their survival from seed to seedling depended on the density of a different specialized natural enemy. Similarly, all strains of natural enemy were likewise equivalent within any simulation. Across all simulations, seed production was fixed at 10,000 seeds per adult tree per time-step, spore production at 100,000 spores per adult host tree, and natural enemy infectivity (β, Eqn 18.2) at 0.001.

Simulations were run for different combinations of seed and spore dispersal scales (σ in Eqn 18.1), with each varying independently between 10, 20, 40 and 80 m (1, 2, 4 and 8 cell widths). Each combination of dispersal scales was represented by one simulation of 10,000 time-steps, or 1000 tree generations (given a mortality rate of 10% per time-step). Given the degree of temporal autocorrelation observed within any given simulation, we sampled the simulated landscape every 200 time-steps (20 tree generations) and treated these samples as independent for the purposes of statistical analysis. To avoid transient effects of initial conditions, we discarded data from the first 1999 time-steps. We chose this threshold because species richness always plateaued between 500 and 1500 time-steps.

Analyses

We tracked three effects of natural enemies across the simulations: the average proportion of seeds killed by natural enemies, the strength of population-level density dependence, and the average species richness at equilibrium. Because natural enemy attack exclusively affected the survival of seeds to the juvenile stage (Eqn 18.2), the proportion of seeds killed by natural enemies was measured as

$$1 - \frac{\text{the number of seeds surviving to become juveniles}}{\text{the number of seeds produced}}$$

The mean proportion of seeds killed, the average equilibrium species richness and their confidence intervals were calculated from the values recorded every 200 time-steps between time 2000 and time 10,000. The strength of population-level density dependence was evaluated by analysing the relationship between births per adult and adult abundance (Fig. 18.3). In particular, we used the slope of the relationship between births per adult and adult relative abundance in the vicinity of the equilibrium abundance as our measure of density dependence (see Fig. 18.3 for details). Because all species are biologically equivalent in these simulations, all have the same equilibrium abundance within a given simulation, and thus this relationship is the same whether one species is tracked over multiple time-steps as it changes in abundance, or multiple species varying in abundance are examined at one time. We thus aggregated data for multiple species and multiple time-steps, sampling all species every ten time-steps within each sampling period of 200 time-steps (we sampled multiple times per sampling period in order to have large enough sample sizes to reliably calculate a slope for that period). For each response variable, the effects of seed dispersal and natural enemy dispersal and their interaction were analysed using ANOVA, treating each 200-time-step sample as one independent data point.

Tests of robustness

A key simplifying assumption of our main simulations is that adult host plants are the source of natural enemies. In reality, many natural enemies that attack seeds and seedlings are found in high concentrations around adults only because there are many seeds and/or seedlings there, not because the adult plants themselves are an important reservoir. If the source of natural enemy spores or inoculum is not the adult trees, but

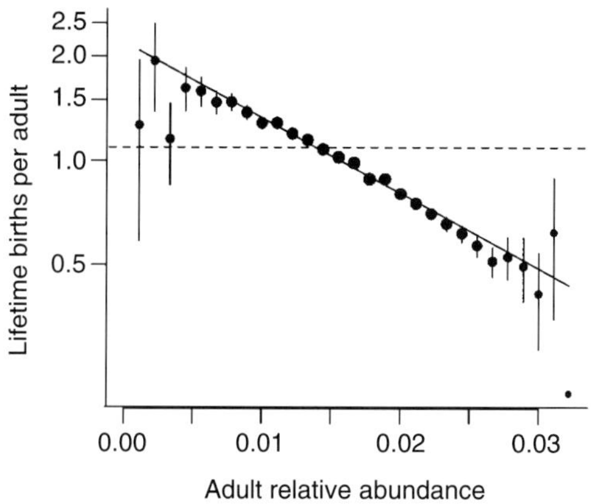

Fig. 18.3. To calculate population-level density dependence we first calculated the average lifetime births per adult for each adult relative abundance (filled circles; vertical lines show 95% confidence intervals). We then regressed log(lifetime births per adult) against adult relative abundance (solid black line), weighting each abundance by the number of occurrences (circle size reflects this weighting). The slope of the regression line is our measure of population-level density dependence; steeper (more negative) slopes indicate stronger density dependence. The horizontal dashed line indicates the birth rate at which the population neither increases nor decreases in abundance. The example here is from a simulation in which seed and natural enemy dispersal scales were both 40 m.

rather the seeds or seedlings previously killed by the natural enemy, then the spatial distribution of sources of natural enemies changes as the seed dispersal distance changes, and we might expect different effects of seed dispersal distances on the maintenance of diversity.

To evaluate such a possibility, we ran simulations in which natural enemies dispersed from dead seeds, rather than from adult trees. Because natural enemy densities can vary considerably within the area of an adult tree and do not immediately spread over such an area, these simulations were run with a finer grid for natural enemies and seeds superimposed over the coarse grid for adults – specifically, there were 25 of these seed and natural enemy cells per adult cell (each effectively 2 m × 2 m). Because tracking natural enemies in this way greatly slowed the simulations, these simulations were done on a smaller grid of 400 (20 × 20) adult cells, over fewer time-steps (1000), with only 20 species, and without immigration or speciation. Seed and natural enemy dispersal scales were 10 or 40 m (1 or 4 adult cell widths). Each killed seed resulted in the production of 100 spores of the natural enemy. A parallel set of simulations was done with natural enemies dispersing from adult trees, and each adult tree producing 100,000 spores. We examined the proportion of seeds killed by natural enemies and the population-level density dependence under each set of dispersal distances within each set of simulations. The effects on species richness could not be evaluated in parallel with the main simulations because there was no immigration in these sets of simulations.

Results

Dispersal and total mortality

As expected, longer seed dispersal distances led to lower total mortality due to natural enemies (Fig. 18.4a). This effect was strongest for the shortest enemy dispersal scales, decreasing as enemy dispersal increased, but remained significant at all the scales examined here. Further, for any given seed dispersal scale, enemy dispersal scales that were relatively similar to the seed dispersal scale resulted in the highest mortality; this effect was most pronounced for longer seed dispersal scales. Seed dispersal scale, enemy dispersal scale, and the interaction of the two, all had significant effects on total mortality, explaining 65.2, 19.2 and 15.5% of the variation, respectively (Table 18.1).

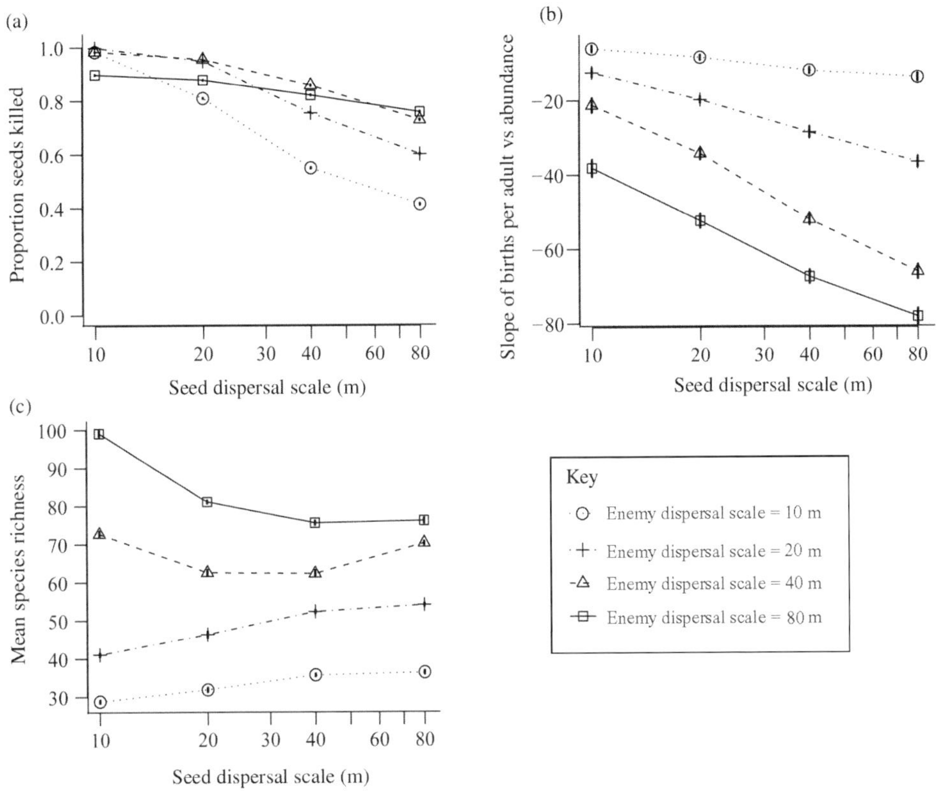

Fig. 18.4. Effects of seed and natural enemy dispersal scales on (a) total seed mortality due to natural enemies, (b) population-level density dependence as quantified by the slope of the relationship of births per adult to adult relative abundance (more negative slopes reflect stronger density dependence) and (c) mean equilibrium species richness maintained in the model community. The points show means across samples for a given set of parameter values; 95% confidence intervals on the means were plotted as vertical bars but are so narrow that they are not visible.

Table 18.1. ANOVA results for effects of seed dispersal scale, enemy dispersal scale, and their interaction on juvenile survival, population-level density dependence and species richness; all effects are significant at $P < 0.0001$, denoted by ***.

		F-ratio		
Independent variable	*df*	Juvenile survival	Density dependence	Species richness
Seed dispersal	3	118375.1 ***	1048.5 ***	225.3 ***
Enemy dispersal	3	34862.9 ***	2953.9 ***	19895.5 ***
Seed dispersal × Enemy dispersal	9	9386.5 ***	86.8 ***	620.9 ***
Error	624			

Dispersal and population-level density dependence

Longer seed dispersal distances also led to stronger population-level density dependence (Fig. 18.4b). That is, there was a stronger decrease in recruitment with adult density when seed dispersal scales were longer. This effect was strongest for the longest enemy dispersal scales and more modest as enemy dispersal decreased, but remained significant at all the scales examined here. Longer enemy dispersal scales also resulted in stronger density dependence, an effect that was more pronounced at longer seed dispersal scales, and was stronger overall than the effect of seed dispersal scales. Seed dispersal scale, enemy dispersal scale, and the interaction of the two, all had significant effects on density dependence, explaining 23.5, 66.1 and 5.8% of the variation, respectively (Table 18.1).

Dispersal and equilibrium species richness

The effects of seed dispersal scale on the equilibrium species richness maintained depended on enemy dispersal scale (Fig. 18.4c). For long enemy dispersal scales, species richness was highest at short seed dispersal scales; in contrast, for short enemy dispersal scales, species richness was highest at long seed dispersal scales. The effects of enemy dispersal scale alone were straightforward and much stronger: for any given seed dispersal scale, increased enemy dispersal scale always increased species richness. Seed dispersal scales, enemy dispersal scale, and the interaction of the two, all had significant effects on species richness, explaining 1.0, 89.7 and 8.4% of the variation, respectively (Table 18.1).

While the proportion of seeds killed by natural enemies and the strength of population-level density dependence were both correlated with species richness as expected, the influence of seed dispersal on species richness cannot be predicted from the influence of seed dispersal on either of these variables (Fig. 18.5). The proportion of seeds killed was positively correlated with species richness overall ($r = 0.24$, $n = 640$), and shorter seed dispersal always increased the proportion of seeds killed, but did not necessarily increase species richness (Fig. 18.5a). Population-level density dependence was negatively correlated with species richness ($r = -0.72$, $n = 640$), and longer seed dispersal always made this density dependence more negative (indicating stronger population regulation), but did not necessarily increase species richness (Fig. 18.5b). The effects of enemy dispersal on species richness also could not be predicted from the effects on the proportion of seeds killed, but could be predicted from the effects on population-level density dependence. That is, increased enemy dispersal distances sometimes increased and sometimes decreased the proportion of seeds killed, yet always made population-level density dependence more negative and always increased species richness.

Robustness of results to the source of natural enemies

The qualitative effects of seed dispersal scale and natural enemy dispersal scale on the proportion of seeds killed and population-level density dependence were the same whether natural enemies dispersed from adult trees or from killed seeds (Fig. 18.6). In both cases, the proportion of seeds killed was higher at the lower seed dispersal distance and at enemy dispersal scales closer to seed dispersal scales, although both effects were smaller when natural enemies dispersed from killed seeds (Fig. 18.6a,b). Population-level density dependence was consistently stronger under the

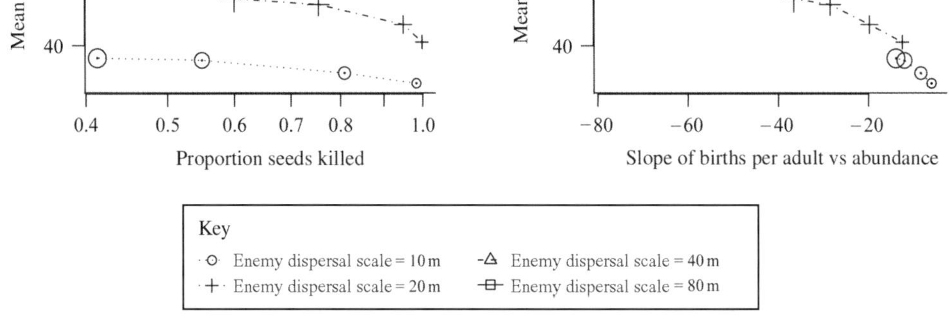

Fig. 18.5. The relationships of species richness to (a) the proportion of seeds killed and (b) the population-level density dependence across simulations varying in seed and enemy dispersal scales. Symbol size increases with seed dispersal scale, while symbol and line type reflect enemy dispersal scale as shown in the key. 95% confidence intervals are not shown, as they are generally smaller than the size of the symbols.

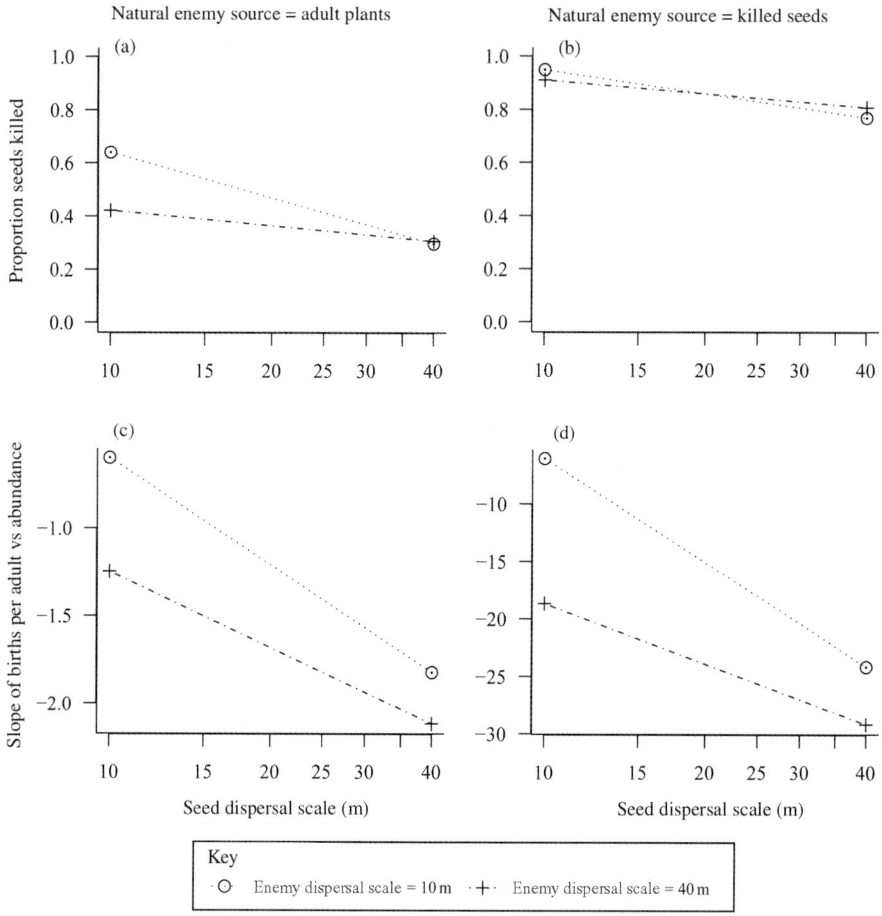

Fig. 18.6. The effects of seed and natural enemy dispersal scales on (a, b) total seed mortality due to natural enemies and (c, d) population-level density dependence in simulations in which the source of natural enemy spores is either (a, c) adult plants of the host species or (b, d) seeds of the host species recently killed by natural enemies.

longer seed dispersal distance and longer natural enemy dispersal distance in both sets of simulations (Fig. 18.6c,d). Overall, density dependence was much stronger when natural enemies dispersed from killed seeds.

Discussion

The complex effects of dispersal on species richness

The results of this study demonstrate that the effects of dispersal distances on the diversity-maintaining influence of natural enemy interactions are complex. In particular, seed dispersal distances have opposite effects on

effectively imposed a minimum establishment distance from any existing adult (parent, other conspecific or heterospecific), which is perhaps more realistic for a closed-canopy forest; under these conditions shorter dispersal distances always reduced density dependence.

Strong local resource competition in tropical forests and many other plant communities prevents the pronounced clumping that appeared in the model of Adler and Muller-Landau (2005), and thus precludes the diversity-enhancing effect of short dispersal found there. Large trees (> 30 cm in diameter) of all species combined in a Panamanian tropical forest have a spatial pattern that is the opposite of clumped: it is more regular than a random distribution (H.C. Muller-Landau, 2004, unpublished results; data available at http://ctfs.si.edu/datasets/bci/). To implicitly represent this local resource competition and capture the effects on community spatial pattern, we used a discrete space model with regularly spaced adults in this study. The convention of a regular rectangular grid with one adult tree per cell is common in forest models, and indeed in spatial models of plant communities in general (Hubbell, 2001; Chave *et al.*, 2002). Discrete- and continuous-space models have various advantages and disadvantages in general (Durrett and Levin, 1994; Bolker, 2004). The particular choice of continuous formulation in our previous study enabled analytical treatment (Adler and Muller-Landau, 2005). Furthermore, different models are appropriate for different communities. The results of Adler and Muller-Landau (2005) are likely to apply well to communities in which the spatial scales of seed and natural enemy dispersal are shorter than the spatial scales of resource competition, and in which strong overall clumping of adults can and does develop; desert plant communities may be a good example. The results from the current study should better predict effects in closed-canopy communities such as tropical forests.

Robustness of the results

Our model is obviously a simplification of the complexities of plant and natural enemy dynamics in real communities. Some of the simplifying assumptions are intended to narrow our focus to the question at hand (e.g. the assumption that all species are identical); others are made to speed up simulations to reasonable timescales (e.g. the size of the grid). Many of these assumptions have quantitative effects on the results; however, the qualitative patterns that we observe are robust to more realistic assumptions regarding the sources of natural enemies and many other aspects of plant–enemy interactions.

In this chapter we have presented results demonstrating that the qualitative effects of seed and natural enemy dispersal distances on diversity maintaining forces are the same whether natural enemies disperse from adult host plants or recently killed seeds. The effects on species richness could not be evaluated because of the shorter time span of the runs and the concomitant lack of inclusion of immigration, but we expect that these would also be qualitatively the same, because species richness is driven by effects of population-level density dependence and seed dispersal distance alone. The similarity in the qualitative effects reflects the fact that under Gaussian seed dispersal, seed densities are considerably higher under adult trees and decline with increasing distance so that, in practice, the areas under adult trees are the main source of natural enemies even when these are explicitly produced by dead seeds. If seed dispersal patterns were such that there was no concentration of seeds under parent trees, then the two scenarios would be expected to diverge.

While we did not conduct simulations relaxing other assumptions, we can infer their effects from theory and the results of other studies. All other things being equal, a larger model landscape or a higher immigration rate will support higher species richness for all dispersal scales (Chave *et al.*, 2002), but will not affect the direction of changes with dispersal scale. An increase in natural enemy infectivity will increase the proportion of seeds killed, but will leave qualitative patterns unchanged (Adler and Muller-Landau, 2005). If species of plant vary in their competitive abilities, susceptibility to natural enemy attack, or seed or natural enemy dispersal scales, then their reproductive rates will equilibrate at different relative abundances; however, each species' reproductive rate will still be negatively density-dependent, and thus, as long as a species' equilibrium abundance is non-zero, natural enemy interactions will help maintain it in the community. If natural enemies are specialized at a higher taxonomic level than species, for example at the generic level as seems most common for insects and pathogens (Gilbert and Sousa, 2002; Novotny and Basset, 2005), then the impacts on diversity depend on how effects of a given natural enemy vary across species. Effects that are identical across all species in a genus would contribute to maintenance of generic diversity, but not to the maintenance of species diversity within genera. In the more realistic case of unequal effects across multiple affected plant species and by multiple shared natural enemies, the natural enemies contribute to the maintenance of species diversity the degree that different species are most limited by different natural enemies, i.e. different factors, as stated in classic competition theory.

Implications for understanding effects of changes in seed dispersal patterns

The effects of seed dispersal distances on natural enemy interactions in this study shed light on the potential impacts of anthropogenic changes in seed dispersal patterns for population viability and community diversity. In considering the implications, we must take into consideration that these changes affect different species to differing degrees (with some species completely unaffected), in communities where seed dispersal already varies among species. Thus, we cannot simply extrapolate from differences between simulations in which seed dispersal distances of all species are higher or lower – we must take variation among species into account.

The first direct effect of reduced seed dispersal distances on interactions with natural enemies is elevated mortality due to specialized natural enemies concentrated around parents (Janzen, 1970). Because these reductions in seed dispersal affect some, but not all, of the species in a community, the affected species will be at a relative disadvantage and will be expected to decrease in abundance. Such changes in abundance alone can lead to profound changes in community structure. However, the most important question from a conservation perspective is whether these declines will continue to local extinction. The many density-dependent factors that stabilize plant populations and communities have the potential to prevent such extinctions by ensuring that, at some lowered abundance, the reproductive rate of the affected species will be high enough for it to maintain itself (Muller-Landau, 2007). These factors include not only interactions with natural enemies, but also habitat niche differentiation and temporal partitioning of varying environmental conditions, among others (Chesson, 2000; Barot, 2004).

Unfortunately, the results shown here suggest that reduced seed dispersal not only increases mortality due to natural enemies and thereby decreases survival at any given abundance, it also reduces the strength of the population-level density dependence of mortality due to specialized natural enemies. Shorter seed dispersal scales led seeds to continue to encounter their natural enemies at high rates even when the abundance of their conspecifics in the overall landscape was very low. This implies that species whose seed dispersal is decreased will derive less benefit from rarity, and are more likely to continue to decrease in abundance to extinction. On the other hand, species whose seed dispersal is enhanced are likely to not only experience elevated survival but also increased density dependence. Thus, their increase is more likely to be checked before they achieve monodominance.

To what degree are these predictions regarding the effects of changed seed dispersal on density dependence and diversity sensitive to the details of our model? Incorporation of additional complexity or changed functional forms for natural enemy dynamics could result in quite different influences of changed seed dispersal distances on density dependence and diversity. For example, if natural enemies have a minimum population size larger than that supported by a single host individual, then natural enemies and the mortality they induce could locally disappear before the host reaches extinction. This could elevate host reproductive rates sufficiently to facilitate persistence, or even increases to a level at which the natural enemy reappears. Similarly, relaxing the degree of host specialization of natural enemies, adding dependence of enemy movement distances on host availability, adding seed or spore dormancy, and many other potential changes could have major impacts on how seed dispersal affects density dependence. In general, however, we expect the qualitative result found here to be relatively robust – decreased seed dispersal will decrease the density dependence of natural enemy interactions, and thereby reduce the potential for species declines to stop short of extinction.

Conclusions

Seed dispersal clearly plays an important role in structuring plant interactions with specialized natural enemies. In this chapter we demonstrated the importance of seed dispersal distances to the magnitude and density dependence of mortality due to natural enemies. But dispersal distances are only one facet of seed dispersal patterns (Muller-Landau and Hardesty, 2005) – habitat-specific deposition and clumping of seed rain irrespective of habitat are likely to further affect natural enemy interactions. Furthermore, the specific influences of all these factors depend on the details of natural enemy life history – specificity, longevity, movement distances, reproductive rates, etc. Ultimately, we need to better understand these and other details of plant and natural enemy biology, and incorporate this understanding into mechanistic models in order to elucidate the influences of seed dispersal on natural enemy interactions, and specifically on the density-dependent, and consequently diversity-enhancing, effects of natural enemies on plant populations and communities.

Acknowledgements

We thank Janneke HilleRisLambers, Gene Schupp and an anonymous reviewer for helpful comments that improved the manuscript, and Andrew Dennis, Ronda Green, Gene Schupp and David Westcott for all their hard work in organizing the 2005 *Symposium/Workshop on Frugivory and Seed Dispersal* in Brisbane, and in editing this book. Funding for H.C.M. to prepare this chapter and attend the Symposium/Workshop was provided by the National Science Foundation (DEB No. 0453445 to H.C.M. and OISE No. 0456402 to E.W. Schupp) and the University of Minnesota.

References

Adler, F.R. and Muller-Landau, H.C. (2005) When do localized natural enemies increase species richness? *Ecology Letters* 8, 438–447.

Armstrong, R.A. (1989) Competition, seed predation, and species coexistence. *Journal of Theoretical Biology* 141, 191–195.

Baker, R. (1978) Inoculum potential. In: Horsfall, J.G. and Cowling, E.B. (eds) *Plant Disease: An Advanced Treatise*. Academic Press, New York, pp. 137–157.

Barot, S. (2004) Mechanisms promoting plant coexistence: can all the proposed processes be reconciled? *Oikos* 106, 185–192.

Becker, P., Lee, L.W., Rothman, E.D. and Hamilton, W.D. (1985) Seed predation and the coexistence of tree species: Hubbell's models revisited. *Oikos* 44, 382–390.

Bolker, B. (2004) Continuous-space models for population dynamics. In: Hanski, I. and Gaggioti, O.E. (eds) *Ecology, Genetics, and Evolution of Metapopulations*. Elsevier Science, San Diego, California, pp. 45–69.

Bond, W.J. (1994) Do mutualisms matter: assessing the impact of pollinator and disperser disruption on plant extinction. *Philosophical Transactions of the Royal Society London B: Biological Sciences* 344, 83–90.

Chave, J., Muller-Landau, H.C. and Levin, S.A. (2002) Comparing classical community models: theoretical consequences for patterns of diversity. *American Naturalist* 159, 1–23.

Chesson, P. (2000) Mechanisms of maintenance of species diversity. *Annual Reviews of Ecology and Systematics* 31, 343–366.

Clark, J.S., Macklin, E. and Wood, L. (1998) Stages and spatial scales of recruitment limitation in southern Appalachian forests. *Ecological Monographs* 68, 213–235.

Connell, J.H. (1971) On the roles of natural enemies in preventing competitive exclusion in some marine animals and in rain forest trees. In: den Boer, P.J. and Gradwell, G.R. (eds) *Dynamics of Populations*. Proceedings of the Advanced Study Institute on Dynamics of Numbers in Populations, Oosterbeek, 1970. Centre for Agricultural Publishing and Documentation (PUDOC), Wageningen, The Netherlands, pp. 298–312.

Durrett, R. and Levin, S. (1994) The importance of being discrete (and spatial). *Theoretical Population Biology* 46, 363–394.

Emmons, L.H. (1989) Tropical rain forests: why they have so many species and how we may lose this biodiversity without cutting a single tree. *Orion* 8, 8–14.

Gilbert, G.S. and Sousa, W.P. (2002) Host specialization among wood-decay polypore fungi in a Caribbean mangrove forest. *Biotropica* 34, 396–404.

Gillett, J.B. (1962) Pest pressure, an underestimated factor in evolution. *Systematics Association Publication* 4, 37–46.

Hammond, D.S. and Brown, V.K. (1998) Disturbance, phenology and life-history characteristics: factors influencing distance/density-dependent attack on tropical seeds and seedlings. In: Newbery, D.M., Prins, H.H.T. and Brown, N.D. (eds) *Dynamics of Tropical Communities*. Blackwell Science, Oxford, UK, pp. 51–78.

Howe, H.F. and Smallwood, J. (1982) Ecology of seed dispersal. *Annual Review of Ecology and Systematics* 13, 201–228.

Hubbell, S.P. (1980) Seed predation and the coexistence of tree species in tropical forests. *Oikos* 35, 214–229.

Hubbell, S.P. (2001) *The Unified Neutral Theory of Biodiversity and Biogeography*. Princeton University Press, Princeton, New Jersey.

Hubbell, S.P., Ahumada, J.A., Condit, R. and Foster, R.B. (2001) Local neighbourhood effects on long-term survival of individual trees in a neotropical forest. *Ecological Research* 16, 859–875.

Janzen, D.H. (1970) Herbivores and the number of tree species in tropical forests. *American Naturalist* 104, 501–528.

Muller-Landau, H.C. and Hardesty, B.D. (2005) Seed dispersal of woody plants in tropical forests: concepts, examples, and future directions. In: Burslem, D., Pinard, M. and Hartley, S. (eds) *Biotic Interactions in the Tropics: Their Role in the Maintenance of Species Diversity*. Cambridge University Press, Cambridge, UK, pp. 267–309.

Muller-Landau, H.C., Levin, S.A. and Keymer, J.E. (2003) Theoretical perspectives on evolution of long-distance dispersal and the example of specialized pests. *Ecology* 84, 1957–1967.

Muller-Landau, H.C. (2007) Predicting the long-term effects of hunting on plant species diversity and composition in tropical forests. *Biotropica* 39, 372–384.

Nathan, R. and Casagrandi, R. (2004) A simple mechanistic model of seed dispersal, predation and plant establishment: Janzen–Connell and beyond. *Journal of Ecology* 92, 733–746.

Novotny, V. and Basset, Y. (2005) Review: Host specificity of insect herbivores in tropical forests. *Proceedings of the Royal Society London B: Biological Sciences* 272, 1083–1090.

Pacala, S.W. and Crawley, M.J. (1992) Herbivores and plant diversity. *American Naturalist* 140, 243–260.

Redford, K.H. (1992) The empty forest. *BioScience* 42, 412–422.

Ridley, H.N. (1930) *The Dispersal of Plants Throughout the World.* Lovell Reeve & Co., Ashford, UK.

Schupp, E.W. (1992) The Janzen–Connell model for tropical tree diversity: population implications and the importance of spatial scale. *American Naturalist* 140, 526–530.

Soons, M.B., Nathan, R. and Katul, G.G. (2004) Human effects on long-distance wind dispersal and colonization by grassland plants. *Ecology* 85, 3069–3079.

Wright, S.J. (2003) The myriad consequences of hunting for vertebrates and plants in tropical forests. *Perspectives in Plant Ecology, Evolution and Systematics* 6, 73–86.

Wright, S.J. and Duber, H.C. (2001) Poachers and forest fragmentation alter seed dispersal, seed survival, and seedling recruitment in the palm *Attalea butyraceae*, with implications for tropical tree diversity. *Biotropica* 33, 583–595.

19 Out of One Shadow and into Another: Causes and Consequences of Spatially Contagious Seed Dispersal by Frugivores

C. Kwit, D.J. Levey, S.A. Turner, C.J. Clark and J.R. Poulsen

Introduction

Seed shadows have been estimated and evaluated for many species of trees and despite the variety of distributions that have been used to describe them, their shape is universally leptokurtic (Willson, 1993; J.S. Clark *et al.*, 1998, 1999; Dalling *et al.*, 2002; Greene *et al.*, 2004; C.J. Clark *et al.*, 2005; Uriarte *et al.*, 2005). By definition, therefore, areas close to maternal trees receive greater seed rain than areas farther away. Subsequently, density-dependent mortality (i.e. higher per-capita mortality; Harms *et al.*, 2000) by specialist predators is thought to prevent successful establishment of offspring beneath or near parent trees, making such locations more favourable for heterospecific propagules present in low densities. Thus, the 'scattering' of seeds in low densities away from parent plants, presumably beneath heterospecific trees, is considered essential both to population persistence and to species coexistence (Howe and Miriti, 2000; but see Terborgh *et al.*, 2002; Wright, 2002, for alternative hypotheses). This encapsulates both the population outcome and the de facto community result – species coexistence – of the Janzen–Connell hypothesis (Janzen, 1970; Connell, 1971; D.A. Clark and D.B. Clark, 1984).

Most tests of the Janzen–Connell hypothesis have been pursued in tropical forests, where up to 80% of trees and shrubs have their seeds dispersed by a rich assemblage of vertebrate frugivores (Howe and Smallwood, 1982). Studies have traditionally included monitoring seed and seedling survival, growth and recruitment, and comparing propagule performance in 'high versus low conspecific density' locations, or across a continuum of densities: a common conclusion is that negative density dependence is pervasive (Harms *et al.*, 2000). However, studies based more specifically on Janzen–Connell's theme of 'beneath-parent versus elsewhere' have yielded mixed results (Hammond and Brown, 1998). This contrast in results between studies focusing on density and distance raises the possibility that escape from the parent tree is not necessarily equivalent to escape from mortality. In general, the spatial distribution of dispersed seeds has been overlooked or oversimplified – which is not surprising, given that seed shadows are notoriously hard to quantify. Yet without a clear picture of real dispersal kernels, it is difficult to begin to address the relative importance of distance-dependent and density-dependent seed predation.

The Janzen–Connell hypothesis is particularly appealing because of its simplicity. Its most compelling outcome is the population recruitment curve (PRC), illustrating that the area with the highest probability of seedling establishment occurs at some distance from a parent tree. The PRC is derived from the product of the decreasing density of seeds that are dispersed away from a parent tree and the increasing probability of seed survival away from the parent tree, resulting in a peak recruitment probability at some intermediate distance. This representation of the Janzen–Connell hypothesis poses several limitations, in particular when considering the effects of multiple fruiting trees, generalist predators and frugivore behaviour. The role of additional fruiting trees is difficult to interpret: PRC and dispersal curves are traditionally drawn from a single tree. Summing or averaging curves of conspecifics whose seed shadows overlap is not straightforward, as seeds from one tree can affect density-dependent relationships arising from other parents. Janzen (1970, p. 504) noted that

> the probability that a juvenile plant will be eaten by a *density-responsive* predator is primarily a function of the ecological distance between that juvenile and other juveniles.

The 'ecological distance' between dispersed seeds will be close (and spatially clumped) in areas of seed shadow overlap. Also, when the actions of generalist predators are included, the PRC may become more influenced by heterospecific seed densities, especially when generalist predators act in a density-dependent fashion. The complexities of seed dispersal patterns and the PRC are further increased by the inclusion of frugivore behaviour (Schupp *et al.*, 2002) – a mechanism which the original Janzen–Connell hypothesis acknowledged, but did not explicitly incorporate – which can lead to patchy seed deposition and result in highly clumped seed distributions (at certain spatial scales) that density-dependent predators may be attracted to (see also Muller-Landau and Adler, Chapter 18, this volume).

We suggest that spatial variation in seed deposition and its influence on density-dependent mortality and recruitment have been greatly overlooked within the Janzen–Connell framework. Incorporating their impacts on the Janzen–Connell model is akin to exploring the potential causes and consequences of complex (possibly 'bumpy' or multimodal) or uninterpretable PRCs. This requires an integration of distance and density effects arising from overlapping seed shadows of multiple trees along with heterospecific seed influences. We will argue that this type of approach can begin to help resolve the current debate over the role of Janzen–Connell processes in explaining latitudinal trends in tree species diversity (HilleRisLambers *et al.*, 2002). If the magnitude or strength of density-dependent mortality is a function of the spatial structure (the quantitative representation of the degree of clumping) of dispersed seeds, and frugivores accentuate that structure at lower latitudes, then density-dependent seed mortality may indeed help explain the latitudinal gradient in tree diversity.

Using simulations, we illustrate how the assumed Janzen–Connell relationship between distance from parent and seed density can break down. We emphasize spatial patterns of seed dispersal, the resultant spatial structure of seeds, and the potential consequences for the population and the community. We end our modelling exercise by detailing a special form of dispersal limitation: contagious seed dispersal, which is defined as the patchy deposition of seeds such that some sites receive many seeds and others receive few to none (Schupp *et al.*, 2002). This type of spatially restricted seed dispersal is exemplified by dispersal beneath perches, sleeping sites, roosts, nest locations and latrines (see Wenny, 2001; Schupp *et al.*, 2002), and can include dispersal beneath heterospecific trees, considered by the Janzen–Connell hypothesis to be an inherently safe location for establishment and the promotion of species coexistence. By investigating contagious seed dispersal, we examine the effects of spatial variability in seed dispersal curves on subsequent demographic processes. We explain where such dispersal is likely to take place and suggest potential outcomes for seed survival based on the resulting spatial deposition patterns. In short, we illustrate how contagious seed dispersal relates to and modifies the original Janzen–Connell model and its community-level outcomes. We stress that this is a first attempt at illustrating the important role of contagious seed dispersal and its consequences, and that any conclusions derived from our results are restricted to the various parameters ascribed to our simulations.

Model Simulations

We build our models in a stepwise fashion, starting with a single reproductive tree to illustrate the traditional representation of the Janzen–Connell model. We then introduce conspecific trees to examine the relationship between seed density and distance from the parent plant in a more complex system and to quantify the spatial structure of seeds (i.e. whether seeds are clumped, randomly distributed or regularly distributed, and the scale of observation). Next, we add heterospecific trees to investigate the potential effects of generalist predators with regard to distance from parent trees and to quantify the spatial structure of all seeds available to generalist predators. We end by introducing spatial contagion, simulating seed dispersal to non-fruiting trees, as well as conspecific and heterospecific fruiting trees.

Our simulated trees occur in a hypothetical 1-ha (100 m × 100 m) plot and exhibit equal fecundities and similar seed dispersal distributions (see below). In all but a few cases, seed dispersal of 1000 seeds per adult tree was initially assumed, so as to conform to an exponential probability distribution function (pdf):

$$f(x; \lambda) = \lambda e^{-\lambda x}$$

with a mean dispersal distance, defined by $1/\lambda$, of 10 m. Whenever 100 reproductive trees were included in a given simulation, we limited crop size to 100 seeds per tree due to constraints of computation time. Dispersal of each seed was initially simulated as an isotropic (i.e. random azimuth) process, with distances drawn at random from the distribution. The exponential distribution is a 'simple' and continuously declining special case of the generalized gamma probability density function. Although fundamentally different from other distributions used to describe seed dispersal kernels (e.g. lognormal), this inherently leptokurtic distribution provides an adequate fit for many species (Willson, 1993; C.J. Clark *et al.*, 2005). Spatial simulations were performed in FORTRAN.

We used a variation of traditional univariate Ripley's K analyses (Ripley, 1976; Diggle, 1983), $\boldsymbol{L}(t)$, to detect the relative strength and spatial scales of clumping of seeds (i.e. spatial structure). This is useful due to the effects of spatial scale on the evaluation of density-dependent mortality (Schupp, 1988, 1992), and because some predators may have larger effects on seed mortality patterns (e.g. inflict higher mortality) when seeds are more tightly structured. The $\boldsymbol{L}(t)$ statistic is edge-corrected and makes use of the essence of the Ripley's K function: the expected number of seeds within a given distance t of each focal seed divided by the mean number of such events (seeds) per unit area. In our analyses, a positive value of $\boldsymbol{L}(t)$ indicates a positive spatial association or clustering of seeds at scales less than or equal to a given distance, a value of 0 indicates a random distribution, and negative values indicate a regular distribution. Signs for $\boldsymbol{L}(t)$ in our analyses are the opposite of those reported in other papers in the scientific literature, and this is solely an artefact of our calculations. We obtained 95% confidence envelopes by running 1000 simulations for each scenario of interest. Spatial statistical analyses were performed using FORTRAN.

While we hope that the parameters and their values are biologically relevant, we stress that our simulations are simplistic and that they do not account for variation in parameters. None the less, the simulations clearly illustrate that adult tree densities, the role of generalist predators, and frugivore behaviour – all factors generally overlooked in most discussions of the Janzen–Connell model – probably have strong influences on dispersal patterns and seed survival.

One Tree, One Species

In the first simulation, 1000 seeds are dispersed from a single reproductive tree randomly located within a 1-ha plot (Fig. 19.1a). According to the Janzen–Connell model, seed survival increases with distance from the parent tree, and the shape of the function depends on the dispersal curve. Hence, at some intermediate distance from the parent, the chances of future establishment are highest where the product of seed density (which in this case is analogous to distance) and survival reaches a maximum. The spatial pattern of our one-tree model reflects a clumped spatial pattern of seeds across the entire range of investigated spatial scales (Fig. 19.1a). The highest degree of clumping, as determined by $\boldsymbol{L}(t)$, occurred at scales of $<\sim 20$ m, corresponding to the diameter of the mean seed dispersal distance. It is easy to envision how the smooth (i.e. isotropic) dispersal function, coupled with high degrees of spatial clumping of seeds at scales corresponding to the mean seed dispersal distance, could result in a predictable seed survival function and a smooth and predictable PRC in all directions from the sole parent tree. In fact, this is the only case in which the high degree of spatial clumping is directly associated with distance from a parent tree. Subsequent seed mortality patterns that reflect distance- and density-dependence are possible here if seed pathogens originate near the parent tree, or if seed predators operate efficiently at spatial scales corresponding to high degrees of clumping.

More Than One Tree, One Species

The interpretation of the original Janzen–Connell model becomes more complicated with the addition of other trees. When seeds are dispersed from ten reproductive adult trees located randomly within the 1-ha plot, seed shadows commonly overlap, with seeds from one parent tree landing near other fruiting conspecifics, where seed densities are high (Fig. 19.1b). In this way, the link between escape by distance and escape from high densities of seeds is broken. If seed survival is negatively density-dependent, seed survival and the PRC become difficult to predict, due to the high spatial variation in seed densities. Although Janzen (1970) acknowledged this limitation to the model, it is rarely considered in empirical or theoretical work. The underlying problem is that the PRC is based on a one-dimensional representation of seed dispersal (distance from the parent tree), whereas the actual distribution of seeds is two-dimensional (distance and direction). Because the two-dimensional world of seed distribution contains areas of greater seed density wherever there are dispersal foci (e.g. conspecific fruiting trees; or frugivory hubs, see Carlo *et al.*, Chapter 16, this volume), and because these peaks vary in distance from the parent depending on the direction in which dispersal foci are located, the PRC is expected to be anything but smooth. In addition to complicating the PRC, the overlap of seed shadows could

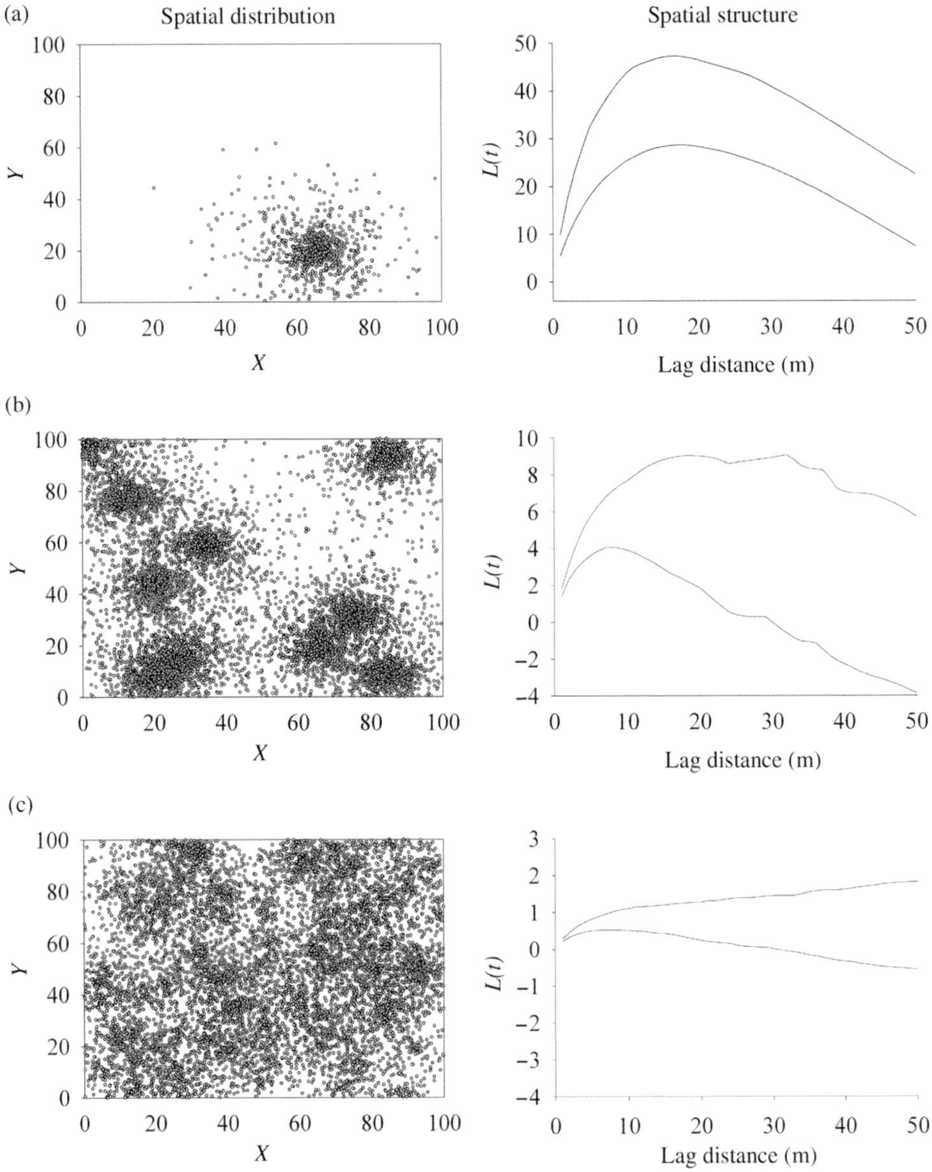

Fig. 19.1. Spatial distributions (one simulation) and resultant spatial structures considering only a single species: (a) 1000 seeds dispersed isotropically (random azimuth) from one hypothetical reproductive tree; (b) 1000 seeds dispersed isotropically from each of ten reproductive trees; and (c) 100 seeds dispersed isotropically from each of 100 reproductive trees. Simulations assume a 100 m × 100 m plot, randomly placed reproductive trees, and distances travelled by seeds from parents drawn from an exponential distribution with mean dispersal distance of 10 m. The resulting spatial structure is based on 95% confidence envelopes obtained from 1000 simulations in which positions of adults changed with each simulation. Clumping is evident at most spatial scales (traditionally referred to as 'lag distance') considered, and is highest at scales <~20 m.

lessen the per-tree area of optimal seed density at which recruitment is most likely to occur (see Fig. 9 in Janzen, 1970). This could lead to a decrease in population recruitment, and if stochastic processes led to population extinction, this would by default endanger coexistence. However, the decreased recruitment in the short term could be offset by higher recruitment in the future, thus ensuring coexistence (Schupp, 1992).

The spatial structure of seeds in this multi-tree model is similar to the '*One tree, one species*' model. Values of $\boldsymbol{L}(t)$ indicate that seeds are clumped over much of the plot, with clumping most evident at scales $<\sim 20$ m (Fig. 19.1b). At scales $>\sim 30$ m, however, the spatial distribution of seeds approaches a random distribution (i.e. confidence envelopes of $\boldsymbol{L}(t)$ include zero). Seed survival in the face of density-dependent predators will be driven, at least in part, by the scale at which the greatest degrees of clumping occurs and whether the seed predators operate at such scales.

When 100 fruiting trees are included in the model, results were similar to those with ten fruiting trees. The spatial distribution of seeds (Fig. 19.1c) resulted in clumped seed structure at scales $<\sim 20$ m. At scales $>\sim 30$ m, the spatial structure of seeds did not differ significantly from random (Fig. 19.1c). However, values of $\boldsymbol{L}(t)$ were smaller than those observed in the ten fruiting tree model, indicating that the degree of clumping was relatively less pronounced.

More Than One Tree, More Than One Species

Generalist predators can complicate the Janzen–Connell model. This is exemplified by the dispersal of seeds from an individual of one species under the canopy of synchronously fruiting heterospecific individuals whose seeds are consumed by a shared predator, and where overall seed densities are still high. For example, in our heuristic model, even if a given parent tree is the only tree of that species in the plot, a fair proportion of its seeds land near and beneath other trees with desirable seeds, just as they did in Figure 19.1b, c. Once again, seed survival is probably not a simple increasing function with distance away from parent plant, and the PRC and survival curves are likely to be bumpy and spatially variable.

To a generalist seed predator, the spatial structure of seeds it can consume is a function of the dispersal patterns of multiple species, and hence may be no different than that described in the '*More than one tree, one species*' model above (see Fig 19.1b,c). Hence, in our models, generalist predators are exposed to high degrees of seed clumping at scales $<\sim 20$ m, and continue to be subject to clumped seed distributions at scales $<\sim 30$ m.

Contagious Seed Dispersal

Focal points

Frugivore behaviour can lead to substantial spatial variation in seed deposition. We first describe contagious seed dispersal to locations not associated with fruit (hereafter, 'focal points'). To accomplish this, we impose 'spheres of influence' around randomly located focal points, such that these focal points (e.g. sleeping sites, display sites, roosts, nests, perches, latrines) attract dispersed seeds. We then simulate seed dispersal with our '*One tree, one species*' model, modifying it so that if a seed is dispersed in the 1-ha plot and lands within 20 m ('sphere of influence') of a focal point, then the seed is 'moved' to the nearest focal point. The seed is then dispersed from the focal point at a distance chosen from an exponential distribution with a mean dispersal distance of 5 m (half that of the original mean dispersal distance); this shorter dispersal distance was chosen to incorporate the fact that time had passed since the original seed ingestion event (pertinent for mobile frugivores), and due to the sedentary nature of frugivores at certain types of such locations. We ran our contagious seed dispersal simulation using ten and 100 randomly located dispersal foci. To investigate the spatial patterns created when there are multiple dispersal foci and reproductive trees in the same plot, we also added spatial contagion to the '*More than one tree, one species*' model. We acknowledge that while this methodology may be representative of frugivores visiting display sites or fruit-processing roosts (see Schupp *et al.*, 2002), it is merely a first step and only one of countless ways in which contagious dispersal can be modelled. Our modelling procedure results in slight increases in the mean seed dispersal distance, but we focus on the resulting changes in seed spatial structure.

The addition of focal points has a complex effect on spatial pattern and structure of seeds, depending on the number of focal points and number of fruiting trees. Regardless of the number of reproductive trees, the inclusion of a few focal points (n = 10) had a greater impact on seed spatial structure than did the inclusion of many focal points (n = 100). In all cases when ten focal points were simulated, the highest degree of spatial clumping was at small spatial scales, typically about half those at which the highest degrees of clumping occurred without focal points. This pattern was least conspicuous in the simulation with one reproductive tree (Fig. 19.2), most dramatic in the simulation with ten reproductive trees (Fig. 19.3), and slightly less evident when 100 trees were included (Fig. 19.4). The inclusion of many focal points essentially washed out patterns of spatial contagion, as the spatial patterns created by the simulation of 100 dispersal foci in plots with one and ten trees resembled those cases in which no focal points were included (Figs 19.2 and 19.3). For the 100-tree model, the scale at which high degrees of seed clumping occurred, when 100 focal points were present, was still notably less than when isotropic dispersal took place (Fig. 19.4).

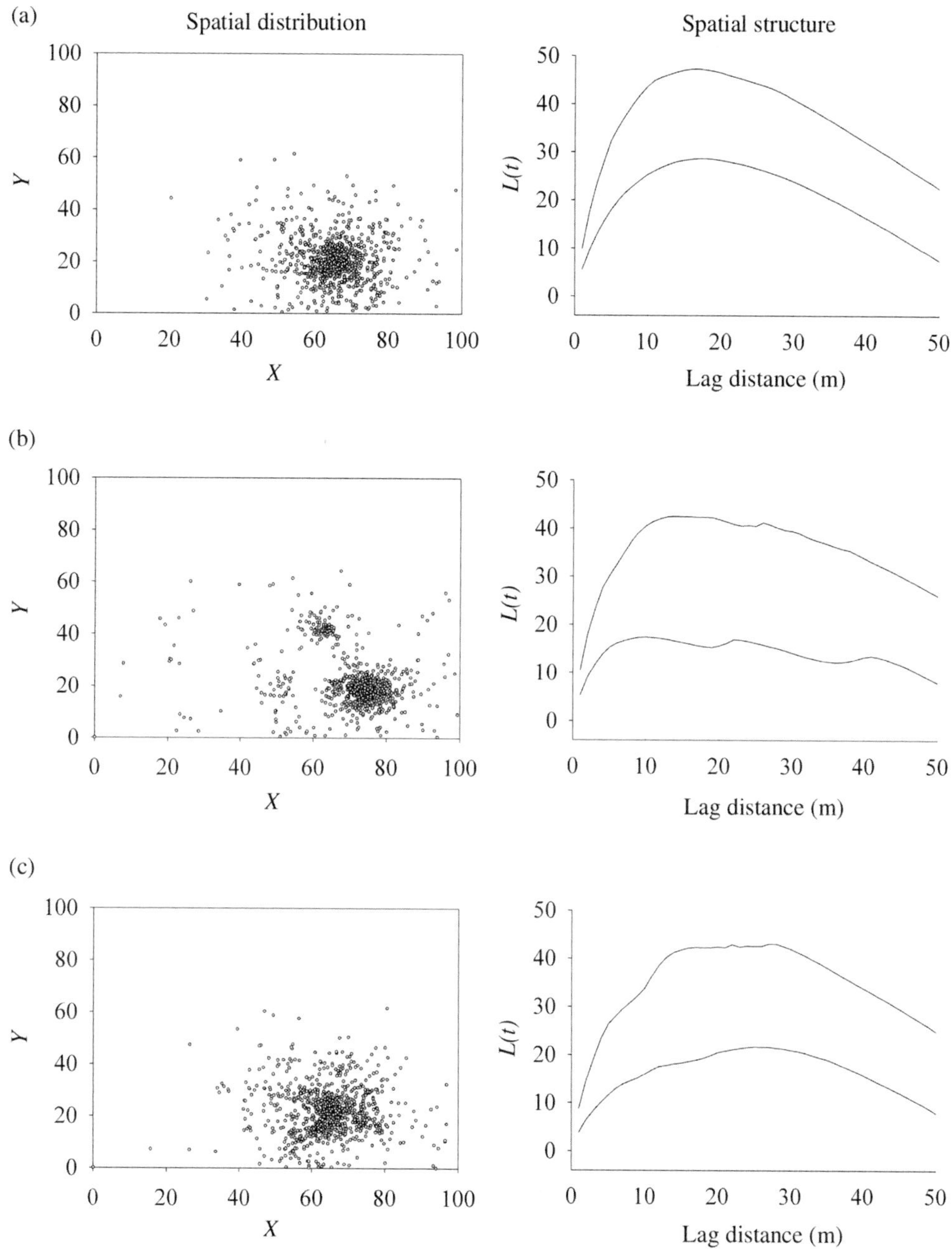

Fig. 19.2. Spatial distributions (one simulation) and resultant spatial structures of 1000 seeds dispersed from a single reproductive adult tree exhibiting: (a) isotropic dispersal; (b) contagious seed dispersal towards ten focal points: and (c) contagious seed dispersal towards 100 focal points. The resulting spatial structure is based on 95% confidence envelopes obtained from 1000 simulations in which positions of adults changed with each simulation.

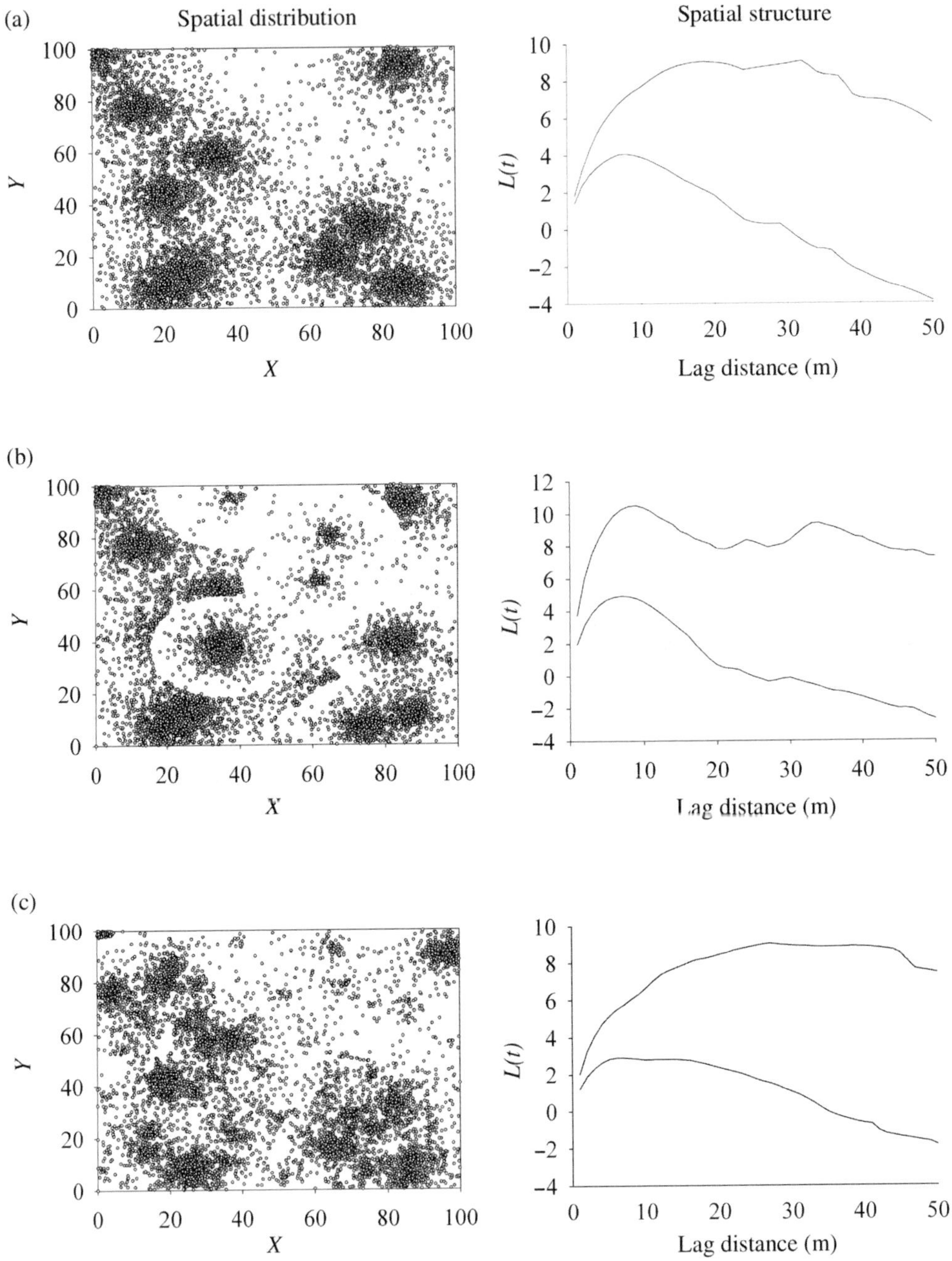

Fig. 19.3. Spatial distributions (one simulation) and resultant spatial structures of 1000 seeds dispersed from each of ten reproductive adult trees exhibiting: (a) isotropic dispersal; (b) contagious seed dispersal towards ten focal points; and (c) contagious seed dispersal towards 100 focal points. The resulting spatial structure is based on 95% confidence envelopes obtained from 1000 simulations in which positions of adults changed with each simulation.

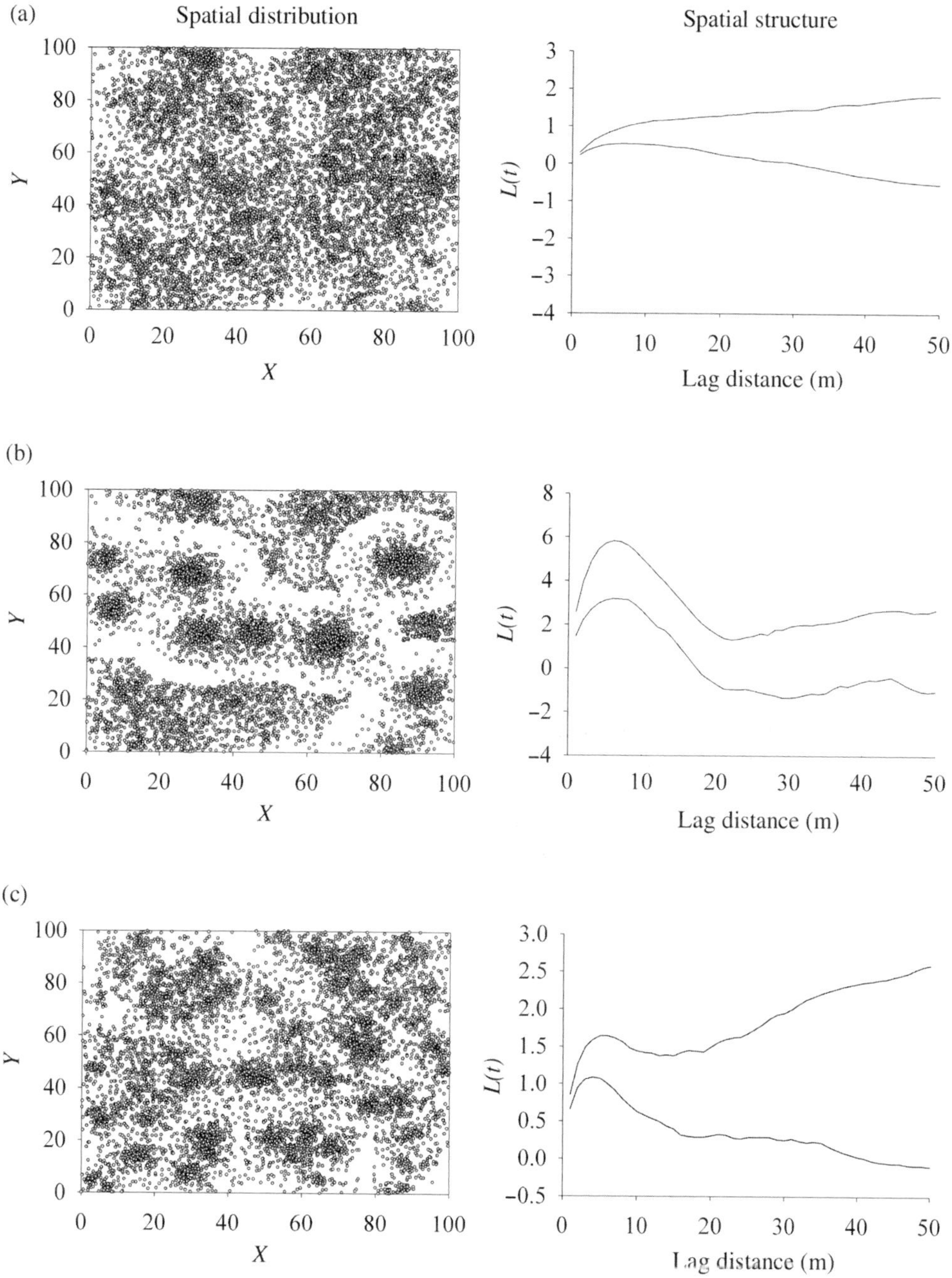

Fig. 19.4. Spatial distributions (one simulation) and resultant spatial structures of 100 seeds dispersed from each of 100 reproductive adult trees exhibiting: (a) isotropic dispersal; (b) contagious seed dispersal towards ten focal points; and (c) contagious seed dispersal towards 100 focal points. The resulting spatial structure is based on 95% confidence envelopes obtained from 1000 simulations in which positions of adults changed with each simulation.

Seed dispersal in systems with focal points can have dramatic effects on seed survival. Focal points can lead to dispersal limitation, as evidenced by the high degree of clumping occurring at smaller spatial scales. The consequences for seed survival are likely to be context-dependent (see Schupp, Chapter 20, this volume), and are most pertinent to rare tree species. For rare species, contagious dispersal may help seeds escape from species-specific, distance-dependent predators, only to subject them to more intense predation pressure from species-specific, density-dependent predators. Generalist predators acting in a density-dependent fashion may also inflict high mortality on seeds of rare tree species.

Fruiting trees

Next we describe contagious seed dispersal beneath conspecific fruiting trees, a situation in which fruit drives the movements and time budgets of frugivores. Here, we impose 'spheres of influence' around fruiting trees. We simulate dispersal by ten conspecific reproductive adults following our '*More than one tree, one species*' simulation, but if a seed disperses in the 1-ha plot and lands within 20 m of another fruiting conspecific, then the seed is 'moved' to the nearest conspecific, from which it is dispersed at a distance picked from an exponential distribution with a mean dispersal distance of 5 m. We repeat the simulation with 100 conspecific adults acting as dispersal foci.

The spatial pattern and the spatial structure of seeds are significantly altered when seeds are dispersed contagiously towards fruiting trees. In our simulations, seeds were clumped at higher densities under conspecific fruiting trees (Fig. 19.5), as some seeds that normally would have been scattered in lower densities away from parent trees are re-routed to high-density patches beneath conspecifics. Field studies have found evidence for this form of contagion (Alcántara *et al.*, 2000; Jordano and Schupp, 2000). Contagion around conspecific trees probably attracts species-specific distance- and density-dependent seed predators; this in itself may result in lower probabilities of population persistence than when contagious dispersal to conspecifics does not occur. This type of contagion may also contribute to lower probabilities of population persistence if fewer seeds land in the areas of 'optimum' seed density (at which the product of it and the probability of survival (see Janzen, 1970) reaches a peak), or if fewer seeds make their way to favourable recruitment sites that are unpredictable in space (Schupp *et al.*, 2002). This is a case in which dispersal limitation is visually apparent, despite seeds being dispersed away from parent trees. On the other hand, such contagion may lead to increased persistence, due to more seeds landing in low densities where recruitment may be most favoured. The exact outcome will depend on the complex interactions between patterns of seed dispersal and the myriad of factors affecting recruitment.

The spatial structure of the seed distribution reflects dispersal limitation through the increased magnitude of clumping at slightly smaller scales of observation (Fig. 19.5) than when seed dispersal was isotropic (see Fig.

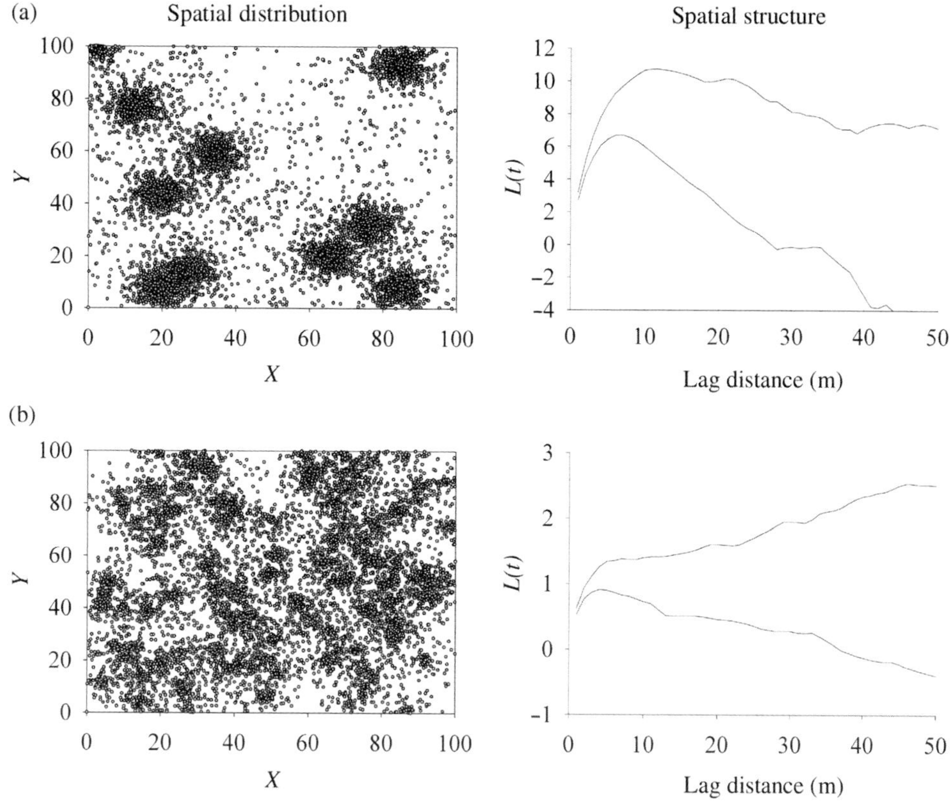

Fig. 19.5. Spatial distributions (one simulation) and resultant spatial structures of spatially contagious seed dispersal beneath: (a) ten; and (b) 100 fruiting adult trees. This was accomplished by assigning 20 m 'spheres of influence' around the adult trees and 'drawing in' dispersed seeds from other parents. First, seed distances were drawn from an exponential distribution with mean dispersal distance of 10 m, and were dispersed isotropically from parent trees (1000 seeds per tree in (a); 100 seeds per tree in (b)). If a seed dispersed > 5 m from a parent tree remained in the plot, and that seed fell within the 20 m sphere of influence of another tree, that seed was moved from its location to the nearest such tree, and then isotropically dispersed from that point at a distance drawn from an exponential distribution with mean dispersal distance of 5 m. The resulting spatial structures, which are based on 95% confidence envelopes obtained from 1000 simulations in which positions of adults changed with each simulation, indicate high degrees of clumping at scales <~10 m.

19.1b,c for comparison). Higher degrees of spatial clumping, based on $\boldsymbol{L}(t)$ values, are now evident at scales <~10 m rather than 20 m. Density-dependent predators operating at such scales are likely to inflict high mortality on seeds. Because clumping is directly tied to locations near conspecific parent trees, distance-dependent predators may also inflict high seed mortality, although the distance-to-parent relationship of the PRC is probably quite complex. The consequences of the clumped spatial structure for survival curves away from parent plants, density-dependent mortality relationships, population persistence and community coexistence remain largely unexplored.

Heterospecific fruiting trees may also act as seed dispersal foci (Schupp *et al.*, 2002; Verdú and García-Fayos, 2003; C.J. Clark *et al.*, 2004; Kwit *et al.*, 2004). For any given parent tree, many of its seeds land in high densities beneath heterospecific trees where species-specific, density-dependent predators are more likely to kill seeds (when found) than if the seeds had been scattered across the landscape. Because of the overall high density of seeds, generalist predators may further decrease the possibility of seedling recruitment beneath the crown of the heterospecific fruiting tree, an area considered as a site for colonization under the Janzen–Connell hypothesis. The reduction in the number of potential recruitment sites may have negative repercussions for population persistence and, ultimately, for the maintenance of species diversity. However, the consequences of contagious dispersal for recruitment depend, in part, on the behaviour of seed predators and their response to clumped seed distributions under heterospecifics versus their response to the scattered low-density seeds away from conspecifics and heterospecifics. If recruitment probabilities are lower under fruiting heterospecifics than at equally distant sites not under fruiting trees, dispersal limitation will be amplified.

Discussion

Contagious seed dispersal represents a fundamental challenge to the Janzen–Connell hypothesis because it severs the relationship between distance from the parent plant and seed density. Contagious dispersal generates a spatial distribution in which seeds occur in high densities in areas far from parent trees – areas envisioned to contain low seed densities according to the traditional Janzen–Connell model. With a greater proportion of the total seeds being dispersed in a clumped distribution, fewer seeds are found in low-density areas.

Contagious seed dispersal will commonly occur in systems where frugivore behaviour alters dispersal patterns of seeds and when trees that share dispersers have overlapping fruiting phenologies (although these are not the only conditions under which it will occur, as high densities of seeds can also be consistently dispersed to non-fruit-related dispersal foci). Because the vast majority of trees and shrubs in species-rich tropical forests have their seeds dispersed by vertebrate frugivores (Howe and Smallwood, 1982), which are a diverse group in such settings as well, we suggest that contagious seed dispersal beneath heterospecific fruiting trees is more likely to occur in tropical than in temperate forests. C.J. Clark *et al.* (2004) noted higher than expected densities of hornbill- and touraco-dispersed seeds beneath heterospecific fruiting trees in a tropical forest. Similar patterns have been predicted in other tropical forests for fruiting trees and shrubs in and along the edges of treefall gaps (Schupp *et al.*, 1989), where densities of fruiting plants and fruit-eating birds are high (Levey, 1988, 1990).

The consequences of spatially contagious seed dispersal from both population and community standpoints should be viewed in a two-dimensional context. The spatial scale of seed clumping, as perceived by specialist and generalist predators, is critical in determining the potential strength of the density-dependent mortality relationship. We expect the combined impact of contagion and predators to have the greatest impact on the strength of the density-dependent mortality in tropical systems. If correct, this would provide the mechanism sought by HilleRisLambers *et al.* (2002) to explain the more pronounced effects of the Janzen–Connell hypothesis in the tropics than in temperate zones. As HilleRisLambers *et al.* (2002) point out; the strength of Janzen–Connell effects may ultimately be responsible for the maintenance of higher species diversity in tropical forests.

Density-dependent mortality is generally recognized as an important process that can influence population abundance and distribution, but the strength of negative density dependence, which determines its influence on population size and species diversity, is driven by the spatial distribution and structure of dispersed seeds, which is not a simple function of distance from parent trees. We argue that for the Janzen–Connell model to explain latitudinal gradients in tree diversity, contagious dispersal must be taken into consideration, and its contribution needs to be clarified.

Future directions

Seed dispersal is the first step in the process that dictates propagule survival, population persistence and, ultimately, species co-existence. While our results highlight how seed dispersers can alter the spatial structure of seeds, they are contingent on the parameters chosen for our simulations. Further investigations into the effects of changes in model parameters (e.g. mean seed dispersal distances, shapes and types of dispersal distributions, changes in the rules governing contagion, etc.), as well as extensions of our models to incorporate responses of enemies to local seed densities (and not solely degrees of clumping per se) and encounter rates, would undoubtedly be informative. Further documentation of density-dependent mortality determining tree recruitment in other forests worldwide will also prove useful. In addition, studies in two areas of research are necessary to bolster our hypothesis that contagious seed dispersal can explain latitudinal gradients in tree species diversity.

1. Documentation of contagious seed dispersal beneath heterospecific fruiting trees

Studies describing and quantifying the occurrence of contagious seed dispersal beneath heterospecific fruiting trees are needed in order to ascertain its prevalence and the resulting spatial structure of seeds. Studies addressing seed dispersal patterns of a single species in diverse and less-diverse systems are needed to document underlying community-level effects (e.g. the effects of the presence of heterospecific fruiting trees on seed dispersal patterns; see Carlo, 2005). Other potential effects not considered in our heuristic model, such as the density and diversity of adult fruiting trees whose seeds share common dispersers, will help in determining how spatial patterns of contagious seed dispersal beneath heterospecific fruiting trees differ across latitudinal gradients.

2. Examination of how spatial patterns of seed deposition affect density-dependent mortality relationships

To determine how the spatial patterns of seeds influence density-dependent mortality and seedling recruitment, we propose an experimental approach. Seeds from fruiting plants could be redistributed under non-fruiting conspecifics in an isotropic leptokurtic distribution, or in a contagiously dispersed manner (as described here). Plots within each distribution could be sampled to compare the strength of density-dependent mortality. Conspecific and heterospecific regeneration could then be followed to assess the effect of seed distributions on population and community recruitment. The role of generalist versus specialist predators could be examined by carefully selecting species to be preyed upon either by generalists or specialists. If contagious dispersal contributes to higher magnitudes of density-dependent mortality in the tropics, these studies will have exciting ramifications for explaining latitudinal gradients in tree species diversity.

Acknowledgements

We are grateful to numerous people whose ideas, contributions, and input led to the development of this chapter; including D. Gorchov, K. Harms, K. Holbrook, P. Jordano, B. Loiselle, S. Russo, E. Schupp, H. Stevens, J. Tewksbury, C. Whelan, J. Wright and two anonymous reviewers. We are indebted for support and assistance from B. Collins, G. Wein, the Savannah River Ecology Laboratory, University of Georgia, and Financial Assistance Award Number DE-FC09-96SR18546 from the US Department of Energy to the University of Georgia Research Foundation. Travel of the senior author to Australia to participate in the Symposium/Workshop was partially supported by National Science Foundation (OISE No. 0456402 to E.W. Schupp).

References

Alcántara, J.M., Rey, P.J., Valera, F. and Sánchez-Lafuente, A.M. (2000) Factors shaping the seedfall pattern of a bird-dispersed plant. *Ecology* 81, 1937–1950.

Carlo, T.A. (2005) Interspecific neighbors change seed dispersal pattern of an avian-dispersed plant. *Ecology* 86, 2440–2449.

Clark, C.J., Poulsen, J.R., Conner, E.F. and Parker, V.T. (2004) Fruiting trees as dispersal foci in a semi-deciduous tropical forest. *Oecologia* 139, 66–75.

Clark, C.J., Poulsen, J.R., Bolker, B.M., Conner, E.F. and Parker, V.T. (2005) Comparative seed shadows of bird-, monkey-, and wind-dispersed trees. *Ecology* 86, 2684–2694.

Clark, D.A. and Clark, D.B. (1984) Spacing dynamics of a tropical rain-forest tree: evaluation of the Janzen–Connell model. *American Naturalist* 124, 769–788.

Clark, J.S., Macklin, E. and Wood, L. (1998) Stages and spatial scales of recruitment limitation in southern Appalachian forests. *Ecological Monographs* 68, 213–235.

Clark, J.S., Silman, M., Kern, R., Macklin, E. and HilleRisLambers, J. (1999) Seed dispersal near and far: patterns across temperate and tropical forests. *Ecology* 80, 1475–1494.

Connell, J.H. (1971) On the role of natural enemies in preventing competitive exclusion in some marine animals and in rain forest trees. In: den Boer, P.J. and Gradwell, G.R. (eds) *Dynamics of Populations*. Centre for Agricultural Publishing and Documentation (PUDOC), Wageningen, The Netherlands, pp. 298–313.

Dalling, J.W., Muller-Landau, H.C., Wright, S.J. and Hubbell, S.P. (2002) Role of dispersal in the recruitment limitation of neotropical pioneer species. *Journal of Ecology* 90, 714–727.

Diggle, P.J. (1983) *Statistical Analysis of Spatial Point Patterns*. Academic Press, London, UK.

Greene, D.F., Canham, C.D., Coates, K.D. and Lepage, P.T. (2004) An evaluation of alternative dispersal functions for trees. *Journal of Ecology* 92, 758–766.

Hammond, D.S. and Brown, V.K. (1998) Disturbance, phenology and life-history characteristics: factors influencing distance/density-dependent attack on tropical seeds and seedlings. In: Newbery, D.M., Prins, H.H.T. and Brown, N.D. (eds) *Dynamics of Tropical Communities*. Blackwell, Oxford, UK, pp. 51–78.

Harms, K.E., Wright, S.J., Calderón, O., Hernández, A. and Herre, E.A. (2000) Pervasive density-dependent recruitment enhances seedling diversity in a tropical forest. *Nature* 404, 493–495.

HilleRisLambers, J., Clark, J.S. and Beckage, B. (2002) Density-dependent mortality and the latitudinal gradient in species diversity. *Nature* 417, 732–735.

Howe, H.F. and Miriti, M.N. (2000) No question: seed dispersal matters. *Trends in Ecology and Evolution* 15, 434–436.

Howe, H.F. and Smallwood, J. (1982) Ecology of seed dispersal. *Annual Review of Ecology and Systematics* 13, 201–228

Janzen, D.H. (1970) Herbivores and the number of tree species in tropical forests. *American Naturalist* 104, 501–528.

Jordano, P. and Schupp, E.W. (2000) Seed disperser effectiveness: the quantity component and patterns of seed rain for *Prunus mahaleb*. *Ecological Monographs* 70, 591–615.

Kwit, C., Levey, D.J. and Greenberg, C.H. (2004) Contagious seed dispersal beneath heterospecific fruiting trees and its consequences. *Oikos* 107, 303–308.

Levey, D.J. (1988) Tropical wet forest treefall gaps and distributions of understory birds and plants. *Ecology* 69, 1076–1089.

Levey, D.J. (1990) Habitat-dependent fruiting behavior of an understory tree, *Miconia centrodesma*, and tropical treefall gaps as keystone habitats for frugivores in Costa Rica. *Journal of Tropical Ecology* 6, 409–420.

Ripley, B.D. (1976) The second-order analysis of stationary point processes. *Journal of Applied Probability* 13, 255–266.

Schupp, E.W. (1988) Seed and early seedling predation in the forest understory and in treefall gaps. *Oikos* 51, 71–78.

Schupp, E.W. (1992) The Janzen–Connell model for tropical tree diversity: population implications and the importance of spatial scale. *American Naturalist* 140, 526–530.

Schupp, E.W., Howe, H.F., Augspurger, C.K. and Levey, D.J. (1989) Arrival and survival in tropical treefall gaps. *Ecology* 70, 562–564.

Schupp, E.W., Milleron, T. and Russo, S.E. (2002) Dissemination limitation and the origin and maintenance of species-rich tropical forests. In: Levey, D.J., Silva, W.R. and Galetti, M. (eds)

Seed Dispersal and Frugivory: Ecology, Evolution and Conservation. CAB International, Wallingford, UK, pp. 19–33.

Terborgh, J., Pitman, N., Silman, M., Schichter, H. and Núñez, V.P. (2002) Maintenance of tree diversity in tropical forests. In: Levey, D.J., Silva, W.R. and Galetti, M. (eds) *Seed Dispersal and Frugivory: Ecology, Evolution and Conservation.* CAB International, Wallingford, UK, pp. 1–17.

Uriarte, M., Canham, C.D., Thompson, J., Zimmerman, J.K. and Brokaw, N. (2005) Seedling recruitment in a hurricane-driven tropical forest: light limitation, density-dependence and the spatial distribution of parent trees. *Journal of Ecology* 93, 291–304.

Verdú, M. and García-Fayos, P. (2003) Frugivorous birds mediate sex-biased facilitation in a dioecious nurse plant. *Journal of Vegetation Science* 14, 35–42.

Wenny, D.G. (2001) Advantages of seed dispersal: a re-evaluation of directed dispersal. *Evolutionary Ecology Research* 3, 51–74.

Willson, M.F. (1993) Dispersal mode, seed shadows, and colonization patterns. *Vegetatio* 107/108, 261–280.

Wright, S.J. (2002) Plant diversity in tropical forests: a review of mechanisms of species coexistence. *Oecologia* 130, 1–14.

20 The Suitability of a Site for Seed Dispersal is Context-dependent

E.W. Schupp

Introduction

Disperser effectiveness, defined as 'the contribution a disperser makes to the future reproduction of a plant' (Schupp, 1993, p. 16), is determined by two critical components of dispersal: the 'quantity' of seeds dispersed and the 'quality' of dispersal provided to each seed. More explicitly, effectiveness is the product of:

1. The number of seeds dispersed (the *quantity component*);
2. The probability that a dispersed seed survives to become a new reproductive adult (the *quality component*).

Quality, and thus overall effectiveness, depends to a large extent on where seeds are deposited – on the biotic and abiotic conditions that the potential new recruits will face (Schupp, 1993). In the original development of the disperser effectiveness framework (Schupp, 1993), I argued that it was critical not to view sites of deposition as simply suitable or unsuitable for recruitment, but rather to view sites as varying continuously in suitability. That is, sites vary continuously in the likely fate of recruits – in the probability that a seed will be consumed by a rodent, attacked by pathogens, or germinate; that a seedling will be grazed by slugs, infected with a fungus, or escape pests and grow rapidly; that a juvenile will be heavily browsed by an elephant, die of drought, or grow to be a reproductive adult; and much more. Although this continuum view of suitability is generally accepted, especially in the context of attempts to estimate disperser effectiveness, the binary view of suitable/unsuitable continues to emerge and influence our thinking. For example, we (e.g. Schupp and Fuentes, 1995; Silman *et al.*, 2003) frequently fall back on the terminology of 'safe sites' which, by implication, are suitable sites that presumably contrast with 'unsafe' or unsuitable sites.

The problems with evaluating site suitability go beyond the contrast between the continuum versus binary views. Both theoretical and empirical work demonstrates that the 'suitability' of a site is in reality very dependent on the context. The consequences for plant recruitment of a particular site type may vary between different life history stages, between years, and between places. For example, if we compare the suitability of shrub canopies relative to open interspaces for plant recruitment, our conclusion may be very different depending on whether we study seed success or seedling success, conduct our study in a wet year or a dry year, or have our experimental plots on a north-facing slope or a south-facing slope. Because the suitability of deposition sites goes to the core of disperser effectiveness, it is critical that we have a clear understanding of site suitability, including its dependence on context, when designing experiments to quantify disperser effectiveness or when evaluating and interpreting the results of others.

In this chapter, my goal is to present an overview of relevant ideas using selected examples. I begin by presenting a very brief overview of the concept of the safe site and discuss why this may have contributed to confusion as well as clarification. I then more thoroughly review context-dependent site suitability.

- *First*, I discuss cases where very simple 'toggles', such as the presence or absence of an enemy, have been shown to alter the relative suitability of alternative sites for recruitment.
- *Second*, I discuss more complex processes affecting the suitability of sites for seed deposition: (i) the balance between interference and facilitation; and (ii) life-stage conflicts, an extension of the idea of seed–seedling conflicts (Schupp, 1995).
- *Third*, I show how all of these aspects interact, creating further complexities that may hinder our ability to quantify disperser effectiveness.
- *Finally*, I conclude with the implications for the design of studies of site suitability.

There are two take-home messages from this chapter. The first is that although it is more difficult to quantify the suitability of sites for dispersal than we have often assumed – and than we would like – acknowledging the difficulties is the first step to overcoming them. The second is that such studies are still important and extremely fruitful lines of research.

The Safe Site Concept

In order to better understand our present inconsistent views on the suitability of sites for dispersal, I believe it is useful to revisit the roots of the 'safe site'. This concept has had an extremely positive influence on plant ecology, but in my view it has also been responsible for generating confusion. Harper (1977) defined the 'safe site' as the very local area, determined by the size of the seed, that provides:

1. The stimuli necessary for breaking seed dormancy;
2. The conditions that are required for germination to proceed;
3. The resources, such as oxygen and water, used in germination;
4. The absence of hazards such as seed consumers, competitors and pre-emergence pathogens.

Key to the original idea is that a safe site is: (i) a *very* small-scale environment immediately surrounding an individual seed that (ii) promotes or allows the emergence and initial establishment of a seedling. As such, it is an all-or-none phenomenon. Given that a seed is present, a seedling either emerges and establishes and the site is 'safe', or a seedling fails to emerge and establish and the site is 'unsafe'. In addition, the safe site, as originally defined, spans only a very brief period of the life and challenges that a recruiting plant will face – the passage through the environmental sieve from a seed in the soil to an established seedling.

Despite a very clear and specific meaning, the concept has very frequently been misused in two ways.

- *It is often applied to an immense area from the perspective of a seed*. For example, for a pioneer species, we might refer to a treefall gap as being a 'safe site' while considering the forest understorey as unsafe, perhaps without actually referring to it as such. From the perspective of a seed, however, a treefall gap presents extraordinary heterogeneity in a multitude of factors that will influence successful establishment: litter depth, light intensity and spectral composition, soil moisture, temperature, pathogen spores, predator activity, and more. Some seeds will produce seedlings and some will fail. The gap is 'safe' for some and 'unsafe' for others. What we really mean in the context of safe sites is that the gap has more safe sites than the understorey so that, on average, a seed landing in a gap has a higher probability of establishing as a seedling. But individual sites are still either safe or not.
- *It is often applied to stages well past initial emergence and establishment*. For example, it may be applied to the survival and growth of established seedlings, or even saplings.

From the perspective of developing a predictive understanding of the consequences of seed dispersal we are unlikely to be interested in understanding safe sites as originally defined. We are concerned, for example, with the probability that a seed falling in a gap will successfully produce a new adult – relative to the probability that a seed in the understorey will – not in whether or not an individual seed germinates and establishes. We are, in fact, concerned with the more expanded view of 'safe sites' that is more common in the literature – the approach that considers the larger spatial and temporal scales of plant recruitment.

But if this more typical, modified view of safe sites is more useful, why am I being so critical? My criticism is not with the approach we are taking to understand recruitment, but rather with the implications that come with our continued use of the term 'safe site'. The major drawback is that 'safe site' *implies* a world of absolutes in which successful seedling establishment explains all. Sites are safe or not, and once a seedling is established we know the success of recruitment. Neither of these is true from the vantage point from which we typically view dispersal, and it is unlikely that Harper believed either to be true. That is, however, what we *imply* when we use the safe site concept as a framework for seed dispersal studies. And I believe that this implication, consciously or unconsciously, can mislead us in our interpretations of site suitability and make it more difficult for us to accept the obvious context-dependence of suitability.

Although I believe that the safe site concept has only limited direct applicability for understanding seed disperser effectiveness and has, in fact, led to confusion, it is very valuable none the less. Empirically, its value lies in understanding the factors affecting the fate of individual seeds, such as the specific requirements for breaking dormancy and germinating; this value is extremely reductionistic. I believe its primary value, however, lies in how it should make us think, rather than as a metric to use in real-world studies. It should force us to learn to think like a seed, and by extension think like a seedling, sapling, etc. Only by seeing the world from that perspective can we hope to understand the suitability of dispersal sites.

Simple 'Toggles' can Alter Suitability

In many cases a simple change in a single driver in the biotic or abiotic environment can alter the relative suitability of distinct patch types within the landscape, perhaps at a relatively fine patch scale such as microhabitats, perhaps at some larger scale. At a larger scale, Fowler (2002) studied the effects of grazing, competition and topographic position on six species of perennial grasses in a west Texas savannah. Without grazing, all six grasses grew better on flat areas than on hill slopes, although one had lower survival on flats. However, with heavy grazing, the four tallest species performed best on slopes. These taller species were most severely affected by grazing, which was especially intense on flatter areas frequented by cattle. As a consequence, for those species vulnerable to grazing, flats switch from being the most suitable sites for establishment in the absence of heavy grazing to being the least suitable sites in the presence of grazing – the most suitable site for dispersal depends on the grazing context.

Silman *et al.* (2003) similarly demonstrated that suitability depends on the presence or absence of a major herbivore. In Cocha Cashu, Peru, white-lipped peccary (*Tayassu pecari*; Tayassuidae) herds were common in the mid-1970s, went extinct regionally in 1978, and recolonized in 1990. After a 12-year absence of peccaries, *Astrocaryum murumuru* (Arecaceae) seedlings increased in density and developed a spatial distribution that was random with respect to cover objects such as logs and tree trunks. Following recolonization by peccaries, seedling density and spatial pattern returned to 1970 conditions, with fewer seedlings that were strongly biased to being near cover objects. Thus, seed and seedling destruction by peccaries not only reduced seedling numbers, but also altered site suitability. In the absence of peccaries the entire understorey is apparently more or less equally suitable for palm seedling establishment; but in the presence of peccaries, sites near cover objects become far more suitable than sites away from such objects. Although suitability did not completely switch, as in the previous example, the relative suitability of sites for dispersal still depended on the herbivory context.

Both of these examples are based on the presence versus the absence of herbivores, but that should not be required. Varying abundances of cattle or peccaries in space or time would still be expected to alter the relative suitability of distinct dispersal sites. Although the changes may be more subtle, the consequences could still be large from the plant's perspective.

Processes other than herbivory can also provide simple toggles altering the relative suitability of dispersal sites. In the eastern Great Basin, Utah, E.W. Schupp, J.M. Gómez and M. Fuentes (1997, unpublished results) sowed seeds of Utah juniper (*Juniperus osteosperma*; Cupressaceae) beneath juniper trees, beneath sagebrush shrubs (*Artemisia tridentata*; Asteraceae), and in open interspaces. Seeds sown in 1994 had seedlings emerging in 1995, 1996 and 1997, while seeds sown in 1995 had seedlings emerging only in 1997. In all four cohorts, seedling emergence differed between microhabitats, but the most suitable microhabitat, based on the proportion of remaining seeds producing a seedling, differed from year to year. In 1995, emergence was much lower beneath shrubs than beneath trees and in open interspaces, which did not differ from each other. In 1996, the pattern reversed, with the shrub microhabitat having the greatest seedling emergence. Finally, in 1997, seedling emergence was much greater beneath trees than in the other microhabitats regardless of year of sowing. This last result suggests that the year-to-year differences were probably driven by climate variability. Whatever the cause, the suitability of a site type was not constant but depended on the context of the year.

More Complex Drivers

The balance between interference and facilitation

In this section I focus on the effects of existing plants on the recruitment of new individuals. The traditional and still dominant perspective is one of interference (Bruno *et al.*, 2003). Due to competition for light and soil resources or to allelopathy, existing plants reduce the growth and/or survival of potential recruits in their vicinity; the interaction is negative for recruits. However, over the last few decades it has become increasingly accepted that such interactions may also result in the facilitation of recruitment (Hunter and Aarssen, 1988). By ameliorating the biotic environment (e.g. reducing herbivory) or the abiotic environment (e.g. improving climatic or soil conditions), existing plants may improve the growth and/or survival of potential recruits in their vicinity; the interaction may indeed be positive for recruits. Such an expanded view of the potential outcomes of plant–plant interactions has implications for ecological theory, including ideas about the relationship between the fundamental and the realized niche, the effects of diversity on invasibility, and more (Bruno *et al.*, 2003).

There are two major insights to be gained from this growing body of work that are especially relevant to understanding the context-dependence of site suitability for dispersal. First, both interference and facilitation can be occurring simultaneously. What we observe is the net effect of the interaction, or the difference between the strength of interference and the strength of facilitation. Second, both the type of facilitation and the net effect of the interaction can vary both spatially and temporally as the biotic and abiotic environments change.

Positive interactions among plants are thought to be most prevalent in environments with high physical stress, where habitat amelioration is likely to be especially beneficial, and in environments with high consumer pressure, where associational defences should be especially beneficial (Bertness and Callaway, 1994; Brooker and Callaghan, 1998; Callaway and Pugnaire, 1999). Because Bertness and Callaway (1994) assumed that the physical stress and the consumer pressure gradients are negatively correlated, they predicted that positive interactions between plants would be most common in environments with either very high or very low physical stress and less common in intermediate environments. However, it is unclear how valid this assumption is. Herbivory may be especially detrimental under physically stressful conditions where lost tissue is difficult to replace. It is likely that a consumer-effect gradient, which may not be equivalent to a consumer-pressure gradient, is more important in determining the balance between interference and facilitation through biotic amelioration.

Independent of the exact relationship between consumer and stress gradients, it is expected that the net effect of existing vegetation on recruitment should vary continuously across the appropriate gradient. Figure 20.1 presents a hypothetical response across a stress gradient, the gradient that I will concentrate on. Similar arguments can be built around the consumer gradient, focusing on the effects of existing vegetation on herbivore-induced mortality. Note that for simplicity the illustrated response is: (i) in the form of a straight line; and (ii) that is symmetrical around the '0' or 'no-net-effect' line, although neither condition is necessary. For example, it could be argued that the net negative effect region might span more of the stress gradient than the net positive effect region.

At the low-stress end of the gradient, the net effect of the interaction is expected to be strongly negative because competition is much greater than any potential facilitative benefits. With increasing stress, the net negative effect weakens as the facilitative benefits increase and competition possibly becomes less important (*sensu* Welden and Slauson, 1986) relative to other forces affecting fitness. At some point facilitation becomes stronger than

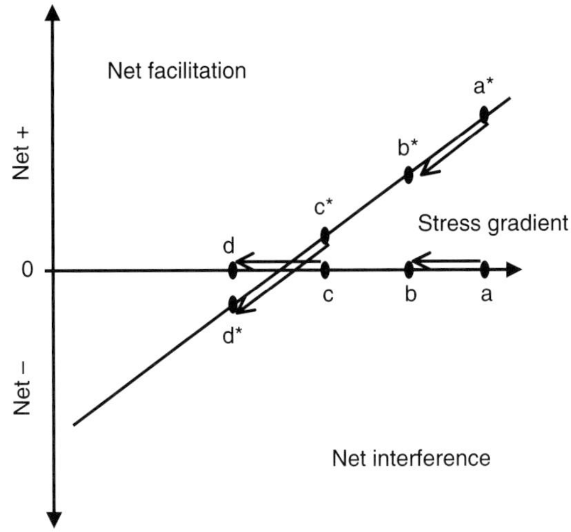

Fig. 20.1. A hypothetical response of the balance between facilitation and interference across an environmental stress gradient where the net effect of an existing plant on recruitment of a new individual shifts from a net negative effect (interference) to a net positive effect (facilitation). Small letters connected by arrows represent temporal or spatial shifts in stress (e.g. a | b might represent a shift from a drought to a non-drought year) while the same letters accompanied by an asterisk (*) represent resultant shifts in the net effect of the interaction. See text for details.

interference and the net effect becomes positive, with the strength of the net positive effect continuing to increase with increasing stress.

The world is not constant, however, and temporal and spatial variability in the environment directly leads to variability in the relative suitability of a given 'type' of dispersal site (e.g. beneath a shrub canopy). For example, the net effect of *A. tridentata* shrubs on *J. osteosperma* recruitment in Tintic Valley, Utah, may shift over time and over very short spatial scales. It is likely that many such shifts occur without altering the rankings of sites, as in 'a' | 'b' (Fig. 20.1). This example might represent a switch from a drought to a non-drought year (temporal variability) or perhaps the difference between a south-facing and a north-facing slope (spatial variability). In this scenario, the most suitable type of site for dispersal (e.g. beneath a sagebrush shrub) does not change, but the relative fitness advantage for potential recruits arriving in that type of site may vary temporally and spatially from slight in some cases to extreme in others; in more stressful microsites or years it is very beneficial (a*), while in less stressful microsites or years it is less beneficial (b*). For example, Maestre *et al.* (2003) experimentally detected simultaneous interference and facilitation by tussock grasses on the survival of shrub seedlings at two sites in south-eastern Spain. Although the net effect was facilitative at both sites throughout the study, net facilitation was stronger at the drier of the two sites, as expected. Similarly, the strength of net facilitation of tree seedling survival by shrubs was greater at drier sites and in drier years than at wetter sites and in wetter years at Mediterranean montane locations in the Sierra Nevada, Spain (Gómez-Aparicio *et al.*, 2004, 2005). In contrast, Tielbörger and Kadmon (2000) reported that the relative advantage of shrubs over open areas for annual plant density and reproduction *increased* with precipitation in the Negev Desert, Israel, contrary to the prediction.

Similar temporal and spatial shifts in the strength of net interference at the opposite end of the gradient are also expected (Goldberg *et al.*, 1999). For example, salt-tolerant fugitive species in salt marshes of New England, USA, were competitively suppressed by existing vegetation, but the strength of the net competitive effect was weaker in hotter, more stressful years, than in cooler, less stressful years (Bertness and Ewanchuk, 2002).

Most interesting is the realization that a temporal or spatial shift in the environment, as in 'c' | 'd' (Fig. 20.1), might shift a net facilitative interaction (c*) to a net interference interaction (d*). Thus, environmental variability may lead to a complete alteration in the rankings of site suitability. In such cases it will be more advantageous to be dispersed to, for example, shrubs in some years and places and to open interspaces in others. In fact, a number of studies using either natural environmental variability or experimental manipulations of precipitation (e.g. Greenlee and Callaway, 1996; Ibáñez and Schupp, 2001) or nocturnal warming (Lloret *et al.*, 2005) have documented shifts from net facilitation of recruitment by existing vegetation under more stressful conditions to net interference under more favourable conditions. In such a system, a study of site suitability conducted during a drought cycle might yield very different conclusions than one conducted during an inter-drought cycle.

Confounding contexts

The context dependencies of the outcomes of plant–plant interactions are extremely varied and complex, making it even more difficult to adequately quantify the suitability of sites for dispersal, much less generalize about site suitability. A limited overview of confounding contexts includes:

- *A shrub is not a shrub; a tree is not a tree.* Not all existing vegetation in the community has the same effect on potential recruits (Callaway and Pugnaire, 1999). Studying the effects of six shrub species on herbaceous biomass in south-eastern Spain, Pugnaire *et al.* (2004) reported net interference by one species, neutrality by another and net facilitation by the remaining four. During primary succession following a volcanic eruption, vegetation patches containing *Salix reinii* (Salicaceae) facilitated recruitment better than patches without *S. reinii* because the willows were a source of ectomychorrizal inocula (Nara and Hogetsu, 2004). Even considering a given potential 'facilitator' species at a given site in a given year, the net effect need not be consistent. In Sonora, Mexico, larger *Olneya tesota* (Fabaceae) trees appear to be better facilitators than smaller trees (Tewksbury and Lloyd, 2001). When *Quercus douglasii* (Fagaceae) fail to access deep water they proliferate roots near the surface and have a net competitive effect on the herbaceous understorey, but when the oaks contact deep water they produce fewer shallow roots and have a net facilitative effect on the understorey (Callaway and Pugnaire, 1999; Callaway *et al.*, 2003).

- *A recruit is not a recruit.* Recruits from different species can respond very differently to the existing vegetation and climatic conditions of a site (Callaway and Pugnaire, 1999). In north-eastern Spain, *Erica multiflora* (Ericaceae) and *Globularia alypum* (Scrophulariaceae) seedlings had distinct responses to canopy cover, climate manipulations and the interaction between canopy and climate (Lloret *et al.*, 2005). Similarly, in the Negev Desert of Israel there was a general trend for an increasing net facilitative effect of shrubs on annuals with increasing rainfall, but the trend was not consistent and the four species of annuals studied all responded somewhat differently (Tielbörger and Kadmon, 2000). Even within a given species, the outcome of the interaction between recruits and existing vegetation can shift ontogenetically from net facilitation for new recruits to net interference for larger plants, as seen with *Ambrosia dumosa* (Asteraceae) in the Mojave Desert, USA (Miriti, 2006).
- *Net effects can change rapidly.* In Australia, artificial shade mimicking an under-canopy environment decreased *Enchylaena tomentosa* (Chenopodiaceae) seedling survival during the cool season but increased survival during the hot summer (Hastwell and Facelli, 2003), suggesting a seasonal switch from net interference to net facilitation. Similarly, although *A. dumosa* shrubs generally facilitated annual plants in the Mojave Desert, there were within-season shifts in the strengths of both positive and negative effects, resulting in a continually varying net effect (Holzapfel and Mahall, 1999).
- *Interactions among environmental variables may be more important than the main effects.* In the succulent Karoo vegetation of South Africa, the effects of shrubs on recruiting seedlings were not straightforward or predictable based on expected responses to a simple gradient of abiotic stress, but were instead the result of a complex interaction between rainfall and soil nutrients (Riginos *et al.*, 2005).
- *There are clear exceptions to the model.* Substantial evidence is accumulating that supports the basic prediction of the model – net facilitation increases with increasing abiotic stress. None the less, whether this pattern is typical in arid and semiarid environments is under debate (see Maestre *et al.*, 2005, 2006; Lortie and Callaway, 2006) and some studies provide obviously contradictory results. As noted above, in the Negev Desert, the net strength of facilitation by shrubs on the density and reproduction of annuals tended to increase with increasing precipitation (Tielbörger and Kadmon, 2000). Similarly, the net effect of *Stipa tenacissima* (Poaceae) tussock grasses on *Pistacia lentiscus*

(Anacardiaceae) seedlings switched from net interference to net facilitation along an increasing rainfall gradient (Maestre and Cortina, 2004). Furthermore, in a preliminary meta-analysis of a broad database, Goldberg *et al.* (1999) found evidence suggesting a trend for the facilitation of survival to be more common in sites with less stress, as indicated by greater standing biomass. In all of these cases, however, although the results were contrary to predictions, the net effects of the interactions still changed with the environment; that is, the net effect was still context-dependent as defined.

Life-stage conflicts

A seed–seedling conflict exists when sites or environmental conditions favourable for seed success may be unfavourable for seedling success (Schupp, 1995). For example, *Prunus mahaleb* (Rosaceae) in southern Spain faces a conflict between conditions suitable for seed survival and conditions suitable for seedling survival. Of nine recognized microhabitats in the study site, 'deep-soil', with abundant herbaceous vegetation and no woody plant cover, had the highest seed survival but the lowest 2-year seedling survival. In complete contrast, 'low spiny shrubs' had the lowest seed survival but the highest 2-year seedling survival (Schupp, 1995). Not only do shrubs provide an abiotic environment that facilitates seedling survival, they also provide an environment favourable to rodent seed predators attempting to avoid their own predators while feeding. A review of the literature at that time suggested that such conflicts are widespread both taxonomically and geographically (Schupp, 1995).

The focus of my original paper (Schupp, 1995) was on seed–seedling conflicts simply because it was developed for a symposium on seed and seedling ecology. None the less, I noted at the time that similar conflicts are likely to extend over successive life history stages and ecological processes, such as seed survival; seed germination; seedling emergence; seedling survival in the face of herbivores, pathogens and falling branches; and more. A critical extension of the concept, then, is 'life-stage conflicts'. Sites are not inherently 'good' or 'bad' for recruitment, but rather 'goodness' is the net outcome, or sum, of a long sequence of potential benefits and detriments associated with different life history stages and ecological processes. *Tachigalia versicolor* (Caesalpinioideae) in Panama still provides an excellent example (Kitajima and Augspurger, 1989); relative to the forest understorey, treefall gaps were detrimental to seed survival due to rodents and to seedling survival due to herbivores, but beneficial to seedling survival due to fewer pathogens and to further growth and survival. The net outcome appears to be a strong overall benefit to a seed landing in a gap, primarily because the very strong reduction in pathogen attack in gaps overwhelms the granivore- and herbivore-generated detriments. But what if the strengths of the different processes acting at the different stages fluctuate spatially or temporally, much as the strengths of interference and facilitation fluctuate?

Figure 20.2 presents a simplified hypothetical example from a semi-arid environment, which considers the suitability of a shrub microhabitat relative to an open interspace microhabitat for the recruitment of a new plant. The black downward-pointing arrows represent a relative disadvantage for recruits beneath shrubs (a detriment), while white upward-pointing arrows represent a relative advantage for recruits beneath shrubs (a benefit). Finally, the width of an arrow represents the relative strength of the effect.

As shown in Fig. 20.2, year 1 is relatively wet and has a relatively large population of rodents (assume a species of plant where rodents are unimportant for seed dispersal and instead are predominantly seed consumers). Under these conditions, shrub microhabitats have considerably greater seed predation due to the concentration of rodent foraging activity there. In contrast, drought, and especially herbivory, kill fewer seedlings beneath shrubs because of facilitation; shrubs marginally ameliorate the moderately stressful abiotic conditions and effectively reduce access to

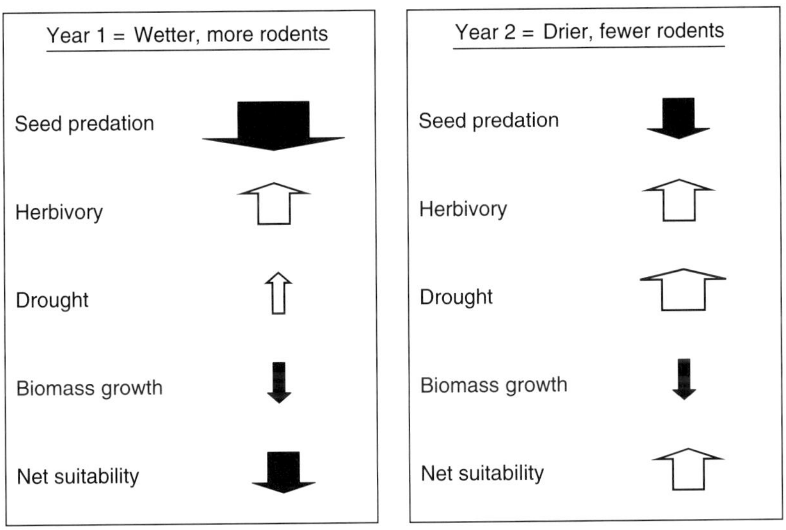

Fig. 20.2. A simplified hypothetical example from a semiarid ecosystem demonstrating the potential for life-stage conflicts to alter net suitability of potential dispersal sites as environmental conditions shift. This figure compares the suitability of a shrub microhabitat relative to an open interspace microhabitat for recruitment of a new plant. Black downward pointing arrows represent a relative disadvantage for recruits beneath shrubs (detriment), while upward pointing arrows represent a relative advantage for recruits beneath shrubs (benefit). The width of an arrow represents the relative strength of the effect. Year 1 is relatively wet and has a relatively large population of rodents, thus it represents an environment with relatively mild abiotic stress but intense seed predation (assume a plant species where rodents are unimportant for seed dispersal and instead are predominantly seed consumers). Year 2 is relatively dry and has a relatively small population of rodents, thus it represents an environment with more extreme abiotic stress but relatively little seed predation. See text for details.

browsing mammals. Lastly, through competition for light and soil resources, shrubs reduce growth and biomass accumulation. The net suitability of the shrub microhabitat relative to the interspace microhabitat, then, is moderately negative, driven predominantly by the strong detriment associated with high seed predation beneath shrubs.

In contrast, year 2 is relatively dry and has a relatively small population of rodents. For simplicity, in this hypothetical system the relative strengths of the herbivory and competition effects are unchanged by this annual shift. However, the relative intensity of seed predation beneath shrubs is substantially reduced. Although seed predation is still greater beneath shrubs than in open interspaces, the large reduction in rodent densities results in a reduction in the relative detriment of the shrub microhabitat driven by seed predation. In contrast, under the drier conditions of year 2, the relative benefit provided by shrubs is much greater than in year 1 because the net effect of facilitation increases with the greater abiotic stress. Due to the large reduction in the detriment associated with seed predation coupled with the large increase in the benefit associated with drought, the net suitability switches from moderately negative in year 1 to moderately positive in year 2.

Although life-stage conflicts have not received the theoretical and empirical attention given to the balance between interference and facilitation, examples continue to be published, reinforcing the idea that the phenomenon is widespread and important. In the Sierra Nevada, south-eastern Spain, *Quercus ilex* (Fagaceae) exhibits a conflict during recruitment; open microhabitats had high acorn survival but very low seedling survival over the dry summer, while beneath-oak microhabitats had very low acorn survival but relatively high seedling survival (Gómez, 2004). As a consequence, fluctuations in the strengths of seed consumption and physical stress, both of which are common, may lead to a switch in the relative suitability of potential recruitment microhabitats. In some cases, whether a conflict even exists may vary with environmental conditions. *Liquidambar styraciflua* (Hamamelidaceae) in South Carolina, USA, had greater seedling emergence but less seedling survival and growth in pit microsites than on mound microsites when exposed to full sunlight, but the conflict disappeared under shade, where mounds were the best microsite for all stages (Battaglia *et al.*, 2000). In the same forest, some potential recruits may face a conflict while others do not.

Putting it All Together

The variety of processes contributing to context-dependence in the suitability of sites for dispersal are not really independent. In fact, as several examples already reviewed have shown, the balance between interference and facilitation may be a prime force in creating a life-stage conflict (e.g. Gómez, 2004). I conclude by presenting overviews of three examples of the interrelatedness of these concepts and the complexities these bring to any attempt to understand plant recruitment and the suitability of sites for dispersal.

Recruitment of *Pinus sylvestris* (Pinaceae) in the Sierra Nevada, south-eastern Spain

Overall, shrubs appear to be the best place for pine recruitment, at least through the sapling stage (Castro *et al.*, 2004). Based on experimental and observational data, seedling emergence and seedling and sapling survival were all far greater beneath shrub canopies than in open interspaces, demonstrating a net facilitative effect of shrubs on emergence and survival, at least during the years of this study. For the few survivors, however, seedling growth was greater in open interspaces than beneath shrubs, demonstrating net interference of growth by shrubs and the existence of a life-stage conflict. Although it seems that the survival advantage outweighs the growth disadvantage in the shrub microhabitat, the long-term fitness consequences of increased growth in the open are not known.

Recruitment of *Cryptantha flava* (Boraginaceae) in eastern Utah, USA

This example, summarized from Peek and Forseth (2003) and Casper (1994, 1996), shows even greater complexities. Mortality of *Cryptantha* was always greater in open interspaces than beneath shrubs, but was only 1.2-fold greater in a less stressful year and 2.3-fold greater in a drier year. Thus, although there was net facilitation of survival in both years, the strength of facilitation varied greatly with annual climatic variation. In contrast, in the wetter year, plants in open interspaces and beneath shrubs were the same size, but in the drier year they were smaller beneath shrubs. Thus, with increasing stress there was a suggestion of a switch from a net neutral effect of shrubs on growth to a net interference effect, contrary to expectations. To further complicate the picture, larger plants may be more susceptible to future drought. Lastly, plants beneath shrubs had less leaf N% but appeared to live longer and have reduced reproduction. Despite the many documented significant microhabitat effects, the natural distribution of *Cryptantha* was independent of microhabitats, suggesting that the effects might cancel each other out. These results suggest that microhabitat might have few demographic consequences in this system, but might still have considerable consequences for individual fitness.

Recruitment of *Cercocarpus ledifolius* (Rosaceae) in northern Utah, USA

The last example, from Russell and Schupp (1998), Schultz *et al.* (1996) and Ibáñez and Schupp (2001, 2002), again reveals a complex web of effects influencing recruitment, obscuring any simple attempt to determine the consequences of patterns of dispersal. Results of a 2-year experiment manipulating water, herbivory and microhabitat demonstrated that a reduction in climatic stress can switch a net facilitative effect to a net interference effect. It appears that, in most years, seedling survival over the first summer will be substantially greater beneath adult *C. ledifolius* trees than beneath shrubs or in open interspaces, but that in unusually wet years survival will be substantially greater in open interspaces than beneath shrubs or trees. Interestingly, the three-way microhabitat × water supplementation × herbivore protection interaction had the greatest impact on patterns of recruitment, suggesting an interaction between stressors, as in Riginos *et al.* (2005); simultaneously relieving water stress and herbivory had a much greater impact on patterns of survival than expected from the response to one stress at a time. Life-stage conflicts were also evident in *C. ledifolius* recruitment. *Cercocarpus ledifolius* litter decreased seedling emergence but increased seedling survival. As reported in Battaglia *et al.* (2000), conflicts may even switch on and off with changing environmental conditions. Because seed predation was greater in open interspaces than beneath shrubs or trees, there is the potential for a switch from no conflict most years to a strong seed–seedling conflict in unusually wet years, with greater seed mortality but greater seedling survival in the open. Lastly, observational data strongly suggest that even though tree microhabitats appear to be most favourable for initial recruitment in most years and open microhabitats appear to be most favourable in the rare, unusually wet years, it is the shrub microhabitat that might ultimately be most favourable for new adult recruitment as competition with trees increases over time as recruits grow (see Miriti, 2006) and periodic drought eliminates plants in the open. With a little literary license, the '*Tortoise and the Hare*' may hold an important lesson for plant ecologists; minimal but steady advantages may in many cases ultimately outweigh inconsistent bouts of substantial advantages. But as noted by A.J. Dennis (Australia, 2006, personal communication), sometimes in the real world the hare wakes up and wins the race.

Implications for the Evolution of Dispersal Systems

The realization that the suitability of a given type of site for recruitment is frequently context-dependent has implications for the evolution of specialist versus generalist seed dispersal strategies. If the most suitable site type for recruitment can vary from year-to-year or place-to-place, selection for directed dispersal (*sensu* Howe and Smallwood, 1982) by specialists that preferentially deposit seeds in the 'most suitable sites' for recruitment should be reduced, simply because the 'most suitable sites' are neither constant nor predictable. The highest quality dispersal may be provided by one disperser species or disperser functional group (see Dennis and Westcott, Chapter 9, this volume) in some years and places but another disperser species or disperser functional group in other years and places, resulting in fluctuating selection favouring the maintenance of diffuse disperser networks that widely disperse seeds. Even rare spatial and temporal switches in the identity of the highest-quality disperser could dilute selection for specialization, especially in long-lived plants where the occasional individual that recruits under 'atypical' conditions may survive and contribute to the population for decades, centuries, or even millennia.

Implications for the Design and Interpretation of Research

Although quantifying or generalizing about the quality of seed dispersal is far more complex than we have generally acknowledged, it is not a hopeless endeavour. Extreme context-dependence of site suitability can easily turn a simple exercise into a challenge, but not an insurmountable challenge. It is critical to accept this challenge in both the design of experimental studies and the interpretation of results.

The major lesson is simply a strong reinforcement of what we already know. A study on site suitability conducted on a single life history stage in a single year at a single site with limited environmental variation is going to yield only very limited ecological insight. For example, in a constantly changing world, what you observe one year may be very different from what you will observe in another (e.g. Adler *et al.*, 2006). While such results have intrinsic value, we must be cautious with interpretations and generalizations. And we must begin to design studies that can tell us more.

There are also some specific lessons that can help us design better studies, including, but not limited to:

- *Expand the range of life history stages studied.* Seed survival studies can be extremely valuable, but if combined with studies of seedling growth and survival they can be far more valuable. Also they may give a very different answer to the question 'What is the most suitable site for recruitment?'
- *Replicate experiments over a range of environmental conditions* relevant to the species of interest. The range explored will depend on the ecological system, but possibilities include: across elevational or latitudinal gradients representing climatic variation, across herbivory gradients, across distinct soil types, or perhaps even across communities. Such an approach can greatly help us to understand how sensitive results are to spatial environmental variation. Are results fairly consistent, or are they highly context-dependent?
- *Similarly, replicate experiments across years.* Three are better than two, four are better than three, but any replication is better than none. As noted above, results may vary greatly from year-to-year, but they may also be fairly consistent. We need to know how sensitive results are to temporal environmental variation.
- *Experimentally manipulate the environment.* With a combination of rain-out shelters and water supplementation experiments, we can more easily

explore the responses of a system to a wide range of precipitation. If herbivory is a key component of the system, use a series of exclosures to selectively restrict access to particular grazers or design a realistic clipping experiment to simulate different levels or types of herbivory. You can alter the temperature, manipulate competitors, and more. The key is not to haphazardly manipulate anything you can think of, but to understand your system sufficiently to know the most likely critical variables and how they should be manipulated in order to obtain a broad yet realistic insight.

Recognize the difficulties and be optimistic. No study will ever be perfect. But we can still strive to make our work as complete and as meaningful as possible.

Acknowledgements

The ideas presented in this chapter have been influenced by my interactions, at various levels, with many students, postdoctoral fellows, and other colleagues, especially, and in no particular order, S. Russell, I. Ibáñez, P. Jordano, J.M. Gómez, M. Fuentes, J. Boettinger, P.-M. Forget, J. Chambers, S. Vander Wall, R.M. Callaway, T. Milleron, and many others. Thanks to all for the productive working relations, discussions and general inspiration. And thanks to A.J. Dennis for his encouragement to think in more detail about the evolutionary implications of context-dependence. The preparation of this chapter has been supported by the Ecology Center and the Utah Agricultural Experiment Station (UAES), Utah State University, Logan, Utah, USA (approved as UAES Journal Paper No. 7772). Travel to Australia to participate in the Symposium/Workshop was partially supported by the National Science Foundation (OISE No. 0456402 to E.W.S.).

References

Adler, P.B., HilleRisLambers, J., Kyriakidis, P.C., Guan, Q. and Levine, J.M. (2006) Climate variability has a stabilizing effect on the coexistence of prairie grasses. *Proceedings of the National Academy of Sciences USA* 103, 12793–12798.

Battaglia, L.L., Fore, S.A. and Sharitz, R.R. (2000) Seedling emergence, survival and size in relation to light and water availability in two bottomland hardwood species. *Journal of Ecology* 88, 1041–1050.

Bertness, M.D. and Callaway, R.M. (1994) Positive interactions in communities. *Trends in Ecology and Evolution* 9, 191–193.

Bertness, M.D. and Ewanchuk, P.J. (2002) Latitudinal and climate-driven variation in the strength of biological interactions in New England salt marshes. *Oecologia* 132, 392–401.

Brooker, R.W. and Callaghan, T.V. (1998) The balance between positive and negative plant interactions and its relationship to environmental gradients: a model. *Oikos* 81, 196–207.

Bruno, J.F., Stachowicz, J.J. and Bertness, M.D. (2003) Inclusion of facilitation into ecological theory. *Trends in Ecology and Evolution* 18, 119–125.

Callaway, R.M. and Pugnaire, F.I. (1999) Facilitation in plant communities. In: Pugnaire, F.I. and Valladares, F. (eds) *Handbook of Functional Plant Ecology*. Marcel Dekker, New York, pp. 623–648.

Callaway, R.M., Pennings, S.C. and Richards, C.L. (2003) Phenotypic plasticity and interactions among plants. *Ecology* 84, 1115–1128.

Casper, B.B. (1994) Postdispersal seedling competition and the evolution of single-seededness in *Cryptantha flava*. *Evolution* 48, 1377–1382.

Casper, B.B. (1996) Demographic consequences of drought in the herbaceous perennial *Cryptantha flava*: effects of density, associations with shrubs, and plant size. *Oecologia* 106, 144–152.

Castro, J., Zamora, R., Hódar, J.A. and Gómez, J.M. (2004) Seedling establishment of a boreal tree species (*Pinus sylvestris*) at its southernmost distribution limit: consequences of being in a marginal Mediterranean habitat. *Journal of Ecology* 92, 266–277.

Fowler, N.L. (2002) The joint effects of grazing, competition, and topographic position on six savanna grasses. *Ecology* 83, 2477–2488.

Goldberg, D.E., Rajaniemi, T., Gurevitch, J. and Stewart-Oaten, A. (1999) Empirical approaches to quantifying interaction intensity: competition and facilitation along productivity gradients. *Ecology* 80, 1118–1131.

Gómez, J.M. (2004) Importance of microhabitat and acorn burial on *Quercus ilex* early recruitment: non-additive effects on multiple demographic processes. *Plant Ecology* 172, 287–297.

Gómez-Aparicio, L., Zamora, R., Gómez, J.M., Hodar, J.A., Castro, J. and Baraza, E. (2004) Applying plant facilitation to forest restoration: a meta-analysis of the use of shrubs as nurse plants. *Ecological Applications* 14, 1128–1138.

Gómez-Aparicio, L., Gómez, J.M., Zamora, R. and Boettinger, J.L. (2005) Canopy vs. soil effects of shrubs facilitating tree seedlings in Mediterranean montane ecosystems. *Journal of Vegetation Science* 16, 191–198.

Greenlee, J.T. and Callaway, R.M. (1996) Abiotic stress and the relative importance of interference and facilitation in montane bunchgrass communities in western Montana. *American Naturalist* 148, 386–396.

Harper, J.L. (1977) *Population Biology of Plants*. Academic Press, London.

Hastwell, G.T. and Facelli, J.M. (2003) Differing effects of shade-induced facilitation on growth and survival during the establishment of a chenopod shrub. *Journal of Ecology* 91, 941–950.

Holzapfel, C. and Mahall, B.E. (1999) Bidirectional facilitation and interference between shrubs and annuals in the Mojave Desert. *Ecology* 80, 1747–1761.

Howe, H.F. and Smallwood, J. (1982) Ecology of seed dispersal. *Annual Review of Ecology and Systematics* 13, 201–228.

Hunter, A.F. and Aarssen, L.W. (1988) Plants helping plants. *Bioscience* 38, 34–40.

Ibáñez, I. and Schupp, E.W. (2001) Positive and negative interactions between environmental conditions affecting *Cercocarpus ledifolius* seedling survival. *Oecologia* 129, 543–550.

Ibáñez, I. and Schupp, E.W. (2002) Effects of litter, soil surface conditions, and microhabitat on *Cercocarpus ledifolius* Nutt. seedling emergence and establishment. *Journal of Arid Environments* 52, 209–221.

Kitajima, K. and Augspurger, C.K. (1989) Seed and seedling ecology of a monocarpic tropical tree, *Tachigalia versicolor*. *Ecology* 70, 1102–1114.

Lloret, F., Peñuelas, J. and Estiarte, M. (2005) Effects of vegetation canopy and climate on seedling establishment in Mediterranean shrubland. *Journal of Vegetation Science* 16, 67–76.

Lortie, C.J. and Callaway, R.M. (2006) Re-analysis of meta-analysis: support for the stress-gradient hypothesis. *Journal of Ecology* 94, 7–16.

Maestre, F.T. and Cortina, J. (2004) Do positive interactions increase with abiotic stress? A test from a semi-arid steppe. *Proceedings of the Royal Society of London B* 271 (Suppl.), S331–S333.

Maestre, F.T., Bautista, S. and Cortina, J. (2003) Positive, negative, and net effects in grass–shrub interactions in Mediterranean semiarid grasslands. *Ecology* 84, 3186–3197.

Maestre, F.T., Vallardes, F. and Reynolds, J.F. (2005) Is the change of plant–plant interactions with abiotic stress predictable? A meta-analysis of field results in arid environments. *Journal of Ecology* 93, 748–757.

Maestre, F.T., Vallardes, F. and Reynolds, J.F. (2006) The stress-gradient hypothesis does not fit all relationships between plant–plant interactions and abiotic stress: further insights from arid environments. *Journal of Ecology* 94, 17–22.

Miriti, M.N. (2006) Ontogenetic shift from facilitation to competition in a desert shrub. *Journal of Ecology* 94, 973–979.

Nara, K. and Hogetsu, T. (2004) Ectomycorrhizal fungi on established shrubs facilitate subsequent seedling establishment of successional plant species. *Ecology* 85, 1700–1707.

Peek, M.S. and Forseth, I.N. (2003) Microhabitat dependent responses to resource pulses in the aridland perennial, *Cryptantha flava*. *Journal of Ecology* 91, 457–466.

Pugnaire, F.I., Armas, C. and Valladares, F. (2004) Soil as a mediator in plant–plant interactions in a semi-arid community. *Journal of Vegetation Science* 15, 85–92

Riginos, C., Milton, S.J. and Wiegand, T. (2005) Context-dependent interactions between adult shrubs and seedlings in a semi-arid shrubland. *Journal of Vegetation Science* 16, 331–340.

Russell, S.K. and Schupp, E.W. (1998) Effects of microhabitat patchiness on patterns of seed dispersal and seed predation of *Cercocarpus ledifolius* (Rosaceae). *Oikos* 81, 434–443.

Schultz, B.W., Tausch, R.J. and Tueller, P.T. (1996) Spatial relationships among young *Cercocarpus ledifolius* (curlleaf mountain mahogany). *Great Basin Naturalist* 56, 261–266.

Schupp, E.W. (1993) Quantity, quality and the effectiveness of seed dispersal by animals. *Vegetatio* 107/108, 15–29.

Schupp, E.W. (1995) Seed–seedling conflicts, habitat choice and patterns of plant recruitment. *American Journal of Botany* 82, 399–409.

Schupp, E.W. and Fuentes, M. (1995) Spatial patterns of seed dispersal and the unification of plant population ecology. *EcoScience* 2, 267–275.

Silman, M.R., Terborgh, J.W and Kiltie, R.A. (2003) Population regulation of a dominant rain forest tree by a major seed predator. *Ecology* 84, 431–438.

Tewksbury, J.J. and Lloyd, J.D. (2001) Positive interactions under nurse-plants: spatial scale, stress gradients and benefactor size. *Oecologia* 127, 425–434.

Tielbörger, K. and Kadmon, R. (2000) Temporal environmental variation tips the balance between facilitation and interference in desert plants. *Ecology* 81, 1544–1553.

Welden, C.W. and Slauson, W.L. (1986) The intensity of competition versus its importance: an overlooked distinction and some implications. *Quarterly Review of Biology* 61, 23–44.

21 Mycorrhizal Plants and Vertebrate Seed and Spore Dispersal: Incorporating Mycorrhizas into the Seed Dispersal Paradigm

T.C. THEIMER AND C.A. GEHRING

Introduction

Animals may positively influence the recruitment of mycorrhizal plants, not only by acting directly as seed dispersers, moving seeds to locations where mycorrhizal inoculum is present, but also by acting indirectly as spore dispersers of the symbiotic mycorrhizal fungi that colonize these plants' roots. Many of the plant families which use vertebrates as important seed dispersers are mycorrhizal (e.g. Fagaceae, Pinaceae, Juglandaceae, Lauraceae), and most plants are capable of forming some sort of mycorrhizal relationship (Trappe, 1987). Given that mycorrhizas can significantly affect plant establishment and performance (Smith and Read, 1997), the way in which seed placement affects the potential for that symbiosis to develop may be an important, albeit often overlooked, component of the qualitative aspect of disperser effectiveness (*sensu* Schupp, 1993). In addition, numerous vertebrate animals, ranging from marsupial potoroos (Claridge *et al.*, 1992) to elephants (Paugy *et al.*, 2004) to some terrestrial birds (Reddell *et al.*, 1997) and many rodents (Maser and Maser, 1988; Janos *et al.*, 1995; Reddell *et al.*, 1997; Terwilliger and Pastor, 1999; Mangan and Adler, 2000, 2002; Pyare and Longland, 2001; Colgan and Claridge, 2002) have been reported to deposit viable mycorrhizal spores in their faeces. In spite of the potential for these animals to affect the plant–mycorrhiza symbiosis, the relative importance of vertebrate spore dispersal, from the plants' viewpoint, remains largely unknown. In a review of mammal spore dispersal published 10 years ago, Johnson (1996) pointed out that no study had yet linked animal-dispersed spores to mycorrhizal infection of plants in the field, and the same holds true today.

The importance of vertebrate seed and spore dispersal to mycorrhizal plants will depend on the types of plants and fungi forming the symbiosis, how dependent the host plants are on their fungal symbionts, and how rare the opportunities for obtaining those symbionts are. Generally, mycorrhizal fungi are divided into aseptate fungi in the Glomeromycota (arbuscular mycorrhizas (AM)) and septate fungi in the Ascomycetes and Basidiomycetes (including ectomycorrhiza (EM), ectendomycorrhiza, arbutoid, monotropoid, ericoid and orchid mycorrhizas) (Smith and Read, 1997). Arbuscular mycorrhizas and ectomycorrhizas are the most widely distributed and studied types of mycorrhizal fungi, so we will focus on these. Arbuscular mycorrhizas are associated with a wide array of plant species, and are found in most herbaceous plants and most tropical and some temperate tree species. Ectomycorrhizas are symbionts of a smaller range of host-plant families, almost all of which are woody perennials. In some cases, such as with members of the Fagaceae and Myrtaceae, these trees may be colonized by both AM and EM fungi, either simultaneously or at different life stages. Although the mycorrhizal symbiosis is most often viewed as a mutualism, with the fungi improving host-plant access to soil resources and providing protection from some pathogens in exchange for photosynthate (Smith and Read, 1997), in some cases the relationship may be more dynamic, potentially shifting from mutualistic to parasitic, depending on environmental context (Johnson *et al.*, 1997). Some plants may be obligately dependent on mycorrhizal fungi (the expansion of introduced pines in the southern hemisphere was argued to be limited initially by the absence of an appropriate mycorrhizal inoculum, for example (Richardson *et al.*, 1994)), while other plants appear to be facultatively so; but for a large number of plants the level of mycorrhizal dependency is unknown.

Colonization of plants by both AM and EM fungi can occur through spores or other resistant propagules (hereafter referred to simply as spores) or via colonized roots or hyphal networks in the soil. Arbuscular mycorrhizas form relatively large, asexual spores (10–600 μm in diameter) either singly, in aggregations, or in sporocarps below the soil surface, so that spore dispersal is largely constrained to movement through the soil by biota or water. The potential for vertebrate animals to act as important longer-distance spore dispersers has been suggested in several studies that confirmed the presence of viable AM spores in the faeces of birds and mammals (Maser and Maser, 1988; Claridge *et al.*, 1992; Janos *et al.*, 1995; Reddell *et al.*, 1997; Mangan and Adler, 2000, 2002; Colgan and Claridge, 2002). Although some species in one family of AM fungi (Gigasporaceae) apparently colonize new root systems primarily from spores (Hart and Reader, 2002), those of another family (Glomaceae) are able to colonize plants readily from hyphae, and hyphal production of AM fungi in the soil can be substantial (1 g of grassland soil may contain 100 m of AM hyphae (Miller *et al.*, 1995)). Most EM fungi form spore-producing fungal bodies above-ground (epigeous), where the wind can disperse spores; but approximately one-quarter of EM species form underground fruiting bodies (hypogeous) (Smith and Read, 1997). Many hypogeous fungi produce a strong scent that attracts mycophagous mammals and this, along with other characteristics, has led to the hypothesis that these fungi evolved these traits to increase spore dispersal by mammals (Maser *et al.*, 1978; Johnson, 1996). EM fungi also produce enormous quantities of hyphae, which can function as an important source of EM inoculum for neighbouring plants. In addition, both AM and EM fungi are capable of forming common mycorrhizal networks that link the root systems of individual plants below-ground and may allow for resource exchange (Newman, 1988; Simard *et al.*, 2002). These networks can be critical for seedling establishment in some habitats (Nara, 2006).

Traditionally, the symbiotic relationship between plants and fungi has been considered to be relatively general for AM fungi, due to the range of fungi that often colonize the same species of plant and the enormous diversity of plants that can be colonized by a single species of fungus. Likewise, between 5000 and 6000 species of fungi have been estimated to form EM associations, and many of these apparently form associations with a wide range of host plants (Molina *et al.*, 1992). This general lack of specificity, combined with the widespread occurrence of mycorrhizal plants and the potential for plants to be inoculated from infected plant roots and fungal hyphae as well as from spores, suggests that inoculum would rarely be limited, and therefore animal dispersal of seeds or spores would be important only in a limited range of circumstances. However, recent research has shown preferential associations between certain plant and fungal taxa (e.g. Weber *et al.*, 2005) and demonstrated circumstances in which inoculum is limited (e.g. Lovelock *et al.*, 2003), challenging these traditional views of the mycorrhizal symbiosis in ways that increase the potential for animal seed and spore dispersal to have important effects on plant recruitment and performance.

In this chapter, we first describe the different ways in which vertebrate dispersers could potentially interact with spores and seeds in order to have favourable impacts on mycorrhizal plants. We then review the recent literature to explore the conditions under which mycorrhizal inoculum might be limited enough to make vertebrate seed and spore dispersal important to mycorrhizal plants. We conclude by suggesting how current and future studies of vertebrate seed dispersal might incorporate mycorrhizal fungi into the potential benefits that plants derive from the actions of dispersers. Although invertebrate animals may also have significant impacts on mycorrhizal spore dispersal (e.g. Lilleskov and Bruns, 2005), and some of the ideas presented here may apply to them, our focus throughout is on vertebrate animals and, within that, primarily on birds and mammals.

Roles of Vertebrate Seed and Spore Dispersers and Their Potential Impact on Mycorrhizal Plants

In seed dispersal systems, disperser effectiveness has been viewed in terms of both quantity of seeds dispersed and the quality of treatment provided to each dispersed seed; a major determinant of the quality of dispersal is the suitability of the site at which the seed arrives (Schupp, 1993; see

Schupp, Chapter 20, this volume). For mycorrhizal plants, the qualitative component of dispersal effectiveness could depend on whether the microsite where seeds arrive contains the appropriate mycorrhizal symbionts. In this context, the role of any one species of animal could be described in two ways: first, in terms of the propagules it disperses and, second, in terms of the microsite to which it delivers those propagules. Species that carry seeds but not inoculum would act solely as *seed dispersers*, those that carry inoculum but not seeds would act as what we term *spore dispersers*, and those that transport both seeds and inoculum would act as what we term *dual dispersers* (see Fig. 21.1). The relative importance of these behaviours to plants depends on the microsite into which seeds and inoculum are dispersed (Fig. 21.1). If microsites containing the appropriate inoculum are relatively rare, then a disproportionate dispersal of seeds into these sites would be a form of *directed seed dispersal* (*sensu* Howe and Smallwood, 1982). Note that directed seed dispersal could be carried out by animals that are either seed dispersers or dual dispersers, as defined above, but in the latter case only the dispersal of the seeds they carry is important. For example, a dual disperser could carry both seeds and inoculum to a microsite where inoculum is already present. In terms of plant establishment, the important action of the dual disperser would be that of directed seed dispersal, as the movement of spores by the dual disperser would be unnecessary, given that the microsite already contained the appropriate inoculum.

By analogy with directed seed dispersal, the movement of spores or other sources of inoculum (hyphae or roots) to microsites where host plant seeds arrive by other means but where the appropriate inoculum is lacking could be considered what we term *directed spore dispersal*. This role could be played by any animal that disproportionately deposits inoculum in sites lacking it, including some birds that apparently carry spores, hyphae or other inocula incidentally (Reddell *et al*., 1997), or by a variety of mycophagous vertebrate animals that pass viable spores through the gut (Maser and Maser, 1988; Janos *et al*., 1995; Reddell *et al*., 1997; Mangan and Adler, 2000, 2002; Pyare and Longland, 2001; Colgan and Claridge, 2002). As with directed seed dispersal, directed spore dispersal could be carried out by an animal that was either a spore disperser or a dual disperser.

Some microsites will lack both symbiotic partners, and movement of inoculum and seeds to these areas is a scenario that we term 'complementary seed and spore dispersal', or simply *complementary dispersal*. Complementary dispersal could occur either sequentially, when spores and seeds are deposited at a microsite at different times, or simultaneously. Sequential complementary dispersal could occur either through independent visitation by a spore-dispersing and a seed-dispersing species (a microsite frequented by a spore-dispersing mammal that is also the site of seed deposition by a seed-dispersing bird as a hypothetical example) or by sequential visitation by a dual disperser that deposits seeds and spores at the same site at different times (a scatter-hoarding rodent that buried seeds at one time and defecated spores at another). Complementary dispersal could also occur simultaneously

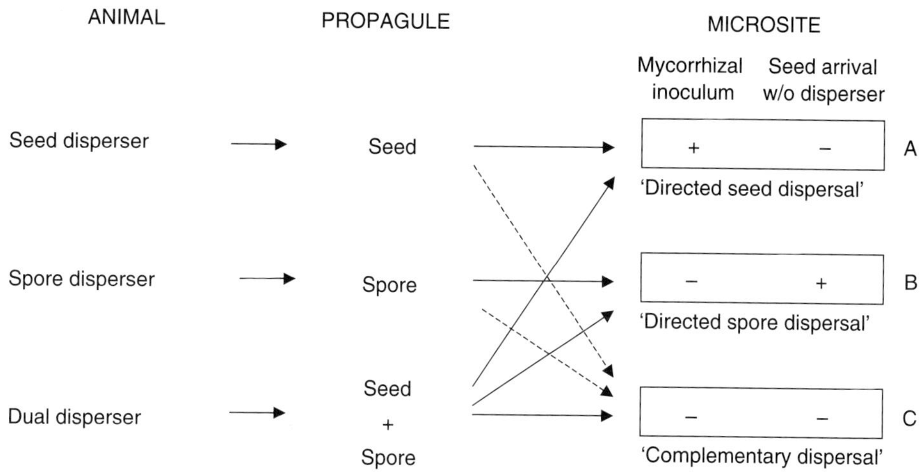

Fig. 21.1. The role of animal dispersers of seeds of a mycorrhizal plant and the spores of that plant's mutualistic mycorrhizal fungi can be viewed in terms of: (i) which propagules are dispersed; (ii) whether the microsite in which those propagules are deposited already has appropriate mycorrhizal inoculum present (+) or not (−); and (iii) whether seeds would arrive at that microsite in the absence of animals (+) or not (−). Different species of animals could act either as seed dispersers by dispersing only seeds, spore dispersers by dispersing only spores, or as dual dispersers by dispersing both seeds and spores. Dispersal of seeds of mycorrhizal plants to microsites that have appropriate mycorrhizal inoculum but where seeds would not arrive in the absence of animal dispersal (Microsite A) would be a form of *directed seed dispersal* and could be achieved by seed dispersers or dual dispersers. Dispersal of viable mycorrhizal spores to microsites lacking inoculum, where appropriate host plant seeds arrive by other means (Microsite B) we term *directed spore dispersal* and could be accomplished by either spore dispersers or dual dispersers. Dispersal of both seeds and spores to microsites lacking both (Microsite C) is a process we term *complementary dispersal* and could be accomplished by a seed disperser and spore disperser acting in conjunction (dotted arrows) or by a dual disperser alone.

if a dual disperser deposited both propagule types at the same site at the same time. We hypothesize that simultaneous complementary dispersal would be most likely in two groups of dispersers. The first group are vertebrates that feed on fungi, and therefore potentially carry fungal spores in the oral cavity, and also carry seeds in their mouths while dispersing them. Most obvious in this respect are scatter-hoarding rodents, some of which could potentially carry both seeds and sporocarps in their cheek pouches, but other animals that eat fungi and scatter-hoard seeds or fruits, such as the musky rat-kangaroo (*Hypsiprymnodon moschatus*; Potoroidae) (Dennis, 2002, 2003), could also act in this way. The second group would be animals that ingest both fungal sporocarps and seeds and then pass viable seeds and viable spores through the digestive tract to be deposited together in the same faecal mass. Large mammals are most likely to play this role, including the white-tailed deer (*Odocoileus virginianus*; Cervidae), which have been shown to disperse a variety of seeds over relatively long distances (Myers *et al.*, 2004) and which have also been implicated as dispersers of mycorrhizal spores (Ashkannejhad and Horton, 2006). Simultaneous complementary dispersal would be more likely when fungal fruiting and seed production coincided. For example, in the south-western US pine forests, the greatest abundance and diversity of sporocarps occurs in the autumn (States and Gaud, 1997), coinciding with seed production of mycorrhizal ponderosa (*Pinus ponderosa*; Pinaceae) and pinyon (*Pinus edulis*; Pinaceae) pines. As a result, scatter-hoarding rodents could potentially interact with fungal fruiting bodies while they were scatter-hoarding seeds. Mycorrhizal mutualisms have been argued to be relatively unstable through evolutionary time, in part due to the lack of a mechanism for linking mycorrhizal and plant-host propagules (Hibbett *et al.*, 2000). The potential role of dual dispersers is intriguing in this regard, as they probably collect spores and seeds from a restricted area, thereby increasing the probability of carrying host-plant seeds and their associated mycorrhizas, and if they deposited seeds and spores simultaneously, they could strengthen the link between symbiotic partners.

The relative importance of simultaneous versus sequential complementary dispersal would depend in part on the specific plant and fungal species interacting, and how long their respective propagules could survive independently. Both EM and AM fungi are highly dependent on plant hosts for energy (Smith and Read, 1997), but mycorrhizal fungal spores may be able to survive in the soil for relatively extended periods (Kjøller and Bruns, 2003) and some species may be resistant to disturbances such as fire (Izzo *et al.*, 2006). Plants vary in their dependency on mycorrhizal fungi, but for those that are dependent, seed size may be an important factor in determining how rapidly colonization must occur. For example, plants that produce seedlings with small seed stores and/or high nutrient requirements may require relatively rapid mycorrhizal colonization for survival, and therefore a tight temporal coupling of spore and seed delivery; while seedlings with larger seeds and lower nutrient demands may not (Allsopp and Stock, 1995; Siqueira *et al.*, 1998). As a result of this variation in plant dependency and spore longevity, sequential dispersal could result in effective establishment of the symbiosis in some cases, while in others simultaneous deposition would be necessary.

Conditions Potentially Limiting Inoculum and Increasing the Importance of Animal Seed and Spore Dispersal

The importance of directed seed dispersal, directed spore dispersal and complementary dispersal depends upon the background availability of mycorrhizal inoculum in the sites where animals deposit seeds and spores. Mycorrhizal inoculum could be limiting for a plant due to the complete absence of any mycorrhizal inoculum, the absence of the appropriate major type of inoculum (e.g. AM inoculum required by an AM plant is absent even though EM inoculum is present), or the absence of specific mycorrhizal species that are most beneficial to that plant. In each case, limitation may be brought about through a variety of natural and anthropogenic causes (Fig. 21.2). A complete absence of inoculum would require that successful establishment of the mutualism would be coupled with directed spore dispersal or complementary dispersal, whereas patchy distribution of fungal inoculum could be overcome by either of these two processes or through directed seed dispersal into existing inoculum patches (Fig. 21.2).

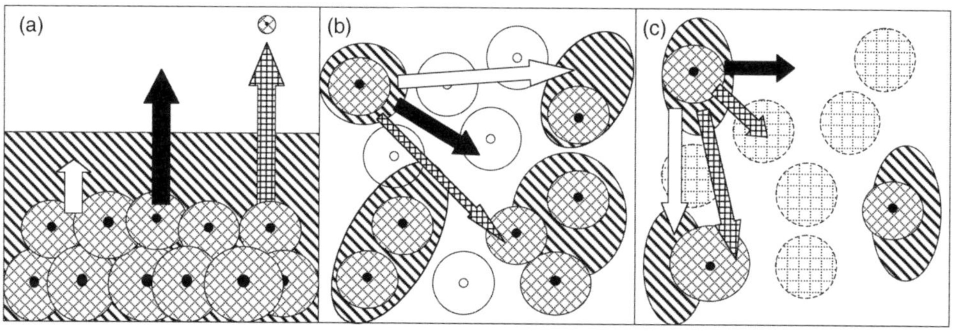

Fig. 21.2. Three hypothetical conditions under which mycorrhizal inoculum could be limiting and therefore directed seed dispersal, directed spore dispersal or complementary dispersal are potentially more important. Small circles represent plant stems and the larger circles around them represent the root and/or hyphal network surrounding each stem. Plants can be of two types, mycorrhizal (black circles with hatched root/hyphal network) or non-mycorrhizal (open circles, open root networks). Diagonally hatched areas represent locations where seeds would arrive in the absence of animal seed dispersers, while open areas represent locations lacking both inoculum and seed input in the absence of animals. (a) An area lacking any mycorrhizal inoculum, as might occur under primary succession, would require directed spore dispersal from existing sources on the periphery of the uncolonized area into areas where seeds would arrive without animal seed dispersal (open arrow) or complementary dispersal to areas where neither inoculum nor seeds arrive by other means (solid arrow). Once a mycorrhizal plant colonized the open area, seeds carried to the vicinity of that plant could be a form of directed seed dispersal to a microsite with inoculum already present (hatched arrow). (b) Plant assemblages containing both mycorrhizal and non-mycorrhizal plant species could have inoculum 'gaps' in areas dominated by root systems of non-mycorrhizal plants. Inoculum limitation again could be overcome through directed spore dispersal to areas lacking inoculum but where seeds arrive without animal dispersal (open arrow) or complementary dispersal of seeds and spores to areas lacking both (solid arrow). Directed seed dispersal could be important if seeds are selectively placed in areas with appropriate inoculum but where seeds have a low probability of arriving by other means (hatched arrow). (c) In plant assemblages in which all plants have similar mycorrhizas, inoculum could be limited in areas where thinning and/or fire have removed plants (dashed, hatched circles), destroyed inoculum and/or created gaps in the root/hyphal network (open areas). Inoculum limitation could be overcome through directed spore dispersal to areas where seeds would arrive without animal dispersal (open arrow) or through complementary dispersal to areas lacking inoculum where seeds would be unlikely to arrive without animal dispersal (solid arrow). Directed seed dispersal would be important when seeds were dispersed to areas with appropriate inoculum, either around living trees or in sites where trees had been removed or killed, if inoculum sources in those areas were still viable (hatched arrows).

Mycorrhizal inoculum is most likely to be completely absent in sites undergoing primary succession or recovering from severe disturbance (Fig. 21.2a). For example, volcanic activity can result in large areas devoid of mycorrhizal inoculum (Allen, 1987; Nara, 2006). Animals could affect the rate of colonization of these areas by dispersing spores, seeds, or both. For example, animal transport of mycorrhizal spores has been suggested as being important for the mycorrhizal colonization of volcanic fields of Mount St Helens, USA (Allen and McMahon, 1988), the retreating front of glaciers (Cazares and Trappe, 1994), and sand dune habitats (Ashkannejhad and Horton, 2006). To our knowledge, the potential for animals to facilitate the initial colonization of such sites through simultaneous complementary dispersal has never been investigated, although it would seem to be a potentially important mechanism. Once areas have been colonized by isolated mycorrhizal plants, directed seed dispersal by animals to the vicinity of these plants could encourage further colonization. The plant seeds dispersed into these inoculum patches need not be of the same species as the established plant in order to realize the benefit of directed seed dispersal, as long as the seed was from a mycorrhiza-dependent plant compatible with the fungal inoculum present. For example, Nara and Hogetsu (2004) reported that the successful establishment by three species of trees on volcanic areas of Mt Fuji was facilitated by the presence of an initial species of tree that colonized the area with mycorrhizal inoculum appropriate for all three species.

Many ecosystems contain a mixture of plant species varying in the presence or type of mycorrhizal associations they form, resulting in patchy availability of mycorrhizal inoculum (Fig. 21.2b) (Kovacic *et al.*, 1984; Weber *et al.*, 2005). For example, in pinyon–juniper woodlands, the species co-dominate and one forms EM associations (pinyon: e.g. *Pinus edulis*; Pinaceae), while the other forms AM associations (juniper: e.g. *Juniperus monosperma*; Cupressaceae). In areas of pinyon–juniper woodland where pinyon pine density was low, less than 10% of pinyon seedlings became colonized by EM fungi (Haskins and Gehring, 2005). Given that, under certain environmental conditions, pinyon pine seedlings have enhanced survival under nurse plants (Drivas and Everett, 1988; Callaway *et al.*, 1996; Chambers, 2001; Mueller *et al.*, 2005; but see Schupp, Chapter 20, this volume) and animal seed dispersers such as rodents often scatter-hoard pinyon seeds near shrubs or trees (Vander Wall, 1997; Hollander and Vander Wall, 2004), these animals could act as a means of directed seed dispersal if they placed seeds under pinyons or other EM plants with compatible mycorrhizal inoculum. If these rodents also transported EM spores they would be dual dispersers, and if they dispersed seeds and spores under juniper canopies they could act as a means of complementary dispersal. Simultaneous complementary dispersal, perhaps via spores and seeds carried together in rodents' cheek pouches, would be the most effective means of establishing the symbiosis in areas lacking the appropriate inoculum.

Within either of the two major types of mycorrhizal fungi (AM versus EM), the mycorrhizal mutualism has traditionally been viewed as having relatively low host specificity; however, recent work has shown that in some cases the species of fungus is important as well (Kiers *et al.*, 2000; Lovelock *et al.*, 2003). As a result, if mycorrhizal inocula are present in microhabitats that seed-dispersing animals favour, or if the spores they transport differ from other fungal species in their relative benefit to plants, then the species of fungus that animals interact with becomes crucially important. Although thousands of species of fungi form EM associations with plants (Molina *et al.*, 1992), the biology of only a small fraction of these fungi is known in detail. It is clear, however, that species of EM fungi vary in their environmental tolerance (e.g. Swaty *et al.*, 2004), energetic costs to the host plant (e.g. Durall *et al.*, 1994), and the type of resources they take up from the soil, and thus the benefits they provide (e.g. Finlay *et al.*, 1992). Furthermore, some EM taxa are recognized as host specialists and others as host generalists (Molina *et al.*, 1992), and in some cases EM diversity has been positively associated with plant nutrient uptake and productivity (Baxter and Dighton, 2001; Jonsson *et al.*, 2001). Arbuscular mycorrhizal fungi are taxonomically less diverse than EM fungi, with approximately 160 described species that show little apparent host specificity. However, recent studies indicate that different AM fungus isolates are more beneficial to some host species than to others (see, e.g. Gange *et al.*, 1993; Kiers *et al.*, 2000). Furthermore, AM fungal communities respond significantly to host-plant species identity (e.g. Sieverding, 1989; Johnson *et al.*, 1992; Bever *et al.*, 1996; Lovelock *et al.*, 2003).

Thus, although plants may have access to abundant AM or EM inoculum, the diversity or species composition of this inoculum could vary significantly across the landscape, leading to heterogeneity in the suitability of sites for mycorrhizal infection and seedling success. The increasing evidence that fungus identity and fungal diversity may impact plant performance suggests that directed dispersal of seeds or spores by animals may be important, not only in providing plants access to the appropriate type of inoculum (e.g. AM versus EM), but also by influencing the diversity and species composition of the inoculum that plants encounter. When we assessed spore diversity and soil inoculum potential inside and outside vertebrate exclosures in a tropical rainforest in northern Queensland, Australia, we found significantly reduced spore diversity in areas from which terrestrial vertebrates had been excluded for 3 years compared with nearby open plots (Gehring *et al.*, 2002). The spatial distribution of AM fungi such as *Glomus rubiforme* (Glomaceae), which form sporocarps attractive to mycophagus animals, was significantly more restricted when terrestrial vertebrates were excluded; a pattern not evident for AM species that did not produce sporocarps, further implicating the importance of spore dispersal by mycophagous mammals. The reduced spore diversity and abundance in exclosures translated into reduced root colonization of plants grown in exclosure soil cores under glasshouse conditions (Gehring *et al.*, 2002) and in seedlings germinating naturally in exclosures in the field (T.C. Theimer and C.A. Gehring, 2005, unpublished results), indicating that the remaining sources of inoculum were not sufficient to make up for the losses caused by terrestrial vertebrate exclusion. One potential reason for this effect may have been that our Australian rainforest plot contained significant numbers of tree species that were non-mycorrhizal (Gehring and Connell, 2006). Such a system would exhibit a patchier distribution of fungal inocula in the absence of vertebrate spore dispersal, as areas dominated by non-mycorrhizal species would create gaps in the fungal hyphal network (Fig. 21.2b).

Management practices that alter the relative abundance of mycorrhizal fruiting bodies, roots and hyphae might also have the potential to change the importance of seed and spore dispersal by animals (Fig. 21.2c). In several important EM-dominated forests, thinning and burning of tree stands potentially alters both rooting density (and therefore inoculum potential) and diversity and abundance of mycorrhizal types. For example, Lazaruk *et al.* (2005) found that *Picea glauca* (Pinaceae) stands subjected to partial tree cuts had EM colonization and richness that did not differ from unharvested stands, while clear-cut and burned areas were lower in both. Likewise, stand-replacing wildfire can potentially destroy inoculum of some mycorrhizal types (e.g. Bruns *et al.*, 2002), while prescribed fire can result in decreased mycorrhizal biomass and potentially reduce inoculum potential or alter species composition (e.g. Stendell *et al.*, 1999; Tuininga and Dighton, 2004; Hart *et al.*, 2005). In the light of these studies, directed spore dispersal and complementary dispersal may be more important in systems where fire and tree thinning negatively affect mycorrhizal inoculum potential.

Linking Mycorrhizal Spore and Seed Dispersal Studies: Challenges and Opportunities

Currently, no studies that we are aware of have evaluated the potential for mycorrhizal inoculum to be one of the benefits derived from directed dispersal of seeds to specific microsites. Likewise, studies of mycorrhizal spore dispersal by vertebrates have focused primarily on identifying the spores carried in faeces and their ability to inoculate plants after gut passage. Although these studies have documented the potential for animals both to disperse viable spores and to vary in their potential effectiveness as spore dispersers (e.g. Pyare and Longland, 2001), they have not elucidated whether that dispersal is really important to plants or fungi. We argue that future seed and spore dispersal studies could help to answer that question by considering the following critical aspects of the plant–animal–fungus interaction.

As has been stressed throughout this chapter, a critical step in determining whether animal seed and spore dispersal is important to plants is to determine whether appropriate inoculum is limiting. Inoculum limitation could be investigated by assessing mycorrhizal inoculum potential across microsites where seeds arrive either by animal dispersal or through other mechanisms. Although removing soil cores from selected microsites and growing plants in these cores under glasshouse conditions may be an efficient first step, soil coring could sever hyphae and roots that may act as sources of inoculation under field conditions. As a result, the planting of seeds or seedlings to selected microsites would more effectively assess the natural probability of inoculation. Planting to microsites in the field is commonly used to assess disperser effectiveness in seed dispersal studies and examining these seedlings for mycorrhizal colonization would be a valuable addition.

Assessing the potential of microsites for inoculation may be confounded if animal spore dispersers maintain inoculum in some microsites where it would normally be lacking. As a result, determining the importance of animals as spore dispersers may require exclosure experiments that can tease apart the relative contribution of animals in maintaining inoculum potential. In many cases, experimental designs used to investigate the importance of animals as seed dispersers could also be used to address their importance as spore dispersers, if plants grown in animal-proof exclosures were monitored for mycorrhizal colonization.

Given the increasing evidence that host specificity may be important in mycorrhizal interactions as cited above, inoculum limitation need not require the complete absence of inoculum for animal dispersal to be important. A critical question is whether the benefits that a plant receives from a fungus with spores transported by animals, or by inoculum present in microsites where seeds are deposited by animals, are greater than the benefits received from fungi with spores that are naturally more widely distributed. Initial steps along this research pathway could include testing plant responses to fungal spores demonstrated to be transported by vertebrates (often hypogeous EM fungi or sporocarpic AM fungi) versus plant response to spores from sympatric fungi that would be predicted to disperse spores relatively broadly without dispersal by vertebrate animals (epigeous EM fungi or non-sporocarpic AM fungi).

Any study assessing the importance of animals as agents of mycorrhizal inoculation should acknowledge that the benefit of mycorrhizal fungi to plants, and therefore the importance of directed seed or spore dispersal or complementary dispersal, may vary among plants and with environmental context (e.g. Jonsson *et al.*, 2001; see Schupp Chapter 20, this volume). The cost to the plant of maintaining fungal mutualists is primarily a loss of photosynthate, while benefits include enhanced mineral and water uptake and protection from pathogens. As a result, plants in environments with abundant nutrients and water, or in low-light conditions where photosynthate is limited and the demand for nutrients and water are minimal, may receive little net benefit from the mutualism (e.g. Son and Smith, 1988; Gehring, 2003). Under extreme conditions, the association has been argued to shift more toward parasitism than mutualism (e.g. Johnson *et al.*, 1997). In contrast, conditions that increase the demand for water and nutrients, such as high-light environments, increased competition among plants, or drought, could make mycorrhizal colonization more beneficial and therefore increase the importance of animal dispersers. Thus, mycorrhizal inoculation, like other microsite characteristics, may have very different impacts on plants, depending on a suite of environmental conditions that vary in space and time (see Schupp, Chapter 20, this volume).

Conclusions

Animals can impact on the symbiosis between plants and mycorrhizal fungi by acting as seed dispersers, spore dispersers or dual dispersers of both seeds and spores. Animals acting in one of these roles could positively affect the recruitment of mycorrhizal plants: (i) through *directed seed dispersal* to microsites containing the appropriate mycorrhizal inoculum; (ii) through *directed spore dispersal* to microsites lacking inoculum but where seeds arrive by other means; or (iii) by *complementary dispersal* of both seeds and spores to microsites lacking both. Although none of these roles have been investigated in detail in terms of how they affect development of the mycorrhizal symbiosis, especially intriguing is the potential for dual dispersers to tightly link the dispersal of seeds and mycorrhizal fungi through simultaneous deposition of both propagules. Likewise, animal dispersal of spores and seeds to microsites where neither would arrive by other means (complementary dispersal) may represent an important, but under-investigated, means of establishing the symbiotic relationship. Critical to a fundamental role for animal dispersal in the plant–fungus interaction is whether appropriate mycorrhizal inoculum is limiting, and we suggest that areas in which inoculum is reduced due to natural or anthropogenic causes represent prime sites for studying the importance of animal seed and spore dispersal. Finally, several fundamental questions could be addressed by modifying current or future seed dispersal studies to assess:

1. The probability of mycorrhizal infection in the absence of animal dispersers;
2. The relative benefits received from animal-influenced inoculum versus other sources under varying environmental conditions.

The answers to these questions will move the field from the current, initial phase of establishing the potential for strong interaction to a fuller understanding of when and if strong interactions actually exist.

Acknowledgements

W. Longland and an anonymous reviewer provided valuable comments on an earlier draft, E. Schupp greatly improved the conceptual basis and clarity of this chapter. We thank A. Dennis, R. Green, E. Schupp and D. Westcott for organizing the Symposium that stimulated this contribution. Travel of the senior author to Australia to participate in the Symposium/Workshop was partially supported by the National Science Foundation (OISE No. 0456402 to E.W. Schupp). The junior author served at the National Science Foundation during the preparation of this chapter and acknowledges that support.

References

Allen, M.F. (1987) Re-establishment of mycorrhiza on Mount St. Helens: migration vectors. *Transactions of the British Mycological Society* 88, 413–417.

Allen, M.F. and McMahon, J.A. (1988) Direct VA mycorrhizal inoculation of colonizing plants by pocket gophers (*Thommomys talpoides*) on Mount St. Helens. *Mycologia* 80, 754–756.

Allsopp, N. and Stock, W.D. (1995) Relationships between seed reserves, seedling growth and mycorrhizal response on 14 related shrubs (Rosidae) from a low-nutrient environment. *Functional Ecology* 9, 248–254.

Ashkannejhad, S. and Horton, T.H. (2006) Ectomycorrhizal ecology under primary succession: interactions involving *Pinus contorta*, suilloid fungi and deer. *New Phytologist* 169, 345–354.

Baxter, J.W. and Dighton, J. (2001) Ectomycorrhizal diversity alters growth and nutrient acquisition of grey birch (*Betula populifolia*) seedlings in host–symbiont culture conditions. *New Phytologist* 152, 139–149.

Bever, J.D., Morton, J.B., Antonovics, J. and Schultz, P.A. (1996) Host-dependent sporulation and species diversity of arbuscular mycorrhizal fungi in a mown grassland. *Journal of Ecology* 84, 71–82.

Bruns, T.D., Tan, J., Bidartondo, M.I., Szaro, T.M. and Redecker, D. (2002) Survival of *Suillus pungens* and *Amanita francheti* ectomycorrhizal genets was rare or absent after a stand-replacing wildfire. *New Phytologist* 155, 517–523.

Callaway, R.M., DeLucia, E.H., Moore, D., Nowak, R. and Schlesinger, W.H. (1996) Competition and facilitation: contrasting effects of *Artemisia tridentata* on desert vs. montane pines. *Ecology* 77, 2130–2141

Cazares, E. and Trappe, J.M. (1994) Spore dispersal of ectomycorrhizal fungi on a glacier forefront by mammal mycophagy. *Mycologia* 86, 507–510.

Chambers, J.C. (2001) *Pinus monophylla* establishment in an expanding *Pinus–Juniperus* woodland: environmental conditions, facilitation, and interacting factors. *Journal of Vegetation Science* 12, 27–40

Colgan, W. and Claridge, A.W. (2002) Mycorrhizal effectiveness of *Rhizopogon* spores recovered from faecal pellets of small forest dwelling mammals. *Mycological Research* 106, 314–320.

Claridge, A.W., Tanton, M.T., Seebeck, J.H., Cork, S.J. and Cunningham, R.B. (1992) Establishment of ectomycorrhizae on the roots of two species of *Eucalyptus* from fungal spores contained in the faeces of the long-nosed potoroo (*Potorous tridactylus*). *Australian Journal of Ecology* 17, 207–217.

Dennis, A.J. (2002) The diet of the musky rat-kangaroo, *Hypsiprymnodon moschatus*, a rainforest specialist. *Wildlife Research* 29, 209–219.

Dennis, A.J. (2003) Scatterhoarding by musky rat-kangaroos, *Hypsiprymnodon moschatus*, a tropical rain-forest marsupial from Australia: implications for seed dispersal. *Journal of Tropical Ecology* 19, 619–627.

Drivas, E.P. and Everett, R.L. (1988) Water relations characteristics of competing singleleaf pinyon seedlings and sagebrush nurse plants. *Forest Ecology and Management* 23, 27–37.

Durall, D.M., Marshall, J.D., Jones, M.D., Crawford, R. and Trappe, J.M. (1994) Morphological changes and photosynthate allocation in ageing *Hebeloma crustuliniforme* (Bull.) Quel. and *Laccaria bicolor* (Maire) Orton mycorrhizas of *Pinus ponderosa* Dougl. ex Laws. *New Phytologist* 127, 719–724.

Finlay, R.D., Frostegard, A. and Sonnerfeldt, A.M. (1992) Utilization of organic and inorganic nitrogen sources by ectomycorrhizal fungi in pure culture and in symbiosis with *Pinus contorta* Dougl. ex Loud. *New Phytologist* 120, 105–115.

Gange, A.C., Brown, V.K. and Sinclair, G.S. (1993) Vesicular–arbuscular mycorrhizal fungi: a determinant of plant community structure in early succession. *Functional Ecology* 7, 616–622.

Gehring, C.A. (2003) Growth responses to arbuscular mycorrhizas by rain forest seedlings vary with light intensity and tree species. *Plant Ecology* 167, 127–139.

Gehring, C.A. and Connell, J.H. (2006) The occurrence and potential importance of arbuscular mycorrhizal fungi in tree seedlings of two Australian rainforests. *Mycorrhiza* 16, 89–98.

Gehring, C.A., Wolf, J.E. and Theimer, T.C. (2002) Terrestrial vertebrates promote arbuscular mycorrhizal fungal diversity and inoculum potential in a rainforest soil. *Ecology Letters* 5, 540–548.

Hart, M.M. and Reader, R.J. (2002) Taxonomic basis for variation in the colonization strategy of arbuscular mycorrhizal fungi. *New Phytologist* 153, 335–344.

Hart, S.E., Classen, A.T. and Wright, R.J. (2005) Long-term interval burning alters fine root and mycorrhizal dynamics in a ponderosa pine forest. *Journal of Applied Ecology* 42, 752–761.

Haskins, K.E. and Gehring, C.A. (2005) Evidence for mutualist limitation: the impacts of conspecific density on the mycorrhizal inoculum potential of woodland soils. *Oecologia* 145, 123–131.

Hibbett, D.S., Luz-Beatriz, G. and Donoghue, M.S. (2000) Evolutionary instability of ectosymbiosis in basidiomycetes. *Nature* 407, 506–508.

Hollander, J.L. and Vander Wall, S.B. (2004) Effectiveness of six species of rodents as dispersers of single-leaf pinon pine (*Pinus monophylla*). *Oecologia* 138, 57–65.

Horton, T.R. and Bruns, T.D. (1998) Multiple-host fungi are the most frequent and abundant ectomycorrhizal types in a mixed stand of Douglas fir (*Pseudotsuga menziesii*) and bishop pine (*Pinus muricata*). *New Phytologist* 139, 331–339.

Howe, F.H. and Smallwood, J. (1982) Ecology of seed dispersal. *Annual Review of Ecology and Systematics* 13, 210–227.

Izzo, A.D., Canright, M. and Bruns, T.D. (2006) The effects of heat treatments on ectomycorrhizal resistant propagules and their ability to colonize bioassay seedlings. *Mycological Research* 110, 196–202.

Janos, D.P., Sahley, C.T. and Emmons, L.H. (1995) Rodent dispersal of vesicular–arbuscular mycorrhizal fungi in Amazonian Peru. *Ecology* 76, 1852–1858.

Johnson, C.N. (1996) Interactions between mammals and ectomycorrhizal fungi. *Trends in Ecology and Evolution* 11, 503–507.

Johnson, N.C., Tilman, D. and Wedin, D. (1992) Plant and soil controls on mycorrhizal fungal communities. *Ecology* 73, 2034–2042.

Johnson, N.C., Graham, J.H. and Smith, F.A. (1997) Functioning of mycorrhizal associations along the parasitism–mutualism continuum. *New Phytologist* 135, 575–585.

Jonsson, L.M., Nilsson, M.C., Wardle, D.A. and Zackrisson, O. (2001) Context dependent effects of ectomycorrhizal species richness on tree seedling productivity. *Oikos* 93, 353–364.

Kiers, E.T., Lovelock, C.E., Krueger, E.L. and Herre, E.A. (2000) Differential effects of tropical arbuscular mycorrhizal fungal inocula on root colonization and tree seedling growth: implications for tropical forest diversity. *Ecology Letters* 3, 106–113.

Kjøller, R. and Bruns, T.D. (2003) *Rhizopogon* spore bank communities within and among California pine forests. *Mycologia* 95, 603–613.

Kovacic, D.A., St John, T.V. and Dyer, M.I. (1984) Lack of vesicular arbuscular mycorrhizal inoculum in a ponderosa pine forest. *Ecology* 65, 1755–1759.

Lazaruk, L.W., Kernaghan, G., Macdonald, S.E. and Khasa, D. (2005) Effects of partial cutting on the ectomycorrhizae of *Picea glauca* forests in northwestern Alberta. *Canadian Journal of Forest Research* 35, 1442–1454.

Lilleskov, E.A. and Bruns, T.D. (2005) Spore dispersal of a resupinate ectomycorrhizal fungus, *Tomentella sublilacina*, via soil food webs. *Mycologia* 97, 762–769.
Lovelock, C.E., Anderson, K.A. and Morton, J.B. (2003) Arbuscular mycorrhiza communities in tropical forests are affected by host tree species and environment. *Oecologia* 135, 268–279.
Mangan, S.A. and Adler, G.H. (2000) Consumption of arbuscular mycorrhizal fungi by terrestrial and arboreal small mammals in a Panamanian cloud forest. *Journal of Mammalogy* 81, 563–570.
Mangan, S.A. and Adler, G.H. (2002) Seasonal dispersal of arbuscular mycorrhizal fungi by spiny rats in a neotropical rainforest. *Oecologia* 131, 587–597.
Maser, C. and Maser, Z. (1988) Interactions among squirrels, mycorrhizal fungi and coniferous forests in Oregon. *Great Basin Naturalist* 48, 358–369.
Maser, C., Trappe, J.M. and Nussbaum, R.A. (1978) Fungal–small mammal interrelationships with emphasis on Oregon coniferous forests. *Ecology* 59, 799–809.
Miller, R.M., Reinhardt, D.R. and Jastrow, J.D. (1995) External hyphal production of vesicular–arbuscular mycorrhizal fungi in pasture and tallgrass prairie communities. *Oecologia* 103,17–23
Molina, R., Massicotte, H. and Trappe, J.M. (1992) Specificity phenomenon in mycorrhizal symbiosis: community–ecological consequences and practical limitations. In: Allen, M.F. (ed.) *Mycorrhizal Functioning*. Chapman and Hall, London, UK, pp. 357–424.
Mueller, R.C., Scudder, C.M., Porter, M.E., Trotter, R.T. III, Gehring, C.A. and Whitham, T.G. (2005) Differential mortality in response to severe drought: implications for long-term vegetation shifts. *Journal of Ecology* 93, 1085–1093.
Myers, J.A., Vellend, M., Gardescu, S. and Marks, P.L. (2004) Seed dispersal by white-tailed deer: implications for long-distance dispersal, invasion and migration of plants in eastern North America. *Oecologia* 139, 35–44.
Nara, K. (2006) Ectomycorrhizal networks and seedling establishment during early primary succession. *New Phytologist* 169, 169–178.
Nara, K. and Hogetsu, T. (2004) Ectomycorrhizal fungi on established shrubs facilitate subsequent seedling establishment of successional plant species. *Ecology* 85, 1700–1707.
Newman, E.I. (1988) Mycorrhizal links between plants: their functioning and ecological significance. *Advances in Ecological Research* 18, 243–270.
Paugy, M., Baillon, F., Chevalier, D. and Duponnois, R. (2004) Elephants as dispersal agents of mycorrhizal spores in Burkina Faso. *African Journal of Ecology* 421, 225–227.
Pyare, S. and Longland, W.S. (2001) Patterns of ectomycorrhizal–fungi consumption by small mammals in remnant old-growth forests of the Sierra Nevada. *Journal of Mammalogy* 82, 681–689.
Reddell, P., Spain, A.V. and Hopkins, M. (1997) Dispersal of spores of mycorrhizal fungi in scats of native mammals in tropical forests of northeastern Australia. *Biotropica* 29, 184–192.
Richardson, D.M., Williams, P.A. and Hobbs, R.J. (1994) Pine invasions in the southern hemisphere: determinants of spread and invadability. *Journal of Biogeography* 21, 511–527.
Schupp, E.W. (1993) Quantity, quality and the effectiveness of seed dispersal by animals. *Vegetatio* 107/108, 15–29.
Sieverding, E. (1989) Ecology of VAM fungi in tropical agrosystems. *Agriculture Ecosystems Environment* 29, 369–390.
Simard, S.W., Durall, D. and Jones, M. (2002) Carbon and nutrient fluxes within and between mycorrhizal plants. In: van der Heijden, M.G.A. and Sanders, I. (eds) *Mycorrhizal Ecology*. Springer, Berlin, Germany. pp. 33–74.

Smith, S.E. and Read, D.J. (1997) *Mycorrhizal Symbiosis*, 2nd edn. Academic Press, London.

Son, S.L. and Smith, S.E. (1988) Mycorrhizal growth responses: interactions between photon irradiance and phosphorus nutrition. *New Phytologist* 108, 305–314.

States, J.S. and Gaud, W.S. (1997) Ecology of hypogeous fungi associated with ponderosa pine. I. Patterns of distribution and sporocarp production in some Arizona forests. *Mycologia* 89, 712–721.

Stendell, E.R., Horton, T.R. and Bruns, T.D. (1999) Early effects of prescribed fire on the structure of the ectomycorrhizal fungus community in a Sierra Nevada ponderosa pine forest. *Mycological Research* 103, 1353–1359.

Swaty, R.L., Deckert, R.J., Whitham, T.G. and Gehring, C.A. (2004) Ectomycorrhizal abundance and community composition shifts with drought: predictions from tree rings. *Ecology* 85, 1072–1084.

Terwilliger, J. and Pastor, J. (1999) Small mammals, ectomycorrhizae and conifer succession in beaver meadows. *Oikos* 85, 83–94.

Trappe, J.M. (1987) Phylogenetic and ecologic aspects of mycotrophy in the angiosperms from an evolutionary standpoint. In: Safir, G.R. (ed.) *Ecophysiology of VA Mycorrhizal Plants.* CRC Press, Boca Raton, Florida, pp. 5–25.

Tuininga, A.R. and Dighton, J. (2004) Changes in ectomycorrhizal communities and nutrient availability following prescribed burns in two upland pine–oak forests in the New Jersey pine barrens. *Canadian Journal of Forest Research* 34, 1755–1765.

Vander Wall, S.B. (1997) Dispersal of single leaf pinyon pine *Pinus monophylla* by seed caching rodents. *Journal of Mammalogy* 78, 181–191.

Weber, A., Karst, J., Gilbert, B. and Kimmins, J.P. (2005) *Thuja plicata* exclusion in ectomycorrhiza-dominated forests: testing the role of inoculum potential of arbuscular mycorrhizal fungi. *Oecologia* 143, 148–156.

22 The Influence of Seed Source, Habitat and Fungi on *Cecropia* Seed Survival in Two Neotropical Forests

R.E. Gallery, J.W. Dalling, B.T. Wolfe and A.E. Arnold

Introduction

Seed dispersal confers two fundamental advantages to plants: an escape from sources of mortality that are concentrated around parents, and an increased probability of colonizing suitable habitats or microhabitats (Howe and Smallwood, 1982). For pioneer species in old-growth forest, dispersal is especially important for recruitment, since pioneers primarily germinate and establish in treefall gaps. Since gaps are ephemeral and their locations are largely unpredictable (see Hubbell and Foster, 1986), recruitment success by pioneers increases with fecundity and with the range and uniformity of seed dispersal (Murray, 1988; Dalling and John, 2007). However, life history trade-offs can weaken selection for traits that maximize dispersal. For example, small seeds can be produced in greater numbers but have lower probabilities of establishment (Smith and Fretwell, 1974; Dalling and Hubbell, 2002). Furthermore, small-seeded species may be limited to recruiting in specialized microsites such as those that lack leaf litter or maintain favourable soil moisture conditions (Metcalfe and Grubb, 1997; Engelbrecht *et al.*, 2006).

For pioneer species with limited dispersal in space, recruitment success can be increased by maintaining soil seed banks that disperse seeds over time. Seeds of some species can persist in a viable state in the soil for up to several decades (Dalling *et al.*, 1997; Murray and Garcia-C., 2002). These seeds germinate in response to cues that reflect favourable conditions for seedling establishment, such as high red : far-red ratios of light irradiance (Vázquez-Yanes and Smith, 1982; Vázquez-Yanes *et al.*, 1990), elevated soil moisture and nitrate concentrations (Daws *et al.*, 2002), and fluctuating soil temperatures (Vázquez-Yanes, 1974; Vázquez-Yanes and Orozco-Segovia, 1982). Seed persistence may be a critical trait for some pioneers with very low fecundity or with establishment requirements that are rare, such as unusually large and infrequent disturbances. For tropical pioneers with broader establishment requirements, seed banks can increase recruitment success by buffering against inter-annual variations in seed production and unpredictable germination opportunities (e.g. Cheke *et al.*, 1979; Hall and Swaine, 1980; Putz and Appanah, 1987).

Demographic effects of fungi in seed banks

If recruitment of pioneers is largely dependent upon even transient seed banks, then understanding the spatial distribution of seed mortality in the soil may be important for predicting seedling establishment. When seeds land on the soil surface, rodents, ants and beetles may be important seed predators (e.g. Levey and Byrne, 1993; Kaspari, 1996; Fornara and Dalling, 2005). Fungi account for some mortality on the soil surface (Jones, 1994) but once seeds are incorporated into the soil, infection by fungi and other microorganisms may become the dominant source of seed mortality. Declines in seed viability in the soil have often been attributed to infection by pathogenic fungi (Baskin and Baskin, 1998; Dalling *et al.*, 1998b; Alvarez-Buylla and Martínez-Ramos, 1990).

Within the genus *Cecropia*, seed mortality from fungal infection appears to vary among species. For example, Murray and Garcia-C. (2002) found that *Cecropia polyphlebia* (Urticaceae) seeds are largely resistant to fungal attack in lower montane, rain-forest soils, while Dalling *et al.* (1998b) found that up to 50% of *Cecropia insignis* mortality in the soil could be attributed to fungal pathogens in a moist semideciduous tropical forest. However, the degree to which different seed characteristics or environmental conditions influence these patterns remains unknown.

Seeds from the same mother plant may also differ in their susceptibility to fungi. For example, variation in nutritional status during seed development or damage to the seed coat during dispersal can influence intrinsic susceptibility to infection (Wulff, 1986; Fenner and Thompson, 2005, and references therein; see Levey *et al.*, Chapter 2, this volume). Seeds may also harbour asymptomatic infections by endophytic fungi (fungi that colonize and live within living plant tissues without causing disease), which accumulate via contagious spread (i.e. horizontal transmission; Arnold, 2002) or by vertical transmission from maternal plants (e.g. Bose, 1947). There is some evidence that endophytes infect *Cecropia* seeds prior to dispersal (Gallery *et al.*, 2007), although the frequency of these infections is still being determined. Fungal colonization at the pre-dispersal stage may affect post-dispersal seed survival in the soil, as endophytic colonization has been shown to confer increased resistance to subsequent infection by pathogens (Arnold *et al.*, 2003).

Mortality from fungal infection appears to be especially important for small-seeded species with thin fruit or seed walls (Crist and Friese, 1993; Fenner and Thompson, 2005). In temperate and subtropical grassland communities, fungicide treatments increased the survival of buried seeds by

10–30% (Lonsdale 1993; Leishman *et al.*, 2000; Blaney and Kotanen, 2002). Fungicide treatments also increased survival of *Betula papyrifera* (white birch, Betulaceae) seeds in the forest understorey in Ontario, Canada; 8% of seeds germinated in the control (water) treatment compared with 32% in the fungicide treatment (O'Hanlon-Manners and Kotanen, 2004). Survival of *Miconia argentea* (Melastomataceae) and *Cecropia insignis* seeds buried for 6 months in mature forest on Barro Colorado Island (BCI), Panama, increased from 5–10% for untreated seeds to 50% for seeds treated with a broad-spectrum fungicide – Captan (Dalling *et al.*, 1998b). Similar increases in survival were found in a greenhouse experiment at La Selva, Costa Rica, when *Cecropia insignis* seeds potted in forest soil were treated with the fungicide Benomyl (R.E. Gallery, 2002, unpublished results). However, these studies did not address the spatial distribution of fungi in soils or identify the fungal species involved in seed mortality.

More recent experiments have begun to address the diversity and spatial structure of seed-infecting fungal communities in tropical forests. Gallery *et al.* (2007) examined seed survival and fungal infection of seeds of four *Cecropia* species (*C. insignis*, *C. longipes*, *C. obtusifolia* and *C. peltata*) buried in common gardens below crowns of *C. insignis* trees at BCI. Seed survival varied among burial sites and *Cecropia* species, with the lowest survival found in *C. insignis*. The question remains whether these patterns reflect variation in fungus-induced seed mortality.

Results from fungal culturing and direct PCR of dead seeds show that soil-incubated *Cecropia* seeds are infected with a highly diverse assemblage of fungi (primarily Ascomycota) composed of a few common and many rare species (Gallery *et al.*, 2007). Randomization tests coupled with Jaccard's index of similarity were used to determine whether fungal community composition was more strongly influenced by burial sites (distinct tree crowns) or by the species identity of the buried seeds. Although the most common genotypes of fungi were cultivated from all sites and from all *Cecropia* species, seeds shared more fungal species when buried beneath the same crown than when below different crowns. Furthermore, the communities of fungi isolated from seeds were more similar within a *Cecropia* species than among species.

While providing a first resolution of the spatial and host-associated patterns of fungal infection in this tropical pioneer genus, these results fail to resolve the importance of host-generalist versus host-specific fungi in determining seed mortality patterns. If the common, host-generalist fungi are predominantly responsible for seed mortality, then opportunities for escape from pathogens via seed dispersal may be limited. On the other hand, if locally distributed fungi and/or fungi with strong host preferences contribute significantly to seed mortality, then the probability of a seed surviving fungal attack will depend greatly on where the seed lands, thereby contributing to spatial variation in recruitment success.

In this chapter, we describe two experiments that examine spatial variation in the survival of *Cecropia* seeds in the seed bank. The objective of this study was to assess the relative importance of the intrinsic traits of *Cecropia* species, including provenances and maternal seed sources, the heterogeneity in seed-infecting fungal communities (Gallery *et al.*, 2007), and environmental conditions on the survival of seeds in the soil. First, we use reciprocal transplant experiments between two provenances of co-occurring *Cecropia* species to broadly test the relative effects of environment and seed source on seed survival. We then examine local variation in survivorship within the dispersal range of individual *Cecropia* trees and ask whether dispersal away from habitat conditions associated with *Cecropia* crowns (i.e. high conspecific seed bank density and litter cover) increases seed survival in the soil. We also test whether seeds experience lower survival below their parent crown compared with below conspecific crowns. Finally, we present new data highlighting fungi occurring in soil-incubated *C. insignis* seeds at the La Selva Biological Station, Costa Rica, and compare these results to a larger data set from Barro Colorado Island, Panama (Gallery *et al.*, 2007; A.E. Arnold, 2006, unpublished results). We use these data to discuss the potential for local and regional-scale differences in fungal communities to influence *Cecropia* seed survival.

Methods

Study sites and species

We examined the survival of *Cecropia* seeds in soils at two neotropical sites, Barro Colorado Island (BCI), Panama (9°9′ N, 79°51′ W) and La Selva Biological Station, Costa Rica (10°26′ N, 83°59′ W). These two sites are separated by 475 km and experience distinctly different rainfall and seasonality. The semideciduous forest at BCI experiences an intense, 4-month dry season from January to May, which accounts for less than 10% of its annual precipitation (annual mean = 2600 mm; Windsor, 1990). In contrast, La Selva's aseasonal, wet forest receives approximately 4000 mm of rain annually (Sanford *et al.*, 1994) with no less than 100 mm of rain in any month. Despite these differences, old-growth forests at these sites share many species of vascular plants (Gentry, 1990).

The study species, *Cecropia insignis* Liebm. and *C. obtusifolia* Bertol. (Urticaceae; Sytsma *et al.*, 2002) are dioecious pioneer trees common in lowland moist and wet forests in Central America (Holdridge *et al.*, 1971; Croat, 1978). *Cecropia insignis* is one of the most common species of pioneer tree at both study sites and one of the most abundant species in the seed bank (Putz, 1983; Young *et al.*, 1987; Dalling *et al.*, 1995, 1997; Dupuy and Chazdon, 1998). Although *C. obtusifolia* is more abundant in younger forests at BCI and La Selva (Croat, 1978; Hammel, 1986), its seeds are still regularly found in the seed bank of mature forest (Young *et al.*, 1987; Dalling *et al.*, 1995, 1997).

At both sites, *C. obtusifolia* fruits nearly continuously, with peak fruiting during February–August, while *C. insignis* fruits in April–July, with peak fruit production in early June (Croat, 1978; Milton, 1991). Thus the peak of seed production for BCI and La Selva overlaps with the period of the highest frequency of gap formation, which generally correlates with the wet season at BCI (May–December) and/or exceptionally wet periods in both forests (Hartshorn, 1978; Brokaw, 1982; Garwood, 1983). Although *Cecropia* seeds are widely dispersed by bats, birds and monkeys, below-crown seed rain is also very high (e.g. 64,000 *C. insignis* seeds/m^2/year below fruiting individuals at BCI; Dalling *et al.*, 1998b). Seeds of both species are small (mean fresh mass = 0.6 mg; Dalling *et al.*, 1997; Gallery *et al.*, 2007). Like most common small-seeded species that do not fruit all-year-round, seeds of *C. insignis* and *C. obtusifolia* tend to be seasonally abundant in the seed bank, with most seeds surviving less than a year (Alvarez-Buylla and Martínez-Ramos, 1990; Dalling *et al.*, 1997, 1998a).

General protocol

We conducted two experiments to investigate potential sources of variation in spatial heterogeneity of seed survival in the seed bank. In each we used a common protocol based on a reciprocal burial design, whereby seeds from different maternal sources were buried under conspecific crowns. Seeds were collected from seed traps or directly from the crowns of *C. insignis* and *C. obtusifolia* individuals (maternal sources) at both BCI and La Selva. Immediately after collection, seeds were removed from infructescences and rinsed in a 10% Clorox® bleach (0.5% sodium hypochlorite) solution for 2 min to remove surface contaminants. Seeds were then surface-dried under sterile conditions in a darkroom and sorted by maternal source into lots of 30, each of which was mixed with 10 g of sterilized forest soil (autoclaved at 115°C for 2 h) and enclosed in a nylon mesh bag (0.5 mm mesh size). Four seed bags from each maternal source were buried in soil at a depth of 3 cm in one 3 m × 3 m plot below each focal crown. After 5 months of incubation (July–December, which encompasses the greatest seasonal fluctuation in seed bank densities of *Cecropia*; Dalling *et al.*, 1998b), bags were recovered and germination trials were conducted for 8 weeks to determine the percentage of seeds surviving. Bag contents (seeds and soil) were transferred into Petri dishes lined with sterile filter paper, watered with filtered water, sealed with Parafilm, and placed in ambient-air growth-houses (mean maximum = 30.8°C ± 1.0°C; mean minimum = 23.9°C ± 0.8°C) with 30% full sunlight and high red : far-red irradiance to induce germination (see Dalling *et al.*, 1998b). Seed survival was measured as the proportion of 30 seeds that germinated (radicle and cotyledon emergence), adjusted for the initial viability, which was determined at the onset of the experiment with a subsample of 100 seeds from each maternal source.

Experiment 1: Is seed source or environment a better predictor of seed survival?

Previous results showed differences in seed survival among four sympatric *Cecropia* species when buried in common gardens below *C. insignis* crowns at BCI (Gallery *et al.*, 2007). In that study *C. insignis* seeds experienced the lowest survival, with foreign seeds (from La Selva) suffering lower survival than local seeds (from BCI). Seed survival also differed significantly between crowns. These differences may have resulted from intrinsic seed characteristics that varied among *Cecropia* species, from small-scale spatial differences and/or host affinities of fungal communities capable of infecting *Cecropia* seeds, and from small-scale variations in environmental conditions (e.g. soil temperature and moisture, litter cover) that influence seed–fungal interactions in the soil.

For the present experiment, we established common gardens by burying seed bags concurrently at BCI and La Selva. We used seeds from six individuals (maternal sources) of *C. insignis* from BCI and five from La Selva, and four *C. obtusifolia* from both BCI and La Selva. We used an incomplete reciprocal design where seeds were buried:

1. Below their maternal crown and two other randomly chosen conspecifics at their 'home' site (e.g. BCI seeds at BCI); or
2. Below three randomly chosen conspecifics at the 'away' site (e.g. BCI seeds at La Selva) (n = 11 for *C. insignis*, n = 8 for *C. obtusifolia*).

Using the general methods described above, seed bags were field-incubated from July to December 2001.

This experiment allowed us to examine two co-occurring *Cecropia* species with similar seed traits (e.g. seed mass, seed longevity in soil, susceptibility to infection by particular fungi; Gallery *et al.*, 2007) and determine the relative importance of seed source and environmental conditions on seed survival. Transplanting seeds over a large geographical scale (475 km between BCI and La Selva), but still within the species' natural range, tested whether seed source (local or novel provenance) affected seed survival. In turn, transplants below different crowns within sites tested whether seeds suffer lower survival below their own parent's crown. This result may be expected if the fungal communities in a below-crown area are locally adapted to the genotypes of seed that they most commonly encounter (see Gilbert, 2005); however, evidence for such local adaptation in natural systems is lacking (but see Parker, 1985). Alternatively, similar survival of different seed sources buried below the same crown and/or site-specific patterns of seed survival would suggest that extrinsic factors play a large role in determining where and for how long a seed will survive.

Experiment 2: How does local dispersal affect seed fate?

Limited seed dispersal implies that the highest densities of seeds and their specialist natural enemies should be found below crowns of reproductive adults (Janzen, 1970; Connell, 1971). For *Cecropia*, seed densities in the soil decline logarithmically with increasing distance from fruiting crowns (Alvarez-Buylla and Martinez-Ramos, 1990; Dalling *et al.*, 1998a). Annually replenished seed banks below reproductive crowns may create a positive feedback with fungal inoculum resulting in overall high inoculum densities in high seed bank areas (e.g. Bever, 1994). Therefore, seed mortality due to fungal infection may be highest in the seed-dense areas below reproductive crowns.

Seed survival may also depend on other environmental conditions associated with tree crowns. For example, litter cover may influence seed survival indirectly by changing soil moisture and nutrient conditions that promote fungal activity and infection potential (Agarwal and Sinclair, 1987; Agrios, 1997), and/or change community composition of fungi and micro-invertebrates in soil and litter (Facelli and Pickett, 1991). Results from large-scale litter manipulation plots in Panama revealed higher saprophytic fungal biomass (hyphal length) in the top 2 cm of soil under litter-addition plots than under litter-removal plots (A. Vincent, Panama, 2006, personal communication). If similar fungi infect both *Cecropia* leaf and seed tissue (as found by A.E. Arnold, 2006, unpublished results) then the inoculum provided by *Cecropia* leaf litter may overwhelm any potentially density-dependent effects that arise from seed-to-seed transmission of fungi in the soil.

We addressed this issue by conducting a second experiment in July 2003, in which seed bags were buried:

1. Below female *C. insignis* crowns;
2. In plots at least 50 m from *C. insignis* trees;
3. Below male *C. insignis* crowns.

This approach allowed us to test for the effect of high versus low *C. insignis* litter and potential pathogens associated with adult trees on seed survival (e.g. Gilbert, 1995; Packer and Clay, 2000) independent of the effects of high versus low seed bank density.

We buried *C. insignis* seeds from six BCI and six La Selva trees (n = 12) in 3 m × 3 m plots:

1. Below six fruiting (female) *C. insignis* crowns;
2. At six locations 50 m away from each of those fruiting individuals;
3. Below six flowering (male) *C. insignis* crowns at least 50 m from the nearest fruiting *C. insignis*.

All seed sources were buried in all plots. To address the temporal aspect of seed survival, a subsample of seed bags (432; from the six BCI seed sources × 4 bags/source × 18 plots) was removed after 1 month and the remaining bags (864; from the six BCI and six La Selva sources) were left to incubate for 5 months.

Natural *C. insignis* seed bank densities were measured during the first week of August, which corresponds with the end of the fruiting season and peak *C. insignis* densities in the seed bank (Dalling *et al.*, 1998b). Densities were measured by coring 235 cm^3 (10 cm diameter × 3 cm depth) of soil at nine randomly chosen locations within the plots below each of the *C. insignis* crowns. Soil cores were spread thinly (0.5 cm) over sand flats (3 cm deep) in an ambient-air growth-house and watered regularly. Seedlings were tagged upon emergence and grown until they could be positively identified; no seedlings died before they could be identified (protocol follows Dalling *et al.*, 1997).

Leaf litter cover was measured at ten randomly chosen locations within each plot. To correspond with seed bank measurements, litter cover was measured during the last week of July using a point intercept method (Elzinga *et al.*, 2001). A 0.5-m-long galvanized steel rod (2 mm diameter) was pushed through the litter and the number of times a leaf touched the rod was recorded.

A reduced version of the experiment was conducted concurrently at La Selva (n = 12). Subsamples for seed lots for the BCI experiment were used, but plots were established only: (i) below six fruiting female *C. insignis* crowns; and (ii) 50 m away from those individuals. No seed bank density or litter thickness measurements were taken.

Data analyses

The percentage of viable seeds that germinated after 5 months of soil incubation was analysed using factorial analysis of variance (ANOVA) with type III sums of squares (proc mixed; SAS, 2003). Prior to analysis, germination data were logit transformed to approximate normality. Below-crown plots were nested within site and maternal sources of seeds were nested within provenance. Maternal source was treated as a random effect and was used to test provenance. Because the four bags per maternal source incubated at a given below-crown plot could not be considered independent (Gallery *et al.*, 2007), mean germination of the four bags was used in each model.

Results and Discussion

Experiment 1: Environment is a better predictor of seed survivorship than seed source

Seed survival at BCI was significantly different from that at La Selva for both species. *C. insignis* seeds experienced higher survival when buried at BCI than at La Selva ($F_{(1, 17)}$ = 24.41, P < 0.001; Fig. 22.1a) while the opposite was true for *C. obtusifolia* ($F_{(1, 7)}$ = 19.53, P = 0.0031; Fig. 22.1b). BCI and La Selva experience distinctly different seasonality and total amounts of rainfall, but the effects, if any, that these factors have on *Cecropia* seed banks remain equivocal.

Within each site, seed survival differed significantly among crowns for both *Cecropia* species. Seed survival under particular crowns ranged from 6.8% (± 6.6% SE) to 61.9% (± 8.0%) for *C. insignis* ($F_{(9, 17)}$ = 1.92, P = 0.11), and from 9.9% (± 6.3%) to 79.3% (± 8.3%) for *C. obtusifolia* ($F_{(6, 7)}$ = 4.00, P = 0.046). If survival was determined only by species-level traits

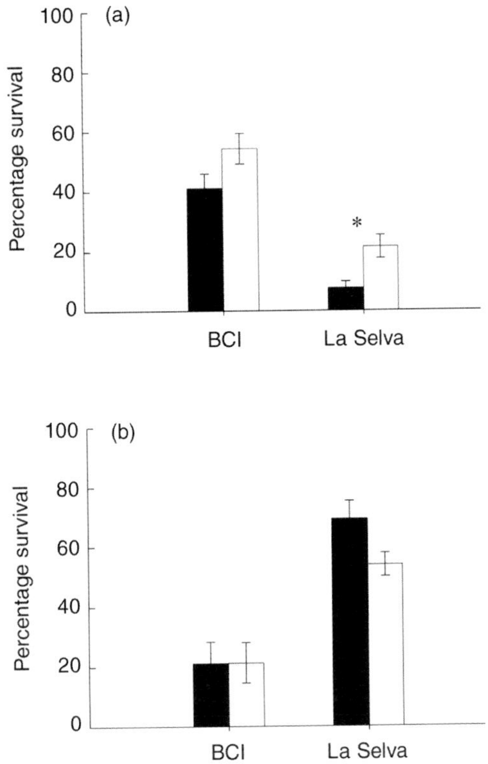

Fig. 22.1. Percentage seed survival of (a) *C. insignis* and (b) *C. obtusifolia* seeds in plots below crowns of fruiting (female) conspecific trees at BCI and La Selva. Bars show differences in seed survival between provenances; dark bars represent the mean (± 1 SE) percentage germination of BCI seed sources, and white bars represent the mean (± 1 SE) percentage germination of La Selva seed sources. (a) *C. insignis*: La Selva seed sources (white bars) experienced higher survival than BCI seed sources (dark bars) at both sites (*$P < 0.05$). Survival was higher at BCI than at La Selva regardless of seed source. (b) *C. obtusifolia*: La Selva (white bars) and BCI (dark bars) seed sources experienced similar survival at both sites. Survival was higher at La Selva than at BCI regardless of seed source.

then we would expect similar survival within a species regardless of where seeds are buried. Instead, these results suggest that seed survival is also a consequence of the location to which seeds are dispersed and incorporated into the seed bank (see also Fenner and Thompson, 2005). Therefore, even with a specific type of dispersal site (e.g. beneath a conspecific crown), differences in local habitat conditions and/or variation in biotic conditions (including fungal communities) contribute greatly to variation in seed survival in seed banks, and thus to the suitability of a site for dispersal (see Schupp, Chapter 20, this volume).

The effect of provenance was significant for *C. insignis* ($F_{(1, 17)} = 5.29$, $P = 0.035$), with La Selva seed sources experiencing slightly higher survival than BCI seed sources at both sites. Provenance was not significant for

C. obtusifolia ($F_{(1, 7)} = 0.39$, $P = 0.55$). The relatively similar survival between BCI and La Selva seed sources for both species at both BCI and La Selva suggests that rare or 'foreign' *Cecropia* seed sources are unlikely to have a substantial survival advantage (Fig. 22.1a,b). However, it remained unclear whether source effects are important at smaller (local) scales. To explore this, we assessed survival for seeds buried beneath their mothers versus other conspecific crowns. We found that seeds buried beneath their mothers did not suffer lower survival than seeds buried under other conspecific female crowns (results similar at BCI and La Selva for both *C. insignis* ($t = 0.93$, $df = 1, 17$, $P = 0.36$; linear contrasts within ANOVA) and *C. obtusifolia* ($t = 0.10$, $df = 1, 7$, $P = 0.92$)). Thus, maternal sources of seed on a small or local scale did not appear to be important in shaping seed survival.

Previous results showed that *C. insignis* and *C. obtusifolia* do not differ in the incidence of fungal infection when buried in common gardens (Gallery *et al.*, 2007). Given that these species experience similar infection rates, the particular fungi infecting the seeds may account for the different survival patterns within forest communities. Our current research examining the local- and geographic-scale differences in communities of fungi that infect *Cecropia* seeds will enable us to determine whether different fungi are responsible for the different seed survival patterns at these two sites.

Experiment 2: Dispersal away from *C. insignis* crowns increases seed survival

At BCI, seed survival was high after 1 month of burial and did not differ for *C. insignis* seeds buried beneath female crowns (80.1% ± 3.1%), below male crowns (78.4% ± 4.0%), or 50 m away from female crowns (78.5% ± 4.1%). After 5 months of burial, however, survival was lower for seeds buried below female or male *C. insignis* compared with 50 m away (Fig. 22.2a; $t = 5.25$, $df = 2, 16$, $P < 0.001$; linear contrasts within ANOVA). At La Selva, survival was also lower for seeds buried for 5 months below female *C. insignis* compared with 50 m away (Fig 22.2b; $t = 4.23$, $df = 1, 20$, $P < 0.001$). This pattern is consistent with results from a similar seed-incubation experiment in a tropical forest in Los Tuxtlas, Mexico: *C. obtusifolia* seed survival at sites 10–28 m from the nearest fruiting *C. obtusifolia* was higher than that of seeds incubated below crowns (Alvarez-Buylla and Martínez-Ramos, 1990). Seed survival did not differ significantly below female and male *C. insignis* at BCI (Fig 22.2a; $t = 1.03$, $df = 2, 16$, $P = 0.32$). Together, these results suggest that, regardless of parental/non-parental status of seed sources, conditions below the crowns of *Cecropia* are generally unfavourable for conspecific seed survival.

Similar to the results from Expt 1, the effect of provenance was not significant for *C. insignis* ($F_{(1, 10)} = 0.16$, $P = 0.70$). Focusing on seed survival below female *C. insignis*, seeds did not experience lower survival below their maternal crowns compared with below other conspecifics ($F_{(1, 87)} = 2.03$, $P = 0.16$). These results suggest that seeds dispersed to areas below conspecific crowns are at an equal disadvantage to seeds falling

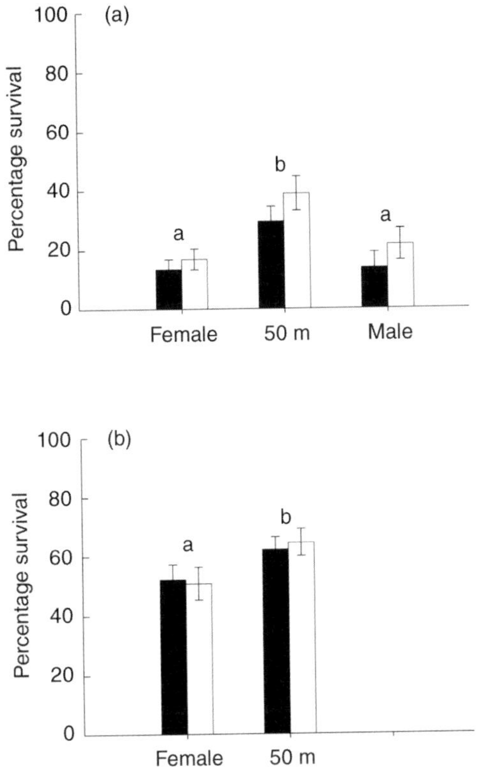

Fig. 22.2. Percentage seed survival of *C. insignis* seeds after 5 months of incubation in plots: (i) below six crowns of fruiting (female) *C. insignis* trees; (ii) six plots 50 m from those fruiting individuals; and (iii) below six crowns of flowering (male) *C. insignis* trees at least 50 m from the nearest fruiting *C. insignis* at BCI (a). Seeds were not buried below male *C. insignis* trees at La Selva (b). Bars show differences in seed survival between provenances, which were not significant in any burial plot in either site. Dark bars represent the mean (± 1 SE) percentage germination of six BCI seed sources; white bars represent the mean (± 1 SE) percentage germination of six La Selva seed sources. Different letters show significant ($P < 0.05$) differences in survival among burial plots at BCI and La Selva.

below their maternal crown. They therefore fail to provide evidence for feedback mechanisms that select against common host genotypes either at single-individual or single-population scales.

Experiment 2: Effects of seed density and litter inputs on seed survival

Seed bank densities varied below crowns, but, on average, were 2.5-fold higher below females (446 ± 55 seeds/m^2) than below males (170 ± 59 seeds/m^2; two-sample *t*-test, $t = 1.44$, $df = 1, 10$, $P < 0.05$). Of the 54 cores taken below male crowns, 44% yielded viable *C. insignis* seeds. In contrast, 76% of the 54 cores taken below female crowns contained viable *C. insignis* seeds. As noted above, seed survival below female (high seed bank density) and male (low seed bank density) *C. insignis* at BCI was not significantly different (Fig 22.2a), suggesting that seed survival is not primarily driven by factors that are sensitive to local seed density (see also Dalling *et al.*, 1998b).

At BCI, we found that the litter layer was patchier and overall litter thickness was higher below *Cecropia* crowns relative to 50 m away from crowns. Percentage cover under *C. insignis* crowns ranged from 85% (male) to 90% (female), relative to 72% at sites 50 m away. Furthermore, 16% of the ground sampled in the 50-m plots was covered with a layer of three or more leaves, compared with 41% of plots below female crowns and 33% of plots below male crowns. The litter composition below male and female *C. insignis* trees comprised primarily *C. insignis* leaves, whereas no *C. insignis* leaves were encountered in any plots 50 m away (B.T. Wolfe and R.E. Gallery, 2003, unpublished results). In some cases, litter thickness reached up to nine *Cecropia* leaves below male and female crowns. Mean litter cover below male and female *C. insignis* crowns was similar (1.8 ± 0.19 leaves below females; 2.2 ± 0.21 leaves below males; t = 0.66, df = 2, 15, P = 0.51) and greater than litter cover in plots 50 m from *Cecropia* crowns (1.0 ± 0.20 leaves; t = 1.8, df = 2, 15, P < 0.05, linear contrasts within ANOVA), suggesting a relationship between the presence of *Cecropia* litter and survival of *Cecropia* seeds. Similarities in litter composition below male and female *Cecropia* trees may therefore play a role in explaining the similar survival rates of seeds below crowns.

Leaf litter could directly influence seed survival below *C. insignis* individuals if fungi that are present in *Cecropia* leaves contribute to the inoculum that infects seeds. In tropical forests, pathogen transmission from parent to offspring through deposits of aerial spores or leaves has been suggested with seedlings (Gilbert, 1995; Mancini *et al.*, 2001), but not with seeds. Direct transmission is not necessary given that, in the absence of a suitable host or under unfavourable conditions, many fungi are capable of long-term persistence in soil (e.g. *Fusarium*, *Rhizoctonia*) either as saprophytes or as mycelia and sclerotia (masses of mycelium) in dead parent material, or as resting or other types of spores (Bruehl, 1987; Agrios, 1997).

Genotype comparisons with an existing database of endophyte sequence data indicate that several fungal taxa isolated from *C. insignis* seeds show high sequence affinity (99%) at a fast-evolving locus (internal transcribed spacer region) to isolates obtained as endophytes from living leaf tissue and as saprophytes from leaf litter (A.E. Arnold, 2006, unpublished results). Studies of leaf-litter and wood-decay fungi in Panama (e.g. Cornejo *et al.* 1994; Ferrer and Gilbert, 2003) suggest that saprophytes are non-randomly distributed with regard to host species, and thus may demonstrate differential host affinity. Given that the litter composition below male and female *C. insignis* trees was composed primarily of *C. insignis* leaves, the observed spatial structure of fungal communities below individual *Cecropia* crowns (Gallery *et al.*, 2007) could be largely influenced by leaf litter fungi.

Linking *Cecropia* Seed Survival to Fungal Community Composition

Cecropia seed-infecting fungi

To fully understand the ecological importance of seed-infecting fungi for recruitment of pioneer species, it is critical to assess the abundance, species richness, and species composition of endophytes, saprophytes and pathogens associated with seeds. Our previous studies of canopy-collected and soil-incubated seeds of four sympatric *Cecropia* species at BCI indicated that taxonomically diverse communities of fungi infect *Cecropia* seeds and have potentially important impacts on seed persistence in the soil seed bank (Gallery *et al.*, 2007). At the time, we lacked comparable data to determine whether fungal communities are similar at La Selva.

Here we examined representative seed-infecting fungi from surface-sterilized *C. insignis* seeds incubated in soil below six *C. insignis* crowns for 1 month at La Selva, Costa Rica, in 2003. After surface sterilization, seeds were plated on 2% malt extract agar (MEA) in sterile Petri dishes. MEA is a general medium that has previously been shown to encourage growth by diverse microfungi (see, e.g. Fröhlich and Hyde, 1999), although it probably underestimates the number of fungi present. From 250 randomly chosen seeds, 32 distinct fungal morphotypes emerged in culture, with seven morphotypes accounting for 72% of the total isolates. This pattern of relative abundance is consistent with seed-infecting fungi from other tropical forests (Gallery *et al.*, 2007) and temperate ecosystems (Schafer and Kotanen, 2004) and with that of other diverse groups of tropical fungi, for which a few species are very common but most are rare (e.g. Fröhlich and Hyde, 1999; Arnold *et al.*, 2000; Gilbert *et al.*, 2002).

Preliminary molecular analyses focused on 43 representative isolates that comprised rare and common morphotypes. Methods for DNA extraction, PCR, sequencing and sequence assembly followed Arnold *et al.* (2007). We extracted total genomic DNA from living cultures on MEA plates using a modified phenol : chloroform method. We used diluted DNA (1:10 dilution of extraction yields) in PCR to amplify the ITS (internal transcribed spacer) and 5.8s regions of the nuclear ribosomal repeats using primers ITS1F and ITS4 or LR3 (White *et al.*, 1990; see also http://www.lutzonilab.net), which are effective for diverse Ascomycota, Basidiomycota and Zygomycota. PCR products were cleaned and sequenced for both forward and reverse reads, and all contigs were verified manually before submission to BLAST searches of the NCBI GenBank database for provisional identification at higher taxonomic levels.

The identified morphotypes include representatives of diverse orders of Ascomycota, including Eurotiales, Chaetothyriales, Pleosporales, Sordariales, Hypocreales, Diaporthales and Phyllachorales (Table 22.1). These orders are frequently represented among endophytic fungi inhabiting leaves of trees at our study sites (Arnold and Lutzoni, 2007) and also contain numerous pathogens and saprophytic species. Several orders containing putatively saprophytic Basidiomycota were also recovered. High affinity matches based on BLAST searches of GenBank, coupled with phylogenetic analysis of isolate sequences among top matches (Arnold, 2002; Arnold and Lutzoni, 2007) indicate that diverse species of Ascomycota, including *Rhinocladiella*, *Botryosphaeria*, *Nectria*, *Fusarium*, *Chaetomium*, *Alternaria* and *Colletotrichum* occur within seeds of *C. insignis* (Table 22.1). Several genotypes were associated preferentially with viable or inviable seeds, providing a basis for future experimental analyses (Table 22.1).

These data are largely congruent with findings from a larger sampling of 220 fungi from *Cecropia* seeds at BCI (Gallery *et al.*, 2007). In that study, species of *Nectria*, *Fusarium* and *Chaetomium* were especially common, with

Table 22.1. Identities of 43 representative *Cecropia* seed-infecting fungi based on high affinity matches from BLAST searches (in GenBank), coupled with phylogenetic analysis of isolate sequences among top matches. Fungi were isolated from *Cecropia insignis* seeds following soil incubation in the forest understorey of the La Selva Biological Station, Costa Rica. Data indicate number of isolates sharing that genotype, and the number of times a genotype was recovered from germinated (viable) or ungerminated (inviable) seeds. Taxa that are of uncertain placement at the ordinal or family levels are listed as I.S. (*incertae sedis*). Cases in which top BLAST matches were to unidentified sequences are marked N/I (not identified). Taxon names are given for N/I sequences if placement was confirmed by phylogenetic analysis. Ordinal names marked with asterisks are Basidiomycota; all others are Ascomycota. ITS sequences are available on request from the authors.

				Seed Status	
Order	Family	Genus	Isolates	Viable	Inviable
Chaetothyriales	Herpotrichiellaceae	*Rhinocladiella*	7	1	6
Aphyllophorales*	Corticiaceae	*Athelia*	4	–	4
Diaporthales	Valsaceae	*Phomopsis*	3	–	3
N/I, I.S. Ascomycota	N/I, Botryosphaeriaceae	N/I, *Botryosphaeria*	3	–	3
Diaporthales	Valsaceae	*Diaporthe*	2	–	2
Diaporthales	Valsaceae	N/I, *Diaporthe/Phomopsis*	2	–	2
Eurotiales	Trichocomaceae	*Penicillium*	2	–	2
Hypocreales	Mitosporic hypocreales	*Fusarium*	2	1	1
N/I, Pleosporales	N/I, Pleosporaceae	N/I, *Alternaria*	2	1	1
Phyllachorales	Phyllachoraceae	*Colletotrichum*	2	2	–
Pleosporales	Pleosporaceae	*Curvularia*	2	–	2
Agaricales*	Pleurotaceae	*Pleurotus*	1	1	–
Aphyllophorales*	Polyporaceae	*Ceriporiopsis*	1	–	1
Eurotiales	Trichocomaceae	*Emericella*	1	1	–
Hypocreales	Clavicipitaceae	N/I, Unknown	1	1	–
Hypocreales	Nectriaceae	*Nectria*	1	1	–
I.S. Ascomycota	Botryosphaeriaceae	*Botryosphaeria*	1	–	1
I.S. Ascomycota	Dothioraceae	*Aureobasidium*	1	–	1
I.S. Ascomycota	Mycosphaerellaceae	*Cladosporium*	1	–	1
I.S. Ascomycota	Botryosphaeriaceae	N/I, *Botryosphaeria*	1	1	–
Saccharomycetales	Saccharomycetaceae	*Debaryomyces*	1	–	1
Sordariales	Chaetomiaceae	*Chaetomium*	1	–	1
Tremellales*	I.S., Tremellales	*Cryptococcus*	1	–	1

Phomopsis, *Botryosphaeria* and other taxa isolated at lower frequencies. Genotype comparisons show that despite the recovery of some genera at both La Selva and BCI, only five genotypes recovered at La Selva were also represented at BCI (comparisons based on 99% ITS sequence similarity). Shared genotypes included the most common *Fusarium* genotype at BCI and a genotype that does not match named sequences in GenBank, but which corresponds in phylogenetic analyses to *Rhinocladiella*. However, 64% of genotypes recovered from BCI and La Selva have been found only once, and thus could not be compared across sites. Among the genotypes found more than once, 85% were specific only to BCI or La Selva. Together, these observations indicate that distinct fungal communities capable of infecting *Cecropia* seeds occur at each site. The large differences in seed survival rates observed between sites (Expt 1) may therefore indicate that the frequency with which seeds encounter pathogenic fungi also varies between these forests.

Conclusions

Taken together, the experiments described in this chapter demonstrate that environmental conditions in the seed bank have a potentially greater influence on *Cecropia* seed survival than maternal sources or intrinsic seed characteristics. Contrary to the expectation based on local adaptation of fungal communities to their hosts, we found that seeds did not experience lower survival in their local site (BCI versus La Selva), and that survival was not lower for seeds buried below their maternal crowns as opposed to other conspecific crowns. Many abiotic factors probably influence *Cecropia* seed survival in soil and need to be examined in greater detail. For example, in temperate and tropical systems, significant sources of seed mortality have been attributed to the direct effects of fluctuations in soil moisture (Schafer and Kotanen, 2003) and temperature (Lonsdale, 1993), and to the indirect effects of these factors on microbial activity.

Cecropia seeds dispersed away from fruiting crowns and incorporated into the seed bank have higher survival than seeds that are incorporated into soils below conspecific crowns (fruiting or not). We examined two conditions associated with *Cecropia* below-crown sites and found that *Cecropia* litter cover may have a greater effect on seed survival than high *Cecropia* seed bank density. Litter cover can create moist microhabitats and increase saprophytic fungal biomass in general, which may negatively affect the survival of seeds in soil. Our data also show that many fungi associated with live leaf tissue and litter infect *Cecropia* seeds, suggesting that litter could even provide a direct inoculum source of fungi capable of infecting seeds.

Directions for future research

The specificity of fungal–seed interactions and the ecological factors influencing the activity and diversity of these fungi are only just beginning to be examined. Fungi recovered from *Cecropia* seeds incubated in the soil are likely to include seed pathogens and parasites, saprophytes and endophyte species that may be harmless or beneficial to a particular host. While most common fungal genotypes isolated appear to be widespread generalists, Gallery *et al.* (2007) found some evidence that other genotypes are either very patchily distributed or demonstrate affinities for particular hosts. In a study using similar experimental protocols, Schafer and Kotanen (2004) found evidence for both generalist and specialist seed pathogens of four co-occurring grasses in temperate meadows. Further, they found that grass seeds differed in their susceptibility to a range of fungi they commonly encounter in soil. These species-specific interactions among seeds and fungi suggest fungi have the potential to differentially limit recruitment of susceptible hosts. However, to determine the community-wide effects of fungi on a particular host, it will be necessary to identify the dominant seed pathogens and determine their spatial distribution in soils.

We have identified putative *C. insignis* seed pathogens by considering seed viability associated with particular fungal isolates (Table 22.1; Gallery *et al.*, 2007). The next step is to determine whether these isolates are responsible for killing their host seeds. Traditional approaches in plant pathology, which focus on satisfying Koch's postulates (Agrios, 1997), argue for the importance of re-inoculation studies to determine causality. By infecting freshly collected, asymptomatic seeds with our focal isolates, assessing seed survival, and re-isolating the inoculants from live or dead seeds, we hope to identify at least some of the fungal pathogens of *Cecropia* seeds in Panama and Costa Rica. These results will provide the framework for developing future hypotheses aimed at testing species-specific interactions among fungi and seeds in tropical soils.

Acknowledgements

We thank the organizers of the *Fourth International Symposium/Workshop on Frugivory and Seed Dispersal* for the opportunity to contribute to this volume: Andrew Dennis, Ronda Green, Gene Schupp and David Westcott. Travel of the senior author to Australia to participate in the Symposium/Workshop was partially supported by the National Science Foundation (OISE No. 0456402 to E.W. Schupp). This research was supported by NSF DEB-0343953 and DEB-0516564 to J.W.D. and A.E.A. with additional support from an NSF Postdoctoral Fellowship in Microbial Biology to A.E.A. (DEB-0200413), a Mellon STRI-OTS Comparative Research Grant to J.W.D. and R.E.G., and the UIUC Undergraduate Mentoring in Environmental Biology Program (UMEB) to B.T.W. We thank the Smithsonian Tropical Research Institute and the Organization for Tropical Studies for providing facilities, logistic support, and permission to conduct this research at Barro Colorado Island and La Selva Biological Station. We thank Deborah and David Clark for use of their *Cecropia* TREES database at La Selva. The TREES project is funded by NSF's LTREB programme (DEB-0129038) and the TEAM project of Conservation International with support from the Gordon and Betty Moore Foundation. We also thank François Lutzoni for logistic support at Duke University, and Evelyn Sanchez, Olman Paniagua and Gilbreth Hurtado for field assistance. Finally, we wish to thank Eugene Schupp and two anonymous reviewers for valuable comments.

References

Agarwal, V.K. and Sinclair, J.B. (1987) *Principles of Seed Pathology*, 2nd edn. CRC Press, Boca Raton, Florida.

Agrios, G.N. (1997) *Plant Pathology*, 4th edn. Academic Press, New York.

Alvarez-Buylla, E. and Martínez-Ramos, R. (1990) Seed bank versus seed rain in the regeneration of a tropical pioneer tree. *Oecologia* 84, 314–325.

Arnold, A.E. (2002) Neotropical fungal endophytes: diversity and ecology. PhD dissertation, University of Arizona, Tucson, Arizona.

Arnold, A.E. and Lutzoni, F. (2007) Diversity and host range of foliar fungal endophytes: are tropical leaves biodiversity hotspots? *Ecology* 88, 541–549.

Arnold, A.E., Maynard, A., Gilbert, G.S., Coley, P.D. and Kursar, T.A. (2000) Are tropical fungal endophytes hyperdiverse? *Ecology Letters* 3, 267–274.

Arnold, A.E., Mejía, L., Kyllo, D., Rojas, E., Maynard, Z. and Herre, E.A. (2003) Fungal endophytes limit pathogen damage in a tropical tree. *Proceedings of the National Academy of Sciences USA* 100, 15649–15654.

Arnold, A.E., Henk, D.A., Eells, R.L., Lutzoni, F. and Vilgalys, R. (2007) Diversity and phylogenetic affinities of foliar fungal endophytes in loblolly pine inferred by culturing and environmental PCR. *Mycologia* (in press).

Baskin, C.C. and Baskin, J.M. (1998) *Seeds: Ecology, Biogeography, and Evolution of Dormancy and Germination.* Academic Press, San Diego, California.

Bever, J.D. (1994) Feedback between plants and their soil communities in an old field community. *Ecology* 75, 1965–1977.

Blaney, C.S. and Kotanen, P.M. (2002) Persistence in the seed bank: an experimental comparison of native and alien plants. *Écoscience* 9, 509–517.

Bose, S.R. (1947) Hereditary (seed-borne) symbioses in *Casuarina equisetifolia. Nature* 159, 152–154.

Brokaw, N.V.L. (1982) Treefalls: frequency, timing and consequences. In: Rand, A.S., Windsor, D.M. and Leigh, E.G. Jr (eds) *The Ecology of a Tropical Forest: Seasonal Rhythms and Long-term Changes.* Smithsonian Institution Press, Washington, DC, pp. 101–108.

Bruehl, G.W. (1987) *Soilborne Plant Pathogens*. Macmillan Press, New York.

Cheke, A.S., Nanakorn, W. and Yankoses, C. (1979) Dormancy and dispersal of seeds of secondary forest species under the canopy of a primary tropical rain forest in northern Thailand. *Biotropica* 11, 88–95.

Connell, J.H. (1971) On the role of natural enemies in preventing competitive exclusion in some marine animals and in rain forest trees. In: Den Boer, P.J. and Gradwell, G. (eds) *Dynamics of Populations.* PUDOC, Wageningen, The Netherlands, pp. 298–312.

Cornejo, F.J., Varela, A. and Wright, S.J. (1994) Tropical forest litter decomposition under seasonal drought: nutrient release, fungi and bacteria. *Oikos* 70, 183–190.

Crist, T.O. and Friese, C.F. (1993) The impact of fungi on soil seeds: implications for plants and granivores in semiarid shrub-steppe. *Ecology* 74, 2231–2239.

Croat, T.B. (1978) *Flora of Barro Colorado Island.* Stanford University Press, Stanford, California.

Dalling, J.W. and Hubbell, S.P. (2002) Seed size, growth rate and gap microsite conditions as determinants of recruitment success for pioneer species. *Journal of Ecology* 90, 557–568.

Dalling, J.W. and John, R. (2007) Seed limitation and the coexistence of pioneer species. In: Carson, W.P. and Schnitzer, S. (eds) *Tropical Forest Community Ecology.* Blackwell Scientific, Oxford, UK (in press)

Dalling, J.W., Swaine, M.D. and Garwood, N.C. (1995) Effect of soil depth on seedling emergence in tropical soil seed-bank investigations. *Functional Ecology* 9, 119–121.

Dalling, J.W., Swaine, M.D. and Garwood, N.C (1997) Soil seed bank community dynamics in seasonally moist lowland tropical forest, Panama. *Journal of Tropical Ecology* 13, 659–680.

Dalling, J.W., Hubbell, S.P. and Silvera, K. (1998a) Seed dispersal, seedling establishment and gap partitioning among tropical pioneer trees. *Journal of Ecology* 86, 674–689.

Dalling, J.W., Swaine, M.D. and Garwood, N.C. (1998b) Dispersal patterns and seed bank dynamics of pioneer trees in moist tropical forest. *Ecology* 79, 564–578.

Daws, M.I., Burslem, D.F.R.P., Crabtree, L.M., Kirkman, P., Mullins, C.E. and Dalling, J.W. (2002) Differences in seed germination responses may promote coexistence of four sympatric *Piper* species. *Functional Ecology* 16, 258–267.

Dupuy, J.M. and Chazdon, R.L. (1998) Long-term effects of forest regrowth and selective logging on the seed bank of tropical forests in NE Costa Rica. *Biotropica* 30, 223–237.

Elzinga, C.L., Salzer, D.W., Willoughby, J.W. and Gibbs, J.P. (2001) *Monitoring Plant and Animal Populations.* Blackwell Science, Malden, Massachusetts, pp. 178–186.

Engelbrecht, B.M.J., Dalling, J.W., Pearson, T.R.H., Wolf, R.L., Galvez, D.A., Koehler, T., Ruiz, M.C. and Kursar, T.A. (2006) Short dry spells in the wet season increase mortality of tropical pioneer seedlings. *Oecologia* 148, 258–269.

Facelli, J.M. and Pickett, S.T.A. (1991) Plant litter: its dynamics and effects on plant community structure. *Botanical Review* 57, 1–32.

Fenner, M. and Thompson, K. (2005) *The Ecology of Seeds.* Cambridge University Press, Cambridge, UK.

Ferrer, A. and Gilbert, G.S. (2003) Effect of tree host species on fungal community composition in a tropical rain forest in Panama. *Diversity and Distributions* 9, 455–468.

Fornara, D.A. and Dalling, J.W. (2005) Post-dispersal removal of seeds of pioneer species from five Panamanian forests. *Journal of Tropical Ecology* 21, 79–84.

Fröhlich, J. and Hyde, K.D. (1999) Biodiversity of palm fungi in the tropics: are global fungal diversity estimates realistic? *Biodiversity and Conservation* 8, 977–1004.

Gallery, R.E., Dalling, J.W. and Arnold, A.E. (2007) Diversity, host affinity, and distribution of seed-infecting fungi: a case study with neotropical *Cecropia. Ecology* (in press)

Garwood, N.C. (1983) Seed germination in a seasonal tropical forest in Panama: a community study. *Ecological Monographs* 53, 159–181.

Gentry, A.H. (1990) Floristic similarities and differences between southern Central America and Upper and Central Amazonia. In: Gentry, A.H. (ed.) *Four Neotropical Rainforests.* Yale University Press, New Haven, Connecticut, pp. 141–157.

Gilbert, G.S. (1995) Rain forest plant diseases: the canopy-understory connection. *Selbyana* 16, 75–77.

Gilbert, G.S. (2005) The dimensions of plant disease in tropical forests. In: Burslem, D.R.F.P., Pinard, M.A. and Hartley, S. (eds) *Biotic Interactions in the Tropics.* Cambridge University Press, Cambridge, UK, pp. 141–164.

Gilbert, G.S., Ferrer, A. and Carranza, J. (2002) Polypore fungal diversity and host density in a moist tropical forest. *Biodiversity and Conservation* 11, 947–957.

Hall, J.B. and Swaine, M.D. (1980) Seed stocks in Ghanaian forest soils. *Biotropica* 12, 256–263.

Hammel, B.A. (1986) The vascular flora of La Selva Biological Station, Costa Rica, Cecropiaceae. *Selbyana* 9, 192–195.

Hartshorn, G.S. (1978) Tree falls and tropical forest dynamics. In: Tomlinson, P.B. and Zimmermann, M.H. (eds) *Tropical Trees as Living Systems.* Cambridge University Press, Cambridge, UK, pp. 617–638.

Holdridge, L.R., Grenke, W.C., Hatheway, W.H., Liang, T. and Tosi, J.R. Jr (1971) *Forest Environments in Tropical Life Zones.* Pergamon Press, New York, USA.

Howe, H.F. and Smallwood, J. (1982) Ecology of seed dispersal. *Annual Review of Ecology and Systematics* 13, 201–228.

Hubbell, S.P. and Foster, R.B. (1986) Canopy gaps and the dynamics of a neotropical forest. In: Crawley, M.J. (ed.) *Plant Ecology.* Blackwell Scientific, Oxford, UK, pp. 77–96.

Janzen, D.H. (1970) Herbivores and the number of species in tropical forests. *American Naturalist* 104, 501–528.

Jones, M.B. (1994) Secondary seed removal by ants, beetles, and rodents in a neotropical moist forest. MS thesis, University of Florida, Gainesville, Florida.

Kaspari, M. (1996) Testing resource-based models of patchiness in four neotropical litter ant assemblages. *Oikos* 76, 443–454.

Leishman, M.R., Masters, G.J., Clarke, I.P. and Brown, V.K. (2000) Seed bank dynamics: the role of fungal pathogens and climate change. *Functional Ecology* 14, 293–299.

Levey, D.J. and Byrne, M.M. (1993) Complex ant–plant interactions: rain forest ants as secondary dispersers and post-dispersal seed predators. *Ecology* 74, 1802–1812.

Lonsdale, W.M. (1993) Losses from the seed bank of *Mimosa pigra*: soil microorganisms vs. temperature fluctuations. *Journal of Applied Ecology* 30, 654–660.

Mancini, F., Apetorgbor, M., Cobbinah, J. and Ragazzi, A. (2001) Potential fungal pathogens on seeds and seedlings of *Milicia excelsa* of three ecological zones in Ghana. *Journal of Plant Disease and Protection* 108, 31–38.

Metcalfe, D.J. and Grubb, P.J. (1997) The ecology of very small-seeded shade-tolerant trees and shrubs in lowland rain forest in Singapore. *Functional Ecology* 11, 215–221.

Milton, K. (1991) Leaf change and fruit production in six neotropical Moraceae species. *Journal of Ecology* 79, 1–26.

Murray, K.G. (1988) Avian seed dispersal of three neotropical gap-dependent plants. *Ecological Monographs* 58, 271–298.

Murray, K.G. and Garcia-C., J.M. (2002) Contributions of seed dispersal and demography to recruitment limitation in a Costa Rican cloud forest. In: Levey, D.J., Silva, W.R. and Galetti, M. (eds) *Seed Dispersal and Frugivory: Ecology, Evolution, and Conservation.* CAB International, Wallingford, UK, pp. 323–338.

O'Hanlon-Manners, D.L. and Kotanen, P.M. (2004) Evidence that fungal pathogens inhibit recruitment of a shade-intolerant tree, white birch (*Betula papyrifera*), in understory habitats. *Oecologia* 140, 650–653.

Packer, A. and Clay, K. (2000) Soil pathogens and spatial patterns of seedling mortality in a temperate tree. *Nature* 404, 278–281.

Parker, M.A. (1985) Local population differentiation for compatibility in an annual legume and its host-specific fungal pathogen. *Evolution* 39, 713–723.

Putz, F.E. (1983) Treefall pits and mounds, buried seeds, and the importance of soil disturbance to pioneer trees on Barro Colorado Island, Panama. *Ecology* 64, 1069–1074.

Putz, F.E. and Appanah, S. (1987) Buried seeds, newly dispersed seeds, and the dynamics of a lowland forest in Malaysia. *Biotropica* 19, 326–333.

Sanford, R.L. Jr, Paaby, P., Luvall, J.C. and Phillips, E. (1994) Climate, geomorphology, and aquatic systems. In: McDade, L.A., Bawa, K.S., Hespenheide, H.A. and Hartshorn, G.S. (eds) *La Selva: Ecology and Natural History of a Neotropical Rain Forest.* University of Chicago Press, Chicago, Illinois, pp. 19–33.

SAS (2003) *Statistical Analysis Software, v. 9.1.* SAS Institute, Cary, North Carolina.

Schafer, M. and Kotanen, P.M. (2003) The influence of soil moisture on losses of buried seeds to fungi. *Acta Oecologica* 24, 255–263.

Schafer, M. and Kotanen, P.M. (2004) Impacts of naturally-occurring soil fungi on seeds of meadow plants. *Plant Ecology* 175, 19–35

Smith, C.C. and Fretwell, S.D. (1974) The optimum balance between size and number of offspring. *American Naturalist* 108, 499–506.

Sytsma, K.J., Morawetz, J., Pires, J.C., Nepokroeff, M., Conti, E., Zjhra, M., Hall, J.C. and Chase, M.W. (2002) Urticalean rosids: circumscription, rosid ancestry, and phylogenetics based on *rbcL*, *trnL-F*, and *ndhF* sequences. *American Journal of Botany* 89, 1531–1546.

Vázquez-Yanes, C. (1974) Studies on the germination of seeds of *Ochroma lagopus* Swartz. *Turrialba* 24, 176–179.

Vázquez-Yanes, C. and Orozco-Segovia, A. (1982) Seed germination of a tropical rain forest pioneer tree (*Heliocarpus donnell-smithii*) in response to diurnal fluctuation of temperature. *Physiologia Plantarum* 56, 295–298.

Vázquez-Yanes, C. and Smith, H. (1982) Phytochrome control of seed germination in the tropical rain forest pioneer trees *Cecropia obtusifolia* and *Piper auritum* and its ecological significance. *New Phytologist* 92, 477–485.

Vázquez-Yanes, C., Orozco-Segovia, A., Rincon, E., Sanchez-Coronado, M.E., Huante, P., Toledo, J.R. and Barradas, V.L. (1990) Light beneath the litter in a tropical forest: effect on seed germination. *Ecology* 71, 1952–1958.

White, T.J., Bruns, T., Lee, S. and Taylor, J. (1990) Amplification and direct sequencing of fungal ribosomal RNA genes for phylogenetics. In: Innis, M.A., Gelfand, D.H., Sninsky, J.J. and White, T.J. (eds) *PCR Protocals: A Guide to Methods and Applications.* Academic Press, San Diego, Carolina, pp. 315–322.

Windsor, D.M. (1990) Climate and moisture variability in a tropical forest: long term records from Barro Colorado Island, Panama. *Smithsonian Contributions to the Earth Sciences.* Smithsonian Institute Press, Washington, DC, p. 144.

Wulff, R.D. (1986) Seed size variation in *Desmodium paniculatum*. I. Factors affecting seed size. *Journal of Ecology* 74, 87–97.

Young, K.R., Ewel, J.J. and Brown, B.J. (1987) Seed dynamics during forest succession in Costa Rica. *Vegetatio* 71, 157–174.

23 Determinants of Tree Species Distributions: Comparing the Roles of Dispersal, Seed Size and Soil Specialization in a Bornean Rainforest

S.E. Russo, M.D. Potts, S.J. Davies and S. Tan

Introduction

Understanding the factors influencing plant species distributions in local communities is critical to explaining how these communities are structured. This is evidenced by the prominence of spatial variation in many hypotheses of species diversity, such as the niche differentiation (Macarthur and Levins, 1967; Tilman and Pacala, 1993), dispersal limitation (Levin, 1974; Hurtt and Pacala, 1995) and natural enemies hypotheses (Janzen, 1970; Connell, 1971). From an ecological perspective, spatial distribution patterns (dispersion) are extremely important for sedentary organisms such as plants because they can mediate species coexistence, which influences the maintenance of species diversity. From an evolutionary perspective, a plant's location affects its fitness and the selection pressures it is exposed to, which influence lineage diversification and the origins of species diversity.

Three important factors that influence the dispersion of plants are *seed dispersal* (Nathan and Muller-Landau, 2000; Russo and Augspurger, 2004), *seed size* (Westoby *et al.*, 1996) and *niche specialization* (Tilman and Kareiva, 1997). However, there is little consensus as to the relative magnitude of their effects and how they interact in determining the dispersion of plants in different communities. Through seed size–number trade-offs and interactions with dispersers, dispersal mode and seed size shape spatial patterns of seed deposition, which establish the initial template of offspring dispersion. The pattern of seed deposition determines how many offspring 'lottery tickets' are played in different environments (Dalling *et al.*, 2002; Schupp *et al.*, 2002; Russo and Augspurger, 2004). Over evolutionary time, spatial patterns of seed dispersal can thereby influence the frequency with which a species experiences novel selection regimes as well as gene flow between populations (Herrera, 2002; Holt *et al.*, 2004).

The size of a plant's seed can influence both its dispersal and the abiotic and biotic environments in which it is most successful (Baker, 1972; Grubb, 1977) due to trade-offs between size and either competitive (Levins and Culver, 1971; Tilman, 1994) or establishment (Dalling and Hubbell, 2002; Coomes and Grubb, 2003) ability. For example, compared with smaller seeds, larger seeds often have greater establishment probability in soils with low mineral nutrient availability and in shaded environments (Westoby *et al.*, 2002). Niche specialization after the seedling stage can modify post-dispersal offspring dispersion (Wada and Ribbens, 1997; Dalling and Hubbell, 2002). Functional traits governing resource acquisition determine the environments in which species are most successful (Reich *et al.*, 1995; Wright *et al.*, 2001). Hence, over ecological timescales, seed dispersal, seed size and niche specialization affect dispersion, whereas over evolutionary timescales, they affect the selection pressures and gene flow that facilitate evolution in novel environments (Van Tienderen, 1991; Westoby *et al.*, 1996; Herrera, 2002; Holt *et al.*, 2004).

The research reported in this chapter was conducted on a 52-ha forest dynamics plot (Lambir) in species-rich, mixed dipterocarp rainforest in Borneo. At Lambir, spatial variations in the distribution of tree species are strongly influenced by variations in soil chemistry and topography (Davies *et al.*, 2005). Soils in Lambir range from sandy loams, which are sandstone-derived, nutrient-poor and well drained; to clays, which are shale-derived, more nutrient-rich and less well drained (Lee *et al.*, 2002). The distributions of most species of tree were significantly aggregated on at least one of four distinct soil types identified in Lambir (Davies *et al.*, 2005). Such edaphic niche specialization is widespread and consistent among species of tree in forests across Borneo, making it likely to be a species-level trait (Ashton, 1964; Brunig, 1974; Baillie *et al.*, 1987; Potts *et al.*, 2002). Variation in soil moisture and fertility may be two important factors underlying this floristic variation among soil types (Palmiotto *et al.*, 2004) and may have influenced the evolution of species to different edaphic niches.

Local adaptation of a population to a novel environment depends on many characteristics of the population and traits in question (Holt *et al.*, 2002). One important factor is gene flow between populations in different environments (Wright, 1943; Lenormand, 2002). Furthermore, only a narrow range of successful colonization rates may facilitate adaptive evolution in a novel sink environment (Holt *et al.*, 2004). Gene flow in plants is partly mediated by colonization effected first by immigration and then successful establishment, processes that are influenced by seed dispersal mode, seed size and environmental niche specialization (Gibson and Wheelwright, 1995; Stanton and Galen, 1997; Jordano and Godoy, 2000; see Hardesty, Chapter 12, this volume). Thus, at Lambir, whether a species of tree is a soil specialist or a generalist may depend on its dispersal mode and seed size. In this study we tested this hypothesis by comparing ecological and evolutionary patterns of covariation among dispersal mode, seed size and soil specialization at Lambir. By arraying these species-level traits on a supertree phylogeny of the species of tree, in this community, we assessed whether divergences in dispersal mode and seed size were associated with divergences in edaphic niche. We also sought to dissect the effects of these factors as determinants of the distributions of species of tree at small and large spatial scales in this rainforest tree community.

Methods

Study site and data sets

Lambir Hills National Park, Sarawak, Malaysia, (4°11′ N, 114°01′ E) encompasses 6800 ha of lowland mixed dipterocarp forest and has the highest tree species diversity recorded in the Palaeotropics (Ashton and Hall, 1992; Lee *et al.*, 2002). The Park receives *c.*3000 mm of rainfall annually, with all months averaging > 100 mm (Watson, 1985). Animal seed dispersers include primates, birds, small arboreal and terrestrial mammals, and large terrestrial mammals (Shanahan and Debski, 2002). In 1991, a 52-ha research plot (hereafter, Lambir) was established in the Park to monitor all woody plants ≥ 1 cm diameter at breast height (dbh). Methods for this project followed similar studies coordinated by the Centre for Tropical Forest Science (Condit, 1998; Ashton *et al.*, 1999). All trees ≥ 1 cm dbh excluding palms were tagged, mapped, identified, and their diameters measured to the nearest 1 mm. A full description of the floristic composition and stand structure of Lambir is in Lee *et al.* (2002).

The soils and geomorphology of Lambir have been previously described (Lee *et al.*, 2002). Soils are Typic and Inceptic Hapludults and related Dystrudepts (Soil Survey Staff, 1999). They range from coarse loams that are sandstone-derived, leached, nutrient-depleted and well drained, with up to 10 cm of raw humus on the surface (referred to in previous publications as *humult*); to clays that are shale-derived, less nutrient-depleted and less well drained, with little raw humus (referred to in previous publications as *udult*; Lee *et al.*, 2002). In a previous study (Davies *et al.*, 2005), four soil types in Lambir were distinguished based on variations in soil nutrients (total C, N and P, and exchangeable K, Ca and Mg) and elevation at a 20 × 20 m scale (Fig. 23.1). Ranked in ascending order of fertility and moisture, these soil types are here referred to as: sandy loam (30.64 ha in area in the plot), loam (7.36 ha), fine loam (10.80 ha) and clay (3.20 ha), which correspond to habitats *A*, *B*, *C* and *D*, respectively, in Davies *et al.* (2005). Lambir is topographically heterogeneous, and disturbance rates vary among the different soil types (Palmiotto, 1998). The loam and fine loam have steeper average slopes than the sandy loam and clay (35.2° and 20.4° versus 17.1° and 18.0°, respectively). Consequently, they have more frequent and larger gaps resulting from landslips, as suggested by a map of the locations of the fastest growing 2% of all stems (Fig. 23.1f).

Davies *et al.* (2005) identified the tree species in Lambir showing spatial distribution patterns biased with respect to these four soil types using a Poisson cluster model (Plotkin *et al.*, 2000). Species were classified as either

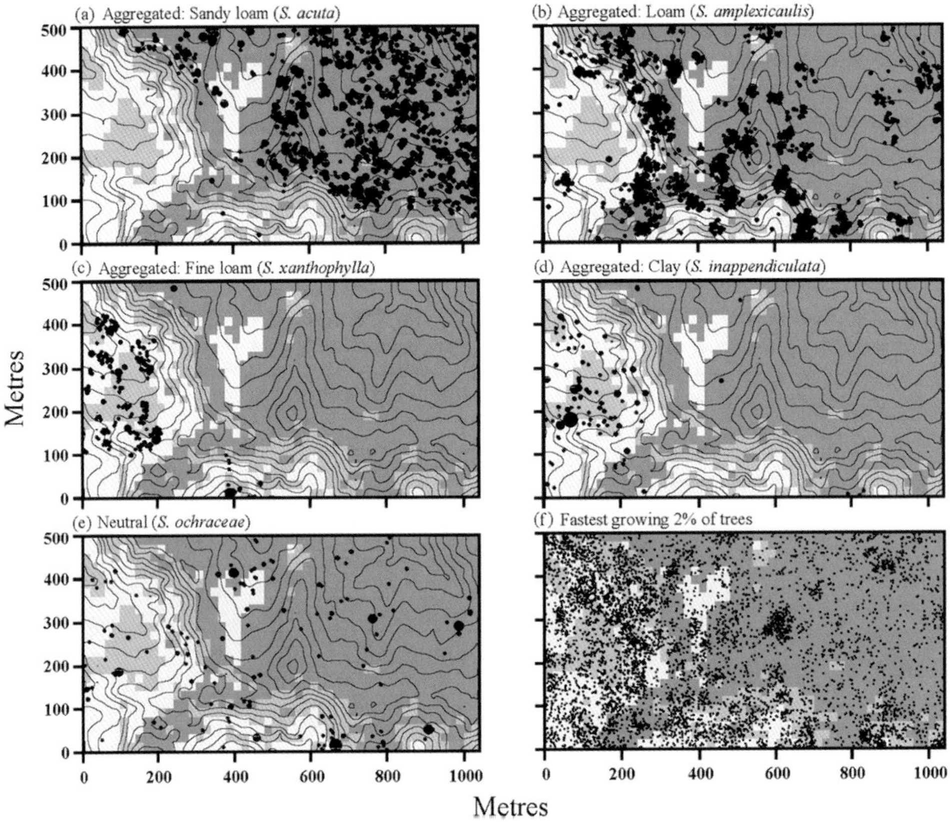

Fig. 23.1. Distributions of five species of *Shorea* (Dipterocarpaceae) that are significantly aggregated on (a) sandy loam, (b) loam, (c) fine loam, or (d) clay soils or (e) with a neutral distribution, in the Lambir plot, Borneo. In parts a–e, circles are scaled according to the diameter of the tree. Locations of the fastest growing 2% of all trees in the Lambir plot are also shown (f). Shading indicates soil types: darkest grey, sandy loam; light grey, loam; white, fine loam; and lightest grey, clay.

aggregated on, repelled from, or having a neutral distribution with respect to each soil type. Among the 764 species tested, 73% had distributions significantly aggregated on one or two soil types (Davies *et al.*, 2005). Here we focus on these species plus those with a neutral distribution with respect to soil (no aggregation or repulsion; 13% of species). For simplicity, for species that were found to be aggregated on two soil types (11% of all species; Davies *et al.*, 2005), we designated the soil type with greater tree density as that of aggregation. Thus, we investigated five possible species-aggregation patterns in this study: species aggregated (specialists) on *sandy loam* (285 species), on *loam* (66 species), on *fine loam* (152 species) and on *clay* (57 species), and species showing *neutral* patterns (generalists: 100 species). Five species of *Shorea* (Dipterocarpaceae) showing these distribution patterns are illustrated in Fig. 23.1a–e.

The results of the Poisson cluster model analysis indicated that at larger spatial scales, distributions of tree species within Lambir are strongly determined by edaphic variation (Davies *et al.*, 2005). Visual inspection of species' distributions within the plot (e.g. Fig. 23.1a–e) none the less indicated substantial interspecific variation in dispersion at smaller spatial scales. Here we use spatial statistics (see Statistical analyses section) to test the hypothesis that interspecific variation in small-scale dispersion can be explained by dispersal mode, seed size and niche specialization.

We collected data on primary dispersal mode and seed length on as many as possible of the 764 species for which soil specialization was categorized (308 species in 119 genera and 47 families). The seed was defined as the diaspore with the dispersal morphology removed. Seed weight may be a preferred metric of seed size but we were able to collect seed lengths for more species, allowing stronger tests of covariation between seed size, dispersal mode and soil specialization. We collected data from diverse sources, including measurements of seeds in herbarium specimens, seed measurements reported in the literature, and data from other researchers (see Acknowledgements).

Based on fruit morphology, personal observations, and reports in the literature, we categorized species into four primary dispersal modes: (i) ballistic (i.e. explosive) dispersal; (ii) gravity (i.e. unassisted) dispersal; (iii) dispersal by animals; and (iv) dispersal by wind. Although more finely divided dispersal categories would be preferable, for most species we have little detailed information about species-specific dispersal mechanisms or the distributions of dispersal distances of seeds. In addition, it is likely that many species in each primary dispersal category are secondarily dispersed by small mammals, but insufficient data exist to account for this process.

We constructed a phylogenetic hypothesis of the evolutionary relationships among the 308 species in our analyses, based on phylogenetic hypotheses for these taxa published in the literature. This supertree phylogeny (Sanderson *et al.*, 1998) was assembled using the software Phylomatic (Webb and Donoghue, 2004) and is based on the most recent Angiosperm Phylogeny Group tree, with generic relationships within families based on other published phylogenies. All species within genera are represented as polytomies (unresolved within-genus relationships), which limits our ability to test patterns of trait variation among species within genera. Because the final phylogenetic tree was a composite from multiple sources, branch lengths could not be estimated. We therefore assumed a speciational model of evolution (branch lengths = 1 across the tree; Ackerly, 2004). Some comparative analyses that we used (i.e. those using the software Discrete; Pagel, 1994) do not accept polytomies. For these, we arbitrarily resolved polytomy relationships into a completely bifurcating tree, and gave the branches added in the process very short branch lengths (equal to 10^{-6}). How polytomies were resolved does not influence the outcomes of analyses when the branch lengths are so short. Species values for dispersal mode, seed size and edaphic niche specialization were arrayed on this phylogeny.

Statistical analyses

For non-phylogenetic analyses (analyses using species values), we used general linear models and type III tests (proc mixed in SAS; SAS Institute, 2000) to test the relationships between continuous variables and between continuous and categorical variables. Standard transformations were used to improve normality of continuous dependent variables. We used Poisson regression to test relationships between categorical variables in which the response variable was counts of species in each cross-classification category (proc genmod in SAS). For all tests of association, as a continuity correction, a value of 1 was added to all cell counts in cross-classification tables containing a zero or when 20% of all counts were < 5.

Tests of phylogenetic trait conservatism evaluate whether trait values of closely related species are more similar (trait evolution is conserved) or more dissimilar (traits have diverged evolutionarily) to each other than what would be expected by chance, or whether species' traits are randomly distributed across the phylogenetic tree (Blomberg *et al.*, 2003). Phylogenetic trait conservatism was tested using the methods implemented in the software Phylocom (Webb *et al.*, 2005). The methods in Phylocom are similar to those for independent contrasts but are modified to handle polytomies by calculating the standard deviation of the descendent trait values at each node from the observed species' values on the phylogeny. Significance is assessed based on 1000 randomizations of species' values across the tips of the phylogeny.

Phylogenetic independent contrasts (PICs) (Felsenstein, 1985), as calculated in the software Phylocom (Webb *et al.*, 2005), were used to test for correlated evolution between two continuous variables and to compare mean contrast values between two groups. In the first case, PICs were analysed using general linear models (proc mixed in SAS) with the intercept forced through the origin. In the second case, PICs were analysed using either Student's *t* or sign tests in two-way comparisons between levels of a factor (e.g. for dispersal modes: wind versus animal dispersal, wind versus ballistic dispersal, wind versus gravity dispersal, and so on). The phylogenetic tree was pruned so as to include only the species appropriate for the particular two-way comparison, and separate tests were run for each comparison.

Correlated evolution between categorical variables was analysed using a likelihood ratio test of whether the transition rates between different character states of a trait on the phylogeny are best modelled as independent of or dependent upon each other, as implemented in the software Discrete (Pagel, 1994). Multistate categorical variables were analysed by transforming them into a series of binary state categorical variables, as described above.

We used the neighbourhood density function (Ω) to analyse species dispersion within soil types (Condit *et al.*, 2000; Dale *et al.*, 2002). This statistic is a modification of the second-order spatial statistic known as Ripley's K (Ripley, 2003) but the value of the statistic is calculated within annuli surrounding a focal tree. Many annuli included areas outside of the

52-ha plot, necessitating an edge correction so that only the area inside the plot was used (Condit *et al.*, 2000).

Dispersion is defined relative to an Ω value of 1, which indicates Poisson random dispersion within the tested annulus; values significantly >1 indicate clumped dispersion, and values <1 indicate over-dispersion (i.e. regularity). Significance was assessed by calculating Ω from 100 realizations of a Poisson random process. The number of times that Ω from the randomized data exceeded the observed value of Ω was tallied and the probability calculated as above. We calculated this statistic for all species with at least 50 saplings across the plot in each of three size classes (1–2 cm, 2–4 cm and 1–5 cm) and within each of three annuli (0–10 m, 10–20 m and 0–20 m). The results from analyses based on 1–5 cm saplings and the 0–20 m annulus are presented here. The results from this subset of data are consistent with the others tested. Because all species were significantly clumped at this scale, Ω can be considered a measure of small-scale clumping.

Results

Phylogenetic trait conservatism

Seed size, dispersal mode, and small-scale clumping intensity (Ω) were highly conserved phylogenetically (Table 23.1). In contrast, soil specialization was randomly distributed phylogenetically, meaning that related species were neither more nor less likely to share edaphic niches than distantly related species. This was true regardless of whether soil specialization categories were broken out individually (four soil specialists plus generalists) or whether all specialists were lumped into one category (specialists versus generalists; Table 23.1).

The random distribution of soil specialization on the phylogeny is illustrated by the fact that 73% of the families and 48% of the genera with > 1 species that were analysed for soil specialization contained at least one specialist and one generalist. Furthermore, 36% of families and 15% of

Table 23.1. Phylogenetic niche conservatism in dispersal mode, seed size and small-scale clumping intensity (Ω) for 1–5 cm saplings at 0–20 m spatial scale, for 308 tree species in a Bornean rainforest. Probabilities are based on randomization of the standard deviations of trait values across the supertree phylogeny of the 308 species. The multistate trait, soil specialization, was transformed into a series of binary traits to meet assumptions of the analysis, and the range of probabilities for possible pairwise comparisons for different binary combinations is given.

Trait	Conserved?	Probability
Dispersal mode	Yes	0.002
Seed size	Yes	0.002
Small-scale clumping intensity	Yes	0.002
Soil specialization	No (random)	0.1118–0.882
Generalists versus specialists	No (random)	0.498

genera with > 4 species had at least one representative of every soil specialist and one generalist. In fact no genus or family with five or more species was composed of only one kind of soil specialist or only of generalists. These results indicate that soil specialization is evolutionarily far more labile than dispersal mode, seed size or small-scale clumping intensity, all of which were more conserved.

Covariation between dispersal mode, seed size and soil specialization

We next tested whether dispersal mode and soil specialization showed significant covariation in ecological and evolutionary contexts. From an ecological perspective, soil specialists and generalists had similar overall frequencies of species in each of the four dispersal modes (Table 23.2). In other words, generalists did not have proportionally more species that were good dispersers (wind- or animal-dispersed), relative to poor dispersers (ballistically or gravity-dispersed). This result held whether soil specialists were all tested separately ($\chi^2 = 12.39$, $\nu = 12$, $P = 0.415$) or grouped together ($\chi^2 = 4.86$, $\nu = 3$, $P = 0.182$). Similarly, from an evolutionary perspective, transitions between different dispersal modes on the phylogenetic tree were not significantly associated with transitions from being a generalist to a specialist species (likelihood ratio test for a model of independent versus dependent trait transitions $\nu = 4$ for all tests; animal versus gravity: $\chi^2 = 3.03$, $P = 0.554$; animal versus ballistic: $\chi^2 = 0.83$, $P = 0.935$; wind versus ballistic: $\chi^2 = 1.72$, $P = 0.787$; wind versus animal: $\chi^2 = 1.44$, $P = 0.838$; wind versus gravity: $\chi^2 = 0.77$, $P = 0.9426$; ballistic versus gravity: $\chi^2 = 2.76$, $P = 0.175$). Furthermore, transitions on the phylogeny from being a well-dispersed species (wind- or animal-dispersed) to being a poorly dispersed species (ballistically or gravity-dispersed) were not associated with transitions from being a generalist to a specialist (likelihood ratio test for a model of independent versus dependent trait transitions; $\chi^2 = 6.984$, $\nu = 4$, $P = 0.137$). Thus, evolutionary transitions between modes of dispersal and soil specialization categories were relatively independent.

Table 23.2. Numbers of tree species in a Bornean rainforest in each cross-classification of dispersal mode and soil specialization categories. There was no significant association ($P > 0.05$) between dispersal mode and soil specialization categories, either with all five soil specialization categories tested separately or with all specialist categories grouped and compared with generalists.

	Soil specialization				
Dispersal mode	Sandy loam	Loam	Fine loam	Clay	Generalist
Ballistic	7	1	4	2	1
Gravity	8	0	1	1	4
Animal	79	20	53	12	24
Wind	18	7	10	7	9

We used a general linear model to test for main effects of both dispersal mode, soil specialization, and their interaction on interspecific variation in seed length. Seed size varied significantly among dispersal modes ($F_{(3, 260)} = 7.57$, $P < 0.001$; Fig. 23.2a) and among soil specialization categories ($F_{(4, 260)} = 7.29$, $P < 0.001$; Fig. 23.2b), but there was no significant interaction between these factors ($F_{(11, 249)} = 0.72$, $P = 0.720$). In other words, species with the same dispersal mode had similar seed sizes across the different soil specialization categories.

When ranked by increasing mean seed length, ballistic-dispersed species had the smallest seeds, followed by animal-, wind- and gravity-dispersed species (Fig. 23.2a). Seed size of wind-dispersed species was somewhat larger than expected, probably due to an effect of species in the Dipterocarpaceae, 83% of which have winged fruits with large seeds (33 of 40 species). Specialists of both the poorest (sandy loam) and the richest (clay) soils and generalists had the largest seeds (Fig. 23.2b). The smallest seeds were of species specializing on the two soils with intermediate fertility and moisture, the loam and fine loam soils.

We used PICs to test for correlated evolution between changes in seed length and either dispersal modes or soil specialization categories, using binary codings of these categorical variables. The results were similar to those of the non-phylogenetic analyses. Among the species contrasted, smaller seeds had an evolutionary association with specialists of loam compared with specialists of sandy loam (17 of 19 contrasts; sign test $P < 0.001$) and compared with specialists of clay (10 of 12 contrasts; sign test $P = 0.039$). No other seed length–soil specialization pairwise combinations were significantly correlated (sign tests, $P > 0.05$), even when specialists of the two richer soils (fine loam and clay) were grouped and compared with specialists of the two poorer soils grouped (sandy loam and loam; 20 of 43 contrasts, sign test, $P = 0.761$). In addition, larger seeds had no significant evolutionary

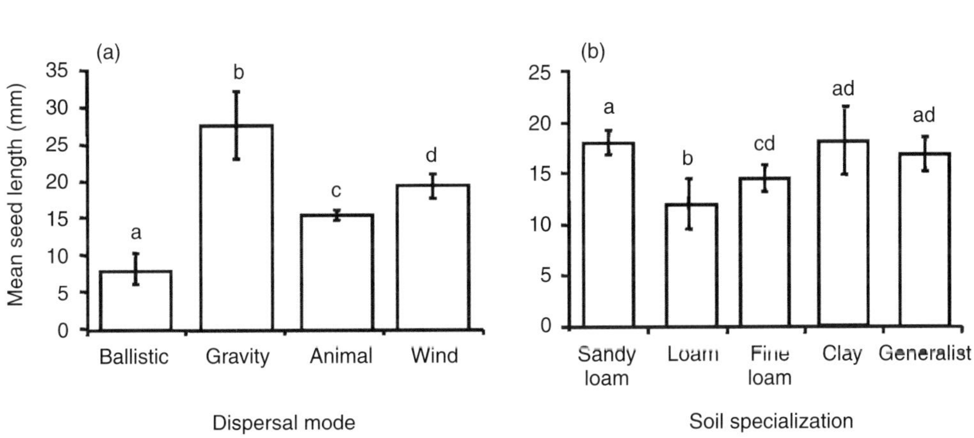

Fig. 23.2. Variation in seed length (mm) among tree species with different (a) dispersal modes and (b) soil specializations in a Bornean rainforest. Error bars are ± 1 SE. Lower-case letters indicate statistically significant differences among means within each panel.

association with being a generalist compared to a specialist (19 of 30 contrasts, sign test; $P = 0.201$).

Among the species contrasted in dispersal mode, larger seeds had an evolutionary association with animal- compared with gravity-dispersed species (3 of 3 contrasts; *t*-test, $P < 0.001$) and weakly for wind- compared with gravity-dispersed species (5 of 5 contrasts; sign test, $P = 0.063$). Other seed length–dispersal mode pairwise combinations were either not significantly correlated (wind versus animal dispersal: 4 of 9 contrasts; and ballistic versus animal dispersal: 1 of 3 contrasts; sign tests $P > 0.05$) or there were too few contrasts to test (wind versus ballistic and gravity versus ballistic dispersal).

Effects of dispersal mode, seed size and soil specialization on small-scale dispersion

The most intensely clumped families (largest values of Ω) were Violaceae (mean $\Omega = 37.6$), which consists of two ballistic-dispersed, understorey tree species in the genus *Rinorea*, followed by the Dipterocarpaceae (mean $\Omega = 23.4$), with 40 gravity- and wind-dispersed species, and lastly, Theaceae (mean $\Omega = 13.1$), with two animal-dispersed, pioneer species in the genus *Adinandra*. The two most clumped species were dipterocarps, *Shorea macrophylla* ($\Omega = 113.28$) and *Dipterocarpus kunstleri* ($\Omega = 109.27$), both wind-dispersed pioneers that are specialists of clay soil. The least clumped species were in the families Anisophylleaceae (mean $\Omega = 1.9$; one animal-dispersed and one gravity-dispersed species) and Bombacaeae (mean $\Omega = 2.1$), Annonaceae (mean $\Omega = 3.0$) and Burseraceae (mean $\Omega = 3.0$), all three of which contain only animal-dispersed species. The gravity-dispersed species in the Anisophylleaceae has quite large seeds that germinate from within the leathery fruit, which suggests that it may be secondarily dispersed by small mammals.

The effects of seed length, dispersal mode and soil specialization on small-scale clumping intensity were tested using a general linear model. All two-way interactions were tested, except the soil specialization × dispersal mode interaction, because some factor combinations had less than two observations. In addition to the significant main effects (dispersal mode: $F_{(3, 216)} = 8.42$, $P < 0.001$; soil specialization: $F_{(4, 216)} = 6.55$, $P < 0.0001$), the only interaction that was significant was that between seed length and dispersal mode (seed length × dispersal mode: $F_{(3, 216)} = 5.15$, $P = 0.002$; seed length × soil specialization: $F_{(4, 212)} = 0.56$, $P = 0.692$).

Variation due to dispersal mode was significant, with animal-dispersed species being the least intensely clumped, wind and gravity-dispersed species being most intensely clumped, and ballistically dispersed species having intermediate clumping intensity (Fig. 23.3a). The intense clumping of wind-dispersed species is somewhat surprising, but can be explained by the importance of the large-seeded Dipterocarpaceae in the wind-dispersal category (80% of wind-dispersed species).

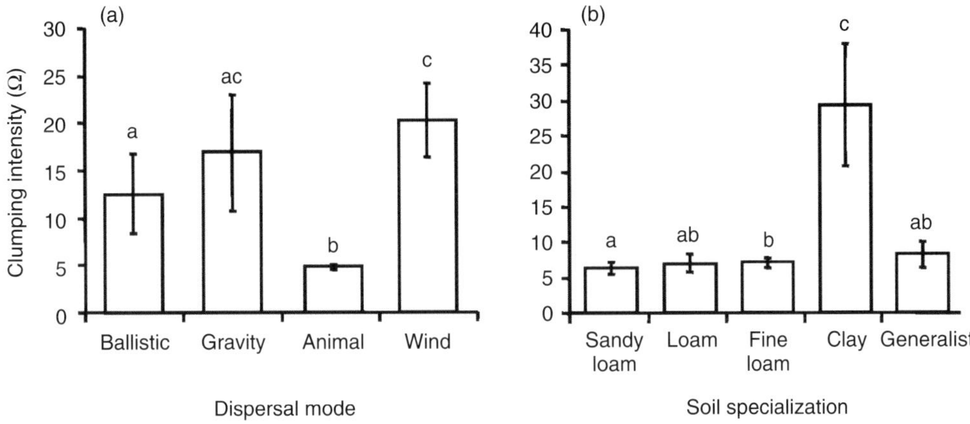

Fig. 23.3. Variation in Ω, the intensity of small-scale (0–20 m) clumping of juveniles (1–5 cm dbh saplings), among tree species with different (a) dispersal modes and (b) soil specializations in a Bornean rainforest. Error bars are ± 1 SE. Lower-case letters indicate statistically significant differences among means within each panel.

By far the most intensely clumped species were among specialists of the clay soil, whereas the least-clumped species were specialists of the two poorer soils, sandy loam and loam soils, and generalists (Fig. 23.3b). This result is partly due to the fact that of the 17 species for which we had seed size and dispersal information that were also clay specialists, seven were from the two most-clumped families, Violaceae and Dipterocarpaceae, and included the two most intensely clumped species on the entire plot. Even after removing species in these two families from the analysis, the clay specialists still remained the most intensely clumped. Thus, these two families were not completely responsible for this significant effect, but rather make the underlying difference of the clay specialists relative to the others even stronger.

Slopes of the relationship between small-scale clumping intensity and seed length were significantly different from zero for all dispersal modes except for ballistically dispersed species (Fig. 23.4a; $b = -0.013$, $SE_b = 0.023$, $P = 0.586$). Animal- and gravity-dispersed species had significantly negative, but shallow, slopes (animal: $b = -0.012$, $SE_b = 0.006$, $P = 0.044$; gravity: $b = -0.042$, $SE_b = 0.013$, $P = 0.001$). In contrast, wind-dispersed species had a significantly positive slope ($b = 0.026$, $SE_b = 0.008$, $P = 0.002$). The negative slopes indicate that animal- and gravity-dispersed species with larger seeds have less intense small scale clumping than do those with smaller seeds, whereas the reverse is true for wind-dispersed species.

Some analyses of small-scale clumping intensity based on PICs differed from those based on species values. Among the species contrasted, increased clumping was not an evolutionary pattern associated with specialist, compared with generalist, species (17 of 26 contrasts; sign test, $P = 0.169$), parallel to analyses based on species values. Based on PICs, species with

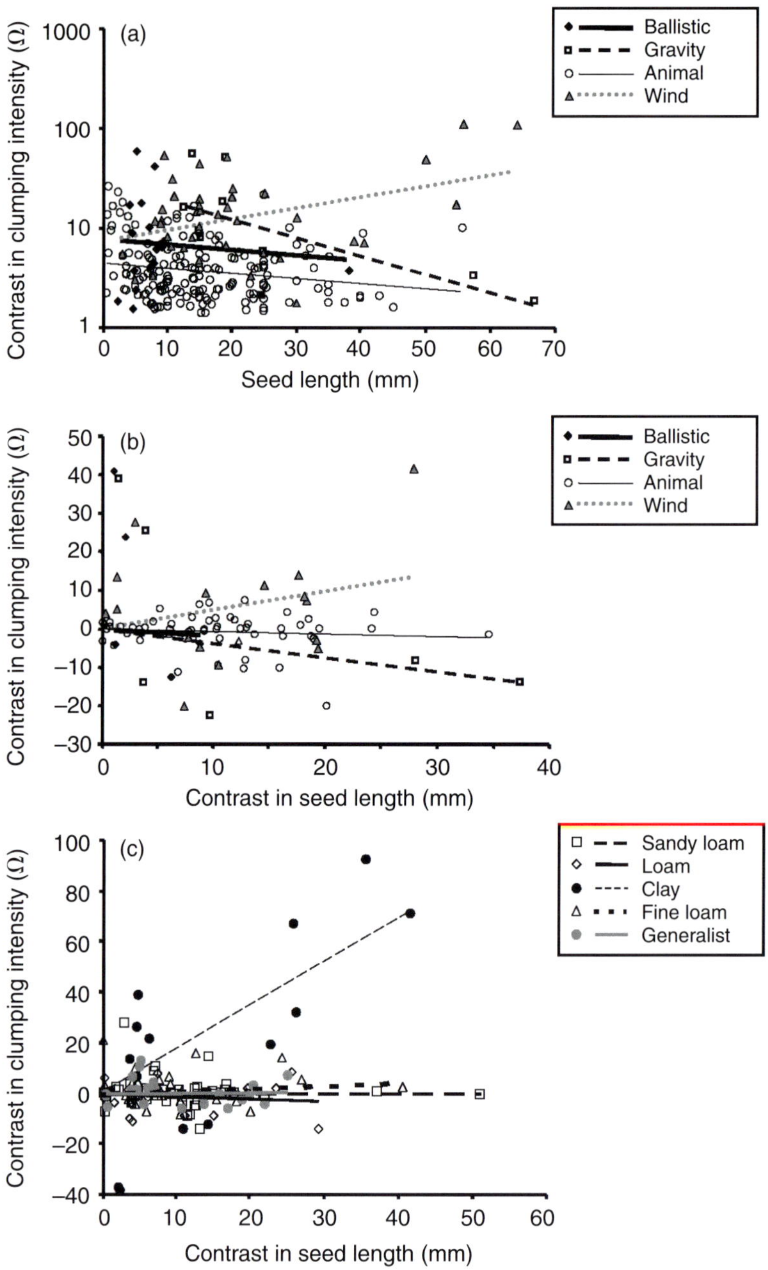

Fig. 23.4. Variation in Ω, the intensity of small-scale (0–20 m) clumping of juveniles (1–5 cm dbh saplings) as a function of seed length (mm) for tree species in a Bornean rainforest, distinguishing dispersal mode (a, non-phylogenetic analysis; and b, phylogenetic analysis) and soil specialization (c, phylogenetic analysis). Lines are best-fit lines based on ordinary least-squares regression. Non-phylogenetic analyses were based on species' values; phylogenetic analyses were based on independent contrasts with the intercept forced through the origin (see Methods for details).

different dispersal modes varied in their relationship between seed length and small-scale clumping intensity (Fig. 23.4b). Slopes of the relationship between PICs in small-scale clumping intensity and seed length were not different from zero for ballistically dispersed ($b = -0.189$, $SE_b = 0.964$, $P = 0.845$), gravity-dispersed ($b = -0.374$, $SE_b = 0.222$, $P = 0.096$) and animal-dispersed species ($b = -0.038$, $SE_b = 0.114$, $P = 0.742$). In contrast, wind-dispersed species had a significantly positive slope ($b = 0.486$, $SE_b = 0.188$, $P = 0.012$), as was found in analyses based on species values.

In contrast to analyses based on species values, there was significant variation among soil specialists in the relationship between PICs for seed length and for small-scale clumping intensity (Fig. 23.4c). This effect was due to the significant positive slope for clay specialists ($b = 1.742$, $SE_b = 0.141$, $P < 0.001$). For all other soil specialists, there was no significant relationship (sandy loam: $b = -0.008$, $SE_b = 0.102$, $P = 0.936$; loam: $b = -0.113$, $SE_b = 0.177$, $P = 0.526$; fine loam: $b = 0.106$, $SE_b = 0.141$, $P = 0.453$, generalists: $b = 0.014$, $SE_b = 0.171$, $P = 0.933$).

Discussion

We found no strong ecological or evolutionary relationships between the probability of being a soil specialist or generalist and either dispersal mode or seed size. Dispersal modes occurred with equal frequency among the five different soil specialization categories, and seed sizes of species in the same dispersal mode were consistent across soil specialization categories, findings that are consistent with patterns from neotropical forests (Hammond and Brown, 1995). In addition, there were no significant associations between evolutionary transitions between different dispersal modes and soil specialization on the phylogenetic supertree of the species in this Bornean rainforest community. Therefore, generalists were not significantly better dispersers than specialists. Using these broad categories to estimate 'good' versus 'poor' dispersing species is clearly inferior to having more detailed information on the actual spatial patterns of seed dispersal, but we lack such information in this forest.

There was significant variation in seed size among soil specialization categories. None the less, the patterns we observed in this Bornean rainforest were inconsistent with the hypotheses that:

1. Larger seeds enhance establishment on soils with low fertility and moisture, restricting smaller-seeded species to richer soils;
2. Larger seeds enhance establishment on any soil type, facilitating evolution of a generalist strategy;
3. Smaller seeds enhance dispersal, facilitating evolution of a generalist strategy.

Although specialists of the poorest soil (sandy loam) and generalists had the largest seeds, they were not alone, as specialists of the richest soil (clay) also had large seeds. Specialists of loam and fine loam had the smallest seed sizes, which are often observed among tropical tree species with fast growth rates (Bazzaz and Pickett, 1980). These two soils, particularly loam, are located on steep slopes with greater frequency of disturbance from landslips. Consequently, they are likely to have greater light availability, which may favour species with faster growth responses. Hence, seed size variation among soil specialization categories may be better explained by light availability and disturbance on the different soil types, rather than soil fertility and moisture, which is consistent with similar analyses in New World tropical forests (Foster and Hanson, 1985; Hammond and Brown, 1995).

Tree species with different dispersal modes varied in seed size. Wind-dispersed seeds are generally smaller than seeds having other dispersal modes (Hammond and Brown, 1995; Westoby *et al.*, 1996). In contrast, as a group, wind-dispersed trees in this community had the largest seed sizes. This surprising result is simply because the dipterocarps, which comprise a majority of species in this forest, are large-seeded, yet most have wings for dispersal by wind (Webber, 1934).

Dispersal mode and seed size were highly conserved on this phylogeny of species in this forest community, as has been observed in temperate floras (Lord *et al.*, 1995). In contrast, soil specialization was evolutionarily labile. Consistent with this result, many families and genera contained representatives of multiple soil specialist and generalist species, but this was not true of dispersal mode. Taken together, these results indicate that tree species' occurrence in, and possibly diversification into, different edaphic niches in this forest may not depend strongly on dispersal mode or seed size. These results raise the possibility that edaphic niche specialization may have had a central role in the more recent diversification of these Bornean tree species, as has been found for the Burseraceae in Amazonia (Fine *et al.*, 2005). In contrast, divergences at deeper nodes in the phylogeny are more associated with differences among lineages in life history traits that are often viewed as more conserved, such as seed size and dispersal mode. A large-scale phylogenetic analysis of seed size variation also found that deeper divergences were associated with large shifts in seed size (Moles *et al.*, 2005). However, studies in a temperate conifer community and an African tropical forest found that variation in seed size was randomly distributed on the phylogenies of species in those communities (Veech *et al.*, 2000; Zanne *et al.*, 2005), suggesting that in some communities there may be ecological sorting to minimize similarity in seed size among species, which may be important for community assembly. In contrast, in this Bornean rainforest, ecological sorting via edaphic niche specialization (Russo *et al.*, 2005) appears to play a more prominent role.

Tree species dispersion within the 52-ha Lambir plot was determined by different mechanisms operating at different ecological scales. At larger spatial scales, distributions are largely determined by edaphic variation (Davies *et al.*, 2005). At small spatial scales of 0–20 m, clumping intensity of juveniles was significantly associated with variation in dispersal mode, seed size and soil specialization, but most strongly by dispersal mode and seed size. The fact that small-scale clumping intensity was strongly phylogenetically conserved in this community may be due to its strong association with dispersal mode and seed size, which were likewise very well-conserved traits on this phylogeny.

Spatial extent of soil types in the plot may affect our estimates of clumping intensity of soil specialists, especially since sample size requirements for spatial statistics necessitated using species with at least 50 stems on the plot. For example, specialists of the soil with the smallest area, clay, were by far the most intensely clumped. It is clear that area does not explain all soil-related variation in clumping intensity because clay specialists were over four times more intensely clumped than specialists of loam, which has only twice the area of clay and is distributed in a narrow band through the plot.

For gravity- and animal-dispersed species, larger seeds were associated with less intense clumping of juveniles. This negative relationship between clumping and seed size may be explained by at least a few processes.

- Small mammals that secondarily disperse seeds by scatter- or larder-hoarding them may prefer larger seeds (Jansen *et al.*, 2002), which would reduce clumping intensity.
- Depending on predator size, larger seeds may experience higher predation rates, especially near parents, which would also reduce clumping, although there is evidence both for and against larger-seeded species suffering greater predation (Kollmann *et al.*, 1998; Coomes and Grubb, 2003; Moles and Westoby, 2004).
- If different species of trees in this community experience a strong trade-off between seed size and number, species with larger seeds may produce fewer seeds, which could reduce clumping of their seed shadows relative to more fecund, smaller-seeded species, particularly if fruits with smaller seeds are more often multi-seeded. The reductions in potential recruitment that large-seeded species suffer (due to seed size–number trade-offs) may be greater than the establishment advantages gained from having larger seeds (Grubb and Coomes, 1997; Coomes and Grubb, 2003; Moles and Westoby, 2004).

For animal-dispersed species, the size of the effect of seed size on clumping intensity was relatively small. This result may be a consequence of the fact that many large-gaped dispersal agents often co-disperse large numbers of both large and small seeds, which could therefore be similar in their spatial patterns of seed deposition (Julliot, 1996; Whitney *et al.*, 1998; Galetti *et al.*, 2000; McConkey, 2000; Poulsen *et al.*, 2002; see Stevenson, Chapter 15, this volume; see Dennis and Westcott, Chapter 9, this volume). Thus, for animal-dispersed species, variation in seed deposition patterns due to seed size alone may be more strongly affected by seed size–number trade-offs (i.e. seed limitation) than by effects of seed size on dispersal agent behaviour (Muller-Landau *et al.*, 2002). In addition, neither negative relationship remained significant after incorporating phylogeny.

In contrast, for wind-dispersed species, seed size explained more variation in small-scale clumping intensity, which increased with increasing seed size. This finding is consistent with larger seeds being harder to disperse well by wind because of the tendency towards higher wing loading and faster terminal velocities with increasing seed size (Augspurger and Franson, 1987). Of course, larger wind-dispersed species may also be preferentially secondarily dispersed, have higher predation rates, and be subject to the same kinds of seed size–number trade-offs as species with other dispersal modes, but the positive slope suggests that the effects of poorer wind dispersal may be relatively greater.

Based on species values, the relationship between clumping intensity and seed size did not vary among soil specialists and generalists. On the other hand, in analyses with PICs, clay specialists showed steeply increasing clumping intensity with seed size. PICs of clay specialists were the highest values in the analysis. Six of the 17 clay specialists were dipterocarps, including the three most extreme PICs, suggesting that taxon sampling strongly influenced the dramatic increase in clumping with seed size on clay. Indeed, for all dipterocarps, there was a significant increase in small-scale clumping with seed size ($b = 0.026$, $\text{SE}_b = 0.009$, $n = 40$, $P = 0.011$, $r^2 = 0.16$), whereas this relationship was not different from zero for all other families with at least 10 species (Euphorbiaceae, $n = 40$; Myristicaceae, $n = 14$; Annonaceae, $n = 13$; Sapotaceae, $n = 12$; Moraceae, $n = 11$). Although a likely explanation for the intense small-scale clumping of dipterocarps is poor wind dispersal due to their large seeds, this is not the only possible explanation, particularly given reports of dipterocarp fruits being uplifted over the canopy (Webber, 1934). Because all dipterocarps are ectomycorrhizal (Alexander, 1989), spatial distributions of mycorrhizae may also play a role (see Theimer and Gehring, Chapter 21, this volume).

Conclusions and Future Directions

This study demonstrated that different mechanisms operate at different spatial scales in shaping the distribution patterns of tree species. Our analyses suggest that smaller-scale dispersion of juveniles is more strongly controlled by variation in dispersal mode and seed size, whereas larger-scale variation is more influenced by edaphic variation in this Bornean rainforest. Further tests of this hypothesis will include spatial analyses conducted at a larger set of spatial scales and sapling size classes. In addition, the role of small-scale variation in light, soil fertility and moisture is unaccounted for, and may also contribute to variation in smaller-scale dispersion.

We also found evidence that, in this rainforest tree community, species' occurrence in, and possibly diversification into, edaphic niches are relatively independent of constraints from dispersal mode and seed size. Because this preliminary conclusion is an explicit statement of trait evolution, a more robust test would use fully resolved phylogenies of several monophyletic groups of tree species combined with data on their comparative ecologies.

Acknowledgements

We thank the organizers of the *Frugivory and Seed Dispersal Symposium/ Workshop* for their invitation to contribute our research. We thank the Forestry Corporation of Sawarak, Malaysia, for permission to conduct research in Lambir Hills National Park. Peter S. Ashton, Rhett Harrison, Kim McConkey, Richard Saunders, Michael Shanahan and Campbell Webb graciously shared data on seed lengths and dispersal modes. We are indebted to Laura Lukas, who tirelessly gathered data in the Harvard University Herbaria collections, and to Emily Wood who facilitated that data collection. We appreciate comments from Peter S. Ashton and Andrew Marshall on earlier drafts of this chapter. Travel of the senior author to Australia to participate in the Symposium/Workshop was partially supported by the National Science Foundation (OISE #0456402 to E.W. Schupp).

References

Ackerly, D.D. (2004) Adaptation, niche conservatism, and convergence: comparative studies of leaf evolution in the California chaparral. *American Naturalist* 163, 654–671.

Alexander, I. (1989) Mycorrhizas in tropical forest. In: Proctor, J. (ed.) *Mineral Nutrients in Tropical Forest and Savanna Ecosystems.* Blackwell Scientific, Oxford, UK, pp. 169–188.

Ashton, P.S. (1964) Ecological studies in the mixed dipterocarp forests of Brunei State. *Oxford Forestry Memoir* 25, 1–75.

Ashton, P.S. and Hall, P. (1992) Comparisons of structure among mixed dipterocarp forests of north-western Borneo. *Journal of Ecology* 80, 459–481.

Ashton, P.S., Boscolo, M., Liu, J. and LaFrankie, J.V. (1999) A global programme in interdisciplinary forest research: the CTFS perspective. *Journal of Tropical Forest Science* 11, 180–204.

Augspurger, C.K. and Franson, S.E. (1987) Wind dispersal of artificial fruits varying in mass, area, and morphology. *Ecology* 68, 27–42.

Baillie, I.C., Ashton, P.S., Court, M.N., Anderson, J.A.R., Fitzpatrick, E.A. and Tinsley, J. (1987) Site characteristics and the distribution of tree species in mixed dipterocarp forest on Tertiary sediments in central Sarawak, Malaysia. *Journal of Tropical Ecology* 3, 201–220.

Baker, H.G. (1972) Seed weight in relation to environmental conditions in California. *Ecology* 53, 997–1010.

Bazzaz, F.A. and Pickett, S.T.A. (1980) Physiological ecology of tropical succession: a comparative review. *Annual Review of Ecology and Systematics* 11, 287–310.

Blomberg, S.P., Garland, T.E. Jr and Ives, A.R. (2003) Testing for phylogenetic signal in comparative data: behavioural traits are more labile. *Evolution* 57, 717–745.

Brunig, E.F. (1974) *Ecological Studies in the Kerangas Forests of Sarawak and Brunei.* Borneo Literature Bureau, Kuching, Malaysia.

Condit, R. (1998) *Tropical Forest Census Plots: Methods and Results from Barro Colorado Island, Panama and a Comparison with Other Plots.* Springer, Berlin, Germany.

Condit, R., Ashton, P.S., Baker, P., Bunyavejchewin, S., Gunatilleke, S., Gunatilleke, N., Hubbell, S.P., Foster, R.B., Itoh, A., LaFrankie, J.V., Lee, H.S., Losos, E., Manokaran, N., Sukumar, R. and Yamakura, T. (2000) Spatial patterns in the distribution of tropical tree species. *Science* 288, 1414–1418.

Connell, J.H. (1971) On the role of natural enemies in preventing competitive exclusion in some marine animals and in rain forest trees. In: den Boer, P.J. and Gradwell, G.R. (eds)

Proceedings of the Advanced Study Institute on Dynamics of Numbers in Populations: Osterbeek, The Netherlands, 7–18 September 1970. Centre for Agricultural Publishing and Documentation (PUDOC), Wageningen, The Netherlands, pp. 298–312.

Coomes, D.A. and Grubb, P.J. (2003) Colonization, tolerance, competition and seed-size variation within functional groups. *Trends in Ecology and Evolution* 18, 283–291.

Dale, M.R.T., Dixon, P., Fortin, M.-J., Legendre, P., Myers, D.E. and Rosenberg, M.S. (2002) Conceptual and mathematical relationships among methods for spatial analysis. *Ecography* 25, 558–577.

Dalling, J.W. and Hubbell, S.P. (2002) Seed size, growth rate and gap microsite conditions as determinants of recruitment success for pioneer species. *Journal of Ecology* 90, 557–568.

Dalling, J.W., Muller-Landau, H.C., Wright, S.J. and Hubbell, S.P. (2002) Role of dispersal in the recruitment limitation of neotropical pioneer species. *Journal of Ecology* 90, 714–727.

Davies, S.J., Tan, S., LaFrankie, J.V. and Potts, M.D. (2005) Soil-related floristic variation in the hyperdiverse dipterocarp forest in Lambir Hills, Sarawak. In: Roubik, D.W., Sakai, S. and Hamid, A. (eds) *Pollination Ecology and Rain Forest Diversity, Sarawak Studies.* Springer, New York, pp. 22–34.

Felsenstein, J. (1985) Phylogenies and the comparative method. *American Naturalist* 125, 1–15.

Fine, P.V.A., Daly, D.C., Muñoz, G.V., Mesones, I. and Cameroni, K.M. (2005) The contribution of edaphic heterogeneity to the evolution and diversity of Burseraceae trees in the western Amazon. *Evolution* 59, 1464–1478.

Foster, S.A. and Hanson, C.H. (1985) The relationship between seed size and establishment conditions in tropical woody plants. *Ecology* 66, 773–780.

Galetti, M., Laps, R. and Pizo, M.A. (2000) Frugivory by toucans (Ramphastidae) at two altitudes in the Atlantic Forest of Brazil. *Biotropica* 32, 842–850.

Gibson, J.P. and Wheelwright, N.T. (1995) Genetic structure in a population of a tropical tree *Ocotea tenera* (Lauraceae): influence of avian seed dispersal. *Oecologia* 103, 49–54.

Grubb, P.J. (1977) The maintenance of species-richness in plant communities: the importance of the regeneration niche. *Biological Reviews* 52, 107–145.

Grubb, P.J. and Coomes, D.A. (1997) Seed mass and nutrient content in nutrient-starved tropical rainforest in Venezuela. *Seed Science Research* 7, 269–280.

Hammond, D.S. and Brown, V.K. (1995) Seed size of woody plants in relation to disturbance, dispersal, soil type in wet neotropical forests. *Ecology* 76, 2544–2561.

Herrera, C.M. (2002) Seed dispersal by vertebrates. In: Herrera, C.M. and Pellmyr, O. (eds) *Plant–Animal Interactions: An Evolutionary Approach.* Blackwell Science, Oxford, UK, pp. 185–210.

Holt, R.D., Gomulkiewicz, R. and Barfield, M. (2002) The phenomenology of niche evolution via quantitative traits in a 'black-hole' sink. *Proceedings of the Royal Society London B* 270, 215–224.

Holt, R.D., Knight, T.M. and Barfield, M. (2004) Allee effects, immigration, and the evolution of species' niches. *American Naturalist* 163, 253–262.

Hurtt, G.C. and Pacala, S.W. (1995) The consequences of recruitment limitation: reconciling chance, history, and competitive differences between plants. *Journal of Theoretical Biology* 176, 1–12.

Jansen, P.A., Bartholomeus, M., Bongers, F., Elzinga, J.A., den Ouden, J. and Van Wieren, S.E. (2002) The role of seed size in dispersal by a scatter-hoarding rodent. In: Levey, D.J., Silva, W.R. and Galetti, M. (eds) *Seed Dispersal and Frugivory: Ecology, Evolution, and Conservation.* CABI International, Wallingford, UK, pp. 209–225.

Janzen, D.H. (1970) Herbivores and the number of tree species in tropical forests. *American Naturalist* 104, 501–528.

Jordano, P. and Godoy, J.A. (2000) RAPD variation and population genetic structure in *Prunus mahaleb* (Rosaceae), an animal-dispersed tree. *Molecular Ecology* 9, 1293–1305.

Julliot, C.M. (1996) Fruit choice by red howler monkeys (*Alouatta seniculus*) in a tropical rain forest. *American Journal of Primatology* 40, 261–282.

Kollmann, J., Coomes, D.A. and White, S.M. (1998) Consistencies in post-dispersal seed predation of temperate fleshy-fruited species among seasons, years and sites. *Functional Ecology* 12, 683–690.

Lee, H.S., Davies, S.J., LaFrankie, J.V., Tan, S., Yamakura, T., Itoh, A., Ohkubo, T. and Ashton, O. (2002) Floristic and structural diversity of mixed dipterocarp forests in Lambir Hills National Park, Sarawak, Malaysia. *Journal of Tropical Forest Science* 14, 379–400.

Lenormand, T. (2002) Gene flow and the limits to natural selection. *Trends in Ecology and Evolution* 17, 183–189.

Levin, S.A. (1974) Dispersion and population interactions. *American Naturalist* 108, 207–228.

Levins, R. and Culver, D. (1971) Regional coexistence of species and competition between rare species. *Proceedings of the National Academy of Sciences* 68, 1246–1248.

Lord, J., Westoby, M. and Leishman, M. (1995) Seed size and phylogeny in six temperate floras: constraints, niche conservatism, and adaptation. *American Naturalist* 146, 349–364.

Macarthur, R. and Levins, R. (1967) The limiting similarity, convergence, and divergence of coexisting species. *American Naturalist* 101, 377–385.

McConkey, K.R. (2000) Primary seed shadow generated by gibbons in the rain forests of Barito Ulu, Central Borneo. *American Journal of Primatology* 52, 13–29.

Moles, A.T. and Westoby, M. (2004) Seedling survival and seed size: a synthesis of the literature. *Journal of Ecology* 92, 372–383.

Moles, A.T., Ackerly, D.D., Webb, C.O., Tweddle, J.C., Dickie, J.B. and Westoby, M. (2005) A brief history of seed size. *Science* 307, 576–580.

Muller-Landau, H.C., Wright, S.J., Calderón, O., Hubbell, S.P. and Foster, R.B. (2002) Assessing recruitment limitation: concepts, methods and case-studies from a tropical forest. In: Levey, D.J., Silva, W.R. and Galetti, M. (eds) *Seed Dispersal and Frugivory: Ecology, Evolution, and Conservation.* CABI International, Wallingford, UK. pp. 35–54.

Nathan, R. and Muller-Landau, H.C. (2000) Spatial patterns of seed dispersal, their determinants and consequences for recruitment. *Trends in Ecology and Evolution* 15, 278–285.

Pagel, M. (1994) Detecting correlated evolution on phylogenies: a general method for the comparative analysis of discrete characters. *Proceedings of the Royal Society London B: Biological Sciences* 255, 37–45.

Palmiotto, P.A. (1998) The role of specialization in nutrient-use efficiency as a mechanism driving species diversity in a tropical rain forest. PhD thesis, Yale University, New Haven, Connecticut.

Palmiotto, P.A., Davies, S.J., Vogt, K.A., Ashton, M.S., Vogt, D.J. and Ashton, P.S. (2004) Soil-related habitat specialization in dipterocarp rain forest tree species in Borneo. *Journal of Ecology* 92, 609–623.

Plotkin, J.B., Potts, M.D., Leslie, N., Manokaran, N., LaFrankie, J. and Ashton, P.S. (2000) Species-area curves, spatial aggregation, and habitat specialization in tropical forests. *Journal of Theoretical Biology* 207, 81–99.

Potts, M.D., Ashton, P.S., Kaufman, L.S. and Plotkin, J.B. (2002) Habitat patterns in tropical rain forests: a comparison of 105 plots in northwest Borneo. *Ecology* 83, 2782–2797.

Poulsen, J.R., Clark, C.J., Connor, E.F. and Smith, T.B. (2002) Differential resource use by primates and hornbills: implications for seed dispersal. *Ecology* 83, 228–240.

Reich, P.B., Ellsworth, D.S. and Uhl, C. (1995) Leaf carbon and nutrient assimilation and conservation in species of differing successional status in an oligotrophic Amazonian forest. *Functional Ecology* 9, 65–76.

Ripley, B.D. (2003) *Statistical Analysis of Spatial Point Patterns.* Oxford University Press, Oxford, UK.

Russo, S.E. and Augspurger, C.K. (2004) Aggregated seed dispersal by spider monkeys limits recruitment to clumped patterns in *Virola calophylla*. *Ecology Letters* 7, 1058–1067.

Russo, S.E., Davies, S.J., King, D.A. and Tan, S. (2005) Soil-related performance variation and distributions of tree species in a Bornean rain forest. *Journal of Ecology* 93, 879–889.

Sanderson, M.J., Purvis, A. and Henze, C. (1998) Phylogenetic supertrees: assembling the trees of life. *Trends in Ecology and Evolution* 13, 105–109.

SAS Institute (2000) *The SAS System Release 8.1.* SAS Institute, Cary, North Carolina.

Schupp, E.W., Milleron, T. and Russo, S.E. (2002) Dissemination limitation and the origin and maintenance of species-rich tropical forests. In: Levey, D.J., Silva, W.R. and Galetti, M. (eds) *Seed Dispersal and Frugivory: Ecology, Evolution, and Conservation.* CAB International, Wallingford, UK. pp. 19–34.

Shanahan, M. and Debski, I. (2002) Vertebrates of Lambir Hills National Park, Sarawak. *Malayan Nature Journal* 56, 103–118.

Soil Survey Staff (1999) *Soil Taxonomy*, 2nd edn. Agricultural Handbook 436. US Department of Agriculture, Washington, DC.

Stanton, M.L. and Galen, C. (1997) Life on the edge: adaptation versus environmentally mediated gene flow in the snow buttercup, *Ranunculus adoneus. American Naturalist* 150, 143–178.

Tilman, D. (1994) Competition and biodiversity in spatially structured habitats. *Ecology* 75, 2–16.

Tilman, D. and Kareiva, P. (eds) (1997) *Spatial Ecology: The Role of Space in Population Dynamic and Interspecific Interactions.* Princeton University Press, Princeton, New Jersey.

Tilman, D. and Pacala, S. (1993). The maintenance of species richness in plant communities. In: Ricklefs, R.E. and Schluter, D. (eds) *Species Diversity in Ecological Communities.* University of Chicago Press, Chicago, Illinois. pp. 13–25.

Van Tienderen, P.H. (1991) Evolution of generalists and specialists in spatially heterogeneous environments. *Evolution* 45, 1317–1331.

Veech, J.A., Charlet, D.A. and Jenkins, S.H. (2000) Interspecific variation in seed mass and the co-existence of conifer species: a null model test. *Evolutionary Ecology Research* 2, 353–363.

Wada, N. and Ribbens, E. (1997) Japanese maple (*Acer palmatum* var. *Matsumurae,* Aceraceae) recruitment patterns: seeds, seedlings, and sapling in relation to conspecific adult neighbours. *American Journal of Botany* 84, 1294–1300.

Watson, H. (1985) *Lambir Hills National Park: Resource Inventory with Management Recommendations.* National Parks and Wildlife Office, Forest Department, Kuching, Sarawak, Malaysia.

Webb, C., Ackerly, D. and Kembel, S. (2005) *Phylocom: Software for the Analysis of Community Phylogenetic Structure and Character Evolution* [available at http://www.phylodiversity.net/phylocom/].

Webb, C.O. and Donoghue, M.J. (2004) Phylomatic: tree assembly for applied phylogenetics. *Molecular Ecology Notes* 5, 181–183.

Webber, M.L. (1934) Fruit dispersal. *Malayan Forester* 3, 18–19.

Westoby, M., Leishman, M. and Lord, J. (1996) Comparative ecology of seed size and dispersal. *Philosophical Transactions of the Royal Society London B: Biological Sciences* 351, 1309–1318.

Westoby, M., Falster, D.S., Moles, A.T., Vesk, P.A. and Wright, I.J. (2002) Plant ecological strategies: some leading dimensions of variation between species. *Annual Review of Ecology and Systematics* 33, 125–159.

Whitney, K.D., Fogiel, M.F., Lamperti, A.M., Holbrook, K.M., Stauffer, D.J., Hardesty, B.D., Parker, V.T. and Smith, T.B. (1998) Seed dispersal by *Ceratogymna* hornbills in the Dja Reserve, Cameroon. *Journal of Tropical Ecology* 14, 351–371.

Wright, I.J., Reich, P.B. and Westoby, M. (2001) Strategy shifts in leaf physiology, structure and nutrient content between species of high- and low-rainfall and high- and low-nutrient habitats. *Functional Ecology* 15, 423–434.

Wright, S. (1943) Isolation by distance. *Genetics* 28, 114–138.

Zanne, A.E., Chapman, C.A. and Kitajima, K. (2005) Evolutionary and ecological correlates of early seedling morphology in East African trees and shrubs. *American Journal of Botany* 92, 972–978.

IV Management Implications and Conservation

RONDA J. GREEN AND ANDREW J. DENNIS

Introduction

Humans have been modifying ecosystems for thousands of years, but with increasing speed in recent decades, causing changes in species abundance (including extinctions), community structures and ecological processes. Seed dispersal is one process that has been disrupted, with potentially far-reaching consequences. For example, Böhning-Gaese (Chapter 25) concludes that the loss of seed dispersers can have far-reaching effects on spatial distribution and genetic structure of plant populations as well as on the composition of plant communities. This conclusion is based on extensive comparisons between South Africa and Madagascar, the latter region harbouring a far lower diversity of dispersal agents. Similarly, Silva *et al.* (Chapter 26) conclude that, despite some resilience in plant–frugivore networks, the extinction of endangered animals, especially large-bodied frugivores, can lead to reduced plant species richness and simplified ecological communities. This conclusion is drawn from models employing a large data set detailing the complex interaction networks in the Atlantic forests of Brazil.

There is ample reason for concern about the future of dispersers. Sekercioglu *et al.* (2004) estimate that at least 25% of the world's frugivorous birds are extinction-prone, making this one of the most vulnerable guilds of birds. They also point out that some dispersers have no comparable guild members to substitute for them should they decline. Similarly, medium and large mammals, including many that disperse large seeds, are subject to high risk (see Corlett, Chapter 24; Wright, Chapter 28). Indeed, the problem of large-seeded fruits lacking dispersers is a recurring theme in this section, as well as elsewhere in the literature (e.g. Green, 1993; McConkey and Drake, 2002; Dennis *et al.*, 2005). Pizo (Chapter 29) demonstrates that most dispersal in fragments of Atlantic forest in Brazil is carried out by small generalist frugivores, a restricted subset of what would have originally been dispersing seeds in continuous forests. Pizo concurs with Wright (Chapter 28) that such small generalists may dominate future seed dispersal patterns, ultimately dictating much regional vegetation composition and structure.

Given the complexities of ecosystem function, some may feel it is simpler to replace lost dispersers with human efforts to maintain vegetation communities than to attempt to maintain, attract or reintroduce natural dispersal agents. However, Hougner *et al.* (2006) concluded that protecting the dispersers is substantially cheaper than investing in humans taking over their role. This conclusion was based on a seed-hoarding animal and its role with a single conspicuous forest tree species in Europe. In less accessible areas, and with rare or cryptic species of plant, the relative value of protecting dispersal agents is likely to be much greater.

To use resources more efficiently and to make conservation efforts most effective, we need continued research in many directions while simultaneously putting our current knowledge into practice (see Green, Chapter 27). This should include studies in relatively intact areas to set benchmarks and understand processes, and in areas subjected to various degrees and types of disturbance to quantify the consequences for the integrity of dispersal systems. There are many rare and threatened plants as well as potential dispersers for which we know very little regarding seed dispersal (see Green, Chapter 27). There is also a tendency for researchers to focus on trees, rainforests and resident frugivores, leaving much to be discovered and understood regarding other habitats and the roles of migratory animals and of seed predators that may disperse some seeds (see Bronstein *et al.*, Chapter 7; Green, Chapter 27; Pizo, Chapter 29).

Aids to predicting where the greatest risks to biodiversity might lie are important for directing conservation efforts. Corlett (Chapter 24) asks whether seed dispersers or pollinators are in most urgent need of conservation effort in Asia. He concludes that large seed dispersers should demand our most immediate action, and that efforts to enhance habitat for seed dispersers are likely to do the same for pollinators. Silva *et al.* (Chapter 26) demonstrate the utility of analysing complex networks to predict the risks of species loss. The use of such analyses in different habitats and geographical regions might help to point to some of the more urgent targets for conservation effort.

Collectively, the authors of this section have highlighted the far-reaching consequences to the process of seed dispersal brought about by loss of diversity in seed-dispersing animals. They have presented large data sets from a range of geographical regions and use a range of approaches to support their conclusions. They have also demonstrated the application of theory in practical conservation settings. A simple conclusion to be drawn from the chapters in this section is that, despite resilience, the loss or decline of seed dispersers is likely to result in a cascading loss of diversity. Direct and active management, potentially even including relocation of large seeds (see Wright, Chapter 28), and reintroduction of dispersers (see, e.g. Corlett, Chapter 24), will be required to alleviate these losses.

References

Dennis, A.J., Lipsett-Moore, G.J., Harrington, G.N., Collins, E. and Westcott, D.A. (2005) Seed predation, seed dispersal and habitat fragmentation: does context make a difference in tropical Australia. In: Forget, P.-M., Lambert, J.E., Hulme, P.E. and Vander Wall, S.B. (eds) *Seed Fate: Predation, Dispersal and Seedling Establishment.* CAB International, Wallingford, UK, pp. 117–135.

Green, R.J. (1993) Avian seed dispersal in and near subtropical rainforests. *Wildlife Research* 20, 535–557.

Hougner, C., Colding, J. and Söderqvist, T. (2006) Economic valuation of a seed dispersal service in the Stockholm National Urban Park, Sweden. *Ecological Economics* 59, 364–374.

McConkey, K.R. and Drake, D.R. (2002) Extinct pigeons and declining bat populations: are large seeds still being dispersed in the tropical Pacific? In: Levey, D.J., Silva, W.R. and Galetti, M. (eds) *Seed Dispersal and Frugivory: Ecology, Evolution and Conservation.* CAB International, Wallingford, UK, pp. 381–395.

Sekercioglu, H., Daily, G.C. and Ehrlich, P.R. (2004) Ecosystem consequences of bird declines. *Proceedings of the National Academy of Sciences of the USA* 101, 18042–18047.

24 Pollination or Seed Dispersal: Which Should we Worry About Most?

R.T. Corlett

Introduction

A search of the Internet finds many sites and publications, targeted at both scientists and the general public, that deal with the 'pollination crisis', but none that even mention a 'dispersal crisis'. Part of this difference reflects the importance of pollination for agricultural crops, while no major crop depends on natural seed dispersal, but many authors also suggest that there is – or soon could be – a pollination crisis in natural ecosystems and that there is therefore a need for the active conservation of pollination interactions (e.g. Buchmann and NabSEhan, 1996; Allen-Wardell *et al.*, 1998; Kearns *et al.*, 1998; Steffan-Dewenter *et al.*, 2005). The major reasons for this concern about pollination systems are the dependence of most species of plant on pollination by animals, especially insects, and the evidence that some pollinator populations (most clearly, of exotic honeybees; Ghazoul, 2005a) have experienced dramatic declines in recent decades. A priori, plants that depend on the services of animals are expected to be more vulnerable than those that do not, because the vulnerabilities of the mutualist are added to those of the plant itself.

However, the same argument applies equally to seed dispersal, which is also, at least in the tropics, usually dependent on animals, many of which have experienced dramatic population declines (Corlett, 1998). Indeed, the increased interest in both pollination and seed dispersal in the tropics over the past two decades has been partly in response to fears that failures of these mutualisms may both accelerate the erosion of biodiversity in disturbed and fragmented forests, and slow its recovery when human pressures are reduced (Bond, 1994). Plants may persist in fragmented and degraded landscapes for decades or centuries, but if their animal mutualists are extinct and were essential for reproduction, these plants must be counted among the 'living dead', with no prospect of contributing to future generations (Janzen, 2001). In theory, failed mutualisms could lead to a cascade of extinctions affecting multiple species (Kearns and Inouye, 1997; but see also Memmott *et al.*, 2004).

There has been a tendency in the ecology and conservation literature to treat pollination and seed dispersal as almost interchangeable (as easily demonstrated by a search on *Google Scholar* for 'pollination and seed dispersal'). Yet these are very different processes, usually carried out by very different animals, so the implications of their vulnerability for conservation management are also very different. Managing a reserve to maintain populations of pollinating insects (e.g. *Apis* bees) will be very different from managing the same reserve to maintain populations of seed-dispersing vertebrates (e.g. monkeys or fruit pigeons).

This is not to deny the close evolutionary and ecological relationships between flowering and fruiting (Primack, 1987). Apart from the obvious morphological continuity between flowers and fruits, the phenologies of the two events are linked, although the constraints imposed by the phenology of one event on the other are less strict in the tropics and subtropics, where fruit development times can vary widely within a genus, than in more seasonal climates (Primack, 1987; Corlett, 1993). The two mutualisms are also linked through the influence of seed dispersal mechanisms on local plant densities and thus the distances moved by pollinators: limited seed dispersal (e.g. by wind or rodents, or following the loss of dispersal agents) may lead to clumping of adults and thus limited pollen dispersal (Hardy *et al.*, 2006). At the level of the individual plant, inter-annual variation in pollination success influences fruit crop size (Jordano and Schupp, 2000), which in turn can influence between-plant variation in fruit removal in complex ways (e.g. Davidar and Morton, 1986; Osada, 2005; Saracco *et al.*, 2005). More generally, a complete failure of pollination renders any worries about failure of seed dispersal superfluous. However, except in those rare cases where pollination and seed dispersal are carried out by the same species (e.g. Kelly *et al.*, 2004), the most likely proximate causes of pollination failure will be different from the most likely causes for dispersal failure, even if the ultimate causes (e.g. habitat loss or degradation) may be the same. It is therefore reasonable to treat them as more or less independent.

The basic aim of this chapter, therefore, is to ask the questions: Which mutualism should conservationists worry about most and where should conservation effort be targeted? I have concentrated on tropical forests because dependence on animals for both pollination and seed dispersal is particularly high in these forests and the threats from human activities particularly severe. No tropical forests are more threatened than those of South-east Asia and no tropical fauna more endangered (Sodhi *et al.*, 2004). Many of the examples will therefore come from two recent reviews of, respectively, flower visitors and pollination, and frugivory and seed dispersal in the Oriental region (i.e. tropical and subtropical Asia) (Corlett, 1998, 2004) updated with reference to more recent literature (Table 24.1), but some of the gaps in our knowledge for this region will be filled, where

Table 24.1. The major taxonomic groups of pollinators and seed dispersal agents in the Oriental region, modified from Corlett (1998, 2004). Importance: +, minor, to ++++, very important.

Pollinators		Seed dispersal agents	All fruits	Large fruits (> 20 mm diameter)
Social bees (*Apis, Trigona*)	++++	Fruit pigeons (*Ducula, Ptilinopus*)	++++	++++
Beetles	+++	Hornbills (Bucerotidae)	++++	++++
Non-social bees	++	Fruit bats (Pteropodidae)	++++	+++
Flies	++	Primates	++++	+++
Birds	++	Barbets (Megalaimidae)	++++	++
Butterflies and moths	+	Civets (Viverridae)	++++	++
Bats	+	Broadbills (*Calyptomena*)	+++	++
Wasps	+	Starlings (Sturnidae)	+++	+
Thrips	+	Bulbuls (Pycnonotidae)	++++	
		Babblers (Sylviidae)	+++	
		Pheasants (Phasianidae)	++	+
		Other passerines	++	+
		Large terrestrial herbivores (deer, elephants, rhinoceroses, tapirs)	++	++
		Other birds and mammals	+	+

possible, by reference to studies elsewhere. Ideally, we would look at the vulnerabilities of the pollination and seed dispersal systems separately for each species of plant or, better still, at entire plant–pollinator and plant–disperser networks, but there are very few areas of the tropics (or, indeed, outside the tropics) where we have enough information to do this (but see Silva *et al.*, Chapter 26, this volume). This review aims instead to derive generalizations that can be applied in conservation planning and management where such detailed information is lacking.

Why Worry?

One assumption underlying most concerns for tropical mutualisms is that the animal partners are likely to be more vulnerable to anthropogenic pressures than the plants they pollinate or disperse (e.g. Memmott *et al.*, 2004). This assumption seems inherently reasonable. Plants are typically much longer-lived than their animal mutualists, and thus pass less frequently through the vulnerable juvenile stages, and individual plants are sensitive only to changes in their immediate vicinity, while their mutualists are affected by changes over a much larger area. This is expected to make plants less vulnerable to fragmentation effects than animals are (Corlett and Turner, 1997). Moreover, where a species is directly targeted for exploitation, the harvesting of plants is usually confined to mature individuals, leaving juveniles to continue the population, while the harvesting of animals is typically much less selective. Patterns of extinction in human-modified landscapes provide empirical support for the greater vulnerability of animals. In tropical Asia, for example, both Singapore and Hong Kong have retained highly diverse forest floras while losing a large proportion of their faunas (Corlett and Turner, 1997; Brook *et al.*, 2003; Hau *et al.*, 2005). It is not, however, necessary for animals to be more vulnerable than plants to cause problems for mutualism. If the vulnerabilities of the mutualists are different, then the sum of these vulnerabilities will be greater than either alone. This could happen, for example, if the animal partner was vulnerable to exploitation and the plant to the microclimatic consequences of fragmentation.

Persistence of Plant Populations Without Their Mutualists

The longer that an animal-dependent plant population can persist after the local extinction of its mutualists, the greater the chance that it will eventually be rescued by the spontaneous or managed reappearance of these mutualists (or of novel replacements). For pollination, the rather limited data set for tropical and subtropical forests suggests a high incidence of dioecy (16–31%) in tree species and a predominance of outcrossing in hermaphrodites, but with many hermaphrodites also retaining at least some capacity to set seeds without pollinator visits (Richards, 1997; Adam and Williams, 2001; Morales and Galetto, 2003). Loss of pollinators will therefore prevent seed production in many, but not all, species. Cessation of seed production will lead to local extinction of the plant population within limits set by the maximum lifespan of individual plants (and the seed bank, if present), but long-lived and clonally propagating plants may persist for centuries without setting seeds.

The degree of dependence of animal-dispersed plants on their dispersal agents is less obvious, since the seeds in fruits that remain beneath the parent plant may have a non-zero chance of establishment (e.g. Chapman and Chapman, 1995, 1996; Robertson *et al.*, 2006; see also Traveset *et al.*, Chapter 4, this volume). If undispersed seeds can grow into adults, a plant population that has lost all its dispersal mutualists could persist beyond the lifespan of individual plants (Kelly *et al.*, 2004). There is insufficient data available to assess how widespread *in situ* population persistence without seed dispersal is in natural or human-impacted landscapes, but it could potentially reduce the medium-term vulnerability of plants to disperser losses. In the highly degraded landscape of Hong Kong, for example, 35 species of Fagaceae (*Castanopsis*, *Lithocarpus* and *Quercus* s.l.) persist in scattered small populations despite the apparent loss of their dispersal agents (including probably a forest rat, *Leopoldamys edwardsi*: Chung and Corlett, 2006). On a much longer timescale, the 'fruits the gomphotheres ate' have apparently persisted for millennia in the Neotropics without their megafaunal dispersers (Janzen and Martin, 1982; see Donatti *et al.*, Chapter 5, this volume).

Mutualist Limitation of Plant Population Sizes

If plant populations are limited by the activity of pollinators and/or seed dispersal agents in intact natural ecosystems, then any decline in the activity of either is likely to have an eventual impact on population size. If, instead, plant population sizes are limited by the availability of resources for seed production or by the carrying capacity of the habitat, then a decline in the activity of mutualists may have no impact. Pollinator or disperser limitation would, ideally, be assessed directly by the experimental manipulation of pollinator and disperser populations, but the only available evidence is less direct.

Experimental hand-pollinations suggest that pollen limitation of fruit set is not only widespread (> 50% of plant species studied in some populations or years; Burd, 1994; Knight *et al.*, 2005) but of high magnitude (a weighted average 75% higher fruit set than controls after pollen supplementation; Knight *et al.*, 2005). Pollen limitation could, in most cases, be reduced by an increase in pollinator visits, so these results can be interpreted as evidence for pollinator limitation. However, most studies have been done in disturbed habitats, where pollinator abundance may have already been reduced, and the ability of plants to reallocate resources among flowers or between years casts some doubt on the lifetime significance of studies that involve hand-pollination of flowers on only part of a plant and/or in a single year (Ashman *et al.*, 2004). Whole-plant, whole-lifetime pollen supplementation studies have only been done on monocarpic plants, although several studies of polycarpic plants have extended over multiple years. There have also been very few studies that have looked at the demographic consequences of pollen limitation (Ashman *et al.*, 2004). While the evidence is incomplete, it does appear that more pollinator visits will usually mean more seeds in pollen-limited plants. Moreover, pollen limitation may not only reduce seed set, but also offspring quality, by reducing pollen competition and thus the selectivity among gametes, but this has rarely been assessed (Colling *et al.*, 2004).

Experiments that add seeds to unoccupied habitats suggest that seed limitation of local recruitment may be a general phenomenon in many ecosystems, including tropical forests (Turnbull *et al.*, 2000; Makana and Thomas, 2004; Svenning and Wright, 2005). This seed limitation could result from insufficient seed production, inadequate seed dispersal, or both. Indirect evidence suggests that the non-uniform distribution of seeds is a major component of seed limitation in at least some species (e.g. Dalling *et al.*, 1998, 2002). However, this dispersal limitation could in turn result from either the 'quantity' or 'quality' components of dispersal (Schupp, 1993), and the quality component – the probability that a dispersed seed will recruit a new reproductive individual – will not, in general, be influenced by the abundance of dispersal agents. The clearest evidence for disperser (as opposed to dispersal) limitation therefore comes from the proportion of a fruit crop that is not consumed by potential dispersal agents, and this has rarely been systematically assessed in tropical forests. There is some evidence to suggest, however, that seed dispersal can be limited by the activity of dispersal agents even in relatively intact ecosystems, at least in some species and years (e.g. Herrera, 1984; Chapman and Chapman, 1996).

The evidence that plant populations in intact communities are limited by the abundance and activity of animal mutualists is mostly indirect and open to alternative interpretations, but it does suggest that recruitment of at least some species of plant could be sensitive to any decline in population of their pollinators or dispersal agents. The evidence is not sufficient, however, to suggest which mutualism might be more vulnerable to such declines.

The Magnitude of the Task

Kelly *et al.* (2004) point out that flowers are not only more numerous than fruits on the same individual plant – often greatly so – but also shorter-lived, so the rate at which flowers must be pollinated for maximum fruit set may be several orders of magnitude greater than the rate at which fruits must be consumed for maximum seed dispersal. They argue that this makes pollination more difficult, and thus more vulnerable to disruption. While this argument is convincing for the rare cases in which pollination and seed dispersal are carried out by the same animal (the bellbird, *Anthornis melanura*, in the mistletoes studied by Kelly *et al.*, 2004), the densities of the (mostly invertebrate) pollinators of a plant typically far exceed those of its (mostly vertebrate) dispersal agents, so the numerical magnitude of the task facing each individual pollinator is unlikely to be greater than that facing individual dispersal agents. However, the ability of ripe fruits of many species to persist on the plant for more than a month (Tang *et al.*, 2005) may make dispersal mutualisms less vulnerable than pollination mutualisms to human impacts that result in reduced mutualist density, simply because a fruit can wait longer than a flower.

Specialization Versus Generalization

All else being equal, the vulnerability of a species of plant to the loss of animal mutualists will be increased by specialization; that is, dependence on one or a few animal species. Theoreticians have argued on a variety of grounds that the potential for tight co-evolution between plants and their pollinators is greater than between plants and seed dispersal agents (Wheelwright and Orians, 1982; Herrera, 2002). Unlike seed dispersal, some degree of specificity among flower visitors is essential for effective pollination, which requires that pollen is moved to a conspecific flower (Pellmyr, 2002). This specificity is generally ensured by 'payment on delivery' of pollen, which gives the plant considerable control over pollinator movements. Frugivores, by contrast, must be 'paid in advance', which makes it almost impossible to direct seed dispersal agents to particular targets, thus reducing the advantages of specialization (although the dispersal of mistletoe seeds to appropriate-sized twigs by flowerpeckers is a striking exception; Reid, 1991). Moreover, it is much easier for a plant to mechanically restrict access by flower visitors (usually invertebrates and usually smaller than the flower) to floral rewards than to restrict access by frugivores (usually vertebrates and usually much larger than the fruit) to fruit rewards: hence the amazing morphological diversity of flowers and the relative uniformity of fruits (Herrera, 2002).

The apparent uniformity of fruits may, however, have been exaggerated by the focus in frugivory studies on a few easily measured fruit traits: size, colour, presence or absence of an inedible covering, and crude nutrient content (typically, lipids, total sugars and nitrogen). Given the relative size and strength of frugivores, specialization is more likely to be mediated by chemistry than morphology (Cipollini and Levey, 1997; see also Levey *et al.*, Chapter 2, this volume). There is evidence that frugivores can detect small differences in fruit chemistry (e.g. Schaefer *et al.*, 2003a) and several recent studies have shown that the rate of fruit removal across species is positively related to nutrient levels and negatively to chemical defences (Schaefer *et al.*, 2003b; Tsang and Corlett, 2005). This trade-off between attraction of frugivores and defence against pests and pathogens may be a major factor underlying the observed rankings of fruit preferences (Herrera, 1998; Carlo *et al.*, 2003), but it is not clear how often it leads to specialization by frugivores.

Although theory suggests that flowers have both the means and motivation for specialization, most recent reviews have emphasized the relative rarity of specialization and the commonness of generalization in pollinator systems (e.g. Waser *et al.*, 1996; Olesen and Jordano, 2002). Fenster *at al.* (2004), however, point out that most plants are pollinated by a small subset of the available pollinators, with this subset of effective pollinators often forming a recognizable 'functional group', even if they are not taxonomically related. In the Oriental region, the best-documented examples of extreme specialization (i.e. a single major pollinator) involve extreme floral morphology (e.g. Kato *et al.*, 1993), nocturnal flowering (e.g. Rao *et al.*, 2001), deception (e.g. Dakwale and Bhatnagar, 1997), or unusual rewards, such as oil (e.g. Li, 1997) and especially brood sites. Provision of brood sites is the basis of the two known examples of obligate 1:1 relationships in the Oriental region: between *Ficus* species (Moraceae) and their pollinating wasps, and between *Glochidion* spp. (and related Phyllanthaceae) and their pollinating moths (Kawakita *et al.*, 2004).

Specialization in dispersal mutualisms is generally assumed to be relatively low, and there are no known examples of obligate plant–disperser relationships (Herrera, 2002). Jordano (1987) found that the average number of interactions per plant species was around 50% higher (6.0 versus 4.1) in disperser than pollination systems, although the variance was high and the methodologies used in the studies reviewed were necessarily different. In general, the number of animal dispersers per species of plant declines with increasing fruit size, due to gape and/or weight limitations, except for soft, multi-seeded fruits from which dispersers can remove seed-containing pieces (Corlett, 1998). Large, single-seeded fruits have the fewest dispersers – both recorded and potential – in the Oriental region, but the number of observed dispersers is always > 1 in published studies from sites with intact disperser faunas (e.g. Hamann and Curio, 1999; Kitamura *et al.*, 2002, 2004, 2006). Pulp chemistry has also been shown to limit disperser diversity in non-tropical systems (Herrera, 2002) and the best-documented tropical example of a fruit with single effective disperser, *Balanites wilsoniana*, which is consumed largely by African elephants, is both very large (9 × 6 cm) and potentially toxic (Cochrane, 2003).

It is important to note, however, that published frugivory data for the tropics is biased towards the most attractive fruit (such as *Ficus* spp.) and thus to detecting generalization, while the pollination data, with some exceptions (e.g. Momose *et al.*, 1998; Devy and Davidar, 2003), is biased towards 'unusual' flower types (e.g. orchids and aroids), and thus to detecting specialization. If dispersal biologists documented frugivory and seed dispersal for the least attractive fruits (with infrared-triggered cameras?) and pollination biologists recorded all visitors and potential pollinators for open access flowers, it is possible that our views on the relative specialization of the two mutualisms would change.

An additional problem is that the Oriental data sets for both mutualisms consist largely of observations of flower visitation and fruit consumption, rather than pollination and seed dispersal, and there is little direct evidence for the effectiveness of either flower visitors in pollination or frugivores in seed dispersal. Comparisons of the two literatures for the Oriental region (Corlett, 1998, 2004) suggest that theoretical expectations may have biased pollination biologists towards recognizing one or a few 'legitimate' pollinators, and dispersal biologists towards assuming that all frugivory equals seed dispersal, with equally little evidence.

Human Impacts on Mutualists

Exploitation

The only pollinators that are directly exploited to a significant extent in the Oriental region are the *Apis* bees (Roubik, 1989; Corlett, 2004). The long-term survival of one species, *A. laboriosa*, may be threatened by overexploitation, and populations of other species are being at least locally diminished. Some species of stingless bees are also exploited locally for the medicinal properties of their honey, but the impact appears minor. Stingless bees may also be impacted indirectly by selective logging of their nest trees (Eltz *et al.*, 2003). The highly nectarivorous dawn bat, *Eonycteris spelaea*, roosts in large colonies in caves, where they are vulnerable to quarrying, but it is not clear to what extent its services as a pollinator can be substituted by other, more tolerant, pteropodid bats (Corlett, 2004).

In striking contrast to the pollinators, *all* the major dispersal agents of large fruits in the Oriental region – large birds, large fruit bats, primates, civets and terrestrial frugivores – are directly exploited by hunters, and many species have now been eliminated from most of their natural ranges (Corlett, 1998, 2002; Robinson and Bennett, 2000). Kitamura *et al.* (2005) estimated that about one-third of the 105 plant species with a diameter of 10 cm or above at breast height (1.3 m) in a 4-ha plot at Khao Yai, Thailand, are dispersed largely or only by large frugivores that are vulnerable to hunting. Hamann and Curio (1999) estimated that 60% of late-successional tree species in a submontane forest in the Philippines would lose all dispersal agents if large avian frugivores (hornbills and fruit pigeons) were hunted to extinction, as they have been in most of the Philippines. This pattern appears to be general throughout the region (Corlett, 1998, 2002). This widespread loss of an entire functional group of seed dispersal agents has no obvious parallel in pollination networks. Moreover, the species lost include not only specialist frugivores (such as elephants and rhinoceroses), whose deletion is unlikely to propagate far through the plant–disperser network, but also generalists (such as many primates) whose loss may influence numerous species (see Silva *et al.*, Chapter 26, this volume).

Habitat loss, degradation and fragmentation

It is often difficult to separate the impacts of habitat degradation from those of hunting for large vertebrates in the Oriental region, but habitat effects are clearer for smaller, unexploited vertebrates and invertebrates (Corlett, 2000). The 'catastrophic extinctions' following deforestation in Singapore included many large vertebrates, but also a range of smaller birds and mammals, and at least one-third of the recorded butterfly species (Brook *et al.*, 2003). Among Oriental birds, babblers (Sylviidae) are particularly sensitive to forest disturbance and are particularly poor recolonizers (Ginny-Ng *et al.*, 2005; Peh *et al.*, 2005). Most babblers seem to be at least partly frugivorous (Corlett, 1998), but these generally small-gaped birds are unlikely to be the sole dispersers for any species of plant. Similarly, there is no evidence to suggest that any species of plant depends on the vulnerable forest interior butterflies for pollination (Corlett, 2004), but the high extinction rate in the best-studied invertebrate group suggests that specialist pollinators in other groups may also be vulnerable. The best Oriental evidence that this is the case comes from the Lambir Hills National Park, Sarawak, where the 1998 ENSO droughts in northern Borneo caused the local extinction of the species-specific fig-wasp pollinators for all 25 dioecious species of fig, with recolonization of the semi-isolated reserve taking >6 months for most species (Harrison, 2001).

For pollination mutualisms, at least, changes in the abundance of pollinators at the landscape scale may be less important than localized declines in pollinator activity as a result of reduced plant density and/or habitat fragmentation (Ghazoul, 2005a,b). Isolated plants or plant populations may suffer reduced fecundity and/or seed quality because pollinators are unable or unwilling to cross the distance to them from the nearest conspecifics. Impacts on offspring quality are much more difficult to assess than impacts on seed set, but could result from increased inbreeding or from reduced pollen competition when less pollen is deposited on each stigma (Colling *et al.*, 2004). Some Oriental pollinators, such as birds, bats, social bees, large non-social bees, and the fig-wasp pollinators of monoecious figs, may move hundreds or thousands of metres and are willing to cross matrix habitats between forest fragments, but the distances travelled by small, non-social bees, beetles, flies, and the pollinators of dioecious figs may be shorter (Corlett, 2004).

The problem is somewhat different for dispersal mutualisms, since, unlike pollen, seeds do not have to be moved to a conspecific stigma. However, plants must still attract enough dispersal agents to remove the fruit crop and, in the long term, an ability to disperse seeds between fragments provides an alternative mechanism for continued gene flow. Distances travelled by vertebrate dispersal agents are generally similar to those travelled by invertebrate pollinators (Corlett, 1998). Moreover, large frugivores in the Oriental region may be more mobile than elsewhere in the tropics as an evolutionary response to the synchronous supra-annual fruiting cycles of many woody species (Primack and Corlett, 2005). The mobility of understorey frugivores may be constrained, however, by unwillingness to cross open areas, so understorey plants in isolated fragments may suffer reduced dispersal. On the other hand, small and medium-sized non-forest frugivorous birds may be more mobile than their forest counterparts and thus provide dispersal services between widely separated fragments for plants bearing fruits with diameters that do not exceed their gape limits (Weir and Corlett, 2007).

There have been no studies of fragmentation effects on pollination and plant recruitment in the Oriental region, but studies elsewhere provide many examples of reduced pollination or seed set (Aizen *et al.*, 2002; Burgess *et al.*, 2006) and a few of reduced recruitment (Cordeiro and Howe, 2003) that can be attributed to mutualist declines, as well as a few examples of positive effects (9 of 85 species reviewed; Burgess *et al.*, 2006). The impacts are species- and situation-specific, and there is as yet no evidence for the predicted 'cascade' of extinctions affecting multiple species (Kearns and Inouye, 1997).

Ashworth *et al.* (2004) suggest that the widespread asymmetry of plant–pollinator mutualisms, with specialist plants linked with generalist animals, and vice versa (Bascompte *et al.*, 2003), explains why specialist plants appear to be no more vulnerable than generalists in fragments (Aizen *et al.*, 2002). The same argument should apply to plant–disperser networks, which are also highly asymmetrical (Bascompte *et al.*, 2003; Silva *et al.*, Chapter 26, this volume). More generally, nesting of mutualist networks would be expected to make them less vulnerable to random disturbances, since the most vulnerable individual species are likely to be specialists, and the generalists they interact with have multiple alternative partners. On the other hand, nested networks may be particularly vulnerable to disturbances affecting generalists (such as their selective exploitation, as noted above), since such disturbances may propagate through many species in the network (see Silva *et al.*, Chapter 26, this volume). It is also important to note both that nesting is not perfect in real networks and that the current qualitative network models capture only part of the complexity of plant–animal relationships.

Invasive species

The only record of an exotic pollinator established in the wild in the Oriental region is *Apis mellifera* on the Bonin (Ogasawara) islands, where the feral colonies are supported largely by exotic plants (Kato *et al.*, 1999). Ghazoul (2002, 2004) suggests that invasive alien plant species may be an unrecognized threat to pollination mutualisms through their impacts on the foraging behaviour of native pollinators. In disturbed dry dipterocarp forest in Thailand, the invasive American shrub, *Chromolaena odorata*, was visited preferentially by the butterfly pollinators of the native tree, *Dipterocarpus obtusifolius*, although fruit set was not reduced. Butterfly abundance was not increased by the availability of extra resources in this case, but the nest density of stingless bees was greatly increased in Bornean rainforest where they had access to pollen from crops near the forest (Eltz *et al.*, 2002).

Although non-native birds, including partly frugivorous babblers, bulbuls and laughingthrushes, are established in many parts of the Oriental region, they are almost all Oriental species that have been moved around the region as cagebirds (e.g. Leven and Corlett, 2004; Yap and Sodhi, 2004). There is no evidence to suggest that they are causing any disruption to dispersal mutualisms. The impact of invasive plants on native plant–disperser mutualisms has not been studied in the region, but the mostly anecdotal evidence from elsewhere suggests that, as with pollination, impacts could be positive (by subsidizing native frugivore populations), negative (by replacing natives in the diet), or neutral (Buckley *et al.*, 2006).

The survivors: pollinators and dispersers in highly degraded landscapes

Direct exploitation, habitat loss and degradation, and exotic invasions tend to occur together in human-dominated landscapes, with a cumulative impact on the availability of mutualists. Hong Kong (22° N, 114° E) provides a ‘worst case scenario’ for the survival of mutualists, with 7 million people in a land area of 1100 km^2 that has already seen at least 7000 years of continuous human occupation (Dudgeon and Corlett, 2004). Both pollination and seed dispersal are overwhelmingly dominated by generalists, with *Apis cerana* visiting 86% and dominating on 55% of 83 woody species studied (Corlett, 2001) and two bulbuls (*Pycnonotus* spp.)

accounting for the majority of dispersed seeds (Au *et al*., 2006). Tolerance of matrix habitats (non-forest in this case) is a good predictor of animal persistence in degraded landscapes (Laurance *et al*., 2002), and the surviving pollinators (R.T. Corlett, personal observation) and seed dispersal agents (Weir and Corlett, 2007) in Hong Kong are apparently all capable of crossing gaps of hundreds of metres between habitat patches. It is therefore mutualist extinctions at the landscape level, rather than lack of mobility of the survivors, that is the major potential threat to the Hong Kong flora. There is no evidence that pollination failure is threatening any plant species in Hong Kong. Hong Kong still supports 24 species of native fig (Corlett, 2006) and eight species of *Glochidion*, all of which set seed, suggesting that these obligate 1:1 relationships are not especially vulnerable to disruption. By contrast, mostly anecdotal evidence suggests that dispersal failure has reduced some tree species to small, isolated populations, apparently maintained only by *in situ* recruitment (Corlett, 2002).

Singapore (1° N, 104° E) has been more recently deforested than Hong Kong and still has some substantial primary forest fragments, although these have lost a large proportion of their fauna (Brook *et al*., 2003). As in Hong Kong, there have been no observations that suggest pollination failures, while decomposing large fruits under fruiting plants are a conspicuous feature of the surviving forest fragments (R.T. Corlett, personal observation). Further evidence for dispersal failures comes from the almost total absence of tree species with large, single-seeded fruits from secondary forests on abandoned cultivation (Corlett, 2002) and the under-representation of the family Myristicaceae, which is dispersed by large birds in South-east Asia, in recruitment in the largest surviving primary forest fragment (Ercelawn *et al*., 1998).

Which mutualism should conservationists worry about most?

This review suggests that in the Oriental region, and probably elsewhere in the tropics, dispersal mutualisms are in general more vulnerable than pollination mutualisms. This is largely because the entire functional group of dispersers of large, single-seeded fruits is extremely vulnerable to direct exploitation (Table 24.2). Some plant populations can probably persist longer without dispersal agents than without pollinators, because unpollinated flowers set no seeds while undispersed seeds may germinate and grow (Kelly *et al*., 2004), but it is not known how general this ability is (Robertson *et al*., 2006).

Evidence that pollination mutualisms are likely to be, in general, less vulnerable to human impacts than dispersal mutualisms is no reason for complacency about pollination. The impact of pollinator declines is less easily detected than dispersal failures, so the problem may have been underestimated. Moreover, declines in offspring quality as a result of changes in pollinator abundance or behaviour are probably impossible to detect without time-consuming, species-by-species experiments.

Table 24.2. The relative vulnerability of pollination and seed dispersal mutualisms to human impacts in the Oriental region.

	Pollination	Seed dispersal
Can plant populations persist without mutualists?	Sometimes	Often?
Does mutualist activity limit plant population sizes?	Often?	Sometimes?
How specialized are plant–mutualist relationships?	Relatively specialized	Relatively generalized
How vulnerable are the animal mutualists to direct exploitation?	Low vulnerability	High vulnerability
How vulnerable are the animal mutualists to habitat degradation?	Varied	Varied
How vulnerable are the mutualisms to invasive species?	Low vulnerability	Low vulnerability

Can the decline and loss of mutualists be mitigated?

The ideal mitigation would be the reintroduction of missing pollinators or dispersal agents, or the reinforcement of depleted populations, but this may be limited, in practice, by many factors, including inadequate habitat, continued exploitation, the problems of sourcing animals, and the general difficulties of reintroductions even under ideal conditions. There have been too few tropical forest reintroductions – apparently none of pollinators – for a direct assessment of the likelihood of success, but a large-gaped frugivorous bird, *Penelope superciliaris*, has been captive-bred and reintroduced to forests in south-eastern Brazil (Pereira and Wajntal, 1999; Pizo, Chapter 29, this volume), and primates, ungulates and birds are over-represented among reintroduction projects globally (Seddon *et al.*, 2005), so there is some relevant experience to build on. Moreover, large vertebrate extinctions are often a result of hunting, rather than lack of suitable habitat, and reducing hunting pressure is usually easier than increasing habitat quantity or quality. However, although no Oriental frugivore or pollinator is known to be globally extinct, at least 20 species of large frugivores are currently classified as endangered (including species of hornbill, fruit pigeon, flying fox, primates and rhinoceros; IUCN, 2006) and most lack viable captive populations that could be used for reintroduction.

Where the major problem is insufficient habitat to maintain a viable population of mutualists, an alternative might be the maintenance of a semi-wild population, with supplementary feeding, provision of nesting sites, or other support. This is a potential strategy for many Oriental vertebrate dispersal agents and at least some pollinators, including some *Apis* bees and those species of non-social bees that will use artificial nests (e.g. Klein *et al.*, 2002). There is also considerable experience with manipulating stingless bee colonies in the Neotropics (Quezada-Euán *et al.*, 2001) that may be transferable to the Oriental region.

In regions where key mutualists are globally extinct (such as many islands in the Pacific; e.g. Meehan *et al.*, 2005), a potential alternative to the reintroduction of missing taxa would be the introduction of novel substitutes, but it would be very hard to justify this approach in the Oriental region, where all known mutualists appear to be still extant. Unless or until reintroductions become practical, the only realistic alternative, therefore, is human-assisted pollination or seed dispersal. Artificial seed dispersal will usually be much easier than artificial pollination in tropical forests because, except for epiphytes and hemi-epiphytes, the targets for seed dispersal are on the ground surface and thus easily accessible, while the targets for pollination are spread through a large and mostly inaccessible volume. Moreover, the complex structure of many flowers – particularly those associated with specialization – can make hand-pollination a slow and difficult task. The relative ease of dispersing seeds is partly counterbalanced by a relatively low success rate, as a result of seed predation and other post-dispersal processes. Planting nursery-raised seedlings avoids most of these problems, although it causes additional disturbance and could potentially introduce soil-borne weeds and pathogens. Overall, however, it appears that although disperser loss is more common than pollinator loss, it is also more easily mitigated.

Conservation management of threatened mutualisms

Ideally, all protected areas would be managed so as to support viable populations of all pollinators and dispersers. Indeed, viable populations of mutualists may not be enough to maintain all mutualisms, since depleted but still viable populations of mutualists in a 'half-empty forest' may visit only their most preferred flowers or fruits (Redford and Feinsinger, 2001). The target may therefore need to be set higher, at maintaining natural population densities. In practice, this diffuse target is rendered meaningless in the tropics by our ignorance of most pollinator and disperser mutualisms, and by the limited resources available for any form of management. Limited resources need to be targeted at clearly defined and potentially solvable problems. The evidence reviewed in this chapter suggests that maintaining viable populations of large frugivores at near their natural population densities should be a management target for all protected areas in the Oriental region, although it is a target that few, if any, reserves currently meet.

Meeting this target will require the complete suppression of hunting in protected areas. Although the most highly preferred prey of Oriental hunters are typically pigs (*Sus* spp.) (Robinson and Bennett, 2000), which are not significant seed dispersal agents, most other prey are, and restricting hunting to pigs is likely to be more difficult than stopping it altogether. Even in the absence of hunting, many protected areas in the region are too small to support long-term, viable populations of large frugivores. Where contiguous unprotected habitat is available, for example as timber production forest, then the protection of frugivore populations should be extended beyond the margins of the reserve. More generally, corridors linking protected areas and habitat patches, buffer zones around protected areas, and the reforestation of deforested land within protected areas are all likely to help maintain vulnerable frugivore populations. Although these management options are already viewed as a general 'good thing', targeting them at the preservation of viable populations of specific dispersal agents should both help with their detailed design and with the persuasion of decision makers.

The disperser mutualisms of least concern are those involving small and medium-sized passerines (such as bulbuls) and small fruit bats (particularly *Cynopterus* spp.), since these occur in all habitats (Corlett, 2002). By contrast, the reduction or elimination of large frugivore populations is so widespread in the Oriental region that it would be prudent to start mitigation now in protected areas that have lost entire functional groups of dispersers, rather than waiting until declines in plant populations become visible. Reintroduction of missing dispersal agents should be considered wherever practical, but unless or until this is possible, the best available options will usually be either artificial dispersal of seeds or planting of seedlings. Which is cheapest will be species- and situation-specific, depending on whether the higher cost of producing seedlings is justified by the greater success rate. A variety of genetic strategies are possible, but it will usually be preferable to use seeds from the nearest available sources unless inbreeding is already likely to be a problem. The inclusion of under-dispersed species in the planting (or direct seeding) mix for reforestation is already a widely recommended strategy (e.g. Martínez-Garza and Howe, 2003; Lamb *et al.*, 2005).

As discussed above, it will be difficult to target mitigation specifically at pollination. Theory and some empirical evidence suggest that a focus on the preservation of the generalist pollinators that pollinate most flowers, including those of most specialists, would be a useful general strategy (Memmott *et al.*, 2004), as well as being far more practical than identifying and managing populations of numerous specialist pollinators. Similarly, it may be useful to identify and protect generalist plants that support many specialist pollinators. In the Oriental region, the ubiquity of the generalist bees, *Apis cerana* and *Xylocopa* spp., in disturbed landscapes (Corlett, 2004) suggests that mutualisms involving these species are of least concern. This may not be true of other generalist pollinators, particularly the large *Apis* species (*A. dorsata* and *A. laboriosa*), but probably also some stingless bees (*Trigona* spp.). Both large *Apis* species form dense nest aggregations at preferred sites that are reoccupied after seasonal (or more prolonged) periods of absence, so protection of these sites is crucial. Stingless bees may also benefit from targeted protection of their nesting sites (Eltz *et al.*, 2003). More generally, the habitat management prescriptions suggested for preserving dispersal mutualisms are likely to be at least as beneficial for pollinators.

Information Gaps: What do we Need to Know?

This review has identified a number of gaps in the information needed for informed conservation management of mutualisms in the region. We need a lot more information on the natural history of pollination and dispersal mutualisms, but not a continuation of the current scattershot approach to collecting it. The diversity of tropical forests makes it difficult to generalize from case studies of individual plant or animal species, particularly when these are selected on the basis of accessibility and/or unusualness. The target should be the elucidation of entire plant–pollinator and plant–disperser networks at a range of sites, representing the major forest types and a variety of human impacts, including isolated forest fragments, logged forests and secondary forests on abandoned cultivation. Currently, the most detailed community-level pollination studies available are those in the tropical lowland rainforest at Lambir, Sarawak (Momose *et al.*, 1998), and the mid-elevation rainforest at Kakachi, in the western Ghats, India (Devy and Davidar, 2003, 2006). The most ambitious attempt at a community-level dispersal study is that of Kitamura *et al.* (2002), in the tropical seasonal forest at Khao Yai National Park, Thailand. Additional work at these sites and the initiation of long-term, community-level studies at additional sites would, in most cases, be more valuable than one-off studies at new sites, however well designed.

With regard to seed dispersal, the most obvious natural history gap concerns the fate of fallen fruits (Corlett, 1998; Yasuda *et al.*, 2005). A better understanding of what happens to fruits that fall below the canopy of the parent plant would contribute to answering two simple but extremely important questions identified in this chapter:

1. What proportion of all fruits (including those on the ground) are removed by potential dispersal agents?
2. Can the seeds in fruits that remain undispersed below the parent plant grow into adults?

It would be particularly valuable to answer both of these questions for species of plants with a range of fruit characteristics across a gradient of human impacts.

With regard to pollination, the most important information gap is probably the effect of habitat fragmentation on pollinator movements (Ghazoul, 2005b). Superficial observations suggest that most flowers attract abundant potential pollinators in even the most degraded landscapes (Corlett, 2001, 2004), but almost nothing is known about interplant movements by pollinators in such landscapes, and thus the consequences of landscape degradation for seed set and offspring quality.

Finally, there is an urgent need for well-documented experimental reintroductions of locally extinct dispersal agents. Experiments with pollinators would also be of interest, although these currently appear to be less urgent. Given the very high risk of failure in even the best-planned reintroduction project, these experiments should initially be carried out with surplus individuals of species with secure wild or captive populations. There is also a need for experiments with the maintenance of semi-wild populations of mutualists at sites where the available habitat is too small and/or degraded to support a viable wild population. Would it be practical, for example, to employ some of Asia's large surplus of domesticated elephants as dispersal agents and general ecosystem engineers in the many protected areas in the region that have lost their entire megafauna?

Acknowledgements

This chapter has benefited greatly from comments on an earlier draft made by Andrew Dennis and Dave Kelly, and from discussions with many of the participants at the *Frugivores and Seed Dispersal Workshop* in 2005.

References

Adam, P. and Williams, G. (2001) Dioecy, self-compatibility and vegetative reproduction in Australian subtropical rainforest trees and shrubs. *Cunninghamia* 7, 89–100.

Aizen, M.A., Ashworth, L. and Galetto, L. (2002) Reproductive success in fragmented habitats: do compatibility systems and pollination specialization matter? *Journal of Vegetation Science* 13, 885–892.

Allen-Wardell, G., Bernhardt, P., Bitner, R., Burquez, A., Buchmann, S., Cane, J., Cox, P.A., Dalton, V., Feinsinger, P., Ingram, M., Inouye, D., Jones, C.E., Kennedy, K., Kevan, P., Koopowitz, H., Medellin, R., Medellin-Morales, S., Nabhan, G.P., Pavlik, B., Tepedino, V., Torchio, P. and Walker, S. (1998) The potential consequences of pollinator declines on the conservation of biodiversity and stability of food crop yields. *Conservation Biology* 12, 8–17.

Ashman, T.L., Knight, T.M., Steets, J.A., Amarasekare, P., Burd, M., Campbell, D.R., Dudash, M.R., Johnston, M.O., Mazer, S.J., Mitchell, R.J., Morgan, M.T. and Wilson, W.G. (2004) Pollen limitation of plant reproduction: ecological and evolutionary causes and consequences. *Ecology* 85, 2408–2421.

Ashworth, L., Aguilar, R., Galetto, L. and Aizen, M.A. (2004) Why do pollination generalist and specialist plant species show similar reproductive susceptibility to habitat fragmentation? *Journal of Ecology* 92, 717–719.

Au, A.Y.Y., Corlett, R.T. and Hau, B.C.H. (2006) Seed rain into upland plant communities in Hong Kong, China. *Plant Ecology* 186, 13–22.

Bascompte, J., Jordano, P., Melian, C.J. and Olesen, J.M. (2003) The nested assembly of plant–animal mutualistic networks. *Proceedings of the National Academy of Sciences of the USA* 100, 9383–9387.

Bond, W.J. (1994) Do mutualisms matter? Assessing the impact of pollinator and disperser disruption on plant extinction. *Philosophical Transactions of the Royal Society London B: Biological Sciences* 344, 83–90.

Brook, B.W., Sodhi, N.S. and Ng, P.K.L. (2003) Catastrophic extinctions follow deforestation in Singapore. *Nature* 424, 420–423.

Buchmann, S.G. and Nabhan, G.P. (1996) *The Forgotten Pollinators*. Island Press, Washington, DC.

Buckley, Y.M., Anderson, S., Catterall, C.P., Corlett, R.T., Engel, T., Gosper, C.R., Nathan, R., Richardson, D.M., Setter, M., Spiegel, O., Vivian-Smith, G., Voigt, F., Weir, J. and

Westcott, D.A. (2006) Management of plant invasions mediated by frugivore interactions. *Journal of Applied Ecology* 43, 848–857.

Burd, M. (1994) Bateman's principle and plant reproduction: the role of pollen limitation in fruit and seed set. *Botanical Review* 60, 83–139.

Burgess, V.J., Kelly, D., Robertson, A.W. and Ladley, J.J. (2006) Positive effects of forest edges on plant reproduction: literature review and case study of bee visitation to flowers of *Peraxilla tetrapetala* (Loranthaceae). *New Zealand Journal of Ecology* 30, 179–190.

Carlo, T.A., Collazo, J.A. and Groom, M.J. (2003) Avian fruit preferences across a Puerto Rican forested landscape: pattern consistency and implications for seed removal. *Oecologia* 134, 119–131.

Chapman, C.A. and Chapman, L.J. (1995) Survival without dispersers: seedling recruitment under parents. *Conservation Biology* 9, 675–678.

Chapman, C.A. and Chapman, L.J. (1996) Frugivory and the fate of dispersed and non-dispersed seeds of six African tree species. *Journal of Tropical Ecology* 12, 491–504.

Chung, K.P.S. and Corlett, R.T. (2006) Rodent diversity in a highly degraded tropical landscape: Hong Kong, South China. *Biodiversity and Conservation* 15, 4521–4532.

Cipollini, M.L. and Levey, D.J. (1997) Secondary metabolites of fleshy vertebrate-dispersed fruits: adaptive hypotheses and implications for seed dispersal. *American Naturalist* 150, 346–372

Cochrane, E.P. (2003) The need to be eaten: *Balanites wilsoniana* with and without elephant seed-dispersal. *Journal of Tropical Ecology* 19, 579–589.

Colling, G., Reckinger, C. and Matthies, D. (2004) Effects of pollen quantity and quality on reproduction and offspring vigor in the rare plant *Scorzonera humilis* (Asteraceae) *American Journal of Botany* 91, 1774–1782.

Cordeiro, N.J. and Howe, H.F. (2003) Forest fragmentation severs mutualism between seed dispersers and an endemic African tree. *Proceedings of the National Academy of Sciences of the USA* 100, 14052–14056.

Corlett, R.T. (1993) Reproductive phenology of Hong Kong shrubland. *Journal of Tropical Ecology* 9, 501–510.

Corlott, R.T. (1998) Frugivory and seed dispersal by vertebrates in the Oriental (Indomalayan) Region. *Biological Reviews* 73, 413–448.

Corlett, R.T. (2000) Environmental heterogeneity and species survival in degraded tropical landscapes. In: Hutchings, M.J., John, E.A. and Stewart, A.J.A. (eds) *The Ecological Consequences of Environmental Heterogeneity.* Blackwell Science, Oxford, UK, pp. 333–355.

Corlett, R.T. (2001) Pollination in a degraded tropical landscape: a Hong Kong case study. *Journal of Tropical Ecology* 17, 155–161.

Corlett, R.T. (2002) Frugivory and seed dispersal in degraded tropical East Asian landscapes. In: Levey, D.J., Silva, W.R. and Galetti, M. (eds) *Seed Dispersal and Frugivory: Ecology, Evolution and Conservation.* CAB International, Wallingford, UK, pp. 451–465.

Corlett, R.T. (2004) Flower visitors and pollination in the Oriental (Indomalayan) Region. *Biological Reviews* 79, 497–532.

Corlett, R.T. (2006) Figs (*Ficus*, Moraceae) in urban Hong Kong, South China. *Biotropica* 38, 116–121.

Corlett, R.T. and Turner, I.M. (1997) Long-term survival in tropical forest remnants in Singapore and Hong Kong. In: Laurance, W.F. and Bierregaard, R.O. (eds) *Tropical Forest Remnants: Ecology, Management and Conservation of Fragmented Communities.* University of Chicago Press, Chicago, Illinois, pp. 333–345.

Dakwale, S.R. and Bhatnagar, S. (1997) Midge-trapping behaviour and floral biology of *Theriophonum crenatum* Blume. Rumph. (Araceae). *Current Science* 73, 606–609.

Dalling, J.W., Hubbell, S.P. and Silvera, K. (1998) Seed dispersal, seedling establishment and gap partitioning among tropical pioneer trees. *Journal of Ecology* 86, 674–689.

Dalling, J.W., Muller-Landau, H.C., Wright, S.J. and Hubbell, S.P. (2002) Role of dispersal in the recruitment limitation of neotropical pioneer species. *Journal of Ecology* 90, 714–727.

Davidar, P. and Morton, E.S. (1986) The relationship between fruit crop sizes and fruit removal rates by birds. *Ecology* 67, 262–265.

Devy, M.S. and Davidar, P. (2003) Pollination systems of trees in Kakachi, a mid-elevation wet evergreen forest in Western Ghats, India. *American Journal of Botany* 90, 650–657.

Devy, M.S. and Davidar, P. (2006) Breeding systems and pollination modes of understorey shrubs in a medium elevation wet evergreen forest, southern Western Ghats, India. *Current Science* 90, 838–842.

Dudgeon, D. and Corlett, R.T. (2004) *The Ecology and Biodiversity of Hong Kong.* Joint Publishing, Hong Kong.

Eltz, T., Bruhl, C.A., van der Kaars, S. and Linsenmair, K.E. (2002) Determinants of stingless bee nest density in lowland dipterocarp forests of Sabah, Malaysia. *Oecologia* 131, 27–34.

Eltz, T., Bruhl, C.A., Imiyabir, Z. and Linsenmair, K.E. (2003) Nesting and nest trees of stingless bees (Apidae: Meliponini) in lowland dipterocarp forests in Sabah, Malaysia, with implications for forest management. *Forest Ecology and Management* 172, 301–313.

Ercelawn, A.C., LaFrankie, J.V., Lum, S.K.Y. and Lee, S.K. (1998) Short-term recruitment of trees in a forest fragment in Singapore. *Tropics* 8, 105–115.

Fenster, C.B., Armbruster, W.S., Wilson, P., Dudash, M.R. and Thomson, J.D. (2004) Pollination syndromes and floral specialization. *Annual Review of Ecology Evolution and Systematics* 35, 375–403.

Ghazoul, J. (2002) Impact of logging on the richness and diversity of forest butterflies in a tropical dry forest in Thailand. *Biodiversity and Conservation* 11, 521–541.

Ghazoul, J. (2004) Alien abduction: disruption of native plant–pollinator interactions by invasive species. *Biotropica* 36, 156–164.

Ghazoul, J. (2005a) Buzziness as usual? Questioning the global pollination crisis. *Trends in Ecology and Evolution* 20, 367–373.

Ghazoul, J. (2005b) Pollen and seed dispersal among dispersed plants. *Biological Reviews* 80, 413–443.

Ginny-Ng, S.L., Sharim-Senik, M., Zubaid, A. and Davison, G.W.H. (2005) A comparison of the bird fauna of a forest fragment and contiguous forest: lessons for avian conservation. *Malayan Nature Journal* 57, 209–220.

Hamann, A. and Curio, E. (1999) Interactions among frugivores and fleshy fruit trees in a Philippine submontane rainforest. *Conservation Biology* 13, 766–773.

Hardy, O.J., Maggia, L., Bandou, E., Breyne, P., Caron, H., Chevallier, M.-H., Doligez, A., Dutech, C., Kremer, A., Latouche-Hallé, C., Troispoux, V., Veron, V. and Degen, B. (2006) Fine-scale genetic structure and gene dispersal inferences in 10 neotropical tree species. *Molecular Ecology* 15, 559–571.

Harrison, R.D. (2001) Drought and the consequences of El Niño in Borneo: a case study of figs. *Population Ecology* 43, 63–75.

Hau, B.C.H., Dudgeon, D. and Corlett, R.T. (2005) Beyond Singapore: Hong Kong and Asian biodiversity. *Trends in Ecology and Evolution* 20, 281–282.

Herrera, C.M. (1984) A study of avian frugivores, bird-dispersed plants, and their interaction in Mediterranean scrublands. *Ecological Monographs* 54, 1–24

Herrera, C.M. (1998) Long-term dynamics of Mediterranean frugivorous birds and fleshy fruits: a 12-year study. *Ecological Monographs* 68, 511–538.

Herrera, C.M. (2002) Seed dispersal by vertebrates. In: Herrera, C.M. and Pellmyr, O. (eds) *Plant–Animal Interactions: An Evolutionary Approach.* Blackwell Science, Oxford, UK, pp. 185–208.

IUCN (2006) *2006 IUCN Red List of Threatened Species* [available at http://www.iucnredlist.org, downloaded on 4 June 2006].

Janzen, D.H. (2001) Latent extinction: the living dead. In: Leven, S.A. (ed.) *Encyclopedia of Biodiversity, Volume 3.* Academic Press, San Diego, California, pp. 689–699.

Janzen, D.H. and Martin, P.S. (1982) Neotropical anachronisms: the fruits the gomphotheres ate. *Science* 215, 19–27.

Jordano, P. (1987) Patterns of mutualistic interactions in pollination and seed dispersal: connectance, dependence asymmetries, and coevolution. *American Naturalist* 129, 657–677.

Jordano, P. and Schupp, E.W. (2000) Seed disperser effectiveness: the quantity component and patterns of seed rain for *Prunus mahaleb. Ecological Monographs* 70, 591–615.

Kato, M., Itino, T. and Nagamitsu, T. (1993) Melittophily and ornithophily of long-tubed flowers in Zingiberaceae and Gesneriaceae in West Sumatra. *Tropics* 2, 129–142.

Kato, M., Shibata, A., Yasui, T. and Nagamasu, H. (1999) Impact of introduced honeybees, *Apis mellifera*, upon native bee communities in the Bonin (Ogasawara) Islands. *Researches on Population Ecology* 41, 217–228.

Kawakita, A., Takimura, A., Terachi, T., Sota, T. and Kato, M. (2004) Cospeciation analysis of an obligate pollination mutualism: have *Glochidion* trees (Euphorbiaceae) and pollinating *Epicephala* moths (Gracillariidae) diversified in parallel? *Evolution* 58, 2201–2214.

Kearns, C.A. and Inouye, D.W. (1997) Pollinators, flowering plants, and conservation biology. *BioScience* 47, 297–307.

Kearns, C.A., Inouye, D.W. and Waser, N.M. (1998) Endangered mutualisms: the conservation of plant–pollinator interactions. *Annual Review of Ecology and Systematics* 29, 83–112.

Kelly, D., Ladley, J.J. and Robertson, A.W. (2004) Is dispersal easier than pollination? Two tests in New Zealand Loranthaceae. *New Zealand Journal of Botany* 42, 89–103.

Kitamura, S., Yumoto, T., Poonswad, P., Chuailua, P., Plongmai, K., Maruhashi, T. and Noma, N. (2002) Interactions between fleshy fruits and frugivores in a tropical seasonal forest in Thailand. *Oecologia* 133, 559–572.

Kitamura, S., Suzuki, S., Yumoto, T., Poonswad, P., Chuailua, P., Plongmai, K., Noma, N., Maruhashi, T. and Suckasam, C. (2004) Dispersal of *Aglaia spectabilis*, a large-seeded tree species in a moist evergreen forest in Thailand. *Journal of Tropical Ecology* 20, 421–427.

Kitamura, S., Suzuki, S., Yumoto, T., Chuailua, P., Plongmai, K., Poonswad, P., Noma, N., Maruhashi, T. and Suckasam, C. (2005) A botanical inventory of a tropical seasonal forest in Khao Yai National Park, Thailand: implications for fruit–frugivore interactions. *Biodiversity and Conservation* 14, 1241–1262.

Kitamura, S., Suzuki, S., Yumoto, T., Poonswad, P., Chuailua, P., Plongmai, K., Maruhashi, T., Noma, N. and Suckasam, C. (2006) Dispersal of *Canarium euphyllum* (Burseraceae), a large-seeded tree species, in a moist evergreen forest in Thailand. *Journal of Tropical Ecology* 22, 137–146.

Klein, A.M., Steffan-Dewenter, I., Buchori, D. and Tscharntke, T. (2002) Effects of land-use intensity in tropical agroforestry systems on coffee flower-visiting and trap-nesting bees and wasps. *Conservation Biology* 16, 1003–1014.

Knight, T.M., Steets, J.A., Vamosi, J.C., Mazer, S.J., Burd, M., Campbell, D.R., Dudash, M.R., Johnston, M.O., Mitchell, R.J. and Ashman, T.-L. (2005) Pollen limitation of plant reproduction: pattern and process. *Annual Review of Ecology, Evolution and Systematics* 36, 467–497.

Laurance, W.F., Lovejoy, T.E., Vasconcelos, H.L., Bruna, E.M., Didham, R.K., Stouffer, P.C., Gascon, C., Bierregaard, R.O., Laurance, S.G. and Sampaio, E. (2002) Ecosystem decay of Amazonian forest fragments: a 22-year investigation. *Conservation Biology* 16, 605–618.

Lamb, D., Erskine, P.D. and Parrotta, J.A. (2005) Restoration of degraded tropical forest landscapes. *Science* 310, 1628–1632.

Leven, M.R. and Corlett, R.T. (2004) Invasive birds in Hong Kong, China. *Ornithological Science* 3, 43–55.

Li, J. (1997) On the systematics of *Thladiantha* (Cucurbitaceae). *Acta Botanica Yunnanica* 19, 103–127.

Makana, J.R. and Thomas, S.C. (2004) Dispersal limits natural recruitment of African mahoganies. *Oikos* 106, 67–72.

Martínez-Garza, C. and Howe, H.F. (2003) Restoring tropical diversity: beating the time tax on species loss. *Journal of Applied Ecology* 40, 423–429.

Meehan, H.J., McConkey, K.R. and Drake, D.R. (2005) Early fate of *Myristica hypargyraea* seeds dispersed by *Ducula pacifica* in Tonga, Western Polynesia. *Austral Ecology* 30, 374–382.

Memmott, J., Waser, N.M. and Price, M.V. (2004) Tolerance of pollination networks to species extinctions. *Proceedings of the Royal Society London B: Biological Sciences* 271, 2605–2611.

Momose, K., Yumoto, T., Nagamitsu, T., Kato, M., Nagamasu, H., Sakai, S., Harrison, R.D., Itioka, T., Hamid, A.A. and Inoue, T. (1998) Pollination biology in a lowland dipterocarp forest in Sarawak, Malaysia. I. Characteristics of the plant–pollinator community in a lowland dipterocarp forest. *American Journal of Botany* 85, 1477–1501.

Morales, C.L. and Galetto, L. (2003) Influence of compatibility system and life form on plant reproductive success. *Plant Biology* 5, 567–573.

Olesen, J.M. and Jordano, P. (2002) Geographic patterns in plant–pollinator mutualistic networks. *Ecology* 83, 2416–2424.

Osada, N. (2005) Influence of between-year variation in the density of *Rhus trichocarpa* fruits on the removal of fruit by birds. *Plant Ecology* 176, 195–202.

Peh, K.S.H., de Jong, J., Sodhi, N.S., Lim, S.L.H. and Yap, C.A.M. (2005) Lowland rainforest avifauna and human disturbance: persistence of primary forest birds in selectively logged forests and mixed-rural habitats of southern Peninsular Malaysia. *Biological Conservation* 123, 489–505.

Pellmyr, O. (2002) Pollination by animals. In: Herrera, C.M. and Pellmyr, O. (eds) *Plant–Animal Interactions: An Evolutionary Approach.* Blackwell Science, Oxford, UK, pp. 157–184.

Pereira, S.L. and Wajntal, W. (1999) Reintroduction of guans of the genus *Penelope* (Cracidae, Aves) in reforested areas in Brazil: assessment by DNA fingerprinting. *Biological Conservation* 87, 31–38.

Primack, R.B. (1987) Relationships among flowers, fruits, and seeds. *Annual Review of Ecology and Systematics* 1, 409–430.

Primack, R.B. and Corlett, R.T. (2005) *Tropical Rain Forests: An Ecological and Biogeographical Comparison.* Blackwell, Oxford, UK.

Quezada-Euán, J.J.G., May-Itzá, W.D.J. and Gonzalez-Acereto, J.A. (2001) Meliponiculture in Mexico: problems and perspective for development. *Bee World* 82, 160–167.

Rao, S.P., Atluri, J.B. and Reddi, C.S. (2001) Intermittent mass blooming, midnight anthesis and rockbee pollination in *Pterocarpus santalinus* (Fabaceae). *Nordic Journal of Botany* 21, 271–276.

Redford, K.H. and Feinsinger, P. (2001) The half-empty forest: sustainable use and the ecology of interactions. In: Reynolds, J.D., Mace, G.M., Redford, K.H. and Robinson, J.G. (eds) *Conservation of Exploited Species.* Cambridge University Press, Cambridge, UK, pp. 370–399.

Reid, N. (1991) Coevolution of mistletoes and frugivorous birds? *Australian Journal of Ecology* 16, 457–469.

Richards, A.J. (1997) *Plant Breeding Systems.* Chapman and Hall, London.

Robertson, A.W., Trass, A., Ladley, J.J. and Kelly, D. (2006) Assessing the benefits of frugivory for seed germination: the importance of the deinhibition effect. *Functional Ecology* 20, 58–66.

Robinson, J.G. and Bennett, E.L. (2000) *Hunting for Sustainability in Tropical Forests.* Columbia University Press, New York.

Roubik, D.W. (1989) *Ecology and Natural History of Tropical Bees.* Cambridge University Press, Cambridge, UK.

Saracco, J.F., Collazo, J.A., Groom, M.J. and Carlo, T.A. (2005) Crop size and fruit neighbourhood effects on bird visitation to fruiting *Schefflera morototoni* trees in Puerto Rico. *Biotropica* 37, 81–87.

Schaefer, H.M., Schmidt, V. and Bairlein, F. (2003a) Discrimination abilities for nutrients: which difference matters for choosy birds and why? *Animal Behaviour* 65, 531–541.

Schaefer, H.M., Schmidt, V. and Winkler, H. (2003b) Testing the defence trade-off hypothesis: how contents of nutrients and secondary compounds affect fruit removal. *Oikos* 102, 318–328.

Schupp, E.W. (1993) Quantity, quality and the effectiveness of seed dispersal by animals. *Plant Ecology* 107–108, 15–29

Seddon, P.J., Soorae, P.S. and Launay, F. (2005) Taxonomic bias in reintroduction projects. *Animal Conservation* 8, 51–58.

Sodhi, N.S., Koh, L.P., Brook, B.W. and Ng, P.K.L. (2004) Southeast Asian biodiversity: an impending disaster. *Trends in Ecology and Evolution* 19, 654–660.

Steffan-Dewenter, I., Potts, S.G. and Packer, L. (2005) Pollinator diversity and crop pollination services are at risk. *Trends in Ecology and Evolution* 20, 651–652.

Svenning, J.C. and Wright, S.J. (2005) Seed limitation in a Panamanian forest. *Journal of Ecology* 93, 853–862.

Tang, A.M.C., Corlett, R.T. and Hyde, K.D. (2005) The persistence of ripe fleshy fruits in the presence and absence of frugivores. *Oecologia* 142, 232–237.

Tsang, A.C.W. and Corlett, R.T. (2005) Reproductive biology of the *Ilex* species (Aquifoliaceae) in Hong Kong, China. *Canadian Journal of Botany* 83, 1645–1654

Turnbull, L.A., Crawley, M.J. and Rees, M. (2000) Are plant populations seed-limited? A review of seed sowing experiments. *Oikos* 88, 225–238.

Waser, N.M., Chittka, L., Price, M.V., Williams, N. and Ollerton, J. (1996) Generalization in pollination systems, and why it matters. *Ecology* 77, 1043–106

Weir, J.E.S. and Corlett, R.T. (2007) How far do birds disperse seeds in the degraded tropical landscape of Hong Kong, China? *Landscape Ecology* 22, 131–140.

Wheelwright, N.T. and Orians, G.H. (1982) Seed dispersal by animals: contrasts with pollen dispersal, problems of terminology, and constraints on coevolution. *American Naturalist* 119, 402–413.

Yap, C.A.M. and Sodhi, N.S. (2004) Southeast Asian invasive birds: ecology, impact and management. *Ornithological Science* 3, 57–67

Yasuda, M., Miura, S., Ishii, N., Okuda, T. and Hussein, N.A. (2005) Fallen fruits and terrestrial vertebrate frugivores: a case study in a lowland tropical rainforest in Peninsula Malaysia. In: Forget, P.-M., Lambert, J.E., Hulme, P.E. and Vander Wall, S.B. (eds) *Seed Fate: Predation, Dispersal and Seedling Establishment.* CAB International, Wallingford, UK, pp. 151–174.

25 Do Seed Dispersers Matter? A Biogeographical Approach

K. Böhning-Gaese

Introduction

One of the key questions currently receiving attention in ecology concerns the relationship between biodiversity and ecosystem functioning (Schulze and Mooney, 1994; Loreau *et al.*, 2002; Hooper *et al.*, 2005). An important functional relationship in many ecosystems is the interaction between frugivorous animals and animal-dispersed plants. The loss, fragmentation and degradation of forests, together with hunting, can lead to changes in frugivore assemblages. For example, in many tropical forests in South-east Asia, large-bodied frugivores such as hornbills are driven to local extinction (see Corlett, Chapter 24, this volume). Similarly, declines in, or the local extinction of, frugivorous animals have been observed in fragmented forests in Spain (Santos and Tellería, 1994), Brazil (Da Silva and Tabarelli, 2000), Tanzania (Cordeiro and Howe, 2003) and Australia (Moran *et al.*, 2004). This raises the question of how changes in frugivore assemblages will affect seed dispersal, tree populations and plant communities?

Changes in frugivore assemblages are expected to have far-reaching consequences for tree species and communities, at least in theory (Howe, 1984; Terborgh, 1986; Bond, 1995). A first consequence of a depauperate frugivore community might be low rates of seed dispersal, with effects on the spatial distribution of seeds, seedlings and eventually trees. In addition, seeds carry genetic information. Thus, changes in the spatial distribution of seeds and seedlings might have consequences for the genetic structure of plant populations. Further effects of low rates in seed dispersal can be expected for the reproductive success of trees. Seed dispersal has been shown to increase the number of regenerating seedlings through escape from enemies and colonization of, or directed dispersal to, suitable habitats (Howe, 1986; Wenny, 2001). A depauperate frugivore assemblage might therefore lead to declines in the numbers of seedlings, saplings and, eventually, the abundance of a tree species. Finally, consequences for the composition of tree communities can be expected if some species of trees suffer less from the loss of seed dispersers than others.

In spite of these theoretical links from seed dispersers to seed dispersal rates, and to tree populations and communities, changes in the frugivore assemblage might not necessarily lead to the anticipated changes in plant populations (Bond, 1995). For example, seed dispersal rates might not decline in spite of some seed dispersers going extinct if the remaining species visit the trees more frequently, or if other species replace the lost species (Farwig *et al.*, 2006). The spatial distribution of trees might not change if it is mainly determined by the spatial distribution of suitable habitat patches (Jeltsch *et al.*, 1996; Wiegand *et al.* 2000). No effects on the genetic structure might be seen if gene flow is more prevalent through pollen than through seeds (Ellstrand, 1992; Petit *et al.*, 2005). The abundance of the trees might remain the same if a decline in the seedling stage is compensated by higher survival rates at other stages in the life cycle of the plant (Harper, 1977; Wang and Smith, 2002; Howe and Miriti, 2004).

To understand the consequences of the loss of seed dispersers for tree populations and communities, rigorous tests are needed. The most rigorous approaches to understanding the relationship between biodiversity and ecosystem functioning are experimental. Scientists have experimentally designed ecosystems with varying species composition, or they have experimentally removed species from ecological communities and studied the consequences for ecosystem functioning (Naeem *et al.*, 1994, Loreau *et al.*, 2002; Diaz *et al.*, 2003). These studies have provided convincing evidence for the importance of biodiversity for ecosystem functioning (Loreau *et al.*, 2002; Hooper *et al.*, 2005). Furthermore, in these systems it has been possible to discover whether the effects of biodiversity were caused by species richness per se or by the presence of certain functionally important species or groups. However, only a limited number of ecosystems can be studied in this way – for example, microcosms or annual plants – and the conclusions are limited to small spatial and temporal scales (Loreau *et al.*, 2002; Hooper *et al.* 2005). When studying seed dispersal systems, it is not possible to manipulate the species composition of frugivore communities for practical and ethical reasons (an exception might be ants). Thus, an alternative approach to understanding the relationship between biodiversity and ecosystem functioning is to use a 'natural experiment' (Begon *et al.*, 2006); for example, the comparison between two biogeographical regions that differ in frugivore assemblages.

A particularly interesting system for such a comparative approach is continental Africa and Madagascar. The diversity of frugivorous birds is markedly different in the two areas, with a generally high diversity in Africa and a depauperate frugivore community in Madagascar, where important frugivorous bird groups such as hornbills, turacos and barbets are missing (Fleming *et al.*, 1987; Langrand, 1990; Goodman and Ganzhorn, 1997). Madagascar appears never to have had a high diversity of frugivorous birds; there is no evidence of frugivorous birds going extinct, either in historical or in earlier times (Goodman and Ganzhorn, 1997). I took such a biogeographical approach and compared seed dispersal systems between Oribi Gorge Nature Reserve in South Africa and Kirindy Forest/Centre de Formation Professionelle Forestière de Morondava (CFPF) in Madagascar. The general patterns of avian frugivore diversity in Africa and Madagascar are reflected by the avian frugivore communities at the study sites. At Oribi Gorge NR, 14 (out of a total of 226) species of birds are mainly frugivorous, whereas at Kirindy Forest/CFPF only four species (out of 114) feed mainly on fruit. In contrast, lemurs play an important role in seed dispersal in Madagascar (Scharfe and Schlund, 1996; Dew and Wright, 1998; Ganzhorn *et al.*, 1999; Bollen *et al.*, 2004), while primates are not as diverse and are less important seed dispersers in South Africa (Frost, 1980; Knight and Siegfried, 1983). At Oribi Gorge NR, three species of primates are found, all of them at least partly frugivorous; Kirindy Forest/CFPF has eight species of lemurs, all mainly or partly frugivorous.

In the review presented in this chapter, I provide an overview of the results from a comparison of seed dispersal systems at these two sites in South Africa and Madagascar. Studies were conducted at the level of tree populations, comparing two species in the genus *Commiphora*: *C. harveyi* in South Africa and *C. guillauminii* in Madagascar, and at the level of the tree community, comparing the tree species between Oribi Gorge NR and Kirindy Forest/CFPF. At the level of tree populations I studied seed dispersal, focusing on tree visitors, seed dispersers and seed dispersal rates, the spatial distribution of seedlings and trees, and the spatial genetic structure of the tree populations at the local and biogeographical scale. At the level of tree communities I compared seed dispersal syndromes between the two sites. All work is based on a collaborative research effort involving a large number of students and collaborators (see Acknowledgements). In the discussion I assess the limitations and advantages of such a comparative, biogeographical approach.

Study System

The studies were carried out between 1992 and 2003 on the KwaZulu–Natal south coast in South Africa and in western Madagascar (see map, Fig. 25.1). In South Africa, the study site was Oribi Gorge NR, a 1850-ha nature reserve located 110 km south of Durban and 21 km inland from Port Shepstone. The reserve is situated in a 280-m-deep gorge whose steep slopes are covered with coastal scarp forest; it is characterized by a subtropical climate with an average rainfall of 1176 mm (Glen, 1996). The study site in Madagascar was the Kirindy Forest/CFPF, a 10,000-ha forestry concession located 60 km north of Morondava. It is characterized by a tropical climate with an average annual rainfall of 730 mm (Sorg and Rohner, 1996) and the vegetation is classified as dry deciduous forest.

The population studies were conducted with *Commiphora harveyi* in South Africa and *C. guillauminii* in Madagascar (Burseraceae). The comparison of

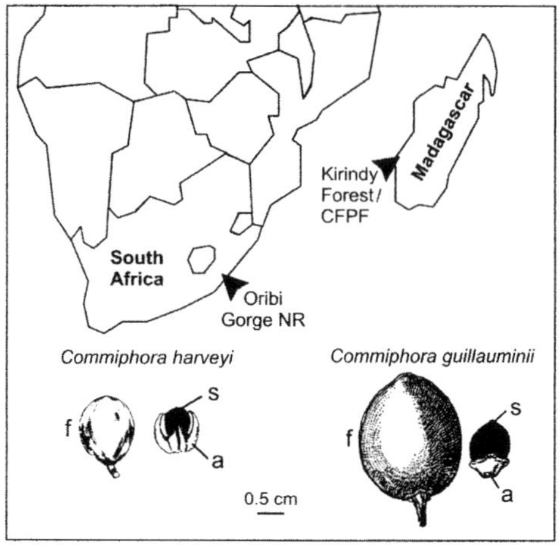

Fig. 25.1. Map of study sites (arrows), study tree species and their corresponding fruits in South Africa and Madagascar; (f) whole fruit with outer coverings, (s) seed, (a) aril (redrawn after de la Bathie, 1946 and van der Walt, 1973). Adapted from Bleher and Böhning-Gaese (2001).

the same species between both sites was not possible because all *Commiphora* species found in Madagascar are endemic to Madagascar. Both species are common canopy trees at the respective study sites, growing up to 20 m in height. Both species are dioecious and have small, inconspicuous flowers. The pollination system of the two species is rather similar, with similar numbers and a comparable assembly of species of insects visiting the trees. Mean flower visitation rates and fruit set are very low (Farwig *et al.*, 2004; Voigt *et al.*, 2005).

The South African species had a median number of 1879 fruits per tree (range 324–15,231, n = 15 trees), while the Malagasy species had 747 fruits per tree (range 345–4591, n = 12 trees). Both species have morphologically similar, typically bird-dispersed, roundish fruits consisting of an unpalatable greenish-reddish fleshy outer covering that splits into two halves to expose a single diaspore (see Fig. 25.1). The diaspore consists of a brilliant black seed (South Africa: 7.1 × 5.7 × 5.1 mm, n = 20; Madagascar: 12.2 × 8.0 × 6.3 mm, n = 20), which is partly enveloped by a red fleshy aril (Fig. 25.1). Arils in the South African species are cup-shaped with four lobes (van der Walt, 1986); whereas in Madagascar the lobes are absent (de la Bathie, 1946) (Fig. 25.1). Both arils are rich in lipids (South Africa 60.1%, Madagascar 81.1%; F.A. Voigt, 2000, unpublished data) and have a similar fraction of proteins (South Africa 3.1%, Madagascar 3.4%; F.A. Voigt, 2000, unpublished data).

Secondary seed dispersal by ants played a role in Madagascar but not in South Africa. However, in Madagascar, seeds were moved only for relatively short distances and seedling establishment profited much less from seed dispersal by ants than by birds (Böhning-Gaese *et al.*, 1995, 1996, 1999; Voigt *et al.*, 2002).

Seed dispersal by frugivorous birds and primates

The two species of *Commiphora* in South Africa and Madagascar have similar fruits and very similar seed sizes and can therefore be used as a model system to test the influence of frugivore diversity on seed dispersal rates. If seed dispersal is influenced by the frugivore assemblage, I would expect higher seed dispersal rates in South Africa than in Madagascar (Fig. 25.2).

To test this hypothesis, observations were conducted on 15 trees in South Africa and 12 trees in Madagascar (Böhning-Gaese *et al.*, 1995, 1999; Bleher and Böhning-Gaese, 2000, 2001). Each tree was observed for 2 days, with observations conducted in blocks from sunrise to 10:00 hours,

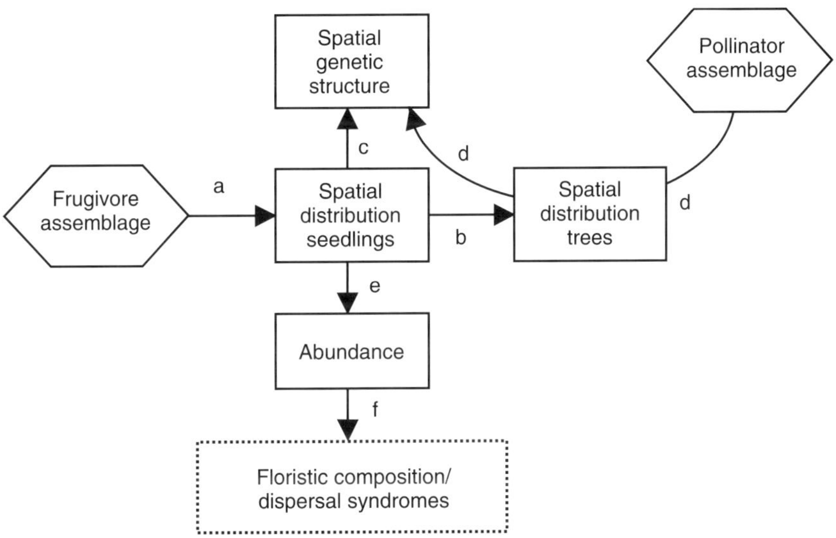

Fig. 25.2. Flow diagram demonstrating the potential consequences of a depauperate frugivore assemblage in Madagascar (diamond) for tree populations (rectangles with solid lines) and the tree community (rectangle with dashed line). A depauperate assemblage of frugivorous birds can lead, through low seed dispersal rates and short seed transport distances, to a clumped spatial distribution of (a) seedlings and (b) trees. Furthermore, the clumped spatial distribution of seedlings (c) and short pollen transport distances, the latter caused by the clumped spatial distribution of trees (d), result in low gene flow and high spatial genetic structure. On the other hand, the more clumped spatial distribution of seedlings can lead, through lower survival of seedlings and saplings (e) to a lower abundance of the tree. In the long term this can result in the local extinction of the tree and of other trees that have fruits with bird-dispersal syndromes (f). Thus, it is possible that the floristic composition of the forest changes and the frequency of bird-dispersal syndromes declines in favour of primate-dispersal syndromes (assuming that trees with primate-dispersal syndromes are well-dispersed and have high abundance and a high probability of local survival). The flow diagram is restricted to the patterns that have been investigated in the present study and assumes that other factors (spatial distribution of suitable establishment sites, competition, predation, herbivory, pollinator assemblage, etc.) do not override the effects of seed dispersal.

from 10:00 to 14:30 hours, and from 14:30 hours to sunset (total number of observation hours in South Africa 345 h, in Madagascar 312 h). In addition, night observations were conducted at both sites. For information on seed dispersal rates, fruit traps were established under tree crowns of the observed trees in South Africa and Madagascar.

All visiting species, both in South Africa and Madagascar, were diurnal. At both sites mainly birds were observed visiting the trees (South Africa: 96.4% of the visitors; Madagascar: 99.1%). In South Africa 15 species of birds and one primate were documented visiting trees; in Madagascar six species of birds and one lemur. In terms of dispersal, in South Africa 12 species of birds and one primate dispersed seeds away from the tree crowns, with the crowned hornbill (*Tockus alboterminatus*; Bucerotidae) being the most important disperser (34.0% of the dispersed seeds). Other species that dispersed seeds were the red-billed woodhoopoe (*Phoeniculus purpureus*; Phoeniculidae, 29.6% of the dispersed seeds), black-collared barbet (*Lybius torquatus*; Ramphastidae, 10.1%) and the red-fronted tinker barbet (*Pogoniulus pusillus*; Ramphastidae, 8.3%). Most species dispersed seeds by swallowing the seed and aril. In Madagascar, only three species of animals dispersed seeds. The main disperser was the lesser vasa parrot (*Coracopsis nigra*; Psittacidae, 87.1% of the dispersed seeds) and the two other species were the crested drongo (*Dicrurus forficatus*; Dicruridae, 10.4%) and the Madagascar bulbul (*Hypsipetes madagascariensis*; Picnonotidae, 2.5%). Lesser vasa parrots harvested a large number of fruits. However, they nibbled off the arils without destroying the seeds and dropped most of the seeds directly under the crown. Only some seeds were dispersed away from the crown when the parrots left the trees while still carrying a seed in their beaks. The mean proportion of fruits per tree which were handled by visitors was similar at both sites (South Africa: 90.1%; Madagascar: 93.2%). In contrast, the total percentage of seeds dispersed away from the parent tree showed a highly significant difference (South Africa: 70.8% of the seeds; Madagascar: 7.9%).

Spatial distribution of seedlings and trees

If seed dispersal away from the parent tree was higher in South Africa than in Madagascar, I would have expected seedlings and trees to have a less clumped distribution in South Africa than in Madagascar (Fig. 25.2). The spatial distribution of seedlings was determined along a 1290-m-long transect in South Africa and 11 transects of 250 m length, spread across the study site (total 2750 m), in Madagascar (Böhning-Gaese *et al.*, 1999; Bleher and Böhning-Gaese, 2001). For each seedling the distance to the closest female *Commiphora* tree was measured. Seedling distributions differed markedly between South Africa and Madagascar. In South Africa the median distance of a seedling to the closest female *Commiphora* tree was 21.1 m, in Madagascar it was 0.9 m (Fig. 25.3). Distinguishing between seedlings < 1 year and > 1 year of age (Fig. 25.3) demonstrated that the

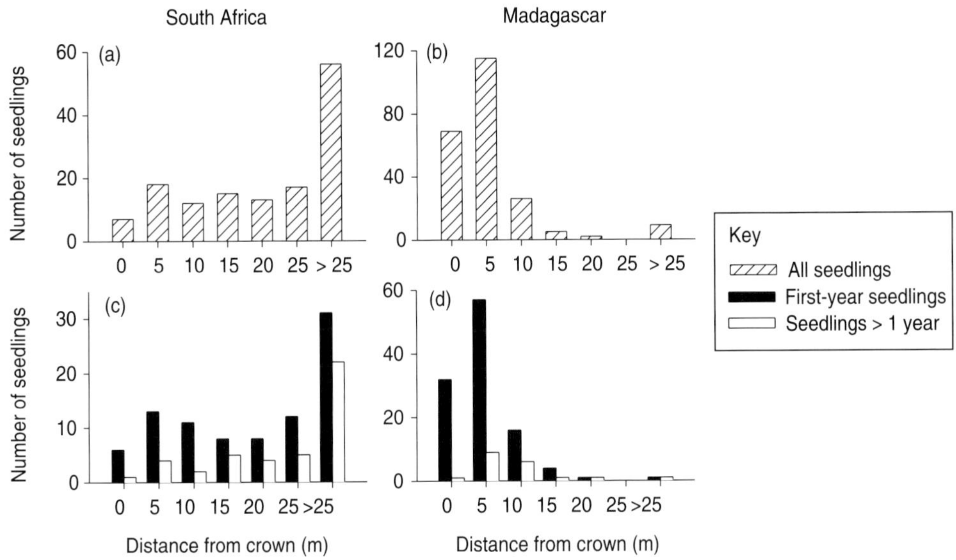

Fig. 25.3. Spatial distribution of all seedlings (a, b) and of first-year seedlings and seedlings older than 1 year (c, d) found in relation to distance from the nearest tree crown for the South African and Malagasy study sites. Adapted from Bleher and Böhning-Gaese (2001).

proportion of older seedlings was particularly low in the Malagasy species among the seedlings growing at high densities under and up to 5 m distance from a crown (Fig. 25.3). This suggests that the high density of seedlings under and close to crowns might lead to particularly high seedling mortality.

However, the distance of a seedling from the nearest female tree is a poor measure of the true seed dispersal distance (Nathan and Muller-Landau, 2000). In a spatially explicit computer model, Bleher *et al.* (2002) simulated seed dispersal and seedling establishment using data derived from the field studies. Simulations were run under a South African and a Malagasy scenario. The results revealed that under the South African scenario the actual median seedling-to-parent tree distance (corresponding to dispersal distance) was 64 m, while in Madagascar it remained at 0.9 m. Thus, the actual dispersal distances must be even higher for the South African species as compared with the distances measured between a seedling and the closest female tree.

The spatial distribution of trees was determined by the T-square method (Ludwig and Reynolds, 1988). The spatial distribution of trees revealed a significantly uniform tree distribution of the South African *Commiphora* species (clumping index $C = 0.36$) and a significantly clumped tree distribution in Madagascar ($C = 0.67$). The use of the same spatially explicit computer model demonstrated that the mean seed dispersal distance was one of the most significant factors (besides tree density) influencing the spatial distribution of trees (Bleher *et al.*, 2002). Thus, the low rate of seed dispersal away from the parent tree in Madagascar could be connected with a much more clumped spatial distribution of seedlings and trees.

Genetic structure

The more clumped spatial distribution of seedlings and trees in the Malagasy than the South African *Commiphora* species might also lead to a higher spatial genetic structure of the trees (Fig. 25.2). However, in plants, genes are not only transported through seeds but also through pollen, and the distances that both seeds and pollen are moved influences the spatial genetic structure. Pollen transport distances are not known for the two *Commiphora* species. However, it appears possible that pollen is transported over shorter distances in the Malagasy than in the South African species. Both species have similar pollinator assemblages. However, the Malagasy species grows in dense clumps of trees. If pollen transport distances are influenced by nearest-neighbour distances, gene flow through pollen might occur over shorter distances in the Malagasy than in the South African species (Hardy *et al.*, 2006). Thus, poor seed dispersal in the Malagasy species might lead to a higher spatial genetic structure; *first*, directly through short seed dispersal distances and, *second*, indirectly through short pollen transport distances between neighbours in dense clumps of trees (Hardy *et al.*, 2006).

To study the genetic structure of the tree populations, F.A. Voigt *et al.* (2006, unpublished results) sampled the leaf material of 158 trees in South Africa and of 136 individuals in Madagascar. Trees were sampled at different spatial scales to test whether the spatial genetic structure expected at small spatial scales also extended to larger spatial scales up to the geographical ranges of the two species. Altogether, trees were sampled from 15 sample sites (at least 8 km apart) in South Africa and from 12 sample sites in Madagascar (also at least 8 km apart). At each sample site, trees were sampled in 1–5 subgroups, each within a circle with a diameter of 300 m (small spatial scale). Second, within each sample site, subgroups were pooled to 1–4 groups each within a circle with a diameter of 3 km (medium spatial scale). At the largest spatial scale, trees were pooled within each sample site (large spatial scale).

The genetic structure was evaluated using amplified fragment length polymorphism (AFLP). The genetic variance was studied at the small and medium spatial scales using Wright's hierarchical *F*-statistics (Wright, 1978; Weir and Cockerham, 1984). Thereby, F.A. Voigt *et al.* (2006, unpublished results) defined all trees within a circle with a diameter of 300 m (subgroups, small spatial scale) or of 3 km (groups, medium spatial scale) as populations and nested them into sample sites. To calculate the genetic differentiation at the largest spatial scale, each sample site was defined as one population and an analysis of molecular variance (AMOVA) was conducted. Additionally, the spatial genetic structure was assessed at different geographical scales using spatial autocorrelation analyses. F.A. Voigt *et al.* (2006, unpublished results) calculated kinship coefficients for each pair of individuals using the program SPaGeDi (Hardy and Vekemans, 2002). The kinship coefficient was defined according to Hardy and Vekemans (2002) as the ratio of differences of probabilities of identity in state (Rousset 2002) and plotted against the geographical distance separating the individuals.

The results demonstrated that both species had similar levels of molecular variation in terms of number of polymorphic loci. At the smallest spatial scale, the hierarchical F-statistics revealed that the differentiation among subgroups within sample sites was higher in Madagascar ($F_{ST} = 0.08$, $P < 0.05$) than in South Africa ($F_{ST} = 0.026$, $P < 0.05$). Similarly, at the medium spatial scale, the F_{ST} value among groups within sample sites was higher in Madagascar ($F_{ST} = 0.069$, $P < 0.05$) than in South Africa ($F_{ST} = 0.014$, P = n.s.). Thus, at the small and medium spatial scales, populations within sample sites were more different in Madagascar than in South Africa. Both patterns supported the initial hypothesis. In contrast, the genetic differentiation among sample sites, i.e. at the largest spatial scale, was higher in the South African *Commiphora* species ($F_{ST} = 0.16$, $P < 0.05$) than in the Malagasy species ($F_{ST} = 0.053$, $P < 0.05$). The spatial autocorrelation analysis revealed that in the Malagasy species the kinship coefficient dropped almost to zero in the distance class of up to 3 km, and in the South African species in the distance class of up to 30 km. Thus, trees in Madagascar were related to each other within distances of up to 3 km, but in South Africa up to 30 km.

Floristic composition of the forest

The next step to test is the link between seed dispersal, seedling establishment, the abundance of trees and the floristic composition of the forest (Fig. 25.2). Given the small number of older seedlings growing under and close to the crown of the Malagasy species (Fig. 25.3), it might be possible that the poor seed dispersal in this species eventually limits its abundance and that other species of trees usually dispersed by birds are also limited in abundance. On the other hand, species of trees in Madagascar with fruits with typical primate-dispersal syndromes are well dispersed by lemurs (Scharfe and Schlund, 1996; Dew and Wright, 1998; Ganzhorn *et al.*, 1999; Bollen *et al.*, 2004). If seed dispersal has consequences for the abundance of these trees, one might expect that trees with primate-dispersal syndromes should be comparatively frequent. Thus, one might expect that seed dispersal has consequences for the abundance of the species and, through local extinction of poorly dispersed and survival of well dispersed species, eventually on the floristic composition of the forest (Fig. 25.2). Such a test is very difficult to conduct. However, it is at least possible to test for the expected outcome. Thus, it is possible to test whether there are more tree species with typical primate-dispersal syndromes in Madagascar, and more species with bird-dispersal syndromes in South Africa.

I tested this hypothesis by comparing the fruit-dispersal syndromes of the tree species found at the two study sites (Voigt *et al.*, 2004). In general, fruit syndromes are defined according to the colour, size and husk thickness of the fruits. Birds are assumed to prefer different colours than primates do (for example red, blue and black fruits versus green and brown fruits) and to favour smaller fruits with thinner husks than primates do. Since no generally accepted characterization of fruits as 'bird-fruits' or 'primate-fruits' exists, Voigt *et al.* (2004) worked with five different fruit classification systems from different regions (Janson, 1983, Neotropics; Knight and Siegfried, 1983, South Africa; Gautier-Hion *et al.*, 1985, Central Africa; Dew and Wright, 1998, Madagascar; and an average of these four systems). Data were used from 214 species of trees from Oribi Gorge NR and 133 from Kirindy Forest/CFPF. All analyses were conducted at the species and genus level and controlled for potential phylogenetic effects using Mantel tests (Böhning-Gaese and Oberrath, 1999).

Voigt *et al.* (2004) found proportionally more tree species that had fruits with bird-colours in South Africa and proportionally more species that had fruits with primate-colours in Madagascar (Fig. 25.4). The differences were significant in four of the five classification systems. No

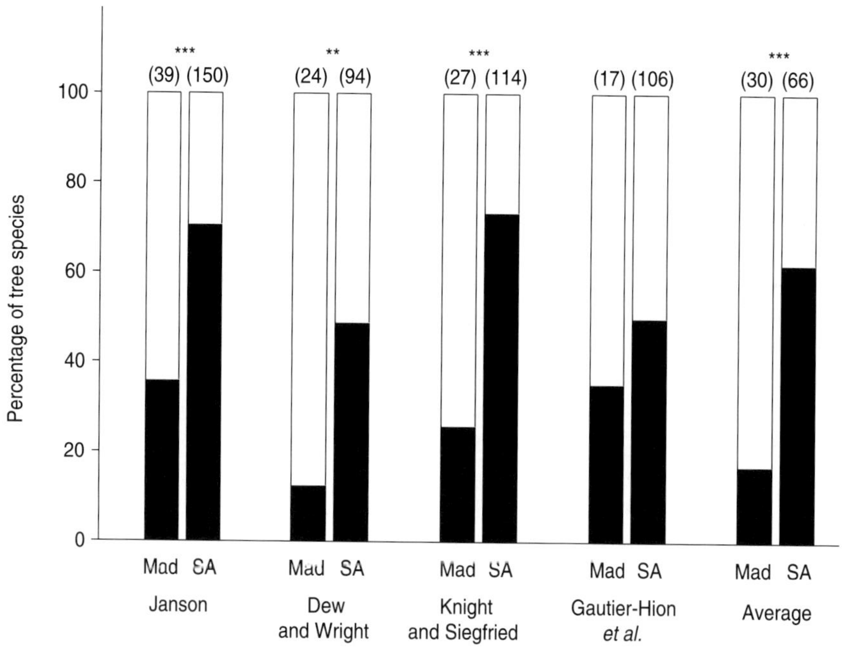

Fig. 25.4. Proportion of tree species with bird- and primate-colours in Madagascar (Mad) and South Africa (SA) according to Janson (1983, $\chi^2 = 16.1$), Dew and Wright (1998, $\chi^2 = 10.5$), Knight and Siegfried (1983, $\chi^2 = 21.4$), Gautier-Hion *et al.* (1985, $\chi^2 = 1.3$) and average classification ($\chi^2 = 17.1$). χ^2 values are based on χ^2 tests with 1 *df.* ***P* < 0.01; ****P* < 0.001; bird-colours = black bars; primate-colours = white bars. Numbers in brackets are the numbers of tree species included in the respective analyses.

significant difference for fruit size was found between the South African and the Malagasy study sites (fruit size in mm, log-transformed: South Africa – mean 1.09 ± 0.28 SD, n = 138 species; Madagascar – mean 1.17 ± 0.28 SD, n = 53; t-test, $t_{(189)} = 1.79$, $P = 0.07$). Similarly, husk thickness did not differ between the two study sites (percentage of species with no or a thin husk versus species with a thick husk: South Africa 89.9% versus 10.1%, $n_{total} = 148$; Madagascar 85.7% versus 14.3%, $n_{total} = 56$; χ^2 test, $\chi^2 = 0.70$, $df = 1$, $P = 0.40$). The results were the same at the species and genus level and phylogenetic effects on fruit colour, size and husk thickness were weak or absent (Mantel tests: $t < 3.45$, $R^2 < 0.27\%$).

Discussion

Comparing seed dispersal systems between South Africa and Madagascar demonstrated that the diversity of fruit-eating birds was correlated with seed dispersal rates, the spatial distribution of seedlings and trees, the genetic structure of the trees at local spatial scales, and the fruit characteristics of the forest. With a depauperate avian frugivore community in Madagascar, the seeds of the Malagasy *Commiphora* species were dispersed by basically one species of bird (and that inefficiently), had a more clumped spatial distribution of seedlings and trees and a higher spatial genetic structure at scales up to 3–30 km. Furthermore, the Malagasy tree community was dominated by trees with primate-fruits, and the South African community by species with bird-fruits.

In this non-experimental, biogeographical approach, it is not possible to distinguish whether the correlation between frugivore diversity and seed dispersal rates was caused by an effect of species richness per se, or by the presence of functionally important species. The study sites in South Africa and Madagascar differed in both species richness and seed dispersal efficiency of the dominant tree visitors. Furthermore, no data are available on the quality of seed dispersal (*sensu* Schupp, 1993). Nevertheless, the difference in primary seed dispersal rates between the two *Commiphora* species is unusually large (70.8% versus 7.9%). Thus, the two species represent two, rather extreme, values in seed dispersal rates. Thus, in spite of these limitations, it is possible to evaluate whether the profound differences in seed dispersal rates can be connected with the spatial distribution and genetic structure of the trees and the composition of the tree communities.

In this study the clumped spatial distribution of seedlings and trees in the Malagasy species was explained by poor seed dispersal. However, the spatial distribution of seedlings and trees is also influenced by the spatial distribution of suitable establishment sites, by intraspecific competition, seed predation, herbivory and the gap dynamics of the canopy trees (Harper, 1977; Schupp, 1993; Wang and Smith, 2002; Howe and Miriti, 2004). Little is known about the actual seed transport distances, the microhabitats that the seeds are deposited in, and about post-dispersal processes that might influence seedling establishment and growth in the two *Commiphora* species.

Seeds of the Malagasy species are secondarily dispersed by ants (see Study System) but seedling establishment profited much less from seed dispersal by ants than by birds (Böhning-Gaese *et al.*, 1995, 1999; Voigt *et al.*, 2002). Seed predation by rodents of seeds under the canopy and close to ant colonies is high in Madagascar (K. Böhning-Gaese, 1993, 1994, unpublished data). Unfortunately, comparable data are not available for the South African species. Furthermore, comparative data on the spatial distribution of suitable establishment sites, herbivory or gap dynamics are not available.

The poor seed dispersal rate in the Malagasy *Commiphora* was connected with a higher spatial genetic structure up to distances of 3–30 km. It is important to note, however, that the genetic structure of trees is influenced by a complex interplay of factors, including genetic variation, gene flow through pollen, local population density, habitat distribution and the history of the populations (Hartl, 1980; Sork *et al.*, 1999; Hamrick, 2004). Genetic variation is similar in the two species and should not bias the results. Pollen transport distances are not known for the two *Commiphora* species. It is possible that pollen dispersal distances in the Malagasy species are shorter than in the South African species, because pollen donors are often the nearest neighbours and the trees are standing in dense clumps (Hardy *et al.*, 2006). However, this remains to be investigated. The local population density is high in both species and should have no effect on the results. However, habitat distribution is very different for the two species. The habitat of the South African *Commiphora* species has a more patchy distribution (mostly northerly slopes of deep gorges) than that of the Malagasy species (formerly more or less continuous dry tropical forest). This patchy habitat distribution should have no influence on the genetic structure within populations but might lead to little gene flow among populations and might explain why the South African species has higher levels of differentiation among sample sites. The population histories of the two species are not known.

Finally, the differences found in fruit colours between South Africa and Madagascar could reflect the differences first in the proportion of the main disperser groups (many birds and few primates in Africa versus many lemurs and few birds in Madagascar) and, second, in the colour vision ability of the primates (Old Word monkeys are trichromatic, lemurs are dichromatic and unable to see in the red spectrum). At present it is not possible to tell which of those two factors is more important. The fact that the difference in fruit size and husk thickness was not significant between South Africa and Madagascar is probably caused by the limited usefulness of these fruit traits to distinguish between bird- and primate-fruits. However, the patterns in fruit colours might be explained also by alternative factors, for example abiotic conditions or the unique biogeographical history of Madagascar and its lineages. To test for the effect of abiotic conditions necessitates demonstrating that abiotic conditions such as temperature or rainfall influence the evolution of fruit colours. To test for historic effects requires an understanding of the phylogenetic relationships of all the South African and Malagasy species used in the analysis. Neither test can be addressed so far because the necessary data are missing.

Thus, a general limitation of this comparative, biogeographical approach is that it is not possible to demonstrate without any doubt that the differences in the seed dispersal systems found for the South African and Malagasy study sites are caused only by seed dispersers. Using a biogeographical approach, it is always possible to formulate alternative hypotheses that explain the described patterns. However, in the case of the present study system, these hypotheses are so far speculative and, to my knowledge, not supported by data. In contrast, the frugivore assemblage and seed dispersal rate could be consistently connected with almost all the patterns and offer the most parsimonious explanation for these patterns. This strongly suggests that seed dispersers matter and that seed dispersal is the overriding factor influencing the spatial distribution and genetic structure of the trees as well as the fruit characteristics of the forests. However, further studies are necessary to test whether some of the assumptions are met and to evaluate alternative hypotheses.

The great advantage of this biogeographical approach and this specific study system is that frugivore assemblages (birds versus lemurs) and seed dispersal rates (70.8% versus 7.9%) vary to a degree rarely found in other systems. Thus, the two systems mark the most extreme consequences that a difference in seed dispersers can have on tree populations and communities. Furthermore, the two study sites have been separated for a long time period and, thus, the long-term consequences, especially on the spatial distribution and genetic structure of trees and the composition of tree communities, can be investigated. If habitat fragmentation, hunting or other environmental factors lead to strong declines in frugivore diversity, such as found between the South African and Malagasy study sites, and if this biogeographical comparison can be used to make temporal predictions, this implies that the loss of seed dispersers has far-reaching effects on the spatial distribution and genetic structure of tree populations and on the composition of tree communities.

Acknowledgements

This review is based on 13 years of field and laboratory work and collaborations with a multitude of persons and institutions. I would especially like to thank the co-authors of the publications on which this chapter is based: R. Arafeh, F. Bairlein, B. Bleher, J. Burkhardt, N. Farwig, J. Fietz, B. Gaese, J. Ganzhorn, E.-M. Griebeler, S. Jung, M. Kraemer, R. Oberrath, S. Rabemanantsoa, E. Randrianirina, J. Schmid, D. Schwab, M. Verhaagh and F. Voigt. R. Zahner and two anonymous reviewers read an earlier version of the manuscript and suggested many improvements. I thank the 'Commission Tripartite' of the Malagasy Government and the Ministère pour la Production Animale et des Eaux et Forêts for their permission to work in Madagascar, the Centre de Formation Professionelle Forestière de Morondava for permission to work on their forestry concession, and the KwaZulu–Natal Nature Conservation Service in South

Africa for their permission to work in Oribi Gorge Nature Reserve. The studies were funded by Deutscher Akademischer Austauschdienst, Deutsche Forschungsgemeinschaft (Bo 1221/3-1, 7-1,2; Ga 342/3-1,2) and by the H. and E. Walter Foundation.

References

Begon, M., Townsend, C.R. and Harper, J.L. (2006) *Ecology*, 4th edn. Blackwell, Oxford, UK.

Bleher, B. and Böhning-Gaese, K. (2000) Seed dispersal by birds in a South African and a Malagasy *Commiphora* species. *Ecotropica* 6, 43–54.

Bleher, B. and Böhning-Gaese, K. (2001) Consequences of frugivore diversity for seed dispersal, seedling establishment and the spatial pattern of seedlings and trees. *Oecologia* 129, 385–394.

Bleher, B., Oberrath, R. and Böhning-Gaese, K. (2002) Seed dispersal, breeding system, tree density and the spatial pattern of trees: a simulation approach. *Basic and Applied Ecology* 3, 115–123.

Böhning-Gaese, K. and Oberrath, R. (1999) Phylogenetic effects on morphological, life-history, behavioural and ecological traits of birds. *Evolutionary Ecology Research* 1, 347–364.

Böhning-Gaese, K., Gaese, B.H. and Rabemanantsoa, S.B. (1995) Seed dispersal by frugivorous tree visitors in the Malagasy tree species *Commiphora guillaumini. Ecotropica* 1, 41–50.

Böhning-Gaese, K., Burkhardt, J.F. and Schmid, J. (1996) Seed dispersal in the tree *Commiphora guillaumini*: a combination of ornithochory and myrmecochory in a dry tropical forest in western Madagascar. In: Ganzhorn, J.U. and Sorg, J.P. (eds) *Ecology and Economy of a Tropical Dry Forest in Madagascar. Primate Report* 46 (1: Special Issue), 305–310

Böhning-Gaese, K., Gaese, B.H. and Rabemanantsoa, S.B. (1999) Importance of primary and secondary seed dispersal in the Malagasy tree *Commiphora guillaumini. Ecology* 80, 821–832.

Bollen, A., Van Elsacker, L. and Ganzhorn, J.U. (2004) Tree dispersal strategies in the littoral forest of Sainte Luce (SE-Madagascar). *Oecologia* 139, 604–616.

Bond, W.J. (1995) Assessing the risk of plant extinction due to pollinator and disperser failure. In: Lawton, J.H. and May, R.M. (eds) *Extinction Rates.* Oxford University Press, Oxford, UK, pp. 131–147.

Cordeiro, N.J. and Howe, H.F. (2003) Forest fragmentation severs mutualism between seed dispersers and an endemic African tree. *Proceedings of the National Academy of Sciences USA* 100, 14052–14056.

Da Silva, J.M.C. and Tabarelli, M. (2000) Tree species impoverishment and the future flora of the Atlantic forest of northeast Brazil. *Nature* 404, 72–74.

de la Bathie, H.P. (1946) Flore *de Madagascar et des Comores. 106e Famille: Burseracées.* Tananarive, Madagascar.

Dew, J.L. and Wright, P. (1998) Frugivory and seed dispersal by primates in Madagascar's eastern rainforest. *Biotropica* 30, 425–437.

Diaz, S., Symstad, A.J., Chapin, F.S., Wardle, D.A. and Huenneke, L.F. (2003) Functional diversity revealed by removal experiments. *Trends in Ecology and Evolution* 18, 140–146.

Ellstrand, N.C. (1992) Gene flow by pollen: implications for plant conservation genetics. *Oikos* 63, 77–86.

Farwig, N., Randrianirina, E.F., Voigt, F.A., Kraemer, M. and Böhning-Gaese, K. (2004) Pollination ecology of the dioecious tree *Commiphora guillauminii* in Madagascar. *Journal of Tropical Ecology* 20, 307–316.

Farwig, N., Böhning-Gaese, K. and Bleher, B. (2006) Enhanced seed dispersal of *Prunus africana* in fragmented and disturbed forests? *Oecologia* 147, 238–252.

Fleming, T.H., Breitwisch, R. and Whiteside, G.H. (1987) Patterns of tropical vertebrate frugivore diversity. *Annual Review of Ecology and Systematics* 18, 91–109.

Frost, P.G.H. (1980) Fruit–frugivore interactions in a South African coastal dune forest. In: Nöhring, R. (ed.) *Acta XVII: Congress International Ornithology.* Deutsche Ornithologische Gesellschaft, Berlin, pp. 1179–1184.

Ganzhorn, J.U., Fietz, J., Rakotovao, E., Schwab, D. and Zinner, D. (1999) Lemurs and the regeneration of dry deciduous forest in Madagascar. *Conservation Biology* 13, 794–804.

Gautier-Hion, A., Duplantier, J.M., Quris, R., Feer, F., Sourd, C., Decoux, J.P., Doubost, L., Emmons, L., Erard, C., Hecketsweiler, P., Moungazi, A., Roussilhon, C. and Thiollay, J.M. (1985) Fruit-characters as a basis of fruit choice and seed dispersal in a tropical forest vertebrate community. *Oecologia* 65, 324–337.

Glen, H.F. (1996) A description of the vegetation of Oribi Gorge Nature Reserve, Natal. Part I. *Trees in South Africa* 46, 18–27.

Goodman, S.M. and Ganzhorn, J.U. (1997) Rarity of figs (*Ficus*) on Madagascar and its relationship to a depauperate frugivore community. *Revue d'Ecologie–la Terre et la Vie* 52, 321–329.

Hamrick, J.L. (2004) Response of forest trees to global environmental changes. *Forest Ecology and Management* 197, 323–335.

Hardy, O.J. and Vekemans, X. (2002) SPaGeDi: a versatile computer program to analyse spatial genetic structure at the individual or population levels. *Molecular Ecology Notes* 2, 618–620.

Hardy, O.J., Maggia, L., Bandou, E., Breyne, P., Caron, H., Chevallier, M.-H., Doligez, A., Dutech, C., Kremer, A., Latouche-Hallé, C., Troispoux, V., Veron, V. and Degen, B. (2006) Fine-scale genetic structure and gene dispersal inferences in 10 neotropical tree species. *Molecular Ecology* 15, 559–571.

Harper, J.L. (1977) *Population Biology of Plants.* Academic Press, San Diego, California.

Hartl, D. (1980) *Principles of Population Genetics.* Sinauer Associates, Sunderland, Massachusetts.

Hooper, D.U., Chapin, F.S., Ewel, J.J., Hector, A., Inchausti, P., Lavorel, S., Lawton, J.H., Lodge, D.M., Loreau, M., Naeem, S., Schmid, B., Setala, H., Symstad, A.J., Vandermeer, J. and Wardle, D.A. (2005) Effects of biodiversity on ecosystem functioning: a consensus of current knowledge. *Ecological Monographs* 75, 3–35.

Howe, H.F. (1984) Implications of seed dispersal by animals for tropical reserve management. *Biological Conservation* 30, 261–281.

Howe, H.F. (1986) Seed dispersal by fruit-eating birds and mammals. In: Murray, D.R. (ed.) *Seed Dispersal.* Academic Press, Sydney, Australia, pp. 123–189

Howe, H.F. and Miriti, M.N. (2004) When seed dispersal matters. *BioScience* 54, 651–660.

Janson, C.H. (1983) Adaptation of fruit morphology to dispersal agents in Neotropical forest. *Science* 219, 183–184.

Jeltsch, F., Milton, S.J., Dean, W.R.J. and Van Rooyen, N. (1996) Tree spacing and coexistence in semiarid savannas. *Journal of Ecology* 84, 583–595.

Knight, R.S. and Siegfried, W.R. (1983) Inter-relationship between type, size and colour of fruits and dispersal in Southern African trees. *Oecologia* 56, 405–412.

Langrand, O. (1990) *Birds of Madagascar.* Yale University Press, New Haven, Connecticut.

Loreau, M., Naeem, S. and Inchausti, P. (2002) *Biodiversity and Ecosystem Functioning: Synthesis and Perspectives.* Oxford University Press, Oxford, UK.

Ludwig, J.A. and Reynolds, J.F. (1988) *Statistical Ecology.* Wiley and Sons, New York.

Moran, C., Catterall, C.P., Green, R.J. and Olsen, M.F. (2004) Functional variation among frugivorous birds: implications for rainforest seed dispersal in a fragmented subtropical landscape. *Oecologia* 141, 584–595.

Naeem, S., Thompson, L.J., Lawler, S.P., Lawton, J.H. and Woodfin, R.M. (1994) Declining biodiversity can alter the performance of ecosystems. *Science* 368, 734–737.

Nathan, R. and Muller-Landau, H. (2000) Spatial patterns of seed dispersal, their determinants and consequences for recruitment. *Trends in Ecology and Evolution* 15, 278–285.

Petit, R.J., Duminil, J., Fineschi, S., Hampe, A., Salvini, D. and Vendramin G.G. (2005) Comparative organization of chloroplast, mitochondrial and nuclear diversity in plant populations. *Molecular Ecology* 14, 689–701.

Rousset, F. (2002) Inbreeding and relatedness coefficients: what do they measure? *Heredity* 88, 371–380.

Santos, T. and Tellería, J.L. (1994) Influence of forest fragmentation on seed consumption and dispersal of Spanish juniper *Juniperus thurifera*. *Biological Conservation* 70, 129–134.

Scharfe, F. and Schlund, W. (1996) Seed removal by lemurs in a dry deciduous forest of western Madagascar. In: Ganzhorn, J.U. and Sorg, J.P. (eds) *Ecology and Economy of a Tropical Dry Forest in Madagascar. Primate Report* 46 (1: Special Issue), 295–304.

Schulze, E.-D. and Mooney, H.A. (1994) *Biodiversity and Ecosystem Function.* Springer, Berlin, Germany.

Schupp, E.W. (1993) Quantity, quality and the effectiveness of seed dispersal by animals. *Vegetatio* 107/108, 15–29.

Sorg, J.P. and Rohner, U. (1996) Climate and tree phenology of the dry deciduous forest of the Kirindy Forest. In: Ganzhorn, J.U. and Sorg, J.P. (eds) *Ecology and Economy of a Tropical Dry Forest in Madagascar. Primate Report* 46 (1: Special Issue), 57–80.

Sork, V.L., Nason, J., Campbell, D.R. and Fernandez, J.F. (1999) Landscape approaches to historical and contemporary gene flow in plants. *Trends in Ecology and Evolution* 14, 219–224.

Terborgh, J. (1986) Community aspects of frugivory in tropical forests. In: Estrada, A. and Fleming, T.H. (eds) *Frugivores and Seed Dispersal.* Sinauer Associates, Sunderland, Massachusetts, pp. 371–384.

van der Walt, J.J.A. (1973) The South African species of *Commiphora. Bothalia* 11, 53–102.

van der Walt, J.J.A. (1986) Burseraceae. *Flora of Southern Africa* 18, 5–34.

Voigt, F.A., Burkhardt, J.F., Verhaagh, M. and Böhning-Gaese, K. (2002) Regional differences in ant community structure and consequences for secondary dispersal of Commiphora seeds. *Ecotropica* 8, 59–66.

Voigt, F.A., Bleher, B., Fietz, J., Ganzhorn, J.U., Schwab, D. and Böhning-Gaese, K. (2004) A comparison of morphological and chemical fruit traits between two sites with different frugivore assemblages. *Oecologia* 141, 94–104.

Voigt, F.A., Jung, S., Farwig, N. and Böhning-Gaese, K. (2005) Low fruit set in a dioecious tree: pollination ecology of *Commiphora harveyi* in South Africa. *Journal of Tropical Ecology* 21, 179–188.

Wang, B.C. and Smith, T.B. (2002) Closing the seed dispersal loop. *Trends in Ecology and Evolution* 17, 379–385.

Weir, B.S. and Cockerham, C.C. (1984) Estimation of *F*-statistics for the analysis of population structure. *Evolution* 38, 1358–1370.

Wenny, D.G. (2001) Advantages of seed dispersal: a re-evaluation of directed dispersal. *Evolutionary Ecology Research* 3, 51–74.

Wiegand, K., Jeltsch, F. and Ward, D. (2000) Do spatial effects play a role in the spatial distribution of desert-dwelling *Acacia raddiana*? *Journal of Vegetation Science* 11, 473–484.

Wright, S. (1978) *Evolution and the Genetics of Populations, Vol. 4. Variability Within and Among Natural Populations.* University of Chicago Press, Chicago, Illinois.

26 Investigating Fragility in Plant–Frugivore Networks: A Case Study of the Atlantic Forest in Brazil

W.R. Silva, P.R. Guimarães Jr, S.F. dos Reis and P. Guimarães

Introduction

Mutualistic interactions among free-living species, such as in pollination and seed dispersal by animals, have been a major factor in the evolution of flowering plants and animals. The recent disruption of these interactions caused by human activities may bring about a considerable loss of biodiversity (Buchmann and Nabhan, 1996; Jordano, 2000; Herrera, 2002; Fontaine *et al.*, 2006). An important characteristic of mutualistic interactions among free-living species is that exclusive, pairwise interactions are rare (Thompson, 2005). In contrast, in a given ecological community, each species interacts with a few to dozens of mutualistic partners (Jordano, 1987). As a result, at the community level, mutualistic interactions often emerge as web-like systems (Jordano, 1987; Jordano *et al.*, 2003).

Web-like systems such as food webs, the Internet and human social interactions are traditionally described as graphs or networks (reviewed by Albert and Barabási, 2002). Recently, the notion of complex networks has emerged as a key concept for understanding mutualistic interactions among species in a given community (Olesen and Jordano, 2002; Bascompte *et al.*, 2003; Jordano *et al.*, 2003; Vazquez and Aizen, 2004; Guimarães *et al.*, 2006). Indeed, by applying the network approach, several consistent patterns in the structure of mutualistic ecological interactions have been described, including high heterogeneity in the degree of specialization (Jordano *et al.*, 2003), strong asymmetrical specialization among partners (Bascompte *et al.*, 2003; Vazquez and Aizen, 2004; Guimarães *et al.*, 2006) and an allometric relationship between number of interactions and species richness (Bascompte *et al.*, 2003).

The description of structural patterns of plant–animal networks allows inferences to be made about the robustness of ecological interactions to disturbances (e.g. Jordano *et al.*, 2003). In addition, the network approach also allows theoretical investigation of the vulnerability of networks by means of numerical simulations (reviewed by Albert and Barabási, 2002), an approach recently introduced for mutualistic interactions (Memmott *et al.*, 2004; Fortuna and Bascompte, 2006; Pascual and Dunne, 2006). Here, we define fragility as the vulnerability of ecological systems to disturbances. We investigate the fragility of the largest plant–frugivore network characterized so far (Bascompte *et al.*, 2003): the interactions between vertebrates and fruiting plants in the Parque Estadual Intervales (hereafter PEI network), a reserve that is part of one of the last remaining forest blocks of the endangered Atlantic forest in south-eastern Brazil.

The remainder of the chapter is divided into four sections:

1. In *Plant–Animal Interactions as Networks* we review how to describe plant–animal mutualisms as networks, and quantify the resulting structure.
2. In *Investigating Fragility* we discuss how to infer the fragility of a network to the two general classes of disturbances: disturbances that propagate through the network and extinction of species.
3. In *A Case Study: PEI Network* we investigate different aspects of fragility of PEI network, including the investigation for the first time of how disturbances may cascade through plant–frugivore networks.
4. In the *Discussion* we discuss our results, with special attention to our findings for the studied Atlantic forest.

Plant–Animal Interactions as Networks

In this section we introduce the description of field data of plant–animal interactions in a given community as a complex network. We focus on three basic aspects of the network description:

1. How to gather the data and potential bias (*Data sampling*);
2. The equivalence of binary matrices and graphs (*Graph and matrix representations*);
3. The basic structural aspects of plant–animal networks and their biological interpretation (*Basic structural features*).

Data sampling

Ideally, the network description should include all interactions between all elements of the system. In the case of plant–frugivore networks, this means that every interaction between species of frugivores and plants should be recorded. Methods to record these interactions include all the traditional methods used in studies of seed dispersal; for example, direct observation of animals foraging for fruits, focal observations under fruiting plants, inspection of scats of frugivorous species, records from automatic camera traps placed close to fallen fruits and faeces samples from animals captured using mist nets (see, e.g. Donatti *et al.*, Chapter 5, this volume).

However, it is very difficult to sample all the interactions in an ecological community. Most plant–frugivore networks are indeed subnets of larger networks, in which a specific taxonomic group (e.g. birds or primates) is sampled (see data set from Jordano *et al.*, 2003). Analyses of subnetworks also occur for other web-like systems, and subnets may not have the same properties as entire networks (Stumpf *et al.*, 2005). Nevertheless, analyses of subnets are helping to shed light on important questions about the organization of ecological communities (Olesen and Jordano, 2002; Jordano *et al.*, 2003; Vazquez and Aizen, 2004; Guimarães *et al.*, 2006; Lewinsohn *et al.*, 2006).

It is important to note that, even studying a subnetwork, it is almost certainly true that we are not able to record every interaction. It is clear that future work is needed to allow a more rigorous statistical characterization of ecological networks. So far, the study of mutualistic networks has been based on the assumption that the sampled network is a good approximation of the complete network. This belief is supported by the use of data from long-term field studies (e.g. Bascompte *et al.*, 2003), by complementary analyses that control for potential sampling bias (Guimarães *et al.*, 2006), and by the assumption that non-recorded interactions are usually rare, 'weak' interactions that do not have much importance for the biological processes related to the structure of plant–animal interactions. However, it is worth mentioning that some missed interactions may sometimes reflect the difficulty of recording all possible interactions in fieldwork, especially when sampling forest habitats that may contain species of frugivores with nocturnal and/or secretive behaviour, confined to the understorey or canopy.

Graph and matrix representations

Plant–animal interactions data can be described as a binary, or presence/absence, matrix in which species of fruiting plants are represented by rows and species of frugivores are depicted as columns (the opposite definition is also valid) (Jordano *et al.*, 2003). Quantitative descriptions of plant–animal interactions (number of visits to fruiting trees, number of fruits consumed, etc) are useful for estimating the strength of the interaction, but as this measure is not always available for most of the interactions recorded during the fieldwork, the binary matrix is sufficient as a first approximation. The matrix element, a_{ij}, has a value of 1 if the frugivore species i harvests the fruit of the fruiting plant j and zero if these species do not interact (Jordano *et al.*, 2003). This matrix, depicted in Figure 26.1a, can be interpreted as an adjacency matrix; that is, a matrix that describes a graph or network (Harary, 1969). The basic idea is that rows and columns represent two sets (fruiting plants and frugivores, respectively) of elements called nodes or vertices, and the cells represent the presence or absence of a link (also called edge) connecting two nodes. By definition, in a seed dispersal network there is no direct interaction between two plants or two animals, and for this

(a)

	Animal 1	Animal 2	Animal 3	Animal 4	Animal 5
Plant 1	1	0	1	1	1
Plant 2	1	1	0	1	1
Plant 3	0	1	1	0	1
Plant 4	1	0	0	1	1

(b)

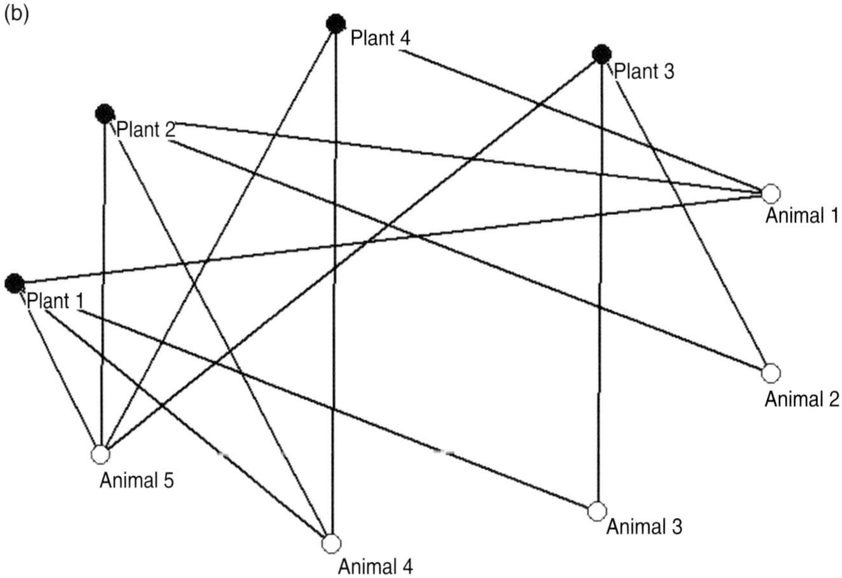

Fig. 26.1. The equivalence between two representations of plant–animal interactions (see text for further details): (a) binary matrix, in which plants are rows, animals are columns and interactions are represented by '1'; (b) network, in which plants are black nodes, animals are white nodes and interactions are represented by links.

reason the network is called bipartite (Fig. 26.1b). The network contains all information described in the original, binary matrix and, as a consequence, both representations are equivalent (Harary, 1969; Lewinsohn *et al.*, 2006).

Basic structural features

The network literature is full of examples of network metrics, whose main function is to describe different structural aspects of the network (see Wasserman *et al.*, 1994). Here we focus on some basic aspects of network structure and their classical measurements.

Size

Size is the most fundamental network feature and is defined as the number of nodes in the network. In plant–frugivore networks it is the number of species.

Components

A component is a maximally connected subnetwork. Therefore, if a network has more than one component, its structure is characterized by at least two totally disconnected subnetworks (Fig. 26.2a,b). If the network describing the interactions between assemblages of plants and frugivores shows more than one component, the ecological dynamics may be independent in each component.

Average degree and connectance

The degree of a node is equal to its number of links. In the context of ecological interactions, the degree can be interpreted as the level of ecological specialization (see Olesen and Jordano, 2002): specialists are those species with few mutualistic partners (few links), whereas generalists are species that interact with many species in the network (many links) (Jordano *et al.*, 2003). Regarding frugivory, the terms 'specialist' and 'generalist' used hereafter are used in a somewhat different sense from that usually found in typical studies dealing with this theme (for comparison see Moermond and Denslow, 1985). The average degree for an ecological network can be interpreted as the average level of ecological specialization (or generalism) of a species in a community. It is important to note that the average degree is a function of network size. An alternative measure for the average level of ecological specialization is connectance, which is the proportion of possible links in the network that are actually established (Fig. 26.2c,d). In fact, connectance is a size-independent measure (Olesen and Jordano, 2002).

Nestedness

A nested network is a bipartite network in which there is a specific pattern of interaction among nodes, in which the nodes with small degrees interact with a proper subset of the nodes that interact with a larger degree node (Fig. 26.2e,f). Bascompte *et al.* (2003) defined the degree of nestedness, N, as $N = (100 - T)/100$, in which T is a standardized measure of how much the binary matrix that describes the network departs from perfect nestedness. To calculate T, the binary matrix is maximally packed (see Atmar and Patterson, 1993, for further details). Then, an isocline of perfect nestedness is calculated and deviations from isocline (i.e. unexpected recorded presences and absences of interactions that deviate from a perfectly nested pattern) are standardized and recorded. The degree of deviation of this isocline is T. In ecological terms, the degree of nestedness

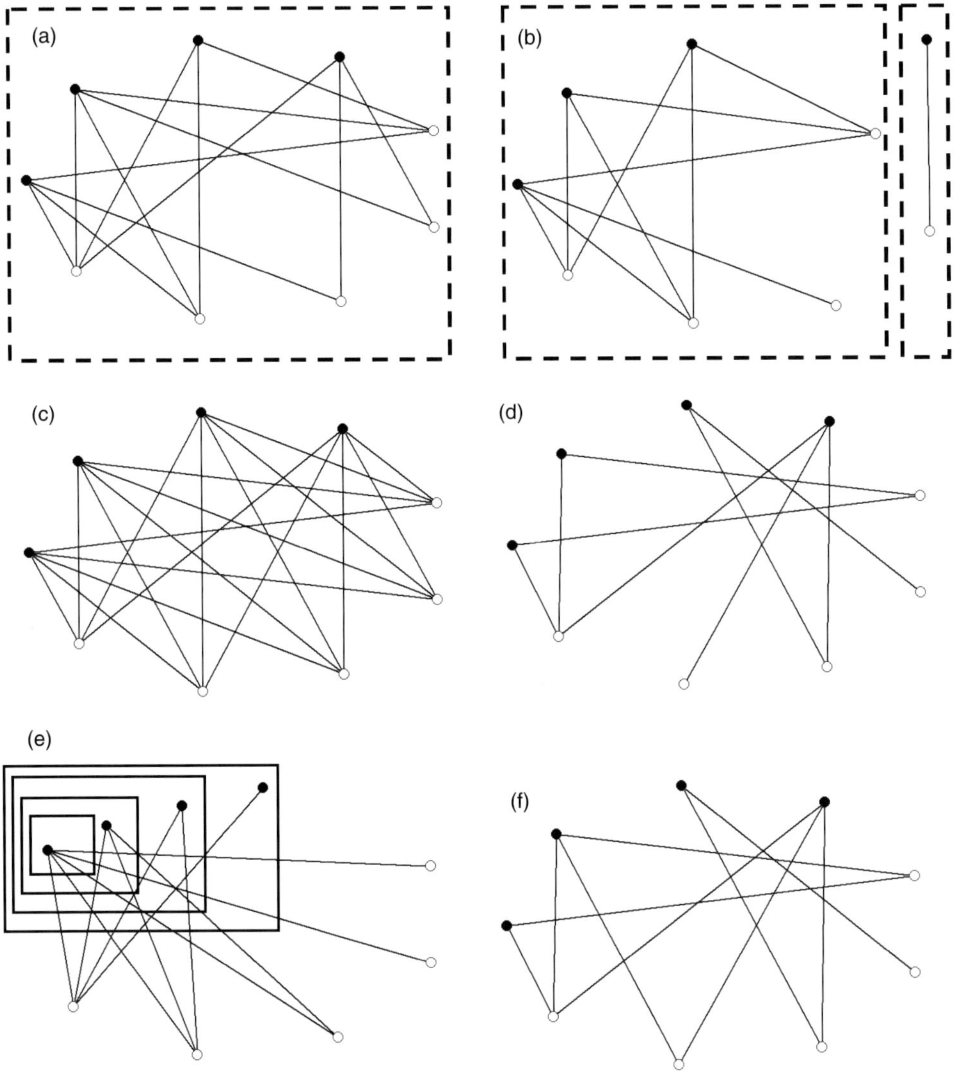

Fig. 26.2. Basic structural aspects of networks: (a) network formed by one component, (b) network formed by two components. Components identified by the dashed rectangles, (c) a network showing high connectance, (d) a network showing low connectance, (e) nested network in which the specialist interacts with a subset (small rectangle) of the partners that also interact with more generalist species (larger rectangles), (f) non-nested network. All networks were drawn using Pajek (see de Nooy *et al.*, 2005).

(N) is a measure of the level of observed asymmetrical specialization: values of N close to 1 therefore indicate strong asymmetrical patterns in specialization (= high degree of nestedness). One of the main discoveries of the network approach to plant–animal mutualisms is that they are often nested (Bascompte *et al.*, 2003; Thompson, 2005; Guimarães *et al.*, 2006).

Investigating Fragility

Ecologists have been concerned about the vulnerability of ecological systems to disturbance for a long time (Pimm, 2002). In network theory, 'fragility' can be defined in many distinct ways (see Pascual and Dunne, 2006). Here, we explore two classes of fragility.

First, a network can be considered fragile if the removal of a node or a group of nodes markedly affects its structure, changing the observed values for one or more network metrics (Fig. 26.3a,b). Examples of the disturbances that can be modelled as node removal include death (in social networks), bankruptcy of companies (in economic networks), and errors in protein synthesis (in molecular networks) (reviewed by Albert and Barabási, 2002). In ecology, node removal can be used to simulate the local extinction of frugivores or/and plants (Memmott *et al.*, 2004; Fortuna and Bascompte, 2006).

Second, a network can be considered fragile if any disturbance easily cascades through the entire network. In this case, the disturbance can be modelled as a diffusion process, in which the disturbance flows from the perturbed node to others (Fig. 26.3c,d). Examples of disturbances that propagate in networks are diseases and technological innovations, in social networks (de Nooy *et al.*, 2005). In ecological networks, an example of

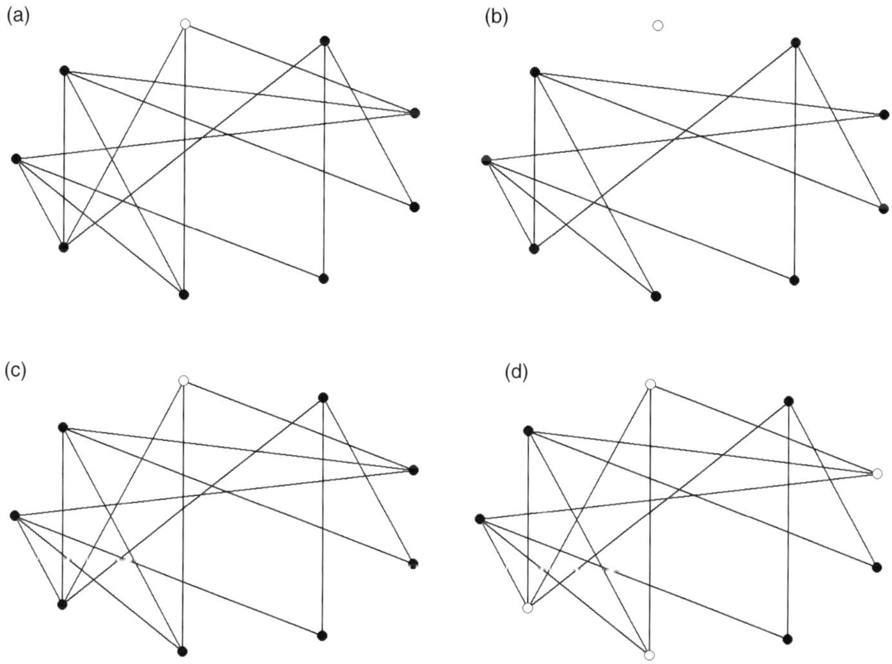

Fig. 26.3. Simulating disturbance in networks: (a) one node is selected (in white) and (b) removed from the network, (c) one node is perturbed (in white) and (d) the disturbance propagates to the nodes that interact with it (in white).

disturbances that can be modelled as diffusion processes are the indirect effects of the increase in abundance of a given species, in which the disturbance (change in abundance) may affect species that directly interact with the perturbed species and may also cascade through the network.

The investigation of fragility in terms of experimental manipulations is very difficult (but see Fontaine *et al.*, 2006) and is certainly limited to species-poor systems. However, the researcher can use at least two broad classes of approach: the inference of fragility based on structural aspects (see *Structure and fragility*) or the simulation of disturbance dynamics (see *Simulating diffusion and extinctions*).

Structure and fragility

Due to simulation and observational studies of complex networks in different scientific areas, we now have many examples of relationships between structural aspects of the network and its vulnerability to disturbances (see Albert and Barabási, 2002). Thus, we can infer aspects of the fragility of a given network by simply calculating and interpreting the values of different network metrics. Below, we describe possible interpretations of the different values for network metrics already introduced in this chapter. It is important to note that much of the interpretation will depend on the particulars of the dynamics of the diffusion or node removal that we assume to occur in the network (see *Simulating diffusion and extinctions*). Here we interpret the values of different metrics by assuming two very simple dynamics: (i) random node removals in which any species has the same probability of going extinct; and (ii) degree-dependent diffusion processes, in which the probability that species i will be affected by a perturbed partner j is inversely proportional to the number of partners of i. In these dynamics, a generalist species is less likely to be affected by changes in the abundance of a partner, whereas extreme specialists will often be affected by changes in the abundance of its partner.

Size

The effects of node removal will probably be weaker in large networks, since the larger the network, the less probable it is that the removal of a single node will change the community-level properties of the network. Thus we can usually assume that large networks are less fragile than smaller networks. Accordingly, the diffusion of disturbances is clearly dependent on the size of a network. In food webs, it is usually assumed that the effects of a disturbance in a given species will only affect species that are connected by a small path of interactions to the perturbed species (Williams *et al.*, 2002). If the network is very large, a disturbance will probably affect a small proportion of the network. In contrast, in smaller networks, some disturbances will easily diffuse through the entire network (but see Pimm, 2002, for a more complete discussion, considering other fragility regimes).

Components

The number of components may not be an important structural aspect for the removal of species, except in extreme situations in which the network is formed by components of few (e.g. two) species. In this case, the extinction of a species may lead to the disconnection of the mutualistic partner. In contrast, the number of components can markedly affect the propagation of disturbances. By definition, there is no path between two components and, as a consequence, a disturbance that affects one component cannot propagate to the other component. In conclusion, networks formed by more than one component are less sensitive to the propagation of disturbances than networks formed by a single component.

Average degree and connectance

In networks with low connectance, the extinction of species may lead to deep structural changes, such as the emergence of a new component (e.g. Fig. 26.3c,d). In contrast, in densely connected networks there are several paths connecting groups of species, and the disconnection of a species or the emergence of new components are less likely. Thus, we can usually assume that densely connected networks are less fragile than sparse networks with regard to disturbances. Accordingly, in networks with low connectance (sparse networks) a randomly selected species will usually have few interactions. As a consequence, species are likely to be more sensitive to the propagation of disturbances. In contrast, in networks that show high connectance, each species will have many interactions and, therefore, it is unlikely that an environmental disturbance that affects only one species will cascade through its ecological partners (but see Pimm, 2002).

Nestedness

The existence of high levels of asymmetrical specialization implies that the fragility of the network markedly depends on the identity of the perturbed species. In a perfectly nested network, all specialists interact with the most generalist species. As a consequence, if the extinct species is a generalist, all the specialists may lose their unique mutualistic partner. In contrast, if a specialist goes extinct, no species loses its mutualistic partners. Similarly, a disturbance that affects a generalist will propagate to all specialists (e.g. Fig. 26.3c,d). In contrast, a disturbance that affects a specialist will rarely propagate to the generalist.

Simulating diffusion and extinctions

The inference of fragility from structural aspects is a powerful tool for predicting the possible consequences of disturbances and extinctions for the network. However, the structure of real networks cannot be completely described using one or a few structural features. Moreover, different metrics can lead to opposite interpretations. For example, a network can be simultaneously large, implying less sensibility to species removal, and sparse, implying higher sensibility to species removal. Is such a large and sparse network fragile? The answer cannot be derived directly from both measures because we still do not understand the extent to which the effects of one topological pattern balance the effects of other structural patterns. To circumvent the above problems, we can adopt a more direct approach by simulating disturbances propagating through the network, or species removal. By performing numerical simulations of disturbances we are able to investigate how fragile the network is to different disturbance dynamics without being limited to the measured structural metrics.

A Case Study: PEI Network

The Atlantic forest and the Parque Estadual Intervales

The Atlantic forest of south-eastern Brazil is among the top five world hotspots of biodiversity and is also one of the most threatened ecosystems in the world (Myers *et al.*, 2000). This forest has been reduced, since colonial times, to 8% of its original extent of more than 1 million km^2 along the Brazilian Atlantic coast (Dean, 1995). Today the largest continuous forest tracts are confined to south-eastern Brazil, a region where 70% of the country's population live (Ministério do Meio Ambiente, 2000). Direct threats to this biome include logging, poaching, wildlife trade, urban and industrial development, and deforestation driven by agriculture and expansion of pastureland (see several chapters in Galindo-Leal and Câmara, 2003).

The Parque Estadual Intervales (PEI) is a 42,000-ha reserve in south-eastern São Paulo state characterized by a mosaic of pristine forest and second-growth vegetation along an altitudinal gradient ranging from 60 to 1100 m (for a detailed description of PEI, see Carvalho *et al.*, 2002). Different sites inside the park were visited monthly from January 1999 to October 2002, for a 3–4 day census at each site. During the censuses, interactions between frugivorous vertebrates and plants were recorded, using several methods: mist netting and searching for seeds in faecal samples (birds and bats), collecting scats in forest trails (large birds and mammals), trapping with folding Sherman traps (opossums) and observation of fruit-eating behaviour (birds and primates). Like many other sites in the Brazilian Atlantic forest, the fauna of frugivorous vertebrates at PEI includes several species of threatened birds and mammals (de Vivo and Gregorin, 2001; Vielliard and Silva, 2001) which are liable to experience severe shifts in abundance, leading to extinction in a medium-term future (Brooks *et al.*, 1999; Chiarello, 2000). Thus, we applied the approach introduced in previous sections of this chapter to investigate and predict the consequences of these disturbances to the organization of the biodiversity of PEI.

Basic structural features and inferences of fragility in the PEI network

The PEI network (Fig. 26.4) is the largest plant–frugivore network recorded to date, with 317 species, making it 1.5 times larger in number of species than the second largest plant–frugivore network, and more than six times higher than the average number of species found in recorded networks (49.1 ± 63.8 species, mean ± SD; data from Bascompte *et al.*, 2003). All the nodes represent individual species: = 207 plant and 110 animal species. As with other typical large plant–frugivore networks (Bascompte *et al.*, 2003), the PEI network is formed by a single but sparse component, showing a small average degree (5.4 ± 6.3 animal partners per plant and 10.2 ± 11.5 plant partners per animal). Consequently, very low connectance was observed (only 4.9%). We also calculated the degree of nestedness, N, of the PEI network using ANINHADO (Guimarães and Guimarães, 2006). The PEI is a highly asymmetrical network, since $N = 0.96$ is very close to perfect nestedness ($N = 1$), indicating that specialists often depend on generalists in the PEI network, corroborating previous analysis with a less complete version of the PEI network (Bascompte *et al.*, 2003).

What can we infer about the fragility of the PEI network based on structural aspects? The large size of the network suggests that it is probably not very sensitive to disturbances, even considering that the network is

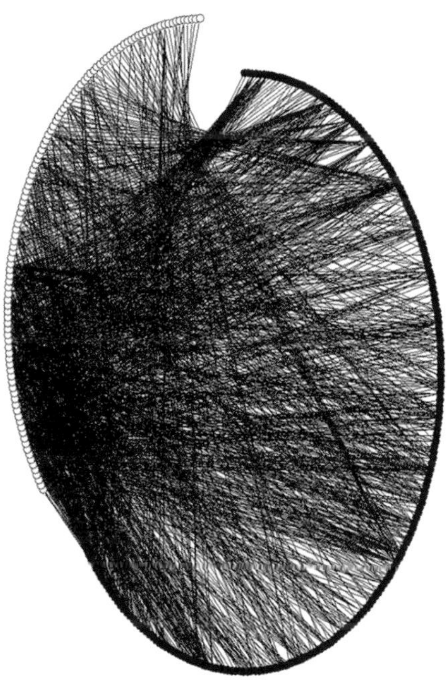

Fig. 26.4. The PEI network, in which plants are represented by black nodes and animals are represented by white nodes. Links represent interactions between frugivores and plants.

formed by a single component. However, the sparse structure and the strong pattern of asymmetrical specialization suggest that the network is potentially vulnerable to disturbances, especially if the perturbed species is a generalist one supporting many species with few interactions. Due to the uncertainties associated with simple analyses of topology, we simulated disturbances in the network and describe our findings in the next two sections. Special attention is given to: (i) the possibility that single-species disturbances propagate through a large proportion of interacting species; and (ii) the consequences of the extinction of already endangered vertebrates.

Simulating disturbances that propagate in the PEI network

We simulated the degree-dependent diffusion processes (n = 1000 simulations). The algorithm used was defined as follows:

1. A species is randomly selected in the network and perturbed;
2. The probability that each partner species directly connected with the perturbed species ('target' species) is affected is equal to $1/k$, where k is the number of partners of the target species;
3. The simulation proceeds until all species are affected or until no new species is perturbed.

Although very simple, these propagation dynamics have some interesting properties: generalist species are less susceptible to being affected by disturbances, while specialists will often be affected by changes in the abundance of their partner. Indeed, an extreme specialist interacting with a single partner will always be affected by disturbance of its partner.

In 58.5% of simulations, the disturbance did not propagate to any species in the network. However, the distribution function that describes the probability to find a simulation in which S species are perturbed by the propagation decays slowly, following a power-law (Fig. 26.5). As a consequence, there is a small, but non-negligible, probability that the disturbances will affect nearly half of the species in the network.

Simulating species extinction in the PEI network

In PEI, 11 species of vertebrates (10%) are considered threatened (adapted from São Paulo state, Law No. 42.838, February 1998), eight species of birds *Tinamus solitarius* (Tinamidae), *Aburria jacutinga* and *Penelope obscura* (Cracidae), *Laniisoma elegans*, *Lipaugus lanioides*, *Procnias nudicollis* and *Pyroderus scutatus* (Cotingidae) and *Tangara peruviana* (Thraupidae), and three species of mammals: *Alouatta guariba* and *Brachyteles arachnoides* (Cebidae) and *Tapirus terrestris* (Tapiridae). As the number of interactions of a species is directly related to its abundance (Jordano, 1987) and as a consequence of strong asymmetrical specialization, we should expect that:

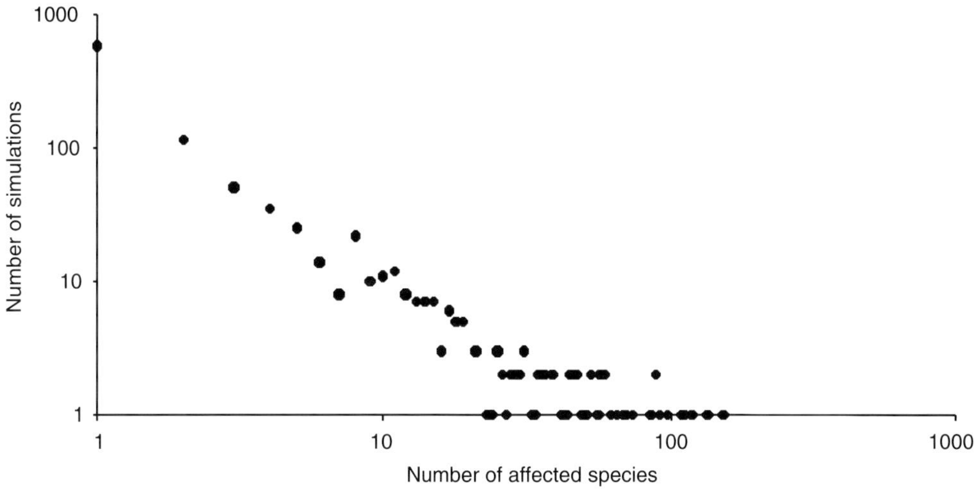

Fig. 26.5. Probability that a disturbance propagates through the PEI network (see text for further details), affecting *n* species. Note that the probability decays as a power-law, indicating that there is a non-negligible probability of a disturbance cascades to a great part of the network.

1. Many endangered species, often rare, will usually show few interactions; **2.** As species with few interactions usually interact with generalists, few plant species will show total seed dispersal collapse (i.e. loss of all seed dispersers) due the extinction of endangered seed dispersers.

The network with and without the endangered species has similar macroscopic structures (Table 26.1). This is a consequence of the relatively small number of endangered animals. The main difference between the networks is that the network without endangered species is formed by two components. However, one component contains only two species (0.7% of the remaining species) and all remaining species are in the other, giant, component.

Only 11 species of plants will lose all their seed dispersers. Although this number represents only 5.3% of the plant richness, we do not know a priori whether this number is too small, too large, or as expected if any set of 11

Table 26.1. Differences in macroscopic properties of the PEI network before and after the extinction of 11 species of endangered vertebrates.

	Size	No. of components	*C* (%)	<*k*> (plants)	<*k*> (animals)	*N*
Before the extinction	307	1	4.92	5.4 ± 6.3	10.2 ± 11.5	0.96
After the extinction	295	2	5.25	5.2 ± 5.9	10.3 ± 11.6	0.96
Difference (%)	3.9	100	6.7	3.7	1.1	0.1

C = connectance, <*k*> = average degree, *N* = degree of nestedness (see text for further details).

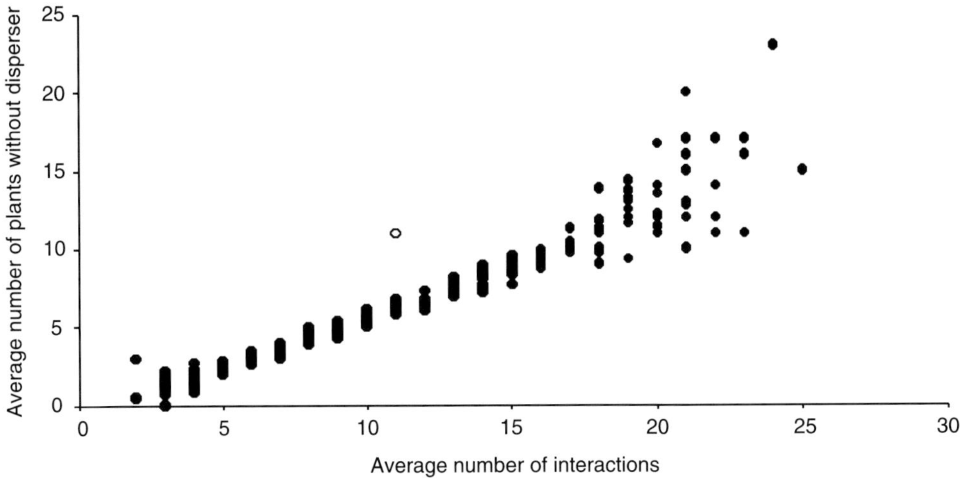

Fig. 26.6. Relationship between the average degree (number of interactions) of 11 animal species and the average number of plants that will lose all seed dispersers with their extinction: random extinction of 11 animal species (black nodes) and the extinction of the endangered animal species (white node).

species of animals goes extinct. We simulated the extinction of random sets of 11 species of animals and recorded the number of plants that lost all their seed dispersers. The results suggest that 11 species of plants without seed dispersers is too high when compared to what is expected by random extinctions ($P = 0.09$). Additionally, we compared the effects of the extinction of endangered species with the extinction of 11 randomly selected seed dispersers but with a similar average number of interactions. In this situation, the seed dispersal collapse of 11 species of plants is twice as high as expected for a random extinction of 11 seed dispersers with a similar average number of interactions (Fig. 26.6), indicating that there is an unexpected set of plants that will be severely affected by the extinction of endangered seed dispersers.

Discussion

In different ecosystems, many plants rely on animals to disperse their seeds (Jordano, 2000). Thus, the disruption of these interactions and other mutualisms may profoundly affect the organization of ecological communities (e.g. Galetti *et al.*, 2006) and indirectly affect the ecological services necessary to sustain humankind (Levin, 1999). We now have a relatively good understanding of how the local extinction of frugivores may affect the regeneration of natural plant populations (e.g. McConkey and Drake, 2002; Galetti *et al.*, 2006). However, we need to develop approaches that allow us to predict, at least in broad terms, the consequences of disturbance processes to the organization of plant–animal mutualisms. Only by enhancing our predictive success will we be able to optimize our conservation efforts and efficiently protect threatened ecological areas.

Until recently, attempts to predict the fate of plant communities after animal extinctions were based on indirect evidence; for example, morphological traits such as seed size and frugivore gap width (e.g. Cardoso da Silva and Tabarelli, 2000). An alternative to these indirect approaches is the analysis of network fragility. In this chapter, we investigate for the first time how disturbances may propagate across species in a plant–frugivore network. Moreover, we also contribute to an enhancement of our understanding of how disturbances that may lead to local extinction of species may affect the structure of plant–animal mutualisms (see also Fontaine *et al.*, 2006; Fortuna and Bascompte, 2006).

The study of the mutualistic networks is now only just beginning (but see Jordano, 1987). Considerable work will be needed before we are able to apply this approach to management plans. It is possible that only more complete descriptions of quantitative, well-studied networks will provide information at the necessary level of detail needed to guide conservation policies. Nevertheless, the study of binary networks has provided useful and interesting insights into how complex systems respond to disturbances; for example, assisting in understanding and controlling how viruses propagate in digital and sexual networks (Albert and Barabási, 2002).

In this chapter, we corroborate the general view that large networks may be robust to disturbance (Albert and Barabási, 2002; Jordano *et al.*, 2003; Memmott *et al.*, 2004; Fortuna and Bascompte, 2006). Indeed, our simulations suggest that, for example, the effects of changes in the abundance of a single species will seldom propagate throughout the entire Atlantic forest network. This is probably a result of the large number of species with few interactions that interact with a few super-generalists. Moreover, the extinction of the endangered dispersers will not markedly affect the large-scale properties of plant–frugivore interactions in this tropical community, an expected consequence of the few species that are endangered in PEI.

Nevertheless, two of the ways in which the PEI network responds to disturbances suggest that plant–frugivore networks may be severely affected in some situations. First, although disturbances probably will not propagate throughout the network, our results also emphasize that there is a non-negligible probability that a considerable proportion – up to half of the species – will be affected by a single-species disturbance. These drastic events, in which nearly half of the species in the community will be affected, are also a consequence of asymmetrical specialization. If a random disturbance affects a super-generalist, this disturbance may propagate through many species in the network, eventually affecting a large proportion of the interacting species. Therefore, we suggest that special attention should be devoted to investigating the vulnerability to extinction of the most-connected species in the Atlantic forest such as the already threatened *Penelope obscura* and the palm tree *Euterpe edulis* (Arecaceae), which is threatened by intense harvesting.

Second, although the extinction of endangered birds and mammals will not lead to dramatic shifts in network structure, an unexpectedly large proportion of plant species will probably show seed dispersal collapse. This result is quite surprising, due to the nested structure of plant–animal mutualisms (Bascompte *et al.*, 2003). However, no plant–frugivore network shows perfect nestedness (asymmetrical specialization) and eventual, unique interactions between species with few interactions will occur. Our results indicate that these unique interactions chiefly involve endangered forest seed dispersers such as large-bodied birds and mammals. Unique interactions are community-level properties that emerge as a result of ecological and evolutionary processes relating plants and large-bodied vertebrates (see Donatti *et al.*, Chapter 5, this volume). Thus, the extinction of endangered animals will not only lead to the reduction of plant richness, but to the simplification of the ecological community and the loss of relevant aspects of biodiversity structure. We suggest that future conservation policies in the Atlantic forest should focus on long-term maintenance of these endangered frugivores and their interacting plants.

Acknowledgements

This study was supported by Foundation MB and BIOTA/FAPESP – The Biodiversity Virtual Institute Program (http://www.biota.org.br/) through Grant 98/05090-6. We are grateful to E.Z.F. Setz, N.O. Leiner, É. Hasui, V.S. Gomes and F.C. Passos for their valuable help with fieldwork. We are also indebted to M.L. Kawazaki, J. Tamashiro, R.A. Oliveira and R. Goldemberg for the difficult task of plant identification. P. Jordano provided helpful comments on the manuscript. The Fundação Florestal do Estado de São Paulo provided the logistic support at PEI. P.R.G. and S.F.R. were supported by FAPESP.

References

Albert, R. and Barabási, A.L. (2002) Statistical mechanics of complex networks. *Reviews in Modern Physics* 74, 47–97.

Atmar, W. and Patterson, B.D. (1993) The measure of order and disorder in the distribution of species in fragmented habitat. *Oecologia* 96, 373–382.

Bascompte, J., Jordano, P., Melián, C.J. and Olesen, J.M. (2003) The nested assembly of plant–animal mutualistic networks. *Proceedings of the National Academy of Sciences of the USA* 100, 9383–9387.

Brooks, T., Tobias, J. and Balmford, A. (1999) Deforestation and bird extinctions in the Atlantic forest. *Animal Conservation* 2, 211–222.

Buchmann, S. and Nabhan, G.P. (1996) *The Forgotten Pollinators*. Island Press/Shearwater Books, Washington DC/Covelo, California.

Cardoso da Silva, J.M. and Tabarelli, M. (2000) Tree species impoverishment and the future flora of the Atlantic forest of northeast Brazil. *Nature* 404, 72–74.

Carvalho, M.P., Cezare, C.H.G., Costa, C.G., da Cunha, R.G.T., Dornelles, S.S., Fialho, M.S., Guix, J.C., Hernández, A., Lazarin, P.C., Llorens, L., Martín, M., Mateos, E., Miquel, C., Petroni, L.M., Pinto, B.S., Pisciotta, K., Sánchez, C., Oliveras, I. and Serra, A. (2002) Description of the study areas. In: Mateos, E., Guix, J.C., Serra, A. and Pisciotta, K. (eds) *Censuses of vertebrates in a Brazilian Atlantic Rainforest Area: The Paranapiacaba Fragment.* University of Barcelona, Barcelona, Spain, pp. 27–48.

Chiarello, A.G. (2000) Density and population size of mammals in remnants of Brazilian Atlantic forest. *Conservation Biology* 14, 1649–1657.

de Nooy, W., Mrvar, A. and Batagelj, V. (2005) *Exploratory Social Network Analysis with Pajek.* Cambridge University Press, Cambridge, UK.

de Vivo, M. and Gregorin, R. (2001) *Mamíferos.* In: Secretaria de Estado do Meio Ambiente, Fundação para a Conservação e a Produção Florestal do Estado de São Paulo. *Intervales.* Fundação Florestal, São Paulo, pp. 116–123.

Dean, W. (1995) *With Broadax and Firebrand: The Destruction of the Brazilian Atlantic Fforest.* University of California Press, Berkeley, California.

Fontaine, C., Dajoz, I., Merigueti, J. and Loreau, M. (2006) Functional diversity of plant–pollinator interaction webs enhances the persistence of plant communities. *PLoS Biology* 4, 1–7.

Fortuna, M.A. and Bascompte, J. (2006) Habitat loss and the structure of plant–animal mutualistic networks. *Ecology Letters* 9, 281.

Galetti, M., Donatti, C.I., Pires, A.S., Guimarães, P.R. and Jordano, P. (2006) Seed survival and dispersal of an endemic Atlantic forest palm: the combined effects of defaunation and forest fragmentation. *Botanical Journal of the Linnean Society* 151, 141–149.

Galindo-Leal, C. and Câmara, I.G. (2003) *The Atlantic forest of South America: Biodiversity Status, Threats and Outlook.* Island Press, Washington, DC.

Guimarães, P.R. and Guimarães, P. (2006) Improving the analyses of nestedness for large sets of matrices. *Environmental Modelling and Software* 21, 1512–1513.

Guimarães, P.R., Rico-Gray, V., dos Reis, S.F. and Thompson, J.N. (2006) Asymmetrical specialization in ant–plant networks. *Proceedings of the Royal Society B: Biological Sciences* 273, 2041–2047.

Harary, F. (1969) *Graph Theory.* Addison-Wesley, Reading, UK.

Herrera, C.M. (2002) Seed dispersal by vertebrates. In: Herrera, C.M. and Pellmyr, O. (eds) *Plant–Animal Interactions. An Evolutionary Approach.* Blackwell Science, Oxford, UK, pp. 185–208.

Jordano, P. (1987) Patterns of mutualistic interactions in pollination and seed dispersal: connectance, dependence asymmetries, and coevolution. *American Naturalist* 129, 657–677.

Jordano, P. (2000) Fruits and frugivory. In: Fenner, M. (ed.) *Seeds: The Ecology of Regeneration in Plant Communities.* CAB International, Wallingford, UK, pp. 125–166.

Jordano, P., Bascompte, J. and Olesen, J.M. (2003) Invariant properties in coevolutionary networks of plant–animal interactions. *Ecology Letters* 6, 69–81.

Levin, S. (1999) *Fragile Dominion.* Perseus Books, Reading, UK.

Lewinsohn, T.M., Prado, P.I., Jordano, P., Bascompte, J. and Olesen, J.M. (2006) Structure in plant–animal interaction assemblages. *Oikos* 113, 174–184.

McConkey, K.R. and Drake, D.R. (2002) Extinct pigeons and declining bat populations: are large seeds still being dispersed in the tropical pacific? In: Levey, D.J., Silva, W.R. and Galetti, M. (eds) *Seed Dispersal and Frugivory: Ecology, Evolution and Conservation.* CAB International, Wallingford, UK, pp. 381–395.

Memmott, J., Waser, N.M. and Price, M.V. (2004) Tolerance of pollination networks to species extinctions. *Proceedings of the Royal Society London B* 271, 2605–2611.

Ministério do Meio Ambiente (2000) *Avaliação e ações prioritárias para a conservação da biodiversidade da Mata Atlântica e Campos Sulinos.* MMA/SBF, Brasília, DF.

Moermond, T.C. and Denslow, J.S. (1985) Neotropical avian frugivores: patterns of behaviour, morphology, and nutrition, with consequences for fruit selection. In: Buckley, P.A., Foster, M.S., Morton, E.S., Ridgely, R.S. and Buckley, F.G. (eds) *Neotropical Ornithology.* Ornithological Monographs No. 36. American Ornithologist Union, Washington, DC, pp. 865–897.

Myers, N., Mittermeier, R.A., Mittermeier, C.G., Fonseca, G.A.B. and Kent, J. (2000) Biodiversity hotspots for conservation priorities. *Nature* 403, 853–858.

Olesen, J.M. and Jordano, P. (2002) Geographic patterns in plant–pollinator mutualistic networks. *Ecology* 83, 2416–2424.

Pascual, M. and Dunne, J.A. (2006) *Ecological networks: linking dynamics in food webs.* Oxford University Press, Oxford, UK.

Pimm, S.L. (2002) *Food Webs.* University of Chicago Press, Chicago, Illinois.

Stumpf, M.P.H., Wiuf, C. and May, R.M. (2005) Subnets of scale-free networks are not scale-free: sampling properties of networks. *Proceedings of the National Academy of Sciences of the USA* 102, 4221–4224.

Thompson, J.N. (2005) *The Geographic Mosaic of Coevolution.* University of Chicago Press, Chicago, Illinois.

Vazquez, D.P. and Aizen, M.A. (2004) Asymmetric specialization: a pervasive feature of plant–pollinator interactions. *Ecology* 85, 1251–1257.

Vielliard, J.M.E. and Silva, W.R. (2001) *Avifauna.* In: Secretaria de Estado do Meio Ambiente, Fundação para a Conservação e a Produção Florestal do Estado de São Paulo. *Intervales.* Fundação Florestal, São Paulo, pp. 124–145.

Wasserman, S., Faust, K., Iacobucci, D. and Granovetter, M. (1994) *Social Network Analysis: Methods and Applications.* Cambridge University Press, Cambridge, UK.

Williams, R.J., Berlow, E.L., Dunne, J.A., Barabasi, A.L. and Martinez, N.D. (2002) Two degrees of separation in complex food webs. *Proceedings of the National Academy of Sciences of the USA* 99, 12913–12916.

27 Refining the Conservation Management of Seed-dispersing Frugivores and their Fruits: Examples from Australia

R.J. Green

Introduction

Research on frugivory and seed dispersal has expanded rapidly over the past couple of decades. So have vegetation loss (e.g. Vesk and MacNally, 2006), the invasion of exotic plants and animals into natural and semi-natural habitats (e.g. Low, 2001), altered fire regimes (e.g. Edwards and Russell-Smith, 2006) and other threats to biodiversity, plus the more recent beginnings of climate change (Hilbert *et al.*, 2001). It is thus timely to explore the application of research into frugivores and seed dispersal to conservation management in such a rapidly changing world.

The call for managers and scientists to work together more closely for conservation goals is often heard at ecology-related conferences, and despite some difficulties (Terborgh, 2004; Fazey and McQuie, 2005), this appeal is likely to become increasingly urgent as environmental problems accelerate. Jarman and Brock (1996) observed that while conservation is not itself a scientific goal, collaboration between scientists and managers can produce management-related questions amenable to scientific method. Underpinning both scientific research and management decisions are sound natural-history data, the importance of which should not be underestimated.

Despite the expanding coverage in the research literature (e.g. Harrington *et al.*, 1997; Moran *et al.*, 2004a), the possible disruption of interactions between fleshy-fruited plants and their dispersers tends to merit only a passing mention in publications on conservation management (e.g. Cropper, 1993; Maunder *et al.*, 2002; NSW NPWS, 2002; Playford and Murray, 2002). The case is similar for the habitat restoration undertaken by the mining industry (Langkamp and Plaisted, 1987), with very little reference in the literature on plans to attract natural dispersers to mine sites. The extensive reports and threatened-species recovery plans published by Australia's federal Department of Environment and Heritage (see http://www.deh.gov.au/biodiversity/threatened) rarely mention seed dispersal apart from occasional rather vague comments on whether a species may be dispersed by ants or birds. A publication on threatened species of northern New South Wales, Australia, (NSW NPWS, 2002), which includes many species of plants with fleshy fruit, scarcely mentions dispersers. As with other literature on restoration or threatened plants, there appears to be little consideration of the possibility that the absence of dispersers could be a risky situation or that plans to entice or reintroduce dispersers could form part of the proposed recovery actions.

Researchers and conservation managers alike tend to work under tight restrictions in time and funding, and such topics as plant–animal interactions can be seen as luxuries unaffordable by conservation managers who have to deal with more urgent problems (Smith and Garnett, 2004). However, precisely because of the limited resources of time and money, it is highly advantageous for conservation managers to have access to sufficient relevant information to direct their efforts as efficiently as possible. Hougner *et al.* (2006) conclude that protecting the dispersers is substantially cheaper than investing in human beings to take over their role. Ideally, managers would know:

1. Whether local mutualisms have been disrupted, resulting in a shortage of appropriate dispersers;
2. Whether these disruptions really matter for biodiversity conservation.

Assuming that some, at least, do matter, it would also be advantageous to know whether it is cost-effective to attract, or reintroduce and subsequently maintain, appropriate dispersers and how to do this in the most efficient way.

In this chapter, I explore these questions in relation to the management of altered landscapes in Australia. In particular, I explore our current knowledge base, including the scientific and other literature, and identify the numerous gaps that hinder effective decision making in relation to frugivores and seed dispersal. My focus on altered landscapes is primarily related to habitat fragments. For brevity, I will pool the main categories of habitat fragments – vegetation remnants, replanted or restored sites, and areas of unassisted regeneration. This is because:

1. They all tend to lack a full quota of species;
2. In practice there is little sharp distinction between these categories, as restoration sites are frequently adjacent to habitat remnants (e.g. Hussey and Wallace, 1993);
3. Over time, both restoration sites and unassisted regeneration may progressively come to resemble each other, as well as remnants.

Determining Whether Mutualisms have been Disrupted

The animals that disperse species of plants at a particular site may be absent through global extinction (see Temple, 1977; Johnson, 2006; Donatti *et al.*,

Chapter 5, this volume), regional extinction (see, e.g. Wright, Chapter 28, this volume), or locally through various factors that impede visitation or persistence (e.g. Laurance, 1994; Warburton, 1997). Warning signs of local disruption include habitat fragmentation, the presence of large-seeded fruits (Crome, 1975c; Stocker and Irvine, 1983; Green, 1993), current or past hunting pressure (see Wright, Chapter 28, this volume) and the presence of rare or threatened plants or frugivores. To know whether dispersers are lacking, we need knowledge of which animals disperse different plant taxa. We are unlikely ever to know all the dispersers of all fruits, but functional class approaches (Moran *et al.*, 2004b; Dennis and Westcott, 2006; also Chapter 9, this volume) offer the potential for making predictions. For example, the absence, low numbers or erratic presence of any functional classes could indicate a lack of dispersal for local plants.

The presence of a disperser or functional class does not guarantee that all plants receive appropriate dispersal services. Although most frugivores eat a variety of fruits, not all are inclined to eat all possible fruits within a locality (Crome, 1975a, 1978; Green, 1993; Eby, 1998; Hall and Richards, 2000; Theimer, 2001; Dennis *et al.*, 2005). For example, a compilation of frugivores feeding on rainforest plant taxa, based on the frequency of published reports (Fig. 27.1a), suggests that fruit-pigeons (Columbidae: Treroninae) are frequent consumers for several families (e.g. Lauraceae and Moraceae; see also Crome, 1975a) but not others (e.g. Sapindaceae); that fruit-bats (Pteropodidae) are unlikely to disperse Lauraceae; and that plants of open habitats are dispersed primarily by honeyeaters (Meliphagidae) and other passerines (Fig 27.1b). This is obviously a very broad comparison, and standardized comparisons within plant families and within genera with both captive and free-ranging frugivores would be needed in order to find out which fruits are likely to be shunned by particular animal taxa. Even where plants are being visited, dispersers are far from equal in the services they provide (Dennis and Westcott, 2006). While in highly fragmented habitats any disperser is probably better than none, an altered subset of the original suite of dispersers can result in quite different patterns of seed rain. This is a problem if a long-term goal is to preserve an ecosystem and its evolutionary processes as close as possible to the original. For example, Howe *et al.* (1981) found low numbers of *Ptilinopus magnificus* and *Lopholaimus antarcticus* (Columbidae: Treroninae), both large-gaped, in small rainforest remnants, compared with extensive rainforest, while *Zosterops lateralis* (Zosteropidae) and some others increased. Moran *et al.* (2004a,b) observed the same pattern but also found that another large-gaped disperser, *Sphecotheres viridis* (Oriolidae), increased in abundance. However, *S. viridis* cannot entirely replace the fruit-pigeons due to differences in the diet of the three species (Fig. 27.2) and also differences in movement patterns and reliability of visitation (Frith, 1982; Pizzey and Knight, 1997; R. Green, personal observation).

Ideally, managers would know how the seeds of target plants are dispersed and whether the relevant agents occur in a target locality. Such information is far from complete. At any rate, a simple list of species of

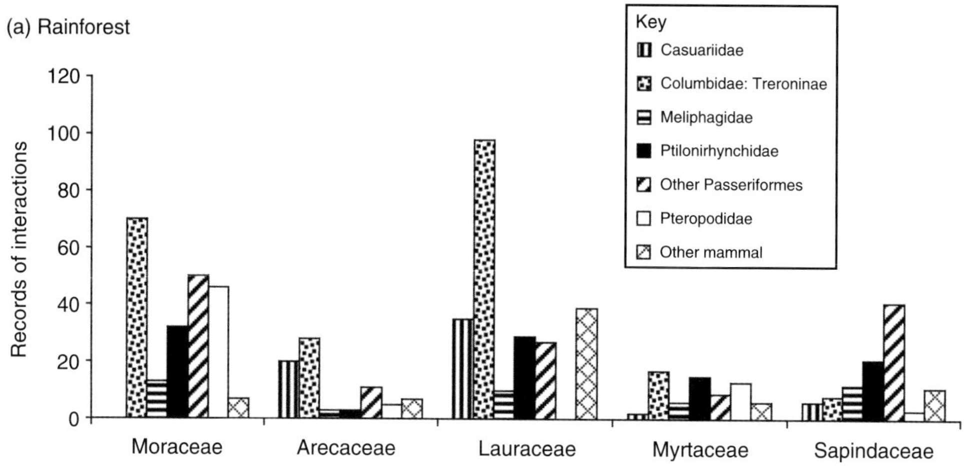

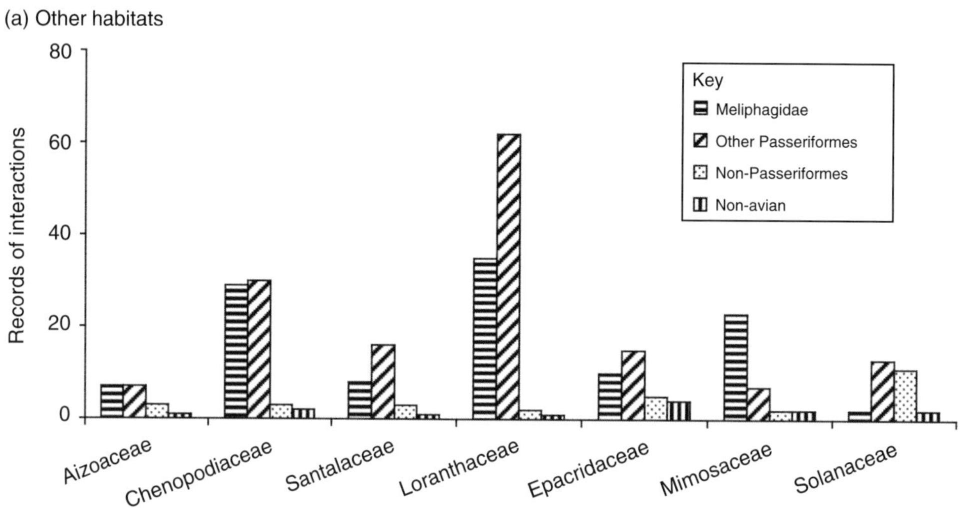

Fig. 27.1. The number of records from the published literature of consumption of fruit by frugivores from species of plants belonging to the respective families (a) in the rainforest and (b) in other habitats.

Literature searched: Barker and Vestjens, 1989, 1990; Bass, 1990; Beland, 1977; Berg, 1975; Berney, 1907a,b; Booth, 1970; Bourke and Austin, 1947; Buchanan, 1989; Butler, 1971; Catterall, 1985; Chisholm, 1944; Church, 1997; Clifford and Drake, 1985; Clifford and Monteith, 1989; Cooper and Cooper, 2004; Crome, 1975a,b,c, 1978; Dennis, 2002; Drew, 1987; Floyd, 1989; Forde, 1986; French, 1990; Friend *et al.*, 1991; Frith, 1952, 1957, 1982; Gannon, 1936; Gilbert, 1935, 1939; Green, 1993; Hall and Richards, 2000; Holland, 1964; Holmes, 1987, 1990; Innes, 1989; Innes and McEvoy, 1992; Johnstone and Burbidge, 1991; Liddy, 1982; Lenz, 1999; Low, 1988; McLennan, 1917; Morris, 1975; Nicholson and Nicholson, 1985, 1988, 1991, 1994, 2000, 2004; Parsons, 1997; Perkins, 1973; Reid, 1987, 1989; Richards, 1990; Roberts and Ingram, 1976; Robertson, 1978; Rogers, 1971, 1977; Rose, 1973; Rowley, 1968; Rowley and Vestjens, 1969; Storr, 1953, 1965; Strahan, 1983; Tadgell, 1936.

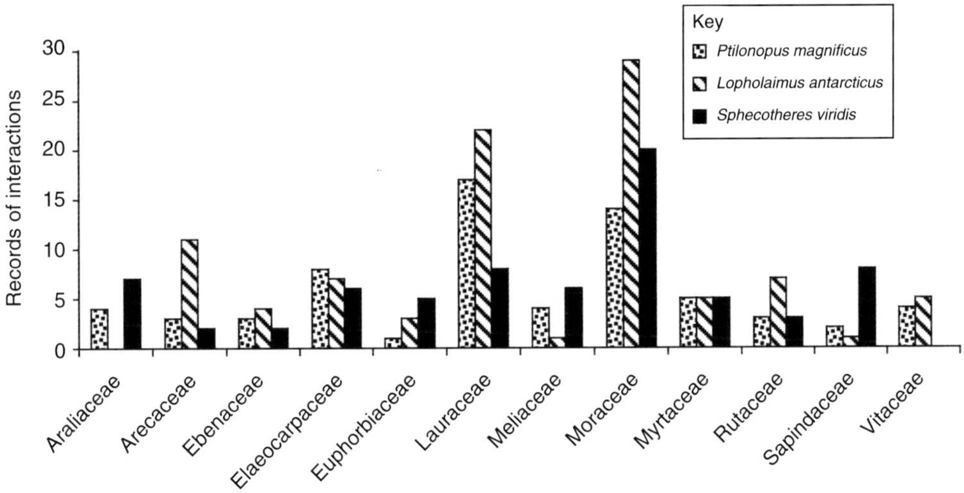

Fig. 27.2. Recorded observations of families of plants visited by the three largest-gaped obligate frugivores of south-eastern Queensland (sources a subset of those from Fig. 27.1).

fleshy-fruited plants and their dispersers that occur in a target locality may not be sufficient for long-term conservation planning. This is because:

1. Dispersers are not necessarily interchangeable (e.g. Poulsen *et al.*, 2002), and those existing in a site may not be the most effective (*sensu* Schupp, 1993);
2. Dispersers present in one season may be absent in others (and thus not available for all fruiting periods);
3. Some of the original plants may be absent from the target community and the absence of their dispersers is thus overlooked;
4. Current dispersers may differ from those present before human-induced changes (important if relatively natural ecological and evolutionary processes are to be preserved in some areas).

When existing information on local plants and their dispersers is lacking, gaps can be filled by observation of consumers using fruiting plants. Much is to be gained by such basic observation, which can be conducted directly (e.g. Green, 1993), using automatic cameras (Yasuda *et al.*, 2005) and by live-trapping to inspect scats (French, 1990) or to detect infrequent or cryptic visitors. It can also be beneficial to observe dispersal and recruitment directly by measuring how much fruit stays on or beneath a mother plant (using tagging or photography), and measuring the distribution and abundance of seedlings or saplings in the vicinity.

Gaps in Knowledge and Sources of Information

In Australia, we know more about fruit eating in rainforests than in other vegetation types, more about trees than other plants, and more about birds than other frugivores (e.g. Crome, 1990; Green, 1993; Moran *et al*., 2004b; Dennis and Westcott, Chapter 9, this volume) but some information does exist for other habitats and taxa (e.g. Berg, 1981; Davidson and Morton, 1981; Clifford and Drake, 1985; Forde, 1986; French, 1991; Green, 1995; Andersen *et al*., 1998). Even the driest of habitats in Australia harbour considerable numbers of fleshy-fruited plants (DEHAA, 1998; Moore, 2005). In moister areas, such as south-east Queensland, the percentages of fleshy-fruited plants amongst the vascular flora are considerable for heaths (9.6–12.4%), open forest (17.3%), freshwater wetlands (8.1%), tidal wetlands (10.4%) and coastal dunes (14.9%) (percentages are of plants recorded by Logan Branch SGAP, 2002, 2005).

In addition to the data in scientific journals, some books on rainforest plants provide information on which fruits are eaten by which frugivores (Nicholson and Nicholson, 1985, 1988, 1991, 1994, 2000, 2004; Floyd, 1989; Cooper and Cooper, 2004), although with no indication of the frequency of interactions or whether consumers are dispersers, fruit thieves or seed predators. However, records are far from complete, even for rainforest. For example, 820 of the 1250 fleshy-fruited species described by Cooper and Cooper (2004) have no consumers recorded and I have been unable to find other records for these in the literature. Details for other habitats are far scarcer. Of over 1000 species of *Acacia* in Australia, many of which have fleshy arils, I can find published reports of dispersers visiting only 13.

Not only are data on interactions between plants and animals lacking, but knowledge of which animals act as dispersers is also incomplete. For example, some Australian lizards are known to eat fruit at least occasionally (Green, 1995; Cogger, 2000) and experienced reptile keepers recommend fruit in the diets of some species (e.g. Weigel, 2004). While several lizards in neighbouring New Zealand are known to act as dispersers (Whitaker, 1978) and I have found potentially viable seeds in the scats of several species of Australian skinks, whether the scat contents represent incidental ingestion or whether lizards contribute significantly to dispersal is unknown. As Australia has a very diverse lizard fauna, many with unknown ecology (Cogger, 2000), their role as dispersers might repay further investigation (see Valido and Olesen, Chapter 6, this volume).

Amongst Australian mammals, fruit-bats have received the most attention (e.g. Eby, 1998; Hall and Richards, 2000) but hoarding rodents are also important dispersers in Australian tropical rainforest (Dennis *et al*., 2005; Theimer, 2005) and may well prove important in other habitats. Australia's most frugivorous marsupial, the musky rat-kangaroo, is also an important disperser in tropical rainforests (Dennis, 2002, 2003) and other marsupials act as dispersers at least occasionally (e.g. Clifford and Drake, 1985; Jarman and Phillips, 1989; M.T. Murphy, Murdoch University, Western Australia, 2005, personal communication). Other marsupials, such as possums, are known to consume fruits. For example, although a 3-year study of squirrel gliders (*Petaurus breviceps*; Petauridae) found no seeds in droppings (Rowston, 1998), Sharpe and Goldingay (1998) observed them feeding on the fruits of mistletoes (Loranthaceae), *Trochocarpa laurina*

(Epacridaceae) and *Acacia concurrens* (Mimosaceae), discarding the seeds while feeding. Whether this simply represents fruit theft or occasionally results in significant dispersal events is not known.

We know little or nothing of the diets of many Australian ants (Shattuck, 2000). However, species in at least nine genera are known to take seeds, some eating only the aril or elaiosome, others eating seeds but also hoarding them (Shattuck, 2000). An *Iridomyrmex* is recorded as carrying elaiosome-bearing seeds up to 180 m from the parent (Whitney, 2002). Australia has many more species of plants with elaiosome-bearing seeds than any other continent (Berg, 1975, 1981), suggesting that they may be far more important than so far recognized, particularly in fire-prone shrublands and woodlands on infertile soils (Berg, 1981; Clifford and Drake, 1981; Milewski and Bond, 1982). Thus, further investigation of lizards, small mammals and ants as seed dispersers (mostly within rather than between fragments) is warranted.

Do Disruptions in Dispersal Matter?

Seed dispersal is a critical process in plant regeneration and persistence in a region and it is fundamental to the structuring and maintenance of plant communities (e.g. Connell, 1971; Howe, 2000; Terborgh *et al.*, 2002). Recent work has indicated that many plants are dispersal-limited in fragmented habitats (Butaye *et al.*, 2002). However, some plants appear less limited by disruptions in seed dispersal than others. For instance, some tolerate growing near parents or siblings, whereas others do not (see Connell, 1971; Penfold and Lamb, 1999; Connell and Green, 2000; Terborgh *et al.* 2002; Butler, 2004) and some have extended seed dormancy, increasing the possibility of dispersal to favourable sites or conditions at later times. On the other hand, some plants are extremely limited by disruptions to dispersal. For example, those that need burial to escape predation (Berg 1975; Hughes and Westoby, 1992; Dennis *et al.*, 2005), particularly when the dispersal agent is susceptible to habitat fragmentation, as is frequently the case (Andersen *et al.*, 1998; Andersen and Morrison, 1998; Gibb and Hochuli, 2002; Ness *et al.*, 2004). Undoubtedly, disruptions to dispersal do matter for some species of plant and in some locations. Consequently, they also matter at the community level, as disruptions to the maintenance of some components of a community will change community structure and long-term trajectories.

Dispersers can contribute to maintaining and enhancing biodiversity (including genetic diversity) in fragmented habitats in three ways:

1. By bringing seeds into fragments from elsewhere;
2. By dispersing seeds within fragments;
3. By dispersing seeds away from fragments to other sites.

The first is often discussed in the applied literature (e.g. White *et al.*, 2004; Neilan *et al.*, 2006) but although much that is relevant to the second is well represented in the theoretical literature (as reference to the Janzen–Connell hypothesis) the possible value of short-distance dispersers within fragments gets far less attention in conservation publications. The third does not get much mention at all, despite there being many threatened plants that occur in habitat fragments and have genetic inbreeding listed as a threatening process (e.g. NSW NPWS, 2002).

It is possible that rarity in some fleshy-fruited species is due to, or at least compounded by, a lack of effective dispersal, especially in the case of species distributed in landscapes with small and isolated fragments of suitable habitat. Of 60 fleshy-fruited Australian species listed as endangered (excluding *Acacia*) only two have known consumers (*Triunia robusta* and *Endiandra cooperana*: Nicholson and Nicholson, 2004; Cooper and Cooper 2004, respectively) and their effectiveness as dispersers is unknown. The scarcity of information may be due to lack of investigation or a genuine lack of dispersal that has contributed to their threatened status.

Fruits with germination inhibitors in their flesh and rapid germination after removal of the flesh appear adapted to germinating after passage through a frugivore, and are probably more likely to suffer from lack of dispersal agents than those with extended dormancy or those that can germinate surrounded by flesh (see Traveset *et al.*, Chapter 4, this volume). While little research has been conducted on this in Australia, numerous examples of such plants can be found in the horticultural and forestry literature (Nicholson and Nicholson, 1985, 1988, 1991, 1994, 2000, 2004; Jones, 1986; Floyd, 1989).

Fleshy-fruited plants, including not only trees but also shrubs, herbs, vines and mistletoes, support many vertebrate and invertebrate animals other than frugivores (Common and Waterhouse, 1981; Recher, 2004). Especially early on in restoration projects, attracting seed dispersers and restoring disrupted dispersal processes can rapidly accelerate biodiversity gains through the introduction of a variety of other local plants and, therefore, resources for other animals. This can also assist seed predators such as the critically endangered Coxen's figparrot (*Cyclopsitta diophthalma coxeni*; Psittacidae; Holmes, 1990).

The value of attracting or reintroducing dispersers that are lacking needs to be weighed against the costs, the urgency of other threats, whether suitable germination conditions and sites are available, and the relative impact of restoring dispersal services. In some cases, direct plantings may be the best solution to maintaining a threatened plant in the short term (ANPC, 1997), but enhancement of natural dispersal systems could still remain a long-term goal. Depending on the context (see Schupp, Chapter 20, this volume), suitable sites for germination may be lacking in altered landscapes. This may make it necessary to restore potential habitat before committing resources to restoring dispersal processes, and may require such action as controlling weeds, which inhibit regeneration in some areas (Fairfax and Fensham, 2000). Whether enticing or reintroducing dispersers is likely to improve the situation also depends partly on whether suitable habitat and sufficient area are available to support them.

Attracting and Maintaining Dispersers

To increase seed dispersal for some plants in habitat fragments, it may be sufficient to attract dispersers that live in or readily move through the landscape matrix. However, others will require additional efforts to attract dispersers that avoid disturbance. Techniques may include planting fruit-bearing trees to attract and support frugivores, buffers to increase the functional size of remnants, extensions to restoration sites, corridors or stepping stones or adding perches to paddocks to provide rest points and dispersal foci (Date *et al.*, 1991; McClanahan and Wolfe, 1993). However, planting to attract dispersers must be done with caution. For example, *Strepera graculina* (Artamidae) were inadvertently encouraged to overwinter in some suburbs by the planting of winter-fruiting plants. They now cause up to 100% mortality of nestlings of small birds in some neighbouring forest remnants (H. Recher, University of New England, New South Wales, 1993, personal communication). In addition, many volant frugivores are likely to disperse weed seeds. A knowledge of which are most likely to bring in troublesome weeds may also influence decisions on where and what to plant in some cases (Loyn and French, 1991; Gosper *et al.*, 2005; Buckley *et al.*, 2006). In situations where it is impossible for dispersers to bring themselves in over long distances, or where they are globally extinct, it may be desirable (with due caution) to reintroduce either the original or substitute dispersers.

Deciding which plants to use to attract dispersers requires knowledge of dietary preferences. Unfortunately, information available on diets frequently gives equal weight to single sightings and frequent usage of a plant by a frugivore. Therefore, I searched the literature for indications of preferences. I used reports of local movements being correlated with diet switching (e.g. Innes, 1989) and which taxa dominated diets when several species were available concurrently (Fig. 27.3). I only used species of birds mentioned in more than five publications that included dietary information and presented it at the family level. Figure 27.3 indicates that some families dominate the diets of some dispersers (e.g. Lauraceae and *P. magnificus*) and that these families are likely to be a good initial choice for attracting those dispersers. It also suggests considerable differences between avian taxa; however, our knowledge gaps on frugivore choice between and within plant families and genera remains enormous.

Attracting frugivores into severely degraded areas can be a slow process, and maintaining them there (throughout the year or for a season) requires a diversity of resources. However, even isolated trees in cleared areas can attract frugivores that bring in seeds of other plants and slowly increase the diversity of resources (e.g. Guevara *et al.*, 1986). The use of artificial perches has been explored in a number of areas (e.g. McDonnell and Stiles, 1983; McClanahan and Wolfe, 1993), and Tinley (Arid Rangelands Recovery Program, Western Australia, 2005, personal communication) demonstrated the functioning of shrubs as perches, and therefore dispersal foci, for seed-dispersing birds in arid regions of

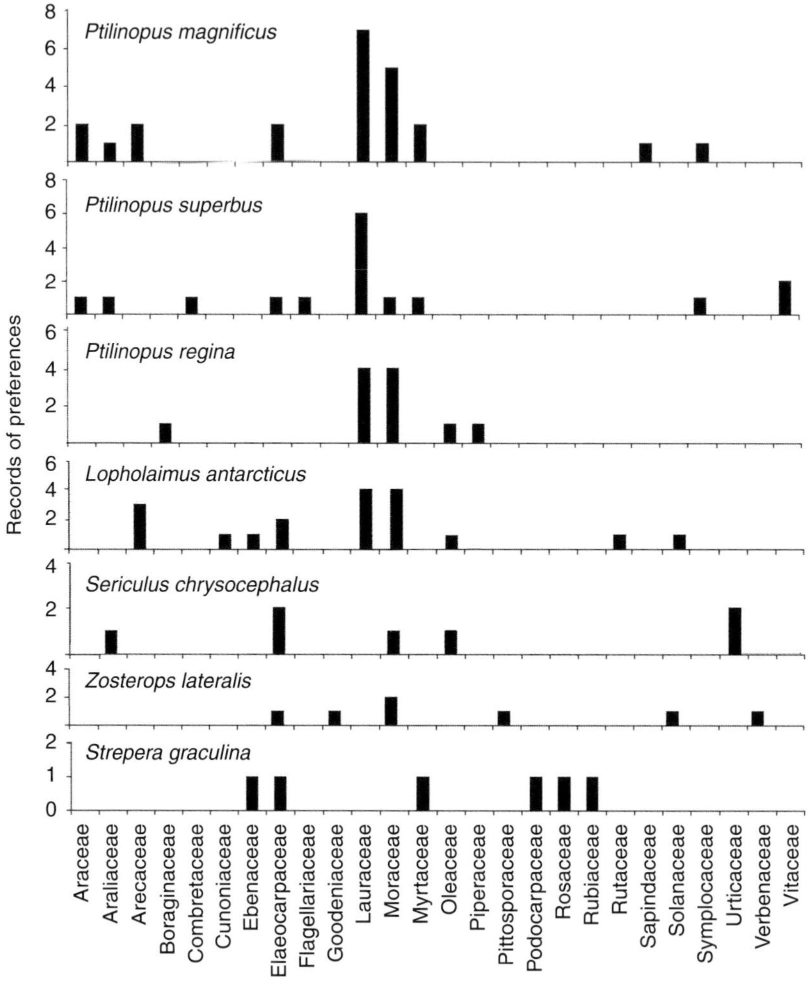

Fig. 27.3. Number of preferences recorded in the literature while wild birds are feeding (e.g. birds recorded as switching very markedly from one fruiting species as another becomes available, or showing the highest percentage of foraging while several plants are fruiting simultaneously) for all potential dispersers with at least five such references in the literature. The vertical axis shows the numbers of times a fleshy-fruited plant within the family was recorded as selected over one or more species of other families.
Literature searched: Catterall, 1985; Crome, 1975a,b, 1978; Drew, 1987; Floyd, 1989; Frith, 1957, 1982; Gannon, 1936; Green, 1993; Innes, 1989; Innes and McEvoy, 1992; Johnstone and Burbidge, 1991; Lenz, 1999; Liddy, 1982; Low, 1988; Rose, 1973; Storr, 1953.

Australia. Even plants generally regarded as environmental weeds can prove useful. Neilan *et al.* (2006) have shown that the introduced tree *Cinnamomum camphori* can be useful foci for the re-establishment of rainforest, recording 180 species of rainforest plants growing under *C. camphori* stands.

While attracting frugivores can be as simple as providing a perch, maintaining resident populations in a degraded area, and therefore year-round dispersal services, requires a more complex solution. Non-mobile or non-migratory dispersers, such as some birds, small mammals and ants, require year-round resources within a local area. To maintain resident, obligate frugivores, in particular, it is necessary to provide nutritional and energy requirements during seasonal shortages, poor years and for breeding. This is a particular challenge if habitat fragments are small. However, it is necessary for maintaining dispersal services for plants that fruit during lean seasons when migratory and nomadic dispersers are absent. These resources could potentially be provided using local fruits that provide the required range of nutrients, but little data are currently available.

Dietary needs can lead to complexities in combinations of foods selected and seasonal changes in selection (e.g. Witmer, 2001) and the generally low protein content of fruits can be a problem for frugivores (e.g. Herbst 1986; Bosque and Pacheco, 2000; Bosque and Calchi, 2003). In Australia, we lack understanding of the details of nutritional needs of frugivores or nutritional contents of fruits and the whole topic is fraught with many complexities (e.g. Martinez del Rio and Restrepo, 1993). Crome (1975a) compared nutrient consumption between Australian frugivores, but detailed information on the physiological requirements of Australian frugivores is sparse. However, some clues can be gleaned from the general physiology of broader taxa, and avicultural, veterinary and wildlife carers' literature (e.g. Henderson, 1997). In some instances this may be matched with data on the chemical content of native fruits (sparse in the seed dispersal literature) from publications on the wild foods industry, medicinal properties of native fruits and military survival texts (e.g. Low, 1988; Brand Miller *et al.*, 1993).

In some areas, the original dispersers are no longer available to be attracted back. For example, the mainland emu, *Dromaius novaehollandia* (Dromidaeidae), an important disperser (Clifford and Monteith, 1989; Rogers *et al.*, 1994; McGrath and Bass, 1999), is now absent from many coastal areas, and the King Island (*D. minor*) and Kangaroo Island emus (*D. baudinianus*) are now extinct (Pizzey and Knight, 1997). Several other Australian frugivores are endangered or already extinct, and others, including the mistletoe-dispersing *Grantiella picta* (Meliphagidae; ACT Government, 1999) are listed as rare or threatened. With the exception of *Casuarius casuarius* (Casuariidae: Crome, 1975c; Stocker and Irvine, 1983), the effects of their low abundance or absence on seed dispersal remain largely uninvestigated. In these situations, reintroduction or the introduction of substitute species could be considered. For example, *D. baudinianus* on Kangaroo Island has now been replaced by the introduction of the mainland species. However, the effects of this substitution on the rest of the ecosystem have not been investigated. While literature on the reintroduction of vertebrate animals is growing (e.g. Kushlan, 1980; Serena, 1994; Richards, 1998) there is much to learn about techniques for different species (including invertebrates), and so it remains an approach to be employed with considerable caution.

Directions for Future Research

It is clear that numerous gaps exist in our knowledge and research base for tackling management issues relating directly to seed dispersal, not least of which is the gap between the questions of general principle that scientists usually ask and the specific, local problems that managers usually tackle. However, we can at least identify issues with managers and incorporate these into research projects that are either directly focused on these or link with more theoretical questions. For example, questions relating to foraging choices, seasonal changes in nutritional needs, interspecific competition, or food as a limiting factor (Newton, 1981) can provide avenues for a marriage between theoretical research and applied outcomes, especially if the results are readily accessible to managers. Other examples include combining the questions of accessibility by different frugivores to different kinds of habitat fragments of varying isolation in different matrices with the theoretical aspects of habitat selection, animal dispersal and movement. In addition, different research areas such as behavioural ecology and physiology could be combined to produce interesting theoretical and applied projects, such as identifying whether the cue for mobile frugivores to leave an area is related to nutrient availability and seasonal requirements.

Conclusions

Suding (2005) writes that:

> Restoration ecology has been likened to battlefield medicine. Actions need to be taken even with imperfect scientific knowledge and practical knowledge develops separately from hard scientific research.

This is likely to remain true for some time, requiring simultaneous conservation action, technique development, applied research and basic research, but hopefully with increasing communication between workers in all fields. While the questions asked by researchers and the needs of restoration or conservation management do not always coincide, there would appear to be much scope for further applied and theoretical research that directly contributes to solving conservation problems. In addition, a readily available and well-publicized, web-based hub or database that aims to create links between and within researchers and managers, and to share data and results, would facilitate an increase in communication and would be a useful tool to allow conservation managers to make informed decisions and access research results.

Atkins (1998) estimated that 14.8% of species of vascular plants in Australia are now threatened, and other continents face similar problems, exacerbated by threats to the animals that disperse them. Intensive management and translocation or seeding of threatened plants and the direct restoration of fragmented landscapes will often be necessary in the short term. However, enhancement of natural dispersal should prove more cost-effective for the long term by providing some degree of protection to a natural ecological and evolutionary process that can provide long-term sustainability to individual plants and communities.

Acknowledgements

Thanks to Andrew Dennis for his assistance in condensing what started out as an over-long and complex manuscript. Thanks also to Dr Carla Catterall and others of Griffith University for supporting my role as Honorary Research Fellow while, amongst other things, working on this chapter, and to Drs Don Butler (Queensland Herbarium), Richard Corlett (University of Hong Kong), Palitha Jayasekara (University of Tokyo) and an anonymous reviewer for comments on the manuscript.

References

ACT Government (1999) Painted honeyeater (*Grantiella picta*): a vulnerable species. *Action Plan No. 19.* Environment ACT, Canberra, Australia.

Andersen, A.N. and Morrison, S.C. (1998) Myrmechory in Australia's seasonal tropics: effects of disturbance on distance dispersal. *Australian Journal of Ecology* 23, 483–491.

Andersen, A.N., Morrison, S., Belbin, L., Ashwath, N. and Brennan, K. (1998) The role of ants in minesite restoration in the Kakadu region of Australia's Northern Territory, with particular reference to their use as bioindicators. *Supervising Scientist Report 130.* Department of Environment and Heritage, Canberra.

ANPC (1997) *Guidelines for the Translocation of Threatened Plants in Australia.* Australian Network for Plant Conservation Working Group, Canberra, Australia.

Atkins, K. (1998) Threatened flora overview. In: Brown, A., Thomson-Dans, C. and Marchant, N. (eds) *Western Australia's Threatened Flora.* Department of Conservation and Land Management, Como, Australia, pp. 11–14.

Barker, R.D. and Vestjens, W.J.M. (1989) *The Food of Australian Birds. Vol. I: Non-Passerines.* CSIRO, Canberra, Australia.

Barker, R.D. and Vestjens, W.J.M. (1990) The Food of Australia Birds. Volume II: Passerines. CSIRO, Canberra, Australia.

Bass, D.A. (1990) Pied currawongs and seed dispersal. *Corella* 14, 24–27.

Beland, P. (1977) Mimicry in orioles in south-eastern Queensland. *Emu* 77, 215–218.

Berg, R.Y. (1975) Myrmechorous plants in Australia and their dispersal by ants. *Australian Journal of Botany* 23, 475–508.

Berg, R.Y. (1981) The role of ants in seed dispersal in Australian lowland heathlands. In: Specht, R.L. (ed.) *Heathlands and Related Shrublands: Analytical Studies.* Elsevier, Amsterdam, pp. 41–50.

Berney, F.L. (1907a) Field notes on birds of the Richmond District, north Queensland. Part IV. *Emu* 6, 106–115.

Berney, F.L. (1907b) Field notes on birds of the Richmond District, north Queensland. Part V. *Emu* 6, 155–159.

Booth, J. (1970) Birds of Fairfax Island (Bunker Group) and Wreck Island (Capricorn Group), Great Barrier Reef. *Sunbird* 1, 85–92.

Bourke, P.A. and Austin, A.F. (1947) Notes on the red-browed Lorilet. *Emu* 46, 286–294.

Bosque, C. and Calchi, R. (2003) Food choice by blue–gray tanagers in relation to protein content. *Comparative Biochemistry and Physiology A: Molecular and Integrative Physiology* 135, 321–327.

Bosque, C. and Pacheco, M.A. (2000) Dietary nitrogen as a limiting nutrient in frugivorous birds. *Revista Chilena de Historia Natural* 73, 441–450.

Brand Miller, J., James, K.W. and Meggiore, P.M.A. (1993) *Tables of Composition of Australian Aboriginal Foods.* Aboriginal Studies Press, Canberra, Australia.

Buchanan, R. (1989) *Bush Regeneration: Recovering Australian Landscape.* TAFE NSW, Sydney, Australia.

Buckley, Y.M., Anderson, S., Catterall, C.P., Corlett, R.T., Engel, T., Gosper, C.R., Nathan, R., Richardson, D.M., Setter, M., Spiegel, O., Vivian-Smith, G., Voigt, F.A., Weir, J.E.S. and Westcott, D.A. (2006) Management of plant invasions mediated by frugivore interactions. *Journal of Applied Ecology* 43, 848–857.

Butaye, J., Jacquemyn, H., Honnay, O. and Hermy, M. (2002) The species pool concept applied to forests in a fragmented landscape: dispersal limitation *versus* habitat limitation. *Journal of Vegetation Science* 13, 27–34.

Butler, D.W. (2004) Seed dispersal syndromes and the distribution of woody plants in south-east Queensland's vine-forests. PhD thesis, University of Queensland, Australia.

Butler, W.H. (1971) Yellow figbird in Western Australia. *Western Australia Naturalist* 12, 21.

Catterall, C.P. (1985) Winter energy deficits and the importance of fruits *versus* insects in a tropical island bird population. *Australian Journal of Ecology* 10, 265–279.

Chisholm, A.H. (1944) An interesting old notebook. *Emu* 43, 281–288.

Church, R.J. (1997) Avian frugivory in a subtropical rainforest: eleven years of observation in Lamington National Park. *Sunbird* 27, 85–97.

Clifford, H.T. and Drake, W.E. (1981) Pollination and dispersal in Eastern Australian heathlands. In: Specht, R.L. (ed.) *Ecosystems of the World. 9B: Heathlands and Related Shrublands.* Elsevier Scientific, Amsterdam, pp. 284–286.

Clifford, H.T. and Drake, W.E. (1985) Seed dispersal by kangaroos and their relatives. *Journal of Tropical Ecology* 1, 373–374.

Clifford, H.T. and Monteith, G.B. (1989) A three-phase seed dispersal mechanism in Australian quinine bush (*Petalostigma pubescens* Domin). *Biotropica* 21, 284–286.

Cogger, H.G. (2000) *Reptiles and Amphibians of Australia.* Reed New Holland, Sydney, Australia.

Common, I.F.B. and Waterhouse, D.F. (1981) *Butterflies of Australia.* Angus and Robertson, London, UK.

Connell, J.H. (1971) On the role of natural enemies in preventing competitive exclusion in some marine animals and in rain forest trees. In: Den Boer, P.J. and Gradwell, G.R. (eds) *Dynamics of Populations.* PUDOC, Wageningen, The Netherlands, pp. 298–312.

Connell, J.H. and Green, P.T. (2000) Seedling dynamics over thirty-two years in a tropical rain forest tree. *Ecology* 81, 568–584.

Cooper, W. and Cooper, W. (2004) *Fruits of the Australian Tropical Rainforest.* Nokomis Editions, Australia.

Crome, F.H.J. (1975a) The ecology of fruit pigeons in Tropical Northern Queensland. *Australian Wildlife Research* 2, 155–185.

Crome, F.H.J. (1975b) Breeding, feeding and status of the Torres Strait Pigeon at Low Isles, north-eastern Queensland. *Emu* 75, 189–198.

Crome, F.H.J. (1975c) Some observations on the biology of the cassowary in northern Queensland. *Emu* 76, 8–14.

Crome, F.H.J. (1978) Foraging ecology of an assemblage of birds in lowland rainforest in northern Queensland. *Australian Journal of Ecology* 3, 195–212.

Crome, F.H.J. (1990) Vertebrates and successions. In: Webb, L.J. and Kikkawa, J. (eds) *Australian Tropical Rainforest: Science, Values, Meaning.* CSIRO Publishing, Melbourne, Australia, pp. 53–64.

Cropper, S. (1993) *Management of Endangered Plants.* CSIRO Publishing, Melbourne, Australia.

Date, E.M., Ford, H.A. and Recher, H.F. (1991) Frugivorous pigeons, stepping stones, and weeds in northern New South Wales. In: Saunders, D.A. and Hobbs, R.J. (eds) *Nature Conservation 2: The Role of Corridors.* Surrey Beatty and Sons, Chipping Norton, Australia, pp. 241–245.

Davidson, D.W. and Morton, S.R. (1981) Myrmecochory in some plants (*F. chenopodiaceae*) of the Australian arid zone. *Oecologia* 50, 357–366.

DEHAA (1998) *A Review of the Simpson Desert Reserve 1988–1998.* Department of Environment, Heritage and Aboriginal Affairs, Adelaide, Australia.

Dennis, A.J. (2002) The diet of the musky rat-kangaroo, *Hypsiprymnodon moschatus*, a rainforest specialist *Wildlife Research* 29, 209–219.

Dennis, A.J. (2003) Scatter-hoarding by musky rat-kangaroos, *Hypsiprymnodon moschatus*, a tropical rain forest marsupial from Australia: implications for seed dispersal. *Journal of Tropical Ecology* 19, 619–627.

Dennis, A.J. and Westcott, D.A. (2006) Reducing complexity in studying seed dispersal at community scales: a functional classification of vertebrate seed dispersers in tropical forests. *Oecologia* 149, 620–634.

Dennis, A.J., Lipsett-Moore, G.J., Harrington, G.N., Collins, E.N. and Westcott, D.A. (2005) Seed predation, seed dispersal and habitat fragmentation: does context make a difference in tropical Australia? In: Forget, P.-M., Lambert, J.E., Hulme, P.E. and Vander Wall, S.B. (eds) *Seed Fate: Predation, Dispersal and Seedling Establishment.* CAB International, Wallingford, UK, pp. 117–135.

Drew, R.A.I. (1987) Reduction in fruit fly (Tephridae: Dacinae) populations in their endemic rainforest habitat by frugivorous vertebrates. *Australian Journal of Zoology* 35, 283–286.

Eby, P. (1998) An analysis of diet specialization in frugivorous *Pteropus poliocephalus* (Megachiroptera) in Australian subtropical rainforest. *Australian Journal of Ecology* 23, 443–456.

Edwards, A. and Russell-Smith, J. (2006) Assessing the effects of fire regimes on fire-sensitive vegetation and habitats of the Arnhem Plateau, tropical northern Australia. *Forest Ecology and Management* 234 (Suppl. 1), S170.

Fairfax, R.J. and Fensham, R.J. (2000) The effect of exotic pasture development on floristic diversity in central Queensland, Australia. *Biological Conservation* 94, 11–21.

Fazey, J. and McQuie, A. (2005) Applying conservation theory in natural areas management. *Ecological Management and Restoration* 6, 147.

Floyd, A.G. (1989) *Rainforest Trees of Mainland South-eastern Australia.* Inkata Press, Melbourne, Australia.

Forde, N. (1986) Relationships between birds and fruits in temperate Australia. In: Ford, H. and Paton, D.C. (eds) *The Dynamic Partnership: The Birds and Plants of Southern Australia.* Government Printer, Adelaide, Australia, pp. 42–58.

French, K. (1990) Evidence for frugivory by birds in montane and lowland forests in southeast Australia. *Emu* 90, 185–189.

French, K. (1991) Characteristics and abundance of vertebrate-dispersed fruits in temperate wet sclerophyll forest in southeastern Australia. *Australian Journal of Ecology* 16, 1–13.

Friend, G.R., Morris, K.D. and McKenzie, N.L. (1991) The mammal fauna of Kimberley rainforests. In: McKenzie, N.L., Johnston, R.B. and Kendrick, P.G. (eds) *Kimberley Rainforests of Australia.* Surrey Beatty and Sons, Chipping Norton, Australia, pp. 393–412

Frith, H.J. (1952) Notes on the pigeons of the Richmond River, NSW. *Emu* 52, 89–99.

Frith, H.J. (1957) Food habits of the topknot pigeon. *Emu* 57, 341–345.

Frith, H.J. (1982) *Pigeons and Doves of Australia.* Rigby, Adelaide, Australia.

Gannon, G.R. (1936) Plants spread by the silvereye. *Emu* 35, 314–316.

Gibb, H. and Hochuli, D.F. (2002) Habitat fragmentation in an urban environment: large and small fragments support different arthropod assemblages. *Biological Conservation* 106, 91–100.

Gilbert, P.A. (1935) The topknot pigeon. *Emu* 35, 301–302.

Gilbert, P.A. (1939) The bower-painting habit of the satin bowerbird (*Ptilonorhynchus violaceus*). *Emu* 39, 209–219.

Gosper, C., Stansbury, C.D. and Vivian-Smith, G. (2005) Seed dispersal of fleshy-fruited invasive plants by birds: contributing factors and management options. *Diversity and Distributions* 11, 549–558.

Green, R.J. (1993) Avian seed dispersal in and near subtropical rainforests. *Wildlife Research* 20, 535–557.

Green, R.J. (1995) Using frugivores for regeneration: a survey of knowledge and problems in Australia. In: Bissonette, J.A. and Krausman, P.R. (eds) *Integrating People and Wildlife for a Sustainable Future*. The Wildlife Society, Bethesda, Maryland, pp. 667–671.

Guevara, S., Purata, S.E. and Van der Maarel, E. (1986) The role of remnant forest trees in tropical secondary succession. *Vegetatio* 66, 77–84.

Hall, L. and Richards, G. (2000) *Flying Foxes: Fruit and Blossom Bats of Australia*. UNSW Press, Sydney, Australia.

Harrington, G.N., Irvine, A.K., Crome, F.H.J. and Moore, L. (1997) Regeneration of large-seeded trees in Australian rainforest fragments: a study of higher order interactions. In: Laurance, W.F., Bierregaard, R.O. (eds) *Tropical Forest Remnants: Ecology, Management and Conservation of Fragmented Communities*. University of Chicago Press, Chicago, Illinois, pp. 292–303.

Henderson, N. (ed.) (1997) *Australian Bird Rehabilitation Manual*. Wildlife Information and Rescue Service Foster Carer Advisory Committee, Forestville, Australia

Herbst, L.H. (1986) The role of nitrogen from fruit pulp in the nutrition of the frugivorous bat *Carollia perspicillata*. *Biotropica* 18, 39–44.

Hilbert, D.W., Ostendorf, B. and Hopkins, M. (2001) Sensitivity of tropical forests to climate change in the humid tropics of north Queensland. *Austral Ecology* 26, 590–603.

Holland, L. (1964) Pigeons of the Woolgoolga District, New South Wales. *Australian Bird Watcher* 4, 43–46.

Holmes, G. (1987) *Avifauna of the Big Scrub Region*. NSW National Parks and Wildlife Service, Sydney, Australia.

Holmes, G. (1990) The biology and ecology of Coxen's figparrot. *RAOU Report no. 65*, Melbourne, Australia.

Hougner, C., Colding, J. and Söderqvist, T. (2006) Economic valuation of a seed dispersal service in the Stockholm National Urban Park, Sweden. *Ecological Economics* 59, 364–374.

Howe, H.F. (2000) No question: seed dispersal matters. *Trends in Ecology and Evolution* 15, 434–436.

Howe, R.W., Howe, T.D. and Ford, H.A. (1981) Bird distributions on small rainforest remnants in New South Wales. *Australian Wildlife Research* 8, 637–651.

Hughes, L. and Westoby, M. (1992) Fate of seeds adapted for dispersal by ants in Australian sclerophyll vegetation. *Ecology* 73, 1285–1299.

Hussey, B.M.J. and Wallace, K.J. (1993) *Managing your Bushland*. Department of Conservation and Land Management, Como, Australia.

Innes, G.J. (1989) Feeding ecology of fruit pigeons in subtropical rainforests of south-eastern Queensland. *Australian Wildlife Research* 16, 365–394.

Innes, G.J. and McEvoy, J. (1992) Feeding ecology of green catbirds (*Ailuroedus crassirostris*) in subtropical rainforests of south-eastern Australia. *Wildlife Research* 19, 317–329.

Jarman, P.J. and Brock, M.A. (1996) Collaboration of science and management in endangered species recovery. In: Stephens, S. and Maxwell, S. (eds) *Back from the Brink: Refining the Threatened Species Recovery Process*. Australian Nature Conservation Agency, ACT, Australia pp. 74–78.

Jarman, P.J. and Phillips, C. (1989) Diets in a community of macropod species. In: Grigg, G. and Hume, I. (eds) *Kangaroos, Wallabies and Rat-Kangaroos*. Surrey Beatty and Sons, Sydney, Australia pp. 143–149.

Johnson, C. (2006) *Australia's Mammal Extinctions: A 50,000 Year History*. Cambridge University Press, Cambridge, UK.

Johnstone, R.E. and Burbidge, A.H. (1991) The avifauna of Kimberley rainforests. In: McKenzie, N.L., Johnstone, R.B. and Kendrick, P.G. (eds) *Kimberley Rainforests of Australia*. Surrey Beatty and Sons, Chipping Norton, Australia, pp. 361–391.

Jones, D.L. (1986) *Rainforest Plants*. Reed Books, Balgowlah, Australia.

Kushlan, J.A. (1980) Reintroduction of indigenous species to natural ecosystems. *Environmental Management* 4, 93–94.

Laurance, W.F. (1994) Rainforest fragmentation and the structure of small mammal communities in tropical Queensland. *Biological Conservation* 69, 23–32.

Langkamp, P. and Plaisted, M. (1987) Appendix 1. Native plant seed usage by the mining industry – a survey. In: Langkamp, P. (ed.) *Germination of Australian Native Plants Seed*. Inkata Press, Melbourne, Australia, pp. 134–181.

Lenz, N. (1999) Evolutionary ecology of the regent bowerbird *Sericulus chrysocephalus*. *Okologie der Vogel* 22 (Suppl.), 1–200.

Liddy, J. (1982) Food of the mistletoebird near Pumicestone Passage, south-eastern Queensland. *Corella* 6, 11–15.

Logan Branch SGAP (2002) *Mangroves to Mountains: A Field Guide to the Native Plants of the Logan–Albert River Catchments*. Logan Branch S.G.A.P. (Qld Region), Copyright Publishing, Brisbane, Australia.

Logan Branch SGAP (2005) *Mangroves to Mountains. Vol. 2: A Field Guide to the Native Plants of South-east Queensland*. Logan Branch S.G.A.P. (Qld Region), Browns Plains, Australia.

Low, T. (1988) *Wild Food Plants of Australia*. Angus and Robertson, Sydney, Australia.

Low, T. (2001) *Feral Future: The Untold Story of Australia's Exotic Invaders*. Penguin Australia, Ringwood, Australia.

Loyn, R.H. and French, K. (1991) Birds and environmental weeds in south-eastern Australia. *Plant Protection Quarterly* 6, 137–149.

Maunder, M., Clubbe, C., Hankamer, C. and Groves, M. (eds) (2002) *Plant Conservation in the Tropics: Perspectives and Practice*. Royal Botanic Gardens, Kew, UK.

Martinez del Rio, C. and Restrepo, C. (1993) Ecological and behavioural consequences of digestion in frugivorous animals. *Vegetatio* 107/108, 205–216.

McClanahan, T.R. and Wolfe, R.W. (1993) Accelerating forest succession in a fragmented landscape: the role of birds and perches. *Conservation Biology* 7, 279–288.

McDonnell, M.J. and Stiles, E.W. (1983) The structural complexity of old field vegetation and the recruitment of bird-dispersed plant species. *Oecologia* 56, 109–116.

McGrath, R.J. and Bass, D. (1999) Seed dispersal by emus on the New South Wales North-east coast. *Emu* 99, 248–252.

McLennan, W. (1917) Northern Australian birds. *Emu* 16, 205–231.

Milewski, A.V. and Bond, W.J. (1982) Convergence of myrmecochory in mediterranean Australia and South Africa. In: Buckley, R. (ed.) *Ant–Plant Interactions in Australia*. Dr W Junk, The Hague, The Netherlands, pp. 651–664

Moore, P. (2005) *A Guide to Plants of Inland Australia*. Reed New Holland, Sydney, Australia.

Moran, C., Catterall, C.P., Green, R.J. and Olsen, M.F. (2004a) Fates of feathered fruit-eaters in fragmented forests. In: D. Lunney (ed.) *Conservation of Australia's Forest Fauna*, 2nd edn. Royal Zoological Society of New South Wales, Mosman, Australia, pp. 699–712.

Moran, C., Catterall, C.P., Green, R.J. and Olsen, M.F. (2004b) Functional variation among frugivorous birds: implications for rainforest seed dispersal in a fragmented subtropical landscape. *Oecologia* 141, 585–595.

Morris, A.K. (1975) The birds of Gosford, Wyong and Newcastle (County of Cumberland). *Australian Birds* 9, 37–76.

Neilan, W., Catterall, C.P., Kanowski, J. and McKenna, S. (2006) Frugivorous birds may facilitate rainforest succession in weed-dominated regrowth in subtropical Australia. *Biological Conservation* 129, 393–407.

Ness, J.H., Bronstein, J.L., Andersen, A.N. and Holland, J.N. (2004) Ant body size predicts dispersal distance of ant-adapted seeds: implications of small-ant invasions. *Ecology* 85, 1244–1250.

Newton, I. (1981) The role of food in limiting bird numbers. In: Klump, H. and Woldendrop, J.W. (eds) *The Integrated Study of Bird Populations*. North Holland Publishing, Amsterdam, pp. 11–30.

Nicholson, N. and Nicholson, H. (1985) *Australian Rainforest Plants*. Terania Rainforest Publishing, The Channon, NSW, Australia.

Nicholson, N. and Nicholson, H. (1988) *Australian Rainforest Plants II*. Terania Rainforest Publishing, The Channon, NSW, Australia.

Nicholson, N. and Nicholson, H. (1991) *Australian Rainforest Plants III*. Terania Rainforest Publishing, The Channon, NSW, Australia.

Nicholson, N. and Nicholson, H. (1994) *Australian Rainforest Plants IV*. Terania Rainforest Publishing, The Channon, Australia.

Nicholson, N. and Nicholson, H. (2000) *Australian Rainforest Plants V*. Terania Rainforest Publishing, The Channon, NSW, Australia.

Nicholson, N. and Nicholson, H. (2004) *Australian Rainforest Plants VI*. Terania Rainforest Publishing, The Channon, NSW, Australia.

NSW NPWS (2002) *Threatened Species of the Upper North Coast of New South Wales: Flora*. NSW National Parks and Wildlife Service, Northern Directorate, Coffs Harbour, Australia.

Penfold, G.C. and Lamb, D. (1999) Species coexistence in an Australian subtropical rain forest: evidence for compensatory mortality. *Journal of Ecology* 87, 316–329.

Perkins, D.L. (1973) Some observations from Fraser Island. *Sunbird* 4, 40–41.

Pizzey, G. and Knight, F. (1997) *The Graham Pizzey and Frank Knight Field Guide to the Birds of Australia*. Angus and Robertson (imprint of HarperCollins), Sydney.

Playford, J. and Murray, R. (2002) *Threatened Plants: Active Bush Care*. Centre for Conservation Biology, University of Queensland, Brisbane, Australia.

Poulsen, J.R., Clark, C.J., Connor, E.F. and Smith, T.B. (2002) Differential resource use by primates and hornbills: implications for seed dispersal. *Ecology* 83, 228–240.

Recher, H.F. (2004) Eucalypt forest birds: the role of nesting and foraging resources in conservation and management. In: Lunney, D. (ed.) *Conservation of Australia's Forest Fauna*, 2nd edn. Royal Zoological Society of New South Wales, Mosman, Australia.

Reid, N. (1987) The mistletoebird and Australian mistletoes: co-evolution or coincidence? *Emu* 87, 130–131.

Reid, N. (1989) Dispersal of mistletoes by honeyeaters and flowerpeckers: components of seed dispersal quality. *Ecology* 70, 137–145.

Richards, G.C. (1990) The spectacled flying-fox, *Pteropus conspicillatus* (Chiroptera: Pteropodidae) in north Queensland. Part 2. Diet, seed dispersal and feeding ecology. *Australian Mammalogy* 13, 25–31.

Richards, J.D. (1998) Return of the natives. *Australian Geographic* 50, 82–97.
Roberts, G.J. and Ingram, G.J. (1976) An annotated list of the land birds of Cooloola. *Sunbird* 7, 1–20.
Robertson, J.S. (1978) Southeast Queensland aspects of the cooperative silvereye project. *Australian Bird-Bander* 9, 51–55.
Rogers, A. (1971) NSW bird report for 1971. *Birds (Sydney)* 7, 89–109.
Rogers, A. (1977) NSW bird report for 1976. *Australian Birds* 11, 81–104.
Rogers, R.W., Butler, D. and Carnell, J. (1994) Dispersal of germinable seeds by emus in semi-arid Queensland. *Emu* 94, 132–134.
Rose, A.B. (1973) Food of some Australian birds. *Emu* 73, 177–183.
Rowley, I. (1968) Unusual feeding technique of a bald coot. *Emu* 67, 295–296.
Rowley, I. and Vestjens, W.J.M. (1969) The comparative ecology of Australian corvids. V. Food. *Wildlife Research* 18, 131–155.
Rowston, C. (1998) The squirrel glider: an autecological study in a fragmented landscape. PhD thesis, Griffith University, Australia.
Schupp, E.W. (1993) Quantity, quality and the effectiveness of seed dispersal by animals. *Vegetatio* 107/108, 15–29.
Serena, M. (ed.) (1994) *Reintroduction Biology of Australian and New Zealand Fauna*. Surrey Beatty and Sons, Chipping Norton, Australia.
Sharpe, D.J. and Goldingay, R.L. (1998) Feeding behaviour of the squirrel glider at Bungawalbin Nature Reserve, north-eastern New South Wales. *Wildlife Research* 25, 243–254.
Shattuck, S.O. (2000) *Australian Ants: Their Biology and Identification*. Monographs on Invertebrate Taxonomy, Vol. 3. CSIRO Publishing, Collingwood, Australia.
Smith, K. and Garnett, S. (2004) Animal–plant interactions: a rainforest conservation manager's perspective. In: Kanowski, J., Catterall, C.P., Dennis, A.J. and Westcott, D.A. (eds) *Animal–Plant Interactions in Conservation and Restoration: Workshop Proceedings*. Cooperative Research Centre for Tropical Rainforest Ecology and Management, Rainforest CRC, Cairns, Australia, pp. 17–19.
Stocker, G.C. and Irvine, A.K. (1983) Seed dispersal by cassowaries (*Casuarius casuarius*) in north Queensland's rainforests. *Biotropica* 15, 170–176.
Storr, G.M. (1953) Birds of the Cooktown and Laura districts, North Queensland. *Emu* 53, 225–248.
Storr, G.M. (1965) The avifauna of Rottnest Island, WA III. Land birds. *Emu* 64, 172–180.
Strahan, R. (ed.) (1983) *The Australian Museum Complete Book of Australian Mammals*. Australian Museum, Sydney, Australia.
Suding, K.N. (2005) Book review: The practice of restoration and the science of ecology. *Trends in Ecology and Evolution* 20, 587–588.
Tadgell, A.J. (1936) Songbirds and palms. *Victorian Naturalist* 53, 74–75.
Temple, S.A. (1977) Plant–animal mutualism: coevolution with dodo leads to near-extinction of plant. *Science* 197, 885–886.
Terborgh, J., Pitman, N., Silman, M., Schichter, H. and Nunez, V.P. (2002) Maintenance of tree diversity in tropical forests. In: Levey, D.J., Silva, W.R. and Galetti, M. (eds) *Seed Dispersal and Frugivory: Ecology, Evolution and Conservation*. CAB International, Wallingford, UK, pp. 1–17.
Terborgh, J. (2004) Reflections of a scientist on the World Parks Congress. *Conservation Biology* 18, 619–620.
Theimer, T.C. (2001) Seed scatterhoarding by white-tailed rats: consequences for seedling recruitment by an Australian rain forest tree. *Journal of Tropical Ecology* 17, 177–189.
Theimer, T.C. (2005) Rodents as conditional mutualists. In: Forget, P.-M., Lambert, J.E., Hulme, P.E. and Vander Wall, S.B. (eds) *Seed Fate: Predation, Dispersal and Seedling Establishment*. CAB International, Wallingford, UK, pp. 283–295.

Vesk, P.A. and MacNally, R. (2006) The clock is ticking: revegetation and habitat for birds and arboreal mammals in rural landscapes of southern Australia. *Agriculture, Ecosystems and Environment* 112, 356–366.

Warburton, N.H. (1997) Structure and conservation of avifauna in isolated rainforest remnants in tropical Australia. In: Laurance, W.F. and Bierregaard, R.O. (eds) *Tropical Forest Remnants: Ecology, Management and Conservation of Fragmented Communities.* University of Chicago Press, Chicago, Illinois, pp. 190–206.

Weigel, J. (2004) *Care of Australian Reptiles in Captivity.* Reptile Keepers Association, Gosford, Australia.

Whitaker, A.H. (1978) The roles of lizards in New Zealand plant reproductive strategies. *New Zealand Journal of Botany* 25, 315–328.

White, E., Tucker, N., Meyers, N. and Wilson, J. (2004) Seed dispersal to revegetated isolated rainforest patches in North Queensland. *Forest Ecology and Management* 192, 409–426.

Whitney, K.D. (2002) Dispersal for distance? *Acacia ligulata* seeds and meat ants *Iridomyrmx viridiaeneus. Austral Ecology* 27, 589–595.

Witmer, C.M. (2001) Nutritional interactions and fruit removal: cedar waxwing consumption of *Viburnum opulus* fruits in spring. *Ecology* 82, 3120–3130.

Yasuda, M., Miura, S., Ishii, N., Okuda, T. and Hussein, N.A. (2005) Fallen fruits and terrestrial vertebrate frugivores: a case study in a lowland tropical rain forest in Peninsular Malaysia. In: Forget, P.-M., Lambert, J.E., Hulme, P.E. and Vander Wall, S.B. (eds) *Seed Fate: Predation, Dispersal and Seedling Establishment.* CAB International, Wallingford, UK, pp. 151–174

28 Seed Dispersal in Anthropogenic Landscapes

S.J. Wright

Introduction

Humans alter the dispersal of seeds through land-use conversion, which changes the mix of habitats through which seeds disperse, and might also alter frugivore abundance and behaviours, and through direct persecution, which further alters the abundances and behaviours of many frugivores. The conversion of formerly continuous natural vegetation cover to human uses often leaves island-like fragments of relatively natural vegetation surrounded by agricultural fields, pastures and towns. Seed dispersal within and among the remaining fragments of natural vegetation will vary with the behaviour and local abundance of seed dispersal agents (Dennis *et al.*, 2005; Ghazoul, 2005). Seed dispersal within a fragment might not occur if the appropriate dispersal agents are absent from the fragment. Seed dispersal between fragments might not occur if the behaviours of the available seed dispersal agents preclude movements across the intervening fields, pastures and towns or, alternatively, might be commonplace if a seed dispersal agent ranges over a wide area and seeks out patches of similar natural vegetation cover.

Humans persecute frugivores for their meat – frugivores are often preferred game species (Corlett, 2007; Peres and Palacios, 2007) – and as agricultural pests. There is a widespread synergism between hunting pressure and habitat fragmentation, with the ease of access leading to greater hunting pressure in fragmented landscapes (Peres, 2001). Reductions in game species and agricultural pests and the widespread suppression of predators that might interfere with domestic animals will have additional indirect effects on the remaining frugivores (Fig. 28.1). As humans reduce both predators and competitors, the remaining non-game species might experience ecological release, their abundances might increase, and the number of seeds they disperse might also increase.

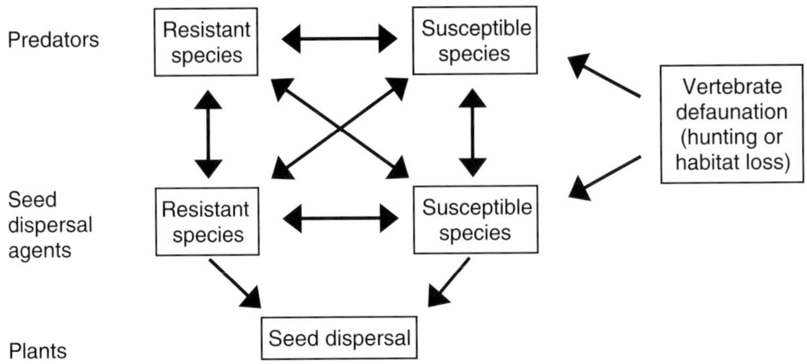

Fig. 28.1. A schematic model to represent interactions between anthropogenic defaunation caused by hunters or habitat loss, vertebrate species that disperse or eat seeds and are susceptible to anthropogenic disturbance or resistant to disturbance, and seed dispersal. The single-headed arrows emerging from anthropogenic defaunation represent direct effects on susceptible species. The double-headed arrows between susceptible and resistant species represent the web of natural interactions that may cause the direct effects of defaunation on susceptible species to cascade to include a much wider range of indirect effects on resistant species and, hence, on seed dispersal. Adapted from Wright (2003).

Both land-use conversion and the direct persecution of frugivores and their predators will alter the seed dispersal patterns of different species of plants, with potentially profound implications for plant population dynamics. Direct persecution of seed dispersal agents and their predators might increase, decrease or have no effect on seed dispersal for different plant species whose seeds are dispersed by animal species whose abundance increases, decreases or remains constant, respectively. Land-use conversion will cause similar interspecific differences. Plants with dispersal agents that are able to move seeds and pollen among habitat fragments will maintain genetic diversity and viable metapopulations (Hamrick, 2004; Bacles *et al.*, 2006), while other plant species whose seeds and pollen fail to disperse among the same habitat fragments will gradually lose genetic diversity and might also fail to re-establish populations after local degeneration (Maina and Howe, 2000; Ghazoul, 2005). Thus, anthropogenic change is likely to alter the relative rankings of dispersal abilities among plant species.

The objective of this review is to develop and evaluate three predictions concerning the types of plants that are likely to maintain seed dispersal and the types that are likely to face reduced seed dispersal in the natural vegetation that remains in anthropogenic landscapes. This is accomplished in three steps.

- The first step is to evaluate the susceptibility of different seed dispersal agents to habitat degradation, hunting, and other forms of direct persecution by humans.

- The second step is to consider case studies where anthropogenic effects on seed dispersal have been evaluated. The case studies are largely drawn from tropical forests because tropical forests have a rich community of plants, frugivores and seed dispersal modes; are among the last remaining biomes where a relatively intact biota of seed dispersal agents remains over large areas, which enables comparisons of seed dispersal in intact and human-altered communities; and are particularly threatened by anthropogenic change today.
- The third and final step is to consider the implications of altered seed dispersal for species of plants present in fragments of natural vegetation and/or after humans remove game and 'pest' species.

The Susceptibility of Primary Seed Dispersal Agents to Anthropogenic Disturbance

Predictions of the *relative* success of seed dispersal in anthropogenic landscapes will be motivated by the differential susceptibility of dispersal agents to anthropogenic disturbance. Primary seed dispersal agents will be partitioned into seven broad categories as follows: bats, small birds (< 300 g body mass), larger birds, non-volant mammals, explosive, water and wind. The four categories for vertebrate-dispersed seeds are necessarily broad because each species of plant with fleshy fruit tends to attract several to many potential seed dispersal agents, and each frugivore tends to consume several to many species with fleshy fruit (see Dennis and Westcott, Chapter 9, this volume; Nathan, Chapter 11, this volume). The four vertebrate-dispersed categories are non-overlapping in that each species of animal appears in a single category; however, a single species of plant might attract animals from several categories. The focus is on fleshy fruits because more than 70% of species from tropical forest have fleshy fruits, and most of the remaining species have dry fruit with seeds dispersed by mechanical means (Willson *et al.*, 1989; Muller-Landau and Hardesty, 2005). The relative impact of anthropogenic disturbance will be evaluated for dispersal agents from these seven categories.

Anthropogenic disturbances will be partitioned into two broad classes: direct persecution and land-use change. Direct persecution will refer to the hunting or trapping of game species for bush-meat or traditional medicines, the collection of live animals for the pet trade, and the eradication of pest species. Land-use change will refer to habitat fragmentation, habitat degradation (e.g. logging of forests), and reductions in total habitat area. There is often a synergism between these two classes of anthropogenic disturbance, with hunting and other forms of direct persecution of animals increasing in fragmented landscapes (Peres, 2001).

Land-use change has the potential to influence six of the seven categories of primary seed dispersal agents, with seed dispersal by explosive means being the exception. Land-use change can alter flooding regimes and hence seed dispersal by water (and fish). Land-use change can also alter the turbulent atmospheric conditions that favour the uplift of seeds and their long-distance dispersal by wind (Nathan *et al.*, 2002; Nathan and Katul, 2005). In particular, when dense forest vegetation is replaced by open fields or pastures, the change in wind fields might enhance seed dispersal by wind. For example, paternity analyses conducted in a chronically deforested Scottish landscape demonstrated that up to 53% of seedlings of the ash tree, *Fraxinus excelsior* (Oleaceae) were derived from seed that had been dispersed more than 3 km by wind (Bacles *et al.*, 2006). Thus, it has been suggested that species with wind-dispersed seeds and both wind- and animal-dispersed pollen are able to maintain high rates of gene flow and genetic diversity across heavily fragmented landscapes (Hamrick, 2004; Bacles *et al.*, 2006); however, these outcomes vary among species and with the severity of habitat fragmentation (Andrén, 1994; Ghazoul, 2005).

Both direct persecution and habitat change have the potential to alter seed dispersal by animals by altering the abundance and behaviours of the animals. Whether seed dispersal is enhanced or reduced will depend on the particular species of animal and the impact on its abundance and behaviour (Dennis *et al.*, 2005). Direct persecution and land-use change will tend to reduce seed dispersal for species of plants with large, animal-dispersed diaspores because there is a strong relationship between the size of an animal species and its susceptibility to most anthropogenic disturbances, and a weaker association between diaspore and frugivore size (see Forget *et al.*, Chapter 1, this volume). Large vertebrates are preferred game species because their large size makes them easy to locate and increases their value for food and their mystique for traditional medicines (Corlett, 2007; Peres and Palacios, 2007). Large vertebrates are also particularly susceptible to hunting as well as to habitat fragmentation because they tend to have long generation times, low reproductive capacities, and large territories or home ranges (Bodmer *et al.*, 1997). Among 4000 non-marine mammal species, the risk of global extinction increases rapidly above 3 kg body mass for these reasons and also because the impact of human population density, hunting, land-use change and other anthropogenic threats increase rapidly with body size (Cardillo *et al.*, 2005). Large vertebrates are among the first species to be exterminated by human activities.

The loss of large vertebrates will impact the dispersal of large seeds. More than 70% of the species of plants in most tropical and subtropical rain forests bear fleshy fruit whose seeds are dispersed by frugivorous birds and mammals (Willson *et al.*, 1989; Muller-Landau and Hardesty, 2005). There is a weak relationship between diaspore and frugivore size among these species. My colleagues and I have determined fresh diaspore mass and assigned diaspores to the seven primary dispersal mode categories described above for 270 species of trees and lianas from Barro Colorado Island, Panama (Muller-Landau and Hardesty, 2005; Wright *et al.*, 2007). Diaspore mass varies by six orders of magnitude among species (Fig. 28.2). Diaspores dispersed by wind tend to be lighter than those dispersed by animals (separate variance t-test = 3.91, $P < 0.001$), and diaspores dispersed *only* by

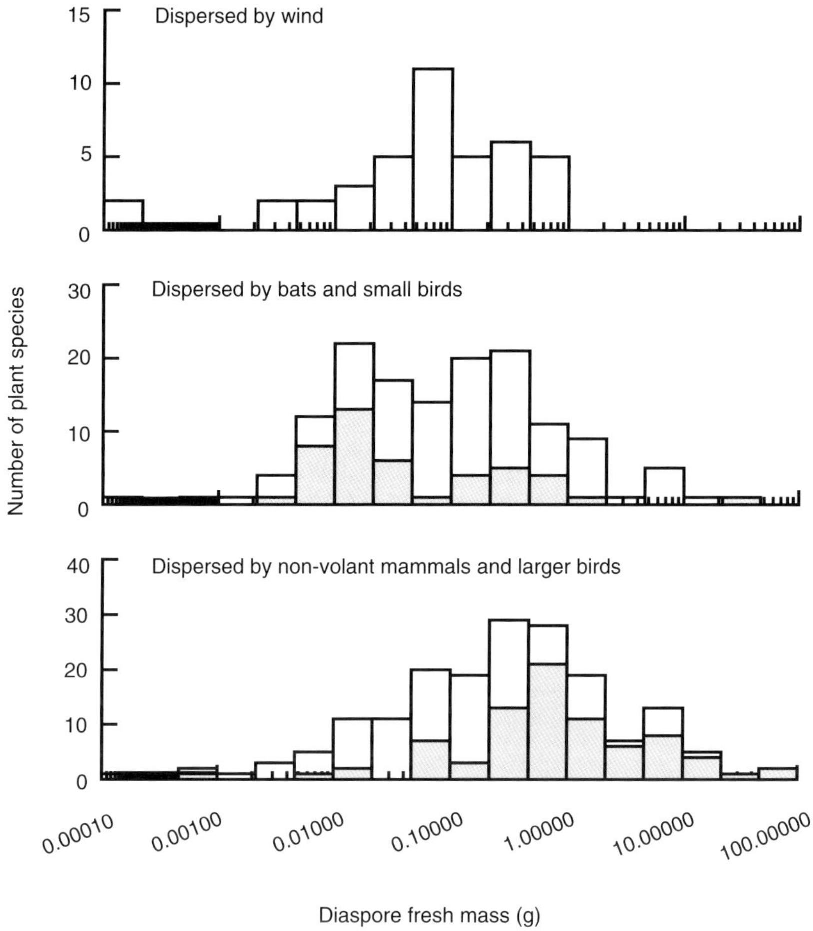

Fig. 28.2. Histograms of diaspore fresh mass for 270 species of plants from Barro Colorado Island, Panama whose primary seed dispersal is by wind (top panel), by bats and small birds with body mass less than 300 g (middle panel), and by non-volant mammals and larger birds (bottom panel). For the species with animal-dispersed seeds (the middle and bottom panels), the shaded portions of the histograms represent species whose primary seed dispersal is limited to the appropriate dispersal agents, while the open portions of the histograms represent 97 species dispersed by both bats/small birds and by non-volant mammals/larger birds. The *t*-tests reported in the text are limited to the shaded portions of the histograms.

bats and small birds (< 300 g) tend to be lighter than those dispersed *only* by non-volant mammals and larger birds (separate variance t test = 2.11, P = 0.037); however, extensive variation in diaspore mass remains within each of these broad categories, and 97 of the 270 species are dispersed by both small and large vertebrates (Fig. 28.2). The strong association between animal size and susceptibility to anthropogenic disturbance and the weaker association between diaspore and frugivore size suggest that anthropogenic disturbance will tend to reduce seed dispersal for species with large diaspores.

Direct persecution and land-use change will tend to have more limited effects on seed dispersal for plants with small-animal-dispersed diaspores because smaller animals tend to be less impacted by anthropogenic change. Smaller animals may even experience ecological release and increase in abundance if anthropogenic disturbances reduce their larger predators and competitors (Figs 28.1 and 28.3). There is clear evidence for ecological release and increased abundance among insular populations that lack competitors and predators found on nearby continents (Wright, 1980). There is similar evidence with respect to hunting, with non-game species increasing in abundance after hunters have removed their competitors and predators (Wright, 2003). In Amazonian forests, for example, medium-sized primates reach their greatest abundance at lightly hunted sites where hunters have reduced populations of the competitively dominant large primates; and the smallest primates reach their greatest abundance at heavily hunted sites where hunters have reduced populations of both medium-sized and large primates (Peres and Dolman, 2000; Peres and Palacios, 2007). Many birds and small mammals are also relatively tolerant of habitat fragmentation because they are able to cross intervening fields

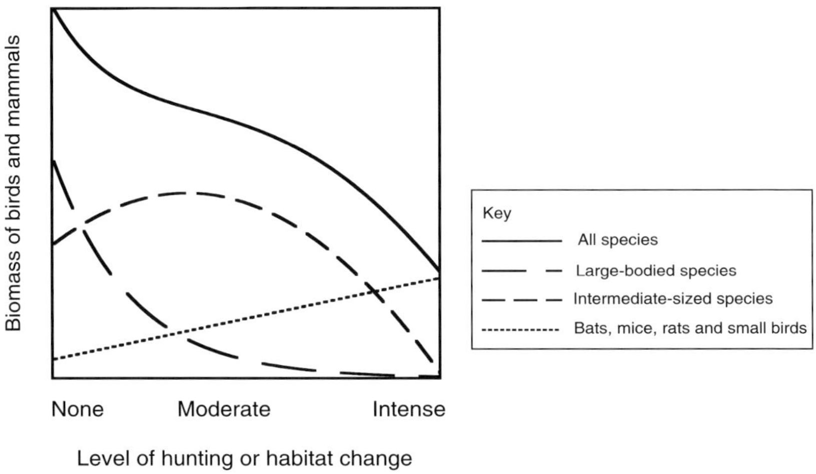

Fig. 28.3. Predicted shifts in the composition of the bird and mammal community with anthropogenic disturbance. Large-bodied species decline rapidly as anthropogenic disturbance increases (see Peres and Palacios, 2007, for an example). Intermediate-sized species might initially increase due to release from larger competitors and predators but then decline as they too are impacted by further increases in the level of anthropogenic disturbance (see Peres and Dolman, 2000, for an example). More speculatively, most small birds, bats and small rodents will escape direct persecution and many of these smaller species will also tolerate habitat fragmentation, so their collective importance might increase steadily with the level of anthropogenic disturbance. Compensation is also likely to be incomplete. so that the total biomass summed over all species declines steadily as large-bodied species become rare and smaller species become more abundant. Adapted from Wright (2003).

and pastures to move between fragments of natural vegetation (Daily *et al.*, 2001, 2003). Thus, to the extent that smaller birds and mammals disperse smaller diaspores (Fig. 28.2), seed dispersal should tend to be maintained, or even increase, for species of plants with smaller diaspores relative to those with larger diaspores.

These considerations motivate three general predictions concerning seed dispersal in anthropogenic landscapes. The predictions are:

1. Both land-use change and direct persecution will tend to reduce the abundances of large-bodied frugivores, and seed dispersal for species of plants with large-animal-dispersed diaspores will decline in anthropogenic landscapes.
2. Land-use change and direct persecution will have mixed, species-specific impacts on smaller-bodied frugivores, and seed dispersal for species of plants with small-animal-dispersed diaspores might increase, decrease or remain unchanged in anthropogenic landscapes, depending on those species-specific impacts.
3. Land-use change will often alter wind fields in ways that favour seed dispersal by wind. Examples include rising air columns over surfaces with reduced albedo and evapotranspiration, and planar air flow over relatively smooth land surfaces where fields and pastures replace tall forests (Nathan and Katul, 2005).

Although seed dispersal might decline absolutely with increasing anthropogenic disturbance for all seed dispersal modes, the decline is predicted to be most severe for species of plants with large-animal-dispersed diaspores, so that plants with wind- and small-animal-dispersed diaspores perform *relatively* better as anthropogenic disturbance increases. The following case studies evaluate these predictions and include examples of species-specific responses to anthropogenic disturbance.

Case Studies

1. Hunting and forest fragmentation in the Neotropics

Widespread forest fragmentation and a 'bush-meat crisis' caused by unsustainable levels of hunting threaten to exterminate many large mammals and birds from many neotropical forests (Peres and Palacios, 2007). The grave consequences of habitat fragmentation and the bush-meat crisis for seed dispersal have been demonstrated in Bolivia, Brazil, Costa Rica, Mexico, Panama and Peru.

Seed dispersal was documented for two coccosoid palms (Arecaceae) with large diaspores for completely protected, lightly hunted, and severely hunted sites separated in some cases by just a few hundred metres in central Panama (Wright *et al.*, 2000). Of the two palms, the diaspore of *Astrocaryum standleyanum* (mean fresh mass = 9.7 g) is highly preferred by many frugivores. Nonetheless, the percentage dispersed away from beneath the parent tree fell from > 95% at fully protected sites to < 5% at the most heavily hunted site. Similar percentages for the less preferred diaspores of *Attalea butyraceae* (Arecaceae; 14.6 g) were about 80% at fully protected sites and < 1% at the most heavily hunted site. Ecologically effective dispersal distances, defined as the distance between the final resting place of the diaspore and the nearest conspecific tree, were subsequently documented for *A. butyraceae* for sites characterized by three levels of hunting pressure and for small island fragments in Gatun Lake (Fig. 28.4). The absolute number of seeds was constant out to 30 m from the nearest reproductive tree at the fully protected site but fell by two orders of magnitude over the same distance at heavily hunted sites and on small-island fragments (Fig. 28.4b). The fruit of both palms, and particularly *A. standleyanum*, are sought

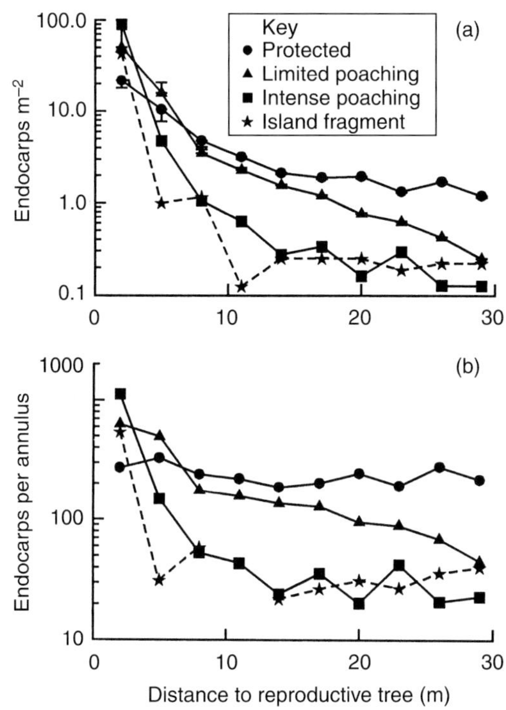

Fig. 28.4. Relationships between (a) the density and (b) the absolute number of endocarps and distance to the nearest reproductive tree for the palm *Attalea butyraceae* for protected and disturbed sites in central Panama. Each endocarp represents one diaspore. The horizontal axis represents distance to the nearest reproductive *A. butyraceae.* The decline in endocarp density with distance was slowest for fully protected sites, intermediate for lightly poached sites, and greatest for severely poached sites and for small island fragments. Absolute numbers of endocarps in 1-m-wide annuli were unaffected by distance at protected sites and declined with distance at disturbed sites (b). Means (± 1 SEM) are plotted in (a). Error bars are smaller than plotting symbols in most cases and are unidirectional in order to minimize clutter (reprinted from Wright and Duber (2001) with permission).

by a wide range of mammal species (Wright *et al.*, 2000). This first case study illustrates how anthropogenic disturbance can cause severe reductions in seed dispersal for species with large diaspores even where their fruits attract many potential seed dispersal agents.

The impact of anthropogenic disturbance for seed dispersal and seed survival has been studied for the genus *Astrocaryum* from three additional neotropical forests, which offers an opportunity to examine the generality of the patterns observed in Panama. In Peru, as in Panama, the number of dispersed seeds of *A. murumuru* fell more rapidly with distance to the nearest seed-bearing conspecific tree at disturbed sites (Wyatt and Silman, 2004). In Bolivia, as in Panama, seed survival and seedling density of *A. murumuru* were greater in an intensively hunted forest than in a lightly hunted forest (Roldán and Simonetti, 2001). In the Atlantic coastal forest of Brazil, seed dispersal of *A. aculeatissimum* was greater for protected sites and lower for defaunated forest fragments, as in Panama and Peru; however, seed survival and seedling density were lower in the defaunated forest fragments in contrast with Panama, Peru and Bolivia (Galetti *et al.*, 2006; Donatti *et al.*, 2007). Anthropogenic disturbance consistently reduced seed dispersal but had site-specific effects on post-dispersal seed survival for three relatively large-seeded species of *Astrocaryum*.

Generality can also be evaluated across species in central Panama and elsewhere. Hunting reduced overall (primary plus secondary) seed dispersal for a third palm species with a large diaspore (fresh mass = 2.71 g), but not for a fourth tree species with a much smaller diaspore (0.15 g) (Beckman and Muller-Landau, 2007). Wright *et al.* (2007) compared plant traits for species in the seedling bank and for the adult trees directly overhead for nine protected and 11 heavily hunted sites in central Panama. Presence in the seedling bank integrates seed dispersal and seed/early seedling survival. None the less, the susceptibility of seed dispersal agents to hunting influenced representation in the seedling bank. Species whose seeds are dispersed by large birds and non-volant mammals were consistently under-represented at sites where these frugivores are heavily hunted, and species whose seeds are dispersed by bats, small birds and mechanical means were consistently over-represented at the same heavily hunted sites. Hunting was also associated with lower numbers of saplings with seeds dispersed by large primates and higher numbers of saplings with seeds dispersed by wind and smaller animals in Peru (Nuñez-Iturri and Howe, 2007). Forest fragmentation tended to reduce seed dispersal for large-seeded species in Costa Rica and Brazil (Pizo, 1997; Guariguata *et al.*, 2000, 2002). Species with large-animal-dispersed diaspores appear later in secondary succession than do species with wind- and small-animal-dispersed diaspores during secondary succession in the Atlantic coastal forests of Brazil (Tabarelli and Peres, 2002). To summarize, hunting, land-use change, and habitat degradation are all associated with reduced seed dispersal among species with large-animal-dispersed diaspores and with *relative* increases among species with wind and small-animal-dispersed diaspores in the Neotropics.

The Panama seedling bank study included one discordant finding. Large-seeded species were consistently over-represented in the seedling bank at hunted sites (Wright *et al.*, 2007). This finding is consistent with the finding that species whose seeds are dispersed by large birds and non-volant mammals are under-represented at those same hunted sites because the relationship between diaspore and frugivore size is weak (Fig. 28.2) and because the survival of dispersed seeds comes into play. Presumably species with large seeds are over-represented because hunters reduce the abundance of predators of large seed and thereby increase the survival of large seeds. The same pattern of increased seedling abundance at hunted sites was observed earlier for *Astrocaryum* and *Attalea* where the nearly complete failure of seed dispersal was more than offset by reductions in seed predation by preferred game species at hunted sites (Wright *et al.*, 2000; Wright and Duber, 2001). The same pattern has also been observed in Mexican forests, where vertebrate defaunation reduces the dispersal of large seeds but increases their subsequent survival so that dense carpets of conspecific seedlings become established around trees with large seeds (Dirzo and Miranda, 1991; Dirzo *et al.*, 2007). The impact of anthropogenic disturbance on seed dispersal is likely to be difficult to detect for the seedling bank and later life stages because the same anthropogenic disturbances also affect seed predators and seedling browsers. However, whenever seed dispersal has been studied directly, there is clear and consistent evidence that anthropogenic disturbance reduces seed dispersal among species with diaspores dispersed by large animals and tends to favour species with wind- and small-animal-dispersed diaspores throughout the Neotropics.

2. Forest fragmentation in Australian rainforests

The Wet Tropics of northern Queensland offer a contrasting example where hunting is not now a factor and the effects of habitat change, specifically habitat fragmentation, can be studied in isolation. Dennis *et al.* (2005) use detailed knowledge of the sensitivity of frugivores and seed predators to habitat fragmentation and of interactions between fruits and frugivores, and seeds and seed predators, to generate three predictions concerning seed dispersal and seed survival in the fragmented landscapes of northern Queensland.

1. The dynamics of species whose seeds are both eaten and dispersed by widespread rodents will be little impacted in forest fragments. These species tend to have woody fruits and large, hard seeds.
2. Plant species whose seeds are dispersed by species of animals that are insensitive to fragmentation, and that are rarely eaten by rodents, will increase in abundance. These species tend to have soft fruits and small seeds.
3. Species whose seeds are dispersed by animals that are rare in fragments, and that are regularly eaten by rodents, will decline in abundance. These species tend to have soft fruits with seeds of variable size and hardness.

These predictions are similar to those generated here, but incorporate seed predation as well as seed dispersal.

White *et al.* (2004) provided an opportunity to evaluate the predictions of Dennis *et al.* (2005) by documenting seedlings present in three 8–11-year-old secondary forest patches that were initiated by purposeful planting. Seedlings of species that were not included among those purposefully planted at each site had to have arrived by long-distance dispersal from a large remnant forest patch that was contiguous with one of the planted forests and 0.6 and 2.0 km distant from the other two planted forests. Table 28.1 compiles immigrant seedlings by seed size and seed dispersal mode. Immigrant species with seeds dispersed by wind, bats and birds occurred in about equal numbers in the one contiguous and two isolated planted forests. In contrast, although sample sizes are small, species with seeds dispersed by non-volant vertebrates were six times more likely to be found in the contiguous planted forest (Table 28.1). Immigrant species with large and intermediate-sized seeds were also virtually absent from the two planted forests distant from the forested reserve (Table 28.1). As predicted here, habitat fragmentation has the least impact on seed dispersal for species with small seeds. Information on fruit traits and seed hardness for the species in White *et al.* (2004) would permit additional tests of the predictions of Dennis *et al.* (2005).

Moran *et al.* (2004) documented the sensitivity of different guilds of frugivorous birds to fragmentation in a subtropical landscape on Australia's central coast, and discuss the implications for seed dispersal there. They distinguish avian guilds defined by gape size and diet. Gape sizes were small, medium and large, enabling birds to swallow fruits of different sizes. Diet categories were: dominated by fruit, mixed including fruits and animals, crushing seeds, or minimal fruit/seeds. Among the potential seed dispersers (excluding seed crushers and those with minimal fruit/seeds in their diet), the guilds that were most sensitive to habitat fragmentation (least likely to be

Table 28.1. Numbers of immigrant species represented as seedlings in three secondary rainforest patches initiated by planting known species in Queensland, Australia. One planted forest was contiguous with a natural forest reserve, and two planted forests were 600 and 2000 m from the reserve. Plant species with dispersal agents from multiple dispersal modes are excluded. Data compiled from Appendix 1 of White *et al.* (2004).

Dispersal mode	Distance to seed source	Seed size		
		< 10 mm	10–20 mm	> 20 mm
Bats and birds	Contiguous	25	7	1
	≥ 600 m	22	1	1
Non-volant mammals and cassowary	Contiguous	1	0	5
	≥ 600 m	1	0	0
Wind	Contiguous	21	3	0
	≥ 600 m	24	0	0

found in small fragments) included species with large gapes and both fruit-dominated and mixed diets, and species with medium gapes and fruit-dominated diets. Species with medium gapes and mixed diets and small gapes and fruit-dominated diets were relatively tolerant of fragmentation; and species with small gapes and mixed diets actually increased in abundance in fragments. There was also one large-gaped species with a fruit-dominated diet that increased in abundance in fragments. Thus the sensitivity of birds and also non-volant mammals to habitat fragmentation should clearly favour dispersal of smaller seeds among forest fragments in tropical and subtropical Australia (Moran *et al.*, 2004; Dennis *et al.*, 2005).

3. Forest fragmentation in Africa

The impact of forest fragmentation on seed dispersal is also clearly related to seed size and seed dispersal mode in African forests. Cordeiro and Howe (2001) documented increased numbers of species of frugivores and increased representation of plant species with animal-dispersed seeds with the size of forest fragments in Tanzania. Two additional studies measured and compared seed dispersal for small forest fragments and nearby blocks of continuous forest, with contrasting outcomes. Cordeiro and Howe (2003) documented decreased visits by frugivores, seed dispersal and seedling density in small forest fragments relative to a nearby, continuous forest block in Tanzania. In contrast, Farwig *et al.* (2006) documented increased visits by frugivores and increased seed dispersal in fragments relative to a nearby, continuous forest block in Kenya. Seed size reconciles these contrasting observations. Cordeiro and Howe (2003) studied a species with a very large seed (90–133 mm long) while Farwig *et al.* (2006) studied a species with a relatively small seed (0.15 g). As in the Neotropics and the Wet Tropics and subtropics of Australia, anthropogenic disturbance favours seed dispersal for species of plants with small seeds.

Implications for Plant Conservation in Anthropogenic Landscapes

It is abundantly clear that humans alter seed dispersal in the natural vegetation that remains in anthropogenic landscapes throughout the tropics. The direct persecution of large frugivores and the sensitivity of large frugivores to habitat change reduce seed dispersal for species with large-animal-dispersed seeds in tropical forests in the Neotropics, Australia and Africa (Wright *et al.*, 2000; Cordeiro and Howe, 2001, 2003; Wright and Duber, 2001; White *et al.*, 2004; Beckman and Muller-Landau, 2007; Nuñez-Iturri and Howe, 2007). The relative lack of direct persecution of smaller species and their relative lack of sensitivity to habitat change creates the potential for increased seed dispersal for species with small-animal-dispersed seeds on the same three tropical continents (Moran *et al.*, 2004; White *et al.*, 2004; Farwig *et al.*, 2006; Wright *et al.*, 2007). The fragmentation of continuous, dense forest also has the potential to increase seed dispersal by wind (Nathan and Katul, 2005; Bacles *et al.*, 2006). The differential impacts of anthropogenic disturbance on seed dispersal is likely to change the species composition of the natural plant communities that remain in tropical landscapes. This change will occur even if seed dispersal is reduced for all species, but the reduction is greater for species whose seeds are dispersed by larger animals.

This potential for change in species composition and genetic structure mediated by change in seed dispersal might be offset during later stages of plant regeneration. Negative density dependence mediated by herbivores or pathogens might stabilize populations that become rare due to a lack of seed input, and delay or prevent their local decline and global extinction (Muller-Landau, 2007). Such negative density dependence might have contributed to the persistence of neotropical plant populations that lack primary seed dispersal agents today and that were presumably once dispersed by megafauna that became extinct more than 10,000 years ago (see Donatti *et al.*, Chapter 5, this volume). Isolated plants can also attract pollinators, and habitat fragments can attract frugivores, ameliorating the loss of genetic variability (Hamrick, 2004). Given these countervailing mechanisms, some policy-makers might require additional evidence of the impact of reduced seed dispersal in order to justify conservation action.

Anthropogenic disturbances have also been short-lived relative to the generation times of many plant species, and we generally lack studies that predate anthropogenic disturbance (Hamrick, 2004; Ghazoul, 2005). Biogeographical comparisons can, however, provide important insights. Böhning-Gaese (Chapter 25, this volume) compared two congeneric species from Madagascar and the South African mainland. Seeds of the Malagasy and South African species are dispersed by 1 versus 12 bird species with just 8% versus 71% of the seeds dispersed, respectively. The consequences for population and genetic structure were clear. Seedlings and trees are aggregated in space in Madagascar but randomly distributed in South Africa. Gene flow occurs over distances up to 3 km versus 30 km and genetic differentiation between local populations is high versus very limited in Madagascar and South Africa, respectively. This biogeographical evidence indicates that anthropogenic disturbance that changes seed dispersal patterns will eventually alter the structure and composition of natural plant populations that remain in anthropogenic landscapes, thus justifying pre-emptive conservation action now.

It is clearly evident that plants with large seeds and with seeds dispersed by large, non-volant mammals will be most severely impacted by anthropogenic disturbance. The relative disadvantage suffered by large-seeded species will be particularly severe during secondary succession and in fragmented landscapes where large seeds might fail to arrive (Table 28.1). In relatively continuous forests where direct persecution has reduced the abundances of most large frugivores and also large seed predators, species of plants with large seeds might increase in abundance but with greatly reduced genetic diversity (Wang *et al.*, 2007). The cumulative disadvantages suffered by large-seeded plants could lead to changes in plant species composition that will be difficult or impossible to reverse. A reduction in the abundances of species whose fruit attract large frugivores will make it difficult to re-establish populations of the frugivores and might even lead to an alternative steady state with forests dominated by species whose seeds are dispersed by wind and small birds and bats but that will not sustain large primates, hornbills, toucans and other charismatic tropical frugivores (Muller-Landau, 2007). Pre-emptive conservation action might focus on moving large seeds to regenerating forests and between forest fragments (Lamb *et al.*, 2005), and even the reintroduction of large frugivores and/or the plants needed to sustain them.

Acknowledgements

I thank Helene Muller-Landau for comments and Mauro Galetti, John Terborgh and Ronda Green for reviews that improved the manuscript.

References

Andrén, H. (1994) Effects of habitat fragmentation on birds and mammals in landscapes with different proportions of suitable habitat: a review. *Oikos* 71, 355–366.

Bacles, C.F.E., Lowe, A.J. and Ennos, R.A. (2006) Effective seed dispersal across a fragmented landscape. *Science* 311, 628–628.

Beckman, N. and Muller-Landau, H.C. (2007) Differential effects of hunting on pre-dispersal seed predation, primary dispersal and secondary seed removal of two tropical tree species. *Biotropica* 39, 328–339.

Bodmer R.E., Eisenberg, J.F. and Redford, K.H. (1997) Hunting and the likelihood of extinction of Amazonian mammals. *Conservation Biology* 11, 460–466.

Cardillo, M., Mace, G.M., Jones, K.E., Bielby, J., Bininda-Emonds, O.R.P., Sechrest, W., Orme, C.D.L. and Purvis, A. (2005) Multiple causes of high extinction risk in large mammal species. *Science* 309, 1239–1241.

Cordeiro, N.J. and Howe, H.F. (2001) Low recruitment of trees dispersed by animals in African forest fragments. *Conservation Biology* 15, 1733–1741.

Cordeiro, N.J. and Howe, H.F. (2003) Forest fragmentation severs mutualism between seed dispersers and an endemic African tree. *Proceedings of the National Academy of Science USA* 100, 14052–14056.

Corlett, R.T. (2007) The impact of hunting on the mammalian fauna of tropical Asian forests. *Biotropica* 39, 292–303.

Daily, G.C., Ehrlich, P.R. and Sánchez-Azofeifa, G.A. (2001) Countryside biogeography: use of human-dominated habitats by the avifauna of southern Costa Rica. *Ecological Applications* 11, 1–13.

Daily, G.C., Ceballos, G., Pacheco, J., Suzan, G. and Sánchez-Azofeifa, A. (2003) Countryside biogeography of neotropical mammals: conservation opportunities in agricultural landscapes of Costa Rica. *Conservation Biology* 17, 1814–1826.

Dennis, A.J., Lipsett-Moore, G., Harrington, G., Collins, E. and Westcott, D.A. (2005) Seed predation, seed dispersal and habitat fragmentation: does context make a difference in tropical Australia? In Forget, P.-M., Lambert, J.E., Hulme, P.E. and Vander Wall, S.B.

(eds) *Seed Fate: Predation, Dispersal and Seedling Establishment.* CAB International, Wallingford, UK, pp. 117–135.

Dirzo, R. and Miranda, A. (1991) Altered patterns of herbivory and diversity in the forest understory: a case study of the possible consequences of contemporary defaunation. In: Price, P.W., Lewinsohn, T.M., Fernandes, G.W. and Benson, W.W. (eds) *Plant–Animal Interactions: Evolutionary Ecology in Tropical and Temperate Regions.* Wiley and Sons, New York, pp. 273–287.

Dirzo, R., Mendoza, E. and Ortíz, P. (2007) Effects of differential mammal defaunation on seed predation patterns in a Mexican tropical rain forest. *Biotropica* 39, 355–362.

Donatti, C.I., Guimarães, P.R. Jr and Galetti, M. (2007) Agoutis and palm fruits: the pervasive effects of defaunation on *Astrocaryum aculeatissimum* in the Atlantic Forest, Brazil. *Biotropica* 39, in press.

Farwig, N., Böhning-Gaese, K. and Bleher, B. (2006) Enhanced seed dispersal of *Prunus africana* in fragmented and disturbed forests? *Oecologia* 147, 238–252.

Galetti, M., Donatti, C.I., Pires, A.S., Guimarães, P.R. Jr and Jordano, P. (2006) Seed survival and dispersal of an endemic Atlantic forest palm: the combined effects of defaunation and forest fragmentation. *Botanical Journal of the Linnean Society* 151, 141–149.

Ghazoul, J. (2005) Pollen and seed dispersal among dispersed plants. *Biological Reviews* 80, 413–433.

Guariguata, M.R., Rosales Adame, J.J. and Finegan, B. (2000) Seed removal and fate in two selectively logged lowland forests with contrasting protection levels. *Conservation Biology* 14, 1046–1054.

Guariguata, M.R., Arias-Le Claire, H. and Jones, G. (2002) Tree seed fate in a logged and fragmented forest landscape, northeastern Costa Rica. *Biotropica* 34, 405–415.

Hamrick, J.L. (2004) Response of forest trees to global environmental changes. *Forest Ecology and Management* 197, 323–335.

Lamb, D., Erskine, P.D. and Parrotta, J.A. (2005) Restoration of degraded tropical forest landscapes. *Science* 310, 1628–1632.

Maina, G.G. and Howe, H.F. (2000) Inherent rarity in community restoration. *Conservation Biology* 14, 1335–1340.

Moran, C., Catterall, C.P., Green, R.J. and Olsen, M.F. (2004) Functional variation among frugivorous birds: implications for rainforest seed dispersal in a fragmented subtropical landscape. *Oecologia* 141, 584–595.

Muller-Landau, H.C. (2007) Integrating theory and data to predict the long-term effects of hunting on plant species composition and diversity. *Biotropica* 39, 372–384.

Muller-Landau, H.C. and Hardesty, B.D. (2005) Seed dispersal of woody plants in tropical forests: concepts, examples, and future directions. In: Burslem, D., Pinard, M. and Hartley, S. (eds) *Biotic Interactions in the Tropics.* Cambridge University Press, Cambridge, UK, pp. 267–309.

Nathan, R. and Katul, G.G. (2005) Foliage shedding in deciduous forests lifts up long-distance seed dispersal by wind. *Proceedings of the National Academy of Sciences USA* 102, 8251–8256.

Nathan, R., Katul, G.G., Horn, H.S., Thomas, S.M., Oren, R., Avissar, R., Pacala, S.W. and Levin, S.A. (2002) Mechanisms of long-distance dispersal of seeds by wind. *Nature* 418, 409–413.

Nuñez-Iturri, G. and Howe, H.F. (2007) Bushmeat and the fate of trees with seeds dispersed by large primates in a lowland rainforest in western Amazonia. *Biotropica* 39, 348–354.

Peres, C.A. (2001) Synergistic effects of subsistence hunting and habitat fragmentation on Amazonian forest vertebrates. *Conservation Biology* 15, 1490–1505.

Peres, C.A. and Dolman, P.M. (2000) Density compensation in neotropical primate communities: evidence from 56 hunted and nonhunted Amazonian forests of varying productivity. *Oecologia* 122, 175–189.

Peres, C.A. and Palacios, E. (2007) Basin-wide effects of game harvest on vertebrate population densities in neotropical forests: implications for animal-mediated seed dispersal. *Biotropica* 39, 304–315.

Pizo, M.A. (1997) Seed dispersal and predation in two populations of *Cabralea canjerana* (Meliaceae) in the Atlantic Forest of southeastern Brazil. *Journal of Tropical Ecology* 13, 559–578.

Roldán, A.I. and Simonetti, J.A. (2001) Plant–mammal interactions in tropical Bolivian forests with different hunting pressures. *Conservation Biology* 15, 617–623.

Tabarelli, M. and Peres, C.A. (2002) Abiotic and vertebrate seed dispersal in the Brazilian Atlantic forest: implications for forest regeneration. *Biological Conservation* 106, 165–176.

Wang, B.C., Leong, M.T., Smith, T.B. and Sork, V.L. (2007) Ecological repercussions of extirpating mammals: reduced seed removal and dispersal of the Afrotropical tree, *Antrocaryon klaineanum* (Anacardiaceae). *Biotropica* 39, 340–347.

White, E., Tucker, N., Meyers, N. and Wilson, J. (2004) Seed dispersal to revegetated isolated rainforest patches in North Queensland. *Forest Ecology and Management* 192, 409–426.

Willson, M.F., Irvine, A.K. and Walsh, N.G. (1989) Vertebrate dispersal syndromes in some Australian and New Zealand plant communities, with geographic comparisons. *Biotropica* 21, 133–147.

Wright, S.J. (1980) Density compensation in island avifaunas. *Oecologia* 45, 385–389.

Wright, S.J. (2003) The myriad effects of hunting for vertebrates and plants in tropical forests. *Perspectives in Plant Ecology, Evolution and Systematics* 6, 73–86.

Wright, S.J. and Duber, H.C. (2001) Poachers and forest fragmentation alter seed dispersal, seed survival, and seedling recruitment in the palm *Attalea butyraceae*, with implications for tropical tree diversity. *Biotropica* 33, 583–595.

Wright, S.J., Zeballos, H., Domínguez, I., Gallardo, M.M., Moreno, M.C. and Ibáñez, R. (2000) Poachers alter mammal abundance, seed dispersal and seed predation in a neotropical forest. *Conservation Biology* 14, 227–239.

Wright, S.J., Hernandéz, A. and Condit, R. (2007) The bush meat harvest alters seedling banks by favoring lianas, large seeds and seeds dispersed by bats, birds and wind. *Biotropica* 39, 363–371.

Wyatt, J.L. and Silman, M.R. (2004) Distance-dependence in two Amazonian palms: effects of spatial and temporal variation in seed predator communities. *Oecologia* 140, 26–35.

29 Frugivory by Birds in Degraded Areas of Brazil

M.A. Pizo

> We can so far take a prophetic glance into futurity as to foretell that it will be the common and widely-spread species, belonging to the larger and dominant groups within each class, which ultimately prevail and procreate new and dominant species.
>
> Charles Darwin, *The Origin of Species* (6th edn)

Introduction

Although Darwin wrote the above words from an evolutionary perspective, in an ecological sense one can observe that in human-induced, altered habitats the common and dominant species frequently influence the composition and dynamics of communities. This is especially true for seed dispersal, a very important step for the recruitment and spatial distribution of plant populations (Cordeiro and Howe, 2003). In areas disturbed by human activities, small, habitat-generalist, frugivorous birds shape the local flora by promoting the dispersal of a variety of small-seeded plant species, including exotic, invasive species (Corlett, 1998; Hamann and Curio, 1999; Renne *et al.*, 2002). Coupled with the general scarcity of large-bodied frugivores capable of dispersing large-seeded plants (Silva and Tabarelli, 2000), the net result is that, over the long run, degraded areas may be dominated by small-seeded plants, as Tabarelli and Peres (2002) reported for the Brazilian Atlantic Forest. Assuming that human impacts on natural habitats will intensify as currently projected, and given the overall abundance of ornithochorous plants in tropical habitats (Fleming, 1991), frugivorous birds thriving in disturbed lands will help to, in the words of Darwin, 'procreate' dominant plant species over the coming decades and centuries.

It seems, therefore, of great interest to know which are the species of birds that eat fruits and disperse seeds in degraded areas. In this chapter, in a manner similar to Corlett's (1998, 2002) treatment of degraded, tropical, east-Asian landscapes, I ask which are these birds in degraded areas of two Brazilian global biodiversity 'hotspots' which have suffered intense and ongoing degradation: the Atlantic Forest and the Cerrado. Of the highly diverse Brazilian Atlantic Forest that once ran from 8° S to 28° S, covering more than 1,000,000 km^2, only 7% remains, and most of what remains is highly degraded (Dean, 1995; Viana *et al.*, 1997). The Cerrado is the second largest Brazilian biome after the Amazon and is composed of a variety of vegetation physiognomies that range from open grasslands to forested landscapes (Oliveira and Marquis, 2002). Agricultural development, mainly during the last four decades, has reduced its area to approximately 20% (Nepstad *et al.*, 1997).

I surveyed the literature looking for studies carried out in an array of disturbed habitats ranging from urban areas to forest fragments that quantitatively assessed two aspects of avian frugivory: number of visits and fruits removed from fruiting plants. As well as the taxonomic composition of this particular group of birds, I also report some of the morphological (body mass, gape width) and behavioural (migratory status) traits that may influence their role as seed dispersers. Comparisons with well-preserved areas permitted a highlighting of the idiosyncrasies of avian frugivory and seed dispersal in Brazilian degraded lands that are, in many aspects, likely to be similar to other disturbed areas around the world.

Methods

I searched the literature (published papers and unpublished theses) looking for studies that provided quantitative measures of birds using fruits from groups of species or individual species of plant. Such measures were of two kinds: the number of feeding visits to fruiting plants performed by each species of bird and/or the number of fruits taken by birds, two important components of the effectiveness of an animal as seed disperser (Schupp, 1993). Although the second parameter is a more direct and reliable measure of the role of a given species as seed disperser, the number of visits is usually positively correlated with the number of seeds dispersed (Schupp, 1993). In fact, for the ten localities in which both measurements were taken, bird visits and number of fruits taken were highly correlated (Spearman rank correlations: mean $r_s \pm$ SD = 0.84 ± 0.09, all $P < 0.001$). Because I was especially interested in the seed dispersal potential of birds, I discarded records of tinamous (Tinamidae), psittacids (Psittacidae), and columbids (Columbidae), which most often act as seed predators rather than seed dispersers. Taken together, these birds were responsible for only 1.4% and 1.2% of bird visits and fruits removed, respectively. I did not exclude from the analysis other granivorous birds (e.g. grassquit *Volatinia jacarina*, rufous-collared sparrow *Zonotrichia capensis*, both Emberizidae) because their role as seed dispersers or seed predators is still uncertain (see below).

I used a broad classification of Atlantic Forest that encompasses Atlantic rainforest and the drier Atlantic semideciduous forest (*sensu* Morellato and Haddad, 2000). For the Cerrado biome, all the studies were made in the forested physiognomy known as *cerradão* (Oliveira-Filho and Ratter, 2002). The definition of degraded areas used here encompass: (i) forest fragments <400 ha; (ii) exotic plantations of *Eucalyptus* (Myrtaceae) and *Pinus* (Pinaceae) usually intermixed with secondary, native vegetation; (iii) rural; and (iv) urban areas. Bird taxonomy follows Sick (1997) and plant taxonomic arrangement follows APG II (2003).

Bird weight and gape width data were obtained from the same studies surveyed for feeding records, especially Argel de Oliveira (1999), supplemented with data from Motta (1990), Magalhães (1999), and from specimens in the collection of the Museu de Zoologia da Universidade de São Paulo (MZUSP). The migratory status of birds was assessed from the literature (e.g. Sick, 1997; del Hoyo *et al.*, 2004) and the experience of the author.

Data analysis

Locality was the unit of analysis. Therefore data from different studies and different species of plants carried out at the same locality were pooled for analysis. Because separating elaenias (*Elaenia* spp.; Tyrannidae) from each other in the field is a difficult task that may lead to errors, I pooled the *Elaenia* species for analysis. As far as is known, all elaenias are similar in their propensity to frugivory (Marini and Cavalcanti, 1998), being represented by two to three species in the studies surveyed.

From each locality I ranked each species of bird in terms of visits to fruiting plants and/or fruits removed and arbitrarily took the top five species for analysis. As I show below, this subset of species was responsible for most visits and fruits taken. Because there were so few Cerrado studies and all of them were made in the forested physiognomy that is frequently contiguous and intermixed with the Atlantic semideciduous forest, I made no attempt to analyse both biomes separately.

Results and Discussion

I found 35 studies carried out in 27 localities in south and south-eastern Brazil (Table 29.1). Most of them were conducted in Atlantic Forest (27 studies, 23 localities), while eight (four localities) were in Cerrado. Localities ranged from 18° to 29° S and 39° to 52° W. Forest fragments were the habitat most studied (16 studies, 13 localities), followed by exotic plantations (eight studies, four localities), rural (eight studies, eight localities), and urban areas (three studies, two localities). Together, these studies involved 1732 h of watching 138 species (46 families; fruit size range: 0.4–60 mm diameter) of ornithochorous plants (including six exotic species). The number of bird species recorded per locality ranged from 5 to 51 for a total of 131 species

Table 29.1. Studies on the consumption of fruits by birds in different kinds of degraded areas in south-south-eastern Brazil. Species of plant marked with an asterisk are exotic.

Study[a]	Biome[b]	Kind of perturbation	Plant species (family) studied
Argel de Oliveira (1999)	AF	Forest fragment	21 spp. (13 families)
Argel de Oliveira and Figueiredo (1996)	AF	Rural area	*Ficus clusiifolia* (Moraceae)
Cazetta *et al.* (2002)[a]	AF	Forest fragment	*Magnolia ovata* (Magnoliaceae)
da Rosa and Marcondes-Machado (2005)	AF	Rural area	*Citharexylum myrianthum* (Verbenaceae)
Fadini and de Marco (2004)	AF	Forest fragment	25 spp. (16 families)
Figueiredo (1996)[b]	AF	Forest fragment	*Ficus* spp. (Moraceae)
Figueiredo *et al.* (1995)[c]	AF	Urban area	*Ficus microcarpa* (Moraceae)
Galetti and Pizo (1996)[b]	AF	Forest fragment	7 spp. (6 families)
Gondim (2001)[a]	AF	Forest fragment	*Trichilia* spp. (Meliaceae)
Gridi-Papp *et al.* (2004)	AF	*Pinus* and secondary forest	*Miconia cinerascens* (Melastomataceae)
			Leandra aff. *sublanata* (Melastomataceae)
Guimarães (2003)	AF	Rural area	*Tapirira guianensis* (Anacardiaceae)
Hasui (1994)	AF	Forest fragment	35 spp. (19 families)
Kindel (1996)	AF	Forest fragment	18 spp. (12 families)
Lombardi and Motta (1993)[c]	AF	Urban area	*Michelia champaca* * (Magnoliaceae)
Lopes (2000)	AF	Forest fragment	4 spp. (4 families)
Manhães *et al.* (2003)	AF	Forest fragment	*Miconia urophylla* (Melastomataceae)
Masteguin and Figueiredo (1995)	AF	Rural area	*Prunus sellowii* (Rosaceae)
Oniki *et al.* (1994)	AF	*Eucalyptus* and forest fragments	*Cecropia* sp. (Urticaceae)
Pizo (2004)	AF	Rural area	16 spp. (13 families)
Scheibler and Melo (2003)	AF	Urban area	*Ligustrum japonicum* * (Oleaceae)
		Rural area	*Ligustrum lucidum* * (Oleaceae)
Silva (1988)	AF	Rural area	*Cereus peruvianus* (Cactaceae)
Silva (2005)	AF	Rural area	*Melia azedarach* * (Meliaceae)
Valente (2001)	AF	*Eucalyptus* and forest fragments	*Alchornea glandulosa* (Euphorbiaceae)
Zimmermann (1996)[d]	AF	Forest fragment	*Alchornea glandulosa* (Euphorbiaceae)
Zimmermann (2001)	AF	Forest fragment	*Trema micrantha* (Cannabaceae)
Zimmermann et al. (2002)[d]	AF	Forest fragment	*Myrsine coriacea* (Myrsinaceae)
Cazetta *et al.* (2002)[e]	CE	*Eucalyptus* and forest fragments	*Magnolia ovata* (Magnoliaceae)
Francisco and Galetti (2001)[e]	CE	*Eucalyptus* and forest fragments	*Myrsine lancifolia* (Myrsinaceae)
Francisco and Galetti (2002a)[e]	CE	*Eucalyptus* and forest fragments	*Ocotea pulchella* (Lauraceae)
Francisco and Galetti (2002b)[e]	CE	*Eucalyptus* and forest fragments	*Davilla rugosa* (Dilleniaceae)
Gondim (2002)	CE	Forest fragment	40 spp. (27 families)
Marcondes-Machado (2002)	CE	Forest fragment	*Miconia rubiginosa* (Melastomataceae)
Melo *et al.* (2003)	CE	Forest fragment	*Faramea cyanea* (Rubiaceae)
Motta and Lombardi (1990)[e]	CE	*Eucalyptus* and forest fragments	*Copaifera langsdorffii* (Fabaceae)

[a] Studies sharing the same superscript letter were carried out at the same locality. [b] Biomes: AF = Atlantic Forest, CE = Cerrado.

(17 families, 112 passeriformes and 19 non-passeriformes; plus 15 species of seed predators as defined above) in 12,903 feeding visits to fruiting plants. All studies except one (Argel de Oliveira, 1999) reported the number of visits by each species of bird. Seventeen studies (11 localities) also reported the number of fruits taken by birds, which totalled 20,948 fruits.

The top five species of birds were on average responsible for 74.0% (range 46.3–100%) of visits and 70.9% (59.5–93.3%) of fruits removed at each locality. Although comparative data taken from well-preserved sites in the Atlantic Forest and the Cerrado are scarce, the available data indicate a small contribution of the five main seed dispersers to the number of visits made to fruiting plants in pristine habitats: Atlantic Forest – 50.6% of visits to *Cabralea canjerana* (Meliaceae) (Pizo, 1997); Cerrado – 46.7% of visits to 19 species of plants (14 families) (Motta, 1990). If this is true then compared with well-preserved habitats, plant species in degraded areas have a greater proportion of their reproductive investment dependent on a small group of bird species (Pizo, 1997).

Fifteen species of birds (plus an unknown number of *Elaenia* species) were ranked among the top five species in terms of feeding visits to fruiting plants and/or fruits removed in at least two localities (Table 29.2). This group of birds is formed solely by Passeriformes, either Oscines (four families, mainly Muscicapidae and Emberizidae) or Suboscines (only one family, Tyrannidae). As expected, all of them but the yellow-legged thrush *Platycichla flavipes* (Muscicapidae, medium sensitivity) have low sensitivity to disturbance according to Stotz *et al.* (1996). Overall, they are widespread species, with many of them ranging from Central America to Argentina (Ridgely and Tudor, 1989, 1994). Based on the percentages of fruits eaten, shown in Table 29.2, some of the birds (e.g. sayaca tanager *Thraupis sayaca*; Emberizidae) are clearly among the selected group of 'super-generalist' species: species that interact with a great number of fruits and form the backbone of the plant–frugivore mutualistic networks in their communities (Jordano *et al.*, 2003).

All birds listed in Table 29.2 are small- to medium-sized birds weighing < 71 g with gape widths < 17 mm (87% with gape widths < 15 mm). This characteristic is similar to that reported by Corlett (2002) for frugivorous birds in degraded areas of east Asia, where he found maximum gape widths of 13–15 mm. The distribution of fruit diameters among ornithochorous fruits of the Atlantic Forest and Cerrado revealed that 20% of species have fruits greater than 15 mm diameter (Fig. 29.1). Although some of these species may be eaten piecemeal by birds and occasionally have secondary terrestrial seed dispersers, they are prone to suffer a lack of primary seed dispersal in degraded areas (Silva and Tabarelli, 2000).

Five of the 15 species of birds, plus some of the *Elaenia* species, are known to be migratory or vagrant (i.e. species with irregular movements). The overall importance of migratory birds as seed dispersers in disturbed areas in southeast Brazil has already been noted (Pizo, 1997). Populations of migratory birds occurring at a given site are especially prone to interannual variation in abundance, which reflects in the number of seeds they disperse

Table 29.2. Species of birds recorded eating fruits in degraded areas of south–south-eastern Brazil. Birds were ranked in decreasing order according to the number of visits to fruiting plants and number of fruits removed. Only birds that appeared among the top five species in two or more localities are shown. A dash indicates that the species did not meet this criterion for a given parameter. Information on gape width and diet are also presented. Within families, birds are arranged in alphabetical order. Species marked with an asterisk are migratory or vagrant species (i.e. species with irregular movements).

				Visits (n = 26 localities)		Number of fruits removed (n = 11 localities)	
Family/Species	Weight (g)	Gape width (mm)	% of species eaten (n =138)	Median rank[a]	Percentage of localities[b]	Median rank[a]	Percentage of localities[b]
Tyrannidae							
Elaenia spp.	16–27.5	9.1–11.2	29.9	10 (20)	23.1	14.5 (10)	27.3
Megarhynchus pitangua	57.3	16.8	8.6	12 (10)	7.7	21 (6)	18.2
Myiodynastes maculatus *	48.0	15.7	15.5	11.5 (14)	7.7	–	–
Pitangus sulphuratus	64.0	14.4	28.2	5 (21)	34.6	7 (8)	–
Tyrannus melancholicus *	41.0	13.2	20.1	7 (19)	23.1	–	–
Muscicapidae							
Platycichla flavipes *	59.7	13.8	2.9	–	–	5 (4)	27.3
Turdus albicollis	67.8	11.0	8.6	5 (7)	11.5	6 (5)	18.2
Turdus amaurochalinus	58.7	9.8	16.1	9 (19)	15.4	8 (10)	18.2
Turdus leucomelas	70.3	10.6	34.5	2 (17)	42.3	2 (7)	36.4
Turdus rufiventris	71.0	13.1	24.1	6 (18)	23.1	4 (8)	45.5
Vireonidae							
Vireo olivaceus *	14.1	7.6	16.7	7 (13)	11.5	–	–
Emberizidae							
Dacnis cayana *	13.3	4.7	29.3	5 (17)	30.8	4 (7)	36.4
Tachyphonus coronatus	28.5	7.4	14.9	10 (9)	7.7	–	–
Tangara cayana	20.0	7.8	31.6	5 (15)	34.6	6.5 (6)	18.2
Thraupis sayaca	36.0	10.4	45.4	3 (21)	61.5	6 (9)	36.4
Zonotrichia capensis	22.1	7.5	6.9	9 (7)	7.7	–	–

[a] Between parentheses are the number of localities where a given species was recorded visiting fruiting plants or removing fruits.
[b] Percentage of localities refers only to the localities where the bird appeared among the top five species in number of visits or fruits consumed.

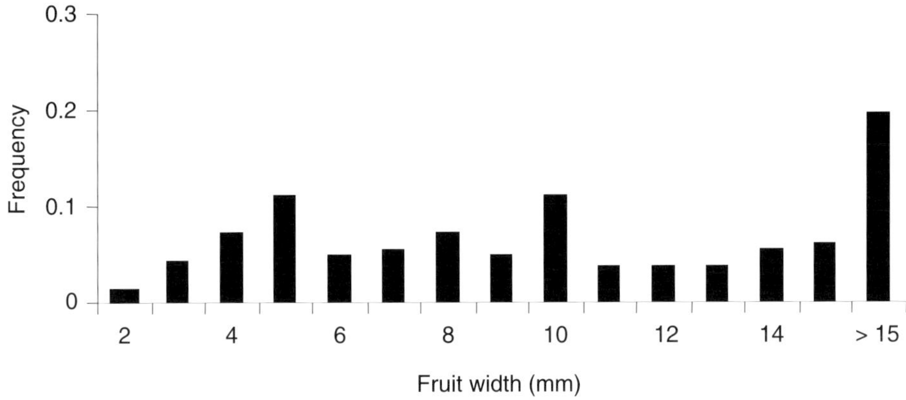

Fig. 29.1. Frequency distribution of fruit diameters for 163 bird-dispersed species (50 families) from the Atlantic Forest and Cerrado (M. Galetti, M.A. Pizo and L.P.C. Morellato, 2006, unpublished results).

(Malmborg and Willson, 1988; Hagan and Johnston, 1992). Therefore, special attention should be paid to these birds in south–south-eastern Brazil because of the possible impact that annual variations in their populations can have on the seed dispersal spectra of plants.

Inspecting the bird species listed in Table 29.2, one notices the absence of toucans (Ramphastidae), manakins (Pipridae) and cotingas (Cotingidae), which are prominent frugivorous birds in undisturbed Atlantic Forest (Silva *et al.*, 2002). Of these, four species of toucans (*Baillonius bailloni*, *Pteroglossus araracari*, *Ramphastos dicolorus*, *R. toco*) were recorded eating fruits in seven of the 27 localities, but their contribution to number of visits or fruits consumed never exceeded 4%, manakins (five species) appeared in eight localities where three species (*Chiroxiphia caudata*, *Manacus manacus* and *Antilophia galeata*) were among the top five species in number of visits, each in a different locality, whereas only one cotinga (*Procnias nudicollis*) was recorded in one locality (Argel de Oliveira, 1999). These figures contrast with assemblages of frugivorous birds recorded in more pristine habitats where they eat fruits from a variety of plant species and are usually important seed dispersers (Galetti *et al.*, 2000; Pizo *et al.*, 2002). Based on 397 bird–fruit interactions recorded in two well-preserved Atlantic Forest sites in south-east Brazil, cotingas and manakins were among the top four most important frugivorous birds (Silva *et al.*, 2002). However, these birds are severely affected by habitat alteration and often disappear from disturbed areas (Willis, 1979), which may result in truncated seed shadows in fragmented habitats (Loiselle and Blake, 2002).

Thrushes (*Turdus* spp., Muscicapidae) and *T. sayaca* should be highlighted for their importance in both number of visits and fruits removed from plants (Table 29.2). They use a variety of habitats in disturbed lands including forest fragments, small forest thickets, isolated trees in pastures, and hedgerows (Pizo, 2004). *Thraupis sayaca*, in particular, may fly for long distances between forest patches (Isler and Isler, 1987), probably making a great contribution to the movement of seeds throughout disturbed landscapes. Especially in Cerrado areas, *Elaenia* spp. are of great importance. These are highly frugivorous birds (Marini and Cavalcanti, 1998) whose relative contribution to seed dispersal probably increases as one moves from forested to more open physiognomies of Cerrado. The other tyrant flycatchers (Tyrannidae) listed in Table 29.2 are of importance in Atlantic Forest and Cerrado because of their abundance, tolerance to disturbance, and broad diets (del Hoyo *et al.*, 2004).

Guans (*Penelope superciliaris* and *P. obscura*; Cracidae) were the largest birds recorded. They were recorded in four of the localities surveyed, contributing with 0.2–4.5% of visits, and figuring among the top five species in number of fruits consumed in a Cerrado area (Gondim, 2002). These species are large enough (weight 950–1200 g; gape width 19–23 mm) to swallow large fruits (> 20 mm diameter) with large seeds (Mikich, 2002). *Penelope superciliaris* is one of the few large avian frugivores able to survive in disturbed Atlantic forests, where it eats the fruits and disperses the seeds of a variety of plant species (Willis, 1979; Mikich, 2002). Moreover, it is a species amenable to captive breeding and reintroduction programmes (Pereira and Wajntal, 1999). Thus, it should be carefully considered in conservation efforts to overcome the usual absence of large avian seed dispersers in disturbed forests (Silva and Tabarelli, 2000).

Although not particularly important in terms of fruits removed from plants, some birds are of special interest. For instance, the euphonias (*Euphonia* spp.; Emberezidae) are the principal vectors of mistletoe seeds of the family Viscaceae (Restrepo *et al.*, 2002), and the patterns of prevalence and spatial distribution of these parasitic plants in disturbed areas are highly influenced by *Euphonia chlorotica* and *E. violaceae*, which were recorded eating fruits in nine localities, being among the top five visitors in one locality each. *Zonothichia capensis* (Emberizidae) and other predominantly granivorous birds (e.g. *Volatinia jacarina*; Emberizidae) are very common in disturbed areas where they frequent habitats not readily used by true frugivorous species (e.g. pastures), and may be important seed vectors to such habitats which otherwise receive few seeds from fleshy fruits (Nepstad *et al.*, 1997; Wijdeven and Kuzee, 2000). What deserves further investigation is the quality of treatment these granivorous birds provide to the seeds they ingest: the proportion of seeds of various species of plants passed intact through the gut and whether they alter the chances of germination success. Similarly, although traditionally regarded as seed predators and not considered in this survey, although they are common in disturbed areas, the role of *Columba* pigeons and *Leptotila* doves (Columbidae) as seed dispersers of small seeds should be carefully investigated, ideally with captive birds (Wheelwright *et al.*, 1984).

Although the survey presented here is geographically restricted, the general pattern (i.e. small birds of the families Tyrannidae, Muscicapidae and Emberizidae as important seed dispersers in disturbed areas) is likely to prevail in other parts of Brazil and elsewhere in the Neotropics. In Amazonia, Silva *et al.* (1996) observed that three Thraupinae (Emberizidae) species, which included two genera highlighted here (*Tachyphonus* and *Thraupis*), were responsible for 70% of movements of frugivorous birds from second-growth forest to abandoned pastures. In a Chaco landscape in Argentina that encompassed logged forest and agricultural areas, *Elaenia parvirostris* (Tyrannidae) and *Turdus amaurochalinus* (Muscicapidae) figured as the most common frugivorous birds (Codesido and Bilenca, 2004). Birds in the family Muscicapidae, particularly thrushes (Turdinae), are indeed prominent seed dispersers in degraded areas of the Old World (Guitián *et al.*, 2000; Corlett, 2002).

In summary, although a great number of frugivorous birds eat fruits in south-southeast Brazil, a few species remove a large proportion of seeds (Pizo, 1997). As observed in similarly degraded areas around the world (Corlett 1998, 2002), small-seeded plant species are favoured due to the overall dominance of small, habitat-generalist birds. Large-seeded plants are, therefore, prone to be limited by low disperser visitation, which ultimately may lead to the limitation of recruitment (Schupp *et al.*, 2002).

A great part of the natural vegetation that 500 years ago covered the Brazilian territory south of the Amazon has been lost or severely degraded by human activities, mainly in the most recent decades (Dean, 1995; Nepstad *et al.*, 1997). Frugivorous birds living in such degraded areas, apart from influencing the composition and structure of the vegetation, also contribute to vegetation restoration (Silva *et al.*, 1996), and the spreading of exotic, invasive plants (Merler *et al.*, 2001). Therefore, it appears to be of great interest to know not only which sorts of birds these are and their morphological characteristics, as reported here, but also to investigate aspects of their ecology important for the dispersal of seeds (e.g. diet, movement rates, habitat use, seed passage time). The knowledge of these aspects will help us to understand the movement of seeds and the dynamics of the vegetation in human-modified areas which are, in much of Brazil and elsewhere, an important repository of our present and future biodiversity.

Acknowledgements

I am grateful to Rodolfo Figueiredo, Carlos Zimmermann and Marilise Krügel, who shared information, Bianca Reinert for measuring some museum skins, and Mauro Galleti, who kindly gave me access to his database of fruit characteristics.

References

APG II (2003) An update of the Angiosperm Phylogeny Group classification for the orders and families of flowering plants: APG II. *Botanical Journal of the Linnean Society* 141, 399–436.

Argel de Oliveira, M.M. (1999) Frugivoria por aves em um fragmento de floresta de restinga no Estado do Espírito Santo, Brasil. PhD thesis, Universidade Estadual de Campinas, Brazil.

Argel de Oliveira, M.M. and Figueiredo, R.A. (1996) Aves que visitam uma figueira isolada em ambiente aberto, Espírito Santo, Brasil. *Iheringia, Série Zoologia* 80, 127–134.

Cazetta, E., Rubim, P., Lunardi, V.O., Francisco, M.R. and Galetti, M. (2002) Frugivoria e dispersão de sementes de *Talauma ovata* (Magnoliaceae) no sudeste brasileiro. *Ararajuba* 10, 199–206.

Codesido, M. and Bilenca, D. (2004) Variación estacional de um ensamble de aves em um bosque subtropical semiárido de Chaco Argentino. *Biotropica* 36, 544–554.

Cordeiro, N.J. and Howe, H.F. (2003) Forest fragmentation severs mutualism between seed dispersers and an endemic African tree. *Proceedings of the National Academy of Sciences* 100, 14052–14056.

Corlett, R.T. (1998) Frugivory and seed dispersal by birds in Hong Kong shrubland. *Forktail* 13, 23–27.

Corlett, R.T. (2002) Frugivory and seed dispersal in degraded tropical east Asian landscapes. In: Levey, D.J., Silva, W.R. and Galetti, M. (eds) *Seed Dispersal and Frugivory: Ecology, Evolution and Conservation.* CAB International, Wallingford, UK, pp. 451–465.

da Rosa, G.A.B. and Marcondes-Machado, L.O. (2005) Frugivoria por aves em *Cytharexylum myrianthum* Cham (Verbenaceae) em áreas de pastagens de Campinas, SP. *Ararajuba* 13, 113–115.

Dean, W. (1995) *With Broadax and Firebrand: The Destruction of the Brazilian Atlantic Forest.* University of California Press, Berkeley, California.

del Hoyo J., Elliott, A. and Christie, D.A. (eds) (2004) *Handbook of the Birds of the World, Vol. 9: Cotingas to Pipits and Wagtails.* Lynx Edicions, Barcelona, Spain.

Fadini, R.F. and de Marco, P. Jr (2004) Interações entre aves frugívoras e plantas em um fragmento de Mata Atlântica de Minas Gerais. *Ararajuba* 12, 97–103.

Figueiredo, R.A. (1996) Vertebrates at neotropical fig species in a forest fragment. *Tropical Ecology* 37, 139–141.

Figueiredo, R.A., Motta, J.C. Jr and Vasconcellos, L.A.S. (1995) Pollination, seed dispersal, seed germination and establishment of seedlings of *Ficus microcarpa*, Moraceae, in southeastern Brazil. *Revista Brasileira de Biologia* 55, 233–239.

Fleming, T.H. (1991) Fruiting plant–frugivore mutualism: the evolutionary theater and the ecological play. In: Price, P.W., Lewinsohn, T.M., Fernandes, G.W. and Benson W.W. (eds) *Plant–Animal Interactions: Evolutionary Ecology in Tropical and Temperate Regions.* John Wiley and Sons, New York, pp. 119–144.

Francisco, M.R. and Galetti, M. (2001) Frugivoria e dispersão de sementes de *Rapanea lancifolia* (Myrsinaceae) por aves numa área de cerrado do Estado de São Paulo, sudeste do Brasil. *Ararajuba* 9, 13–19.

Francisco, M.R. and Galetti, M. (2002a) Aves como potenciais dispersoras de sementes de *Ocotea pulchella* Mart. (Lauraceae) numa área de vegetação de cerrado do sudeste brasileiro. *Revista Brasileira de Botânica* 25, 11–17.

Francisco, M.R. and Galetti, M. (2002b) Consumo de frutos de *Davilla rugosa* (Dilleniaceae) por aves numa área de cerrado em São Carlos, Estado de São Paulo. *Ararajuba* 10, 193–198.

Galetti, M. and Pizo, M.A. (1996) Fruit eating by birds in a forest fragment in southeastern Brazil. *Ararajuba* 4, 71–79.

Galetti, M., Laps, R. and Pizo, M.A. (2000) Frugivory by toucans at two altitudes in the Atlantic Forest of Brazil. *Biotropica* 32, 842–850.

Gondim, M.J.C. (2001) Dispersão de sementes de *Trichilia* spp. (Meliaceae) por aves em um fragmento de mata mesófila semidecídua, Rio Claro, SP, Brasil. *Ararajuba* 9, 101–112.

Gondim, M.J.C. (2002) A exploração de frutos por aves frugívoras em uma área de cerradão no Estado de São Paulo. PhD thesis, Universidade Estadual Paulista, Rio Claro, Brazil.

Gridi-Papp, C.O., Gridi-Papp, M. and Silva, W.R. (2004) Differential fruit consumption of two Melastomataceae by birds in Serra da Mantiqueira, southeastern Brazil. *Ararajuba* 12, 7–13.

Guimarães, M.A. (2003) Frugivoria por aves em *Tapirira guianensis* (Anacardiaceae) na zona urbana do município de Araruama, estado do Rio de Janeiro, sudeste brasileiro. *Atualidades Ornitológicas* 116, 12.
Guitián, J.M., Guitián, P., Munilla, I., Guitián, J., Bermejo, T., Larrinaga, A.R., Navarro, L. and López, B. (2000) Zorzales, Espinos e Serbales: un estudio sobre el consumo de frutos silvestres de las aves migratorias en la costa occidental europea. Universidad de Santiago de Compostela, Santiago de Compostela, Spain.
Hagan, J.M. III and Johnston, D.W. (eds) (1992) *Ecology and Conservation of Neotropical Migrant Landbirds.* Smithsonian Institution Press, Washington, DC.
Hamann, A. and Curio, E. (1999) Interactions among frugivores and fleshy fruit trees in a Philippine submontane rainforest. *Conservation Biology* 13, 766–773.
Hasui, E. (1994) O papel das aves frugívoras na dispersão de sementes em um fragmento de floresta semidecídua secundária em São Paulo, SP. MSc thesis, Universidade de São Paulo, Brazil.
Isler, M.I. and Isler, P.R. (1987) *The Tanagers: Natural History, Distribution and Identification.* Smithsonian Institution Press, Washington, DC.
Jordano, P., Bascompte, J. and Olesen, J.M. (2003) Invariant properties in coevolutionary networks of plant–animal interactions. *Ecology Letters* 6, 69–81.
Kindel, A. (1996) Interações entre plantas ornitocóricas e aves frugívoras na Estação Ecológica de Aracuri, Muitos Capões, RS. MSc thesis, Universidade Federal do Rio Grande do Sul, Brazil.
Loiselle, B.A. and Blake, J.G. (2002) Potential consequences of extinction of frugivorous birds for shrubs of a tropical wet forest. In: Levey, D.J., Silva, W.R. and Galetti, M. (eds) *Seed Dispersal and Frugivory: Ecology, Evolution and Conservation.* CAB International, Wallingford, UK, pp. 397–406.
Lombardi, J.A. and Motta, J.C. Jr (1993) Seeds of the champak, *Michelia champaca* L. (Magnoliaceae), as a food source for Brazilian birds. *Ciência e Cultura* 45, 408–409.
Lopes, R.F. (2000) Frugivoria e dispersão de sementes através da avifauna, em quatro espécies de vegetais na região de Botucatu – SP. MSc thesis, Escola Superior de Agricultura Luiz de Queiroz, Brazil.
Magalhães, J.C.R. (1999) *As aves na Fazenda Barreiro Rico.* Editora Plêiade, São Paulo, Brazil.
Malmborg, P.K. and Willson, M.F. (1988) Foraging ecology of avian frugivores and some consequences for seed dispersal in an Illinois woodlot. *Condor* 90, 173–186.
Manhães, M.A., Assis, L.C.S. and Castro, R.M. (2003) Frugivoria e dispersão de sementes de *Miconia urophylla* (Melastomataceae) por aves em um fragmento de Mata Atlântica secundária em Juiz de Fora, Minas Gerais, Brasil. *Ararajuba* 11, 173–178.
Marcondes-Machado, L.O. (2002) Comportamento alimentar de aves em *Miconia rubiginosa* (Melastomataceae) em fragmento de cerrado, São Paulo. *Iheringia* 92, 97–100.
Marini, M.A. and Cavalcanti, R.B. (1998) Frugivory by *Elaenia* flycatchers. *Hornero* 15, 47–50
Masteguin, M.A. and Figueiredo, R.A. (1995) Consumo de frutos de *Prunus sellowii* Koehne (Rosaceae) por aves em um fragmento florestal em Jundiaí, SP. *Ciência e Natura* 17, 51–56.
Melo, C., Bento, E.C. and Oliveira, P.E. (2003) Frugivory and dispersal of *Faramea cyanea* (Rubiaceae) in cerrado woodyplant formations. *Brazilian Journal of Biology* 63, 75–82.
Merler, J.A., Diuk-Wasser, M.A. and Quintana, R.D. (2001) Winter diet of dusk-legged guan (*Penelope obscura*) at the Paraná river delta region. *Studies on Neotropical Fauna and Environment* 36, 33–38.
Mikich, S.B. (2002) A dieta frugívora de *Penelope superciliaris* (Cracidae) em remanescentes de floresta estacional semidecidual no centro-oeste do Paraná, Brasil, e sua relação com *Euterpe edulis* (Arecaceae). *Ararajuba* 10, 207–217.
Morellato, L.P.C. and Haddad, C.F.B. (2000) Introduction: the Brazilian Atlantic Forest. *Biotropica* 32, 786–792.

Motta, J.C. Jr (1990) Estrutura trófica e composição das avifaunas de três habitats na região central de São Paulo. *Ararajuba* 1, 65–71

Motta, J.C. Jr and Lombardi, J.A. (1990) Aves como agentes dispersores da copaíba (*Copaifera langsdorfii*, Caesalpinaceae) em São Carlos, Estado de São Paulo. *Ararajuba* 1, 105–106.

Nepstad, D.C., Klink, C.A., Uhl, C., Vieira, I.C., Lefebvre, P., Pedlowski, M., Matricardi, E., Negreiros, G., Brown, I.F., Amaral, E., Homma, A. and Walker, R. (1997) Land-use in Amazonia and the Cerrado of Brazil. *Ciência e Cultura* 49, 73–86.

Oliveira, P.S. and Marquis, R.J. (2002) *The Cerrados of Brazil.* Columbia University Press, New York.

Oliveira-Filho, A.T. and Ratter, J.A. (2002) Vegetation physiognomies and wood flora of the cerrado biome. In: Oliveira, P.S. and Marquis R.J. (eds) *The Cerrados of Brazil.* Columbia University Press, New York, pp. 91–120.

Oniki, Y., Melo, T.A. Jr, Scopel, E.T. and Willis, E.O. (1994) Bird use of *Cecropia* (Cecropiaceae) and nearby trees in Espírito Santo state, Brazil. *Ornitologia Neotropical* 5, 109–114

Pereira, S.L. and Wajntal, A. (1999) Reintroduction of guans of the genus *Penelope* (Cracidae, Aves) in reforested areas: assessment by DNA fingerprinting. *Biological Conservation* 87, 31–38.

Pizo, M.A. (1997) Seed dispersal and predation in two populations of *Cabralea canjerana* (Meliaceae) in the Atlantic Forest of southeastern Brazil. *Journal of Tropical Ecology* 13, 559–578.

Pizo, M.A. (2004) Frugivory and habitat use by fruit-eating birds in a fragmented landscape in southeast Brazil. *Ornitologia Neotropical* 15 (suppl.), 117–126.

Pizo, M.A., Silva, W.R., Galetti, M. and Laps, R. (2002) Frugivory in cotingas of the Atlantic Forest of southeast Brazil. *Ararajuba* 10, 177–185.

Renne, I.J., Barrow, W.C. Jr, Randall, L.A.J. and Bridges, W.C. Jr (2002) Generalized avian dispersal syndrome contributes to Chinese tallow tree (*Sapium sebiforum*, Euphorbiaceae) invasiveness. *Diversity and Distributions* 8, 285–295.

Restrepo, C., Sargent, S., Levey, D.J. and Watson, D.M. (2002) The role of vertebrates in the diversification of New World mistletoes. In: Levey, D.J., Silva, W.R. and Galetti, M. (eds) *Seed Dispersal and Frugivory: Ecology, Evolution and Conservation.* CAB International, Wallingford, UK, pp. 83–98.

Ridgely, R.S. and Tudor, G. (1989) *The Birds of South America. Vol. 1: The Oscine Passerines.* Oxford University Press, Oxford, UK.

Ridgely, R.S. and Tudor, G. (1994) *The Birds of South America. Vol. 2: The Suboscine Passerines.* Oxford University Press, Oxford, UK.

Scheibler, D.R. and Melo, T.A. Jr (2003) Frugivory by birds on two exotic *Ligustrum* species (Oleaceae) in Brazil. *Ararajuba* 11, 89–91

Schupp, E.W. (1993) Quantity, quality, and the effectiveness of seed dispersal by animals. *Vegetatio* 107/108, 15–29.

Schupp, E.W., Milleron, T. and Russo, S.E. (2002) Dissemination limitation and maintenance of species-rich tropical forests. In: Levey, D.J., Silva, W.R. and Galetti, M. (eds) *Seed Dispersal and Frugivory: Ecology, Evolution and Conservation.* CAB International, Wallingford, UK, pp. 19–33.

Sick, H. (1997) *Ornitologia Brasileira.* Editora Nova Fronteira, Rio de Janeiro, Brazil.

Silva, J.M.C. and Tabarelli, M. (2000) Tree species impoverishment and the future flora of the Atlantic Forest of northeast Brazil. *Nature* 404, 72–74.

Silva, J.M.C., Uhl, C. and Murray, G. (1996) Plant succession, landscape management, and the ecology of frugivorous birds in abandoned Amazonian pastures. *Conservation Biology* 10, 491–503.

Silva, P.A. (2005) Predação de sementes pelo maracanã-nobre (*Diopsittaca nobilis*, Psittacidae) em uma planta exótica (*Melia azedarach*, Meliaceae) no oeste do Estado de São Paulo, Brasil. *Revista Brasileira de Ornitologia* 13, 183–185.

Silva, W.R. (1988) Ornitocoria em *Cereus peruvianus* (Cactaceae) na Serra do Japi, Estado de São Paulo. *Revista Brasileira de Zoologia* 48, 381–389.

Silva, W.R., De Marco, P. Jr, Hasui, E. and Gomes, V.S.M. (2002) Patterns of fruit–frugivore interactions in two Atlantic forest bird communities of southeastern Brazil: implications for conservation. In: Levey, D.J., Silva, W.R. and Galetti, M. (eds) *Seed Dispersal and Frugivory: Ecology, Evolution and Conservation.* CAB International, Wallingford, UK, pp. 423–435.

Stotz, D.F., Fitzpatrick, J.W., Parker, T.A. III and Moskowitz, D.K. (1996) *Neotropical Birds: Ecology and Conservation.* University of Chicago Press, Chicago, Illinois.

Tabarelli, M. and Peres, C.A. (2002) Abiotic and vertebrate seed dispersal in the Brazilian Atlantic Forest: implications for forest regeneration. *Biological Conservation* 106, 165–176.

Valente, R.M. (2001) Comportamento alimentar de aves em *Alchornea glandulosa* (Euphorbiaceae) em Rio Claro, São Paulo. *Iheringia Série Zooógica* 91, 61–66

Viana, V.M., Tabanez, A.A.J. and Batista, J.L.F. (1997) Dynamics and restoration of forest fragments in the Brazilian Atlantic moist forest. In: Laurance, W.F. and Bierregaard, R.O. Jr (eds) *Tropical Forest Remnants: Ecology, Management, and Conservation of Fragmented Communities.* University of Chicago Press, Chicago, Illinois, pp. 351–365.

Wheelwright, N.T., Haber, W.A., Murray, K.G. and Guindon, C. (1984) Tropical fruit-eating birds and their food plants: a survey of a Costa Rican lower montane forest. *Biotropica* 16, 173–192.

Wijdeven, S.M.J. and Kuzee, M.E. (2000) Seed availability as a limiting factor in forest recovery processes in Costa Rica. *Restoration Ecology* 8, 414–424.

Willis, E.O. (1979) The composition of avian communities in remanescent woodlots in southern Brazil. *Papéis Avulsos de Zoologia* 33, 1–25.

Zimmermann, C.E. (1996) Observações preliminares sobre a frugivoria por aves em *Alchornea glandulosa* (Endl. & Poepp.) (Euphorbiaceae) em vegetação secundária. *Revista Brasileira de Zoologia* 13, 533–538

Zimmermann, C.E. (2001) O uso da grandiúva, *Trema micrantha* Blume (Ulmaceae), na recuperação de áreas degradadas: o papel das aves que se alimentam de seus frutos. *Tangara* 1, 177–182.

Zimmermann, C.E., Santos, D.S., Santos, C.A.K. and Assunção, L.G. (2002) O uso de poleiros naturais para a recuperação de florestas ciliares. In: *Anais do I Simpósio Regional de Mata Ciliar.* Marechal Cândido Rondon, Paraná, pp. 70–75.

Appendix 1

List of plant genera with mean seed dimensions, including length (L), width (W) and roundness (R) and number of species known to be consumed and dispersed by frugivores in four rain forests on different continents (N).

Rainforest	Family	Genus	L (mm)	W (mm)	R	N
Central Africa	Anacardiaceae	*Antrocaryon*	32.2	28.6	0.864	13
		Lannea	6.6	4.3	0.653	2
		Pseudospondias	18.2	10.4	0.572	2
		Spondias	31.5	19.5	0.619	1
		Trichoscypha	30.0	16.3	0.552	3
	Annonaceae	*Annickia*	21.0	11.0	0.528	2
		Annona	9.0	4.7	0.522	1
		Anonidium	40.7	23.6	0.581	1
		Artabotrys	21.3	10.5	0.508	3
		Hexalobus	35.0	20.0	0.571	1
		Isolona	12.5	10.0	0.800	1
		Monanthotaxis	15.0	6.4	0.426	1
		Monodora	18.4	11.7	0.627	2
		Pachypodanthium	14.9	7.2	0.492	3
		Piptostigma	22.5	10.0	0.444	1
		Polyalthia	12.1	11.5	0.947	1
		Uvaria	10.7	6.6	0.623	3
		Uvariastrum	20.0	10.0	0.500	1
		Uvariopsis	16.2	9.5	0.595	2
		Xylopia	13.3	8.6	0.642	7
	Apocynaceae	*Alstonia*	5.0	2.0	0.400	1
		Dictyophleba	23.1	12.9	0.558	1
		Landolphia	26.3	14.8	0.557	2
		Picralima	40.0	20.0	0.500	1
		Rauvolfia	6.8	4.3	0.645	2
	Araliaceae	*Polyscias*	5.0	2.0	0.400	1

Rainforest	Family	Genus	L (mm)	W (mm)	R	N
	Arecaceae	*Borassus*	75.0	65.0	0.867	1
		Elaeis	20.0	14.0	0.700	1
		Eremospatha	20.8	14.9	0.720	1
		Laccosperma	8.0	8.0	1.000	1
		Phoenix	11.0	7.0	0.636	1
		Raphia	53.7	34.3	0.639	1
	Bignonaceae	*Kigelia*	10.0	7.0	0.700	1
	Boraginaceae	*Cordia*	9.0	7.0	0.778	1
		Ehretia	3.3	2.8	0.848	1
	Brassicaceae	*Ritchiea*	10.0	10.0	1.000	1
	Burseraceae	*Canarium*	30.6	12.2	0.398	1
		Dacryodes	29.4	17.7	0.604	5
		Santiria	24.1	16.3	0.676	2
	Celastraceae	*Salacia*	16.2	11.2	0.688	1
	Chrysobalanaceae	*Magnistipula*	45.0	20.0	0.444	1
		Maranthes	55.0	32.5	0.591	1
		Parinari	43.4	33.7	0.766	2
	Clusiaceae	*Allanblackia*	26.0	18.2	0.701	1
		Garcinia	25.8	11.9	0.567	2
		Harungana	3.0	2.0	0.667	1
		Mammea	42.4	32.6	0.768	1
		Pentadesma	42.0	26.0	0.619	1
		Symphonia	25.0	12.5	0.500	1
	Combretaceae	*Combretum*	43.0	24.0	0.558	1
		Terminalia	20.0	12.0	0.600	1
	Connaraceae	*Agelaea*	15.0	6.8	0.453	1
		Cnestis	32.0	23.0	0.719	1
		Paxia	30.7	19.2	0.625	1
		Rourea	15.0	7.4	0.500	2
	Dichapetalaceae	*Dichapetalum*	19.3	10.8	0.566	2
	Ebenaceae	*Diospyros*	24.2	10.9	0.547	7
	Euphorbiaceae	*Alchornea*	6.0	4.0	0.667	1
		Bridelia	7.5	7.5	1.000	1
		Croton	8.4	5.7	0.663	2
		Dichostemma	5.9	5.0	0.839	1
		Discoglypremna	3.4	3.1	0.912	1
		Drypetes	19.9	14.3	0.735	3
		Macaranga	4.0	4.0	1.000	1
		Manniophyton	12.5	9.9	0.792	1
		Neoboutonia	7.0	5.0	0.714	1
		Phyllanthus	4.5	3.9	0.863	2
		Plagiostyles	22.3	15.9	0.712	1
		Ricinodendron	17.3	16.8	0.965	2
		Securinega	2.0	2.0	1.000	1
		Uapaca	19.8	11.5	0.595	5
	Fabaceae	*Acacia*	7.8	5.0	0.634	2
		Albizia	9.5	7.3	0.763	1
		Angylocalyx	32.0	18.0	0.563	1
		Anthonotha	43.8	35.8	0.817	1
		Berlinia	52.0	48.0	0.923	1
		Calpocalyx	12.5	7.5	0.600	1

Rainforest	Family	Genus	L (mm)	W (mm)	R	N
Central Africa *Continued*	Fabaceae *Continued*	*Detarium*	45.7	41.0	0.913	3
		Dialium	10.3	8.3	0.798	6
		Entada	62.0	60.8	0.981	1
		Erythrina	10.5	6.5	0.619	1
		Gilbertiodendron	6.6	5.2	0.788	1
		Millettia	47.0	32.0	0.681	1
		Pachyelasma	25.0	14.0	0.560	1
		Parkia	15.0	10.4	0.691	1
		Pentaclethra	49.3	33.0	0.692	2
		Samanea	9.0	5.0	0.556	1
		Scorodophloeus	22.5	17.5	0.778	1
		Swartzia	11.7	10.7	0.914	1
		Tamarindus	13.2	9.3	0.705	1
		Tetrapleura	9.3	7.3	0.782	2
	Huaceae	*Afrostyrax*	19.7	18.1	0.920	1
	Humiriaceae	*Sacoglottis*	36.0	24.0	0.667	1
	Icacinaceae	*Icacina*	28.1	15.0	0.534	1
		Lavigeria	55.0	40.0	0.727	1
	Irvingiaceae	*Irvingia*	42.2	21.8	0.518	2
		Klainedoxa	43.0	32.0	0.744	1
	Lauraceae	*Beilschmiedia*	35.8	21.8	0.610	3
	Lecythidaceae	*Oubanguia*	20.0	10.0	0.500	1
		Scytopetalum	28.0	10.7	0.382	1
	Linaceae	*Hugonia*	10.9	8.8	0.833	3
	Loganiaceae	*Anthocleista*	2.0	1.0	0.500	4
		Strychnos	17.6	11.1	0.620	4
	Malvaceae	*Bombax*	6.0	5.0	0.833	1
		Cola	27.5	18.9	0.702	6
		Duboscia	8.0	5.0	0.667	2
		Grewia	12.5	7.3	0.595	2
		Leptonychia	10.1	5.1	0.505	1
		Rhodognaphalon	10.5	7.0	0.667	1
		Sterculia	21.8	12.2	0.566	3
	Marantaceae	*Hypselodelphys*	24.5	20.0	0.816	1
		Megaphrynium	12.5	8.6	0.692	2
		Sarcocephalus	1.2	0.8	0.667	1
		Sarcophrynium	12.3	6.7	0.572	2
	Meliaceae	*Guarea*	12.9	10.1	0.780	3
		Trichilia	13.0	7.9	0.605	3
		Turraeanthus	21.0	12.0	0.571	1
	Moraceae	*Antiaris*	9.6	7.9	0.823	1
		Ficus	1.4	1.4	1.000	7
		Milicia	2.0	1.5	0.750	1
		Treculia	12.2	6.7	0.548	1
		Trilepisium	12.5	12.5	1.000	1
	Musaceae	*Ensete*	17.5	15.0	0.857	1
	Myristicaceae	*Coelocaryon*	21.9	12.6	0.575	1
		Pycnanthus	19.7	12.5	0.636	1
		Scyphocephalium	30.0	22.0	0.733	1
		Staudtia	24.8	16.9	0.690	2

Rainforest	Family	Genus	L (mm)	W (mm)	R	N
	Myrtaceae	*Syzygium*	17.7	14.6	0.825	1
	Ochnaceae	*Lophira*	25.8	10.3	0.398	1
	Olacaceae	*Chionanthus*	18.0	11.0	0.611	1
		Coula	23.4	19.8	0.845	1
		Diogoa	22.5	22.5	1.000	1
		Heisteria	13.8	9.5	0.685	2
		Ongokea	24.3	22.1	0.911	1
		Strombosia	21.0	14.4	0.679	2
		Strombosiopsis	25.8	18.2	0.705	1
	Pandaceae	*Panda*	54.0	38.0	0.704	1
	Passifloraceae	*Paropsia*	5.4	3.4	0.634	1
		Passiflora	4.5	2.7	0.593	1
	Phytolaccaceae	*Phytolacca*	2.5	1.5	0.600	1
	Piperaceae	*Piper*	4.0	3.0	0.750	1
	Rhamnaceae	*Maesopsis*	25.5	14.5	0.569	2
		Ventilago	12.0	11.0	0.917	1
	Rhizophoraceae	*Anopyxis*	21.0	16.9	0.802	1
	Rubiaceae	*Atractogyne*	2.0	1.0	0.500	1
		Massularia	5.7	4.2	0.748	1
		Mussaenda	1.0	1.0	1.000	1
		Nauclea	1.6	1.3	0.850	4
		Oxyanthus	4.7	2.2	0.468	1
		Parapentas	17.0	14.0	0.824	1
		Psychotria	5.8	4.0	0.696	1
		Sherbournia	1.9	1.5	0.815	4
		Tarenna	2.9	2.0	0.684	1
		Tricalysia	2.8	2.4	0.857	1
		Vangueria	13.0	5.0	0.385	1
	Ruscaceae	*Dracaena*	18.0	15.3	0.853	1
	Rutaceae	*Teclea*	5.8	2.9	0.504	1
	Salicaceae	*Caloncoba*	5.7	4.4	0.766	2
		Casearia	4.2	4.0	0.952	1
		Dasylepis	8.0	7.0	0.875	1
		Oncoba	7.3	3.8	0.519	2
		Scottellia	5.1	3.6	0.706	1
	Sapindaceae	*Blighia*	19.9	11.3	0.549	2
		Chytranthus	26.4	15.1	0.571	1
		Eriocoelum	17.0	9.0	0.529	1
		Ganophyllum	18.5	10.2	0.551	1
		Lepisanthes	12.0	8.0	0.667	1
		Pancovia	22.5	12.8	0.568	2
	Sapotaceae	*Afrosersalisia*	22.0	10.0	0.455	1
		Baillonella	57.7	33.9	0.587	1
		Chrysophyllum	24.9	13.6	0.540	9
		Manilkara	9.3	5.5	0.591	1
		Mimusops	11.5	9.0	0.783	1
		Omphalocarpum	33.0	21.2	0.642	1
		Pouteria	15.0	10.0	0.667	1
		Synsepalum	15.4	9.0	0.583	3

Rainforest	Family	Genus	L (mm)	W (mm)	R	N
Central Africa *Continued*	Sapotaceae *Continued*	*Tieghemella*	65.0	35.0	0.538	1
		Tridesmostemon	40.0	20.0	0.500	1
		Vitellaria	35.0	25.0	0.714	1
	Solanaceae	*Solanum*	2.0	1.5	0.750	1
	Thymelaeaceae	*Dicranolepis*	8.9	7.3	0.820	1
	Ulmaceae	*Celtis*	6.8	6.5	0.933	2
	Urticaceae	*Myrianthus*	15.7	9.9	0.633	1
	Verbenaceae	*Vitex*	20.3	20.3	0.997	3
	Vitaceae	*Cissus*	14.2	7.7	0.555	3
	Zygophyllaceae	*Balanites*	88.0	47.0	0.534	1
Guianas	Acanthaceae	*Mendoncia*	16.0	6.0	0.375	1
	Anacardiaceae	*Spondias*	28.3	18.5	0.660	2
		Tapirira	11.1	8.4	0.757	2
	Annonaceae	*Annona*	10.3	6.2	0.594	3
		Cremastosperma	11.0	9.0	0.818	1
		Duguetia	16.5	8.2	0.571	3
		Ephedranthus	28.0	14.0	0.500	1
		Guatteria	12.6	7.3	0.571	2
		Oxandra	18.0	8.0	0.444	1
		Pseudoxandra	11.0	11.0	1.000	1
		Rollinia	5.8	3.3	0.561	2
		Unonopsis	11.4	9.1	0.800	5
		Xylopia	6.5	4.5	0.614	2
	Apocynaceae	*Allamanda*	20.0	20.0	1.000	1
		Ambelania	8.0	5.0	0.625	1
		Bonafousia	10.0	4.0	0.400	1
		Couma	10.0	7.0	0.700	1
		Geissospermum	17.5	17.5	1.000	1
		Lacmellea	9.5	7.5	0.801	2
		Landolphia	20.0	15.0	0.750	1
		Parahancornia	13.5	6.8	0.508	2
	Araceae	*Anthurium*	3.1	2.3	0.742	1
		Heteropsis	10.8	7.5	0.697	2
		Monstera	7.0	5.0	0.714	1
		Philodendron	1.7	0.8	0.391	10
		Rhodospatha	2.2	1.3	0.674	2
	Araliaceae	*Oreopanax*	3.0	1.3	0.433	1
		Schefflera	8.8	6.7	0.730	3
	Arecaceae	*Astrocaryum*	28.3	16.5	0.584	2
		Attalea	55.0	28.0	0.509	1
		Bactris	15.2	13.5	0.905	3
		Euterpe	12.5	12.2	0.972	3
		Geonoma	6.5	5.0	0.813	2
		Jessenia	32.0	20.0	0.625	1
		Mauritia	45.0	45.0	1.000	1
		Oenocarpus	13.0	8.0	0.615	2
		Socratea	24.0	17.0	0.708	1
	Boraginaceae	*Cordia*	10.9	7.3	0.689	7

Rainforest	Family	Genus	L (mm)	W (mm)	R	N
	Bromeliaceae	*Aechmea*	3.0	1.0	0.333	1
		Ananas	3.0	1.0	0.333	1
	Burseraceae	*Dacryodes*	18.0	12.5	0.727	2
		Protium	17.4	11.6	0.639	7
		Tetragastris	19.0	11.5	0.616	2
		Trattinnickia	11.0	8.0	0.727	1
	Cactaceae	*Hylocereus*	3.0	2.0	0.667	1
		Pereskia	3.0	2.0	0.667	1
	Canellaceae	*Cinnamodendron*	9.2	8.8	0.951	1
	Caricaceae	*Jacaratia*	8.0	4.0	0.500	1
	Caryocaraceae	*Caryocar*	43.3	31.0	0.688	3
	Celastraceae	*Cheiloclinium*	19.5	11.8	0.601	3
		Maytenus	10.3	6.3	0.650	4
		Peritassa	15.5	7.4	0.477	1
		Salacia	27.0	19.0	0.704	1
		Tontelea	17.5	12.0	0.658	2
	Chrysobalanaceae	*Chrysobalanus*	8.0	8.0	1.000	1
		Couepia	41.4	26.6	0.657	6
		Hirtella	12.0	5.0	0.417	1
		Licania	20.0	15.8	0.776	10
		Parinari	43.8	26.5	0.588	4
	Clusiaceae	*Caraipa*	16.0	10.0	0.625	1
		Clusia	6.0	2.7	0.459	4
		Moronobea	25.0	15.0	0.600	1
		Platonia	42.5	24.5	0.576	1
		Rheedia	20.2	11.2	0.552	3
		Symphonia	20.5	14.5	0.707	1
		Tovomita	18.8	9.3	0.493	2
		Vismia	2.2	1.0	0.415	4
	Combretaceae	*Buchenavia*	25.1	12.9	0.538	2
		Terminalia	20.0	10.5	0.553	2
	Connaraceae	*Connarus*	9.0	9.0	1.000	1
		Rourea	10.0	6.0	0.600	1
	Convolvulaceae	*Dicranostyles*	24.0	14.0	0.583	1
		Maripa	18.8	12.5	0.661	2
	Cucurbitaceae	*Cayaponia*	11.6	9.8	0.878	4
		Gurania	7.0	4.3	0.607	1
		Melothria	3.0	2.3	0.750	1
	Cyclanthaceae	*Asplundia*	1.3	0.6	0.454	3
		Cyclanthus	0.7	0.3	0.429	1
		Evodianthus	1.2	0.4	0.348	1
		Ludovia	1.0	0.7	0.700	1
		Stelestylis	2.0	0.6	0.300	1
		Thoracocarpus	3.1	0.9	0.295	1
	Dilleniaceae	*Davilla*	3.5	3.5	1.000	1
		Doliocarpus	8.9	7.1	0.802	2
	Ebenaceae	*Diospyros*	17.2	9.6	0.569	2
	Elaeocarpaceae	*Sloanea*	14.5	9.5	0.655	2

Rainforest	Family	Genus	L (mm)	W (mm)	R	N
Guianas *Continued*	Euphorbiaceae	*Alchornea*	5.0	4.0	0.800	1
		Alchorneopsis	12.5	7.0	0.560	1
		Conceveiba	8.5	8.5	1.000	1
		Drypetes	18.0	12.0	0.667	2
		Hieronima	3.5	2.5	0.708	2
		Margaritaria	11.0	7.0	0.636	1
		Micrandra	8.9	6.6	0.742	1
		Omphalea	40.0	35.0	0.875	1
		Plukenetia	32.0	23.0	0.719	1
		Sagotia	8.0	4.0	0.500	1
		Sandwithia	8.0	5.0	0.625	1
		Sapium	5.5	4.0	0.727	1
	Fabaceae	*Abarema*	6.5	6.5	1.000	1
		Andira	31.3	31.3	1.000	2
		Bauhinia	10.8	7.3	0.572	2
		Cassia	10.3	7.7	0.748	1
		Clathrotropis	55.0	50.0	0.909	1
		Copaifera	12.0	8.0	0.671	2
		Crudia	33.0	28.0	0.848	1
		Dicorynia	5.0	5.0	1.000	1
		Dioclea	61.0	57.0	0.934	1
		Dipteryx	56.0	32.2	0.574	2
		Dussia	35.7	21.2	0.594	1
		Eperua	54.0	43.5	0.777	2
		Hymenaea	21.5	14.0	0.651	1
		Inga	16.7	9.4	0.576	26
		Parkia	11.8	7.5	0.680	2
		Peltogyne	15.5	13.5	0.876	2
		Pithecellobium	8.5	6.5	0.757	2
		Stryphnodendron	8.0	5.0	0.625	1
		Swartzia	24.5	15.3	0.635	7
		Vouacapoua	49.0	34.0	0.694	1
		Zygia	10.0	3.0	0.300	1
	Goupiaceae	*Goupia*	2.8	1.5	0.545	1
	Heliconiaceae	*Heliconia*	15.0	10.0	0.667	1
	Humiriaceae	*Sacoglottis*	27.5	17.0	0.701	2
	Icacinaceae	*Dendrobangia*	20.0	8.0	0.400	1
		Poraqueiba	25.5	17.0	0.667	1
	Lauraceae	*Cryptocarya*	21.0	20.0	0.952	1
		Nectandra	12.0	9.0	0.750	1
		Ocotea	16.1	11.2	0.703	8
		Persea	11.5	10.0	0.870	1
		Rhodostemonodaphne	30.0	20.0	0.667	1
	Lecythidaceae	*Eschweilera*	22.0	15.9	0.729	12
		Gustavia	16.8	11.5	0.691	2
		Lecythis	28.1	17.6	0.634	8
	Linaceae	*Hebepetalum*	6.0	6.0	1.000	1
	Loganiaceae	*Potalia*	4.8	2.9	0.611	1
		Strychnos	18.9	15.6	0.826	3

Rainforest	Family	Genus	L (mm)	W (mm)	R	N
	Malpighiaceae	*Byrsonima*	7.0	6.0	0.860	2
		Norantea	3.0	1.0	0.333	1
	Malvaceae	*Apeiba*	3.3	2.8	0.857	2
		Bombax	8.0	7.5	0.929	2
		Catostemma	40.0	26.5	0.663	1
		Eriotheca	8.0	7.0	0.875	1
		Guazuma	1.8	1.8	0.992	1
		Quararibea	20.0	11.5	0.600	3
		Sterculia	19.0	14.3	0.766	3
		Theobroma	20.0	16.0	0.800	2
	Marcgraviaceae	*Marcgravia*	1.7	0.8	0.460	3
		Norantea	4.2	1.3	0.310	1
		Souroubea	4.0	1.3	0.313	1
	Melastomataceae	*Bellucia*	0.5	0.5	1.000	1
		Clidemia	0.5	0.3	0.556	1
		Henriettea	1.3	0.9	0.680	1
		Loreya	0.5	0.5	1.000	1
		Miconia	1.0	0.4	0.311	2
		Mouriri	11.8	8.8	0.768	5
	Meliaceae	*Carapa*	32.5	27.5	0.833	2
		Guarea	20.4	13.1	0.642	5
		Trichilia	13.4	6.4	0.581	3
	Menispermaceae	*Abuta*	21.1	13.1	0.620	4
		Anomospermum	32.0	19.0	0.594	1
		Caryomene	48.8	32.3	0.662	1
		Elephantomene	47.0	25.0	0.532	1
		Orthomene	20.0	10.0	0.500	1
	Monimiaceae	*Mollinedia*	15.8	6.5	0.411	1
		Siparuna	25.0	14.0	0.533	2
	Moraceae	*Bagassa*	4.1	2.7	0.653	2
		Batocarpus	11.0	9.9	0.900	1
		Brosimum	10.3	10.0	0.972	4
		Clarisia	28.0	18.0	0.643	1
		Ficus	1.5	1.0	0.731	11
		Helicostylis	8.0	7.0	0.875	1
		Maquira	13.0	9.0	0.692	1
		Naucleopsis	10.0	7.5	0.750	1
		Perebea	9.1	7.1	0.781	2
		Pseudolmedia	6.0	6.0	1.000	1
		Sorocea	9.5	8.5	0.895	1
		Trymatococcus	9.6	9.3	0.971	2
	Myristicaceae	*Iryanthera*	28.2	17.5	0.610	5
		Ostoophlooum	27.0	17.0	0.630	1
		Virola	18.2	13.5	0.764	11
	Myrtaceae	*Eugenia*	20.3	13.4	0.743	6
		Myrcia	12.0	11.0	0.917	1
		Psidium	4.3	3.3	0.765	1
	Nyctaginaceae	*Guapira*	12.0	5.0	0.417	1
	Ochnaceae	*Elvasia*	7.0	1.9	0.271	1

Rainforest	Family	Genus	L (mm)	W (mm)	R	N
Guianas *Continued*	Olacaceae	*Heisteria*	8.3	6.3	0.756	2
		Minquartia	22.0	11.0	0.500	1
	Opiliaceae	*Agonandra*	12.0	10.0	0.833	1
	Passifloraceae	*Dilkea*	14.8	8.7	0.588	1
		Passiflora	6.0	3.3	0.557	3
	Phytolaccaceae	*Phytolacca*	1.3	1.0	0.769	1
	Piperaceae	*Peperomia*	0.8	0.4	0.533	1
		Piper	1.5	1.3	0.849	17
		Pothomorphe	0.5	0.5	1.000	1
	Polygalaceae	*Moutabea*	17.9	11.3	0.635	2
	Polygonaceae	*Coccoloba*	9.0	6.5	0.716	3
	Quiinaceae	*Lacunaria*	16.5	5.5	0.386	2
	Rhamnaceae	*Rhamnidium*	13.0	10.0	0.769	1
		Ziziphus	19.8	12.8	0.646	1
	Rubiaceae	*Amaioua*	3.0	2.0	0.667	1
		Coussarea	12.0	9.5	0.779	2
		Faramea	8.0	5.0	0.700	2
		Guettarda	8.0	5.0	0.625	1
		Isertia	8.5	3.0	0.353	1
		Palicourea	6.0	4.0	0.667	1
		Posoqueria	17.5	17.5	1.000	1
		Psychotria	4.3	3.5	0.828	5
	Rutaceae	*Zanthoxylum*	3.0	2.0	0.667	1
	Salicaceae	*Banara*	1.3	0.8	0.640	1
		Casearia	9.0	5.0	0.556	1
		Laetia	3.5	2.5	0.714	1
		Ryania	4.0	3.0	0.750	1
	Sapindaceae	*Cupania*	12.5	8.5	0.732	2
		Matayba	10.7	7.3	0.691	3
		Paullinia	11.7	9.2	0.827	5
		Pseudima	18.0	18.0	1.000	1
		Talisia	15.0	9.5	0.633	2
	Sapotaceae	*Chrysophyllum*	20.5	11.2	0.550	10
		Diploon	14.0	12.0	0.857	1
		Ecclinusa	21.0	11.8	0.559	4
		Elaeoluma	12.5	10.5	0.826	2
		Manilkara	18.4	8.8	0.490	5
		Micropholis	16.5	7.8	0.477	7
		Pouteria	22.0	12.8	0.609	11
		Pradosia	17.5	11.5	0.657	1
	Scrophulariaceae	*Schlegelia*	3.0	1.0	0.333	1
	Simaroubaceae	*Simarouba*	11.0	6.0	0.545	1
	Smilacaceae	*Smilax*	6.0	5.0	0.833	1
	Solanaceae	*Cestrum*	3.0	1.4	0.467	1
		Lycianthes	4.0	4.0	1.000	1
		Markea	3.3	1.2	0.355	2
		Solanum	2.3	1.8	0.778	10
	Symplocaceae	*Symplocos*	12.0	6.0	0.500	2
	Ulmaceae	*Celtis*	0.9	0.8	0.889	1

Rainforest	Family	Genus	L (mm)	W (mm)	R	N
	Urticaceae	*Cecropia*	1.8	0.7	0.393	2
		Coussapoa	2.3	1.6	0.689	3
		Pourouma	18.0	10.4	0.577	8
	Verbenaceae	*Citharexylum*	11.0	5.0	0.455	1
		Vitex	15.7	9.0	0.573	1
	Violaceae	*Leonia*	15.0	11.0	0.733	1
		Paypayrola	5.0	5.0	1.000	1
Australia	Agavaceae	*Cordyline*	3.7	2.7	0.716	3
	Anacardiaceae	*Euroschinus*	5.0	5.0	1.000	1
		Pleiogynium	27.0	17.0	0.630	1
		Rhus	2.5	2.5	1.000	1
		Semecarpus	21.0	13.5	0.643	1
	Annonaceae	*Cananga*	8.0	5.0	0.625	1
		Cyathostemma	7.0	5.0	0.714	1
		Goniothalamus	19.0	15.0	0.789	1
		Meiogyne	8.0	5.0	0.625	1
		Melodorum	7.0	4.0	0.571	1
		Polyalthia	27.8	16.8	0.601	2
		Pseuduvaria	10.5	10.0	0.952	1
	Apocynaceae	*Alyxia*	14.5	9.5	0.699	2
		Cerbera	63.6	32.6	0.505	2
		Melodinus	8.6	5.0	0.581	1
		Neisosperma	34.4	16.0	0.463	1
	Aquifoliaceae	*Sphenostemon*	12.0	8.0	0.667	1
	Araceae	*Alocasia*	6.0	6.0	1.000	1
		Gymnostachys	10.0	6.0	0.600	1
		Pothos	11.3	5.9	0.524	1
	Araliaceae	*Delarbrea*	12.0	10.0	0.833	1
		Mackinlaya	8.0	4.5	0.563	1
		Motherwellia	6.0	4.0	0.667	1
		Polyscias	4.5	1.9	0.422	3
		Schefflera	4.8	2.3	0.478	2
	Arecaceae	*Archontophoenix*	10.2	9.0	0.901	2
		Calamus	9.0	8.3	0.921	4
		Hydriastele	5.5	5.5	1.000	1
		Licuala	14.0	8.3	0.589	1
		Linospadix	10.1	4.1	0.529	2
		Normanbya	41.4	28.2	0.680	1
		Oraniopsis	20.0	19.3	0.967	1
		Ptychosperma	12.1	6.6	0.541	1
	Asteraceae	*Tragopogon*	3.0	2.5	0.833	1
	Austrobaileyaceae	*Austrobaileya*	17.5	17.5	1.000	1
	Balanopaceae	*Balanops*	12.3	7.2	0.581	1
	Boraginaceae	*Ehretia*	4.0	3.5	0.875	1
	Burseraceae	*Canarium*	13.0	7.5	0.599	4
		Garuga	4.0	4.0	1.000	1
	Celastraceae	*Celastrus*	6.8	3.0	0.444	1
		Siphonodon	27.6	22.8	0.826	1

Rainforest	Family	Genus	L (mm)	W (mm)	R	N
Australia *Continued*	Chrysobalanaceae	*Maranthes*	38.0	20.0	0.526	1
		Parinari	22.5	17.0	0.756	1
	Clusiaceae	*Calophyllum*	21.4	17.4	0.778	2
		Garcinia	8.3	5.3	0.637	1
		Mammea	37.6	35.6	0.945	1
	Combretaceae	*Terminalia*	23.3	13.5	0.582	7
	Cornaceae	*Alangium*	15.0	7.0	0.467	1
	Corynocarpaceae	*Corynocarpus*	35.1	31.7	0.903	1
	Cucurbitaceae	*Diplocyclos*	6.5	3.5	0.538	1
		Mukia	5.5	3.0	0.545	1
		Trichosanthes	11.3	8.0	0.736	2
		Zehneria	3.8	2.7	0.696	1
	Cunoniaceae	*Davidsonia*	34.0	29.9	0.878	1
		Gillbeea	7.0	1.0	0.143	1
		Schizomeria	11.0	9.0	0.833	2
	Ebenaceae	*Diospyros*	9.8	6.5	0.691	6
	Elaeagnaceae	*Elaeagnus*	16.1	6.2	0.385	1
	Elaeocarpaceae	*Aceratium*	17.0	7.2	0.459	4
		Elaeocarpus	16.5	11.6	0.700	17
		Peripentadenia	16.8	13.7	0.811	1
		Sloanea	11.4	7.6	0.695	2
	Erythroxylaceae	*Erythroxylum*	12.0	4.5	0.375	1
	Escalloniaceae	*Abrophyllum*	0.3	0.3	1.000	1
	Euphorbiaceae	*Aleurites*	35.7	32.9	0.923	1
		Antidesma	7.0	5.0	0.714	1
		Breynia	3.9	3.0	0.771	2
		Bridelia	6.0	4.0	0.667	1
		Fontainea	23.0	15.0	0.652	1
		Glochidion	3.6	3.1	0.863	4
		Homalanthus	6.2	4.3	0.699	1
		Mallotus	3.5	2.9	0.846	3
		Omphalea	45.6	42.3	0.926	1
	Fabaceae	*Archidendron*	13.5	10.0	0.741	1
	Flagellariaceae	*Flagellaria*	5.4	5.4	1.000	1
	Goodeniaceae	*Scaevola*	7.3	5.0	0.668	2
	Hemerocallidaceae	*Dianella*	3.5	2.3	0.643	2
	Himantandraceae	*Galbulimima*	11.9	6.7	0.561	1
	Icacinaceae	*Apodytes*	12.0	10.2	0.848	1
		Citronella	17.2	13.3	0.768	2
		Gomphandra	20.5	11.5	0.561	1
		Irvingbaileya	26.6	14.4	0.539	1
		Pennantia	13.0	8.0	0.615	1
	Lauraceae	*Beilschmiedia*	29.6	24.0	0.763	7
		Cinnamomum	15.2	11.2	0.746	2
		Cryptocarya	14.8	11.9	0.835	20
		Endiandra	32.9	20.8	0.627	18
		Litsea	9.3	6.7	0.728	4
		Neolitsea	7.9	6.9	0.866	2
	Lecythidaceae	*Barringtonia*	35.0	27.0	0.771	1

Rainforest	Family	Genus	L (mm)	W (mm)	R	N
	Loganiaceae	*Fagraea*	2.6	1.9	0.741	2
		Strychnos	12.3	10.7	0.885	3
	Loranthaceae	*Amylotheca*	6.0	3.0	0.500	1
		Lysiana	11.0	6.0	0.545	1
	Malvaceae	*Grewia*	7.0	7.0	1.000	1
	Melastomataceae	*Melastoma*	0.5	0.5	1.000	2
		Memecylon	5.0	5.0	1.000	1
	Meliaceae	*Aglaia*	19.1	11.3	0.606	4
		Anthocarapa	16.0	9.0	0.563	1
		Dysoxylum	17.1	11.1	0.653	9
		Melia	10.6	7.1	0.670	1
		Turraea	5.5	3.5	0.636	1
	Menispermaceae	*Hypserpa*	12.3	10.4	0.847	1
		Legnephora	7.8	6.8	0.862	1
		Pachygone	27.0	23.0	0.852	1
		Stephania	4.5	4.0	0.889	1
	Monimiaceae	*Hedycarya*	6.0	5.0	0.833	1
		Levieria	7.0	3.5	0.500	1
		Palmeria	5.0	4.0	0.800	1
		Tetrasynandra	12.0	7.5	0.625	1
	Moraceae	*Ficus*	1.7	1.1	0.698	23
		Maclura	4.5	2.5	0.556	1
		Streblus	4.0	3.0	0.750	1
	Musaceae	*Musa*	4.8	4.1	0.845	1
	Myoporaceae	*Eremophila*	35.0	35.0	1.000	1
	Myristicaceae	*Myristica*	17.3	13.8	0.802	2
	Myrsinaceae	*Ardisia*	4.5	3.9	0.870	2
		Embelia	4.0	4.0	1.000	1
		Maesa	1.0	1.0	1.000	1
		Rapanea	4.0	3.8	0.938	2
	Myrtaceae	*Acmena*	20.8	20.1	0.955	5
		Acmenosperma	11.1	6.3	0.567	1
		Austromyrtus	4.3	3.8	0.900	2
		Decaspermum	4.0	3.0	0.750	1
		Eugenia	9.0	8.4	0.932	1
		Melaleuca	0.5	0.5	1.000	1
		Myrtella	2.0	1.0	0.500	1
		Pilidiostigma	2.3	2.2	0.946	1
		Rhodamnia	3.5	3.1	0.852	3
		Rhodomyrtus	3.4	2.2	0.750	4
		Syzygium	14.2	12.1	0.866	19
	Olacaceae	*Ximenia*	24.0	20.0	0.833	1
	Oleaceae	*Chionanthus*	14.2	9.5	0.670	2
		Jasminum	7.8	6.8	0.871	2
		Notelaea	9.0	8.5	0.944	1
		Olea	12.0	7.0	0.583	1
	Pandanaceae	*Freycinetia*	1.0	1.0	1.000	1
		Pandanus	20.5	6.0	0.338	2

Rainforest	Family	Genus	L (mm)	W (mm)	R	N
Australia *Continued*	Passifloraceae	*Passiflora*	4.0	3.3	0.817	2
	Phytolaccaceae	*Phytolacca*	2.0	2.0	1.000	1
	Piperaceae	*Piper*	4.9	3.2	0.651	2
		Pittosporum	7.7	4.2	0.543	1
	Polygalaceae	*Xanthophyllum*	11.7	10.4	0.888	1
	Proteaceae	*Helicia*	24.3	21.3	0.886	3
	Rhamnaceae	*Alphitonia*	5.7	4.1	0.708	2
		Emmenosperma	5.2	5.2	1.000	1
		Ziziphus	7.5	7.5	1.000	1
	Rhipogonaceae	*Rhipogonum*	7.0	6.0	0.851	1
	Rhizophoraceae	*Carallia*	7.0	6.0	0.857	1
	Rosaceae	*Eriobotrya*	18.0	13.0	0.722	1
		Prunus	24.2	18.1	0.745	2
		Rubus	0.9	0.9	1.000	3
	Rubiaceae	*Aidia*	1.8	1.8	1.000	1
		Antirhea	9.9	5.5	0.558	1
		Atractocarpus	7.0	4.0	0.571	1
		Canthium	4.0	2.8	0.683	2
		Cyclophyllum	8.1	5.1	0.631	1
		Morinda	5.0	3.0	0.619	2
		Nauclea	0.5	0.5	1.000	1
		Psychotria	4.0	2.5	0.625	1
		Timonius	4.0	1.0	0.250	1
	Rutaceae	*Acronychia*	8.6	6.2	0.678	5
		Geijera	12.5	11.1	0.888	1
		Halfordia	5.0	5.0	1.000	1
		Melicope	4.1	2.9	0.701	4
		Sarcomelicope	6.0	4.0	0.667	1
		Zanthoxylum	5.1	4.3	0.843	1
	Salicaceae	*Baileyoxylon*	17.0	17.0	1.000	1
		Scolopia	4.0	3.5	0.875	1
	Santalaceae	*Notothixos*	3.5	1.5	0.414	1
	Sapindaceae	*Alectryon*	7.8	6.3	0.786	2
		Arytera	14.0	8.0	0.571	1
		Castanospora	25.2	21.1	0.839	1
		Cupaniopsis	10.1	5.8	0.579	2
		Diploglottis	27.2	22.5	0.832	3
		Ganophyllum	9.0	8.0	0.889	1
		Guioa	6.9	4.4	0.658	3
		Harpullia	13.1	10.3	0.785	1
		Mischocarpus	10.4	7.2	0.718	5
		Rhysotoechia	13.8	13.3	0.964	1
		Sarcopteryx	8.0	3.0	0.375	1
		Sarcotoechia	13.5	7.0	0.519	1
		Synima	10.8	8.3	0.791	2
		Toechima	14.0	8.1	0.575	1
	Sapotaceae	*Chrysophyllum*	34.0	23.4	0.688	1
		Manilkara	19.0	12.0	0.632	1
		Mimusops	13.0	8.5	0.654	1

Rainforest	Family	Genus	L (mm)	W (mm)	R	N
	Sapotaceae	*Niemeyera*	32.3	28.5	0.900	2
	Continued	*Palaquium*	30.0	16.0	0.533	1
		Pouteria	16.7	9.7	0.539	9
	Smilacaceae	*Smilax*	8.9	6.3	0.709	1
	Solanaceae	*Duboisia*	3.0	3.0	1.000	1
		Physalis	2.0	2.0	1.000	1
		Solanum	1.9	1.7	0.955	4
	Symplocaceae	*Symplocos*	5.0	4.0	0.800	2
	Thymelaeaceae	*Lethedon*	10.0	5.0	0.500	1
		Phaleria	23.0	22.5	0.978	1
	Ulmaceae	*Aphananthe*	6.0	3.0	0.500	1
		Celtis	7.0	7.0	1.000	1
		Trema	2.0	2.0	1.000	2
	Urticaceae	*Dendrocnide*	2.0	1.0	0.500	1
		Pipturus	0.5	0.5	1.000	1
	Verbenaceae	*Callicarpa*	1.5	1.0	0.763	2
		Faradaya	45.0	30.0	0.667	1
		Gmelina	12.1	9.5	0.778	2
		Premna	11.3	10.8	0.900	2
		Viticipremna	6.0	4.0	0.667	1
	Vitaceae	*Cayratia*	4.0	4.0	1.000	1
		Cissus	8.3	5.4	0.683	6
		Leea	4.0	3.0	0.750	1
		Tetrastigma	9.5	5.5	0.579	1
	Zingiberaceae	*Alpinia*	4.3	2.7	0.619	3
		Hornstedtia	3.5	2.5	0.714	1
Thailand	Actinidiaceae	*Saurauia*	0.1	0.1	1.000	1
	Anacardiaceae	*Buchanania*	9.6	6.9	0.719	1
		Choerospondias	18.5	15.1	0.815	1
		Mangifera	47.8	27.7	0.580	1
		Rhus	6.1	4.7	0.771	1
		Spondias	33.4	23.8	0.711	1
	Annonaceae	*Alphonsea*	16.9	10.5	0.620	1
		Artabotrys	18.8	12.9	0.684	1
		Cyathostemma	7.6	5.2	0.686	1
		Dasymaschalon	12.9	4.5	0.351	1
		Desmos	6.7	4.8	0.714	1
		Fissistigma	14.1	8.1	0.572	1
		Miliusa	11.4	7.7	0.681	1
		Platymitra	28.2	16.7	0.595	1
		Polyalthia	19.3	12.8	0.729	3
		Uvaria	12.4	9.6	0.783	3
	Apocynaceae	*Chilocarpus*	7.3	5.4	0.737	1
		Rauvolfia	8.6	5.2	0.606	1
	Aquifoliaceae	*Ilex*	4.0	1.9	0.481	1
	Araceae	*Amorphophallus*	11.0	6.8	0.618	1
		Arisaema	7.1	5.9	0.825	1
		Pothos	13.0	6.3	0.486	1

Rainforest	Family	Genus	L (mm)	W (mm)	R	N
Thailand *Continued*	Arecaceae	*Areca*	27.2	11.0	0.403	1
		Calamus	11.1	9.3	0.842	1
		Livistona	21.8	18.9	0.866	1
	Burseraceae	*Canarium*	35.6	16.6	0.465	1
	Caprifoliaceae	*Lonicera*	4.9	3.2	0.657	1
		Viburnum	10.0	6.5	0.648	1
	Celastraceae	*Bhesa*	15.4	8.1	0.522	1
		Celastrus	7.0	4.4	0.626	1
		Glyptopetalum	15.7	11.0	0.701	1
		Microtropis	10.0	7.6	0.764	1
		Salacia	14.8	12.3	0.832	1
	Chloranthaceae	*Chloranthus*	4.0	3.2	0.803	1
	Clusiaceae	*Garcinia*	25.1	12.8	0.511	2
	Combretaceae	*Terminalia*	23.7	17.2	0.726	1
	Connaraceae	*Rourea*	11.8	7.1	0.634	2
	Convolvulaceae	*Erycibe*	17.2	11.0	0.642	1
	Cornaceae	*Alangium*	9.5	6.1	0.658	2
		Mastixia	22.9	12.6	0.548	1
	Cucurbitaceae	*Trichosanthes*	9.5	6.3	0.702	2
	Daphniphyllaceae	*Daphniphyllum*	10.5	6.1	0.585	1
	Ebenaceae	*Diospyros*	14.8	8.4	0.565	1
	Elaeagnaceae	*Elaeagnus*	33.0	9.9	0.298	1
	Elaeocarpaceae	*Elaeocarpus*	14.8	12.0	0.741	2
		Sloanea	8.8	5.9	0.673	1
	Escalloniaceae	*Polyosma*	20.3	13.3	0.658	1
	Euphorbiaceae	*Antidesma*	4.7	3.8	0.823	1
		Aporusa	6.2	5.0	0.807	2
		Baccaurea	10.4	8.2	0.786	1
		Breynia	4.0	2.6	0.659	2
		Bridelia	6.9	5.1	0.760	3
		Claoxylon	2.9	2.9	0.995	1
		Excoecaria	14.1	12.8	0.906	1
		Glochidion	4.2	3.3	0.807	2
		Hippomane	7.4	4.8	0.654	1
		Macaranga	2.9	2.5	0.881	2
		Mallotus	3.4	3.3	0.986	2
	Flagellariaceae	*Flagellaria*	9.6	8.0	0.833	1
	Gentianaceae	*Fagraea*	2.8	1.5	0.528	1
	Icacinaceae	*Gonocaryum*	45.8	23.9	0.522	1
		Platea	28.7	16.4	0.572	1
	Irvingiaceae	*Irvingia*	30.2	22.9	0.756	1
	Lamiaceae	*Orthosiphon*	5.3	4.5	0.862	2
	Lauraceae	*Actinodaphne*	9.6	7.1	0.742	1
		Beilschmiedia	27.3	13.4	0.495	3
		Cinnamomum	12.5	8.1	0.648	3
		Cryptocarya	13.9	9.6	0.697	2
		Litsea	8.1	6.0	0.745	2
		Neolitsea	11.5	9.2	0.830	2
		Phoebe	12.4	8.2	0.651	2

Rainforest	Family	Genus	L (mm)	W (mm)	R	N
	Loranthaceae	*Helixanthera*	3.0	2.2	0.749	1
		Macrosolen	4.4	3.7	0.837	1
	Magnoliaceae	*Michelia*	7.3	5.6	0.771	2
	Malvaceae	*Muntingia*	0.5	0.3	0.640	1
		Sterculia	14.1	8.9	0.629	2
	Melastomataceae	*Diplectria*	0.1	0.1	1.000	1
		Medinilla	0.1	0.1	1.000	1
		Melastoma	0.7	0.5	0.730	1
		Memecylon	7.0	6.0	0.857	3
	Meliaceae	*Aglaia*	26.4	16.8	0.632	2
		Aphanamixis	14.5	11.0	0.758	1
		Chisocheton	27.1	13.9	0.513	1
		Dysoxylum	15.8	14.3	0.903	2
		Melia	15.3	11.2	0.732	1
		Sandoricum	25.5	15.1	0.594	1
		Walsura	11.2	9.5	0.853	2
	Menispermaceae	*Anamirta*	10.2	8.5	0.832	1
		Cyclea	5.5	5.3	0.958	1
		Diploclisia	16.8	9.8	0.587	1
		Hypserpa	7.7	7.4	0.962	1
		Stephania	5.6	5.0	0.899	1
	Moraceae	*Antiaris*	17.7	14.9	0.842	1
		Artocarpus	11.3	9.4	0.825	2
		Ficus	1.2	1.0	0.866	15
		Maclura	4.7	4.0	0.852	1
		Morus	1.7	1.6	0.967	1
		Trophis	2.6	2.2	0.833	1
	Myristicaceae	*Horsfieldia*	21.2	16.7	0.784	1
		Knema	21.3	13.9	0.653	1
	Myrsinaceae	*Ardisia*	5.5	5.0	0.918	3
		Labisia	4.3	4.0	0.943	1
		Maesa	2.2	2.1	0.938	1
	Myrtaceae	*Decaspermum*	4.1	3.0	0.729	1
		Psidium	3.5	3.0	0.842	1
		Syzygium	14.2	11.7	0.797	5
	Olacaceae	*Schoepfia*	9.8	7.0	0.719	1
	Oleaceae	*Chionanthus*	19.7	9.8	0.494	1
		Jasminum	9.2	5.9	0.648	1
		Ligustrum	7.2	5.0	0.699	1
	Piperaceae	*Piper*	3.6	3.5	0.970	2
	Proteaceae	*Helicia*	37.9	27.2	0.719	1
	Rhamnaceae	*Ziziphus*	11.1	7.7	0.706	2
	Rosaceae	*Prunus*	12.3	10.3	0.874	2
		Rubus	2.2	1.6	0.733	2
	Rubiaceae	*Breonia*	0.1	0.1	1.000	1
		Canthium	16.1	8.5	0.526	2
		Chassalia	6.9	5.3	0.769	1
		Geophila	4.3	3.7	0.858	1
		Ixora	6.0	5.2	0.880	1

Rainforest	Family	Genus	L (mm)	W (mm)	R	N
Thailand *Continued*		*Lasianthus*	4.6	2.5	0.549	12
		Mitragyna	1.2	0.8	0.672	1
		Morinda	6.1	4.4	0.716	1
		Mussaenda	0.8	0.6	0.751	1
		Prismatomeris	7.8	4.7	0.610	1
		Psychotria	5.4	4.8	0.905	3
		Randia	1.7	1.1	0.667	1
	Rutaceae	*Acronychia*	4.0	3.2	0.803	1
		Citrus	12.4	6.9	0.555	1
		Clausena	7.6	6.2	0.834	2
		Euodia	2.7	2.4	0.893	1
		Glycosmis	7.3	6.0	0.815	3
		Melicope	3.7	3.0	0.825	1
		Micromelum	9.7	5.1	0.525	1
		Toddalia	4.7	3.1	0.670	1
	Sabiaceae	*Sabia*	11.3	9.6	0.854	1
	Salicaceae	*Casearia*	5.6	3.9	0.714	4
	Sapindaceae	*Allophylus*	8.4	7.2	0.855	1
		Dimocarpus	12.1	11.4	0.943	1
		Harpullia	20.2	14.2	0.703	1
		Lepisanthes	15.2	6.9	0.456	1
		Mischocarpus	9.5	8.3	0.873	1
		Nephelium	22.4	12.7	0.569	1
		Sapindus	15.7	15.3	0.973	1
	Sapotaceae	*Chrysophyllum*	14.2	10.4	0.732	1
		Palaquium	18.8	10.1	0.538	1
	Simaroubaceae	*Picrasma*	10.4	9.1	0.878	1
	Smilacaceae	*Smilax*	3.5	3.4	0.979	1
	Solanaceae	*Solanum*	2.1	1.8	0.856	2
	Staphyleaceae	*Turpinia*	4.6	4.6	0.994	1
	Symplocaceae	*Symplocos*	5.3	5.2	0.990	1
	Theaceae	*Eurya*	2.1	1.6	0.764	1
		Sladenia	8.0	4.2	0.520	1
		Ternstroemia	18.6	13.5	0.728	1
	Ulmaceae	*Celtis*	6.5	4.8	0.740	1
		Gironniera	5.3	4.8	0.905	1
		Trema	2.6	2.5	0.992	1
	Verbenaceae	*Callicarpa*	1.6	1.2	0.791	1
		Gmelina	15.8	8.2	0.519	1
		Premna	5.2	4.8	0.938	1
	Vitaceae	*Ampelocissus*	10.7	6.0	0.561	1
		Ampelopsis	5.2	4.5	0.856	1
		Cayratia	5.5	4.2	0.765	1
		Cissus	8.9	6.9	0.776	2
		Leea	4.5	3.7	0.820	1
		Tetrastigma	8.6	7.1	0.838	3
		Vitis	5.1	3.7	0.725	1
	Zingiberaceae	*Alpinia*	6.6	5.4	0.809	1
		Hedychium	5.7	4.4	0.785	1

Appendix 2

List of species consuming fruit in four tropical rainforests on different continents showing body masses and how seeds are treated (C – Carry; Ca – Cache; E – Eat; R – Regurgitate; Sp – Spit; Sw – Swallow).

Rainforest	Order	Family	Species	Body mass (g)	Seed treatment
Central Africa	Birds	Musophagidae	*Tauraco persa*	185	Sw-R
		Musophagidae	*Tauraco fasciatus*	275	Sw-R
		Musophagidae	*Tauraco macrorhynchus*	295	Sw-R
		Bucerotidae	*Ceratogymna fistulator*	540	Sw-R
		Bucerotidae	*Ceratogymna cylindricus*	1.348	Sw-R
		Bucerotidae	*Ceratogymna atrata*	2,100	Sw-R
		Bucerotidae	*Buceros bicornis*	3.007	Sw-R
	Bats	Centrarchidae	*Micropterus pusillus*	22–32	Sw-C
		Phyllostomidae	*Micronycteris torquata*	43	Sw-C
		Pteropodidae	*Eidolon helvum*	250–310	Sw-C
		Pteropodidae	*Hypsignathus monstrosus*	275–377	Sw-C
		Pteropodidae	*Epomops buettikoferi*	340–700	Sw-C
	Primates	Cercopithecidae	*Miopithecus talapoin*	1,100-1,371	Sw-Sp
		Cercopithecidae	*Cercopithecus cephus*	2,839-4,132	Sw-Sp
		Cercopithecidae	*Cercopithecus ascanius*	2,924-3,700	Sw-Sp
		Cercopithecidae	*Cercopithecus pogonias*	2,940-4,442	Sw-Sp
		Cercopithecidae	*Cercopithecus mitis*	3,929-5,849	Sw-Sp
		Cercopithecidae	*Cercopithecus nictitans*	4,092-6,715	Sw-Sp
		Cercopithecidae	*Lophocebus albigena*	5,846-7,647	Sw-Sp
		Cercopithecidae	*Colobus guereza*	9,200-13,500	Sw-Sp
		Hominidae	*Pan troglodytes*	35	Sw-Sp
		Hominidae	*Gorilla gorilla*	71,000-175,000	Sw-Sp
	Rodents	Muridae	*Hylomyscus stella*	20	E
		Muridae	*Hylomyscus fumosus*	21	E
		Muridae	*Praomys tullbergi minor*	36	E
		Muridae	*Praomys lukolelae*	48	E
		Muridae	*Hybomys univittatus*	62	E
		Muridae	*Stochomys longicaudatus*	78	E
		Muridae	*Malacomys longipes*	78	E
		Sciuridae	*Heliosciurus rufobrachium*	363	E

Rainforest	Order	Family	Species	Body mass (g)	Seed treatment
Central Africa *Continued*		Sciuridae	*Epixerus ebii*	592	E-Ca
		Sciuridae	*Protoxerus strangeri*	691	E-Ca
		Nesomyidae	*Cricetomys emini*	1,250	E-Ca
		Hystricidae	*Atherurus atherurus*	3,650	E
	Ruminants	Bovidae	*Cephalophus monticola*	4,900	E-R-Sp
		Tragulidae	*Hyemoschus aquaticus*	10,800	E-R-Sp
		Bovidae	*Cephalophus leucogaster*	12,700	E-R-Sp
		Bovidae	*Cephalophus nigrifrons*	13,900	E-R-Sp
		Bovidae	*Cephalophus callipygus*	20,100	E-R-Sp
		Bovidae	*Cephalophus dorsalis*	21,700	E-R-Sp
		Bovidae	*Cephalophus sylvicultor*	68,000	E-R-Sp
	Elephant	Elephantidae	*Loxodonta africana*	5,000,000	E-Sw
Guianas	Bats	Phyllostomidae	*Rhynophylla pumilio*	8.3–9.5	Sw-C
		Phyllostomidae	*Carollia perspicillata*	15.3–15.6	Sw-C
		Phyllostomidae	*Artibeus gnomus*	9.2–10.0	Sw-C
		Phyllostomidae	*Artibeus cinereus*	10.7–12	Sw-C
		Phyllostomidae	*Artibeus obscurus*	33.9–37.7	Sw-C
		Phyllostomidae	*Artibeus jamaicensis*	53.4–56.7	Sw-C
		Phyllostomidae	*Artibeus literatus*	61–70.8	Sw-C
	Birds	Rhamphastidae	*Pteroglossus aracari*	135	Sw-R
		Rhamphastidae	*Selenidera culik*	140	Sw-R-Sp
		Contingidae	*Rupicola rupicola*	185	Sw-R
		Rhamphastidae	*Pteroglossus* spp.	390	Sw-R
		Rhamphastidae	*Rhamphastos tucanus*	700	Sw-R
		Tinamidae	*Tinamus major*	975	E
		Cracidae	*Penelope marail*	1,000	Sw
		Psophiidae	*Psophia crepitans*	1,070	Sw
		Cracidae	*Crax alector*	2,846–3,075	E
	Carnivores	Procyonidae	*Potos flavus*	1,600	Sw
	Marsupials	Didelphidae	*Caluromys philander*	300	Sw
	Primates	Cebidae	*Saimiri sciureus*	500	Sw
		Callitrichidae	*Saguinus midas*	540	Sw
		Cebidae	*Pithecia pithecia*	2,000	E
		Cebidae	*Chiropotes satanas*	3,000	E
		Cebidae	*Cebus apella*	3,100	Sw
		Cebidae	*Alouatta seniculus*	7,350	Sw
		Cebidae	*Ateles paniscus*	10,500	Sw
	Rodents	Cricetidae	*Oryzomys capito*	35	E
		Echimyidae	*Proechimys spp.*	350	E-C
		Dasyproctidae	*Myoprocta exilis*	1,000	E-C
		Dasyproctidae	*Dasyprocta leporina*	3,500	E-C
	Tapir	Tapiridae	*Tapirus terrestris*	250,000	Sw
	Ungulates	Cervidae	*Mazama gouazoubira*	16,500	E-R
		Cervidae	*Mazama americana*	44,000	E-R
Thailand	Birds	Pycnonotidae	*Pycnonotus atriceps*	< 5	Sw
		Pycnonotidae	*Pycnonotus finlaysoni*	< 5	Sw
		Pycnonotidae	*Pycnonotus melanicterus*	< 5	Sw
		Pycnonotidae	*Pycnonotus jocosus*	< 5	Sw
		Pycnonotidae	*Hypsipetes propinquus*	< 5	Sw
		Pycnonotidae	*Hypsipetes flavala*	< 5	Sw
		Pycnonotidae	*Criniger pallidus*	< 5	Sw
		Rhamphastidae	*Megalaima faiostricta*	40	Sw
		Columbidae	*Ducula badia*	500–600	Sw-R
		Bucerotidae	*Anthracoceros albirostris*	700–800	Sw-R
		Bucerotidae	*Anorrhinus austeni*	800–900	Sw-R
		Bucerotidae	*Aceros undulatus*	2,000–2,500	Sw-R
		Bucerotidae	*Buceros bicornis*	2,000–3,000	Sw-R

Rainforest	Order	Family	Species	Body mass (g)	Seed treatment
	Rodents	Sciuridae	*Callosciurus finlaysonii*	300	E-Ca
		Sciuridae	*Ratufa bicolor*	1,400	E-Ca
		Hystricidae	*Hystrix brachyura*	11,000–18,000	E
	Primates	Hylobatidae	*Hylobates lar*	4,000–7,000	Sw
		Hylobatidae	*Hylobates pileatus*	4,000–7,000	Sw
		Cercopithecidae	*Macaca nemestrina*	4,000–9,000	E-Sw
	Ungulates	Cervidae	*Muntiacus muntjak*	20,000–28,000	Sw
		Cervidae	*Cervus unicolor*	185,000–260,000	Sw
	Elephants	Elephantidae	*Elephas maximus*	4,000,000	Sw
Australia	Bat	Pteropodidae	*Pteropus conspicillatus*	700	Sw-C
	Birds	Dicaeidae	*Diceaum hirundinaceum*	9	Sw
		Zosteropidae	*Zosterops lateralis*	10	Sw
		Meliphagidae	*Meliphaga gracilis*	15	R
		Meliphagidae	*Lichenostomus chrysops*	16	R
		Meliphagidae	*Lichenostomus flavus*	22	R
		Meliphagidae	*Xanthotis macleayeana*	25	Sw
		Meliphagidae	*Meliphaga notata*	25	Sw
		Meliphagidae	*Lichenostomus versicolor*	25	R
		Meliphagidae	*Meliphaga lewenii*	33	Sw
		Campephagidae	*Lalage leucomela*	35	Sw
		Meliphagidae	*Lichenostomus frenatus*	36	Sw
		Psittacidae	*Glossopsilla pusilla*	37	E
		Psittacidae	*Psittaculirostris diopthalma*	41	E
		Cracticidae	*Coracina tenuirostris*	50	Sw
		Sturnidae	*Aplonis metallica*	60	Sw
		Cracticidae	*Coracina papuensis*	65	R
		Cracticidae	*Coracina lineata*	70	Sw
		Ptilonorhynchidae	*Prionodura newtoniana*	74	Sw
		Dicruridae	*Dicrurus bracteatus*	80	R
		Columbidae	*Ptilinopus reginae*	85	Sw
		Psittacidae	*Trichoglossus chlorolepidotus*	86	E
		Oriolidae	*Oriolus saggitatus*	96	Sw
		Paradisaeidae	*Ptiloris victoriae*	103	Sw
		Oriolidae	*Oiolus flavocinctus*	108	Sw
		Columbidae	*Ptilinopus superbus*	115	Sw
		Meliphagidae	*Philemon corniculatus*	120	R
		Columbidae	*Chalcophaps indica*	128	E
		Meliphagidae	*Entomyzon cyanotis*	130	Sw
		Oriolidae	*Sphecotheres viridis*	130	Sw
		Psittacidae	*Platycercus elegans*	130	E
		Cracticidae	*Coracina novaehollandiae*	133	Sw
		Psittacidae	*Trichoglossus haematodus*	133	E
		Meliphagidae	*Philemon buceroides*	140	Sw
		Columbidae	*Macropygia ambionensis*	142	E
		Columbidae	*Ptilinopus magnificus*	156	Sw
		Ptilonorhynchidae	*Scenopoeetes dentirostris*	165	Sw
		Cracticidae	*Cracticus mentalis*	170	R
		Ptilonorhynchidae	*Ailuroedus melanotis*	180	Sw
		Cuculidae	*Eudynamis scolopacea*	190	Sw
		Ptilonorhynchidae	*Ptilinorhynchus violaceus*	200	Sw
		Psittacidae	*Alisterus scapularis*	210	E
		Cracticidae	*Strepera graculina*	290	Sw
		Columbidae	*Columba leucomela*	490	E
		Columbidae	*Ducula bicolor*	500	Sw
		Columbidae	*Lopholaimus antarcticus*	500	Sw
		Cacatuidae	*Cacatua galerita*	500	E

Rainforest	Order	Family	Species	Body mass (g)	Seed treatment
Australia		Cuculidae	*Scythrops novaehollandiae*	680	Sw
Continued		Megapodidae	*Alectura lathemi*	2270	E
		Casuaridae	*Casuarius casuarius*	60,000	Sw
	Marsupials	Hypsyprymnodontidae	*Hypsiprymnodon moschatus*	520	C-Ca
		Psuedocheiridae	*Pseudocheirops archeri*	1000	E-Sp
		Psuedocheiridae	*Psuedocheirulus herbertensis*	1100	E-Sp
		Macropodidae	*Thylogale stigmatica*	4500	E-Sp
	Rodents	Muridae	*Melomys cervinipes*	80	E-Ca
		Muridae	*Rattus leucopus*	132	E-Ca
		Muridae	*Rattus fuscipes*	160	E-Ca
		Muridae	*Uromys hadrourus*	190	E-Ca
		Muridae	*Uromys caudimaculatus*	650	E-Ca

References for Appendices 1–2

Central Africa: Gautier-Hion, 1971; Alexandre, 1978; Emmons, 1980; Gautier-Hion *et al.*, 1980; Duplantier, 1982; Sourd, 1983; Dubost, 1984; Gautier-Hion *et al.*, 1985; Tutin and Fernandez, 1985; Sourd and Gautier-Hion, 1986; Debroux, 1988; Feer, 1989; Williamson *et al.*, 1990; Decoux and Erard, 1992; White, 1994; Wrangham *et al.*, 1994; Feer, 1995a,b; Whitney *et al.*, 1998; Lambert, 1999; Voysey *et al.*, 1999; Theuerkauf *et al.*, 2000; Poulsen *et al.*, 2002; Ebigbo, 2005; Richter and Cumming, 2006; C. Lehoucq and C. Erard, Brunoy, 2005, personal communication; J.E. Lambert, Wisconsin, 2005, personal communication; Electronic source – Bats at Bui National Park and Shai Hills <http://mampam.50megs.com/public_html/bvbats.html>.

The Guianas: Guillotin, 1982a,b; van Roosmalen, 1985a,b; Charles-Dominique, 1986; van Roosmalen *et al.*, 1988; Erard and Sabatier, 1989; Erard *et al.*, 1989; Julien-Laferriere, 1989; Théry *et al.*, 1992; Kinzey and Norconk, 1993; Théry and Larpin, 1993; Erard and Théry, 1994; Guillotin *et al.*, 1994; Norconk and Kinzey, 1994; Simmen and Sabatier, 1996; Cockle, 1997; Day, 1997; Norconk *et al.*, 1997; Vié, 1997; Simmons and Voss, 1998; Pack *et al.*, 1999; Galetti *et al.*, 2000; Henry *et al.*, 2000; Julien-Laferrière, 2000; Oliveira and Ferrari, 2000; Gayot *et al.*, 2004; F. Bayart, Brunoy, 2005, personal communication; S. Boinski, Gainesville, 2005, personal communication; K. Holbrook, St Louis, 2005, personal communication; P.A. Jansen, Panama, 2005, personal observation; P.-M. Forget, Brunoy, 2005, personal observation; T. Lobova, New York, 2005, personal communication; Electronic sources – <http://nybg.org/botany/tlobova/mori/batsplants/Checklist/Checklist_Home.htm>; University of Michigan Museum of Zoology <http://animaldiversity.ummz.umich.edu>; Toucans <http://www.summersbirds.com/speciesprofiles-toucans.htm>.

Thailand: Kitamura *et al.*, 2002, 2004.

Australia: Stocker and Irvine, 1983; Strahan, 1995; Baker *et al.*, 1997; Dennis, 1997, 2003; Dennis *et al.*, 2005; A.J. Dennis and D.W. Westcott, 2005, unpublished results.

References

Alexandre, D.Y. (1978) Le rôle disséminateur des éléphants en forêt de Taï, Côte d'Ivoire. *La Terre et la Vie* 32, 47–72.

Baker, G.B., Dettmann, E.B., Scotney, B.T., Hardy, L.J. and Drynan, D.A.D. (1997) *Report on the Australian Bird and Bat Banding Scheme, 1995–1996.* Environment Australia, Canberra.

Charles-Dominique, P. (1986) Inter-relations between frugivorous vertebrates and pioneer plants: *Cecropia*, birds and bats in French Guiana. In: Estrada, A. and Fleming, T.H. (eds) *Frugivores and Seed Dispersal.* Dr. W. Junk, Dordrecht, The Netherlands, pp. 119–135.

Cockle, A. (1997) Modalités de dissémination et d'établissement de lianes de sous-bois (Cylanthaceae et Philodendron) en forêt guyanaise. Doctoral thesis, University of Paris VI, Paris.

Day, R. (1997) Behavioural ecology of the tamarin, *Saguinus midas midas*, in a Guianense primate community. PhD thesis, The Queen's University of Belfast, Belfast.

Debroux, L. (1988) L'aménagement des forêts tropicales fondé sur la gestion des populations d'arbres: l'exemple du moabi (*Baillonella toxisperma* Pierre) dans la forêt du Dja, Cameroun. Doctoral thesis, Universitaire des Sciences Agronomiques de Gembloux, Gembloux, Belgium.

Decoux, J. and Erard, C. (1992) *Tauraco macrorhynchus* and *T. persa* in north-eastern Gabon: behavioral and ecological aspects of their co-existence. In: Bennun, L. (ed.) *Seventh Pan-African Ornithological Congress*, Nairobi, Kenya, pp. 369–380.

Dennis, A.J. (1997) Musky rat-kangaroos, *Hypsiprymnodon moschatus*: cursorial frugivores in Australia's wet tropical rain forests. PhD thesis, James Cook University of North Queensland, Townsville, Australia.

Dennis, A.J. (2003) Scatter-hoarding by musky rat-kangaroos, *Hypsiprymnodon moschatus*, a tropical rain-forest marsupial from Australia: implications for seed dispersal. *Journal of Tropical Ecology* 19, 619–627.

Dennis, A.J., Lipsett-Moore, G.J., Harrington, G., Collins, E. and Westcott, D. (2005) Seed predation, seed dispersal and habitat fragmentation: does context make a difference in tropical Australia? In: Forget, P.-M., Lambert, J., Hulme, P. and Vander Wall, S.B. (eds) *Seed Fate: Predation, Dispersal and Seedling Establishment.* CAB International, Wallingford, UK, pp. 117–135.

Dubost, G. (1984) Comparison of the diets of frugivorous forest ruminants of Gabon. *Journal of Mammalogy* 65, 298–316.

Duplantier, J.-M. (1982) Les rongeurs myomorphes forestiers du nord-est du Gabon: peuplements, utilisation de l'espace et des ressources alimentaires, rôle dans la dispersion et la germination des graines. Université STL Montpellier II, Montpellier, France.

Emmons, L.H. (1980) Ecology and resource partitioning among nine species of African rain forest squirrels. *Ecological Monographs* 50, 31–54.

Ebigbo, N.M. (2005) The role of flying foxes as seed dispersers on the vegetation dynamics in a west African forest–savannah mosaic in Côte d'Ivoire. Ulm University, Ulm, Germany.

Erard, C. and Sabatier, D. (1989) Rôle des oiseaux frugivores terrestres dans la dynamique forestière en Guyane française. *Proceedings of the International Ornithological Congress* 19, 803–815.

Erard, C. and Théry, M. (1994) Frugivorie et ornithochorie en forêt guyanaise: l'exemple des grands oiseaux terrestres et de *Penelope marail*. *Alauda* 62, 27–31.

Erard, C., Théry, M. and Sabatier, D. (1989) Régime alimentaire de *Rupicola rupicola* (Cotingidae) en Guyane française; relations avec la frugivorie et la zoochorie. *Revue d'Ecologie (Terre et Vie)* 44, 47–74.

Feer, F. (1989) Comparaison des régimes alimentaires de *Cephalophus callipygus* et *C. dorsalis*, bovidés sympatriques de la forêt sempervirente africaine. *Mammalia* 53, 563–604.

Feer, F. (1995a) Morphology of fruits dispersed by African forest elephants. *African Journal of Ecology* 33, 279–284.

Feer, F. (1995b) Seed dispersal in African forest ruminants. *Journal of Tropical Ecology* 11, 683–689.

Galetti, M., Laps, R. and Pizo, M.A. (2000) Frugivory by toucans (Ramphastidae) at two altitudes in the Atlantic forest of Brazil. *Biotropica* 32, 842–850.

Gautier-Hion, A. (1971) L'écologie du talapoin du Gabon. *La Terre et la Vie* 25, 427–490.

Gautier-Hion, A., Emmons, L.H. and Dubost, G. (1980) A comparison of the diets of three major groups of primary consumers of Gabon (primates, squirrels and ruminants). *Oecologia* 45, 182–189.

Gautier-Hion, A., Duplantier, J.-M., Quris, R., Feer, F., Sourd, C., Decoux, J., Dubost, G., Emmons, L., Erard, C., Hecketsweiler, P., Moungazi, A., Roussilhon, C. and Thiollay, J.-M. (1985) Fruit characters as a basis of fruit choice, and seed dispersal by a tropical forest vertebrate community. *Oecologia* 65, 324–337.

Gayot, M., Henry, O., Dubost, G. and Sabatier, D. (2004) Comparative diet of the two forest cervids of the genus *Mazama* in French Guiana. *Journal of Tropical Ecology* 20, 31–43.

Guillotin, M. (1982a) Activités et régimes alimentaires: leurs interactions chez *Proechimys cuvieri* et *Oryzomys capito velutinus* (Rodentia) en forêt guyanaise. *Revue d'Ecologie (Terre et Vie)* 36, 337–371.

Guillotin, M. (1982b) Place de *Proechimys cuvieri* (Rodentia, Echimidae) dans les peuplements micromammaliens terrestres de la forêt Guyanaise. *Mammalia* 46, 299–318.

Guillotin, M., Dubost, G. and Sabatier, D. (1994) Food choice and food composition among three major primate species of French Guiana. *Journal of Zoology (London)* 233, 551–579.

Henry, O., Feer, F. and Sabatier, D. (2000) Diet of the lowland tapir (*Tapirus terrestris* L.) in French Guiana. *Biotropica* 32, 364–368.

Julien-Laferriere, D. (1989) *Utilisation de l'espace et des ressources alimentaires chez* Caluromys philander *(Marsupialia-Didelphidae) comparaison avec* Potos flavus *(Eutheria, Procionidae).* Thèse de Doctorat, spécialité Biologie du comportement. Université de Paris XIII, Villetaneuse.

Julien-Laferrière, D. (2000) Frugivory and seed dispersal by kinkajous. In: Bongers, F., Charles-Dominique, P., Forget, P.-M. and Théry, M. (eds) *Nouragues: Dynamics and Plant–Animal Interactions in a Neotropical Rain Forest.* Kluwert Academic, Dordrecht, The Netherlands.

Kinzey, W.G. and Norconk, M.A. (1993) Physical and chemical properties of fruit and seeds eaten by *Pithecia* and *Chiropotes* in Surinam and Venezuela. *International Journal of Primatology* 14, 207–228.

Kitamura, S., Yumoto, T., Poonswad, P., Chuailua, P., Plongmai, K., Maruhashi, T. and Noma, N. (2002) Interactions between fleshy fruits and frugivores in a tropical seasonal forest in Thailand. *Oecologia* 133, 559–572.

Kitamura, S., Shunsuke, S., Yumoto, T., Poonswad, P., Chuailua, P., Plongmai, K., Noma, N., Maruhashi, T. and Suckasam, C. (2004) Dispersal of *Aglaia spectabilis*, a large-seeded tree species in a moist evergreen forest in Thailand. *Journal of Tropical Ecology* 20, 421–427.

Lambert, J.E. (1999) Seed handling in chimpanzees (*Pan troglodytes*) and redtail monkeys (*Cercopithecus ascanius*): implications for understanding hominoid and cercopithecine fruit-processing strategies and seed dispersal. *American Journal of Physical Anthropology* 109, 365–386.

Norconk, M.A. and Kinzey, W.G. (1994) Challenge for neotropical frugivory: travel patterns of spider monkeys and bearded sakis. *American Journal of Primatology* 34, 171–183.

Norconk, M.A., Wertis, C. and Kinzey, W.G. (1997) Seed predation by monkeys and macaws in eastern Venezuela: preliminary findings. *Primates* 38, 177–184.

Oliveira, A.C.M. and Ferrari, S.F. (2000) Seed dispersal by black-handed tamarins, *Saguinus midas niger* (Callitrichinae, Primates): implications for the regeneration of degraded forest habitats in eastern Amazonia. *Journal of Tropical Ecology* 16, 709–716.

Pack, K.S., Henry, O. and Sabatier, D. (1999) The insectivorous–frugivorous diet of the golden-handed tamarin (*Saguinus midas midas*) in French Guiana. *Folia Primatologica* 70, 1–7.

Poulsen, J.R., Clark, C.J., Connor, E.F. and Smith, T.B. (2002) Differential resource use by primates and hornbills: implications for seed dispersal. *Ecology* 83, 228–240.

Richter, H.V. and Cumming, G.S. (2006) Food availability and annual migration of the straw-colored fruit bat (*Eidolon helvum*). *Journal of Zoology* 268, 35–44.

Simmen, B. and Sabatier, D. (1996) Diets of some French Guianan primates: food composition and food choice. *International Journal of Primatology* 17, 661–693.

Simmons, N.B. and Voss, R.S. (1998) Mammals of Paracou, French Guiana: a neotropical rainforest fauna. Part 1. Bats. *Bulletin of the American Museum of Natural History* 237, 1–219.

Sourd, C. (1983) Etudes des modes d'exploitation des ressources fruitières par *Cercopithecus cephus* au cours d'un cycle annuel. Thesis, Université de Rennes I, Rennes, France.

Sourd, C. and Gautier-Hion, A. (1986) Fruit selection by a forest guenons. *Journal of Animal Ecology* 55, 235–244.

Stocker, G.C. and Irvine, A.K. (1983) Seed dispersal by cassowaries *Casuarius casuarius* in north Queensland's rainforests. *Biotropica* 15, 170–176.

Strahan, R. (ed.) (1995) *The Mammals of Australia.* Reed Books/Australian Museum, Chatswood, Australia.

Théry, M. and Larpin, D. (1993) Seed dispersal and vegetation dynamics at a cock-of-the-rock's lek in the tropical forest of French Guiana. *Journal of Tropical Ecology* 9, 109–116.

Théry, M., Erard, C. and Sabatier, D. (1992) Les fruits dans le régime alimentaire de *Penelope marail* (Aves, Cracidae) en Guyane française: frugivorie stricte et sélective? *Revue d'Ecologie (Terre et Vie)* 47, 383–401.

Theuerkauf, J., Waitkuwait, W.E., Guiro, Y., Ellenberg, H. and Porembski, S. (2000) Diet of forest elephants and their role in seed dispersal in the Bossematie Forest Reserve, Ivory Coast. *Mammalia* 64, 447–459.

Tutin, C.E.G. and Fernandez, M. (1985) Food consumed by sympatric populations of *Gorilla g. gorilla* and *Pan t. troglodytes* in Gabon: some preliminary data. *International Journal of Primatology* 6, 27–43.

van Roosmalen, M.G.M. (1985a) *Fruits of the Guiana Flora.* Institute of Systematic Botany, Utrecht University, Utrecht, The Netherlands.

van Roosmalen, M.G.M. (1985b) Habitat preferences, diet, feeding strategy and social organization of the black spider monkey (*Ateles paniscus* Linnaeus 1758) in Surinam. *Acta Amazonica* (Suppl.) 15, 1–238.

van Roosmalen, M.G.M., Mittermeier, R.A. and Fleagle, J.G. (1988) Diet of the northern bearded saki (*Chiropotes satanas chiropotes*): a neotropical seed predator. *American Journal of Primatology* 14, 11–36.

Vié, J.-C. (1997). Ecology and behavior of white-faced sakis (*Pithecia pithecia*) in French Guiana: preliminary results. *XVIth Congress of the International Primatological Society and the XIXth Conference of the American Society of Primatologists, 11–16 August.* University of Wisconsin, Madison, Wisconsin, p. 37.

Voysey, B.C., McDonald, K.E., Rogers, M.E., Tutin, C.E.G. and Parnell, R.J. (1999) Gorillas and seed dispersal in the Lope Reserve, Gabon. I. Gorilla acquisition by trees. *Journal of Tropical Ecology* 15, 23–38.

White, L.J.T. (1994). *Sacoglottis gabonensis* fruiting and the seasonal movements of elephants in the Lopé Reserve, Gabon. *Journal of Tropical Ecology* 10, 121–125.

Whitney, K.D., Fogiel, M.K., Lamperti, A.M., Holbrook, K.M., Stauffer, D.J., Hardesty, B.D., Parker, V.T. and Smith, T.B. (1998) Seed dispersal by *Ceratogymna* hornbills in the Dja Reserve, Cameroon. *Journal of Tropical Ecology* 14, 351–371.

Williamson, E.A., Tutin, C.E.G., Rogers, E.M. and Fernandez, M. (1990) Composition of the diet of lowland gorillas at Lopé in Gabon. *American Journal of Primatology* 21, 265–277.

Wrangham, R.W., Chapman, C.A. and Chapman, L.J. (1994) Seed dispersal by forest chimpanzees in Uganda. *Journal of Tropical Ecology* 10, 355–368.

Appendix 3

Families, species, habitats, fruit size, fruit colour and type of seed disperser of fleshy-fruited species studied at Fazenda Rio Negro, Nhecolândia region, MS, Brazil.

Family	Genus	Species	Habitat	Fruit length (mm)	Fruit width (mm)	Fruit weight (g)	Seed disperser*	Fruit colour
Anacardiaceae	*Spondias*	*lutea*	Semideciduous forest	25.1 ± 1.7	21.7 ± 1.4	6.45 1.4	mammals	yellow
Annonaceae	*Annona*	*dioica*	Cerrado	92.7 ± 4.4	75.1 ± 5.2	250.0 ± 22.7	mammals	green
Annonaceae	*Annona*	*cornifolia*	Cerrado	39.0 ± 7.5	38.0 ± 9.2	23.6 ± 12.7	mammals	orange
Annonaceae	*Unonopsis*	*lindmanii*	Cerrado	9.3 ± 0.2	9.2 ± 0.2	0.6 ± 0.02	birds	orange
Apocynaceae	*Hancornia*	*speciosa*	Semideciduous forest	30.7 ± 3.3	29.8 ± 3.7	15.4 ± 5.3	mammals+rhea	green
Arecaceae	*Acrocomia*	*aculeata*	Semideciduous forest	31.9 ± 2.1	32.1 ± 2.8	19.2 ± 2.7	mammals+rhea	brown
Arecaceae	*Attalea*	*speciosa*	Cerrado	87.7 ± 7.8	50.7 ± 4.1		none	brown
Arecaceae	*Attalea*	*phalerata*	Semideciduous forest	62.7 ± 3.7	34.9 ± 2.9	39.6 ± 6.8	mammals+rhea	orange
Arecaceae	*Bactris*	*glaucescens*	Gallery forest	18.8 ± 3.7	17.7 ± 1.5	2.9 ± 0.8	mammals+rhea	black
Arecaceae	*Copernicia*	*alba*	Semideciduous forest	19.1 ± 1.7	16.4 ± 1.0	2.7 ± 0.5	mammals+rhea	black
Arecaceae	*Syagrus*	*flexuosa*	Semideciduous forest	37.3 ± 2.0	15.9 ± 1.5	6.1 ± 1.7	mammals+tortoise	yellow
Bromeliaceae	*Bromelia*	*balansae*	Cerrado	33.8 ± 2.4	29.2 ± 2.9	16.9 ± 4.4	mammals	yellow
Bursenaceae	*Protium*	*heptaphyllum*	Semideciduous forest	21.0 ± 1.4	15.4 ± 1.1	2.0 ± 0.5	mixed	red/white
Cactaceae	*Cereus*	*peruvianus*	Semideciduous forest	68.6 ± 17.3	47.8 ± 5.3	103.3 ± 40.3	birds	red/white
Caryocaraceae	*Caryocar*	*brasiliense*	Semideciduous forest	60.2 ± 1.3	65.7 ± 3.0	137.2 ± 12.2	mammals	green
Cecropiaceae	*Cecropia*	*pachystachya*	Gallery forest	80.1 ± 8.5	8.0 ± 0.6	1.9 ± 0.3	mixed	green
Chrysobalanaceae	*Couepia*	*uiti*	Gallery forest	44.9 ± 6.2	26.8 ± 3.8	29.2 ± 2.9	mammals	brown
Chrysobalanaceae	*Licania*	*parvifolia*	Cerrado	29.7 ± 3.8	12.5 ± 1.8	2.5 ± 0.7	fish	green
Clusiaceae	*Calophyllum*	*brasiliense*	Gallery forest	25.1 ± 1.8	24.4 ± 1.9	8.0 ± 1.7	bats	green
Connaraceae	*Connarus*	*suberosus*	Cerrado	15.4 ± 1.6	8.9 ± 0.6	0.8 ± 0.09	ants	red/black
Dilleniaceae	*Curatella*	*americana*	Semideciduous forest	11.5 ± 0.9	8.6 ± 0.6	0.9 ± 0.07	birds	red/white
Dilleniaceae	*Doliocarpus*	*dentatus*	Gallery forest	6.7 ± 0.8	5.5 ± 0.6	0.1 ± 0.03	birds	red/white
Ebenaceae	*Diospyros*	*hispida*	Cerrado	42.3 ± 3.5	47.2 ± 2.6	57.3 ± 10.3	mammals	green
Erythroxylaceae	*Erythroxylum*	*anguifugum*	Semideciduous forest	10.0 ± 1.1	6.0 ± 0.7	0.2 ± 0.06	birds	red
Euphorbiaceae	*Sapium*	*haematospermum*	Cerrado	6.4 ± 1.3	4.7 ± 0.6	0.1 ± 0.02	birds	red/green
Fabaceae	*Andira*	*cuyabensis*	Semideciduous forest	40.4 ± 2.3	31.2 ± 0.8	21.8 ± 2.5	bats	green
Fabaceae	*Andira*	*inermis*	Cerrado	46.4 ± 4.9	28.8 ± 2.5		bats	green
Fabaceae	*Dipteryx*	*alata*	Semideciduous forest	48.8 ± 3.6	39.9 ± 2.4	27.4 ± 6.9	mammals	brown
Fabaceae	*Inga*	*uruguensis*	Gallery forest	59.2 ± 14.5	19.0 ± 1.8	9.5 ± 4.3	mammals+fish	brown
Fabaceae	*Inga*	*laurina*	Cerrado	53.1 ± 14.3	21.9 ± 1.6	12.3 ± 5.3	mammals+fish	green
Fabaceae	*Copaifera*	*martii*	Semideciduous forest	18.1 ± 3.3	14.0 ± 1.0	1.4 ± 0.7	birds	white

Flacourtiaceae	*Casearia*	*rupestris*	Semideciduous forest	24.3 ± 4.1	25.9 ± 3.5	9.1 ± 4.4	birds	orange
Guttiferae	*Garcinia*	*brasiliensis*	Gallery forest	27.5 ± 5.9	22.6 ± 2.6	7.9 ± 3.7	mammals+fish	yellow
Hippocrateaceae	*Salacia*	*elliptica*	Gallery forest	53.2 ± 7.3	46.0 ± 5	66.1 ± 21.2	mammals+fish	orange
Lauraceae	*Ocotea*	*diospyrifolia*	Semideciduous forest	9.4 ± 0.8	5.9 ± 0.4	0.2 ± 0.04	birds+fish	red/black
Leguminosae	*Hymenaea*	*stigonocarpa*	Cerrado	97.0 ± 12.5	41.7 ± 4.3	48.5 ± 20.2	mammals	brown
Leguminosae	*Hymenaea*	*courbaril*	Semideciduous forest	108.9 ± 7.9	48.3 ± 3.1		mammals	brown
Leguminosae	*Enterolobium*	*contortisiliquum*	Semideciduous forest	39.1 ± 1.6	78.8 ± 1.6	30.8 ± 3.9	mammals	brown
Loranthaceae	*Psittacanthus*	sp.	Cerrado	26.3 ± 31.9	11.8 ± 1.0	1.8 ± 0.2	birds	green
Malphigiaceae	*Bunchosia*	*paraguariensis*	Cerrado	5.8 ± 0.07	4.8 ± 0.4	0.1 ± 1.5	birds	orange
Malphigiaceae	*Byrsonima*	*orbignyana*	Cerrado	12.6 ± 1.4	12.1 ± 1.6	1.4 ± 1.5	mammals+tortoise	yellow
Malphigiaceae	*Byrsonima*	*verbascifolia*	Cerrado	17.1 ± 1.5	18.1 ± 1.0	4.1 ± 0.6	mammals	yellow
Melastomataceae	*Miconia*	*prasina*	Cerrado	5.7 ± 0.7	7.0 ± 0.4	0.2 ± 0.04	birds	black
Meliaceae	*Trichilia*	*elegans*	Semideciduous forest	7.1 ± 1.1	4.8 ± 0.7		birds	black/red
Moraceae	*Ficus*	*pertusa*	Semideciduous forest	7.1 ± 0.4	7.4 ± 0.5	0.2 ± 0.02	mixed	purple
Moraceae	*Ficus*	*insipida*	Semideciduous forest	21.3 ± 1.4	18.9 ± 2.1	2.3 ± 0.4	mixed	green
Myrtaceae	*Eugenia*	*desynterica*	Semideciduous forest	23.3 ± 7.2	25.8 ± 8.1	9.1 ± 7.8	mammals+tortoise	yellow
Myrtaceae	*Syzigium*	*cumini*	Semideciduous forest	21.3 ± 2.2	17.1 ± 2.4	4.2 ± 1.2	mixed	purple
Myrtaceae	*Psidium*	*nutans*	Semideciduous forest	30.1 ± 5.2	26.4 ± 4.4	13.0 ± 4.6	mixed	yellow
Myrtaceae	Indeterminate1		Gallery forest	19.6 ± 2.4	17.2 ± 1.2	3.4 ± 0.7	birds	orange
Myrtaceae	Indeterminate 2		Cerrado	8.8 ± 1.0	10.8 ± 1.3	0.7 ± 0.2	mixed	red
Ochnaceae	*Ouratea*	*hexasperma*	Semideciduous forest	9.4 ± 0.8	5.9 ± 0.4	0.2 ± 0.04	birds	black/red
Olacaceae	*Ximenia*	*americana*	Semideciduous forest	21.7 ± 2.3	20.9 ± 2.0	6.6 ± 1.4	mammals+tortoise	yellow
Opiliaceae	*Agonandra*	*brasiliensis*	Semideciduous forest	24.8 ± 1.7	24.2 ± 1.9	9.4 ± 2.0	mammals+tortoise	yellow
Rhamnaceae	*Rhamnidium*	*elaeocarpum*	Semideciduous forest	10.9 ± 1.5	7.5 ± 1.0	0.4 ± 0.07	birds	black
Rubiaceae	*Alibertia*	*sessilis*	Cordileira	26.2 ± 2.5	26.1 ± 2.0	11.4 ± 3.2	mixed	black
Rubiaceae	*Genipa*	*americana*	Gallery forest	75.7 ± 5.0	60.0 ± 4.6	130.2 ± 26.7	mixed	brown
Rubiaceae	*Psychotria*	sp.	Gallery forest	12.0 ± 0.81	10.8 ± 1.7	0.8 ± 0.1	birds	white
Rubiaceae	*Psychotria*	*carthagenensis*	Gallery forest	5.9 ± 0.5	5.8 ± 0.5	0.1 ± 0.03	birds	red
Rubiaceae	*Tocoyena*	*formosa*	Gallery forest	48.1 ± 6.9	33.7 ± 5.1	26.9 ± 10.4	mammals+birds	yellow
Rutaceae	*Zanthoxyllum*	*rigidum*	Semideciduous forest	5.1 ± 0.3	3.7 ± 0.15	0.04 ± 0.0	birds	black
Sapindaceae	*Melicoccus*	*lepidopetalus*	Gallery forest	26.5 ± 3.6	20.6 ± 1.6	6.8 ± 1.4	mammals	yellow
Sapindaceae	*Paullinia*	*pinnata*	Cerrado	17.4 ± 1.7	11.2 ± 2.1	1.0 ± 0.4	birds+fish	red/black

Continued

Family	Genus	Species	Habitat	Fruit length (mm)	Fruit width (mm)	Fruit weight (g)	Seed disperser*	Fruit colour
Sapindaceae	*Sapindus*	*saponaria*	Semideciduous forest	18.0 ± 3.0	18.6 ± 3.1	2.8 ± 0.7	bats	brown
Sapindaceae	*Talisia*	*esculenta*	Semideciduous forest	23.8 ± 1.6	21.4 ± 1.3	5.8 ± 1.5	mammals	brown
Sapotaceae	*Chrysophyllum*	*marginatum*	Semideciduous forest	6.7 ± 0.7	6.8 ± 0.5	0.2 ± 0.03	birds	black
Sapotaceae	*Pouteria*	*ramiflora*	Semideciduous forest	41.3 ± 2.1	31.9 ± 2.0	28.1 ± 0.5	mammals	green
Sapotaceae	*Pouteria*	*gardneri*	Semideciduous forest	15.7 ± 3.1	11.9 ± 2.3	1.4 ± 0.4	mammals+tortoise	purple
Simaroubaceae	*Simarouba*	*versicolor*	Cerrado	30.9 ± 0.9	26.1 ± 1.3	12.8 ± 1.1	bats	green
Solanaceae	*Solanum*	*viarum*	Pasture	33.3 ± 1.5	34.5 ± 4.8	15.3 ± 4.8	mammals	green
Sterculiaceae	*Guazuma*	*ulmifolia*	Semideciduous forest	18.0 ± 1.0	19.9 ± 1.1	3.3 ± 0.6	mammals	black
Sterculiaceae	*Sterculia*	*apelata*	Semideciduous forest	25.5 ± 1.3	23.6 ± 0.8	4.9 ± 0.5	birds	red/black
Verbenaceae	*Vitex*	*cymosa*	Semideciduous forest	22.1 ± 3.8	16.4 ± 4.5	4.0 ± 2.9	mixed	black
Vitaceae	*Cissus*	*erosa*	Semideciduous forest	5.1 ± 0.5	5.8 ± 0.6	0.1 ± 0.03	birds+fish	black
Vitaceae	*Cissus*	*spinosa*	Gallery forest	6.8 ± 0.4	8.2 ± 0.6	0.3 ± 0.06	birds+fish	black

*Seed disperser 'mixed' includes mammals, birds and other seed dispersers.

Glossary of Terms Used in Studies of Seed Dispersal

Allele frequency
A measure of the relative frequency at which an allele occurs on a locus within a population.

Allelic richness
The number of alleles detected within a sample or population; a fundamental measure of genetic diversity.

Anemochory
Diaspore dispersal by wind (see van der Pijl, 1972).

Atelechory
The limitation of dispersal to an already occupied and thus obviously suitable site; in a sense, the avoidance of dispersal (see van der Pijl, 1972).

Autochory
Seed dispersal by the plant itself, including ballistic dispersal and creeping diaspores (see van der Pijl, 1972).

Barochory
Seed dispersal by gravity alone (see van der Pijl, 1972).

Contagious seed dispersal
The patchy deposition of seeds, such that some sites receive many seeds and others receive few to none (see Schupp *et al.*, 2002).

Contrast hypothesis
Fruits that are best at attracting seed dispersers are those that are most easily distinguished from their background (see Schmidt *et al.*, 2004).

Defence indication hypothesis
Situations where the production of fruit colour (e.g. anthocyanins) is linked to the production of defensive compounds in leaves (e.g. tannins)

through a common biosynthetic pathway (e.g. phenylpropanoid pathway); fruit colour therefore indicating that defensive chemicals might be present (see Schaefer and Rolshausen, 2006).

Deinhibition
The release of seeds from germination inhibitors such as those that block light or biochemical pathways or control osmotic pressure, after the removal of fruit pulp (Robertson *et al.*, 2006).

Density-dependent effects
The ecological consequences of density. In seed dispersal studies it generally refers to density-dependent seed or seedling mortality, in particular negative density dependence where survival decreases with increasing densities of seeds or seedlings.

Diaspore
The dispersal unit, such as seed, fruit, branches or the entire plant. In many cases, the term 'diaspore' is replaced by 'seed', even if the diaspore is not a seed.

Differential dispersal hypothesis
A situation where co-occuring species display significantly different seed dispersal resulting in different patterns of dispersion, which potentially leads to different spatial arrangements of seedlings and adults influencing forest structure (see Smallwood *et al.*, 1998).

Diplochory
Dispersal by sequential vectors (see van der Pijl, 1972).

Dispersal curve
See '**Dispersal kernal**'

Dispersal kernel
Originally, the probability density function of the post-dispersal density of diaspores relative to their source; it yields the probability (*per-unit-area*) of a diaspore landing at any distance from its source. Recently, the term has been applied more broadly to any probability density function of the distances diaspores are dispersed. See also '**Distance distribution**'.

Dispersal limitation
See '**Dissemination limitation**'

Dispersal vector
An agent (biotic or abiotic, e.g. bird, wind, etc.) that disperses plant seeds or other diaspores.

Disperser effectiveness
The contribution a disperser makes to the future reproduction of a plant (see Schupp, 1993).

Dissemination limitation
When recruitment is limited by the failure to disperse seeds to all suitable dispersal sites or microsites. Largely equivalent to what many call dispersal

limitation, this terminology also acknowledges that recruitment is limited by the dispersal of seeds to unsuitable sites, that there are important delayed consequences to dispersal (see Schupp *et al.*, 2002; Jordano and Godoy, 2002).

Distance distribution
The probability density function of diaspores in relation to their source after dispersal, yielding the probability of a diaspore landing at any distance from its source. See also '**Dispersal kernel**'.

Distance insensitive
A dispersal pattern showing lack of dependence on the distance from the source.

Distance-dependent effects
The ecological consequences of distance. In seed dispersal studies it generally refers to distance-dependent seed or seedling mortality, where mortality increases with proximity to a conspecific adult.

Ecological redundancy
The degree to which two or more species perform the same ecosystem function. Complete redundancy occurs when the remaining species have exact density compensation following the removal of a species (see Walker, 1992).

Endozoochory
Seed dispersal by animals with diaspores carried inside the animal (see van der Pijl, 1972).

Epizoochory
Seed dispersal by animals with the diaspores accidentally carried on the outside of the animal (see van der Pijl, 1972).

Establishment limitation
See '**Microsite limitation**'

Exozoochory
Seed dispersal by animals with the diaspores deliberately carried by the animal (see van der Pijl, 1972).

Frugivore
Generally used to describe organisms that digest the flesh or pulp of fruit (e.g. bacteria, beetles, bears, birds, etc.). Its use sometimes incorporates granivores, which digest seeds.

Fruit flags
See '**Secondary structures in fruit**'

Fruit signals
Methods of communication between plants and animals generally altering the detectability of fruit displays according to the perception of consumers.

Fruit syndrome
A set of fruit traits thought to be associated with particular groups of consumers (e.g. bird-fruit, bat-fruit, etc.; see van der Pijl, 1972).

Fruit types
May be used to refer to fruit syndromes but generally refers to botanical structures (e.g. syconium, drupe, berry, etc.).

Genetic bottleneck
An event that results in killing off or otherwise preventing a significant percentage of a population or species from reproducing resulting in a reduction in genetic diversity for the species.

Genetic diversity
Based upon the average number of alleles per polymorphic locus, the effective number of alleles per locus or Nei's measure of Genetic diversity: $(H_e) = (1-\Sigma p i^2$, where p*i* is the frequency of the *i*th allele).

Genetic drift
The random change of allele frequencies over time in a finite population and through successive generations.

Genetic neighbourhood
The immediate area around an individual that is comprised of a given proportion of related individuals.

Germinability
The ability of a seed to germinate due to some influence (such as passage through a gut or with intact flesh). Can be measured as germination percentage and germination speed.

Germination percentage
The proportion of seeds that germinate after maturation, experimental planting or some treatment (see Robertson *et al.*, 2006).

Germination speed
The length of time a seed takes to germinate from maturation, experimental planting or some treatment (see Robertson *et al.*, 2006).

Germination rate
A confusing term that has been used to refer to either germination percentage or germination speed; a measure of germination success (see Robertson *et al.*, 2006).

Granivore
An organism that consumes seeds or parts of seeds; vertebrate granivores (often called seed predators) frequently also act as seed dispersers.

Haplochory
Dispersal mediated by a single vector, or a particular vector type (see van der Pijl, 1972).

Hydrochory
Seed dispersal by water (see van der Pijl, 1972).

Ichthyochory
Seed dispersal by fishes (see van der Pijl, 1972).

Larder-hoarding
Storage of food items in large caches containing multiple items. Larder-hoarders generally have one or a few large hoards but the term can overlap with the use of scatter-hoard when animals have a large number of small larders (see Vander Wall, 1990).

Life stage conflicts
When the biotic and abiotic conditions associated with a particular type of site are advantageous for some life history stages but detrimental for others. Thus, for example, a site may be highly suitable for seed survival and germination but unsuitable for seedling growth and survival.

Long-distance dispersal
A *proportional definition* assigns long-distance dispersal (LDD) events as those beyond a certain high percentile at the tail of the cumulative (or total) dispersal kernel. An *absolute definition* sets a threshold distance that corresponds to key features of the system of interest (e.g. inter-patch distance).

Mammaliochory
Seed dispersal by mammals (see van der Pijl, 1972).

Megafauna syndrome
Refers to fruits thought to be adapted to dispersal by large-bodied vertebrates (>44kg).

Microsatellite
Tandem repeats of 1–6+ nucleotide length found at high frequency in the nuclear genome. Also known as simple sequence repeats (SSR), variable number tandem repeats (VNTR) or short tandem repeats (STR).

Microsite limitation
Recruitment is limited by the availability of suitable microsites (see Eriksson and Ehrlen, 1992).

Microsite stochasticity
Localized, non-predictable variability in a small environment area or habitat affecting a plant or group of plants.

Myrmecochory
Seed dispersal by ants (see van der Pijl, 1972).

Ornithochory
Seed dispersal by birds (see van der Pijl, 1972).

Phase I dispersal
See '**Primary dispersal**'

Phase II dispersal
See '**Secondary dispersal**'

Polychory
Dispersal by multiple vectors (see van der Pijl, 1972).

Primary dispersal
The stage of seed dispersal where seeds move from the parent plant to a substrate. Primary dispersal may be by frugivorous animals, wind, gravity, and much more (see Chambers and MacMahon, 1994; Vander Wall, 2003).

Quality of dispersal
The quality of dispersal provided to an individual seed, including treatment in mouth and gut, location of deposition, etc. (see Schupp, 1993).

Quantity of dispersal
The number of seeds dispersed by an individual, population, species or community. It is influenced by factors such as the number of visits made, the number of fruits handled per visit, etc. (see Schupp, 1993).

Recruitment limitation
The abundance of a species being below the maximum the environment will support due to limited numbers of recruits (see Muller-Landau *et al.*, 2002).

Safe site
The very local area, determined by the size of the seed, that provides: (i) the stimuli necessary for breaking seed dormancy; (ii) the conditions that are required for germination to proceed; (iii) the resources such as oxygen and water used in germination; and (iv) the absence of hazards such as seed consumers, competitors, and pre-emergence pathogens (see Harper, 1977).

Saurochory
Seed dispersal by reptiles (see van der Pijl, 1972).

Scatter-hoarding
Storage of food items in multiple small and scattered caches. Frequently single but may contain multiple items per cache (see also '**Larder-hoarding**'; Vander Wall, 1990).

Secondary compound
Plant compounds that are not part of a primary metabolic pathway such as photosynthesis, respiration or uptake. Instead, they often function in defence against antagonists or signalling or rewarding mutualists (see Whittaker and Feeny, 1971).

Secondary dispersal
Any movement of seeds following primary dispersal. Secondary dispersal may be by water, wind, animal activity and much more (see Vander Wall 2003).

Secondary metabolite
See '**Secondary compound**'

Secondary structures in fruit
Structures (non-fruit) that increase the detectibility of fruit displays, including for example, bracts, stems or unripe fruit.

Seed fate
Simply, the fate of a seed. However, in seed dispersal studies the term implies passage through a string of processes, such as pre-dispersal predation, treatment and handling by frugivores and granivores, patterns of deposition, the influence of deposition time and site (neighbourhood) and post-dispersal seed and seedling mortality, to list a few (see Forget *et al.*, 2005).

Seed limitation
Recruitment is limited by the availability of seeds (Eriksson and Ehrlen, 1992).

Seed shadow
The spatial distribution of all seeds dispersed from a single plant. Sometimes used to refer to the population-level distribution of seeds, or the sum of all individual seed shadows (see Nathan and Muller-Landau, 2000).

Seedling shadow
The spatial distribution of all seedlings dispersed from a single plant. Sometimes used to refer to the population-level distribution of seedlings, or the sum of all individual seedling shadows.

Source limitation
See '**Seed limitation**'

Spatial structure
The quantitative representation of the degree of clumping of individuals at particular spatial scales. Numerous spatial statistical analyses can detect whether individuals (e.g. seeds) are distributed in a clumped, random, or regular manner. Having 'structure' often implies a non-random distribution.

Suitability of a site for dispersal
The likelihood that a seed dispersed to a particular type of site will survive and produce a new recruit to the population. Site suitability varies continuously and can be very dependent on the ecological context.

Total dispersal kernel
A dispersal kernel (in its broad sense) accounting for the joint contribution of multiple dispersal vectors and/or multiple plant species acting in a particular system. See also '**Dispersal kernel**' and '**Polychory**'.

Contributors

The definitions in this glossary were compiled and contributed by Andrew J. Dennis, Eugene W. Schupp, Ran Nathan, B. Denise Hardesty and Charles Kwit.

References

Chambers, J.C. and MacMahon, J.A. (1994) A day in the life of a seed: movements and fates of seeds and their implications for natural and managed systems. *Annual Review of Ecology and Systematics* 25, 263–292.

Eriksson, O. and Ehrlen, J. (1992) Seed and microsite limitation of recruitment in plant populations. *Oecologia* 91, 360–364.

Forget, P.-M., Lambert, J.E, Hulme, P.E. and Vander Wall, S.B. (eds) (2005) *Seed Fate: Predation, Dispersal and Seedling Establishment.* CAB International, Wallingford, UK.

Harper, J.L. (1977) *Population Biology of Plants.* Academic Press, London, UK.

Jordano, P. and Godoy, J. (2002) Frugivore-generated seed shadows: a landscape view of demographic and genetic effects. In: Levey, D., Silva, W. and Galetti, M. (eds) *Frugivores and Seed Dispersal: Ecology, Evolution, and Conservation.* CAB International, Wallingford, UK, pp. 305–321.

Muller-Landau, H., Wright, S., Calderón, O., Hubbell, S. and Foster, R. (2002) Assessing recruitment limitation: concepts, methods and case-studies from a tropical forest. In: Levey, D., Silva, W. and Galetti, M. (eds) *Seed Dispersal and Frugivory: Ecology, Evolution and Conservation.* CAB International, Wallingford, UK, pp. 35–53.

Nathan, R. and Muller-Landau, H.C. (2000) Spatial patterns of seed dispersal, their determinants and consequences for recruitment. *Trends in Ecology and Evolution* 15, 278–285.

Robertson, A.W., Trass, A., Ladley, J.J. and Kelly, D. (2006) Assessing the benefits of frugivory for seed germination: the importance of the de-inhibition effect. *Functional Ecology* 20, 58–66.

Schaefer, H.M. and Rolshausen, G. (2006) Plants on red alert: do insects pay attention? *BioEssays* 28, 65–71.

Schupp, E.W. (1993) Quantity, quality and the effectiveness of seed dispersal by animals. *Vegetatio* 107/108, 15–29.

Schupp, E., Milleron, T. and Russo, S. (2002) Dissemination limitation and the origin and maintenance of species-rich tropical forests. In: Levey, D., Silva, W. and Galetti, M. (eds) *Seed Dispersal and Frugivory: Ecology, Evolution and Conservation.* CAB International, Wallingford, UK, pp. 19–33.

Schmidt, V., Schaefer, H.M. and Winkler, H. (2004) Conspicuousness, not colour as foraging cue in plant–animal signalling. *Oikos* 106, 551–557.

Smallwood, P.D., Steele, M.A., Ribbens, E. and McShea, W.J. (1998) Detecting the effects of seed hoarders on the distribution of tree species: gray squirrels (*Sciurus carolinensis*) and oaks (*Quercus*) as a model system. In: Steele, M.A., Merritt, J.F. and Zegers, D.A. (eds) *Ecology and Evolutionary Biology of Tree Squirrels.* Virginia Museum of Natural History, Special Publication 6, Martinsville, Virginia, pp. 211–222.

Vander Wall, S.B. (1990) *Food hoarding in animals.* University of Chicago Press, Chicago, Illinois.

Vander Wall, S.B. (2003) Effects of seed size of wind dispersed pines (*Pinus*) on secondary dispersal and caching behaviour of rodents. *Oikos* 100, 25–34.

van der Pijl, L. (1972) *Principles of Dispersal in Higher Plants*, 2nd edn. Springer-Verlag, Berlin, Germany.

Walker, B.H. (1992). Biodiversity and ecological redundancy. *Conservation Biology* 6: 18.23.

Whittaker, R.H. and Feeny, P.P. (1971) Allelochemics: chemical interactions between species. *Science* 171, 757–770.

Index